SOLAR ENGINEERING OF THERMAL PROCESSES

SOLAR ENGINEERING OF THERMAL PROCESSES

Second Edition

JOHN A. DUFFIE
Emeritus Professor of Chemical Engineering

WILLIAM A. BECKMAN
Professor of Mechanical Engineering
Solar Energy Laboratory
University of Wisconsin-Madison

A Wiley-Interscience Publication

JOHN WILEY & SONS, INC.

New York   Chichester   Brisbane   Toronto   Singapore

Library of Congress Cataloging in Publication Data:

Duffie, John A.
    Solar engineering of thermal processes / John A. Duffie, William
A. Beckman. — 2nd ed.
        p. cm.

    "A Wiley-Interscience publication"
    Includes bibliographical references and index.
    ISBN 0-471-51056-4
    1. Solar energy. I. Beckman, William A.   II. Title.
    TJ810.D82 1991   90-25202
    621.47 — dc20   CIP

Printed in the United States of America

10 9 8 7 6 5 4 3 2 1

To
Pat and Sylvia

PREFACE

In the ten years since we prepared the first edition there have been tremendous changes in solar energy science and technology. In the time between 1978 (when we made the last changes in the manuscript of the first edition) and 1991 (when the last changes were made for this edition) thousands of papers have been published, many meetings have been held with proceedings published, industries have come and gone, and public interest in the field has waxed, waned, and is waxing again.

There have been significant scientific and technological developments. We have better methods for calculating radiation on sloped surfaces and modeling stratified storage tanks. We have new methods for predicting the output of solar processes and new ideas on how solar heating systems can best be controlled. We have seen new large-scale applications of linear solar concentrators and salt-gradient ponds for power generation, widespread interest in and adoption of the principles of passive heating, development of low-flow liquid heating systems, and great advances in photovoltaic processes for conversion of solar to electrical energy.

Which of these many new developments belong in a second edition? This is a difficult problem, and from the great spread of new materials no two authors would elect to include the same items. For example, there have been many new models proposed for calculating radiation on sloped surfaces, given measurements on a horizontal surface. Which of these should be included? We have made choices; others might make different choices.

Those familiar with the first edition will note some significant changes. The most obvious is a reorganization of the material into three parts. Part I is on fundamentals, and covers essentially the same materials (with many additions) as the first eleven chapters in the first edition. Part II is on applications and is largely descriptive in nature. Part III is on design of systems, or more precisely on predicting long-term system thermal performance. This includes information on simulations, on *f*-chart, on utilizability methods applied to active and passive systems, and on the solar load ratio method developed at Los Alamos. This section ends with a chapter on photovoltaics and the application of utilizability methods to predicting PV system performance.

While the organization has changed, we have tried to retain enough of the flavor of the first edition to make those who have worked with it feel at home with this one. Where we have chosen to use new correlations, we have included those in the first edition in footnotes. The nomenclature is substantially the same. Many of the figures will be familiar, as will most of the equations. We hope that the transition to this edition will be an easy one.

We have been influenced by the academic atmosphere in which we work, but have also tried to stay in touch with the commercial and industrial world. (Our students who are now out in industry have been a big help to us.) We have taught a course to engineering students at least once each year and have had a steady stream of graduate students in our laboratory. Much of the new material we have included in this edition has been prepared as notes for use by these students, and the selection process has resulted from our assessment of what we thought these students should have. We have also been influenced by the research that our students have done; it has resulted in ideas, developments and methods that have been accepted and used by many others in the field.

We have drawn on many sources for new materials, and have provided references as appropriate. In addition to the specific references, a number of general resources are worthy of note. *Advances in Solar Energy* is an annual edited by K. Böer and includes extensive reviews of various topics; volume 6 appeared in 1990. Two handbooks are available, the *Solar Energy Handbook* edited by Kreider and Kreith and the *Solar Energy Technology Handbook* edited by Dickenson and Cheremisinoff. Interesting new books have appeared, including Iqbal's *Introduction to Solar Radiation*, Rabl's *Active Solar Collectors and Their Applications*, and Hull, Nielsen, and Golding, *Salinity-Gradient Solar Ponds*. The Commission of the European Communities has published an informative series of books on many aspects of solar energy research and applications. There are several journals, including *Solar Energy*, published by the International Solar Energy Society, and the *Journal of Solar Energy Engineering*, published by the American Society of Mechanical Engineers. The June 1987 issue of *Solar Energy* is a cumulative subject and author index to the 2400 papers that have appeared in the first 39 volumes of the journal.

We have aimed this book at two audiences. It is intended to serve as a general source book and reference for those who are working in the field. The extensive bibliographies with each chapter will provide leads to more detailed exploration of topics that may be of special interest to the reader. The book is also intended to serve as a text for university-level engineering courses. There is material here for a two semester sequence, or by appropriate selection of sections it can readily be used for a one semester course. There is a wealth of new problems in Appendix A. A solutions manual is available that includes course outlines and suggestions for use of the book as a text.

We are indebted to students in our classes at Wisconsin and at Borlänge, Sweden who have used much of the text in note form. They have been critics of the best kind, always willing to tell us in constructive ways what is right and what is wrong with the materials. Heidi Burak and Craig Fieschko provided us with very useful critiques of the manuscript. Susan Pernsteiner helped us assemble the materials in useful form.

We prepared the text on Macintosh computers using T/Maker's WriteNow word processor, and set most of the equations with Prescience Company's Expressionist. The assistance of Peter Shank of T/Maker and of Allan Bonadio of Prescience is greatly appreciated. If these pages do not appear as attractive as they might, it should be attributed to our skills with these programs and not to the programs themselves.

Lynda Litzkow prepared the new art work for this edition using MacDraw II. Her assistance and competence have been very much appreciated. Port-to-Print, of Madison, prepared galleys using our disks. The cooperation of Jim Devine and Tracy Ripp of Port-to-Print has been very helpful.

We must again acknowledge the help, inspiration, and forbearance of our colleagues at the Solar Energy Laboratory. Without the support of S. A. Klein and J. W. Mitchell, the preparation of this work would have been much more difficult.

JOHN A. DUFFIE
WILLIAM A. BECKMAN

Madison, Wisconsin
June 1991

PREFACE TO THE FIRST EDITION

When we started to revise our earlier book, Solar Energy Thermal Processes, it quickly became evident that the years since 1974 had brought many significant developments in our knowledge of solar processes. What started out to be a second edition of the 1974 book quickly grew into a new work, with new analysis and design tools, new insights into solar process operation, new industrial developments, and new ideas on how solar energy can be used. The result is a new book, substantially broader in scope and more detailed than the earlier one. Perhaps less than 20 percent of this book is taken directly from *Solar Energy Thermal Processes*, although many diagrams have been reused and the general outline of the work is similar. Our aim in preparing this volume has been to provide both a reference book and a text. Throughout it we have endeavored to present quantitative methods for estimated solar process performance.

In the first two chapters we treat solar radiation, radiation data, and the processing of the data to get it in forms needed for calculation of process performance. The next set of three chapters is a review of some heat transfer principles that are particularly useful and a treatment of the radiation properties of opaque and transparent materials. Chapters 6 through 9 go into detail on collectors and storage, as without an understanding of these essential components in a solar process system it is not possible to understand how systems operate. Chapters 10 and 11 are on system concepts and economics. They serve as an introduction to the balance of the book, which is concerned with applications and design methods.

Some of the topics we cover are very well established and well understood. Others are clearly matters of research, and the methods we have presented can be expected to be out dated and replaced by better methods. An example of this situation is found in Chapter 2; the methods for estimating the fractions of total radiation which are beam and diffuse are topics of current research, and procedures better than those we suggest will probably become available. In these situations we have included in the text extensive literature citations so the interested reader can easily go the the references for further background.

Collectors are at the heart of solar processes, and for those who are starting a study of solar energy without any previous background in the subject, we suggest reading Sections 6.1 and 6.2 for a general description of these unique heat transfer devices. The first half of the book is aimed entirely at development of the ability to calculate how collectors work, and a reading of the description will make clearer the reasons for the treatment of the first set of chapters.

Our emphasis is on solar applications to buildings, as they are the applications developing most rapidly and are the basis of small but growing industry. The same ideas that are the basis of application to buildings also underlie applications to industrial process heat, thermal conversion to electrical energy generation and evaporative processes, which are all discussed briefly. Chapter 15 is a discussion of passive heating, and uses many of the same concepts and calculation methods for estimating solar gains that are developed and used in active heating systems. The principles are the same; the first half of the book develops these principles, and the second half is concerned with their application to active, passive and nonbuilding processes.

New methods of simulation of transient processes have been developed in recent years, in our laboratory and in others. These are powerful tools in the development of understanding of solar processes and in their design, and in the chapters on applications the results of simulation studies are used to illustrate the sensitivity of long-term performance to design variables. Simulation are the basis of the design procedures described in Chapters 14 and 18. Experimental measurements of system performance are still scarce, but in several cases we have made comparisons of predicted and measured performance.

Since the future of solar applications depends on the costs of solar energy systems, we have included a discussion of life cycle ecomonic analysis, and concluded it with a way of combining the many economics parameters in a life cycle saving analysis into just two numbers which can readily be used in system optimization studies. We find the method to be highly useful, but we make no claims for the worth of any of the numbers used in illustrating the method, and each user must pick his own economic parameters.

In order to make the book useful, we have wherever possible given useful relationships in equation, graphical, and tabular form. We have used the recommended standard nomenclature of the journal of *Solar Energy* (**21**, 69, 1978), except for a few cases where additional symbols have been needed for clarity. For example, G is used for irradiance (a rate, W/m^2), H is used for irradiation for a day (an integrated quantity, MJ/m^2), and I is used for irradiation for an hour (MJ/m^2), which can be thought of as an average rate for an hour. A listing of nomenclature appears in Appendix B, and includes page references to discussions of the meaning of symbols where there might be confusion. SI units are used throughout, and Appendix C provides useful conversion tables.

Numerous sources have been used in writing this book. The journal *Solar Energy*, a publication of the international Solar Energy Society, is very useful, and contains a variety of papers on radiation data, collectors of various types, heating and cooling processes, and other topics. Publications of ASME and ASHRAE have provided additional sources. In addition to these journals, there exists a very large and growing body of literature in the form of reports to and by government agencies which are not reviewed in the usual sense but which contain useful information not readily available elsewhere. These materials are not as readily available as journals, but they are referenced where we have not found the material in journals. We also call the reader's attention to *Geliotecknika* (Applied Solar Energy), a journal published by the Academy of Sciences of the USSR which is available in English and the *Revue Internationale d'Heliotechnique*, published by COMPLES in Marseille.

Many have contributed to the growing body of solar energy literature on which we have drawn. Here we note only a few of the most important of them. The work of H.C. Hottel and his colleagues at MIT, that of A. Whillier at MIT continues to be of basic importance. In space heating, the publications of G. O. G. Löf, S. Karaki and their colleagues at Colorado State University provide much of the quantitative information we have on that application.

Individuals who have helped us with the preparation of this book are many. Our graduate students and staff at the Solar Energy Laboratory have provided us with ideas, useful information and reviews of parts of the manuscript. Their constructive comments have been invaluable, and references to their work are included in the appropriate chapters. The help of students in our course on Solar Energy Technology is also acknowledged; the number of errors in the manuscript is substantially lower as a result of their good-natured criticisms.

Critical reviews are imperative, and we are indebted to S. A. Klein for his reading of the manuscript. He has been a source of ideas, a sounding board for a wide variety of concepts, the author of many of the publications on which we have drawn, and a constructive critic of the best kind.

High on any list of acknowledgements for support of this work must be the College of Engineering and the Graduate School of the University of Wisconsin-Madison. The College has provided us with support while the manuscript was in preparation, and the Graduate School made it possible for each of us to spend a half year at the Division of Mechanical Engineering of the Commonwealth Scientific and Industrial Research Organization, Australia, where we made good use of their library and developed some of the concepts of this book. Our Laboratory at Wisconsin has been supported by the National Science Foundation, the Energy Research and Development Administration, and now the Department of Energy, and the research of the Laboratory has provided ideas for the book.

It is again appropriate to acknowledge the inspiration of the late Farrington Daniels. He kept interest in solar energy alive in the 1960s and so helped to prepare for the new activity in the field during the 1970s.

Generous permissions have been provided by many publishers and authors for the use of their tables, drawings and other materials in this book. The inclusion of these material made the book more complete and useful, and their cooperation is deeply appreciated.

A book such as this takes more than authors and critics to bring it into being. Typing and drafting help are essential and we are pleased to not the help of Shirley Quamme and her co-workers in preparing the manuscript. We have been through several drafts of the book which have been typed by our student helpers at the laboratory; it has often been difficult work, and their persistence, skill and good humor have been tremendous.

Not the least, we thank our patient families for their forbearance during the lengthy process of putting this book together.

JOHN A. DUFFIE

WILLIAM A. BECKMAN

Madison, Wisconsin

June 1980

CONTENTS

13. BUILDING HEATING - ACTIVE   514

14. BUILDING HEATING:
PASSIVE AND HYBRID METHODS   555

PART III - THERMAL DESIGN METHODS

Introduction, 667

Part I

FUNDAMENTALS

In Part I, we treat the basic ideas and calculation procedures that must be understood in order to appreciate how solar processes work and how their performance can be predicted. The first five chapters are basic to the material in Chapter 6. In Chapter 6 we develop equations for a collector which give the useful output in terms of the available solar radiation and the losses. An energy balance is developed which says, in essence, that the useful gain is the (positive) difference between the absorbed solar energy and the thermal losses.

The first chapter is concerned with the nature of the radiation emitted by the sun and incident on the earth's atmosphere. This includes geometric considerations, i.e., the direction from which beam solar radiation is received and its angle of incidence on various surfaces and the quantity of radiation received over various time spans. The next chapter covers the effects of the atmosphere on the solar radiation, the radiation data that are available, and how those data can be processed to get the information that we ultimately want – the radiation incident on surfaces of various orientations.

Chapter 3 notes a set of heat transfer problems that arise in solar energy processes and is part of the basis for analysis of collectors, storage units, and other components.

The next two chapters treat interaction of radiation and opaque and transparent materials, i.e., emission, absorption, reflection, and transmission of solar and long wave radiation. These first five chapters lead to Chapter 6, a detailed discussion and analysis of the performance of flat-plate collectors. Chapter 7 is concerned with concentrating collectors and 8 with energy storage in various media. Chapter 9 is a brief discussion of the loads imposed on solar processes and the kinds of information that must be known in order to analyze the process.

Chapter 10 is the point at which the discussions of individual components are brought together to show how solar process *systems* function and how their long-term performance can be determined by simulations. The object is to be able to quantitatively predict system performance; this is the point at which we proceed from components to systems and see how transient system behavior can be calculated.

The last chapter in Part I is on solar process economics. It concludes with a method for combining the large number of economic parameters into two which can be used to optimize thermal design and assess the effects of uncertainties in an economic analysis.

Chapter 1

SOLAR RADIATION

The sun's structure and characteristics determine the nature of the energy it radiates into space. The first major topic in this chapter concerns the characteristics of this energy outside of the earth's atmosphere, its intensity and spectral distribution. We will be concerned primarily with radiation in a wavelength range of 0.25 to 3.0 μm, the portion of the electromagnetic radiation that includes most of the energy radiated by the sun.

The second major topic in this chapter is solar geometry, i.e., the position of the sun in the sky, the direction in which beam radiation is incident on surfaces of various orientations, and shading. The third topic is extraterrestrial radiation on a horizontal surface, which represents the theoretical upper limit of solar radiation available at the earth's surface.

An understanding of the nature of extraterrestrial radiation, the effects of orientation of a receiving surface, and the theoretically possible radiation at the earth's surface is important in understanding and using solar radiation data.

1.1   THE SUN

The sun is a sphere of intensely hot gaseous matter with a diameter of 1.39×10^9 m and is, on the average, 1.5×10^{11} m from the earth. As seen from the earth, the sun rotates on its axis about once every four weeks. However, it does not rotate as a solid body; the equator takes about 27 days and the polar regions take about 30 days for each rotation.

The sun has an effective blackbody temperature of 5777 K.[1] The temperature in the central interior regions is variously estimated at 8×10^6 to 40×10^6 K and the density is estimated to be about 100 times that of water. The sun is, in effect, a continuous fusion reactor with its constituent gases as the "containing vessel" retained by gravitational forces. Several fusion reactions have been suggested to supply the

---

[1] The effective blackbody temperature of 5777 K is the temperature of a blackbody radiating the same amount of energy as does the sun. Other effective temperatures can be defined, for example, that corresponding to the blackbody temperature giving the same wavelength of maximum radiation as solar radiation (about 6300 K).

energy radiated by the sun. The one considered the most important is a process in which hydrogen (i.e., four protons) combines to form helium (i.e., one helium nucleus); the mass of the helium nucleus is less than that of the four protons, mass having been lost in the reaction and converted to energy.

The energy produced in the interior of the solar sphere at temperatures of many millions of degrees must be transferred out to the surface and then be radiated into space. A succession of radiative and convective processes occur with successive emission, absorption, and reradiation; the radiation in the sun's core is in the x-ray and gamma-ray parts of the spectrum, with the wavelengths of the radiation increasing as the temperature drops at larger radial distances.

A schematic structure of the sun is shown in Figure 1.1.1. It is estimated that 90% of the energy is generated in the region of 0 to $0.23R$ (where R is the radius of the sun), which contains 40% of the mass of the sun. At a distance $0.7R$ from the center, the temperature has dropped to about 130,000 K and the density has dropped to 70 kg/m³; here convection processes begin to become important, and the zone from 0.7 to $1.0R$ is known as the **convective zone**. Within this zone the temperature drops to about 5000 K and the density to about 10^{-5} kg/m³.

The sun's surface appears to be composed of granules (irregular convection cells), with dimensions from 1000 to 3000 km and with cell lifetime of a few

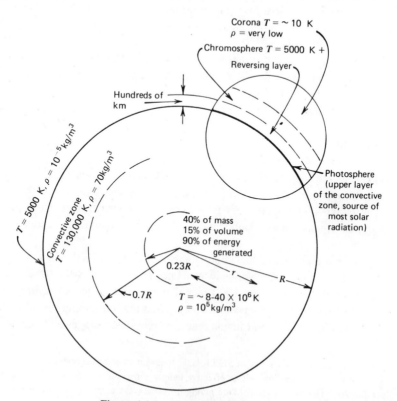

Figure 1.1.1   The structure of the sun.

minutes. Other features of the solar surface are small dark areas called pores, which are of the same order of magnitude as the convective cells, and larger dark areas called sunspots, which vary in size. The outer layer of the convective zone is called the **photosphere**. The edge of the photosphere is sharply defined, even though it is of low density (about 10^{-4} that of air at sea level). It is essentially opaque, as the gases of which it is composed are strongly ionized and able to absorb and emit a continuous spectrum of radiation. The photosphere is the source of most solar radiation.

Outside of the photosphere is a more or less transparent solar atmosphere, observable during total solar eclipse or by instruments that occult the solar disk. Above the photosphere is a layer of cooler gases several hundred kilometers deep called the **reversing layer**. Outside of that is a layer referred to as the **chromosphere**, with a depth of about 10,000 km. This is a gaseous layer with temperatures somewhat higher than that of the photosphere but with lower density. Still further out is the **corona**, a region of very low density and of very high (10^6 K) temperature. For further information on the sun's structure see Thomas (1958) or Robinson (1966).

This simplified picture of the sun, its physical structure, and its temperature and density gradients will serve as a basis for appreciating that the sun does not, in fact, function as a blackbody radiator at a fixed temperature. Rather, the emitted solar radiation is the composite result of the several layers that emit and absorb radiation of various wavelengths. The resulting extraterrestrial solar radiation and its spectral distribution have now been measured by various methods in several experiments; the results are noted in the following two sections.

1.2   THE SOLAR CONSTANT

Figure 1.2.1 shows schematically the geometry of the sun-earth relationships. The eccentricity of the earth's orbit is such that the distance between the sun and the earth

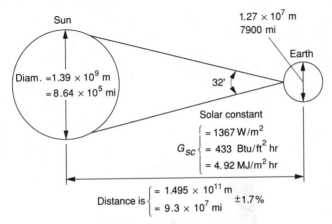

Figure 1.2.1   Sun-earth relationships.

varies by 1.7%. At a distance of one astronomical unit, 1.495×10^{11} m, the mean earth-sun distance, the sun subtends an angle of 32'. The radiation emitted by the sun and its spatial relationship to the earth result in a nearly fixed intensity of solar radiation outside of the earth's atmosphere. The **solar constant**, G_{sc}, is the energy from the sun, per unit time, received on a unit area of surface perpendicular to the direction of propagation of the radiation, at mean earth-sun distance, outside of the atmosphere.

Before rockets and spacecraft, estimates of the solar constant had to be made from ground-based measurements of solar radiation after it had been transmitted through the atmosphere and thus in part absorbed and scattered by components of the atmosphere. Extrapolations from the terrestrial measurements made from high mountains were based on estimates of atmospheric transmission in various portions of the solar spectrum. Pioneering studies were done by C. G. Abbot and his colleagues at the Smithsonian Institution. These studies and later measurements from rockets were summarized by Johnson (1954); Abbot's value of the solar constant of 1322 W/m^2 was revised upward by Johnson to 1395 W/m^2.

The availability of very-high-altitude aircraft, balloons, and spacecraft has permitted direct measurements of solar radiation outside most or all of the earth's atmosphere. These measurements were made with a variety of instruments in nine separate experimental programs. They resulted in a value of the solar constant, G_{sc}, of 1353 W/m^2 with an estimated error of ±1.5%. For discussions of these experiments, see Thekaekara (1976) or Thekaekara and Drummond (1971). This standard value was accepted by NASA [see NASA (1971)] and by the American Society of Testing and Materials.

The data on which the 1353 W/m^2 value was based have been reexamined by Frohlich (1977) and reduced to a new pyrheliometric scale[2] based on comparisons of the instruments with absolute radiometers. Data from Nimbus and Mariner satellites have also been included in the analysis, and as of 1978, Frohlich recommends a new value of the solar constant of G_{sc} = 1373 W/m^2, with a probable error of 1 to 2%. This was 1.5% higher than the earlier value and 1.2% higher than the best available determination of the solar constant by integration of spectral measurements. Additional spacecraft measurements have been made with Hickey et al. (1982) reporting 1373 W/m^2 and Willson et al. (1981) reporting 1368 W/m^2. Measurements from three rocket flights reported by Duncan et al. (1982) were 1367, 1372, and 1374 W/m^2. The World Radiation Center (WRC) has adopted a value of 1367 W/m^2, with an uncertainty of the order of 1%. As will be seen in Chapter 2, uncertainties in most terrestrial solar radiation measurements are an order of magnitude larger than those in G_{sc}. A value of G_{sc} of 1367 W/m^2 (1.960 cal/cm^2 min, 432 Btu/ft^2 hr, or 4.921 MJ/m^{2}hr) is used in this book. [See Iqbal (1983) for more detailed information on the solar constant.]

---

[2] Pyrheliometric scales are discussed in Section 2.2.

1.3 SPECTRAL DISTRIBUTION OF EXTRATERRESTRIAL RADIATION

In addition to the total energy in the solar spectrum (i.e., the solar constant), it is useful to know the spectral distribution of the extraterrestial radiation, that is, the radiation that would be received in the absence of the atmosphere. A standard spectral irradiance curve has been compiled based on high altitude and space measurements. The WRC standard and a 5777 K blackbody spectrum are shown in Figure 1.3.1. Table 1.3.1 provides the same information on the WRC spectrum in numerical form. The average energy $G_{sc,\lambda}$ (in W/m² µm) over small bandwidths centered at wavelength λ is given in the second column. The fraction $f_{0-\lambda}$ of the total energy in the spectrum that is between wavelengths 0 and λ is given in the third column. The table is in two parts, the first at regular intervals of wavelength and the second at even fractions $f_{0-\lambda}$. This is a condensed table; more detailed tables are available. [See Iqbal (1983).]

Example 1.3.1

Calculate the fraction of the extraterrestial solar radiation and the amount of that radiation in the ultraviolet ($\lambda < 0.38$ µm), the visible (0.38 µm $< \lambda < 0.78$ µm), and the infrared ($\lambda > 0.78$ µm) portions of the spectrum.

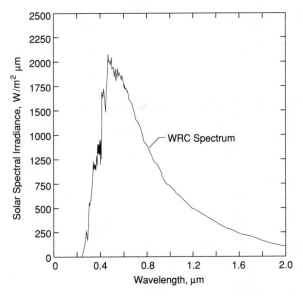

Figure 1.3.1 The WRC standard spectral irradiance curve at mean earth-sun distance.

Table 1.3.1a   Extraterrestrial Solar Irradiance (The WRC Spectrum) in Increments of Wavelength[a]

λ μm	$G_{sc,\lambda}$ W/m² μm	$f_{0-\lambda}$	λ μm	$G_{sc,\lambda}$ W/m² μm	$f_{0-\lambda}$	λ μm	$G_{sc,\lambda}$ W/m² μm	$f_{0-\lambda}$
0.250	13.8	0.002	0.520	1820.9	0.243	0.880	965.7	0.621
0.275	224.5	0.005	0.530	1873.4	0.257	0.900	911.9	0.635
0.300	542.3	0.012	0.540	1873.3	0.271	0.920	846.8	0.648
0.325	778.4	0.023	0.550	1875.0	0.284	0.940	803.8	0.660
0.340	912.0	0.033	0.560	1841.1	0.298	0.960	768.5	0.671
0.350	983.0	0.040	0.570	1843.2	0.311	0.980	763.5	0.683
0.360	967.0	0.047	0.580	1844.6	0.325	1.000	756.5	0.694
0.370	1130.8	0.056	0.590	1782.2	0.338	1.050	668.6	0.720
0.380	1070.3	0.064	0.600	1765.4	0.351	1.100	591.1	0.743
0.390	1029.5	0.071	0.620	1716.4	0.377	1.200	505.6	0.783
0.400	1476.9	0.079	0.640	1693.6	0.401	1.300	429.5	0.817
0.410	1698.0	0.092	0.660	1545.7	0.424	1.400	354.7	0.846
0.420	1726.2	0.104	0.680	1492.7	0.447	1.500	296.6	0.870
0.430	1591.1	0.117	0.700	1416.6	0.468	1.600	241.7	0.890
0.440	1837.6	0.129	0.720	1351.3	0.488	1.800	169.0	0.921
0.450	1995.2	0.143	0.740	1292.4	0.507	2.000	100.7	0.941
0.460	2042.6	0.158	0.760	1236.1	0.526	2.500	49.5	0.968
0.470	1996.0	0.173	0.780	1188.7	0.544	3.000	25.5	0.981
0.480	2028.8	0.187	0.800	1133.3	0.561	3.500	14.3	0.988
0.490	1892.4	0.201	0.820	1089.0	0.577	4.000	7.8	0.992
0.500	1918.3	0.216	0.840	1035.2	0.593	5.000	2.7	0.996
0.510	1926.1	0.230	0.860	967.1	0.607	8.000	0.8	0.999

[a] $G_{sc,\lambda}$ is the average solar irradiance over the interval from the middle of the preceeding wavelength interval to the middle of the following wavelenght interval. For example, at 0.600 μm, 1765.4 W/m² μm is the average value between 0.595 and 0.610 μm.

Table 1.3.1b   Extraterrestrial Solar Irradiance in Equal Increments of Energy

$f_{0-\lambda}$	λ	$f_{0-\lambda}$	λ
0.05	0.364	0.55	0.787
0.10	0.417	0.60	0.849
0.15	0.455	0.65	0.923
0.20	0.489	0.70	1.010
0.25	0.525	0.75	1.114
0.30	0.562	0.80	1.245
0.35	0.599	0.85	1.413
0.40	0.638	0.90	1.658
0.45	0.683	0.95	2.118
0.50	0.731	1.00	∞

Solution

From Table 1.3.1, the fractions of $f_{0-\lambda}$ corresponding to wavelengths of 0.38 and 0.78 μm are 0.064 and 0.544. Thus, the fraction in the ultraviolet is 0.064, the fraction in the visible range is 0.544 − 0.064 = 0.480, and the fraction in the infrared is 1.0 − 0.544 = 0.456. Applying these fractions to a solar constant of 1367 W/m² and tabulating the results, we have:

Wavelength range (μm)	0-0.38	0.38-0.78	0.78-∞
Fraction in range	0.064	0.480	0.456
Energy in range (W/m²)	88	656	623

■

1.4   VARIATION OF EXTRATERRESTRIAL RADIATION

Two sources of variation in extraterrestrial radiation must be considered. The first is the variation in the radiation emitted by the sun. There are conflicting reports in the literature on periodic variations of intrinsic solar radiation. It has been suggested that there are small variations (less than ±1.5%) with different periodicities and variation related to sunspot activities. Willson et al. (1981) report variances of up to 0.2% correlated with the development of sunspots. Others consider the measurements to be inconclusive or not indicative of regular variability. Measurements from Nimbus and Mariner satellites over periods of several months showed variations within limits of ±0.2% over a time when sunspot activity was very low [Frohlich (1977)]. Data of Hickey et al. (1982) over a span of 2.5 years from the Nimbus 7 satellite suggest that the solar constant is decreasing slowly, at a rate of approximately 0.02% per year. See Coulson (1975) or Thekaekara (1976) for further discussion of this topic. For engineering purposes, in view of the uncertainties and variability of atmospheric transmission, the energy emitted by the sun can be considered to be fixed.

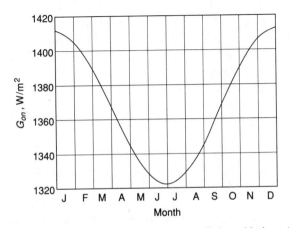

Figure 1.4.1   Variation of extraterrestrial solar radiation with time of year.

Variation of the earth-sun distance, however, does lead to variation of extraterrestrial radiation flux in the range of $\pm$ 3%. The dependence of extraterrestrial radiation on time of year is indicated by Equation 1.4.1 and is shown in Figure 1.4.1.

$$G_{on} = G_{sc}\left(1+ 0.033 \cos \frac{360\,n}{365}\right) \qquad\qquad (1.4.1)$$

where G_{on} is the extraterrestrial radiation, measured on the plane normal to the radiation on the nth day of the year.

1.5   DEFINITIONS

Several definitions will be useful in understanding the balance of this chapter.

Air Mass m The ratio of the mass of atmosphere through which beam radiation passes to the mass it would pass through if the sun were at the zenith (i.e., directly overhead). Thus at sea level, $m = 1$ when the sun is at the zenith, and $m = 2$ for a zenith angle θ_z of 60°. For zenith angles from 0° to 70° at sea level, to a close approximation,

$$m = 1/\cos \theta_z \qquad\qquad (1.5.1)$$

For higher zenith angles, the effect of the earth's curvature becomes significant and must be taken into account. For a more complete discussion of air mass, see Robinson (1966), Kondratyev (1969), or Garg (1982).

Beam Radiation The solar radiation received from the sun without having been scattered by the atmosphere. (Beam radiation is often referred to as direct solar radiation; to avoid confusion between subscripts for direct and diffuse, we use the term beam radiation.)

Diffuse Radiation The solar radiation received from the sun after its direction has been changed by scattering by the atmosphere. (Diffuse radiation is referred to in some meteorological literature as sky radiation or solar sky radiation; the definition used here will distinguish the diffuse solar radiation from infrared radiation emitted by the atmosphere.)

Total Solar Radiation The sum of the beam and the diffuse solar radiation on a surface.[3] (The most common measurements of solar radiation are total radiation on a horizontal surface, often referred to as **global radiation** on the surface.)

Irradiance, W/m² The rate at which radiant energy is incident on a surface, per unit area of surface. The symbol G is used for solar irradiance, with appropriate subscripts for beam, diffuse, or spectral radiation.

---

[3] Total solar radiation is sometimes used to indicate quantities integrated over all wavelengths of the solar spectrum.

Irradiation or **Radiant Exposure**, J/m^2 The incident energy per unit area on a surface, found by integration of irradiance over a specified time, usually an hour or a day. **Insolation** is a term applying specifically to solar energy irradiation. The symbol H is used for insolation for a day. The symbol I is used for insolation for an hour (or other period if specified). The symbols H and I can represent beam, diffuse, or total and can be on surfaces of any orientation.

Subscripts on G, H, and I are as follows: o refers to radiation above the earth's atmosphere, referred to as extraterrestrial radiation; b and d refer to beam and diffuse radiation; T and n refer to radiation on a tilted plane and on a plane normal to the direction of propagation. If neither T nor n appear, the radiation is on a horizontal plane.

Radiosity or **Radiant Exitance**, W/m^2 The rate at which radiant energy leaves a surface, per unit area, by combined emission, reflection, and transmission.

Emissive Power or **Radiant Self-Exitance**, W/m^2 The rate at which radiant energy leaves a surface per unit area, by emission only.

Any of these radiation terms, except insolation, can apply to any specified wavelength range (such as the solar energy spectrum) or to monochromatic radiation. Insolation refers only to irradiation in the solar energy spectrum.

Solar Time Time based on the apparent angular motion of the sun across the sky, with solar noon the time the sun crosses the meridian of the observer.

Solar time is the time used in all of the sun-angle relationships; it does not coincide with local clock time. It is necessary to convert standard time to solar time by applying two corrections. First, there is a constant correction for the difference in longitude between the observer's meridian (longitude) and the meridian on which the local standard time is based.[4] The sun takes 4 minutes to transverse $1°$ of longitude. The second correction is from the equation of time, which takes into account the perturbations in the earth's rate of rotation which affect the time the sun crosses the observer's meridian. The difference in minutes between solar time and standard time is

$$\text{Solar time} - \text{standard time} = 4\left(L_{st} - L_{loc}\right) + E \qquad (1.5.2)$$

where L_{st} is the standard meridian for the local time zone, L_{loc} is the longitude of the location in question (in degrees west) and E is the equation of time (in minutes) from Figure 1.5.1 or Equation 1.5.3[5] [from Spencer (1971), as cited by Iqbal (1983)],

$$E = 229.2(0.000075 + 0.001868\cos B - 0.032077\sin B$$

$$- 0.014615\cos 2B - 0.04089\sin 2B\ ) \qquad (1.5.3a)$$

---

[4] Standard meridians for continental U.S. time zones are: Eastern, $75°W$; Central, $90°W$; Mountain, $105°W$; and Pacific, $120°W$.

[5] All equations use degrees, not radians.

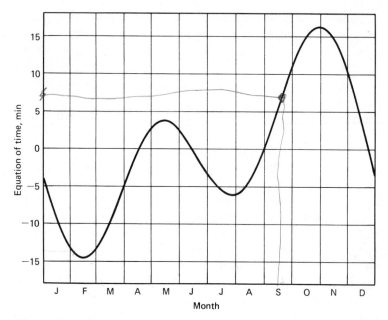

Figure 1.5.1 The equation of time E in minutes, as a function of time of year.

where $\qquad B = (n-1)\dfrac{360}{365}$ $\qquad\qquad\qquad\qquad\qquad$ (1.5.3b)

and n = day of the year. Thus $1 \leq n \leq 365$.

Note that the corrections for equations of time and displacement from the standard meridian are in minutes and that there is a 60 minute difference between daylight saving time and standard time. Time is usually specified in hours and minutes. Care must be exercised in applying the corrections, which can total more than 60 minutes.

Example 1.5.1

At Madison, WI, what is the solar time corresponding to 10:30 AM central time on February 3?

Solution

In Madison, where the longitude is 89.4°, Equation 1.5.2 gives

$$\text{Solar time} = \text{standard time} + 4(90 - 89.4) + E$$
$$= \text{standard time} + 2.48 + E$$

On February 3, $n = 34$, and from Equation 1.5.3 or Figure 1.5.1, E is -13.5 minutes, so the correction to standard time is -11 minutes. Thus 10:30 AM central standard time is 10:19 AM solar time. ■

In this book all times are assumed to be solar times unless indication is given otherwise.

1.6 DIRECTION OF BEAM RADIATION

The geometric relationships between a plane of any particular orientation relative to the earth at any time (whether that plane is fixed or moving relative to the earth) and the incoming beam solar radiation, that is, the position of the sun relative to that plane, can be described in terms of several angles [Benford and Bock (1939)]. Some of the angles are indicated in Figure 1.6.1. The angles are as follows:

ϕ **Latitude**, the angular location north or south of the equator, north positive; $-90° \leq \phi \leq 90°$.

δ **Declination**, the angular position of the sun at solar noon (i.e., when the sun is on the local meridian) with respect to the plane of the equator, north positive; $-23.45° \leq \delta \leq 23.45°$.

β **Slope**, the angle between the plane of the surface in question and the horizontal; $0 \leq \beta \leq 180°$. ($\beta > 90°$ means that the surface has a downward facing component.)

γ **Surface azimuth angle**, the deviation of the projection on a horizontal plane of the normal to the surface from the local meridian, with zero due south, east negative, and west positive; $-180° \leq \gamma \leq 180°$

ω **Hour angle**, the angular displacement of the sun east or west of the local meridian due to rotation of the earth on its axis at 15° per hour, morning negative, afternoon positive.

θ **Angle of incidence**, the angle between the beam radiation on a surface and the normal to that surface.

Additional angles are defined that describe the position of the sun in the sky:

θ_z **Zenith angle**, the angle between the vertical and the line to the sun, i.e., the angle of incidence of beam radiation on a horizontal surface.

α_s **Solar altitude angle**, the angle between the horizontal and the line to the sun, i.e., the complement of the zenith angle.

γ_s **Solar azimuth angle**, the angular displacement from south of the projection of beam radiation on the horizontal plane, shown in Figure 1.6.1. Displacements east of south are negative and west of south are positive.

The declination δ can be found from the equation of Cooper (1969):

$$\delta = 23.45 \sin\left(360\frac{284 + n}{365}\right) \qquad (1.6.1)$$

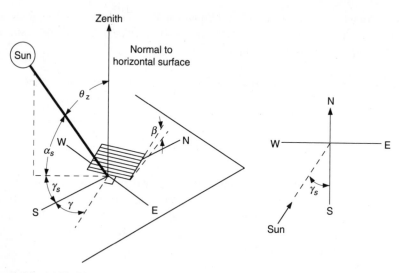

Figure 1.6.1 (a) Zenith angle, slope, surface azimuth angle, and solar azimuth angle for a tilted surface. (b) Plan view showing solar azimuth angle.

Table 1.6.1   Recommended Average Days for Months and Values of n by Months[a]

Month	n for ith Day of Month	For the Average Day of the Month		
		Date	n, Day of Year	δ, Declination
January	i	17	17	−20.9
February	$31 + i$	16	47	−13.0
March	$59 + i$	16	75	−2.4
April	$90 + i$	15	105	9.4
May	$120 + i$	15	135	18.8
June	$151 + i$	11	162	23.1
July	$181 + i$	17	198	21.2
August	$212 + i$	16	228	13.5
September	$243 + i$	15	258	2.2
October	$273 + i$	15	288	−9.6
November	$304 + i$	14	318	−18.9
December	$334 + i$	10	344	−23.0

[a] From Klein (1977)

The day of the year n can be conveniently obtained with the help of Table 1.6.1. Note that declination is a continuous function of time. The maximum rate of change of declination is at the equinoxes, when it is about 0.5°/day. For most engineering calculations, the assumption of an integer n to represent a day results in a satisfactory calculation of declination.

There is a set of useful relationships among these angles. Equations relating the angle of incidence of beam radiation on a surface, θ, to the other angles are

$$
\begin{aligned}
\cos\theta = \; & \sin\delta\sin\phi\cos\beta \\
& - \sin\delta\cos\phi\sin\beta\cos\gamma \\
& + \cos\delta\cos\phi\cos\beta\cos\omega \\
& + \cos\delta\sin\phi\sin\beta\cos\gamma\cos\omega \\
& + \cos\delta\sin\beta\sin\gamma\sin\omega
\end{aligned}
\tag{1.6.2}
$$

and
$$
\cos\theta = \cos\theta_z\cos\beta + \sin\theta_z\sin\beta\,\cos(\gamma_s - \gamma)
\tag{1.6.3}
$$

Example 1.6.1

Calculate the angle of incidence of beam radiation on a surface located at Madison, WI at 10:30 (solar time) on February 13, if the surface is tilted 45° from the horizontal and pointed 15° west of south.

Solution

Under these conditions, n is 44, the declination δ from Equation 1.6.1 is $-14°$, the hour angle ω is $-22.5°$ (15° per hour times 1.5 hours before noon), and the surface azimuth angle γ is 15°. Using a slope β of 45° and the latitude ϕ of Madison of 43°N, Equation 1.6.2 is

$$
\begin{aligned}
\cos\theta = \; & \sin(-14)\sin 43\cos 45 \\
& - \sin(-14)\cos 43\sin 45\cos 15 \\
& + \cos(-14)\cos 43\cos 45\cos(-22.5) \\
& + \cos(-14)\sin 43\sin 45\cos 15\cos(-22.5) \\
& + \cos(-14)\sin 45\sin 15\sin(-22.5)
\end{aligned}
$$

$$
\cos\theta = -0.117 + 0.121 + 0.464 + 0.418 - 0.068 = 0.817
$$

$$
\theta = 35° \qquad\blacksquare
$$

There are several commonly occurring cases for which Equation 1.6.2 is simplified. For fixed surfaces sloped toward the south or north, that is, with a surface azimuth angle γ of 0° or 180° (a very common situation for fixed flat-plate collectors), the last term drops out.

For vertical surfaces, $\beta = 90°$ and the equation becomes

$$
\begin{aligned}
\cos\theta = \; & -\sin\delta\cos\phi\cos\gamma + \cos\delta\sin\phi\cos\gamma\cos\omega \\
& + \cos\delta\sin\gamma\sin\omega
\end{aligned}
\tag{1.6.4}
$$

For horizontal surfaces, the angle of incidence is the zenith angle of the sun, θ_z. Its value must be between $0°$ and $90°$ when the sun is above the horizon. For this situation, $\beta = 0$, and Equation 1.6.2 becomes

$$\cos \theta_z = \cos \phi \cos \delta \cos \omega + \sin \phi \sin \delta \tag{1.6.5}$$

The solar azimuth angle γ_s can have values in the range of $180°$ to $-180°$. For north or south latitudes between $23.45°$ and $66.45°$, γ_s will be between $90°$ and $-90°$ for days less than 12 hours long; for days with more than 12 hours between sunrise and sunset, γ_s will be greater than $90°$ or less than $-90°$ early and late in the day when the sun is north of the east-west line in the northern hemisphere or south of the east-west line in the southern hemisphere. For tropical latitudes, γ_s can have any value when $(\delta - \phi)$ is positive in the northern hemisphere or negative in the southern, e.g., at noon at $\phi = 10°$ and $\delta = 20°$, $\gamma_s = 180°$.

Thus, to calculate γ_s, we must know in which quadrant the sun will be. This is determined by the relationship of the hour angle ω to the hour angle ω_{ew}, when the sun is due west (or east). A general formulation for γ_s, from Braun and Mitchell (1983), is conveniently written in terms of γ_s', a pseudo surface azimuth angle in the first or fourth quadrant:

$$\gamma_s = C_1 \, C_2 \, \gamma_s' + C_3 \left(\frac{1 - C_1 \, C_2}{2} \right) 180 \tag{1.6.6a}$$

where

$$\sin \gamma_s' = \frac{\sin \omega \cos \delta}{\sin \theta_z} \tag{1.6.6b}$$

or

$$\tan \gamma_s' = \frac{\sin \omega}{\sin \delta \cos \omega - \cos \phi \tan \delta} \tag{1.6.6c}$$

$$C_1 = \begin{cases} 1 & \text{if } |\omega| \le \omega_{ew} \\ -1 & \text{if } |\omega| > \omega_{ew} \end{cases} \tag{1.6.6d}$$

$$C_2 = \begin{cases} 1 & \text{if } (\phi - \delta) \ge 0 \\ -1 & \text{if } (\phi - \delta) < 0 \end{cases} \tag{1.6.6e}$$

$$C_3 = \begin{cases} 1 & \text{if } \omega \ge 0 \\ -1 & \text{if } \omega < 0 \end{cases} \tag{1.6.6f}$$

$$\cos \omega_{ew} = (\tan \delta)/(\tan \phi) \tag{1.6.6g}$$

This equation appears cumbersome, but it is easy to use as is illustrated in the next example.

Example 1.6.2

Calculate the zenith and solar azimuth angles for $\phi = 43°$ at **a** 9:30 AM on February 13 and **b** 5:30 AM on July 1.

Solution

a   On February 13 at 9:30, $\delta = -14°$ and $\omega = -37.5°$. From Equation 1.6.5,

$$\cos \theta_z = \cos 43 \cos(-14) \cos(-37.5) + \sin 43 \sin(-14) = 0.398$$

$$\theta_z = 66°$$

From Equation 1.6.6g,

$$\omega_{ew} = \cos^{-1}[\tan(-14)/(\tan 43)] = 105.5°$$

and since $|-37.5| < 105.5$, from Equation 1.6.6d, $C_1 = 1$.

From Equation 1.6.6e, $\phi - \delta = 43 - (-14) = 57$, and $57 > 0$, so $C_2 = 1$.

It is not necessary to calculate C_3 since the second term in Equation 1.6.6a is zero. (C_3 is -1, since $-37.5 < 0$.) Thus Equation 1.6.6a reduces to

$$\gamma_s = \gamma_s' = \sin^{-1} \frac{\sin(-37.5) \cos(-14)}{\sin 66} = -40.3°$$

b   On July 1 at 5:30 AM, $n = 182$, $\delta = 23.1°$, and $\omega = -97.5°$. From Equation 1.6.5,

$$\cos \theta_z = \cos 43 \cos 23.1 \cos(-97.5) + \sin 43 \sin 23.1$$

$$\theta_z = 79.6°$$

From Equation 1.6.6g,

$$\omega_{ew} = \cos^{-1}[(\tan 23.1)/(\tan 43)] = 62.8°$$

Since $|-97.5| > 62.7$, $C_1 = -1$

Since $(43 - 23.1) > 0$, $C_2 = 1$

Since $-97.5 < 0$, $C_3 = -1$

Then from Equation 1.6.6a,

$$\gamma_s = -\sin^{-1} \frac{\sin(-97.5) \cos 23.1}{\sin 79.6} - 180 = 68 - 180 = -112°$$ ∎

Useful relationships for the angle of incidence of surfaces sloped due north or due south can be derived from the fact that surfaces with slope β to the north or south have the same angular relationship to beam radiation as a horizontal surface at an artificial latitude of $(\phi - \beta)$. The relationship is shown in Figure 1.6.2 for the northern hemisphere. Modifying Equation 1.6.5 yields

$$\cos \theta = \cos(\phi - \beta) \cos \delta \cos \omega + \sin(\phi - \beta) \sin \delta \qquad (1.6.7a)$$

For the southern hemisphere modify the equation by replacing $(\phi - \beta)$ by $(\phi + \beta)$, consistent with the sign conventions on ϕ and δ:

$$\cos \theta = \cos(\phi + \beta) \cos \delta \cos \omega + \sin(\phi + \beta) \sin \delta \qquad (1.6.7b)$$

For the special case of solar noon, for the south-facing sloped surface in the northern hemisphere,

$$\theta_{noon} = |\phi - \delta - \beta| \qquad (1.6.8a)$$

and in the southern hemisphere

$$\theta_{noon} = |-\phi + \delta - \beta| \qquad (1.6.8b)$$

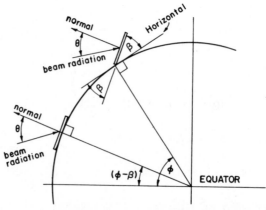

Figure 1.6.2 Section of Earth showing β, θ, ϕ, and $(\phi - \beta)$ for a south-facing surface.

where $\beta = 0$, the angle of incidence is the zenith angle, which for the northern hemisphere is

$$\theta_{z,\text{noon}} = |\phi - \delta| \tag{1.6.9a}$$

and for the southern hemisphere

$$\theta_{z,\text{noon}} = |-\phi + \delta| \tag{1.6.9b}$$

Equation 1.6.5 can be solved for the **sunset hour angle** ω_s, when $\theta_z = 90°$:

$$\cos \omega_s = -\frac{\sin\phi \sin\delta}{\cos\phi \cos\delta} = -\tan\phi \tan\delta \tag{1.6.10}$$

It also follows that the number of daylight hours is given by

$$N = \frac{2}{15}\cos^{-1}(-\tan\phi \tan\delta) \tag{1.6.11}$$

A convenient nomogram for determining day length has been devised by Whillier (1965b) and is shown in Figure 1.6.3. Information on latitude and

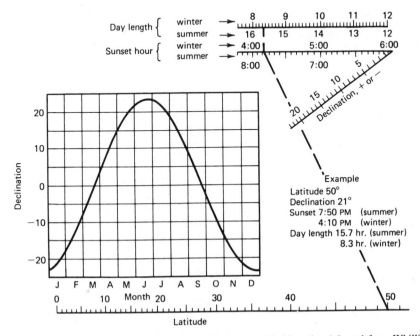

Figure 1.6.3 Nomogram to determine time of sunset and day length. Adapted from Whillier (1965b).

declination leads directly to times of sunrise and sunset and day length, for either hemisphere.

An additional angle of interest is the profile angle of beam radiation on a receiver plane R that has a surface azimuth angle of γ. It is the projection of the solar altitude angle on a vertical plane perpendicular to the plane in question. Expressed another way, it is the angle through which a plane that is initially horizontal must be rotated about an axis in the plane of the surface in question in order to include the sun. The solar altitude angle α_s (i.e., $\angle BAC$), and the profile angle α_p (i.e., $\angle DEF$), for the plane R are shown in Figure 1.6.4. The plane AEDB includes the sun. Note that the solar altitude and profile angle are the same when the sun is in a plane perpendicular to the surface R (e.g., at solar noon for a surface with a surface azimuth angle of $0°$ or $180°$). The profile angle is useful in calculating shading by overhangs. It can be calculated from

$$\tan \alpha_p = \frac{\tan \alpha_s}{\cos(\gamma_s - \gamma)} \tag{1.6.12}$$

Example 1.6.3

Calculate the solar altitude, zenith, and profile angles for a surface facing $25°$ west of south at 4:00 PM solar time on March 16 at a latitude of $43°$.

Solution

The solar altitude angle α_s is a function only of time of day and declination. For March 16, from Equation 1.6.1 (or Table 1.6.1), δ is $-2.4°$. At 4:00 PM, $\omega = 60°$. From Equation 1.6.5, recognizing that $\cos \theta_z = \sin(90 - \theta_z) = \sin \alpha_s$:

$$\sin \alpha_s = \cos 43 \cos(-2.4) \cos 60 + \sin 43 \sin(-2.4) = 0.337$$

$$\alpha_s = 19.7°$$

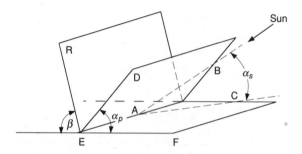

Figure 1.6.4   The solar altitude angle α_s ($\angle BAC$), and the profile angle α_p ($\angle DEF$) for a surface R.

The solar azimuth angle for this time can be calculated with Equation 1.6.6:

$$\tan \gamma'_s = \frac{\sin 60}{\sin 43 \cos 60 - \cos 43 \tan(-2.4)} = 2.330, \quad \gamma'_s = 66.8°$$

$$\cos \omega_{ew} = \tan(-2.4)/\tan 43 = -0.045, \quad \omega_{ew} = 92.6°$$

Thus C_1, C_2, and C_3 are all 1, and $\gamma_s = \gamma'_s = 66.8°$.

The profile angle for the surface with $\gamma = 25°$ is calculated with Equation 1.6.12:

$$\tan \alpha_p = \frac{\tan 19.7}{\cos(66.8 - 25)} = 0.480, \quad \alpha_p = 25.7° \qquad \blacksquare$$

Solar azimuth and altitude angles are tabulated as functions of latitude, declination, and hour angle by the U.S. Hydrographic Office (1940). Information on the position of the sun in the sky is also available with less precision but easy access in various types of charts. Examples of these are the Sun Angle Calculator (1951) and the solar position charts (plots of α_s or θ_z vs. γ_s for various ϕ, δ, and ω) in Section 1.9 and Appendix H. Care is necessary in interpreting information from other sources, since nomenclature, definitions, and sign conventions may vary from those used here.

1.7 ANGLES FOR TRACKING SURFACES

Some solar collectors "track" the sun by moving in prescribed ways to minimize the angle of incidence of beam radiation on their surfaces and thus maximize the incident beam radiation. The angles of incidence and the surface azimuth angles are needed for these collectors. The relationships in this section will be useful in radiation calculations for these moving surfaces. For further information see Eibling et al. (1953) and Braun and Mitchell (1983).

Tracking systems are classified by their motions. Rotation can be about a single axis (which could have any orientation but which in practice is usually horizontal east-west, horizontal north-south, vertical, or parallel to the earth's axis) or it can be about two axes. The following sets of equations (except for Equations 1.7.4) are for surfaces that rotate on axes that are parallel to the surfaces.

For a plane rotated about a horizontal east-west axis with a single daily adjustment so that the beam radiation is normal to the surface at noon each day,

$$\cos \theta = \sin^2 \delta + \cos^2 \delta \cos \omega \qquad (1.7.1a)$$

The slope of this surface will be fixed for each day and will be

$$\beta = |\phi - \delta|$$ (1.7.1b)

The surface azimuth angle for a day will be 0° or 180° depending on the latitude and declination:

If $(\phi - \delta) > 0,\ \gamma = 0$ (1.7.1c)
If $(\phi - \delta) < 0,\ \gamma = 180°$

For a plane rotated about a horizontal east-west axis with continuous adjustment to minimize the angle of incidence,

$$\cos \theta = \left(1 - \cos^2 \delta \sin^2 \omega\right)^{1/2}$$ (1.7.2a)

The slope of this surface is given by

$$\tan \beta = \tan \theta_z |\cos \gamma_s|$$ (1.7.2b)

The surface azimuth angle for this mode of orientation will change between 0° and 180° if the solar azimuth angle passes through ± 90°. For either hemisphere,

If $|\gamma_s| < 90,\ \gamma = 0°$ (1.7.2c)
If $|\gamma_s| > 90,\ \gamma = 180°$

For a plane rotated about a horizontal north-south axis with continuous adjustment to minimize the angle of incidence,

$$\cos \theta = \left(\cos^2 \theta_z + \cos^2 \delta \sin^2 \omega\right)^{1/2}$$ (1.7.3a)

The slope is given by

$$\tan \beta = \tan \theta_z |\cos (\gamma - \gamma_s)|$$ (1.7.3b)

The surface azimuth angle γ will be 90° or –90° depending on the sign of the solar azimuth angle:

If $\gamma_s > 0,\ \gamma = 90°$ (1.7.3c)
If $\gamma_s < 0,\ \gamma = -90°$

For a plane with a fixed slope rotated about a vertical axis, the angle of incidence is minimized when the surface azimuth and solar azimuth angles are equal.

From Equation 1.6.3, the angle of incidence is

$$\cos\theta = \cos\theta_z \cos\beta + \sin\theta_z \sin\beta \qquad (1.7.4a)$$

The slope is fixed, so

$$\beta = \text{constant} \qquad (1.7.4b)$$

The surface azimuth angle is

$$\gamma = \gamma_s \qquad (1.7.4c)$$

For a plane rotated about a north-south axis parallel to the earth's axis with continuous adjustment to minimize θ,

$$\cos\theta = \cos\delta \qquad (1.7.5a)$$

The slope varies continuously and is

$$\tan\beta = \frac{\tan\phi}{\cos\gamma} \qquad (1.7.5b)$$

The surface azimuth angle is

$$\gamma = \tan^{-1}\frac{\sin\theta_z \sin\gamma_s}{\cos\theta' \sin\phi} + 180C_1C_2 \qquad (1.7.5c)$$

where $\qquad \cos\theta' = \cos\theta_z \cos\phi + \sin\theta_z \sin\phi \qquad (1.7.5d)$

$$C_1 = \begin{cases} 0 \;\; \text{if} \; \left(\tan^{-1}\dfrac{\sin\theta_z \sin\gamma_s}{\cos\theta' \sin\phi}\right) + \gamma_s = 0 \\ \\ 1 \;\; \text{otherwise} \end{cases} \qquad (1.7.5e)$$

$$C_2 = \begin{cases} 1 \;\;\; \text{if } \gamma_s \geq 0 \\ -1 \;\; \text{if } \gamma_s < 0 \end{cases} \qquad (1.7.5f)$$

For a plane that is continuously tracking about two axes to minimize the angle of incidence,

$$\cos\theta = 1 \qquad (1.7.6a)$$

$$\beta = \alpha \tag{1.7.6b}$$

$$\gamma = \gamma_s \tag{1.7.6c}$$

Example 1.7.1

Calculate the angle of incidence of beam radiation, the slope of the surface, and the surface azimuth angle for a surface at **a** $\phi = 40°$, $\delta = 21°$, and $\omega = 30°$ (2:00 PM) and **b** $\phi = 40°$, $\delta = 21°$, and $\omega = 100°$ if it is continuously rotated about an east-west axis to minimize θ.

Solution

a   Use Equations 1.7.2 for a surface moved in this way. First calculate the angle of incidence:

$$\theta = \cos^{-1}(1 - \cos^2 21 \sin^2 30)^{1/2} = 27.8°$$

Next calculate θ_z from Equation 1.6.5:

$$\theta_z = \cos^{-1}(\cos 40 \cos 21 \cos 30 + \sin 40 \sin 21) = 31.8°$$

We now need γ_s. Using Equations 1.6.6g, $\cos \omega_{ew} = (\tan 21)/(\tan 40)$ and $\omega_{ew} = 62.8°$. Thus $C_1 = 1, C_2 = 1, C_3 = 1$, and $\gamma_s = \gamma_s'$. So

$$\gamma_s = \sin^{-1} \frac{\sin 30 \cos 21}{\sin 31.8} = 62.4°$$

Then from Equation 1.7.2b

$$\beta = \tan^{-1}(\tan 31.8 \cos 62.4) = 16.0°$$

From Equation 1.7.2c, with $\gamma_s < 90$, $\gamma = 0$.

b   The procedure is the same as in **a**:

$$\theta = \cos^{-1}(1 - \cos^2 21 \sin^2 100)^{1/2} = 66.8°$$

$$\theta_z = \cos^{-1}(\cos 40 \cos 21 \cos 100 + \sin 40 \sin 21) = 83.9°$$

The value of ω_{ew} is still 62.8°, from part **a**. In this case, the constants in Equation 1.6.6a are $C_1 = -1, C_2 = 1$, and $C_3 = 1$.

$$\gamma_s = -\sin^{-1} \frac{\sin 100 \cos 21}{\sin 83.9} + 180 = 112.4°$$

The slope is then

$$\beta = \tan^{-1} \tan 83.9 \, |\cos 112.4| = 74.3°$$

And since $|\gamma_s| > 90$, γ will be 180°. (Note that these results can be checked using Equation 1.6.5.) ■

1.8 RATIO OF BEAM RADIATION ON TILTED SURFACE TO THAT ON HORIZONTAL SURFACE

For purposes of solar process design and performance calculations, it is often necessary to calculate the hourly radiation on a tilted surface of a collector from measurements or estimates of solar radiation on a horizontal surface. The most commonly available data are total radiation for hours or days on the horizontal surface, whereas the need is for beam and diffuse radiation on the plane of a collector.

The geometric factor R_b, the ratio of beam radiation on the tilted surface to that on a horizontal surface at any time, can be calculated exactly by appropriate use of Equation 1.6.2. Figure 1.8.1 indicates the angle of incidence of beam radiation on the horizontal and tilted surfaces. The ratio G_{bT}/G_b is given by[6]

$$R_b = \frac{G_{b,T}}{G_b} = \frac{G_{b,n} \cos \theta}{G_{b,n} \cos \theta_z} = \frac{\cos \theta}{\cos \theta_z} \tag{1.8.1}$$

and $\cos \theta$ and $\cos \theta_z$ are both determined from Equation 1.6.2 (or from equations derived from Equation 1.6.2).

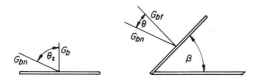

Figure 1.8.1 Beam radiation on horizontal and tilted surfaces.

[6] The symbol G is used in this book to denote rates, while I is used for energy quantities integrated over an hour. The original development of R_b by Hottel and Woertz (1942) was for hourly periods; for an hour (using angles at the midpoint of the hour), $R_b = I_{bT}/I_b$.

Example 1.8.1

What is the ratio of beam radiation to that on a horizontal surface for the surface and time specified in Example 1.6.1?

Solution

Example 1.6.1 shows the calculation for $\cos \theta$. For the horizontal surface, from Equation 1.6.5,

$$\cos \theta_z = \sin(-14) \sin 43 + \cos(-14) \cos 43 \cos(-22.5) = 0.491$$

And from Equation 1.8.1

$$R_b = \frac{\cos \theta}{\cos \theta_z} = \frac{0.818}{0.491} = 1.67 \qquad \blacksquare$$

The optimum azimuth angle for flat-plate collectors is usually $0°$ in the northern hemisphere (or $180°$ in the southern hemisphere). Thus it is a common situation that $\gamma = 0°$ (or $180°$). In this case, Equations 1.6.5 and 1.6.7 can be used to determine $\cos \theta_z$ and $\cos \theta$, respectively, leading in the northern hemisphere, for $\gamma = 0°$, to

$$R_b = \frac{\cos(\phi - \beta) \cos \delta \cos \omega + \sin(\phi - \beta) \sin \delta}{\cos \phi \cos \delta \cos \omega + \sin \phi \sin \delta} \qquad (1.8.2)$$

In the southern hemisphere, $\gamma = 180°$ and the equation is

$$R_b = \frac{\cos(\phi + \beta) \cos \delta \cos \omega + \sin(\phi + \beta) \sin \delta}{\cos \phi \cos \delta \cos \omega + \sin \phi \sin \delta} \qquad (1.8.3)$$

A special case of interest is $R_{b,noon}$, the ratio for south-facing surfaces at solar noon. From Equations 1.6.8a and 1.6.9a, for the northern hemisphere,

$$R_{b,noon} = \frac{\cos|\phi - \delta - \beta|}{\cos|\phi - \delta|} \qquad (1.8.3a)$$

For the southern hemisphere, from Equations 1.6.8b and 1.6.9b,

$$R_{b,noon} = \frac{\cos|-\phi + \delta - \beta|}{\cos|-\phi + \delta|} \qquad (1.8.3b)$$

Hottel and Woertz (1942) pointed out that Equation 1.8.2 provides a convenient method for calculating R_b for the most common cases. They also showed a graphical method for solving these equations. This graphical method has been revised by Whillier (1975), and an adaptation of Whillier's curves are given here. Figures 1.8.2(a–e) are plots of both cos θ_z as a function of ϕ and cos θ as a function of $(\phi - \beta)$ for various dates (i.e., declinations). By plotting the curves for sets of dates having (nearly) the same absolute value of declination, the curves "reflect back" on each other at latitude 0°. Thus each set of curves, in effect, covers the latitude range of –60° to 60°.

As will be seen in later chapters, solar process performance calculations are very often done on an hourly basis. The cos θ_z plots are shown for the midpoints of hours before and after solar noon, and the values of R_b found from them are applied to those hours. (This procedure is satisfactory for most hours of the day, but in hours that include sunrise and sunset, unrepresentative values of R_b may be obtained. Solar collection in those hours is most often zero or a negligible part of the total daily collector output. However, care must be taken that unrealistic products of R_b and beam radiation I_b are not used.)

To find cos θ_z, enter the chart for the appropriate time with the date and latitude of the location in question. Cos θ is found for the same date and latitude by entering with abscissa corresponding to $(\phi - \beta)$. Then R_b is found from Equation 1.8.1. The dates on the sets of curves are shown in two sets, one for north (positive) latitudes and the other for south (negative) latitudes.

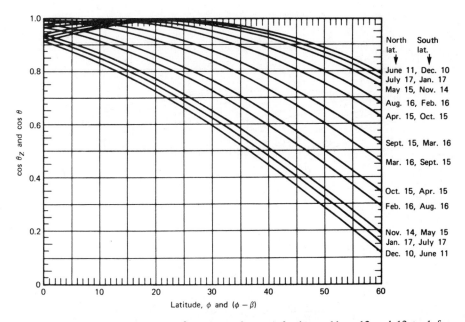

Figure 1.8.2(a) Cos θ vs. $(\phi - \beta)$ and cos θ_z vs. ϕ for hours 11 to 12 and 12 to 1 for surfaces tilted toward the equator. The columns on the right show dates for the curves for north and south latitudes. In south latitudes, use $|\phi|$. Adapted from Whillier (1975).

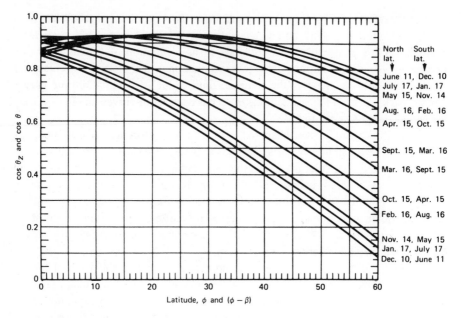

Figure 1.8.2(b) Cos θ vs. $(\phi - \beta)$ and cos θ_z vs. ϕ for hours 10 to 11 and 1 to 2.

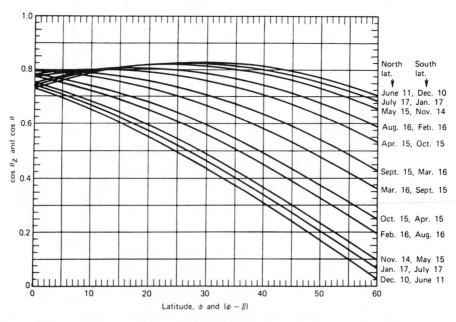

Figure 1.8.2(c) Cos θ vs. $(\phi - \beta)$ and cos θ_z vs. ϕ for hours 9 to 10 and 2 to 3.

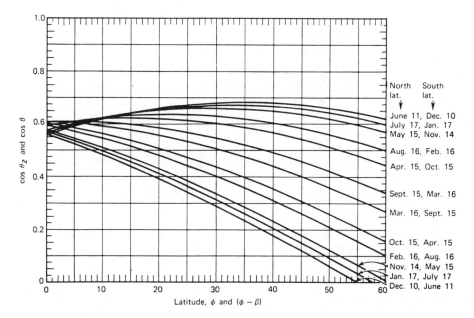

Figure 1.8.2(d)   Cos θ vs. $(\phi - \beta)$ and cos θ_z vs. ϕ for hours 8 to 9 and 3 to 4.

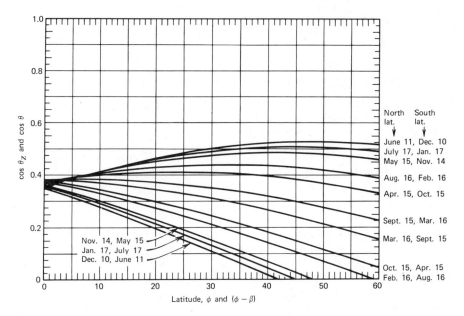

Figure 1.8.2(e)   Cos θ vs. $(\phi - \beta)$ and cos θ_z vs. ϕ for hours 7 to 8 and 4 to 5.

Two situations arise, for positive values or for negative values of $(\phi - \beta)$. For positive values, the charts are used directly. If $(\phi - \beta)$ is negative (which frequently occurs when collectors are sloped for optimum performance in winter or with vertical collectors), the procedure is modified. Determine cos θ_z as before. Determine cos θ from the appropriate absolute value of $(\phi - \beta)$ using the curve for the other hemisphere, that is, with the sign on the declination reversed.

Example 1.8.2

Calculate R_b for a surface at latitude 40°N at a tilt 30° toward the south for the hour 9 to 10 solar time on February 16.

Solution

Use Figure 1.8.2(c) for the hour ±2.5 hours from noon as representative of the hour from 9 to 10. To find cos θ_z, enter at a latitude of 40° for the north latitude date of February 16. Cos $\theta_z = 0.45$. To find cos θ, enter at a latitude of $\phi - \beta = 10°$ for the same date. Cos $\theta = 0.73$. Then

$$R_b = \frac{\cos \theta}{\cos \theta_z} = \frac{0.73}{0.45} = 1.62$$

The ratio can also be calculated using Equation 1.8.2. The declination on February 16 is −13°:

$$R_b = \frac{\cos 10 \cos(-13) \cos(-37.5) + \sin 10 \sin(-13)}{\cos 40 \cos(-13) \cos(-37.5) + \sin 10 \sin(-13)} = 1.61 \qquad ■$$

Example 1.8.3

Calculate R_b for a latitude 40°N at a tilt of 50° toward the south for the hour 9 to 10 solar time on February 16.

Solution

Cos θ_z is found as in the previous example and is 0.45. To find cos θ, enter at an abscissa of 10°, using the curve for February 16 for south latitudes. The value of cos θ from the curve is 0.80. Thus $R_b = 0.80/0.45 = 1.78$. Equation 1.8.2 can also be used:

$$R_b = \frac{\cos(-10) \cos(-13) \cos(-37.5) + \sin(-10) \sin(-13)}{\cos 40 \cos(-13) \cos(-37.5) + \sin 40 \sin(-13)} = 1.80 \qquad ■$$

It is possible, using Equation 1.8.2 or Figures 1.8.2, to construct plots showing the effects of collector tilt on R_b for various times of the year and day. Figure 1.8.3 shows such a plot for a latitude of 40° and a slope of 50°. It illustrates

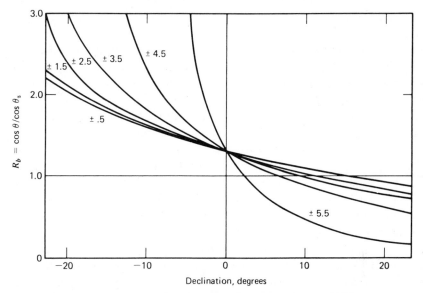

Figure 1.8.3 Ratio R_b for a surface with slope 50° to south at latitude 40° for various hours from solar noon.

that very large gains in incident beam radiation are to be had by tilting a receiving surface toward the equator.

Equation 1.8.1 can also be applied to other than fixed flat-plate collectors. Equations 1.7.1 to 1.7.6 give cos θ for surfaces moved in prescribed ways in which concentrating collectors may move to track the sun. If the beam radiation on a horizontal surface is known or can be estimated, the appropriate one of these equations can be used in the numerator of Equation 1.8.1 for cos θ. For example, for a plane rotated continuously about a horizontal east-west axis to maximize the beam radiation on the plane, from Equation 1.7.2a, the ratio of beam radiation on the plane to that on a horizontal surface at any time is

$$R_b = \frac{\left(1 - \cos^2 \delta \; \sin^2 \omega\right)^{1/2}}{\cos \phi \cos \delta \cos \omega + \sin \phi \sin \delta} \tag{1.8.4}$$

Some of the solar radiation data available are beam radiation on surfaces normal to the radiation, as measured by a pyrheliometer.[7] In this case the useful ratio is beam radiation on the surface in question to beam radiation on the normal surface, R_b' This is simply $R_b' = \cos \theta$, where θ is obtained from Equations 1.7.1 to 1.7.6.

---

[7] Pyrheliometers and other instruments for measuring solar radiation are described in Chapter 2.

1.9   SHADING

Three types of shading problems occur so frequently that methods are needed to cope with them. The first is shading of a collector, window, or other receiver by near-by trees, buildings or other obstructions. The geometries may be irregular, and systematic calculations of shading of the receiver in question may be difficult. Recourse is made to diagrams of the position of the sun in the sky, e.g., plots of solar altitude α_s vs. solar azimuth γ_s, on which shapes of obstructions (shading profiles) can be superimposed to determine when the path from the sun to the point in question is blocked. The second type includes shading of collectors in other than the first row of multirow arrays by the collectors on the ajoining row. The third includes shading of windows by overhangs and wingwalls. Where the geometries are regular, shading is amenable to calculation, and the results can be presented in general form. This will be treated in Chapter 14.

At any point in time and at a particular latitude, ϕ, δ, and ω are fixed. From the equations in Section 1.6, the zenith angle θ_z or solar altitude angle α_s and the solar azimuth angle γ_s can be calculated. A solar position plot of θ_z and α_s vs. γ_s for latitudes of $\pm 45°$ is shown in Figure 1.9.1. Lines of constant declination are labeled by dates of mean days of the months from Table 1.6.1. Lines of constant hour angles labeled by hours are also shown. Plots for latitudes from 0 to $\pm 70°$ are included in Appendix H.

The angular position of buildings, wingwalls, overhangs or other obstructions can be entered on the same plot. For example, as observed by Mazria (1979) and

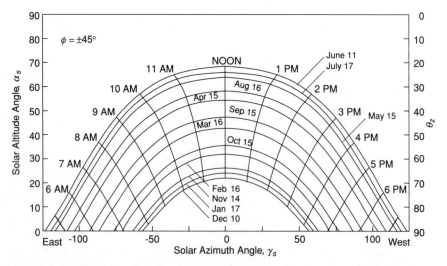

Figure 1.9.1 Solar position plot for $\pm 45°$ latitude. Solar altitude angle and solar azimuth angle are functions of declination and hour angle, indicated on the plots by dates and times. The dates shown are for northern hemisphere; for southern hemisphere use the corresponding dates as indicated in Figure 1.8.2. See Appendix H for other latitudes.

Anderson (1982), if a building or other obstruction of known dimensions and orientation is located a known distance from the point of interest (i.e., the receiver, collector, or window), the angular coordinates corresponding to altitude and azimuth angles of points on the obstruction (the object azimuth angle γ_o and object altitude angle α_o) can be calculated from trigonometric considerations. This is illustrated in Examples 1.9.1 and 1.9.2. Alternatively, measurements of object altitude and azimuth angles may be made at the site of a proposed receiver and the angles plotted on the solar position plot. A variety of instruments are available to measure the angles.

Example 1.9.1

A proposed collector site at S is 10.0 m to the north of a long wall that shades it when the sun is low in the sky. The wall is of uniform height of 2.5 m above the center of the proposed collector area. Show this wall on a solar position chart with **a** the wall oriented east-west and **b** the wall oriented on a southeast to northwest axis displaced 20° from east-west.

a.

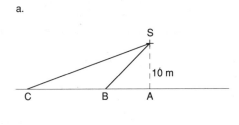

b.

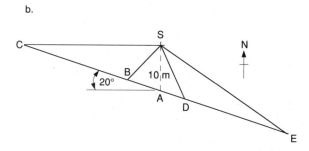

Solution

In each case, we pick several points on the top of the wall to establish the coordinates for plotting on the solar position plot.

a   Take three points indicated by A, B, and C on the diagram with A to the south, and B and C west of A. Points B' and C' are taken to the east of A with the same object altitude angles as B and C and with object azimuth angles changed only in sign.

For point A, the object azimuth γ_{oA} is 0°. The object altitude angle is

$$\tan \alpha_{oA} = 2.5/10, \quad \alpha_{oA} = 14.0°$$

For point B, $SB = \left(10^2 + 10^2\right)^{1/2} = 14.1$ m,

$$\tan \alpha_{oB} = 2.5/14.1, \quad \alpha_{oB} = 10.0°$$
$$\tan \gamma_{oB} = 10/10, \quad \gamma_{oB} = 45.0°$$

For point C, $SC = \left(10^2 + 30^2\right)^{1/2} = 31.6$ m,

$$\tan \alpha_{oC} = 2.5/31.6, \quad \alpha_{oC} = 4.52°$$
$$\tan \gamma_{oC} = 30/10, \quad \gamma_{oC} = 71.6°$$

There are points corresponding to B and C but to the east of A; these will have the same object azimuth angles except with negative signs. The shading profile determined by these coordinates is independent of latitude. It is shown by the solid line on the plot for $\phi = 45°$. Note that at object azimuth angles of 90°, the object distance becomes infinity and the object altitude angle becomes 0°.

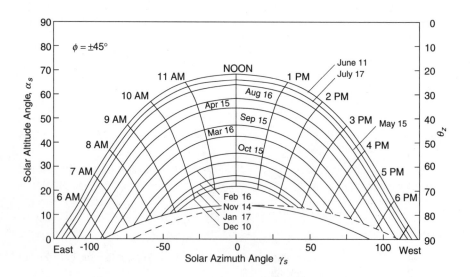

The sun is obscured by the wall only during times shown by cross-hatching on the diagram. The wall does not cast a shadow on S from late March to mid-September. For December 10, it casts a shadow on S before 9:00 AM and after 3:00 PM.

b. The obstruction of the sky does not show east-west symmetry in this case, so five points have been chosen as shown to cover the desirable range. Point A is the same as before, i.e., $\alpha_{oA} = 14.0°$, $\gamma_{oA} = 0°$.

Arbitrarily select points on the wall for the calculation. In this case the calculations are easier if we select values of the object azimuth angle and calculate from them the corresponding distances from the point to the site and the corresponding α_o. In this case we can select values of γ_o for points B, C, D, and E of 45°, 90°, –30°, and –60°.

For point B, with $\gamma_{oB} = 45°$, the distance SB can be calculated from the law of sines:

$$\frac{\sin 70}{SB} = \frac{\sin (180 - 45 - 70)}{10}, \quad SB = 10.4 \text{ m}$$

$$\tan \alpha_{oB} = 2.5/10.4, \quad \alpha_{oB} = 13.5°$$

For the point D, with $\gamma_{oD} = -30°$, the calculation is

$$\frac{\sin 110}{SD} = \frac{\sin(180 - 110 - 30)}{10}, \quad SD = 14.6 \text{ m}$$

$$\tan \alpha_{oD} = 2.5/14.6, \quad \alpha_{oD} = 9.7°$$

The calculations for points C and E give $\alpha_{oC} = 5.2°$ at $\gamma_{oC} = 90°$ and $\alpha_{oE} = 2.6°$ at $\gamma_{oE} = -60.0°$.

The shading profile determined by these coordinates is plotted on the solar position chart for $\phi = 45°$ and is shown as the dashed line. In this case, the object altitude angle goes to zero at azimuth angles of –70° and 110°. In either case, the cross-hatched area under the curves represents the wall and the times when the wall would obstruct the beam radiation are those times (declination and hour angles) in the cross-hatched areas. ■

There may be some freedom in selecting points to be used in plotting object coordinates, and the calculation may be made easier (as in the preceding example) by selecting the most appropriate points. Applications of trigonometry will always provide the necessary information. For obstructions such as buildings, the points selected must include corners or limits that define the extent of obstruction. It may or may not be necessary to select intermediate points to fully define shading. This is illustrated in the following example.

Example 1.9.2

It is proposed to install a solar collector at a level 4.0 m above the ground. A rectangular building 30 m high is located 45 m to the south, has its long dimension on an east-west axis, and has dimensions shown in the sketch. The latitude is 45°.

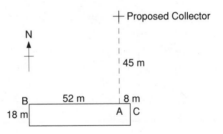

Diagram this building on the solar position plot to show the times of day and year when it would shade the proposed collector.

Solution

Three points that will be critical to determination of the shape of the image are the top near corners and the top of the building directly to the south of the proposed collector. Consider first point A. The object altitude angle of this point is determined by the fact that it is 45 m away and $30 - 4 = 26$ m higher than the proposed collector:

$$\tan \alpha_{oA} = 26/45, \quad \alpha_{oA} = 30.0°$$

The object azimuth angle γ_{oA} is $0°$ as the point A is directly to the south.

For point B, the distance SB is $(45^2 + 52^2)^{1/2} = 68.8$ m. The height is again 26 m. Then

$$\tan \alpha_{oB} = 26/68.8, \quad \alpha_{oB} = 20.7°$$

The object azimuth angle γ_{oB} is

$$\tan \gamma_{oB} = 52/45, \quad \gamma_{oB} = 49.1°$$

The calculation method for point C is the same as for B. The distance SC = $(45^2 + 8^2)^{1/2} = 45.7$ m:

$$\tan \alpha_{oC} = 26/45.7, \quad \alpha_{oC} = 29.6°$$

$$\tan \gamma_{oC} = 8/45, \quad \gamma_{oC} = 10.1°$$

Note again that since point C lies to the east of south, γ_{oC} is by convention negative.

The shading profile of the building can be approximated by joining A and C and A and B by straight lines. A more precise representation is obtained by calculating intermediate points on the shading profile to establish the curve. In this example, an object altitude angle of $27.7°$ is calculated for an object azimuth angle of $25°$.

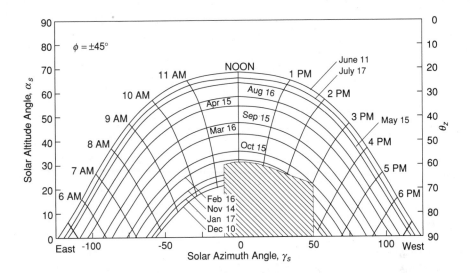

These coordinates are plotted and the outlines of the building are shown on the figure. The shaded area represents the existing building as seen from the proposed collector site. The dates and times when the collector would be shaded from direct sun by the building are evident. ∎

Implicit in the preceding discussion is the idea that the solar position at a point in time can be represented for a point location. Collectors and receivers have finite size, and what one point on a large receiving surface "sees" may not be the same as what another point sees. The problem is often to determine the amount of beam radiation on a receiver. If shading obstructions are far from the receiver relative to its size, so that shadows tend to move over the receiver rapidly and the receiver is either shaded or not shaded, the receiver can be thought of as a point. If a receiver is partially shaded, it can be considered to consist of a number of smaller areas, each of which is shaded or not shaded. Or an integration over the receiver area may be performed to determine shading effects. These integrations have been done for special cases of overhangs and wingwalls.

Overhangs and wingwalls are architectural features that are applied to buildings to shade windows from beam radiation. The solar position charts can be used to determine when points on the receiver are shaded. The procedure is identical to that of Example 1.9.1; the obstruction in the case of an overhang and the times when the point is shaded from beam radiation are the times corresponding to areas above the line. This procedure can be used for overhangs of either finite or infinite length. The same concepts can be applied to wingwalls; the vertical edges of the object in Example 1.9.2 correspond to edges of wingwalls of finite height.

An overhang is shown in cross section in Figure 1.9.2(a) for the most common situation of a vertical window. The projection P is the horizontal distance from the plane of the window to the outer edge of the overhang. The gap G is the vertical

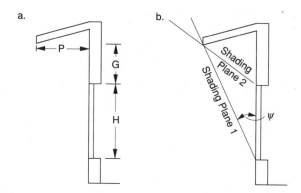

Figure 1.9.2. (a) Cross section of a long overhang, showing projection gap and height. (b) Section showing shading planes.

distance from the top of the window to the horizontal plane that includes the outer edge of the overhang. The height H is the vertical dimension of the window.

The concept of shading planes was introduced by Jones (1980) as a useful way of considering shading by overhangs where end effects are negligible. Two shading planes are labeled in Figure 1.9.2(b). The angle of incidence of beam radiation on a shading plane can be calculated from its surface azimuth angle γ and its slope $\beta =$ $90 + \psi$ by Equation 1.6.2 or equivalent. The angle ψ of shading plane 1 is $\tan^{-1}(P/(G+H))$ and that for shading plane 2 is $\tan^{-1}(P/G)$. Note that if the profile angle α_p is less than $90 - \psi$, the outer surface of the shading plane will "see" the sun and beam radiation will reach the receiver.[8]

Shading calculations are needed when flat-plate collectors are arranged in rows.[9] Normally, the first row is unobstructed, but the second row may be partially shaded by the first, the third by the second, etc. This arrangement of collectors is shown in cross section in Figure 1.9.3.

For the case where the collectors are long in extent so the end effects are negligible, the profile angle provides a useful means of determining shading. As long as the profile angle is greater than the angle CAB, no point on row N will be shaded by row M. If the profile angle at a point in time is CA'B' and is less than CAB, the portion of row N below point A' will be shaded from beam radiation.

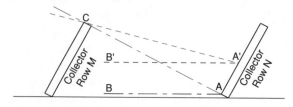

Figure 1.9.3   Section of two rows of a multirow collector array.

---

[8] Use of the shading plane concept will be discussed in Chapters 2 and 14.
[9] See Figure 12.1.2(c) for an example.

Example 1.9.3

A multiple-row array of collectors is arranged as shown in the figure. The collectors are 2.10 m from top to bottom and are sloped at 60° toward the south. At a time when the profile angle (given by Equation 1.6.12) is 25°, estimate the fraction of the area of the collector in row N that will be shaded by the collectors in row M. Assume that the rows are long so end effects are not significant.

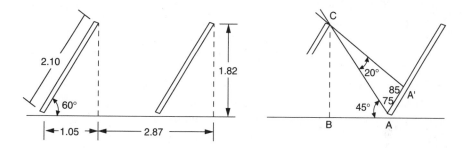

Solution

Referring to the diagram, the angle BAC is $\tan^{-1}(2.87 - 1.05)/1.82 = 45°$, and since α_p is 25°, shading will occur.

The dimension AA' can be calculated:

$$AC = 1.82/\sin 45 = 2.57 \text{ m}$$

$$\angle CAA' = 180 - 45 - 60 = 75°, \text{ and } \angle CA'A = 180 - 75 - 20 = 85°$$

From the law of sines,

$$AA' = \frac{2.57 \sin 20}{\sin 85} = 0.88 \text{ m}$$

The fraction of collector N that is shaded is $0.88/2.10 = 0.42$.   ■

1.10 EXTRATERRESTRIAL RADIATION ON A HORIZONTAL SURFACE

Several types of radiation calculations are most conveniently done using normalized radiation levels, that is, the ratio of radiation level to the theoretically possible radiation that would be available if there were no atmosphere. For these calculations, which are discussed in Chapter 2, we need a method of calculating the extraterrestrial radiation.

At any point in time, the solar radiation incident on a horizontal plane outside of the atmosphere is the normal incident solar radiation as given by Equation 1.4.1 divided by R_b:

$$G_o = G_{sc}\left(1 + 0.033 \cos \frac{360n}{365}\right) \cos \theta_z \qquad (1.10.1)$$

where G_{sc} is the solar constant and n is the day of the year. Combining Equation 1.6.5 for $\cos \theta_z$ with Equation 1.10.1 gives G_o for a horizontal surface at any time between sunrise and sunset.

$$G_o = G_{sc}\left(1 + 0.033 \cos \frac{360n}{365}\right)(\cos \phi \cos \delta \cos \omega + \sin \phi \sin \delta) \qquad (1.10.2)$$

It is often necessary for calculation of daily solar radiation to have the integrated daily extraterrestrial radiation on a horizontal surface, H_o. This is obtained by integrating Equation 1.10.2 over the period from sunrise to sunset. If G_{sc} is in watts per square meter, H_o in joules per square meter is

$$H_o = \frac{24 \times 3600 G_{sc}}{\pi}\left(1 + 0.033 \cos \frac{360n}{365}\right)$$

$$\times \left(\cos \phi \cos \delta \sin \omega_s + \frac{\pi \omega_s}{180} \sin \phi \sin \delta\right) \qquad (1.10.3)$$

where ω_s is the sunset hour angle, in degrees, from Equation 1.6.10.

The monthly mean[10] daily extraterrestrial radiation, $\overline{H}_o$, is a useful quantity. For latitudes in the range +60 to –60 it can be calculated with Equation 1.10.3 using n and δ for the mean day of the month[11] from Table 1.6.1. H_o is plotted as a function of latitude for the northern and southern hemispheres in Figure 1.10.1. The curves are for dates that give the mean radiation for the month and thus show $\overline{H}_o$. Values of H_o for any day can be estimated by interpolation. Exact values of $\overline{H}_o$ for all latitudes are given in Table 1.10.1.

Example 1.10.1

What is H_o, the day's solar radiation on a horizontal surface in the absence of the atmosphere, at latitude 43°N on April 15?

Solution

For these circumstances, $n = 105$ (from Table 1.6.1), $\delta = 9.4°$ (from Equation 1.6.1), and $\phi = 43°$. From Equation 1.6.10

$$\cos \omega_s = -\tan 43 \tan 9.4 \text{ and } \omega_s = 98.9°$$

---

[10] An overbar is used throughout the book to indicate a monthly average quantity.
[11] The mean day is the day having H_o closest to $\overline{H}_o$.

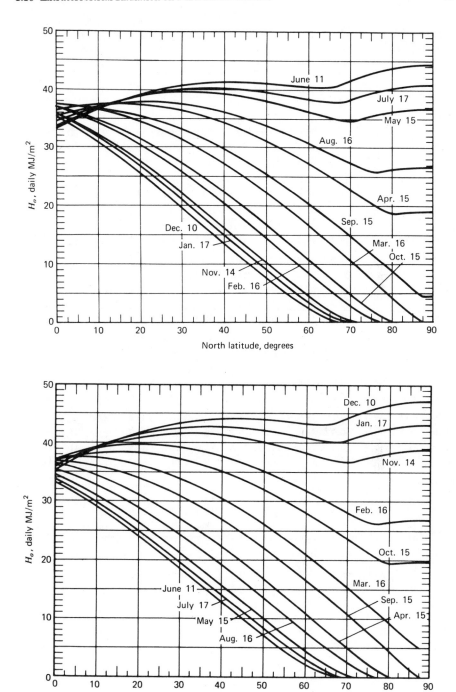

Figure 1.10.1 Extraterrestrial daily radiation on a horizontal surface. The curves are for the mean days of the month from Table 1.6.1.

Table 1.10.1   **Monthly Average Daily Extraterrestrial Radiation, MJ/m^2**

ϕ	Jan	Feb	Mar	Apr	May	Jun	Jul	Aug	Sep	Oct	Nov	Dec
90	0.0	0.0	1.2	19.3	37.2	44.8	41.2	26.5	5.4	0.0	0.0	0.0
85	0.0	0.0	2.2	19.2	37.0	44.7	41.0	26.4	6.4	0.0	0.0	0.0
80	0.0	0.0	4.7	19.6	36.6	44.2	40.5	26.1	9.0	0.6	0.0	0.0
75	0.0	0.7	7.8	21.0	35.9	43.3	39.8	26.3	11.9	2.2	0.0	0.0
70	0.1	2.7	10.9	23.1	35.3	42.1	38.7	27.5	14.8	4.9	0.3	0.0
65	1.2	5.4	13.9	25.4	35.7	41.0	38.3	29.2	17.7	7.8	2.0	0.4
60	3.5	8.3	16.9	27.6	36.6	41.0	38.8	30.9	20.5	10.8	4.5	2.3
55	6.2	11.3	19.8	29.6	37.6	41.3	39.4	32.6	23.1	13.8	7.3	4.8
50	9.1	14.4	22.5	31.5	38.5	41.5	40.0	34.1	25.5	16.7	10.3	7.7
45	12.2	17.4	25.1	33.2	39.2	41.7	40.4	35.3	27.8	19.6	13.3	10.7
40	15.3	20.3	27.4	34.6	39.7	41.7	40.6	36.4	29.8	22.4	16.4	13.7
35	18.3	23.1	29.6	35.8	40.0	41.5	40.6	37.3	31.7	25.0	19.3	16.8
30	21.3	25.7	31.5	36.8	40.0	41.1	40.4	37.8	33.2	27.4	22.2	19.9
25	24.2	28.2	33.2	37.5	39.8	40.4	40.0	38.2	34.6	29.6	25.0	22.9
20	27.0	30.5	34.7	37.9	39.3	39.5	39.3	38.2	35.6	31.6	27.7	25.8
15	29.6	32.6	35.9	38.0	38.5	38.4	38.3	38.0	36.4	33.4	30.1	28.5
10	32.0	34.4	36.8	37.9	37.5	37.0	37.1	37.5	37.0	35.0	32.4	31.1
5	34.2	36.0	37.5	37.4	36.3	35.3	35.6	36.7	37.2	36.3	34.5	33.5
0	36.2	37.4	37.8	36.7	34.8	33.5	34.0	35.7	37.2	37.3	36.3	35.7
−5	38.0	38.5	37.9	35.8	33.0	31.4	32.1	34.4	36.9	38.0	37.9	37.6
−10	39.5	39.3	37.7	34.5	31.1	29.2	29.9	32.9	36.3	38.5	39.3	39.4
−15	40.8	39.8	37.2	33.0	28.9	26.8	27.6	31.1	35.4	38.7	40.4	40.9
−20	41.8	40.0	36.4	31.3	26.6	24.2	25.2	29.1	34.3	38.6	41.2	42.1
−25	42.5	40.0	35.4	29.3	24.1	21.5	22.6	27.0	32.9	38.2	41.7	43.1
−30	43.0	39.7	34.0	27.2	21.4	18.7	19.9	24.6	31.2	37.6	42.0	43.8
−35	43.2	39.1	32.5	24.8	18.6	15.8	17.0	22.1	29.3	36.6	42.0	44.2
−40	43.1	38.2	30.6	22.3	15.8	12.9	14.2	19.4	27.2	35.5	41.7	44.5
−45	42.8	37.1	28.6	19.6	12.9	10.0	11.3	16.6	24.9	34.0	41.2	44.5
−50	42.3	35.7	26.3	16.8	10.0	7.2	8.4	13.8	22.4	32.4	40.5	44.3
−55	41.7	34.1	23.9	13.9	7.2	4.5	5.7	10.9	19.8	30.5	39.6	44.0
−60	41.0	32.4	21.2	10.9	4.5	2.2	3.1	8.0	17.0	28.4	38.7	43.7
−65	40.5	30.6	18.5	7.9	2.1	0.3	1.0	5.2	14.1	26.2	37.8	43.7
−70	40.8	28.8	15.6	5.0	0.4	0.0	0.0	2.6	11.1	24.0	37.4	44.9
−75	41.9	27.6	12.6	2.4	0.0	0.0	0.0	0.8	8.0	21.9	38.1	46.2
−80	42.7	27.4	9.7	0.6	0.0	0.0	0.0	0.0	5.0	20.6	38.8	47.1
−85	43.2	27.7	7.2	0.0	0.0	0.0	0.0	0.0	2.4	20.3	39.3	47.6
−90	43.3	27.8	6.2	0.0	0.0	0.0	0.0	0.0	1.4	20.4	39.4	47.8

Then from Equation 1.10.3, with $G_{sc} = 1367$ W/m^2,

$$H_o = \frac{24 \times 3600 \times 1367}{\pi}\left(1 + 0.033 \cos \frac{360 \times 105}{365}\right)$$

$$\times \left(\cos 43 \cos 9.4 \sin 98.9 + \frac{\pi \times 98.9}{180} \sin 43 \sin 9.4\right)$$

$$= 33.8 \text{ MJ/m}^2$$

From Figure 1.10.1a, for the curve for April, we read $H_o = 34.0$ MJ/m^2, and from Table 1.10.1 we obtain $H_o = 33.8$ MJ/m^2 by interpolation. ∎

It is also of interest to calculate the extraterrestrial radiation on a horizontal surface for an hour period. Integrating Equation 1.10.2 for a period between hour angles ω_1 and ω_2 which define an hour (where ω_2 is the larger),

$$I_o = \frac{12 \times 3600}{\pi} G_{sc}\left(1 + 0.033 \cos \frac{360 \, n}{365}\right)$$

$$\times \left[\cos \phi \cos \delta(\sin \omega_2 - \sin \omega_1) + \frac{\pi (\omega_2 - \omega_1)}{180} \sin \phi \sin \delta\right] \quad (1.10.4)$$

(The limits ω_1 and ω_2 may define a time other than an hour.)

Example 1.10.2

What is the solar radiation on a horizontal surface in the absence of the atmosphere at latitude 43°N on April 15 between the hours of 10 and 11?

Solution

The declination is 9.4° (from the previous example). For April 15, $n = 105$. Using Equation 1.10.4 with $\omega_1 = -30°$ and $\omega_2 = -15°$,

$$I_o = \frac{12 \times 3600 \times 1367}{\pi}\left(1 + 0.033 \cos \frac{360 \times 105}{365}\right)$$

$$\times \left[\cos 43 \cos 9.4 (\sin (-15) - \sin (-30)) + \frac{\pi (-15 - (-30))}{180} \sin 43 \sin 9.4\right]$$

$$= 3.79 \text{ MJ/m}^2 \qquad\qquad ∎$$

The hourly extraterrestrial radiation can also be approximated by writing Equation 1.10.2 in terms of I, evaluating ω at the midpoint of the hour. For the circumstances of Example 1.10.2, the hour's radiation so estimated is 3.80 MJ/m^2. Differences between the hourly radiation calculated by these two methods will be slightly larger at times near sunrise and sunset but are still small. For larger time spans, the differences become larger. For example, for the same circumstances as in Example 1.10.2 but for the two-hour span from 7:00 to 9:00, the use of Equation 1.10.4 gives 4.58 MJ/m², and Equation 1.10.2 for 8:00 gives 4.61 MJ/m².

1.11 SUMMARY

In this chapter we have outlined the basic characteristics of the sun and the radiation it emits, noting that the solar constant, the mean radiation flux density outside of the earth's atmosphere, is 1367 W/m² (within ±1%), with most of the radiation in a wavelength range of 0.3 to 3 μm. This radiation has directional characteristics that are defined by a set of angles that determine the angle of incidence of the radiation on a surface. We have included in this chapter those topics that are based on extraterrestrial radiation and the geometry of the earth and sun. This is background information for Chapter 2, which is concerned with effects of the atmosphere, radiation measurements, and data manipulation.

REFERENCES

Anderson, E. E., *Fundamentals of Solar Energy Conversion*, Addison-Wesley Publishing Co., Reading, MA (1982).

Benford, F. and J. E. Bock, *Trans. of the American Illumination Engineering Soc.*, **34**, 200 (1939). "A Time Analysis of Sunshine."

Braun, J. E. and J. C. Mitchell, *Solar Energy*, **31**, 439 (1983). "Solar Geometry for Fixed and Tracking Surfaces."

Cooper, P I., *Solar Energy*, **12**, 3 (1969). "The Absorption of Solar Radiation in Solar Stills."

Coulson, K. L., *Solar and Terrestrial Radiation*, Academic Press, New York (1975).

Duncan, C. H., R. C. Willson, J. M. Kendall, R. G. Harrison, and J. R. Hickey, *Solar Energy*, **28**, 385 (1982). "Latest Rocket Measurements of the Solar Constant."

Eibling, J. A., R. E. Thomas, and B. A. Landry, Report to the Office of Saline Water, U.S. Department of the Interior (1953). "An Investigation of Multiple-Effect Evaporation of Saline Waters by Steam from Solar Radiation."

Frohlich, C., in *The Solar Output and Its Variation* (O. R. White, ed.), Colorado Associated University Press, Boulder (1977). "Contemporary Measures of the Solar Constant."

Garg, H. P., *Treatise on Solar Energy*, Vol. I, Wiley-Interscience, Chichester (1982).

Hickey, J. R., B. M. Alton, F. J. Griffin, H. Jacobowitz, P. Pelligrino, R. H. Maschhoff, E. A. Smith, and T. H. Vonder Haar, *Solar Energy*, **28**, 443 (1982). "Extraterrestrial Solar Irradiance Variability: Two and One-Half Years of Measurements from Nimbus 7."

Hottel, H. C. and B. B. Woertz, *Trans. ASME*, **64**, 91 (1942). "Performance of Flat-Plate Solar Heat Collectors."

Iqbal, M., *An Introduction to Solar Radiation*, Academic Press, Toronto (1983).

Johnson, F. S., *J. of Meteorology*, **11**, 431 (1954). "The Solar Constant."

Jones, R. E., *Solar Energy*, **24**, 305 (1980). "Effects of Overhang Shading of Windows having Arbitrary Azimuth."

Klein, S. A., *Solar Energy*, **19**, 325 (1977). "Calculation of Monthly Average Insolation on Tilted Surfaces."

Kondratyev, K. Y., *Radiation in the Atmosphere*, Academic Press, New York and London (1969).

Mazria, E., *The Passive Solar Energy Book*, Rondale Press, Emmaus, PA (1979).

NASA SP-8055, National Aeronautics and Space Administration, May (1971). "Solar Electromagnetic Radiation."

Robinson, N. (ed.), *Solar Radiation*, Elsevier, Amsterdam (1966).

Sun Angle Calculator, Libby-Owens-Ford Glass Company (1951).

Spencer, J. W., *Search*, **2** (5), 172 (1971). "Fourier Series Representation of the Position of the Sun."

Thekaekara, M. P. and A. J. Drummond, *National Physical Science*, **229**, 6 (1971). "Standard Values for the Solar Constant and Its Spectral Components."

Thekaekara, M. P., *Solar Energy*, **18**, 309 (1976). "Solar Radiation Measurement: Techniques and Instrumentation."

Thomas, R. N., *Trans. of the Conference on Use of Solar Energy*, **1**, 1, University of Arizona Press (1958). "Features of the Solar Spectrum as Imposed by the Physics of the Sun."

U.S. Hydrographic Office Publications No. 214 (1940). "Tables of Computed Altitude and Azimuth."

Whillier, A., Notes on Solar Energy prepared at McGill University (1965a).

Whillier, A., *Solar Energy*, **9**, 164 (1965b). "Solar Radiation Graphs."

Whillier, A., Personal communication (1975 and 1979).

Willson, R. C., S. Gulkis, M. Janssen, H. S. Hudson, and G. A. Chapman, *Science*, **211**, 700 (1981). "Observations of Solar Irradiance Variability."

Chapter 2

AVAILABLE SOLAR RADIATION

In this chapter we describe instruments for solar radiation measurements, the solar radiation data that are available, and the calculation of needed information from the available data. It is generally not practical to base predictions or calculations of solar radiation on attenuation of the extraterrestrial radiation by the atmosphere, as adequate meteorological information is seldom available. Instead, to predict the performance of a solar process in the future, we use past measurements of solar radiation at the location in question or from a nearby similar location.

Solar radiation data are used in several forms and for a variety of purposes. The most detailed information available is beam and diffuse solar radiation on a horizontal surface, by hours, which is useful in simulations of solar processes. (A few measurements are available on inclined surfaces and for shorter time intervals.) Daily data are often available and hourly radiation can be estimated from daily data. Monthly total solar radiation on a horizontal surface can be used in some process design methods. However, as process performance is generally not linear with solar radiation, the use of averages may lead to serious errors if nonlinearities are not taken into account. It is also possible to reduce radiation data to more manageable forms by statistical methods.

2.1 DEFINITIONS

Figure 2.1.1 shows the primary radiation fluxes on a surface at or near the ground that are important in connection with solar thermal processes. It is convenient to consider radiation in two wavelength ranges.[1]

Solar, or **short-wave, radiation** is radiation originating from the sun, in the wavelength range of 0.3 to 3 μm. In the terminology used throughout this book, solar radiation includes both beam and diffuse components unless otherwise specified.

---

[1] We will see in Chapters 3, 4, and 6 that the wavelength ranges of incoming solar radiation and emitted radiation from flat-plate solar collectors overlap to a negligible extent, and for many purposes the distinction noted here is very useful. For collectors operating at high enough temperatures that there is significant overlap, more precise distinctions are needed.

46

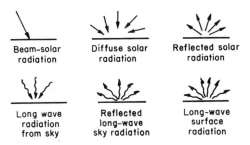

Figure 2.1.1 The radiant energy fluxes of importance in solar thermal processes. Short-wave solar radiation is shown by ⟶. Long-wave radiation is shown by ⤳.

Long-wave radiation is radiation originating from sources at temperatures near ordinary ambient temperatures and thus substantially all at wavelengths greater than 3 μm. Long-wave radiation is emitted by the atmosphere, by a collector, or by any other body at ordinary temperatures. (This radiation, if originating from the ground, is referred to in some literature as "terrestrial" radiation.)

Instruments for measuring solar radiation are of two basic types.

A **pyrheliometer** is an instrument using a collimated detector for measuring solar radiation from the sun and from a small portion of the sky around the sun (i.e., beam radiation) at normal incidence.

A **pyranometer** is an instrument for measuring total hemispherical solar (beam plus diffuse) radiation, usually on a horizontal surface. If shaded from the beam radiation by a shade ring or disc, a pyranometer measures diffuse radiation.

In addition, the terms **solarimeter** and **actinometer** are encountered; a solarimeter can generally be interpreted to mean the same as a pyranometer, and an actinometer usually refers to a pyrheliometer.

In the following sections we discuss briefly the two basic radiation instruments and the pyrheliometric scales that are used in solar radiometry. More detailed discussions of instruments, their use, and the associated terminology are found in Robinson (1966), WMO (1969), Kondratyev (1969), Coulson (1975), Thekaekara (1976), Yellott (1977), and Iqbal (1983). Stewart et al. (1985) review characteristics of pyranometers and pyrheliometers.

2.2 PYRHELIOMETERS AND PYRHELIOMETRIC SCALES

Standard and secondary standard solar radiation instruments are pyrheliometers. The water flow pyrheliometer, designed by Abbot in 1905, was an early standard instrument. This instrument uses a cylindrical blackbody cavity to absorb radiation which is admitted through a collimating tube. Water flows around and over the absorbing cavity, and measurement of its temperature and flow rate provide the means for determining the absorbed energy. The design was modified by Abbot in 1932 to include the use of two thermally identical chambers, dividing the cooling water between them, and heating one chamber electrically while the other is heated by solar radiation; when the instrument is adjusted so as to make the heat produced in the two chambers identical, the electrical power input is a measure of the solar energy absorbed.

Standard pyrheliometers are not easy to use, and secondary standard instruments have been devised that are calibrated against the standard instruments. The secondary standards in turn are used to calibrate field instruments. Robinson (1966) and Coulson (1975) provide detailed discussion and bibliography on this topic. Two of these secondary standard instruments are of importance.

The Abbot silver disc pyrheliometer, first built by Abbot in 1902 and modified in 1909 and 1927, uses a silver disc 38 mm in diameter and 7 mm thick as the radiation receiver. The side exposed to radiation is blackened, and the bulb of a precision mercury thermometer is inserted in a hole in the side of the disc and is in good thermal contact with the disc. The silver disc is suspended on wires at the end of a collimating tube, which in later models has dimensions such that 0.0013 of the hemisphere is "seen" by the detector. Thus any point on the detector sees an aperture angle of 5.7°. The disc is mounted in a copper cylinder, which in turn is in a cylindrical wood box that insulates the copper and the disc from the surroundings. A shutter alternately admits radiation and shades the detector at regular intervals; the corresponding changes in disc temperature are measured and provide the means to calculate the absorbed radiation. A section drawing of the pyrheliometer is shown is Figure 2.2.1.

The other secondary standard of particular importance is the Ångström compensation pyrheliometer, first constructed by K. Ångström in 1893 and modified

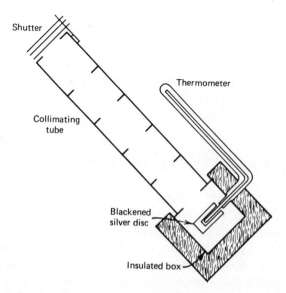

Figure 2.2.1 Schematic section of the Abbot Silver Disc pyrheliometer.

in several developments since then. In this instrument two identical blackened manganin strips are arranged so that either one can be exposed to radiation at the base of collimating tubes by moving a reversible shutter. Each strip can be electrically heated, and each is fitted with a thermocouple. With one strip shaded and one strip exposed to radiation, a current is passed through the shaded strip to heat it to the same temperature as the exposed strip. When there is no difference in temperature, the electrical energy to the shaded strip must equal the solar radiation absorbed by the exposed strip. Solar radiation is determined by equating the electrical energy to the product of incident solar radiation, strip area, and absorptance. After a determination is made, the position of the shutter is reversed to interchange the electrical and radiation heating, and a second determination is made. Alternating the shade and the functions of the two strips compensates for minor differences in the strips such as edge effects and lack of uniformity of electrical heating.

The Ångström instrument serves, in principle, as an absolute or primary standard. However, there are difficulties in applying correction factors in its use, and in practice there are several primary standard Ångström instruments to which those in use as secondary standards are compared.

The Abbot and Ångström instruments are used as secondary standards for calibration of other instruments, and there is a **pyrheliometric scale** associated with each of them. The first scale, based on measurements with the Ångström instrument, was established in 1905 (the Ångström scale of 1905, or AS05). The second, based on the Abbot silver disc pyrheliometer (which was in turn calibrated with a standard water flow pyrheliometer) was established in 1913 (the Smithsonian scale of 1913, or SS13).

Reviews of the accuracy of these instruments and intercomparisons of them led to the conclusions that measurements made on SS13 were 3.5% higher than those on AS05, that SS13 was 2% too high, and that AS05 was 1.5% too low. As a result, the **International Pyrheliometric Scale 1956** (IPS56) was adopted, reflecting these differences. Measurements made before 1956 on the scale AS05 were increased by 1.5%, and those of SS13 were decreased by 2% to correct them to IPS56.

Beginning with the 1956 International Pyrheliometer Comparisons (IPC), which resulted in IPS56, new comparisons have been made at approximately five-year intervals, under World Meteorological Organization auspices, at Davos, Switzerland. As a result of the 1975 comparisons, a new pyrheliometric scale, the World Radiometric Reference (WRR) (also referred to as the Solar Constant Reference Scale, SCRS) was established; it is 2.2% higher than the IPS56 scale. (SS13 is very close to WRR.)

Operational or field instruments are calibrated against secondary standards and are the source of most of the data on which solar process engineering designs must be based. Brief descriptions of two of these, the Eppley Normal Incidence Pyrheliometer (NIP) and the Kipp & Zonen actinometer, are included here. The Eppley NIP is the instrument in most common use in the United States for measuring beam solar radiation, and the Kipp & Zonen instrument is in wide use in Europe. A

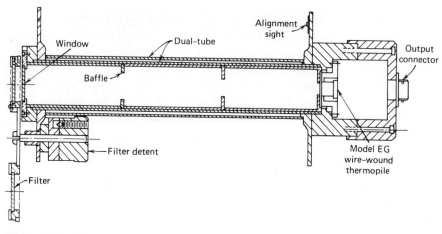

Figure 2.2.2 Cross section of the Eppley Normal Incidence pyrheliometer. Courtesy The Eppley Laboratory.

cross-section of a recent model of the Eppley is shown in Figure 2.2.2. The instrument mounted on a tracking mechanism is shown in Figure 2.2.3. The detector is at the end of the collimating tube, which contains several diaphragms and which is blackened on the inside. The detector is a multijunction thermopile coated with Parsons optical black. Temperature compensation to minimize sensitivity to variations in ambient temperature is provided. The aperture angle of the instrument is 5.7°, so the detector receives radiation from the sun and from an area of the circumsolar sky two orders of magnitude larger than that of the sun.

The Kipp & Zonen actinometer is based on the Linke-Feussner design and uses a 40-junction constantan-manganin thermopile with hot junctions heated by radiation and cold junctions in good thermal contact with the case. In this instrument the assembly of copper diaphragms and case has very large thermal capacity, orders of magnitude more than the hot junctions. On exposure to solar radiation the hot junctions rise quickly to temperatures above the cold junction; the difference in the temperatures provides a measure of the radiation. Other pyrheliometers were designed by Moll-Gorczynski, Yanishevskiy, and Michelson.

The dimensions of the collimating systems are such that the detectors are exposed to radiation from the sun and from a portion of the sky around the sun. Since the detectors do not distinguish between forward-scattered radiation, which comes from the circumsolar sky, and beam radiation, the instruments are, in effect, defining beam radiation. An experimental study by Jeys and Vant-Hull (1976), which utilized several lengths of collimating tubes so that the aperture angles were reduced in step from 5.72° to 2.02°, indicated that for cloudless conditions this reduction in aperture angle resulted in insignificant changes in the measurements of beam radiation. On a day of thin uniform cloud cover, however, with solar altitude angle of less than 32°, as much as 11% of the measured intensity was received from the circumsolar sky between aperture angles of 5.72° and 2.02°. It is difficult to

Figure 2.2.3   An Eppley Normal Incidence pyrheliometer (NIP) on an altazimuth tracking mount. Courtesy The Eppley Laboratory.

generalize from the few data available, but it appears that thin clouds or haze can affect the angular distribution of radiation within the field of view of standard pyrheliometers. The World Meteorological Organization (WMO) recommends that calibration of pyrheliometers only be undertaken on days in which atmospheric clarity meets or exceeds a minimum value.

2.3   PYRANOMETERS

Instruments for measuring total (beam plus diffuse) radiation are referred to as pyranometers, and it is from these instruments that most of the available data on solar radiation are obtained. The detectors for these instruments must have a response independent of wavelength of radiation over the solar energy spectrum. In addition, they should have a response independent of the angle of incidence of the solar radiation. The detectors of most pyranometers are covered with one or two hemispherical glass covers to protect them from wind and other extraneous effects; the covers must be very uniform in thickness so as not to cause uneven distribution of radiation on the detectors. These factors are discussed in more detail by Coulson (1975).

Commonly used pyranometers in the United States are the Eppley and Spectrolab instruments, in Europe the Moll-Gorczynski, in the USSR the Yanishevskiy, and in Australia the Trickett-Norris (Groiss) pyranometer.

The Eppley 180° pyranometer was the most common instrument in the United States. It used a detector consisting of two concentric silver rings; the outer ring was coated with magnesium oxide, which has a high reflectance for radiation in the solar energy spectrum, and the inner ring was coated with Parson's black, which has a very high absorptance for solar radiation. The temperature difference between these rings was detected by a thermopile and was a measure of absorbed solar radiation. The circular symmetry of the detector minimized the effects of the surface azimuth angle on instrument response. The detector assembly was placed in a nearly spherical glass bulb, which has a transmittance greater than 0.90 over most of the solar radiation spectrum, and the instrument response was nearly independent of wavelength except at the extremes of the spectrum. The response of this Eppley was dependent on ambient temperature, with sensitivity decreasing by 0.05 to 0.15%/C [Coulson (1975)]; much of the published data taken with these instruments was not corrected for temperature variations. It is possible to add temperature compensation to the external circuit and remove this source of error. It is estimated that carefully used Eppleys of this type could produce data with less than 5% errors but that errors of twice this could be expected from poorly maintained instruments. The theory of this instrument has been carefully studied by MacDonald (1951).

The Eppley 180° pyranometer is no longer manufactured and has been replaced by other instruments. The Eppley Black and White pyranometer utilizes Parson's-black- and barium-sulfate-coated hot and cold thermopile junctions and has better angular (cosine) response. It uses an optically ground glass envelope and temperature compensation to maintain calibration within ±1.5% over a temperature range of −20 to +40 C. It is shown in Figure 2.3.1.

The Eppley Precision Spectral Pyranometer (PSP) utilizes a thermopile detector, two concentric hemispherical optically ground covers, and temperature compensation

Figure 2.3.1   The Eppley Black and White pyranometer. Courtesy The Eppley Laboratory.

that results in temperature dependence of 0.5% from –20 to +40 C. (Measurements of irradiance in spectral bands can be made by use of bandpass filters; the PSP can be fitted with hemispherical domes of filter glass for this purpose. See Stewart et al. (1985) for information and references.) It is shown in Figure 2.3.2.

The Moll-Gorczynski pyranometer uses a Moll thermopile to measure the temperature difference of the black detector surface and the housing of the instrument. The thermopile assembly is covered with two concentric glass hemispherical domes to protect it from weather and is rectangular in configuration with the thermocouples aligned in a row (which results in some sensitivity to the azimuth angle of the radiation).

Pyranometers are usually calibrated against standard pyrheliometers. A standard method has been set forth in the Annals of the International Geophysical Year [IGY, (1958)], which requires that readings be taken at times of clear skies, with the pyranometer shaded and unshaded at the same time as readings are taken with the pyrheliometer. Shading is recommended to be accomplished by means of a disc held 1 m from the pyranometer with the disc just large enough to shade the glass envelope. The calibration constant is then the ratio of the difference in the output of the shaded and unshaded pyranometer to the output of the pyrheliometer multiplied by the calibration constant of the pyrheliometer and cos θ_z, the angle of incidence of beam radiation on the horizontal pyranometer. Care and precision are required in these calibrations.

It is also possible, as described by Norris (1973), to calibrate pyranometers against a secondary standard pyranometer such as the Eppley precision pyranometer. This secondary standard pyranometer is thought to be good to ±1% when calibrated against a standard pyrheliometer. Direct comparison of the precision Eppley and field instruments can be made to determine the calibration constant of the field instruments.

Figure 2.3.2 The Eppley Precision Spectral pyranometer. Courtesy The Eppley Laboratory.

A pyranometer (or pyrheliometer) produces a voltage from the thermopile detectors that is a function of the incident radiation. It is necessary to use a potentiometer to detect and record this output. Radiation data usually must be integrated over some period of time, such as an hour or a day. Integration can be done by means of planimetry or by electronic integrators. It has been estimated that with careful use and reasonably frequent pyranometer calibration, radiation measurements should be good within ±5%; integration errors would increase this number. Much of the available radiation data prior to 1975 is probably not this good, largely because of infrequent calibration and in some instances because of inadequate integration procedures.

Another class of pyranometers, originally designed by Robitzsch, utilizes detectors that are bimetallic elements heated by solar radiation; mechanical motion of the element is transferred by a linkage to an indicator or recorder pen. These instruments have the advantage of being entirely spring driven and thus require no electrical energy. Variations of the basic design are manufactured by several European firms (Fuess, Casella, and SIAP). They are widely used in isolated stations and are a major source of the solar radiation data that are available for locations outside of Europe, Australia, Japan, and North America. Data from these instruments are generally not as accurate as that from thermopile-type pyranometers.

Another type of pyranometer is based on photovoltaic (solar cell) detectors. Examples are the LI-COR LI-200SA pyranometer and the Yellott Solarimeter. They are less precise instruments than the thermopile instruments and have some limitations on their use. They are also less expensive than thermopile instruments and are easy to use.

The main disadvantage of photovoltaic detectors is their spectrally selective response. Figure 2.3.3 shows a typical terrestrial solar spectrum and the spectral response of a silicon solar cell. If the spectral distribution of incident radiation was

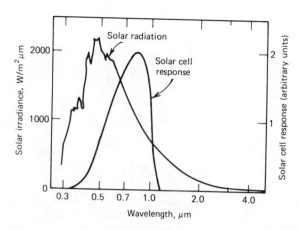

Figure 2.3.3 Spectral distribution of extraterrestrial solar radiation and spectral response of a silicon solar cell. From Coulson, *Solar and Terrestrial Radiation*, Academic Press, New York (1975).

fixed, a calibration could be established that would remain constant; however, there are some variations in spectral distribution[2] with clouds and atmospheric water vapor. LI-COR estimates that the error introduced because of spectral response is ±5% maximum under most conditions of natural daylight and is ±3% under typical conditions.

Photovoltaic detectors have additional characteristics of interest. Their response to changing radiation levels is essentially instantaneous and is linear with radiation. The temperature dependence is ±0.15%/C maximum. The LI-COR instrument is fitted with an acrylic diffuser which substantially removes the dependence of response on the angle of incidence of the radiation. The response of the detectors is independent of its orientation, but reflected radiation from the ground or other surroundings will in general have a different spectral distribution than global horizontal radiation, and measurements on surfaces receiving significant amounts of reflected radiation will be subject to additional errors.

The preceding discussion dealt entirely with measurements of total radiation on a horizontal surface. Two additional kinds of measurements are made with pyranometers: measurements of diffuse radiation on horizontal surfaces and measurement of solar radiation on inclined surfaces.

Measurements of diffuse radiation can be made with pyranometers by shading the instrument from beam radiation. This is usually done by means of a shading ring, as shown in Figure 2.3.4. The ring is used to allow continuous recording of diffuse

Figure 2.3.4 Pyranometer with shading ring to eliminate beam radiation. Courtesy The Eppley Laboratory.

---

[2] This will be discussed in Section 2.6.

radiation without the necessity of continuous positioning of smaller shading devices; adjustments need to be made for changing declination only and can be made every few days. The ring shades the pyranometer from part of the diffuse radiation, and a correction for this shading must be estimated and applied to the observed diffuse radiation [see Drummond (1956, 1964), IGY (1958), or Coulson (1975)]. The corrections are based on assumptions of the distribution of diffuse radiation over the sky and typically are factors of 1.05 to 1.2. An example of shade ring correction factors, to illustrate their trends and magnitudes, is shown in Figure 2.3.5.

Measurements of solar radiation on inclined planes are important in determining the input to solar collectors. There is evidence that the calibration of pyranometers changes if the instrument is inclined to the horizontal. The reason for this appears to be changes in the convection patterns inside the glass dome, which changes the manner in which heat is transferred from the hot junctions of the thermopiles to the cover and other parts of the instrument. The Eppley 180° pyranometer has been variously reported to show a decrease in sensitivity on inversion from 5.5% to no decrease. Norris (1974) measured the response at various inclinations of four pyranometers when subject to radiation from an incandescent lamp source and found correction factors at inclinations of 90° in the range of 1.04 to 1.10. Stewart et al. (1985) plot two sets of data of Latimer (1980), which show smaller correction factors. Figure 2.3.6 shows the set with the greater factors, with the Eppley PSP showing maximum positive effects at $\beta = 90°$ of 2.5% and smaller corrections for Kipp & Zonen instruments. There are thus disagreements of the magnitude of the

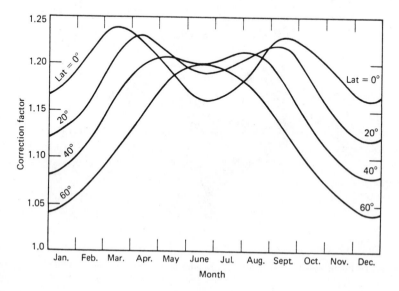

Figure 2.3.5 Typical shade ring correction factors to account for shading of the detector from diffuse radiation. Adapted from Coulson, *Solar and Terrestrial Radiation*, Academic Press, New York (1975).

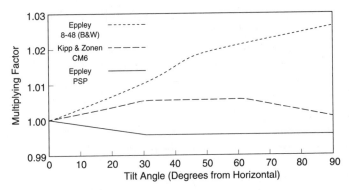

Figure 2.3.6   Effects of inclination of pyranometers on calibration. The instruments are the Eppley PSP, the Eppley 8-48, and the Kipp & Zonen CM6. Adapted from Stewart et al. (1985).

corrections, but for the instruments shown, the corrections are of the order of 1 or 2%.

It is evident from these data and other published results that the calibration of pyranometers is to some degree dependent on inclination and that experimental information is needed on a particular pyranometer in any orientation to adequately interpret information from it.

The Bellani spherical distillation pyranometer is based on a different principle. It uses a spherical container of alcohol which absorbs solar radiation. The sphere is connected to a calibrated condenser receiver tube. The quantity of alcohol condensed is a measure of integrated solar energy on the spherical receiver. Data on the total energy received by a body, as represented by the sphere, are of interest in some biological processes.

2.4   MEASUREMENT OF DURATION OF SUNSHINE

The hours of bright sunshine, that is, the time in which the solar disc is visible, is of some use in estimating long-term averages of solar radiation.[3] Two instruments have been or are widely used. The Campbell-Stokes sunshine recorder uses a solid glass sphere of approximately 10 cm diameter as a lens that produces an image of the sun on the opposite surface of the sphere. A strip of standard treated paper is mounted around the appropriate part of the sphere, and the solar image burns a mark on the paper whenever the beam radiation is above a critical level. The lengths of the burned portions of the paper provide an index of the duration of "bright sunshine." These measurements are uncertain on several counts: the interpretation of what constitutes a burned portion is uncertain, the instrument does not respond to low

---

[3] The relationship between sunshine hours and solar radiation will be discussed in Section 2.7.

levels of radiation early and late in the day, and the condition of the paper may be dependent on humidity.

A photoelectric sunshine recorder, the Foster Sunshine Switch [Foster and Foskett (1953)], is now in use by the United States Weather Service. It incorporates two selenium photovoltaic cells, one of which is shaded from beam radiation and one exposed to it. In the absence of beam radiation, the two detectors indicate (nearly) the same radiation level. When beam radiation is incident on the unshaded cell, the output of that cell is higher than that of the shaded cell. The duration of a critical radiation difference detected by the two cells is a measure of the duration of bright sunshine.

2.5   SOLAR RADIATION DATA

Solar radiation data are available in several forms. The following information about radiation data is important in its understanding and use: whether they are instantaneous measurements (irradiance) or values integrated over some period of time (irradiation) (usually hour or day); the time or time period of the measurements; whether the measurements are of beam, diffuse, or total radiation; the instruments used; the receiving surface orientation (usually horizontal, sometimes inclined at a fixed slope, or normal to the beam radiation); and, if averaged, the period over which they are averaged (e.g., monthly averages of daily radiation).

Most radiation data available are for horizontal surfaces, include both direct and diffuse radiation, and were measured with thermopile pyranometers (or in some cases Robitzsch-type instruments). Most of these instruments provide radiation records as a function of time and do not themselves provide a means of integrating the records. The data were usually recorded in a form similar to that shown in Figure 2.5.1 by recording potentiometers and were integrated graphically. Uncertainties in integration add to uncertainties in pyranometer response; electronic integration is now common.

Two types of solar radiation data are widely available. The first is monthly average daily total radiation on a horizontal surface $\overline{H}$. The second is hourly total radiation on a horizontal surface I for each hour for extended periods such as one or more years. The $\overline{H}$ data are widely available and are given for many stations in Appendix G. The traditional units have been calories per square centimeter; the data in Appendix G are in the more useful megajoules per square meter. These data are available from weather services [e.g., Cinquimani et al. (1978)] and the literature [e.g., from the CEC *European Solar Radiation Atlas* (1984) and Löf et al. (1966a,b)]. The World Meteorological Organization sponsors compilation of solar radiation data at the World Radiation Data Center; these are published in the *Solar Radiation and Radiation Balance Data (The World Network)*.

The accuracy of some of the earlier (pre-1970) data is generally less than desirable, as standards of calibration and care in use of instruments and integration

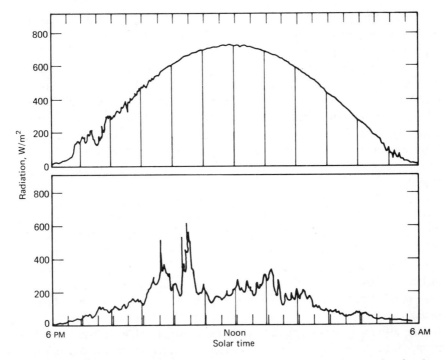

Figure 2.5.1 Total (beam and diffuse) solar radiation on a horizontal surface vs. time for clear and largely cloudy day, latitude 43°, for days near the equinox.

have not always been adequate.[4] Recent laboratory measurements and the averages based thereon are probably good to ±5%. Most of the older average data are probably no better than ±10%, and for some stations a better estimate may be ±20%. Substantial inconsistencies are found in data from different sources for some locations.

A very extensive and carefully compiled monthly average daily solar radiation data base is available for Europe and part of the Mediterranean basin. The *European Solar Radiation Atlas*, Volume I (2nd Edition, 1984), is based on pyranometric data from 139 stations in 29 countries. It includes solar radiation derived from sunshine hours data for 315 stations (with 114 of the stations reporting both) for a total of 340 stations. Ten years of data were used for each station except for a few where data for shorter periods were available. The data and the instruments used to obtain them were carefully evaluated, corrections were made to compensate for instrumental errors, and

---

[4] The SOLMET (1978) program of the United States Weather Service has addressed this problem by careful study of the history of individual instruments and their calibrations and subsequent "rehabilitation" of the data to correct for identifiable errors. The United States data in Appendix G has been processed in this way.

all data are based on the WRR pyrheliometric scale. The Atlas includes[5] tables that show averages, maxima, minima, extraterrestrial radiation, and sunshine hours. Appendix G includes some data from the Atlas.

Average daily solar radiation data are also available from maps that indicate general trends. For example, a world map is shown in Figure 2.5.2 [Löf et al. (1966a,b)].[6] In some geographical areas where climate does not change abruptly with distance (i.e., away from major influences such as mountains or large industrial cities), maps can be used as a source of average radiation if data are not available. However, large-scale maps must be used with care because they do not show local physical or climatological conditions that may greatly affect local solar energy availability.

For calculating the dynamic behavior of solar energy equipment and processes and for simulations of long-term process operation, more detailed solar radiation data (and related meteorological information) are needed. An example of this type of data (hourly integrated radiation, ambient temperature, and wind speed) is shown in Table 2.5.1 for a January week in Boulder, Colorado. Additional information may also be included in these records, such as wet bulb temperature and wind direction.

In the United States there has been a network of stations recording solar radiation on a horizontal surface and reporting it as daily values. Some of these stations also reported hourly radiation. In the 1970s, the U.S. National Oceanic and Atmospheric Administration (NOAA) undertook a program to upgrade the number and quality of the radiation measuring stations, to rehabilitate past data (to account for sensor deterioration, calibration errors, and changes in pyrheliometric scales), and to make these data available (with related meteorological data) on magnetic tapes. In 1978, corrected data tapes of hourly meteorological information (including solar radiation on a horizontal surface based on the Solar Constant Reference Scale) for 26 stations over a period of 23 years became available. These tapes are referred to as the SOLMET tapes and are described in detail in the SOLMET Manual (1978). The monthly average data for the United States stations shown in Appendix G are derived from the SOLMET data.

In the late 1970s, the U.S. federal government funded the development and operation of a national solar radiation network (SOLRAD). Measurements of hourly total horizontal and direct normal radiation were made at the 38 stations that were part of the network. Eleven of the stations also measured diffuse radiation. Data for 1977 to 1980 were checked for quality and are available from the National Climatic Data Center. Funding for much of the program was reduced in 1981, and by 1985 the network was shut down. Since then, some additional funding has become available to upgrade the instrumentation at many of the stations to automate data acquisition and recalibrate pyranometers.

---

[5] Monthly average daily radiation on surfaces other than horizontal are in Volume II of the Atlas.

[6] Figure 2.5.2 is reproduced from deJong (1973), who redrew maps originally published by Löf et al. (1966a). deJong has compiled maps and radiation data from many sources.

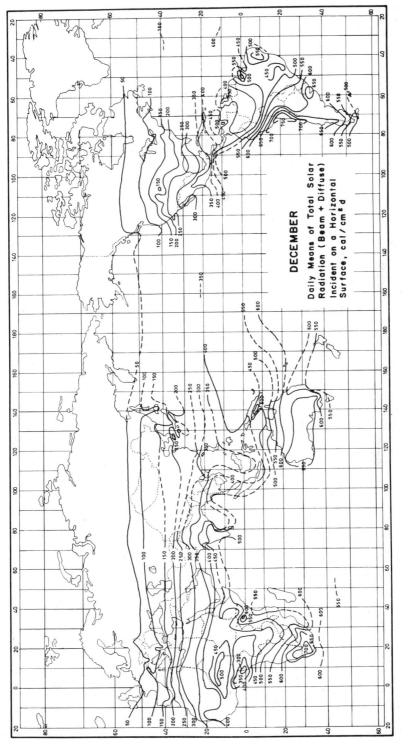

Figure 2.5.2 Average daily radiation on horizontal surfaces for December. Data are in cal/cm², the traditional units. Adapted from deJong (1973) and Löf et al. (1966a).

61

Table 2.5.1 Hourly Radiation for the Hour Ending at the Indicated Time, Air Temperature, and Wind Speed Data for a January Week, Boulder, Colorado.

Day	Hour	I kJ/m^2	T_a C	V m/s	Day	Hour	I kJ/m^2	T_a C	V m/s
8	1	0	−1.7	3.1	8	13	1105	2.8	8.0
8	2	0	−3.3	3.1	8	14	1252	3.8	9.8
8	3	0	−2.8	3.1	8	15	641	3.3	9.8
8	4	0	−2.2	3.1	8	16	167	2.2	7.2
8	5	0	−2.8	4.0	8	17	46	0.6	7.6
8	6	0	−2.8	3.6	8	18	0	−0.6	7.2
8	7	0	−2.2	3.6	8	19	0	−1.1	8.0
8	8	17	−2.2	4.0	8	20	0	−1.7	5.8
8	9	134	−1.1	1.8	8	21	0	−1.7	5.8
8	10	331	1.1	3.6	8	22	0	−2.2	7.2
8	11	636	2.2	1.3	8	23	0	−2.2	6.3
8	12	758	2.8	2.2	8	24	0	−2.2	5.8
9	1	0	−2.8	7.2	9	13	1185	−2.2	2.2
9	2	0	−3.3	7.2	9	14	1009	−1.3	1.7
9	3	0	−3.3	6.3	9	15	796	−0.6	1.3
9	4	0	−3.3	5.8	9	16	389	−0.6	1.3
9	5	0	−3.9	4.0	9	17	134	−2.2	4.0
9	6	0	−3.9	4.5	9	18	0	−2.8	4.0
9	7	0	−3.9	1.8	9	19	0	−3.3	4.5
9	8	4	−3.9	2.2	9	20	0	−5.6	5.8
9	9	71	−3.9	2.2	9	21	0	−6.7	5.4
9	10	155	−3.3	4.0	9	22	0	−7.8	5.8
9	11	343	−2.8	4.0	9	23	0	−8.3	4.5
9	12	402	−2.2	4.0	9	24	0	−8.3	6.3
10	1	0	−9.4	5.8	10	13	1872	2.2	7.6
10	2	0	−10.0	6.3	10	14	1733	4.4	6.7
10	3	0	−8.9	5.8	10	15	1352	6.1	6.3
10	4	0	−10.6	6.3	10	16	775	6.7	4.0
10	5	0	−8.3	4.9	10	17	205	6.1	2.2
10	6	0	−8.3	7.2	10	18	4	3.3	4.5
10	7	0	−10.0	5.8	10	19	0	0.6	4.0
10	8	33	−8.9	5.8	10	20	0	0.6	3.1
10	9	419	−7.2	6.7	10	21	0	0.0	2.7
10	10	1047	−5.0	9.4	10	22	0	0.6	2.2
10	11	1570	−2.2	8.5	10	23	0	1.7	3.6
10	12	1805	−1.1	8.0	10	24	0	0.6	2.7
11	1	0	−1.7	8.9	11	13	138	−5.0	6.7
11	2	0	−2.2	4.9	11	14	96	−3.9	6.7
11	3	0	−2.2	4.5	11	15	84	−4.4	7.6
11	4	0	−2.8	5.8	11	16	42	−3.9	6.3
11	5	0	−4.4	5.4	11	17	4	−5.0	6.3
11	6	0	−5.0	4.5	11	18	0	−5.6	4.5

Day	Hour	I kJ/m^2	T_a C	V m/s	Day	Hour	I kJ/m^2	T_a C	V m/s
11	7	0	−5.6	3.6	11	19	0	−6.7	4.5
11	8	4	−6.1	5.8	11	20	0	−7.8	3.1
11	9	42	−5.6	5.4	11	21	0	−9.4	2.7
11	10	92	−5.6	5.4	11	22	0	−8.9	3.6
11	11	138	−5.6	9.4	11	23	0	−9.4	4.0
11	12	163	−5.6	8.0	11	24	0	−11.1	3.1
12	1	0	−11.7	4.0	12	13	389	−2.2	5.8
12	2	0	−12.8	3.1	12	14	477	−0.6	4.0
12	3	0	−15.6	7.2	12	15	532	2.8	2.2
12	4	0	−16.7	6.7	12	16	461	−0.6	2.2
12	5	0	−16.7	6.3	12	17	33	−1.7	3.1
12	6	0	−16.1	6.3	12	18	0	−4.4	1.3
12	7	0	−17.2	3.6	12	19	0	−7.8	2.7
12	8	17	−17.8	2.7	12	20	0	−7.8	4.0
12	9	71	−13.3	8.0	12	21	0	−8.9	4.9
12	10	180	−11.1	8.9	12	22	0	−10.6	4.9
12	11	247	−7.8	8.5	12	23	0	−12.8	4.9
12	12	331	−5.6	7.6	12	24	0	−11.7	5.4
13	1	0	−10.6	4.0	13	13	1926	5.6	5.4
13	2	0	−10.6	5.4	13	14	1750	7.2	4.5
13	3	0	−10.0	4.5	13	15	1340	8.3	4.9
13	4	0	−11.1	3.1	13	16	703	8.9	4.5
13	5	0	−10.6	3.6	13	17	59	6.7	5.4
13	6	0	−9.4	3.1	13	18	0	4.4	3.6
13	7	0	−7.2	3.6	13	19	0	1.1	3.6
13	8	17	−10.6	4.0	13	20	0	0.0	3.1
13	9	314	−8.3	5.8	13	21	0	−2.2	6.7
13	10	724	−1.7	6.7	13	22	0	2.8	7.2
13	11	1809	1.7	5.4	13	23	0	1.7	8.0
13	12	2299	3.3	6.3	13	24	0	1.7	5.8
14	1	0	−0.6	7.2	14	13	1968	6.7	1.8
14	2	0	−1.1	7.6	14	14	1733	6.7	2.7
14	3	0	−0.6	6.3	14	15	1331	7.2	3.1
14	4	0	−3.9	2.7	14	16	837	6.7	3.1
14	5	0	−1.7	4.9	14	17	96	7.2	2.7
14	6	0	−2.8	5.8	14	18	4	3.3	2.7
14	7	0	−2.8	4.0	14	19	0	0.0	3.6
14	8	38	−5.0	3.1	14	20	0	3.9	5.4
14	9	452	−5.0	4.9	14	21	0	−3.9	3.6
14	10	1110	−1.7	4.5	14	22	0	−3.9	5.8
14	11	1608	2.8	3.1	14	23	0	−6.1	5.4
14	12	1884	3.8	3.6	14	24	0	−6.7	6.3

2.6   ATMOSPHERIC ATTENUATION OF SOLAR RADIATION

Solar radiation at normal incidence received at the surface of the earth is subject to variations due to change in the extraterrestrial radiation as noted in Chapter 1 and to two additional and more significant phenomena, (1) atmospheric scattering by air molecules, water, and dust and (2) atmospheric absorption by O_3, H_2O, and CO_2. Iqbal (1983) reviews these matters in considerable detail.

Scattering of radiation as it passes through the atmosphere is caused by interaction of the radiation with air molecules, water (vapor and droplets), and dust. The degree to which scattering occurs is a function of the number of particles through which the radiation must pass and of the size of the particles relative to λ, the wavelength of the radiation. The path length of the radiation through air molecules is described by the air mass. The particles of water and dust encountered by the radiation depends on air mass and on the time- and location-dependent quantities of dust and moisture present in the atmosphere.

Air molecules are very small relative to the wavelength of the solar radiation, and scattering occurs in accordance with the theory of Rayleigh (i.e., the scattering coefficient varies with λ^{-4}). Rayleigh scattering is significant only at short wavelengths; above $\lambda = 0.6$ μm it has little effect on atmospheric transmittance.

Dust and water in the atmosphere tend to be in larger particle sizes due to aggregation of water molecules and condensation of water on dust particles of various sizes. These effects are more difficult to treat than the effects of Rayleigh scattering by air molecules, as the nature and extent of dust and moisture particles in the atmosphere are highly variable with location and time. Two approaches have been used to treat this problem. Moon (1940) developed a transmission coefficient for precipitable water [the amount of water (vapor plus liquid) in the air column above the observer] that is a function of λ^{-2} and a coefficient for dust that is a function of $\lambda^{-0.75}$. Thus these transmittances are less sensitive to wavelength than is the Rayleigh scattering. The overall transmittance due to scattering is the product of three transmittances, which are three different functions of λ.

The second approach to estimation of effects of scattering by dust and water is by use of Ångström's turbidity equation. An equation for atmospheric transmittance due to aerosols, based on this equation, can be written as

$$\tau_{a,\lambda} = \exp\left(-\beta\lambda^{-\alpha}m\right) \tag{2.6.1}$$

where β is the Ångström turbidity coefficient, α is a single lumped wavelength exponent, and λ is the wavelength in micrometers. Thus there are two parameters, β and α, that describe the atmospheric turbidity and its wavelength dependence; β varies from 0 to 0.4 for very clean to very turbid atmospheres, α depends on the size distribution of the aerosols (a value of 1.3 is commonly used), and β and α will vary with time as atmospheric conditions change.

More detailed discussions of scattering are provided by Fritz (1958) who included effects of clouds, by Thekaekara (1974) in a review, and by Iqbal (1983).

Absorption of radiation in the atmosphere in the solar energy spectrum is due largely to ozone in the ultraviolet and to water vapor and carbon dioxide in bands in the infrared. There is almost complete absorption of short-wave radiation by ozone in the upper atmosphere at wavelengths below 0.29 μm. Ozone absorption decreases as λ increases above 0.29 μm, until at 0.35 μm there is no absorption. There is also a weak ozone absorption band near $\lambda = 0.6$ μm.

Water vapor absorbs strongly in bands in the infrared part of the solar spectrum, with strong absorption bands centered at 1.0, 1.4, and 1.8 μm. Beyond 2.5 μm, the transmission of the atmosphere is very low due to absorption by H_2O and CO_2. The energy in the extraterrestrial spectrum at $\lambda > 2.5$ μm is less than 5% of the total solar spectrum, and energy received at the ground at $\lambda > 2.5$ μm is very small.

The effects of Rayleigh scattering by air molecules and absorption by O_3, H_2O, and CO_2 on the spectral distribution of beam irradiance are shown in Figure 2.6.1 for an atmosphere with $\beta = 0$ and 2 cm of precipitable water, w. The WRC extra-terrestrial distribution is shown as a reference. The Rayleigh scattering is represented by the difference between the extraterrestrial curve and the curve at the top of the shaded areas; its effect becomes small at wavelengths greater than about 0.6 μm. The several absorption bands are shown by the shaded areas.

The effect of air mass is illustrated in Figure 2.6.2, which shows the spectral distribution of beam irradiance for air masses of 0 (the extraterrestrial curve), 1, 2, and 5 for an atmosphere of low turbidity.

The spectral distribution of total radiation depends also on the spectral distribution of the diffuse radiation. Some measurements are available in the ultraviolet and visible portions of the spectrum [see Robinson (1966) and Kondratyev (1969)], which have led to the conclusion that in the wavelength range 0.35 to 0.80

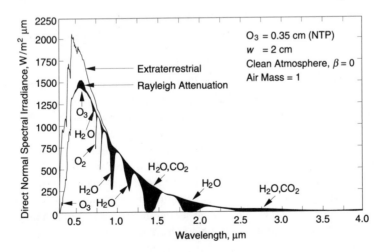

Figure 2.6.1 An example of the effects of Rayleigh scattering and atmospheric absorption on the spectral distribution of beam irradiance. Adapted from Iqbal (1983).

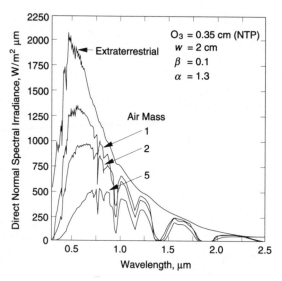

Figure 2.6.2 An example of the spectral distribution of beam irradiance for air masses of 0, 1, 2, and 5. Adapted from Iqbal (1983).

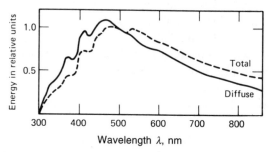

Figure 2.6.3 Relative energy distribution of total and diffuse radiation for a clear sky. From Kondratyev (1965).

μm the distribution of the diffuse radiation is similar to that of the total beam radiation.[7] Figure 2.6.3 shows relative data on spectral distribution of total and diffuse radiation for a clear sky. The diffuse component has a distribution similar to the total but shifted toward the short-wave end of the spectrum; this is consistent with scattering theory which indicates more scatter at shorter wavelengths. Fritz (1958) suggests that the spectrum of an overcast sky is similar to that for a clear sky. Iqbal uses calculated spectral distributions like that of Figure 2.6.4 to show that for typical atmospheric conditions most of the radiation at wavelengths longer than 1 μm is beam, that scattering is more important at shorter wavelengths, and that the spectral distribution of diffuse is dependent on atmospheric conditions.

---

[7] Scattering theory predicts that shorter wavelengths are scattered most, and hence diffuse radiation tends to be at shorter wavelengths. Thus, clear skies are blue.

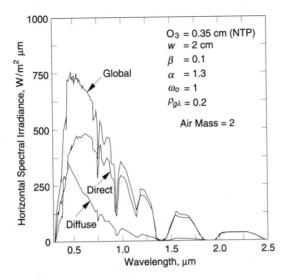

Figure 2.6.4 An example of calculated total, beam, and diffuse spectral irradiances on a horizontal surface for a typical clear atmosphere. Adapted from Iqbal (1983).

For most practical engineering purposes, the spectral distribution of solar radiation can be considered as approximately the same for the beam and diffuse components. It may also be observed that there is no practical alternative; information on atmospheric conditions on which to base any other model is seldom available.

For purposes of calculating properties of materials (absorptance, reflectance, and transmittance) that depend on the spectral distribution of solar radiation, it is convenient to have a representative distribution of terrestrial solar radiation in tabular form. Wiebelt and Henderson (1979) have prepared such tables for several air masses (zenith angles) and atmospheric conditions based on the NASA spectral distribution curves and a solar constant of 1353 W/m². These can be used with little error for most engineering calculations with the more recent value of G_{sc} of 1367 W/m². Table 2.6.1 shows the terrestrial spectrum divided into 20 equal increments of energy, with a mean wavelength for each increment that divides that increment into two equal parts. This table is for a relatively clear atmosphere with air mass 2. It can be used as a typical distribution of terrestrial solar radiation.

In summary, the normal solar radiation incident on the earth's atmosphere has a spectral distribution indicated by Figure 1.3.1. The x-rays and other very short-wave radiation of the solar spectrum are absorbed high in the ionosphere by nitrogen, oxygen, and other atmospheric components. Most of the ultraviolet is absorbed by ozone. At wavelengths longer than 2.5 μm, a combination of low extraterrestrial radiation and strong absorption by CO_2 means that very little energy reaches the ground. Thus, from the view point of terrestrial applications of solar energy, only radiation of wavelengths between 0.29 and 2.5 μm need be considered.

Table 2.6.1   Spectral Distribution of
Terrestrial Beam Radiation at Air Mass 2 and
23 km Visibility, in Equal Increments of
Energy[a]

Energy Band $f_i - f_{i+1}$	Wavelength Range μm	Midpoint Wavelength, μm
0.00 - 0.05	0.300 - 0.434	0.402
0.05 - 0.10	0.434 - 0.479	0.458
0.10 - 0.15	0.479 - 0.517	0.498
0.15 - 0.20	0.517 - 0.557	0.537
0.20 - 0.25	0.557 - 0.595	0.576
0.25 - 0.30	0.595 - 0.633	0.614
0.30 - 0.35	0.633 - 0.670	0.652
0.35 - 0.40	0.670 - 0.710	0.690
0.40 - 0.45	0.710 - 0.752	0.730
0.45 - 0.50	0.752 - 0.799	0.775
0.50 - 0.55	0.799 - 0.845	0.820
0.55 - 0.60	0.845 - 0.894	0.869
0.60 - 0.65	0.894 - 0.975	0.923
0.65 - 0.70	0.975 - 1.035	1.003
0.70 - 0.75	1.035 - 1.101	1.064
0.75 - 0.80	1.101 - 1.212	1.170
0.80 - 0.85	1.212 - 1.310	1.258
0.85 - 0.90	1.310 - 1.603	1.532
0.90 - 0.95	1.603 - 2.049	1.689
0.95 - 1.00	2.049 - 5.000	2.292

[a] From Wiebelt and Henderson (1979).

2.7   ESTIMATION OF AVERAGE SOLAR RADIATION

Radiation data are the best source of information for estimating average incident radiation. Lacking these or data from nearby locations of similar climate, it is possible to use empirical relationships to estimate radiation from hours of sunshine or cloudiness. Data on average hours of sunshine or average percentage of possible sunshine hours are widely available from many hundreds of stations in many countries and are usually based on data taken with Campbell-Stokes instruments. Examples are shown in Table 2.7.1. Cloud cover data (i.e., cloudiness) are also widely available but are based on visual estimates and are probably less useful than hours of sunshine data.

The original Ångström-type regression equation related monthly average daily radiation to clear day radiation at the location in question and average fraction of

Table 2.7.1   Examples of Monthly Average Hours per Day of Sunshine

	Hong Kong	Paris France	Bombay India	Sokoto Nigeria	Perth Australia	Madison Wisconsin
Latitude	22°N	48°N	19°N	13°N	32°S	43°N
Altitude, m	s.l.	50	s.l.	107	20	270
Jan.	4.7	2.1	9.0	9.9	10.4	4.5
Feb.	3.5	2.8	9.3	9.6	9.8	5.7
Mar.	3.1	4.9	9.0	8.8	8.8	6.9
Apr.	3.8	7.4	9.1	8.9	7.5	7.5
May	5.0	7.1	9.3	8.4	5.7	9.1
June	5.3	7.6	5.0	9.5	4.8	10.1
July	6.7	8.0	3.1	7.0	5.4	9.8
Aug.	6.4	6.8	2.5	6.0	6.0	10.0
Sep.	6.6	5.6	5.4	7.9	7.2	8.6
Oct.	6.8	4.5	7.7	9.6	8.1	7.2
Nov.	6.4	2.3	9.7	10.0	9.6	4.2
Dec.	5.6	1.6	9.6	9.8	10.4	3.9
Annual	5.3	5.1	7.4	8.8	7.8	7.3

possible sunshine hours:

$$\frac{\overline{H}}{\overline{H}_c} = a' + b' \, \frac{\overline{n}}{N} \tag{2.7.1}$$

where $\overline{H}$ = monthly average daily radiation on a horizontal surface

$\overline{H}_c$ = average clear sky daily radiation for the location and month in question

a', b' = empirical constants

$\overline{n}$ = monthly average daily hours of bright sunshine

N = monthly average of the maximum possible daily hours of bright sunshine (i.e., the day length of the average day of the month)

A basic difficulty with Equation 2.7.1 lies in the ambiguity of the terms $\overline{n} \, / \, \overline{N}$ and $\overline{H}_c$. The former is an instrumental problem (records from sunshine recorders are open to interpretation). The latter stems from uncertainty in the definition of a clear day. Page (1964) and others have modified the method to base it on extraterrestrial radiation on a horizontal surface rather than on clear day radiation:

$$\frac{\overline{H}}{\overline{H}_o} = a + b \, \frac{\overline{n}}{N} \tag{2.7.2}$$

where $\overline{H}_o$ is the extraterrestrial radiation for the location, averaged over the time period in question, and a and b are constants depending on location. The ratio $\overline{H}/\overline{H}_o$ is termed the clearness index and will be used frequently in later sections and chapters.

Values of $\overline{H}_o$ can be calculated from Equation 1.10.3 using day numbers from Table 1.6.1 for the mean days of the month, or it can be obtained from either Table 1.10.1 or Figure 1.10.1. The average day length $\overline{N}$ can be calculated from Equation 1.6.11 or it can be obtained from Figure 1.6.3, for the mean day of the month as indicated in Table 1.6.1. Löf et al. (1966a) developed sets of constants a and b for various climate types and locations based on radiation data then available. These are given in Table 2.7.2.

Table 2.7.2   Climatic Constants for Use in Equation 2.7.2

Location	Climate[a]	Veg.[b]	Sunshine Hours in Percentage of Possible Range	Avg.	a	b
Albuquerque, NM	BS-BW	E	68-85	78	0.41	0.37
Atlanta, GA	Cf	M	45-71	59	0.38	0.26
Blue Hill, MA	Df	D	42-60	52	0.22	0.50
Brownsville, TX	BS	GDsp	47-80	62	0.35	0.31
Buenos Aires, Arg.	Cf	G	47-68	59	0.26	0.50
Charleston, SC	Cf	E	60-75	67	0.48	0.09
Darien, Manchuria	Dw	D	55-81	67	0.36	0.23
El Paso, TX	BW	Dsi	78-88	84	0.54	0.20
Ely, NV	BW	Bzi	61-89	77	0.54	0.18
Hamburg, Germany	Cf	D	11-49	36	0.22	0.57
Honolulu, HI	Af	G	57-77	65	0.14	0.73
Madison, WI	Df	M	40-72	58	0.30	0.34
Malange, Angola	Aw-BS	GD	41-84	58	0.34	0.34
Miami, FL	Aw	E-GD	56-71	65	0.42	0.22
Nice, France	Cs	SE	49-76	61	0.17	0.63
Poona, India (Monsoon)	Am	S	25-49	37	0.30	0.51
(Dry)			65-89	81	0.41	0.34
Kisangani, Zaire	Af	B	34-56	48	0.28	0.39
Tamanrasset, Algeria	BW	Dsp	76-88	83	0.30	0.43

[a] Climatic classification based on Trewartha's map (1954, 1961), where climate types are:

Af   Tropical forest climate, constantly moist; rainfall all through the year.
Am   Tropical forest climate, monsoon rain; short dry season, but total rainfall sufficient to support rain forest.
Aw   tropical forest climate, dry season in winter.
BS   Steppe or semiarid climate.
BW   Desert or arid climate.
Cf   Mesothermal forest climate; constantly moist; rainfall all through the year.
Cs   Mesothermal forest climate; dry season in winter.
Df   Microthermal snow forest climate; constantly moist; rainfall all through the year.
Dw   Microthermal snow forest climate; dry season in winter.

The following example is based on Madison data (although the procedure is not recommended for a station where there are data) and includes comparisons of the estimated radiation with SOLMET data and estimates for Madison based on the Blue Hill constants (those which might have been used in the absence of constants for Madison).

Example 2.7.1

Estimate the monthly averages of total solar radiation on a horizontal surface for Madison, Wisconsin, latitude 43°, based on the average duration of sunshine hour data of Table 2.7.1.

Solution

The estimates are based on Equation 2.7.2 using constants $a = 0.30$ and $b = 0.34$ from Table 2.7.2. Values of $\overline{H}_o$ are obtained from Figure 1.10.1 and day lengths from Equation 1.6.11, each for the mean days of the month. The desired estimates are obtained in the following table, which shows daily $\overline{H}$ in MJ/m^2. (For

---

Footnotes for Table 2.7.2 (continued)

b Vegetation classification based on Küchler's map, where vegetation types are:

B	Broadleaf evergreen trees.
Bzi	Broadleaf evergreen, shrubform, minimum height 3 feet, growth singly or in groups or patches.
D	Broadleaf deciduous trees.
Dsi	Broadleaf deciduous, shrubform, minimum height 3 feet, plants sufficiently far apart that they frequently do not touch.
Dsp	Broadleaf deciduous, shrubform, minimum height 3 feet, growth singly or in groups or patches.
E	Needleleaf evergreen trees.
G	Grass and other herbaceous plants.
GD	Grass and other herbaceous plants; broadleaf deciduous trees.
GDsp	Grass and other herbaceous plants; broadleaf deciduous, shrubforms, minimum height 3 feet, growth singly or in groups or patches.
M	Mixed broadleaf deciduous and needleleaf evergreen trees.
S	Semideciduous: broadleaf evergreen and broadleaf deciduous trees.
SE	Semideciduous: broadleaf evergreen and broadleaf deciduous trees: needleleaf evergreen trees.

Note: These constants are based on radiation data available before 1966 and do not reflect improvements in data processing and interpretation made since then. The results of estimations for United States stations will be at variance with SOLMET data. It is recommended that these correlations be used only when there are no radiation data available.

comparison, data for Madison from Appendix G are shown, and in the last column estimates of Madison radiation determined by using constants a and b for Blue Hill.)

Month	$\overline{H}_o$ MJ/m^2	$\overline{N}$ hr	$\overline{n}/\overline{N}$	$\overline{H}$, est.[a] MJ/m^2	$\overline{H}$, meas. MJ/m^2	$\overline{H}$, est.[b] MJ/m^2
Jan.	13.2	9.2	0.489	6.2	5.9	7.2
Feb.	18.6	10.3	0.553	9.1	9.1	10.7
Mar.	25.8	11.7	0.590	12.9	12.9	15.4
Apr.	33.4	13.2	0.568	16.5	15.9	19.5
May	39.0	14.5	0.628	20.0	19.8	23.9
June	41.4	15.2	0.665	21.8	22.1	26.2
July	40.1	14.0	0.658	21.0	22.0	25.2
Aug.	35.6	13.8	0.725	19.5	19.4	23.6
Sept.	28.5	12.3	0.699	15.3	14.8	18.5
Oct.	20.7	19.8	0.667	10.9	10.3	13.1
Nov.	14.5	9.5	0.442	6.5	5.7	7.6
Dec.	11.8	8.8	0.443	5.3	4.4	6.2

[a] From SOLMET data.
[b] Using constants for Blue Hill.

The agreement between measured and calculated radiation is reasonably good, even though the constants a and b for Madison were derived from a different data base from the measured data. If we did not have constants for Madison and had to choose a climate close to that of Madison, Blue Hill would be a reasonable choice. The estimated averages using the Blue Hill constants are shown in the last column. The trends are shown, but the agreement is not as good. This is the more typical situation in the use of Equation 2.7.2.

Data are also available on mean monthly cloud cover $\overline{C}$, expressed as tenths of the sky obscured by clouds. Empirical relationships have been derived to relate monthly average daily radiation $\overline{H}$ to monthly average cloud cover $\overline{C}$. These are usually of the form

$$\frac{\overline{H}}{\overline{H}_o} = a'' + b'' \, \overline{C} \tag{2.7.3}$$

Norris (1968) reviews several attempts to develop such a correlation. Bennett (1965) compared correlations of $\overline{H}/\overline{H}_o$ with $\overline{C}$, with $\overline{n}/\overline{N}$, and with a combination of the two variables and found the best correlation to be with $\overline{n}/\overline{N}$, that is, Equation 2.7.2. Cloud cover data are estimated visually, and there is not necessarily a direct relationship between the presence of partial cloud cover and solar radiation at any particular time. Thus there may not be as good a statistical relationship between $\overline{H}/\overline{H}_o$ and $\overline{C}$ as there is between $\overline{H}/\overline{H}_o$ and $\overline{n}/\overline{N}$. Many surveys of solar radiation data [e.g., Bennett (1965) and Löf et al. (1966a,b)] have been based on correlations of radiation with sunshine hour data. However, Paltridge and Proctor (1976) have used cloud cover data to modify clear sky data for Australia and derived therefrom monthly averages of $\overline{H}_o$ which are in good agreement with measured average data.

2.8 ESTIMATION OF CLEAR SKY RADIATION

The effects of the atmosphere in scattering and absorbing radiation are variable with time as atmospheric conditions and air mass change. It is useful to define a standard "clear" sky and calculate the hourly and daily radiation which would be received on a horizontal surface under these standard conditions.

Hottel (1976) has presented a method for estimating the beam radiation transmitted through clear atmospheres which takes into account zenith angle and altitude for a standard atmosphere and for four climate types. The atmospheric transmittance for beam radiation τ_b is G_{bn}/G_{on} (or G_{bT}/G_{oT}) and is given in the form

$$\tau_b = a_0 + a_1 \exp\left(-k/\cos\theta_z\right) \tag{2.8.1a}$$

The constants a_0, a_1, and k for the standard atmosphere with 23 km visibility are found from a_0^*, a_1^*, and k^*, which are given for altitudes less than 2.5 km by

$$a_0^* = 0.4237 - 0.00821(6 - A)^2 \tag{2.8.1b}$$

$$a_1^* = 0.5055 + 0.00595(6.5 - A)^2 \tag{2.8.1c}$$

$$k^* = 0.2711 + 0.01858(2.5 - A)^2 \tag{2.8.1d}$$

where A is the altitude of the observer in kilometers. (Hottel also gives equations for a_0^*, a_1^*, and k^* for a standard atmosphere with 5 km visibility.)

Correction factors are applied to a_0^*, a_1^*, and k^* to allow for changes in climate types. The correction factors $r_0 = a_0/a_0^*$, $r_1 = a_1/a_1^*$, and $r_k = k/k^*$ are given in Table 2.8.1. Thus, the transmittance of this standard atmosphere for beam radiation can be

Table 2.8.1   Correction Factors for Climate Types[a]

Climate Type	r_0	r_1	r_k
Tropical	0.95	0.98	1.02
Midlatitude summer	0.97	0.99	1.02
Subarctic summer	0.99	0.99	1.01
Midlatitude winter	1.03	1.01	1.00

[a] From Hottel (1976).

determined for any zenith angle and any altitude up to 2.5 km. The clear sky beam normal radiation is then

$$G_{cnb} = G_{on} \tau_b \tag{2.8.2}$$

where G_{on} is obtained from Equation 1.4.1. The clear sky horizontal beam radiation is

$$G_{cb} = G_{on} \tau_b \cos \theta_z \tag{2.8.3}$$

For periods of an hour, the clear sky horizontal beam radiation is

$$I_{cb} = I_{on} \tau_b \cos \theta_z \tag{2.8.4}$$

Example 2.8.1

Calculate the transmittance for beam radiation of the standard clear atmosphere at Madison (altitude 270 m) on August 22 at 11:30 AM solar time. Estimate the intensity of beam radiation at that time and its component on a horizontal surface.

Solution

On August 22, n is 234, the declination is 11.4°, and from Equation 1.6.4 the cosine of the zenith angle is 0.846.

The next step is to find the coefficients for Equation 2.8.1. First, the values for the standard atmosphere are obtained from Equations 2.8.1b to 2.8.1d for an altitude of 0.27 km:

$$a_0^* = 0.4237 - 0.00821(6 - 0.27)^2 = 0.154$$

$$a_1^* = 0.5055 + 0.00595(6.5 - 0.27)^2 = 0.736$$

$$k^* = 0.2711 + 0.01858(2.5 - 0.27)^2 = 0.363$$

The climate-type correction factors are obtained from Table 2.8.1 for midlatitude summer. Equation 2.8.1 becomes

$$\tau_b = 0.514(0.97) + 0.736(0.99) \exp\left(-0.363 \times 1.02/0.846\right) = 0.62$$

The extraterrestrial radiation is 1339 W/m^2 from Equation 1.4.1. The beam radiation is then

$$G_{cnb} = 1339 \times 0.62 = 830 \text{ W/m}^2$$

The component on a horizontal plane is

$$830 \times 0.846 = 702 \text{ W/m}^2 \qquad\qquad\qquad \blacksquare$$

It is also necessary to estimate the clear sky diffuse radiation on a horizontal surface to get the total radiation. Liu and Jordan (1960) developed an empirical relationship between the transmission coefficient for beam and diffuse radiation for clear days:

$$\tau_d = \frac{G_d}{G_o} = 0.271 - 0.294\,\tau_b \qquad\qquad (2.8.5)$$

where τ_d is G_d/G_o (or I_d/I_o), the ratio of diffuse radiation to the extraterrestrial (beam) radiation on the horizontal plane. The equation is based on data for three stations. The data used by Liu and Jordan predated that used by Hottel and may not be entirely consistent with it;   until better information becomes available, it is suggested that Equation 2.8.5 be used to estimate diffuse clear sky radiation, which can be added to the beam radiation predicted by Hottel's method to obtain a clear hour's total. These calculations can be repeated for each hour of the day, based on the midpoints of the hours, to obtain a standard clear day's radiation, H_c.

Example 2.8.2

Estimate the standard clear day radiation on a horizontal surface for Madison on August 22.

Solution

For each hour, based on the midpoints of the hour, the transmittances of the atmosphere for beam and diffuse radiation are estimated. The calculation of τ_b is illustrated for the hour 11 to 12 (i.e., at 11:30) in Example 2.8.1, and the beam radiation for a horizontal surface for the hour is 2.53 MJ/m^2 (702 W/m^2 for the hour).

The calculation of τ_d is based on Equation 2.8.5:

$$\tau_d = 0.271 - 0.294(0.62) = 0.089$$

The value of G_{on}, by Equation 1.4.1, is 1339 W/m^2. Then G_o is G_{on} times cos θ_z so that

$$G_{cd} = 1339 \times 0.846 \times 0.089 = 101 \text{ W/m}^2$$

Then the diffuse radiation for the hour is 0.36 MJ/m^2. The total radiation on a horizontal plane for the hour is 2.53 + 0.36 = 2.89 MJ/m^2. These calculations are repeated for each hour of the day. The result is shown in the tabulation, where energy quantities are in MJ/m^2. The beam for the day H_{cb} is twice the sum of column 4, giving 19.0 MJ/m^2. The day's total radiation H_c is twice the sum of column 7, or 22.7 MJ/m^2.

Hours	τ_b	I_{cb} Normal	I_{cb} Horizontal	τ_d	I_{cd}	I_c
11-12, 12-1	0.626	2.99	2.53	0.089	0.36	2.89
10-11, 1-2	0.614	2.93	2.33	0.092	0.35	2.68
9-10, 2-3	0.586	2.80	1.97	0.101	0.34	2.31
8-9, 3-4	0.536	2.56	1.45	0.115	0.31	1.76
7-8, 4-5	0.449	2.14	0.88	0.140	0.27	1.15
6-7, 5-6	0.293	1.39	0.31	0.186	0.20	0.51
5-6, 6-7	0.152	0.73	0.03	0.227	0.04	0.07	■

A simpler method for estimating clear sky radiation by hours is to use data for the ASHRAE standard atmosphere. Farber and Morrison (1977) provide tables of beam normal radiation and total radiation on a horizontal surface as a function of zenith angle. These are plotted in Figure 2.8.1. For a given day, hour-by-hour estimates of I can be made, based on midpoints of the hours.

This method estimates the "clear sky" day's radiation as 10% greater than the Hottel and Liu and Jordan "standard" day method. The difference lies in the definition of a standard (clear) day. While the ASHRAE data are easier to use, the Hottel and Liu and Jordan method provides a means of taking into account climate type and altitude.

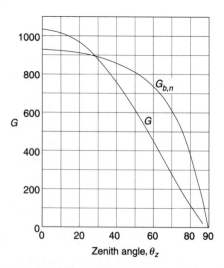

Figure 2.8.1   Total horizontal radiation and beam normal radiation for the ASHRAE standard atmosphere. Data are from Farber and Morrison (1977).

2.9 DISTRIBUTION OF CLEAR AND CLOUDY DAYS AND HOURS

The frequency of occurrence of periods of various radiation levels, for example, of good and bad days, is of interest in two contexts. First, information on the frequency distribution is the link between two kinds of correlations, that of the daily fraction of diffuse with daily radiation and that of the monthly average fraction of diffuse with monthly average radiation. Second, later in this chapter the concept of utilizability is developed; it depends on these frequency distributions.

The monthly average clearness index $\overline{K}_T$ is the ratio[8] of monthly average daily radiation on a horizontal surface to the monthly average daily extraterrestrial radiation. In equation form,

$$\overline{K}_T = \frac{\overline{H}}{\overline{H}_o} \tag{2.9.1}$$

We can also define a daily clearness index K_T as the ratio of a particular day's radiation to the extraterrestrial radiation for that day. In equation form,

$$K_T = \frac{H}{H_o} \tag{2.9.2}$$

An hourly clearness index k_T can also be defined:

$$k_T = \frac{I}{I_o} \tag{2.9.3}$$

The data $\overline{H}$, H, and I are from measurements of total solar radiation on a horizontal surface, that is, the commonly available pyranometer measurements. Values of $\overline{H}_o$, H_o, and I_o can be calculated by the methods of Section 1.10.

If for locations with a particular value of $\overline{K}_T$, the frequency of occurrence of days with various values of K_T is plotted as a function of K_T, the resulting distribution could appear like the solid curve of Figure 2.9.1. The shape of this curve depends on the average clearness index $\overline{K}_T$. For intermediate $\overline{K}_T$ values, days with very low K_T or very high K_T occur relatively infrequently, and most of the days have K_T values intermediate between the extremes. If $\overline{K}_T$ is high, the distribution must be skewed toward high K_T values, and if it is low, the curve must be skewed toward low K_T values. The distribution can be bimodal.

The data used to construct the frequency distribution curve of Figure 2.9.1 can also be plotted as cumulative distribution, that is, as the fraction f of the days that are less clear than K_T versus K_T. In practice, following the precedent of Whillier (1956), the plots are usually shown as K_T versus f. The result is shown as the dashed line in Figure 2.9.1.

---

[8] These ratios were originally referred to by Liu and Jordan (1960) as cloudiness indexes. As their values approach unity with increasing atmospheric clearness, they are also referred to as clearness indexes, the terminology used here.

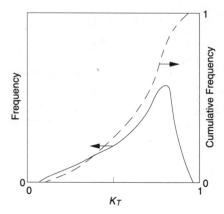

Figure 2.9.1 An example of the frequency of occurrence of days with various clearness indexes and the cumulative frequency of occurrence of those days.

Liu and Jordan found that the cumulative distribution curves are very nearly identical for locations having the same values of $\overline{K}_T$, even though the locations varied widely in latitude and elevation. On the basis of this information, they developed a set of generalized distribution curves of K_T versus f which are functions of $\overline{K}_T$, the monthly clearness index. These are shown in Figure 2.9.2. The coordinates of the curves are given in Table 2.9.1. Thus if a location has a $\overline{K}_T$ of 0.6, 19% of the days will have K_T equal to or less than 0.40.[9]

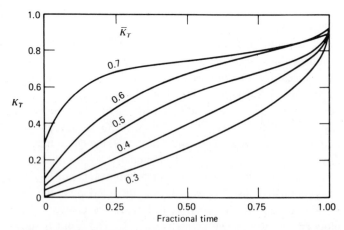

Figure 2.9.2 Generalized distribution of days with various values of K_T as a function of $\overline{K}_T$.

[9] Recent research indicates that there may be some seasonal dependence of these distributions in some locations.

Table 2.9.1 Coordinates of the Liu and Jordan Generalized Monthly K_T Cumulative Distribution Curves

K_T	Value of $f(K_T)$ for $\overline{K}_T =$				
	0.3	0.4	0.5	0.6	0.7
0.04	0.073	0.015	0.001	0.000	0.000
0.08	0.162	0.070	0.023	0.008	0.000
0.12	0.245	0.129	0.045	0.021	0.007
0.16	0.299	0.190	0.082	0.039	0.007
0.20	0.395	0.249	0.121	0.053	0.007
0.24	0.496	0.298	0.160	0.076	0.007
0.28	0.513	0.346	0.194	0.101	0.013
0.32	0.579	0.379	0.234	0.126	0.013
0.36	0.628	0.438	0.277	0.152	0.027
0.40	0.687	0.493	0.323	0.191	0.034
0.44	0.748	0.545	0.358	0.235	0.047
0.48	0.793	0.601	0.400	0.269	0.054
0.52	0.824	0.654	0.460	0.310	0.081
0.56	0.861	0.719	0.509	0.360	0.128
0.60	0.904	0.760	0.614	0.410	0.161
0.64	0.936	0.827	0.703	0.467	0.228
0.68	0.953	0.888	0.792	0.538	0.295
0.72	0.967	0.931	0.873	0.648	0.517
0.76	0.979	0.967	0.945	0.758	0.678
0.80	0.986	0.981	0.980	0.884	0.859
0.84	0.993	0.997	0.993	0.945	0.940
0.88	0.995	0.999	1.000	0.985	0.980
0.92	0.998	0.999		0.996	1.000
0.96	0.998	1.000		0.999	
1.00	1.000			1.000	

Bendt et al. (1981) have developed equations to represent the Liu and Jordan distributions, based on 20 years of data from 90 locations in the United States. The correlation represents the Liu and Jordan curves very well for values of $f(K_T)$ less than 0.9; above 0.9 the correlations overpredict the frequency for given values of the clearness index. The Bendt et al. equations are

$$f(K_T) = \frac{\exp(\gamma K_{T,\min}) - \exp(\gamma K_T)}{\exp(\gamma K_{T,\min}) - \exp(\gamma K_{T,\max})} \tag{2.9.4}$$

where γ is determined from the following equation:

$$\overline{K}_T = \frac{(K_{T,\min} - 1/\gamma)\exp(\gamma K_{T,\min}) - (K_{T,\max} - 1/\gamma)\exp(\gamma K_{T,\max})}{\exp(\gamma K_{T,\min}) - \exp(\gamma K_{T,\max})} \tag{2.9.5}$$

Solving for γ in this equation is not convenient, and Herzog (1985) gives an explicit equation for γ from a curve fit:

$$\gamma = -1.498 + \frac{1.184\xi - 27.182 \exp(-1.5\xi)}{K_{T,\,max} - K_{T,\,min}} \tag{2.9.6a}$$

and

$$\xi = \frac{K_{T,max} - K_{T,min}}{K_{T,max} - \overline{K}_T} \tag{2.9.6b}$$

A value of $K_{T,min}$ of 0.05 was recommended by Bendt et al. Hollands and Huget (1983) recommend that $K_{T,max}$ be estimated from

$$K_{T,max} = 0.6313 + 0.267\overline{K}_T - 11.9(\overline{K}_T - 0.75)^8 \tag{2.9.6c}$$

The universality of the Liu and Jordan distributions has been questioned, particularly as applied to tropical climates. Saunier et al. (1987) propose an alternative expression for the distributions for tropical climates. A brief review of papers on the distributions is included in Knight et al. (1991).

Similar distribution functions have been developed for hourly radiation. Whillier (1953) observed that when the hourly and daily curves for a location are plotted, the curves are very similar. Thus the distribution curves of daily occurrences of K_T can also be applied to hourly clearness indexes. The ordinate in Figure 2.9.2 can be replaced by k_T and the curves will approximate the cumulative distribution of hourly clearness. Thus for a climate with $\overline{K}_T = 0.4$, 0.493 of the hours will have k_T equal to or less than 0.40.

2.10 BEAM AND DIFFUSE COMPONENTS OF HOURLY RADIATION

In this and the following two sections we review methods for estimation of the fractions of total horizontal radiation that are diffuse and beam. The questions of the best methods for doing these calculations are not fully settled. A broader data base and improved understanding of the data will probably lead to improved methods. In each section we review methods that have been published and then suggest one for use. The suggested correlations are in substantial agreement with other correlations, and the set is mutually consistent.

The split of total solar radiation on a horizontal surface into its diffuse and beam components is of interest in two contexts. First, methods for calculating total radiation on surfaces of other orientation from data on a horizontal surface require separate treatments of beam and diffuse radiation (see Section 2.15). Second, estimates of the long-time performance of most concentrating collectors must be based on estimates of availability of beam radiation. The present methods for

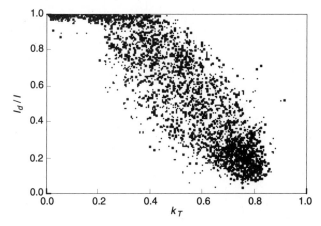

Figure 2.10.1 A sample of diffuse fraction vs. clearness index data from Cape Canaveral, FL. Adapted from Reindl (1988).

estimating the distribution are based on studies of available measured data; they are adequate for the first purpose but less than adequate for the second.

The usual approach is to correlate I_d/I, the fraction of the hourly radiation on a horizontal plane which is diffuse, with k_T, the hourly clearness index. Figure 2.10.1 shows a plot of diffuse fraction I_d/I vs. k_T for Cape Canaveral, FL. In order to obtain I_d/I vs. k_T correlations, data from many locations similar to that shown in Figure 2.10.1 are divided into "bins," or ranges of values of k_T, and the data in each bin are averaged to obtain a point on the plot. A set of these points then is the basis of the correlation. Within each of the bins there is a distribution of points; a k_T of 0.5 may be produced by skies with thin cloud cover, resulting in a high diffuse fraction, or by skies that are clear for part of the hour and heavily clouded for part of the hour, leading to a low diffuse fraction. Thus the correlation may not represent a particular hour very closely, but over a large number of hours it adequately represents the diffuse fraction.

Orgill and Hollands (1977) have used data of this type from Canadian stations, Erbs et al. (1982) have used data from four U. S. and one Australian station, and Reindl et al. (1990a) have used an independent data set from United States and Europe. The three correlations are shown in Figure 2.10.2. They are essentially identical, although they were derived from three separate data bases. The Erbs et al. correlation, Figure 2.10.3, is[10]

---

[10] The Orgill and Hollands correlation has been widely used, produces results that are for practical purposes the same as those of Erbs et al., and is represented by the following equations:

$$\frac{I_d}{I} = \begin{cases} 1.0 - 0.249k_T & \text{for } k_T < 0 \\ 1.557 - 1.84k_T & \text{for } 0.35 < k_T < 0.75 \\ 0.177 & \text{for } k_T > 0.75 \end{cases}$$

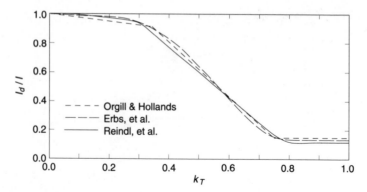

Figure 2.10.2   The ratio I_d/I as a function of hourly clearness index, k_T, showing the Orgill and Hollands (1977), Erbs et al. (1982), and Reindl et al. (1990) correlations.

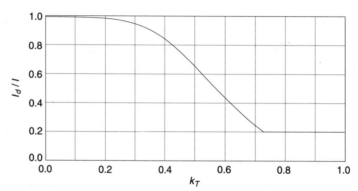

Figure 2.10.3   The ratio I_d/I as a function of hourly clearness index, k_T, from Erbs et al. (1982).

$$\frac{I_d}{I} = \begin{cases} 1.0 - 0.09k_T & \text{for } k_T \leq 0.22 \\[2ex] \begin{aligned} & 0.9511 - 0.1604k_T + 4.388k_T^2 \\ & \quad - 16.638\ k_T^3 + 12.336\ k_T^4 \end{aligned} & \text{for } 0.22 < k_T \leq 0.80 \\[2ex] 0.165 & \text{for } k_T > 0.80 \end{cases} \qquad (2.10.1)$$

For values of k_T greater than 0.8, there are very few data. Some of the data that are available show increasing diffuse fraction as k_T increases above 0.8. This apparent rise in the diffuse fraction is probably due to reflection of radiation from clouds during times when the sun is unobscured but when there are clouds near the path from the sun to the observer. The use of a diffuse fraction of 0.165 is recommended in this region.

In a related approach described by Boes (1975), values of I_d/I from correlations are modified by a restricted random number that adds a statistical variation to the correlation.

2.11 BEAM AND DIFFUSE COMPONENTS OF DAILY RADIATION

Studies of available daily radiation data have shown that the average fraction which is diffuse, H_d/H, is a function of K_T, the day's clearness index. The original correlation of Liu and Jordan (1960) is shown in Figure 2.11.1; the data were for Blue Hill, Massachusetts. Also shown on the graph are plots of data for Israel from Stanhill (1966), for New Delhi from Choudhury (1963), for Canadian stations from Ruth and Chant (1976) and Tuller (1976), for Highett (Melbourne), Australia, from Bannister (1969), and from Collares-Pereira and Rabl (1979a) for four U.S. stations. There is some disagreement, with differences probably due in part to instrumental difficulties such as shading ring corrections and possibly in part due to air mass and/or seasonal effects. The correlation by Erbs (based on the same data set as is Figure 2.10.2) is shown in Figure 2.11.2. A seasonal dependence is shown; the spring, summer, and fall data are essentially the same, while the winter data shows somewhat lower diffuse fractions for high values of K_T. The season is indicated by

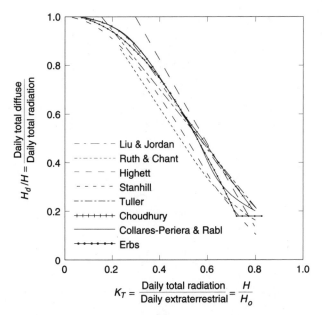

Figure 2.11.1 Correlations of daily diffuse fraction with daily clearness index. Adapted from Klein and Duffie (1978).

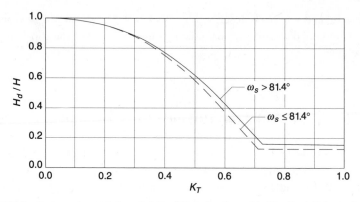

Figure 2.11.2 Suggested correlation of daily diffuse fraction with K_T. From Erbs et al. (1982).

the sunset hour angle ω_s. Equations representing this set of correlations are as follows[11]:

For $\omega_s < 81.4°$

$$\frac{H_d}{H} = \begin{cases} \begin{aligned} & 1.0 - 0.2727\,K_T + 2.4495\,K_T^2 \\ & \quad - 11.9514\,K_T^3 + 9.3879\,K_T^4 \end{aligned} & \text{for } K_T < 0.715 \\[2ex] 0.143 & \text{for } K_T \geq 0.715 \end{cases}$$

(2.11.1a)

and for $\omega_s \geq 81.4°$

$$\frac{H_d}{H} = \begin{cases} \begin{aligned} & 1.0 + 0.2832\,K_T - 2.5557\,K_T^2 \\ & \quad + 0.8448 K_T^3 \end{aligned} & \text{for } K_T < 0.722 \\[2ex] 0.175 & \text{for } K_T \geq 0.722 \end{cases}$$

(2.11.1b)

[11] The Collares-Pereira and Rabl correlation is

$$\frac{H_d}{H} = \begin{cases} 0.99 & \text{for } K_T \leq 0.17 \\[2ex] \begin{aligned} & 1.188 - 2.272 K_T + 9.473 K_T^2 \\ & \quad - 21.865 K_T^3 + 14.648 K_T^4 \end{aligned} & \text{for } 0.17 < K_T < 0.75 \\[2ex] -0.54 K_T + 0.632 & \text{for } 0.75 < K_T < 0.80 \\[2ex] 0.2 & \text{for } K_T \geq 0.80 \end{cases}$$

Example 2.11.1

The day's total radiation on a horizontal surface for St. Louis, Missouri (latitude 38.6°) on September 3 is 23.0 MJ/m^2. Estimate the fraction and amount that is diffuse.

Solution

For September 3, the declination is 7°. From Equation 1.6.10, the sunset hour angle is 95.6°. From Equation 1.10.3, the day's extraterrestrial radiation is 33.3 MJ/m^2. Then

$$K_T = \frac{H}{H_o} = \frac{23.0}{33.3} = 0.69$$

From Figure 2.11.2 or Equation 2.11.1b, H_d/H is 0.26, so an estimated 26% of the day's radiation is diffuse. The diffuse energy is $0.26 \times 23.0 = 6.0$ MJ/m^2. ∎

2.12 BEAM AND DIFFUSE COMPONENTS OF MONTHLY RADIATION

Charts similar to Figures 2.11.1 and 2.11.2 have been derived to show the distribution of monthly average daily radiation into its beam and diffuse components. In this case, the monthly fraction that is diffuse, $\overline{H}_d/\overline{H}$, is plotted as a function of monthly average clearness index, $\overline{K}_T$ $(= \overline{H}/\overline{H}_o)$. The data for these plots can be obtained from daily data in either of two ways. First, monthly data can be plotted by summing the daily data diffuse and total radiation. Second, as shown by Liu and Jordan, a generalized daily H_d/H versus K_T curve can be used with a knowledge of the distribution of good and bad days (the cumulative distribution curves of Figure 2.9.2) to develop the monthly average relationships.

Figure 2.12.1 shows several correlations of $\overline{H}_d/\overline{H}$ versus $\overline{K}_T$. The curves of Page (1964) and Collares-Pereira and Rabl (1979a) are based on summations of daily total and diffuse radiation. The original curve of Liu and Jordan (modified to correct for a small error in $\overline{H}_d/\overline{H}$ at low $\overline{K}_T$) and those labeled Highett, Stanhill, Choudhury, Ruth and Chant, and Tuller are based on daily correlations by the various authors (as in Figure 2.11.1) and on the distribution of days with various $\overline{K}_T$ as shown in Figure 2.9.2. The Collares-Pereira and Rabl curve in Figure 2.12.1 is for their all-year correlation; they found a seasonal dependence of the relationship which they expressed in terms of the sunset hour angle of the mean day of the month. There is significant disagreement among the various correlations of Figure 2.12.1. Instrumental problems and atmospheric variables (air mass, season, or other) may contribute to the differences.

Erbs et al. developed monthly average diffuse fraction correlations from the daily diffuse correlations of Figure 2.11.2. As with the daily correlations, there is a

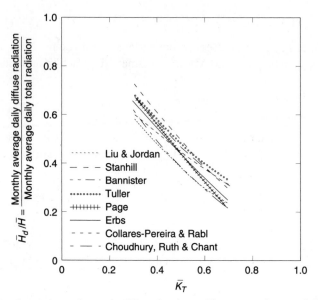

Figure 2.12.1 Correlations of average diffuse fractions with average clearness index. Adapted from Klein and Duffie (1978).

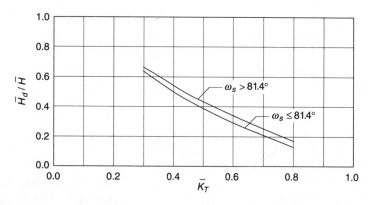

Figure 2.12.2 Suggested correlation of $\overline{H}_d/\overline{H}$ vs. $\overline{K}_T$ and ω_s. Adapted form Erbs et al. (1982).

seasonal dependence; the winter curve lies below the other, indicating lower moisture and dust in the winter sky with resulting lower fractions of diffuse. The dependence of $\overline{H}_d/\overline{H}$ on $\overline{K}_T$ is shown for winter and for the other seasons in Figure 2.12.2. Equations for these correlations are as follows[12]:

[12] The Collares-Pereira and Rabl correlation, with ω_s in degrees, is

$$\frac{\overline{H}_d}{\overline{H}} = 0.775 + 0.00606(\omega_s - 90) - \left[0.505 + 0.00455(\omega_s - 90)\right] \cos\left(115\overline{K}_T - 103\right)$$

For $\omega_s \leq 81.4°$ and $0.3 \leq \overline{K}_T \leq 0.8$

$$\frac{\overline{H}_d}{\overline{H}} = 1.391 - 3.560\overline{K}_T + 4.189\overline{K}_T^2 - 2.137\overline{K}_T^3 \tag{2.12.1a}$$

and for $\omega_s > 81.4°$ and $0.3 \leq \overline{K}_T \leq 0.8$

$$\frac{\overline{H}_d}{\overline{H}} = 1.311 - 3.022\overline{K}_T + 3.427\overline{K}_T^2 - 1.821\overline{K}_T^3 \tag{2.12.1b}$$

Example 2.12.1

Estimate the fraction of the average June radiation on a horizontal surface that is diffuse in Madison, Wisconsin.

Solution

From Appendix G, the June average daily radiation, $\overline{H}$, for Madison is 22.1 MJ/m². From Equation 1.10.3, for June 11 (the mean day of the month, $n = 162$, from Table 1.6.1), when the declination is 23.1°, H_o is 41.8 MJ/m². Thus $\overline{K}_T = 22.1/41.8 = 0.53$. From Equation 1.6.10, $\omega_s = 113.4°$. Then, using either Equation 2.12.1b or the upper curve from Figure 2.12.2, $\overline{H}_d/\overline{H} = 0.40$. ∎

2.13 ESTIMATION OF HOURLY RADIATION FROM DAILY DATA

When hour-by-hour (or other short-time base) performance calculations for a system are to be done, it may be necessary to start with daily data and estimate hourly values from daily numbers. As with the estimation of diffuse from total radiation, this is not an exact process. For example, daily total radiation values in the middle range between clear day and completely cloudy day values can arise from various circumstances, such as intermittent heavy clouds, continuous light clouds, or heavy cloud cover for part of the day. There is no way to determine these circumstances from the daily totals. However, the methods presented here work best for clear days, and those are the days that produce most of the output of solar processes (particularly those processes that operate at temperatures significantly above ambient). Also, these methods tend to produce conservative estimates of long-time process performance.

Statistical studies of the time distribution of total radiation on horizontal surfaces through the day, using monthly average data for a number of stations, have led to generalized charts of r_t, the ratio of hourly total to daily total radiation, as a function of day length and the hour in question:

$$r_t = \frac{I}{H} \tag{2.13.1}$$

Figure 2.13.1 shows such a chart, adapted from Liu and Jordan (1960) and based on Whillier (1956, 1965) and Hottel and Whillier (1958). The hours are designated by

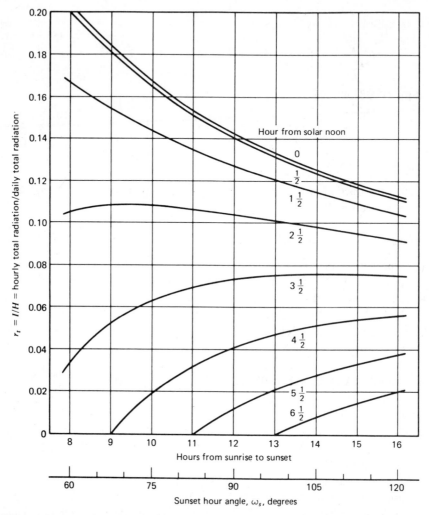

Figure 2.13.1 Relationship between hourly and daily total radiation on a horizontal surface as a function of day length. Adapted form Liu and Jordan (1960).

the time for the midpoint of the hour, and days are assumed to be symmetrical about solar noon. A curve for the hour centered at noon is also shown. Day length can be calculated from Equation 1.6.11, or it can be estimated from Figure 1.6.3. Thus, from a knowledge of day length (a function of latitude ϕ and declination δ) and daily total radiation, the hourly total radiation for symmetrical days can be estimated.

A study of New Zealand data by Benseman and Cook (1969) indicates that the curves of Figure 2.13.1 represent the New Zealand data in a satisfactory way. Iqbal (1979) used Canadian data to further substantiate these relationships. The figure is based on long-term averages and is intended for use in determining averages of hourly radiation. Whillier (1956) recommends that it be used for individual days only if they are clear days. Benseman suggests that it is adequate for individual days, with best results for clear days and increasingly uncertain results as daily total radiation decreases. It is applied to individual days.

The curves of Figure 2.13.1 are represented by the following equation from Collares-Pereira and Rabl (1979a):

$$r_t = \frac{\pi}{24}(a + b \; \cos \; \omega)\frac{\cos \; \omega - \cos \; \omega_s}{\sin \; \omega_s - \frac{\pi \omega_s}{180} \cos \; \omega_s} \tag{2.13.2a}$$

The coefficients a and b are given by

$$a = 0.409 + 0.5016 \; \sin(\omega_s - 60) \tag{2.13.2b}$$

$$b = 0.6609 - 0.4767 \; \sin(\omega_s - 60) \tag{2.13.2c}$$

In these equations ω is the hour angle in degrees for the time in question (e.g., the midpoint of the hour for which the calculation is made), and ω_s is the sunset hour angle.

Example 2.13.1

What is the fraction of the average January daily radiation that is received at Melbourne, Australia, in the hour between 8:00 and 9:00?

Solution

For Melbourne, $\phi = -38°$. From Table 1.6.1 the mean day for January is the 17th. From Equation 1.6.1 the declination is $-20.9°$. From Equation 1.6.11 the day length is 14.3 hours. From Figure 2.13.1, using the curve for 3.5 hours from solar noon, at a day length of 14.3 hours, approximately 7.8% of the day's radiation will be in that hour. Or Equation 2.13.2 can be used; with $\omega_s = 107°$ and $\omega = -52.5°$, the result is $r_t = 0.076$. ∎

Example 2.13.2

The total radiation for Madison on August 23 was 31.4 MJ/m². Estimate the radiation received between 1 and 2 PM.

Solution

For August 23, $\delta = 11°$, and ϕ for Madison is 43°. From Figure 1.6.3, sunset is at 6:45 PM and day length is 13.4 hours. Then from Figure 2.13.1, at day length of 13.4 hours and mean of 1.5 hours from solar noon, the ratio hourly total to daily total r_t is 0.118. The estimated radiation in the hour from 1 to 2 PM is then 3.7 MJ/m². (The measured value for that hour was 3.47 MJ/m².) ∎

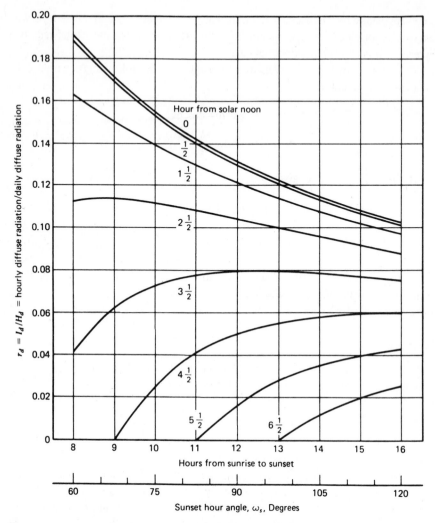

Figure 2.13.2 Relationship between hourly diffuse and daily diffuse radiation on a horizontal surface as a function of day length. Adapted form Liu and Jordan (1960).

Figure 2.13.2 shows a related set of curves for r_d, the ratio of hourly diffuse to daily diffuse radiation, as a function of time and day length. In conjunction with Figure 2.11.2, it can be used to estimate hourly averages of diffuse radiation if the average daily total radiation is known:

$$r_d = \frac{I_d}{H_d} \tag{2.13.3}$$

The curves of Figure 2.13.2, the ratio of hourly to daily extraterrestrial radiation, can be represented by the following equation from Liu and Jordan (1960).

$$r_d = \frac{\pi}{24} \frac{\cos \omega - \cos \omega_s}{\sin \omega_s - \dfrac{\pi \omega_s}{180} \cos \omega_s} \tag{2.13.4}$$

Example 2.13.3

From Appendix G, the average daily June total radiation on a horizontal plane in Madison is 22.1 MJ/m^2. Estimate the average diffuse, the average beam, and the average total radiation for the hours 10 to 11 and 1 to 2.

Solution

The mean daily extraterrestrial radiation $\overline{H}_o$ for June for Madison is 41.7 MJ/m^2 (from Table 1.10.1 or Equation 1.10.3 with $n = 162$), $\omega_s = 113°$, and the day length is 15.1 hours (from Equation 1.6.11). Then (as in Example 2.12.1), $\overline{K}_T = 0.53$. From Equation 2.12.1, $\overline{H}_d/\overline{H} = 0.40$, and the average daily diffuse radiation is 0.40 × 22.1 = 8.84 MJ/m^2. Entering Figure 2.13.2 for an average day length of 15.1 hours and for 1.5 hours from solar noon, we find $r_d = 0.102$. (Or Equation 2.13.3 can be used with $\omega = 22.5°$ and $\omega_s = 113°$ to obtain $r_d = 0.102$.) Thus the average diffuse for those hours is 0.102 × 8.84 = 0.90 MJ/m^2.

From Figure 2.13.1 (or from Equations 2.13.1 and 2.13.2) from the curve for 1.5 hours from solar noon, for an average day length of 15.1 hours, $r_t = 0.108$, and average hourly total radiation is 0.108 × 22.1 = 2.38 MJ/m^2. The average beam radiation is the difference between the total and diffuse, or 2.38 − 0.90 = 1.48 MJ/m^2. ∎

2.14 RADIATION ON SLOPED SURFACES

We turn next to the general problem of calculation of radiation on tilted surfaces when only the total radiation on a horizontal surface is known. For this we need the directions from which the beam and diffuse components reach the surface in question. Section 1.8 dealt with the geometric problem of the direction of beam radiation. The direction from which diffuse radiation is received, i.e., its distribution over the sky dome, is a function of conditions of cloudiness and atmospheric clarity, which are highly variable. Some data are available, for example, from Kondratyev (1969) and Coulson (1975). Figure 2.14.1, from Coulson, shows profiles of diffuse radiation across the sky as a function of angular elevation from the horizon in a plane that includes the sun. The data are for clear sky and for smog conditions.

Clear day data such as that in Figure 2.14.1 have led to a description of the diffuse radiation as being composed of three parts. The first is an **isotropic** part, received uniformly from all of the sky dome. The second is **circumsolar diffuse**, resulting from forward scattering of solar radiation and concentrated in the part of the

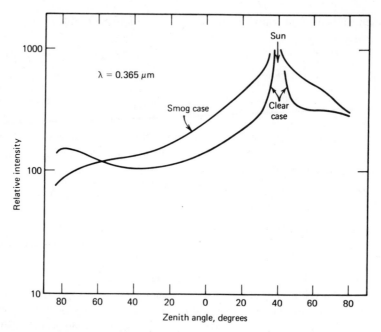

Figure 2.14.1 Relative intensity of solar radiation (at $\lambda = 0.365$ μm) as a function of elevation angle in the principal plane that includes the sun, for Los Angeles, for clear sky and for smog. Adapted from Coulson, *Solar and Terrestrial Radiation,* Academic Press, New York (1975).

sky around the sun. The third, referred to as **horizon brightening**, is concentrated near the horizon, and is most pronounced in clear skies. Figure 2.14.2 shows schematically these three parts of the diffuse radiation.

The angular distribution of diffuse is to some degree a function of the reflectance ρ_g (the albedo) of the ground. A high reflectance (such as that of fresh snow, with ρ_g approximately 0.7) results in reflection of solar radiation back to the sky, which in turn may be scattered to account for horizon brightening.

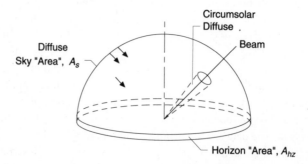

Figure 2.14.2 Schematic of the distribution of diffuse radiation over the sky dome, showing the circumsolar and horizon brightening components added to the isotropic component. Adapted from Perez et al. (1988).

Sky models, in the context used here, are mathematical representations of the diffuse radiation. When beam and reflected radiation are added, they provide the means of calculating radiation on a tilted surface from measurements on the horizontal. Many sky models have been devised. A review of some of them is provided by Hay and McKay (1985). Since 1985, others have been developed. For purposes of this book, three of the most useful of these models are presented. The isotropic model is in Section 2.15, and two anisotropic models are in Section 2.16. The differences among them are in the way they treat the three parts of the diffuse radiation.

It is necessary to know or to be able to estimate the solar radiation incident on tilted surfaces such as flat-plate collectors, windows, or other passive system receivers. The incident solar radiation is the sum of a set of radiation streams including beam radiation, the three components of diffuse radiation from the sky, and radiation reflected from the various surfaces "seen" by the tilted surface. The total incident radiation on this surface can be written as[13]

$$I_T = I_{T,b} + I_{T,d,iso} + I_{T,d,cs} + I_{T,d,hz} + I_{T,refl} \qquad (2.14.1)$$

where the subscripts *iso, cs, hz,* and *refl* refer to the isotropic, circumsolar, horizon, and reflected radiation streams.

For a surface (a collector) of area A_c, the total incident radiation can be expressed in terms of the beam and diffuse radiation on the horizontal surface and the total radiation on the surfaces that reflect to the tilted surface. The terms in Equation 2.14.1 become

$$A_c I_T = I_b R_b A_c + I_{d,iso} A_s F_{s-c} + I_{d,cs} R_b A_c + I_{d,hz} A_{hz} F_{hz-c}$$

$$+ \sum_i I_i \rho_i A_i F_{i-c} \qquad (2.14.2)$$

The first term is the beam contribution. The second is the isotropic diffuse term which includes the product of sky area A_s (an undefined area) and the radiation view factor from the sky to the collector F_{s-c}. The third is the circumsolar diffuse, which is treated as coming from the same direction as the beam. The fourth term is the contribution of the diffuse from the horizon from a band with another undefined area A_{hz}. The fifth term is the set of reflected radiation streams from the buildings, fields, etc., to which the tilted surface is exposed. The symbol *i* refers to each of the reflected streams: I_i is the solar radiation incident on the *i*th surface, ρ_i is the diffuse reflectance of that surface, and F_{i-c} is the view factor from the *i*th surface to the tilted surface. It is assumed that the reflecting surfaces are diffuse reflectors; specular reflectors require a different treatment.

---

[13] This and following equations are written in terms of *I* for an hour. They could also be written in terms of *G*, the irradiance.

In general, it is not possible to calculate the reflected energy term in detail, to account for buildings, trees, etc., the changing solar radiation incident on them, and their changing reflectances. Standard practice is to assume that there is one surface, a horizontal, diffusely reflecting ground, large in extent, contributing to this term. In this case, I_i is simply I and ρ_i becomes ρ_g, a composite "ground" reflectance.

Equation 2.14.2 can be rewritten in a useful form by interchanging areas and view factors (since the view factor reciprocity relation requires that, for example, $A_s F_{s-c} = A_c F_{c-s}$). This eliminates the uncertain areas A_s and A_{hz}. The area A_c appears in each term in the equation and cancels. The result is an equation that gives I_T in terms of parameters that can be determined either theoretically or empirically:

$$I_T = I_b R_b + I_{d,iso} F_{c-s} + I_{d,cs} R_b + I_{d,hz} F_{c-hz} + I \rho_g F_{c-g} \qquad (2.14.3)$$

This equation, with variations, is the basis for methods of calculating I_T that are presented in the following sections.

When I_T has been determined, the ratio of total radiation on the tilted surface to that on the horizontal surface can be determined. By definition,

$$R = \frac{\text{Total radiation on the tilted surface}}{\text{Total radiation on a horizontal surface}} = \frac{I_T}{I} \qquad (2.14.4)$$

Many models have been developed, of varying complication, as the basis for calculating I_T. The differences are largely in the way that the diffuse terms are treated. The simplest model is based on the assumptions that the beam radiation predominates (when it matters) and that the diffuse (and ground-reflected radiation) are effectively concentrated in the area of the sun. Then $R = R_b$ and all radiation is treated as beam. This leads to substantial overestimation of I_T, and the procedure is not recommended. Preferred methods are given in the following two sections.

2.15   RADIATION ON SLOPED SURFACES - ISOTROPIC SKY

It can be assumed [as suggested by Hottel and Woertz (1942)] that the combination of diffuse and ground-reflected radiation is isotropic. With this assumption, the sum of the diffuse from the sky and the ground-reflected radiation on the tilted surface is the same regardless of orientation, and the total radiation on the tilted surface is the sum of the beam contribution calculated as $I_b R_b$ and the diffuse on a horizontal surface, I_d. This represents an improvement over the assumption that all radiation can be treated as beam, but better methods are available.

An improvement on these models, the **isotropic diffuse** model, was derived by Liu and Jordan (1963). The radiation on the tilted surface was considered to include three components: beam, isotropic diffuse, and solar radiation diffusely reflected from the ground. The third and fourth terms in Equation 2.14.3 are taken as zero as all diffuse is assumed to be isotropic. A surface tilted at slope β from the horizontal has a view factor to the sky $F_{c\text{-}s}$ which is given by $(1 + \cos \beta)/2$. (If the diffuse radiation is isotropic, this is also R_d, the ratio of diffuse on the tilted surface to that on the horizontal surface.) The surface has a view factor to the ground $F_{c\text{-}g}$ of $(1 - \cos \beta)/2$, and if the surroundings have a diffuse reflectance of ρ_g for the total solar radiation, the reflected radiation from the surroundings on the surface will be $I\rho_g(1 - \cos \beta)/2$. Thus Equation 2.14.3 is modified to give the total solar radiation on the tilted surface for an hour as the sum of three terms:

$$I_T = I_b R_b + I_d\left(\frac{1 + \cos \beta}{2}\right) + I\rho_g\left(\frac{1 - \cos \beta}{2}\right) \tag{2.15.1}$$

and by the definition of R,

$$R = \frac{I_b}{I}R_b + \frac{I_d}{I}\left(\frac{1 + \cos \beta}{2}\right) + \rho_g\left(\frac{1 - \cos \beta}{2}\right) \tag{2.15.2}$$

Example 2.15.1

Using the isotropic diffuse model, estimate the beam, diffuse, and ground-reflected components of solar radiation and the total radiation on a surface sloped 60° toward the south at a latitude of 40°N for the hour 9 to 10 AM on February 20. Here $I = 1.04$ MJ/m^2 and $\rho_g = 0.60$.

Solution

For this hour, $I_o = 2.31$ MJ/m^2, so $k_T = 1.04/2.31 = 0.45$. From the Erbs correlation, Equation 2.10.1, $I_d/I = 0.757$. Thus

$$I_d = 0.757 \times 1.04 = 0.787 \text{ MJ/m}^2$$

$$I_b = 0.243 \times 1.04 = 0.253 \text{ MJ/m}^2$$

The hour angle ω for the midpoint of the hour is −37.5°. The declination δ is −11.6°. Then

$$R_b = \frac{\cos(40 - 60)\cos(-11.6)\cos(-37.5) + \sin(40 - 60)\sin(-11.6)}{\cos 40 \cos(-11.6)\cos(-37.5) + \sin 40 \sin(-11.6)}$$

$$= \frac{0.799}{0.466} = 1.71$$

Equation 2.15.1 gives the three radiation streams and the total:

$$I_T = 0.253 \times 1.71 + 0.787\left(\frac{1 + \cos 60}{2}\right) + 1.04 \times 0.60\left(\frac{1 - \cos 60}{2}\right)$$

$$= 0.433 + 0.590 + 0.156 = 1.18 \text{ MJ/m}^2$$

Thus the beam contribution is 0.434 MJ/m², the diffuse is 0.590 MJ/m², and the ground-reflected is 0.156 MJ/m². The total radiation on the surface for the hour is 1.18 MJ/m². There are uncertainties in these numbers, and while they are carried to 0.001 MJ in intermediate steps for purposes of comparing sky models, they are certainly no better than 0.01.   ∎

2.16   RADIATION ON SLOPED SURFACES - ANISOTROPIC SKY

The isotropic diffuse model (Equation 2.15.1) is easy to understand, is conservative (i.e., it tends to underestimate I_T), and makes calculation of radiation on tilted surfaces easy. However, improved models have been developed which take into account the circumsolar diffuse and/or horizon brightening components on a tilted surface that are shown schematically in Figure 2.16.1. Hay and Davies (1980) estimate the fraction of the diffuse that is circumsolar and consider it to be all from the same direction as the beam radiation; they do not treat horizon brightening. Reindl et al. (1990b) add a horizon brightening term to the Hay and Davies model, as proposed by Klucher (1979), giving a model to be referred to as the HDKR model. Skartveit and Olseth (1986, 1987a,b) develop methods for estimating the beam and diffuse distribution and radiation on sloped surfaces starting with monthly average radiation. Perez et al. (1987, 1988) treat both circumsolar diffuse and horizon brightening in some detail in a series of models of varying complexity.

The Hay and Davies model is based on the assumption that all of the diffuse can be represented by two parts, the isotropic and the circumsolar. Thus all but the fourth term in Equation 2.14.3 are used. The diffuse radiation on a tilted collector is written as

$$I_{d,T} = I_{T,d,iso} + I_{T,d,cs} \tag{2.16.1}$$

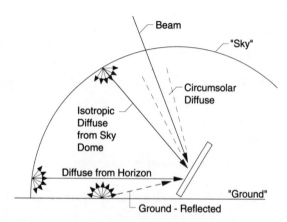

Figure 2.16.1   Beam, diffuse, and ground-reflected radiation on a tilted surface.

and
$$I_{d,T} = I_d\left[(1 - A_i)\left(\frac{1 + \cos \beta}{2}\right) + A_i R_b\right] \qquad (2.16.2)$$

where A_i is an **anisotropy index** which is a function of the transmittance of the atmosphere for beam radiation,

$$A_i = \frac{I_{bn}}{I_{on}} = \frac{I_b}{I_o} \qquad (2.16.3)$$

The anisotropy index determines a portion of the horizontal diffuse which is to be treated as forward scattered; it is considered to be incident at the same angle as the beam radiation. The balance of the diffuse is assumed to be isotropic. Under clear conditions, the A_i will be high, and most of the diffuse will be assumed to be forward scattered. When there is no beam, A_i will be zero, the calculated diffuse is completely isotropic, and the model becomes the same as Equation 2.15.1.

The total radiation on a tilted surface is then

$$I_T = (I_b + I_d A_i) R_b + I_d (1 - A_i)\left(\frac{1 + \cos \beta}{2}\right) + I \rho_g\left(\frac{1 - \cos \beta}{2}\right) \qquad (2.16.4)$$

The Hay and Davies method for calculating I_T is not much more complex than the isotropic model, and leads to slightly higher[14] estimates of radiation on the tilted surface. Reindl et al. (1990a) and others indicate that the results obtained with this model are an improvement over the isotropic model. However, it does not account for horizon brightening. Temps and Coulson (1977) account for horizon brightening on clear days by applying a correction factor of $[1 + \sin^3(\beta/2)]$ to the isotropic diffuse. Klucher (1979) modified this correction factor by a modulating factor F so that it has the form $[1 + F \sin^3(\beta/2)]$ to account for cloudiness.

Reindl et al. (1990b) have modified the Hay and Davies model by the addition of a term like that of Klucher. The diffuse on the tilted surface is

$$I_{d,T} = I_d\left\{(1 - A_i)\left(\frac{1 + \cos \beta}{2}\right)\left[1 + f \sin^3\left(\frac{\beta}{2}\right)\right] + A_i R_b\right\} \qquad (2.16.5)$$

where A_i is as defined by Equation 2.16.3, and

$$f = \sqrt{I_b/I} \qquad (2.16.6)$$

When the beam and ground-reflected terms are added, the HDKR model (the Hay, Davies, Klucher, Reindl model) results. The total radiation on the tilted surface is

---

[14] Recalculation of Example 2.15.1 with Equation 2.16.4 leads to I_T of 1.26 MJ/m², about 7% higher than the isotropic assumption.

then

$$I_T = (I_b + I_d A_i) R_b + I_d(1 - A_i) \left(\frac{1 + \cos \beta}{2}\right) \left[1 + f \sin^3\left(\frac{\beta}{2}\right)\right]$$

$$+ I \rho_g \left(\frac{1 - \cos \beta}{2}\right) \tag{2.16.7}$$

Example 2.16.1

Do Example 2.15.1 using the HDKR model.

Solution

From Example 2.15.1, $I = 1.04$ MJ/m^2, $I_b = 0.253$ MJ/m^2, $I_d = 0.787$ MJ/m^2, $I_o = 2.31$ MJ/m^2, and $R_b = 1.71$. From Equation 2.16.3,

$$A_i = 0.253/2.31 = 0.110$$

The modulating factor f, from Equation 2.16.6, is

$$f = \sqrt{0.253/1.04} = 0.493$$

Then from Equation 2.16.7,

$$I_T = (0.253 + 0.787 \times 0.110)1.71$$

$$+ 0.787 (1 - 0.110)\left(\frac{1 + \cos 60}{2}\right)\left[1 + 0.493 \sin^3\left(\frac{60}{2}\right)\right]$$

$$+ 1.04 \times 0.60\left(\frac{1 - \cos 60}{2}\right)$$

$$= 0.581 + 0.557 + 0.156 = 1.29 \text{ MJ/m}^2$$

In this example, the correction factor to the diffuse to account for horizon brightening is 1.06, and the total estimated radiation on the tilted surface is 9% more than that estimated by the isotropic model.[15] ∎

---

[15] In Chapter 5 we will multiply each of the radiation streams by transmittance and absorptance factors which are functions of the angle of incidence of those streams on collectors. Thus it is generally necessary to calculate each of the streams independently. The differences among the various models may become more significant when these factors are applied.

The Perez et al. model (1988) is based on a more detailed analysis of the three diffuse components. The diffuse on the tilted surface is given by

$$I_{d,T} = I_d \left[(1 - F_1) \left(\frac{1 + \cos \beta}{2} \right) + F_1 \frac{a}{b} + F_2 \sin \beta \right] \tag{2.16.8}$$

where F_1 and F_2 are circumsolar and horizon brightness coefficients and a and b are terms that account for the angles of incidence of the cone of circumsolar radiation (Figure 2.16.1) on the tilted and horizontal surfaces. The circumsolar radiation is considered to be from a point source at the sun. The terms a and b are

$$a = \max \left[0, \cos \theta \right] \tag{2.16.9}$$
$$b = \max \left[\cos 85, \cos \theta_z \right]$$

With these definitions, a/b becomes R_b for most hours when collectors will have useful outputs.

The brightness coefficients F_1 and F_2 are functions of three parameters that describe the sky conditions, the zenith angle θ_z, a clearness ε, and a brightness Δ, where ε is a function of the hour's diffuse radiation I_d and normal incidence beam radiation I_n. It is given by

$$\varepsilon = \frac{\dfrac{I_d + I_n}{I_d} + 5.535 \times 10^{-6} \theta_z^3}{1 + 5.535 \times 10^{-6} \theta_z^3} \tag{2.16.10}$$

where θ_z is in degrees, and

$$\Delta = m \frac{I_d}{I_{on}} \tag{2.16.11}$$

where m is the air mass (Equation 1.5.1) and I_{on} is the extraterrestrial normal incidence radiation (Equation 1.4.1), written in terms of I. Thus these parameters are all calculated from data on total and diffuse radiation (i.e., the data that are used in the computation of I_T).

The brightness coefficients F_1 and F_2 are functions of statistically derived coefficients for ranges of values of ε; a recommended set of these coefficients is shown in Table 2.16.1. The equations for calculating F_1 and F_2 are

$$F_1 = \max \left[0, \left(f_{11} + f_{12}\Delta + \frac{\pi \theta_z}{180} f_{13} \right) \right] \tag{2.16.12}$$

$$F_2 = \left(f_{21} + f_{22}\Delta + \frac{\pi \theta_z}{180} f_{23} \right) \tag{2.16.13}$$

Table 2.16.1   Brightness Coefficients for Perez et al. Anisotropic Sky[a]

Range of ε	f_{11}	f_{12}	f_{13}	f_{21}	f_{22}	f_{23}
0 - 1.065	−0.196	1.084	−0.006	−0.114	0.180	−0.019
1.065 - 1.230	0.236	0.519	−0.180	−0.011	0.020	−0.038
1.230 - 1.500	0.454	0.321	−0.255	0.072	−0.098	−0.046
1.500 - 1.950	0.866	−0.381	−0.375	0.203	−0.403	−0.049
1.950 - 2.800	1.026	−0.711	−0.426	0.273	−0.602	−0.061
2.800 - 4.500	0.978	−0.986	−0.350	0.280	−0.915	−0.024
4.500 - 6.200	0.748	−0.913	−0.236	0.173	−1.045	0.065
6.200 - ↑	0.318	−0.757	0.103	0.062	−1.698	0.236

[a] From Perez et al. (1988).

This set of equations allows calculation of all of the three diffuse components on the tilted surface. It remains to add the beam and ground-reflected contributions. The total radiation on the tilted surface includes five terms: the beam, the isotropic diffuse, the circumsolar diffuse, the diffuse from the horizon, and the ground-reflected term (which parallel the terms in Equation 2.14.3):

$$I_T = I_b R_b + I_d (1 - F_1)\left(\frac{1 + \cos \beta}{2}\right) + I_d F_1 \frac{a}{b}$$

$$+ I_d F_2 \sin \beta + I \rho_g \left(\frac{1 - \cos \beta}{2}\right) \tag{2.16.14}$$

Equations 2.16.8 through 2.16.14, with Table 2.16.1, constitute a working version of the Perez et al. model. Its use is illustrated in the following example.

Example 2.16.2

Do Example 2.15.1 using the Perez et al. method.

Solution

From Example 2.15.1: $I_o = 2.31$ MJ/m^2, $I = 1.04$ MJ/m^2, $I_b = 0.253$ MJ/m^2, $I_d = 0.787$ MJ/m^2, $\cos \theta = 0.799$, $\theta = 37.0°$, $\cos \theta_z = 0.466$, $\theta_z = 62.2°$, and $R_b = 1.71$.

To use Equation 2.16.14, we need a, b, ε, and Δ in addition to the quantities already calculated:

$$a = \max[0, \cos 37.0] = 0.799$$

$$b = \max[\cos 85, \cos 62.2] = 0.466$$

and $a/b = 0.799/0.466 = 1.71$ (the same as R_b in Example 2.15.1)

Next calculate Δ. The air mass m, from Equation 1.5.1, is

$$m = 1/\cos 62.2 = 1/0.466 = 2.144$$

We also need I_{on}. Use Equation 1.4.1 with $n = 51$:

$$I_{on} = 4.921\left[1 + 0.033 \cos \frac{360 \times 51}{365}\right] = 5.025 \text{ MJ/m}^2$$

From the defining equation for Δ, Equation 2.16.11,

$$\Delta = 0.787 \times 2.144/5.025 = 0.336$$

We next calculate ε from Equation 2.16.10. Thus $I_n = I_b/\cos\theta_z = 0.253/\cos 62.2 = 0.544 \text{ MJ/m}^2$, and

$$\varepsilon = \frac{\dfrac{0.787 + 0.544}{0.787} + 5.535 \times 10^{-6}(62.2)^3}{1 + 5.535 \times 10^{-6}(62.2)^3} = 1.30$$

With this we can go to the table of coefficients needed in the calculation of F_1 and F_2. These are, for the third ε range,

$$f_{11} = 0.454 \qquad f_{12} = 0.321 \qquad f_{13} = -0.255$$

$$f_{21} = 0.072 \qquad f_{22} = -0.098 \qquad f_{23} = -0.046$$

So

$$F_1 = \max[0, (0.454 + 0.321 \times 0.336 + 62.2\pi(-0.255)/180)] = 0.285$$

$$F_2 = 0.072 + (-0.098) \times 0.336 + 62.2\pi(-0.046)/180 = -0.011$$

We now have everything needed to use Equation 2.16.14 to get the total radiation on the sloped surface:

$$I_T = 0.253 \times 1.71 + 0.787(1 - 0.285)\left(\frac{1 + \cos 60}{2}\right) + 0.787 \times 0.285 \times 1.71$$

$$- 0.011 \times 0.787 \sin 60 + 1.04 \times 0.60\left(\frac{1 - \cos 60}{2}\right)$$

$$= 0.433 + 0.422 + 0.384 - 0.007 + 0.156 = 1.39 \text{ MJ/m}^2$$

This is about 8% higher than the result of the HDKR model and about 18% higher than the isotropic model for this example. ∎

The next question is which of these models for total radiation on the tilted surface should be used. The isotropic model is the simplest, gives the most conservative estimates of radiation on the tilted surface, and has been widely used. The HDKR model is almost as simple to use as the isotropic and produces results that are closer to measured values. For surfaces sloped toward the equator, these models are suggested. The Perez model is more complex to use and generally predicts slightly higher total radiation on the tilted surface; it is thus the least conservative of the three methods. It agrees the best by a small margin with measurements.[16] For surfaces with γ far from 0° in the northern hemisphere or 180° in the southern hemisphere, the Perez model is suggested. (In examples to be shown in later chapters, the isotropic and HDKR methods will be used, as they are more amenable to hand calculation.)

2.17   RADIATION AUGMENTATION

It is possible to increase the radiation incident on an absorber by use of planar reflectors. In the models discussed in Sections 2.15 and 2.16, ground-reflected radiation was taken into account in the last term, with the ground assumed to be a horizontal diffuse reflector infinite in extent, and there was only one term in the summation in Equation 2.15.2. With ground reflectances normally of the order of 0.2 and low collector slopes, the contributions of ground-reflected radiation are small. However, with ground reflectances of 0.6 to 0.7 typical of snow and with high slopes,[17] the contribution of reflected radiation of surfaces may be substantial. In this section we show a more general case of the effects of diffuse reflectors of finite size.

Consider the geometry sketched in Figure 2.17.1. Consider two intersecting planes, the receiving surface c (i.e., a solar collector or passive absorber) and a diffuse reflector r. The angle between the planes is ψ. The angle ψ is 180° – β if the reflector is horizontal, but the analysis is not restricted to a horizontal reflector. The length of the assembly is m. The other dimensions of the collector and reflector are n and p, as shown.

If the reflector is horizontal, Equation 2.15.3 becomes

$$I_T = I_b R_b + I_d F_{c\text{-}s} + I_r \rho_r F_{c\text{-}r} + I \rho_g F_{c\text{-}g} \tag{2.17.1}$$

where $F_{c\text{-}s}$ is again $(1 + \cos \beta)/2$. The view factor $F_{r\text{-}c}$ is obtained from Figure

---

[16] The HDKR method yields slightly better results than either the isotropic model or the Perez model in predicting utilizable radiation when the critical radiation levels are significant. See Sections 2.20 to 2.22 for notes on utilizable energy.

[17] At a slope of 45°, a flat surface sees 85% sky and 15% ground. At a slope of 90°, it sees 50% sky and 50% ground.

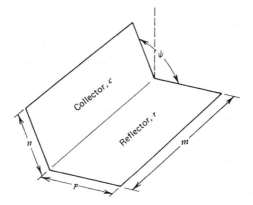

Figure 2.17.1 Geometric relationship of an energy receiving surface c and reflecting surface r.

2.17.2, $F_{c\text{-}r}$ is obtained from the reciprocity relationship $A_c F_{c\text{-}r} = A_r F_{r\text{-}c}$, and $F_{c\text{-}g}$ can be obtained from the summation rule, $F_{c\text{-}s} + F_{c\text{-}r} + F_{c\text{-}g} = 1$. The view factor $F_{r\text{-}c}$ is shown in Figure 2.17.2 as a function of the ratios n/m and p/m for ψ of 90°, 120°, and 150°.

Example 2.17.1

A vertical window receiver in a passive heating system is 3.0 m high and 6.0 m long. There is deployed in front of it a horizontal, diffuse reflector of the same length extending out 2.4 m. What is the view factor from the reflector to the window? What is the view factor from the window to the reflector? What is the view factor from the window to the ground beyond the reflector?

Solution

For the given dimensions, n/m is $3.0/6.0 = 0.5$, p/m is $2.4/6.0 = 0.4$, and from Figure 2.17.2a, the view factor $F_{r\text{-}c}$ is 0.27.

The area of the window is 18.0 m², and the area of the reflector is 14.4 m². From the reciprocity relationship, $F_{c\text{-}r} = (14.4 \times 0.27)/18.0 = 0.22$.

The view factor from window to sky, $F_{c\text{-}s}$, is $(1 + \cos 90)/2$, or 0.50. The view factor from collector to ground is then $1 - (0.50 + 0.22) = 0.28$. ∎

If the surfaces c and r are very long in extent (i.e., m is large relative to n and p, as might be the case with long arrays of collectors for large-scale solar applications), Hottel's "crossed-string" method gives the view factor as

$$F_{r\text{-}c} = \frac{n + p - s}{2p} \qquad\qquad (2.17.2)$$

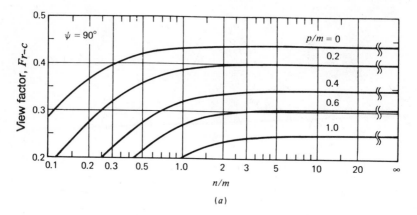

(a)

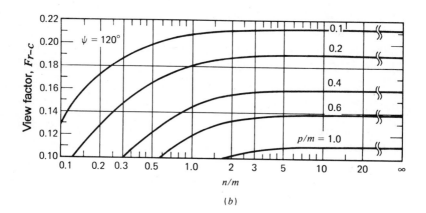

(b)

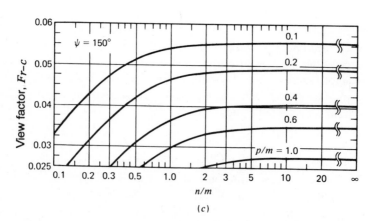

(c)

Figure 2.17.2 View factor F_{r-c} as a function of the relative dimensions of the collecting and reflecting surfaces. Adapted from Hamilton and Morgan (1952).

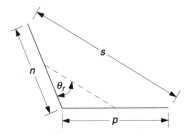

Figure 2.17.3   Section of reflector and collector surfaces.

where s is the distance from the upper edge of the collector to the outer edge of the reflector, measured in a plane perpendicular to planes c and r, as shown in Figure 2.17.3. This can be determined from

$$s = \left(n^2 + p^2 - 2np \cos \psi\right)^{1/2} \qquad (2.17.3)$$

[For a collector array as in Example 2.17.1 but very long in extent, $s = (3.0^2 + 2.4^2)^{0.5} = 3.84$ m and $F_{r-c} = (3 + 2.4 - 3.84)/4.8 = 0.33$.]

It is necessary to know the incident radiation on the plane of the reflector. The beam component is calculated by use of R_{br} for the orientation of the reflector surface. The diffuse component must be estimated from the view factor F_{r-s}. For any orientation of the surface r,

$$F_{r-s} + F_{r-c} + F_{r-g} = 1 \qquad (2.17.4)$$

where the view factors are from surface r to sky, to surface c, and to ground. The view factor F_{r-c} is determined as noted above and F_{r-g} will be zero for a horizontal reflector and will be small for collectors that are long in extent. Thus as a first approximation, $F_{r-s} = 1 - F_{r-c}$ for many practical cases (where there are no other obstructions).

There remains the questions of the angle of incidence of radiation reflected from surface r on surface c. As an approximation, an average angle of incidence can be taken as that of the radiation from the midpoint of surface r to the midpoint of surface c, as shown in Figure 2.17.3.[18] The average angle of incidence θ_r is given by

$$\sin \theta_r = \frac{p \sin \psi}{s} \qquad (2.17.5)$$

The total radiation reflected from surface r with area A_r to surface c with area A_c if

---

[18] As the reflector area becomes very large, the angle of incidence becomes that given by the ground-reflectance curve of Figure 5.4.1, where the angle ψ between the reflector and the collector is β, the abscissa on the figure.

r has a diffuse reflectance of ρ_r is

$$A_c I_r = [I_b R_{br} + (1 - F_{r-c})I_d]\rho_r A_r F_{r-c} \qquad (2.17.6)$$

Example 2.17.2

A south-facing vertical surface is 4.5 m high and 12 m long. It has in front of it a horizontal diffuse reflector of the same length which extends out 4 m. The reflectance is 0.85. At solar noon, the total irradiance on a horizontal surface is 800 W/m² of which 200 is diffuse. The zenith angle of the sun is 50°. Estimate the total radiation on the vertical surface and the angle of incidence on that part of the total that is reflected from the diffuse reflector.

Solution

Here we have an irradiance, the instantaneous radiation, instead of the hourly values of the examples in Section 2.15, so the solution will be in terms of G rather than I.

First estimate F_{r-c} from Figure 2.17.2. At $n/m = 4.5/12 = 0.38$ and $p/m = 4/12 = 0.33$, $F_{r-c} = 0.28$. The total radiation on the reflector is the beam component, 600 W/m², plus the diffuse component, which is $G_d F_{r-s}$ or $G_d(1 - F_{r-c})$. The radiation reflected from the reflector that is incident on the vertical surface is estimated by Equation 2.17.6:

$$G_r = [600 + 200(1 - 0.28)]\frac{0.85 \times 48 \times 0.28}{4.5 \times 12} = 160 \text{ W/m}^2$$

The beam component on the vertical surface is obtained with R_b, which is cos 40/cos 50 = 1.19. Then $G_{bT} = 600 \times 1.19 = 715$ W/m². The diffuse component from the sky on the vertical surface is estimated as

$$G_{dT} = 200\frac{1 + \cos 90}{2} = 100 \text{ W/m}^2$$

The total radiation on the vertical surface (neglecting reflected radiation from ground areas beyond the reflector) is the sum of the three terms:

$$G_T = 160 + 715 + 100 = 975 \text{ W/m}^2$$

An average angle of incidence of the reflected radiation on the vertical surface is estimated with Equation 2.17.5:

$$s = (4.0^2 + 4.5^2)^{0.5} = 6.02 \text{ m}$$

and $\sin \theta_r = 4.5 \sin 90/6.02 = 0.75$

$\theta_r = 49°$ ∎

The contributions of diffuse reflectors may be significant, although they will not result in large increments in incident radiation. In the preceding example, the contribution is approximately 160 W/m². If the horizontal surface in front of the vertical plane were ground with $\rho_g = 0.2$, the contribution from ground-reflected radiation would have been $0.2 \times 800(1 - \cos 90)/2$, or 80 W/m².

It has been pointed out by McDaniels et al. (1975), Grassie and Sheridan (1977), Chiam (1981,1982), and others that a specular reflector can have more effect in augmenting radiation on a collector than a diffuse reflector.[19] Hollands (1971) presents a method of analysis of some reflector-collector geometries, and Bannerot and Howell (1979) show effects of reflectors on average radiation on surfaces. The effects of reflectors that are partly specular and partly diffuse are treated by Grimmer et al. (1978). The practical problem is to maintain high specular reflectance, particularly on surfaces that are facing upward. Such surfaces are difficult to protect against weathering and will accumulate snow in cold climates.

2.18 BEAM RADIATION ON MOVING SURFACES

Sections 2.15 to 2.17 have dealt with estimation of total radiation on surfaces of fixed orientation, such as flat-plate collectors or windows. It is also of interest to estimate the radiation on surfaces that move in various prescribed ways. Most concentrating collectors utilize beam radiation only and move to "track" the sun. This section is concerned with the calculation of beam radiation on these planes, which move about one or two axes of rotation. The tracking motions of interest are described in Section 1.7, and for each the angle of incidence is given as a function of the latitude, declination, and hour angle.

At any time the beam radiation on a surface is a function of G_{bn}, the beam radiation on a plane normal to the direction of propagation of the radiation:

$$G_{bT} = G_{bn} \cos \theta \qquad (2.18.1)$$

where $\cos \theta$ is given by equations in Section 1.7 for various modes of tracking of the collector. If the data that are available are for beam normal radiation, this equation is the correct one to use. Note that as with other calculations of this type, Equation 2.18.1 can be written for an hour, in terms of I rather than G, and the angles calculated for the midpoint of the hour.

Example 2.18.1

A concentrating collector is continuously rotated on a polar axis, i.e., an axis that is parallel to the earth's axis of rotation. The declination is 17.5°, and the beam normal solar radiation for an hour is 2.69 MJ/m². What is I_{bT}, the beam radiation on the aperture of the collector?

---

[19] See Chapter 7 for discussion of specular reflectors.

Solution

For a collector continuously tracking on a polar axis, cos θ = cos δ (Equation 1.7.5a). Thus

$$I_{bT} = I_{bn} \cos \theta = 2.69 \cos 17.5 = 2.57 \text{ MJ/m}^2 \qquad \blacksquare$$

If radiation data on a horizontal surface are used, the R_b concept must be applied. If the data are for hours (i.e., I), the methods of Section 2.10 are used to estimate I_b, and R_b is determined from its definition (Equation 1.8.1) using the appropriate equation for cos θ. If daily data are available (i.e., H), estimates of hourly beam must be made using the methods of Sections 2.10, 2.11, and 2.13. This is illustrated in the next example.

Example 2.18.2

A cylindrical concentrating collector is to be oriented so that it rotates about a horizontal east-west axis so as to constantly minimize the angle of incidence and thus maximize the incident beam radiation. It is to be located at 35°N latitude. On April 13, the day's total radiation on a horizontal surface is 22.8 MJ/m². Estimate the beam radiation on the aperture (the moving plane) of this collector for the hour 1 to 2 PM.

Solution

For this date, δ is 8.67°, ω_s is 96.13°, ω = 22.5°, H_o is 35.1 MJ/m², K_T is 22.8/35.1 = 0.65, and from Figure 2.11.2, H_d/H is 0.34. Thus H_d is 7.75 MJ/m². From Figure 2.13.1 or Equation 2.13.1, r_t is 0.121, and from Figure 2.13.2 or Equation 2.13.2, r_d is 0.115. Thus

$$I = 22.8 \times 0.121 = 2.76 \text{ MJ/m}^2$$

and $\qquad I_d = 7.75 \times 0.115 = 0.89 \text{ MJ/m}^2$

and by difference, $I_b = 1.87$ MJ/m².

Next, calculate R_b from the ratio of Equations 1.7.2a and 1.6.5:

$$R_b = \frac{\left[1 - \cos^2(8.67) \sin^2(22.5)\right]^{1/2}}{\cos 35 \cos 8.67 \cos 22.5 + \sin 35 \sin 8.67} = \frac{0.926}{0.835} = 1.11$$

and $\qquad I_{bT} = I_b R_b = 1.87 \times 1.11 = 2.1 \text{ MJ/m}^2 \qquad \blacksquare$

The uncertainties in these estimations of beam radiation are greater than those associated with estimations of total radiation, and the use of pyrheliometric data is preferred if they are available.

2.19 AVERAGE RADIATION ON SLOPED SURFACES - ISOTROPIC SKY

In Section 2.15, the calculation of total radiation on sloped surfaces from measurements on a horizontal surface was discussed. For use in solar process design procedures,[20] we also need the monthly average daily radiation on the tilted surface. The procedure for calculating $\overline{H}_T$ is parallel to that for I_T, that is, by summing the contributions of the beam radiation, the components of the diffuse radiation, and the radiation reflected from the ground. The state of development of these calculation methods for $\overline{H}_T$ is not as satisfactory as that for I_T.

The first method is that of Liu and Jordan (1962) as extended by Klein (1977), which has been widely used. If the diffuse and ground-reflected radiation are each assumed to be isotropic, then, in a manner analogous to Equation 2.15.1, the monthly mean solar radiation on an unshaded tilted surface can be expressed as

$$\overline{H}_T = \overline{H}\left(1 - \frac{\overline{H}_d}{\overline{H}}\right)\overline{R}_b + \overline{H}_d\left(\frac{1 + \cos\beta}{2}\right) + \overline{H}\rho_g\left(\frac{1 - \cos\beta}{2}\right) \tag{2.19.1}$$

and

$$\overline{R} = \frac{\overline{H}_T}{\overline{H}} = \left(1 - \frac{\overline{H}_d}{\overline{H}}\right)\overline{R}_b + \frac{\overline{H}_d}{\overline{H}}\left(\frac{1 + \cos\beta}{2}\right) + \rho_g\left(\frac{1 - \cos\beta}{2}\right) \tag{2.19.2}$$

where $\overline{H}_d/\overline{H}$ is a function of $\overline{K}_T$, as shown in Figure 2.12.2.

The ratio of the average daily beam radiation on the tilted surface to that on a horizontal surface for the month is $\overline{R}_b$ which is equal to $\overline{H}_{bT}/\overline{H}_b$. It is a function of transmittance of the atmosphere, but Liu and Jordan suggest that it can be estimated by assuming that it has the value which would be obtained if there were no atmosphere. For surfaces that are sloped toward the equator in the northern hemisphere, that is, for surfaces with $\gamma = 0°$,

$$\overline{R}_b = \frac{\cos(\phi - \beta)\cos\delta\sin\omega_s' + (\pi/180)\,\omega_s'\,\sin(\phi - \beta)\sin\delta}{\cos\phi\cos\delta\sin\omega_s + (\pi/180)\omega_s\sin\phi\sin\delta} \tag{2.19.3a}$$

where ω_s' is the sunset hour angle for the tilted surface for the mean day of the month, which is given by

$$\omega_s' = \min\begin{bmatrix}\cos^{-1}(-\tan\phi\tan\delta) \\ \cos^{-1}(-\tan(\phi - \beta)\tan\delta)\end{bmatrix} \tag{2.19.3b}$$

where "min" means the smaller of the two items in the brackets.

---

[20] See Part III.

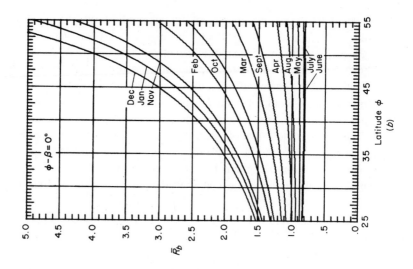

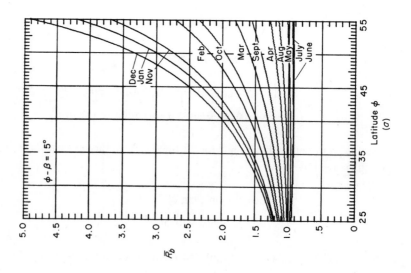

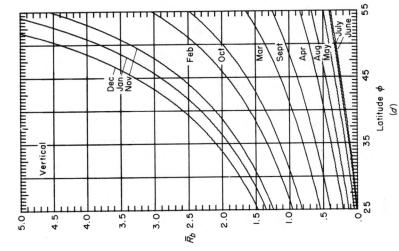

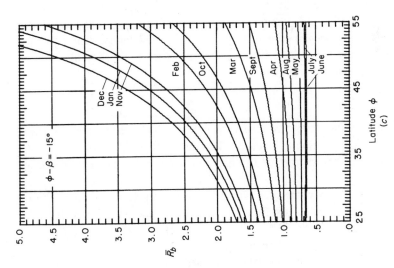

Figure 2.19.1 Estimated $\overline{R}_b$ for surfaces facing the equator as a function of latitude for various ($\phi - \beta$), by months. (a) ($\phi - \beta$) = 15°; (b) ($\phi - \beta$) = 0°; (c) ($\phi - \beta$) = −15°; (d) β = 90°. For the southern hemisphere, interchange months as shown on Figure 1.8.2. From Beckman et al. (1977).

For surfaces in the southern hemisphere sloped toward the equator, with $\gamma = 180°$, the equations are

$$\bar{R}_b = \frac{\cos(\phi + \beta)\cos\delta\sin\omega_s' + (\pi/180)\,\omega_s'\,\sin(\phi + \beta)\sin\delta}{\cos\phi\cos\delta\sin\omega_s + (\pi/180)\omega_s\sin\phi\sin\delta} \qquad (2.19.4a)$$

and

$$\omega_s' = \min \left[\begin{array}{c} \cos^{-1}(-\tan\phi\tan\delta) \\ \cos^{-1}\left(-\tan\left(\phi + \beta\right)\tan\delta\right) \end{array} \right] \qquad (2.19.4b)$$

The numerator of Equation 2.19.3a or 2.19.4a is the extraterrestrial radiation on the tilted surface, and the denominator is that on the horizontal surface. Each of these is obtained by integration of Equation 1.6.2 over the appropriate time period, from true sunrise to sunset for the horizontal surface and from apparent sunrise to apparent sunset on the tilted surface. For convenience, plots of $\bar{R}_b$ as a function of latitude for various slopes for $\gamma = 0°$ (or $180°$ in the southern hemisphere) are shown in Figure 2.19.1, and corresponding tables of $\bar{R}_b$ are in Appendix D. These values of $\bar{R}_b$ can be used for surface azimuth angles of $0°$ (or $180°$) $\pm 15°$ with little error.

The following example illustrates the kind of calculations that will be used in estimating monthly radiation on collectors as part of heating system design procedures.

Example 2.19.1

A collector is to be installed in Madison, latitude $43°$, at a slope of $60°$ to the south. Average daily radiation data are shown in Appendix G. The ground reflectance is 0.2 for all months except December and March ($\rho_g = 0.4$) and January and February ($\rho_g = 0.7$). Using the isotropic diffuse assumption, estimate the monthly average radiation incident on the collector.

Solution

The calculation is detailed below for January, and the results for the year are indicated in a table. The basic equation to be used is Equation 2.19.1. The first steps are to obtain $\bar{H}_d/\bar{H}$ and $\bar{R}_b$. The ratio $\bar{H}_d/\bar{H}$ is a function of $\bar{K}_T$ and can be obtained from Equation 2.12.1 or Figure 2.12.2.

For the mean January day, the 17th, from Table 1.6.1, $n = 17$, $\delta = -20.9°$. The sunset hour angle is calculated from Equation 1.6.10 and is $69.1°$. With $n = 17$ and $\omega_s = 69.1°$, from Equation 1.10.3 (or Figure 1.10.1 or Table 1.10.1), $\bar{H}_o$ is 13.36 MJ/m². Then $\bar{K}_T = 5.85/13.36 = 0.44$.

The Erbs correlation, Equation 2.12.1a, is used to calculate $\bar{H}_d/\bar{H}$ from $\bar{K}_T$ and ω_s and gives $\bar{H}_d/\bar{H} = 0.45$. The calculation of $\bar{R}_b$ requires the sunset hour angle on the sloped collector. From Equations 2.19.3

$$\cos^{-1}[-\tan(43 - 60)\tan(-20.8)] = 96.7°$$

The angle ω_s was calculated as 69.1° and is less than 96.7°, so $\omega_s' = 69.1°$. Then

$$\bar{R}_b = \frac{\cos(-17)\,\cos(-20.9\,\sin 69.1 + (\pi \times 69.1/180)\,\sin(-17)\,\sin(-20.9)}{\cos 43\,\cos(-20.9)\,\sin 69.1 + (\pi \times 69.1/180)\,\sin 43\,\sin(-20.9)} = 2.79$$

The equation for $\bar{H}_T$, Equation 2.19.1, can now be solved:

$$\bar{H}_T = 5.85(1 - 0.45)2.79 + 5.85 \times 0.45\left(\frac{1 + \cos 60}{2}\right) + 5.85 \times 0.7\left(\frac{1 - \cos 60}{2}\right)$$

$$= 12.0 \text{ MJ/m}^2$$

The results for the 12 months are shown in the table. Energy quantities are in megajoules per square meter. The effects of sloping the receiving plane 60° to the south on the average radiation (and thus on the total radiation through the winter season) are large indeed. The monthly results are shown in the table. The $\bar{H}_T$ values are shown to a tenth of a megajoule per square meter. The last place is uncertain due to the combined uncertainties in the data and the correlations for $\bar{H}_d/\bar{H}$ and $\bar{R}$. It is difficult to put limits of accuracy on them; they are probably no better than ±10%.

Month	$\bar{H}$	$\bar{H}_o$	$\bar{K}_T$	$\bar{H}_d/\bar{H}$	$\bar{R}_b$	ρ_g	$\bar{H}_T$
Jan.	5.85	13.36	0.44	0.46	2.79	0.7	11.9
Feb.	9.13	18.80	0.49	0.41	2.04	0.7	15.5
Mar.	12.89	26.01	0.50	0.43	1.42	0.4	15.8
Apr.	15.88	33.75	0.47	0.46	0.96	0.2	14.5
May	19.79	39.39	0.50	0.43	0.71	0.2	15.4
June	22.11	41.74	0.53	0.40	0.62	0.2	15.9
July	21.96	40.52	0.54	0.39	0.66	0.2	16.3
Aug.	19.39	33.88	0.54	0.39	0.84	0.2	16.6
Sep.	14.75	28.77	0.51	0.42	1.21	0.2	15.8
Oct.	10.34	20.89	0.50	0.40	1.81	0.2	14.9
Nov.	5.72	14.61	0.39	0.51	2.56	0.2	9.6
Dec.	4.42	11.90	0.37	0.54	3.06	0.4	8.5

■

2.20 AVERAGE RADIATION ON SLOPED SURFACES - THE K-T METHOD

An alternative approach to calculation of average radiation on sloped surfaces has been developed by Klein and Theilacker (1981). It is a bit more cumbersome to use but shows improved results over the isotropic method when compared with hourly calculations based on many years of radiation data. The method is first outlined below in a form restricted to surfaces facing the equator and then in general form for surfaces of any orientation. As with Equations 2.19.1 and 2.19.2, it is based on the assumption that both diffuse and ground-reflected radiation streams are isotropic.

The long-term value of $\overline{R}$ can be calculated by integrating G_T and G from sunrise to sunset for all days over many years of data for a single month and summing. (For example, data for all days in January for 10 years should represent the long-term average for January.)

$$\overline{R} = \frac{\displaystyle\sum_{day\,=\,1}^{N} \int_{t_{sr}}^{t_{ss}} G_T \, dt}{\displaystyle\sum_{day\,=\,1}^{N} \int_{t_{sr}}^{t_{ss}} G \, dt} \tag{2.20.1}$$

The denominator is $N\overline{H}$. To evaluate the numerator, it is convenient to replace G_T by I_T and exchange the order of the integration and summation. Using Equation 2.15.1, the radiation at any time of the day (i.e., for any hour) for N days is

$$N\overline{I}_T = N\left[\left(\overline{I} - \overline{I}_d\right)R_b + \overline{I}_d\left(\frac{1 + \cos\beta}{2}\right) + \overline{I}\rho_g\left(\frac{1 - \cos\beta}{2}\right)\right] \tag{2.20.2}$$

where the $\overline{I}$ and $\overline{I}_d$ are long-term averages of the total and the diffuse radiation, obtained by summing the values of I and I_d over N days for each particular hour and dividing by N. Equation 2.20.1 then becomes[21]

$$\overline{R} = \frac{\displaystyle\int_{t_{sr}}^{t_{ss}} \left[\left(\overline{I} - \overline{I}_d\right)R_b + \overline{I}_d\left(\frac{1 + \cos\beta}{2}\right) + \overline{I}\rho_g\left(\frac{1 - \cos\beta}{2}\right)\right] dt}{\overline{H}} \tag{2.20.3}$$

Equations 2.13.1 and 2.13.3 define the ratios of hourly to daily total and hourly to daily diffuse radiation, and Equations 2.13.2 and 2.13.4 relate r_t and r_d to time ω and sunset hour angle ω_s. Combining these with Equation 2.20.3 leads, for the case of south-facing surfaces in the northern hemisphere, to

$$\overline{R} = \frac{\cos(\phi - \beta)}{d\,\cos\phi}\left[\left(a - \frac{\overline{H}_d}{\overline{H}}\right)\left(\sin\omega_s' - \frac{\pi\omega_s'}{180}\cos\omega_s''\right)\right.$$

$$\left. + \frac{b}{2}\left(\frac{\pi\omega_s'}{180} + \sin\omega_s'(\cos\omega_s' - 2\cos\omega_s'')\right)\right]$$

$$+ \frac{\overline{H}_d}{\overline{H}}\left(\frac{1 + \cos\beta}{2}\right) + \rho_g\left(\frac{1 - \cos\beta}{2}\right) \tag{2.20.4a}$$

---

[21] The development of this equation assumes that the day length does not change during the month.

where ω_s' is again given by

$$\omega_s' = \min \begin{bmatrix} \cos^{-1}(-\tan\phi\tan\delta) \\ \cos^{-1}[-\tan(\phi-\beta)\tan\delta] \end{bmatrix} \qquad (2.20.4b)$$

and $$\omega_s'' = \cos^{-1}[-\tan(\phi-\beta)\tan\delta] \qquad (2.20.4c)$$

Also, a and b are given by Equations 2.13.2b and 2.13.2c, and d is given by

$$d = \sin\omega_s - \frac{\pi\,\omega_s}{180}\cos\omega_s \qquad (2.20.4d)$$

Equations 2.20.4 can be used in the southern hemisphere for north-facing surfaces by substituting $(\phi+\beta)$ for $(\phi-\beta)$.

Example 2.20.1

Redo Example 2.19.1 for the month of January, using the Klein-Theilacker method.

Solution

For January, from Example 2.19.1, $\overline{H}_o = 13.36$ MJ/m^2, $\overline{H}_d/\overline{H} = 0.45$, and for the mean day of the month ($n = 17$), $\omega_s = \omega_s' = 69.1°$. For the mean day,

$$a = 0.409 + 0.5016\sin(69.1 - 60) = 0.488$$

$$b = 0.6609 - 0.4767\sin(69.1 - 60) = 0.586$$

$$d = \sin 69.1 - \frac{\pi \times 69.1}{180}\cos 69.1 = 0.504$$

$$\omega_s'' = \cos^{-1}[-\tan(43 - 60)\tan(-20.9)] = 96.7°$$

Using Equation 2.20.4a,

$$\overline{R} = \frac{\cos(43 - 60)}{0.504\cos 43}\left[(0.488 - 0.45)\left(\sin 69.1 - \frac{\pi \times 69.1}{180}\cos 96.7\right)\right.$$

$$\left. + \frac{0.586}{2}\left(\frac{\pi \times 69.1}{180} + \sin 69.1(\cos 69.1 - 2\cos 96.7)\right)\right]$$

$$+ 0.45\left(\frac{1 + \cos 60}{2}\right) + 0.7\left(\frac{1 - \cos 60}{2}\right) = 1.441 + 0.338 + 0.175 = 1.95$$

So monthly average radiation on the collector would be

$$\overline{H}_T = \overline{H}\,\overline{R} = 5.85 \times 1.95 = 11.4 \text{ MJ/m}^2$$

The estimated $\bar{H}_T$ in this month is about 5% less than that predicted by Equation 2.19.2.

■

Klein and Theilacker have also developed a more general form for the equations that is valid for any surface azimuth angle γ. If γ is not 0° (or 180°), the times of sunrise and sunset on the sloped surface will not be symmetrical about solar noon, and the limits of integration for the numerator of Equations 2.20.1 and 2.20.3 will have different absolute values. The equation for $\bar{R}$ is

$$\bar{R} = D + \frac{\bar{H}_d}{H}\left(\frac{1+\cos\beta}{2}\right) + \rho_g\left(\frac{1-\cos\beta}{2}\right) \tag{2.20.5a}$$

$$D = \begin{cases} \max\{0, G(\omega_{ss}, \omega_{sr})\} & \text{if } \omega_{ss} \geq \omega_{sr} \\ \\ \max\{0, [G(\omega_{ss}, -\omega_s) + G(\omega_s, \omega_{sr})]\} & \text{if } \omega_{sr} > \omega_{ss} \end{cases} \tag{2.20.5b}$$

$$\begin{aligned} G(\omega_1, \omega_2) = \frac{1}{2d}\Bigg[&\left(\frac{bA}{2} - a'B\right)(\omega_1 - \omega_2)\frac{\pi}{180} \\ &+ (a'A - bB)(\sin\omega_1 - \sin\omega_2) - a'C(\cos\omega_1 - \cos\omega_2) \\ &+ \left(\frac{bA}{2}\right)(\sin\omega_1\cos\omega_1 - \sin\omega_2\cos\omega_2) \\ &+ \left(\frac{bC}{2}\right)(\sin^2\omega_1 - \sin^2\omega_2)\Bigg] \end{aligned} \tag{2.20.5c}$$

$$a' = a - \frac{\bar{H}_d}{H} \tag{2.20.5d}$$

The integration of Equation 2.20.3 starts at sunrise on the sloped surface or a horizontal plane, whichever is latest. The integration ends at sunset on the surface or the horizontal, whichever is earliest. The sunrise and sunset hour angles for the surface are determined by letting $\theta = 90°$ in Equation 1.6.2. This leads to a quadratic equation, giving two values of ω (which must be within $\pm \omega_s$). The signs on ω_{sr} and ω_{ss} depend on the surface orientation:

$$|\omega_{sr}| = \min\left[\omega_s, \cos^{-1}\frac{AB + C\sqrt{A^2 - B^2 + C^2}}{A^2 + C^2}\right] \tag{2.20.5e}$$

$$\omega_{sr} = \begin{cases} -|\omega_{sr}| \text{ if } (A > 0 \text{ and } B > 0) \text{ or } (A \geq B) \\ +|\omega_{sr}| \text{ otherwise} \end{cases}$$

$$|\omega_{ss}| = \min\left[\omega_s, \cos^{-1}\frac{AB - C\sqrt{A^2 - B^2 + C^2}}{A^2 + C^2}\right] \tag{2.20.5f}$$

$$\omega_{ss} = \begin{cases} +|\omega_{ss}| \text{ if } (A > 0 \text{ and } B > 0) \text{ or } (A \geq B) \\ -|\omega_{ss}| \text{ otherwise} \end{cases}$$

where $\qquad A = \cos \beta + \tan \phi \cos \gamma \sin \beta$ (2.20.5g)

$\qquad\qquad B = \cos \omega_s \cos \beta + \tan \delta \sin \beta \cos \gamma$ (2.20.5h)

$\qquad\qquad C = \dfrac{\sin \beta \sin \gamma}{\cos \phi}$ (2.20.5i)

This method of calculating $\overline{R}$ by Equations 2.20.5 works for all surface orientations and all latitudes (including negative latitudes, for the southern hemisphere). It is valid whether the sun rises or sets on the surface twice each day (e.g., on north-facing surfaces when δ is positive) or not at all. Its use is illustrated in the next example.

Example 2.20.2

What is $\overline{H}_T$ for the collector of Example 2.19.1, but with $\gamma = 30°$, for the month of January, estimated by Equations 2.20.5?

Solution

A logical order of the calculation is to obtain A, B, and C, then ω_{sr} and ω_{ss}, and then G, D, $\overline{R}$, and $\overline{H}_T$ (i.e., work backward through Equations 2.20.5). Using data from the previous examples,

$\qquad\qquad A = \cos 60 + \tan 43 \cos 30 \sin 60 = 1.199$

$\qquad\qquad B = \cos 69.1 \cos 60 + \tan(-20.9) \sin 60 \cos 30 = -0.108$

$\qquad\qquad C = \dfrac{\sin 60 \sin 30}{\cos 43} = 0.592$

Next calculate ω_{sr}, the sunrise hour angle with Equation 2.20.5e It will be the minimum of 69.1° and

$$\cos^{-1} \frac{1.199(-0.108) + 0.592 \sqrt{1.199^2 - (-0.108)^2 + 0.592^2}}{1.199^2 + 0.592^2} = 68.3°$$

$\qquad |\omega_{sr}| = \min(69.1, 68.3) = 68.3°$

Since $A > B$, $\omega_{sr} = -68.3°$.
$\qquad$ The sunset hour angle is found next. From Equation 2.20.5f,

$$\cos^{-1} \frac{-0.129 - 0.789}{1.788} = 120.9°$$

Then $\qquad |\omega_{ss}| = \min (69.1, 120.9) = 69.1°$

Since $A > B$, $\omega_{ss} = 69.1°$.

We next calculate G. Since $\omega_{ss} > \omega_{sr}$, $D = \max(0, G(\omega_{ss}, \omega_{sr}))$. From Equation 2.20.5d, with $a = 0.488$ (from Example 2.20.1),

$$a' = 0.488 - 0.450 = 0.038$$

From Equation 2.20.5c, with $b = 0.586$ and $d = 0.504$ and with $\omega_1 = \omega_{ss} = 69.1°$ and $\omega_2 = \omega_{sr} = -68.3°$,

$$G(\omega_{ss}, \omega_{sr}) = \frac{1}{2 \times 0.504}\left\{\left[\frac{0.586 \times 1.199}{2} - 0.038(-0.108)\right][69.1 - (-68.3)]\frac{\pi}{180}\right.$$

$$+\left[0.038 \times 1.199 - 0.586(-0.108)\right][(\sin 69.1 - \sin(-68.3)]$$

$$-0.038 \times 0.592 \,[\cos 69.1 - \cos(-68.3)]$$

$$+\frac{0.586 \times 1.199}{2}[\sin 69.1 \cos 69.1 - \sin(-68.3) \cos(-68.3)]$$

$$\left.+\frac{0.586 \times 0.592}{2}[\sin^2 69.1 - \sin^2(-68.3)]\right\} = 1.28$$

So $D = \max(0, 1.28) = 1.28$.

Then $\bar{R}$, by Equation 2.20.5a, is

$$\bar{R} = 1.28 + 0.45\left(\frac{1 + \cos 60}{2}\right) + 0.7\left(\frac{1 - \cos 60}{2}\right) = 1.80$$

$$\bar{H}_T = \bar{H}\,\bar{R} = 5.85 \times 1.80 = 10.5 \text{ MJ/m}^2 \qquad\blacksquare$$

The uncertainties in estimating radiation on surfaces sloped to the east or west of south are greater than those for south-facing surfaces. Greater contributions to the daily totals occur early and late in the days when the air mass is greater, when atmospheric transmission is less certain, and when instrumental errors in measurements made on a horizontal plane may be larger than when the sun is nearer the zenith. For surfaces with surface azimuth angles more than 15° from south (or north, in the southern hemisphere), the Klein and Theilacker method illustrated in Example 2.20.2 is recommended.

Table 2.20.1 shows a comparison of the results of the monthly calculations for Examples 2.19.1 and 2.20.1. In the winter months, the Liu and Jordan (isotropic) method applied monthly (Equations 2.19.5) indicates the higher radiation. The situation is reversed in the summer months.

Studies of calculation of average radiation on tilted surfaces have been done which account for anisotropic diffuse by other methods. Herzog (1985) has

Table 2.20.1. $\overline{H}_T$ and $\overline{R}$ by the Liu and Jordan (Isotropic) Method and by the Klein and Theilacker Method, from Examples 2.19.1 and 2.20.1 (Madison, $\beta = 60^0$ and $\gamma = 0^0$)

Month	Liu & Jordan		Klein & Theilacker	
	$\overline{R}$	$\overline{H}_T$, MJ/m^2	$\overline{R}$	$\overline{H}_T$, MJ/m^2
Jan.	2.03	11.9	1.95	11.4
Feb.	1.69	15.5	1.64	15.0
Mar.	1.23	15.8	1.21	15.6
Apr.	0.91	14.5	0.93	14.8
May	0.78	15.4	0.80	15.8
June	0.72	15.9	0.74	16.4
July	0.74	16.3	0.77	16.9
Aug.	0.86	16.6	0.88	17.1
Sep.	1.07	15.8	1.06	15.6
Oct.	1.44	14.9	1.38	14.3
Nov.	1.69	9.6	1.60	9.2
Dec.	1.92	8.5	1.81	8.0

developed a correction factor to the Klein and Theilacker method to account for anisotropic diffuse. Page (1986) presents a very detailed discussion of the method used in compiling the tables of radiation on inclined surface that are included in Volume II of the *European Solar Radiation Atlas*. These tables show radiation on surfaces of 9 orientations including surfaces facing all compass points; the tables and the method used to compute them are designed to provide useful information for daylighting and other building applications beyond those of immediate concern in this book.

2.21   EFFECTS OF RECEIVING SURFACE ORIENTATION

The methods outlined in the previous sections for estimating average radiation on surfaces of various orientations can be used to show the effects of slope and azimuth angle on total energy received on a surface on a monthly, seasonal, or annual basis. (Optimization of collector orientation for any solar process that meets seasonally varying energy demands, such as space heating, must ultimately be done taking into account the time dependence of these demands. The surface orientation leading to maximum output of a solar energy system may be quite different from the orientation leading to maximum incident energy.)

To illustrate the effects of the receiving surface slope on monthly average daily radiation, the methods of Section 2.19 have been used to estimate $\overline{H}_T$ for surfaces of several slopes, for values of $\phi = 45°$, $\gamma = 0°$, and ground reflectance 0.2. $\overline{H}_T$ is a function of $\overline{H}_d/\overline{H}$, which in turn is a function of the average clearness index $\overline{K}_T$; the illustration is for $\overline{K}_T = 0.50$, constant through the year, a value typical of many

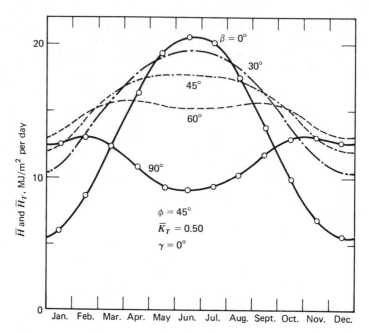

Figure 2.21.1 Variation in estimated average daily radiation on surfaces of various slopes as a function of time of year for a latitude of 45°, $\overline{K}_T$ of 0.50, surface azimuth angle of 0°, and a ground reflectance of 0.20.

temperate climates. Figure 2.21.1 shows the variations of $\overline{H}_T$ (and $\overline{H}$) through the year and shows the marked differences in energy received by surfaces of various slopes in summer and winter.

Figure 2.21.2(a) shows the total annual energy received as a function of slope and indicates a maximum at approximately $\beta = \phi$. The maximum is a broad one, and the changes in total annual energy are less than 5% for slopes of 20° more or less than the optimum. Figure 2.21.2(a) also shows total "winter" energy, taken as the total energy for the months of December, January, February, and March, which would represent the time of the year when most space heating loads would occur. The slope corresponding to the maximum estimated total winter energy is approximately 60°, or $\phi + 15°$. A 15° change in the slope of the collector from the optimum means a reduction of approximately 5% in the incident radiation. The dashed portion of the winter total curve is estimated assuming that there is substantial snow cover in January and February that results in a mean ground reflectance of 0.6 for those 2 months. Under this assumption, the total winter energy is less sensitive to slope than with $\rho_g = 0.2$. The vertical surface receives 8% less energy than does the 60° surface if $\rho_g = 0.6$, and 11% less if $\rho_g = 0.2$.

Calculation of total annual energy for $\phi = 45°$, $\overline{K}_T = 0.50$, and $\rho_g = 0.20$ for surfaces of slopes 30° and 60° are shown as a function of surface azimuth angle in

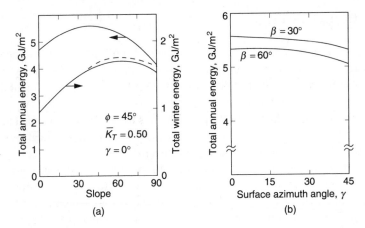

Figure 2.21.2 (a) Variation of total annual energy and total winter (December to March) energy as a function of surface slope for a latitude of 45°, $\overline{K}_T$ of 0.50, and surface azimuth angle of 0°. Ground reflectance is 0.20 except for the dashed curve where it is taken as 0.60 for January and February. (b) Variation of total annual energy with surface azimuth angle for slopes of 30° and 60°, latitude of 45°, $\overline{K}_T$ of 0.50, and ground reflectance of 0.20.

Figure 2.21.2(b). Note the expanded scale. The reduction in annual energy is small for these examples, and the generalization can be made that facing collectors 10° to 20° east or west of south should make little difference in the annual energy received. (Not shown by annual radiation figures is the effect of azimuth angle γ on the diurnal distribution of radiation on the surface. Each shift of γ of 15° will shift the daily maximum of available energy by roughly an hour toward morning if γ is negative and toward afternoon if γ is positive. This could affect the performance of a system for which there are regular diurnal variations in energy demands on the process.) Note that there is implicit in these calculations the assumption that the days are symmetrical about solar noon.

Similar conclusions have been reached by others, for example, Morse and Czarnecki (1958), who estimated the relative total annual beam radiation on surfaces of variable slope and azimuth angle.

From studies of this kind, general "rules of thumb" can be stated. For maximum annual energy availability, a surface slope equal to latitude is best. For maximum summer availability, slope should be approximately 10° to 15° less than the latitude. For maximum winter energy availability, slope should be approximately 10° to 15° more than the latitude. The slopes are not critical; deviations of 15° result in reduction of the order of 5%. The expected presence of a reflective ground cover such as snow leads to higher slopes for maximizing wintertime energy availability. The best surface azimuth angles for maximum incident radiation are 0° in the northern hemisphere or 180° in the southern hemisphere, that is, the surfaces should face the equator. Deviations in azimuth angles of 10° or 20° have small effect on total annual energy availability.

2.22　UTILIZABILITY

In this and the following two sections the concepts of utilizability are developed. The basis is a simple one: if only radiation above a critical or threshold intensity is useful, then we can define a radiation statistic, called utilizability, as the fraction of the total radiation that is received at an intensity higher than the critical level. We can then multiply the average radiation for the period by this fraction to find the total utilizable energy. In these sections we define utilizability and show for any critical level how it can be calculated from radiation data or estimated from $\overline{K}_T$.

In this section we present the concept of monthly average hourly utilizability (the ϕ concept) as developed by Whillier (1953) and Hottel and Whillier (1958). Then in Section 2.23 we show how Liu and Jordan (1963) generalized Whillier's ϕ curves. In Section 2.24 we show an extension of the hourly utilizability to monthly average daily utilizability (the $\overline{\phi}$ concept) by Klein (1978). Collares-Pereira and Rabl (1979a,b) independently extended hourly utilizability to daily utilizability. Evans et al. (1982) have developed a modified and somewhat simplified general method for calculating monthly average daily utilizability.

In Chapter 6 we develop in detail an energy balance equation to represent the performance of a solar collector. The energy balance says, in essence, that the useful gain at any time is the difference between the solar energy absorbed and the thermal losses from the collector. The losses depend on the difference in temperature between the collector plate and the ambient temperature and on a heat loss coefficient. Given a coefficient, a collector temperature, and an ambient temperature (i.e., a loss per unit area), there is a value of incident radiation that is just enough so that the absorbed radiation equals the losses. This value of incident radiation is the **critical radiation level**, I_{Tc}, for that collector operating under those conditions.

If the incident radiation on the tilted surface of the collector I_T is equal to I_{Tc}, all of the absorbed energy will be lost and there will be no useful gain. If the incident radiation exceeds I_{Tc}, there will be useful gain and the collector should be operated. If I_T is less than I_{Tc}, no useful gain is possible and the collector should not be operated. The utilizable energy for any hour is thus $(I_T - I_{Tc})^+$, where the superscript + indicates that the utilizable energy can be zero or positive but not negative.

The fraction of an hour's total energy that is above the critical level is the utilizability for that particular hour:

$$\phi_h = \frac{(I_T - I_{Tc})^+}{I_T} \tag{2.22.1}$$

where ϕ_h can have values from zero to unity. The hour's utilizability is the ratio of the shaded area $(I_T - I_{Tc})$ to the total area (I_T) under the radiation curve for the hour as shown in Figure 2.22.1. (Utilizability could be defined on the basis of rates, i.e., using G_T and G_{Tc}, but as a practical matter, radiation data are available on an hourly basis and that is the basis in use.)

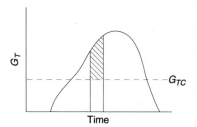

Figure 2.22.1 G_T vs. time for a day. For the hour shown, I_T is the area under the G_T curve. I_{Tc} is the area under the constant critical radiation level curve.

The utilizability for a single hour is not useful. However, utilizability for a particular hour for a month of N days (e.g., 10 to 11 in January) in which the hour's average radiation is $\bar{I}_T$ is useful. It can be found from

$$\phi = \frac{1}{N} \sum_{}^{N} \frac{(I_T - I_{Tc})^+}{\bar{I}_T} \tag{2.22.2}$$

The monthly average utilizable energy for the hour is the product $\bar{I}_T \phi$. The calculation can be done for individual hours (10 to 11, 11 to 12, etc.) for the month and the result summed to get the month's utilizable energy. If the application is such that the conditions of critical radiation level and incident radiation are symmetrical about solar noon, the calculations can be done for hour-pairs (e.g., 10 to 11 and 1 to 2, or 9 to 10 and 2 to 3) and the amount of calculations halved.

Given hourly average radiation data by months and a critical radiation level, the next step is to determine ϕ. This is done by processing the hourly radiation data I_T [as outlined by Whillier (1953)] as follows:

For a given location, hour, month, and collector orientation, plot a cumulative distribution curve of $I_T / \bar{I}_T$. An example for a vertical south-facing surface at Blue Hill, MA, for January is shown in Figure 2.22.2 for the hour-pair 11 to 12 and 12 to 1. This provides a picture of the frequency of occurrence of clear, partly cloudy, or cloudy skies in that hour for the month. For example, for the hour-pair of Figure 2.22.2, for f of 0.20, 20% of the days have radiation that is less than 10% of the average, and for f of 0.80, 20% of the days have radiation in that hour-pair that exceeds 200% of the average.

A dimensionless critical radiation is defined as

$$X_c = \frac{I_{Tc}}{\bar{I}_T} \tag{2.22.3}$$

An example is shown as the horizontal line in Figure 2.22.2 where $X_c = 0.75$ and $f_c = 0.49$. The shaded area represents the monthly utilizability, that is, the fraction of the monthly energy that is above the critical level. Integrating hourly utilizability over

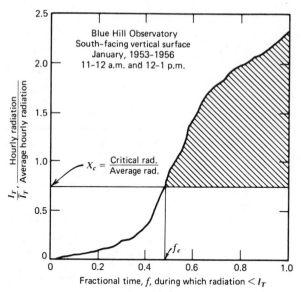

Figure 2.22.2 Cumulative distribution curve for hourly radiation on a south-facing vertical surface in Blue Hill, MA. Adapted from Liu and Jordan (1963).

all values of f_c gives ϕ for that critical radiation level:

$$\phi = \int_{f_c}^{1} \phi_h \, df \tag{2.22.4}$$

As the critical radiation level is varied, f_c varies, and graphical integrations of the curve give utilizability ϕ as a function of critical radiation ratio X_c. An example derived from Figure 2.22.2 is shown in Figure 2.22.3.

Whillier (1953) and later Liu and Jordan (1963) have shown that in a particular location for a 1-month period, ϕ is essentially the same for all hours. Thus, although the curve of Figure 2.22.3 was derived for the hour-pair 11 to 12 and 12 to 1, it is useful for all hour-pairs for the vertical surface at Blue Hill.

The line labeled "limiting curve of identical days" in Figure 2.22.3 would result from a cumulative distribution curve that is a horizontal line at a value of the ordinate of 1.0 in Figure 2.22.2. In other words, every day of the month looks like the average day. The difference between the actual ϕ curve and this limiting case represents the error in utilizable energy that would be made by using a single average day to represent a whole month.

Example 2.22.1

Calculate the utilizable energy on a south-facing vertical solar collector in Blue Hill, MA, for the month of January when the critical radiation level on the collector is 1.07 MJ/m². The averages of January solar radiation on a vertical surface are 1.52, 1.15, and 0.68 MJ/m² for the hour-pairs 0.5, 1.5, and 2.5 hours from solar noon.

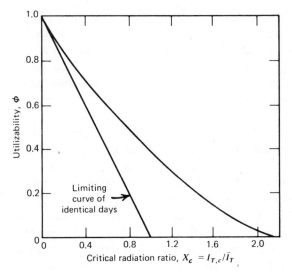

Figure 2.22.3 Utilizability curve derived by numerically integrating Figure 2.22.2. Adapted from Liu and Jordan (1963).

Solution

For the hour-pair 11 to 12 and 12 to 1, the dimensionless critical radiation ratio X_c is

$$X_c = \frac{1.07}{1.52} = 0.70$$

and the utilizability, from Figure 2.22.3, is 0.54. The utilizable energy on the collector during this hour is

$$\bar{I}_T\phi = 1.52 \times 0.54 = 0.82 \text{ MJ/m}^2$$

For the hour-pair 10 to 11 and 1 to 2 the value of X_c is 0.93, ϕ is 0.43, and $\bar{I}_T\phi$ is 0.49. For the hour-pair 9 to 10 and 2 to 3, X_c is 1.57, ϕ is 0.15, and $\bar{I}_T\phi$ is 0.10. The average utilizable energy for the month of January is then

$$N\sum_{\text{hrs}} \bar{I}_T\phi = 31 \times 2(0.82 + 0.49 + 0.10) = 87.5 \text{ MJ/m}^2 \qquad \blacksquare$$

2.23 GENERALIZED UTILIZABILITY

We now have a way of calculating ϕ for specific locations and specific orientations. For most locations the necessary data are not available, but it is possible to make use of the observed statistical nature of solar radiation to develop generalized ϕ curves that depend only on $\overline{K}_T$, latitude, and collector slope. As noted above, ϕ curves are nearly independent of the time of day (i.e., the curves for all hour-pairs are essentially the same). It was observed in early studies [e.g., Whillier (1953)] that ϕ curves based on daily totals of solar radiation are also nearly identical to hourly ϕ curves. It is possible to generate ϕ curves from average hourly values of radiation using the methods of Section 2.13 to break daily total radiation into hourly radiation. However, it is easier to generate ϕ curves from daily totals, and this is the procedure to be described here.

The radiation data most generally available are monthly average daily radiation on horizontal surfaces. Thus, with $\overline{K}_T$ and the long-term distribution of days having particular values of K_T from Figure 2.9.2, it is possible to generate sequences of days that represent the long-term average distribution of daily total radiation. The order of occurrence of the days is unknown, but for ϕ curves the order is irrelevant.

For each of these days, the daily total radiation on an inclined collector can be estimated by a procedure similar to that in Section 2.19 for monthly average radiation. For a particular day, the radiation on a tilted surface, using the isotropic[22] diffuse assumption, can be written as[23]

$$H_T = (H - H_d)\overline{R}_b + H_d\left(\frac{1 + \cos \beta}{2}\right) + H\rho_g\left(\frac{1 - \cos \beta}{2}\right) \tag{2.23.1}$$

where the monthly average conversion of daily beam radiation on a horizontal surface to daily beam radiation on an inclined surface, $\overline{R}_b$, is used rather than the value for the particular day since the exact date within the month is unknown. The value of $\overline{R}_b$ is found from Equations 2.19.3a or its equivalent. If we divide by the monthly average extraterrestrial daily radiation, $\overline{H}_o$, and introduce K_T' based on $\overline{H}_o$ (i.e., $K_T' = H/\overline{H}_o$), Equation 2.23.1 becomes

$$\frac{H_T}{\overline{H}_o} = K_T'\left[\left(1 - \frac{H_d}{H}\right)\overline{R}_b + \frac{H_d}{H}\left(\frac{1 + \cos \beta}{2}\right) + \rho_g\left(\frac{1 - \cos \beta}{2}\right)\right] \tag{2.23.2}$$

The ratio H_d/H is the daily fraction of diffuse radiation and can be found from Figure 2.11.2 (or Equation 2.11.1) as a function of K_T'. Therefore, for each of the days selected from the generalized distribution curve, Equation 2.23.2 can be used to estimate the radiation on a tilted surface. The average of all the days yields the long-term monthly average radiation on the tilted surface. The ratio $H_T/\overline{H}_T$ can then be found for each day. The data for the whole month can then be plotted in the form of a cumulative distribution curve, as illustrated in Figure 2.22.2. The ordinate will be daily totals rather than hourly values, but as has been pointed out, the shape of the

---

[22] Other assumptions for distribution of the diffuse could be used.

[23] Section 2.19 is concerned with monthly average daily radiation on a tilted surface. Here we want the average radiation on an inclined surface for all days having a particular value of K_T.

two curves are nearly the same. Finally, integration of the frequency distribution curve yields a utilizability curve as illustrated in Figure 2.22.3. The process is illustrated in the following example.

Example 2.23.1

Calculate and plot utilizability as a function of the critical radiation ratio for a collector tilted 40° to the south at a latitude of 40°. The month is February and $\overline{K}_T$ is 0.5.

Solution

Since the only radiation information available is $\overline{K}_T$, it will be necessary to generate a ϕ curve from the generalized $\overline{K}_T$ frequency distribution curves. Twenty days, each represented by a K_T from Figure 2.9.2 at $\overline{K}_T = 0.5$, are given in the following table. (Twenty days from the generalized distribution curves are sufficient to represent a month.) For any day with daily total horizontal radiation H and daily diffuse horizontal radiation H_d, the ratio of daily radiation on a south-facing tilted surface to extraterrestrial horizontal radiation is found from Equation 2.23.2. For the condition of this problem, $\overline{R}_b$ is 1.79 from Equation 2.19.3a. The view factors from the collector to the sky and ground are $(1 + \cos \beta)/2 = 0.88$ and $(1 - \cos \beta)/2 = 0.12$, respectively. The ground will be assumed to be covered with snow so that ρ_g is 0.7. Equation 2.23.2 reduces to

$$\frac{H_T}{\overline{H}_o} = K_T' \left[1.87 - 0.91 \left(\frac{H_d}{H} \right) \right]$$

For each day in the table, H_d/H is found from Figure 2.11.1 (or Equation 2.11.1) using the corresponding value of K_T. The results of these calculations are given in columns 2 through 4. The average of column 4 is 0.721. Column 5, the ratio of daily total radiation on a tilted surface H_T to the monthly average value $\overline{H}_T$ is calculated by dividing each value in column 4 by the average value. Column 5 is plotted in the first figure that follows as a function of the day since the data are already in ascending order. The integration, as indicated in this figure, is used to determine the utilizability ϕ. The area under the whole curve is 1.0. The area above a particular value of $H_T/\overline{H}_T$ is the fraction of the month's radiation that is above this level. For $H_T/\overline{H}_T$ equal to 1.2, 13% of the radiation is above this level. The utilizability is plotted in the second figure. Although daily totals were used to generate this figure, the hourly ϕ curves will have nearly the same shape. Consequently, the curve can be used in hourly calculations to determine collector performance as illustrated in Example 2.22.1.

1	2	3	4	5
Day	K_T	H_d/H	$H_T/\overline{H}_o$	$H_T/\overline{H}_T$
1	0.08	0.99	0.078	0.11
2	0.15	0.99	0.145	0.20
3	0.21	0.95	0.211	0.29
4	0.26	0.92	0.269	0.37
5	0.32	0.87	0.345	0.48
6	0.36	0.82	0.405	0.56
7	0.41	0.76	0.483	0.67
8	0.46	0.68	0.576	0.80
9	0.49	0.62	0.640	0.89
10	0.53	0.55	0.726	1.01
11	0.57	0.47	0.822	1.14
12	0.59	0.43	0.872	1.21
13	0.61	0.39	0.924	1.28
14	0.63	0.36	0.972	1.35
15	0.65	0.33	1.020	1.41
16	0.67	0.30	1.070	1.48
17	0.69	0.27	1.121	1.55
18	0.72	0.24	1.189	1.65
19	0.74	0.23	1.229	1.70
20	0.79	0.21	1.326	1.84
			Average = 0.721	

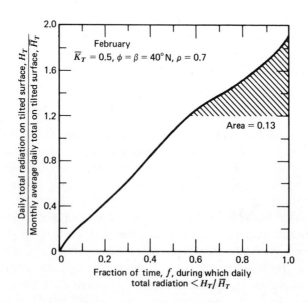

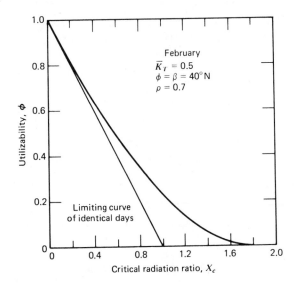

In the preceding example, a ϕ curve was generated from knowledge of the monthly average solar radiation and the known statistical behavior of solar radiation. For some purposes it is necessary to know monthly average hourly utilizability. If this information is needed, the method described in Section 2.13 and illustrated in Example 2.13.3 can be used to determine monthly average hourly radiation from knowledge of monthly average daily radiation (i.e., $\overline{K}_T$).

For each hour or hour-pair, the monthly average hourly radiation incident on the collector is given by

$$\overline{I}_T = \left(\overline{H}r_t - \overline{H}_d r_d\right)R_b + \overline{H}_d r_d\left(\frac{1+\cos\beta}{2}\right) + \overline{H}\rho_g r_t\left(\frac{1-\cos\beta}{2}\right) \qquad (2.23.3)$$

or by dividing by $\overline{H}$ and introducing $\overline{H} = \overline{K}_T\overline{H}_o$,

$$\overline{I}_T = \overline{K}_T\overline{H}_o\left[\left(r_t - \frac{\overline{H}_d}{\overline{H}}r_d\right)R_b + \frac{\overline{H}_d}{\overline{H}}r_d\left(\frac{1+\cos\beta}{2}\right) + \rho_g r_t\left(\frac{1-\cos\beta}{2}\right)\right] (2.23.4)$$

The ratios r_t and r_d are found from Figures 2.13.1 and 2.13.2 for each hour-pair.

Example 2.23.2

Estimate the utilizability for the conditions of Example 2.23.1 for the hour-pair 11 to 12 and 12 to 1. The critical radiation level is 1.28 MJ/m^2.

Solution

At a latitude of 40°N in February the monthly average daily extraterrestrial radiation is 20.5 MJ/m^2 and the declination for the average day of the month is $-13.0°$. The sunset hour angle and the day length of February 16, the mean day of the month, are 78.9° and 10.5 hours, respectively. The monthly average ratio $\overline{H}_d/\overline{H}$ is 0.39 from Figure 2.12.2, and $\omega = 7.5°$. The ratios r_t and r_d from Figures 2.13.1 and 2.13.2 are 0.158 and 0.146. For the mean day in February and from Equation 1.8.2, R_b is 1.62. Then $\bar{I}_T$ from Equation 2.23.4 is

$$\bar{I}_T = 0.5 \times 20.5[(0.158 - 0.39 \times 0.146)1.62 + 0.39 \times 0.146 + 0.88$$

$$+ 0.7 + 0.158 + 0.12] = 2.33 \text{ MJ/m}^2$$

The critical radiation rate for this hour-pair is

$$X_c = \frac{I_{Tc}}{\bar{I}_T} = \frac{1.28}{2.33} = 0.55$$

From the figure of Example 2.23.1, ϕ is 0.50. The utilizable energy for the month for this hour-pair is

$$UE = 2.33 \times 0.50 \times 2 \times 28 = 65.2 \text{ MJ/m}^2 \qquad \blacksquare$$

Liu and Jordan (1963) have generalized the calculations of Example 2.23.1. They found that the shape of the ϕ curves was not strongly dependent on the ground reflectance or the view factors from the collector to the sky and ground. Consequently, they were able to construct a set of ϕ curves for a fixed value of $\overline{K}_T$. The effect of tilt was taken into account by using the monthly average ratio of beam radiation on a tilted surface to monthly average beam radiation on a horizontal surface $\overline{R}_b$ as a parameter. The generalized ϕ curves are shown in Figures 2.23.1 for values of $\overline{K}_T$ of 0.3, 0.4, 0.5, 0.6, and 0.7. The method of constructing these curves is exactly like Example 2.23.1, except that the tilt used in their calculations was 47° and the ground reflectance was 0.2. A comparison of the ϕ curve from Example 2.23.1, in which the tilt was 40° and the ground reflectance was 0.7 with the generalized ϕ curve for $\overline{K}_T = 0.5$ and $\overline{R}_b = 1.79$, shows that the two are nearly identical.

With the generalized ϕ curves, it is possible to predict the utilizable energy at a constant critical level by knowing only the long-term average radiation. This procedure was illustrated (for one hour-pair) in Example 2.23.2. Rather than use the ϕ curve calculated in Example 2.23.1, the generalized ϕ curve could have been used. The only additional calculation is determining $\overline{R}_b$ so that the proper curve can be selected. In Example 2.23.2, $X_c = 0.55$. From Table D-7 in Appendix D, $\overline{R}_b = 1.79$. Figure 2.23.1c is used to obtain ϕ; it is approximately 0.50.

It is convenient for computations to have an analytical representation of the utilizability function. Clark et al. (1983) have developed a simple algorithm to represent the generalized ϕ functions. Curves of ϕ vs. X_c derived from long-term

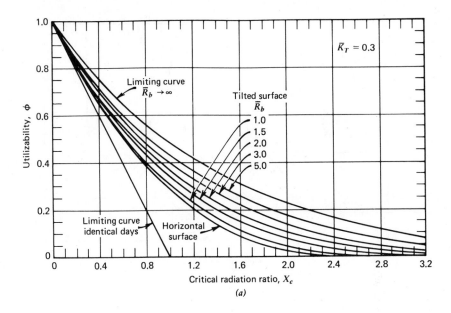

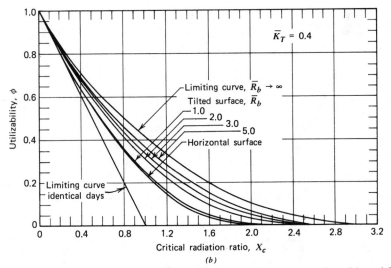

Figure 2.23.1 Generalized ϕ curves for south-facing surfaces. Adapted from Liu and Jordan (1963).

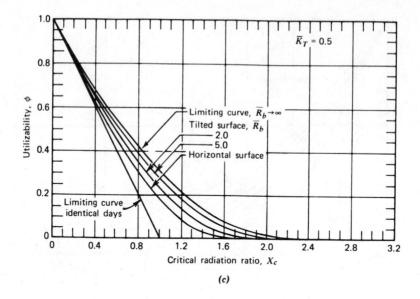

(c)

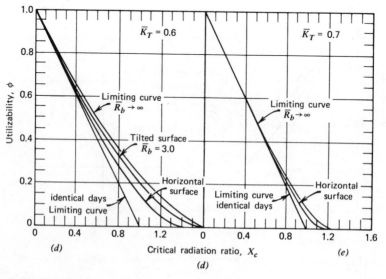

(d)

Figure 2.23.1 (continued)

weather data are represented by

$$\phi = \begin{cases} 0 & \text{if } X_c \geq X_m \\ (1 - X_c/X_m)^2 & \text{if } X_m = 2 \\ \left| g - \left[g^2 + (1 + 2g)(1 - X_c/X_m)^2 \right]^{1/2} \right| & \text{otherwise} \end{cases} \quad (2.23.5a)$$

where $g = (X_m - 1)/(2 - X_m)$ (2.23.5b)

$$X_m = 1.85 + 0.169 \overline{R}_h/\overline{k}^2 - 0.0696 \cos \beta/\overline{k}^2 - 0.981 \overline{k}/\cos^2 \delta \quad (2.23.5c)$$

The monthly average hourly clearness index $\overline{k}_T$ is defined as

$$\overline{k}_T = \frac{\overline{I}}{\overline{I}_o} \quad (2.23.6)$$

It can be estimated using Equations 2.13.2 and 2.13.4:

$$\overline{k}_T = \frac{\overline{I}}{\overline{I}_o} = \frac{r_t}{r_d} \frac{\overline{H}}{\overline{H}_o} = \frac{r_t}{r_d} \overline{K}_T = \overline{K}_T(a + b \cos \omega) \quad (2.23.7)$$

where a and b are given by Equations 2.13.2b and 2.13.2c.

The remaining term in Equation 2.23.5 is $\overline{R}_h$, the ratio of monthly average hourly radiation on the tilted surface to that on a horizontal surface.

$$\overline{R}_h = \overline{I}_T/\overline{I} = \overline{I}_T / \left(\overline{H} r_t \right) \quad (2.23.8)$$

Example 2.23.3

Repeat Example 2.23.2 using the Clark et al. equations.

Solution

The calculations to be made are $\overline{R}_h$, $\overline{k}$, X_m, X_c, g, and finally ϕ. Intermediate results from Example 2.23.2 that are useful here are $\overline{I}_T = 2.33$ MJ/m^2; $r_t = 0.158$; $\omega_s = 78.9°$; $\omega = 7.5°$; and $X_c = 0.549$:

$$\overline{R}_h = \frac{\overline{I}_T}{\overline{I}} = \frac{\overline{I}_T}{r_t \overline{H}} = \frac{2.33}{0.158 \times 20.5 \times 0.50} = 1.44$$

To calculate $\overline{k}_T$ we need the constants a and b in Equation 2.23.7:

$$a = 0.409 + 0.5016 \sin(78.9 - 60) = 0.571$$

and $b = 0.6609 - 0.4767 \sin(78.9 - 60) = 0.506$

Thus $\qquad \bar{k}_T = 0.50(0.571 + 0.506 \cos 7.5) = 0.536$

Next calculate X_m with Equation 2.23.5c:

$$X_m = 1.85 + 0.169 \frac{1.44}{0.536^2} - 0.0696 \frac{\cos 40}{0.536^2} - \frac{0.981 \times 0.536}{\cos^2(-13)} = 1.942$$

The last steps are to calculate g and ϕ with Equations 2.23.5b and 2.23.5a:

$$g = (1.942 - 1)/(2 - 1.942) = 16.24$$

Then $\qquad \phi = \left| 16.24 - \left[16.24^2 + (1 + 2 \times 16.24)\left(1 - \frac{0.549}{1.942}\right)^2 \right]^{1/2} \right| = 0.52$

This is nearly the same ϕ as that from Example 2.23.2. ∎

The ϕ charts graphically illustrate why a single average day should not be used to predict system performance under most conditions. The difference in utilizability as indicated by the limiting curve of identical days and the appropriate ϕ curve is the error that is incurred by basing performance on an average day. Only if $\bar{K}_T$ is high or if the critical level is very low do all ϕ curves approach the limiting curve. For many situations the error in using one average day to predict performance is substantial.

The ϕ curves must be used hourly, even though a single ϕ curve applies for a given collector orientation, critical level, and month. This means that three to six hourly calculations must be made per month if hour-pairs are used. For surfaces facing the equator, where hour-pairs can be used, the concept of monthly average daily utilizability, $\bar{\phi}$, provides a more convenient way of calculating useful energy. However, for processes that have critical radiation levels that vary in repeatable ways through the days of a month and for surfaces that do not face the equator, the generalized ϕ curves must be used for each hour.

2.24 DAILY UTILIZABILITY

The amount of calculations in the use of ϕ curves led Klein (1978) to develop the concept of monthly average daily utilizability, $\bar{\phi}$. This daily utilizability is defined as the sum for a month (over all hours and days) of the radiation on a tilted surface that is above a critical level divided by the monthly radiation. In equation form,

$$\bar{\phi} = \frac{\sum\limits_{\text{days}} \sum\limits_{\text{hours}} (I_T - I_{Tc})^+}{\bar{H}_T N} \qquad\qquad (2.24.1)$$

where the critical level is similar to that used in the ϕ concept.[24] The monthly average utilizable energy is then the product $\overline{H}_T N \overline{\phi}$. The concept of daily utilizability is illustrated in Figure 2.24.1. Considering either of the two sequences of days, $\overline{\phi}$ is the ratio of the sum of the shaded areas to the total areas under the curves.

The value of $\overline{\phi}$ for a month depends on the distribution of hourly values of radiation in the month. If it is assumed that all days are symmetrical about solar noon and that the hourly distributions are as shown in Figures 2.13.1 and 2.13.2, then $\overline{\phi}$ depends on the distribution of daily total radiation, that is, on the relative frequency of occurrence of below average, average, and above average daily radiation values.[25] Figure 2.24.1 illustrates this point. The days in the top sequence are all average days; for the low critical radiation level represented by the solid horizontal line, the shaded areas show utilizable energy, whereas for the high critical level represented by the dotted line, there is no utilizable energy. The bottom sequence shows three days of varying radiation with the same average as before; utilizable energy for the low critical radiation level is nearly the same as for the first set, but there is utilizable energy above the high critical level for the nonuniform set of days and none for the

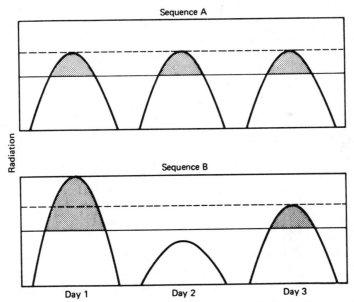

Figure 2.24.1 Two sequences of days with the same average radiation levels on the plane of the collector. From Klein (1978).

[24] The critical level for $\overline{\phi}$ is based on monthly average "optical efficiency" and temperatures rather than on values for particular hours. This will be discussed in Chapter 21.

[25] Klein assumed symmetrical days in his development of $\overline{\phi}$. It can be shown that departure from symmetry within days (e.g., if afternoons are brighter than mornings) will lead to increases in $\overline{\phi}$; thus a $\overline{\phi}$ calculated from the correlations of this section is somewhat conservative.

uniform set. Thus the effect of increasing variability of days is to increase $\bar{\phi}$, particularly at high critical radiation levels.

The monthly distribution of daily total radiation is a unique function of $\bar{K}_T$, as shown by Figure 2.9.2. Thus the effect of daily radiation distribution on $\bar{\phi}$ is related to a single variable, $\bar{K}_T$.

Klein has developed correlations for $\bar{\phi}$ as a function of $\bar{K}_T$ and two variables, a geometric factor $\bar{R}/R_n$ and a dimensionless critical radiation level $\bar{X}_c$. The symbol $\bar{R}$ is the monthly ratio of radiation on a tilted surface to that on a horizontal surface, $\bar{H}_T/\bar{H}$, and is given by Equation 2.19.2, and R_n is the ratio for the hour centered at noon of radiation on the tilted surface to that on a horizontal surface for an average day of the month. Equation 2.15.2 can be rewritten for the noon hour, in terms of $r_{d,n} H_d$ and $r_{t,n} H$, as

$$R_n = \left(\frac{I_T}{I}\right)_n = \left(1 - \frac{r_{d,n} H_d}{r_{t,n} H}\right) R_{b,n} + \left(\frac{r_{d,n} H_d}{r_{t,n} H}\right)\left(\frac{1 + \cos \beta}{2}\right) + \rho_g \left(\frac{1 - \cos \beta}{2}\right)$$

$$(2.24.2)$$

where $r_{d,n}$ and $r_{t,n}$ are obtained from Figures. 2.13.1 and 2.13.2 using the curves for 0 hours from solar noon or from Equations 2.13.2 and 2.13.4.

Note that R_n is calculated for a day that has the day's total radiation equal to the monthly average daily total radiation, i.e., a day in which $H = \bar{H}$. (R_n is *not* the monthly average value of R at noon.) The calculation of R_n is illustrated in Example 2.24.1.

A monthly average critical radiation level $\bar{X}_c$ is defined as the ratio of the critical radiation level to the noon radiation level on a day of the month in which the day's radiation is the same as the monthly average. In equation form,

$$\bar{X}_c = \frac{I_{Tc}}{r_{t,n} R_n \bar{H}}$$

$$(2.24.3)$$

Klein obtained $\bar{\phi}$ as a function of $\bar{X}_c$ for various values of $\bar{R}/R_n$ by the following process. For a given $\bar{K}_T$, a set of days was established that had the correct long-term average distribution of values of K_T (i.e., that match the distributions of Section 2.9). (This is the process illustrated in Example 2.23.1.) The radiation in each of the days in a sequence was divided into hours using the correlations of Section 2.13. These hourly values of beam and diffuse radiation were used to find the total hourly radiation on a tilted surface, I_T. Critical radiation levels were then subtracted from these I_T values and summed according to Equation 2.24.1 to arrive at values of $\bar{\phi}$. •

The $\bar{\phi}$ curves calculated in this manner are shown in Figures 2.24.2(a–e) for values of $\bar{K}_T$ of 0.3 to 0.7. These curves can be represented by the equation

$$\bar{\phi} = \exp\left\{\left[a + b\left(R_n/\bar{R}\right)\right]\left[\bar{X}_c + c\bar{X}_c^2\right]\right\}$$

$$(2.24.4a)$$

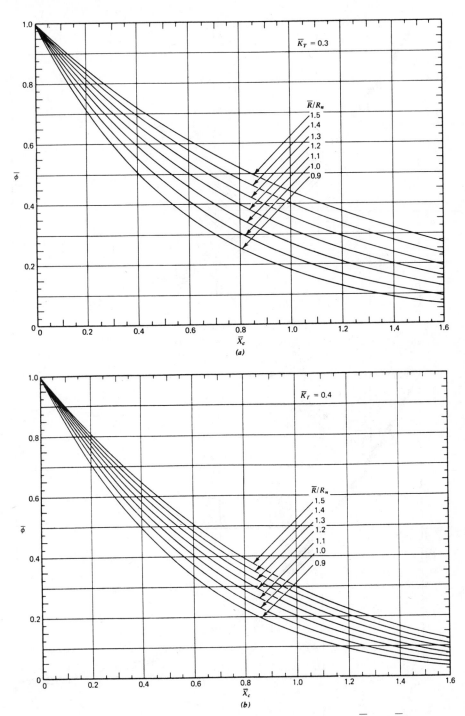

Figure 2.24.2 Monthly average daily utilizability as a function of $\overline{X}_c$ and $\overline{R}/R_n$.

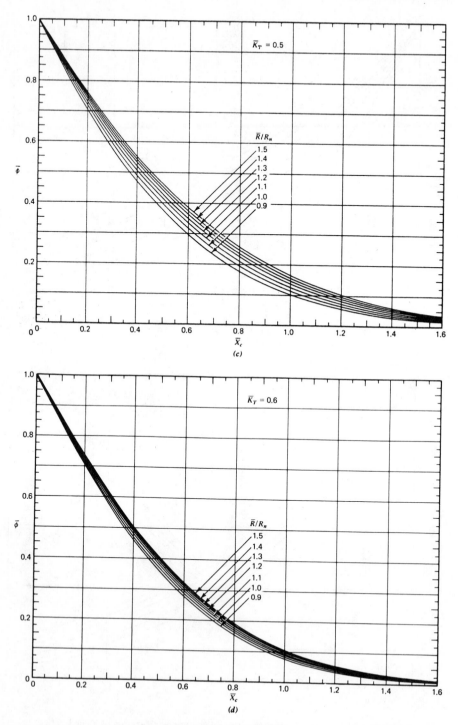

Figure 2.24.2 (continued)

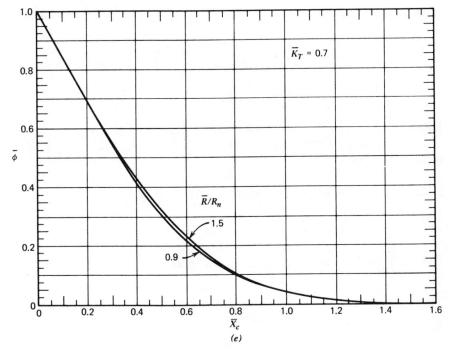

Figure 2.24.2 (continued)

where

$$a = 2.943 - 9.271\overline{K}_T + 4.031\overline{K}_T^2 \qquad (2.24.4b)$$

$$b = -4.345 + 8.853\overline{K}_T - 3.602\overline{K}_T^2 \qquad (2.24.4c)$$

$$c = -0.170 - 0.306\overline{K}_T + 2.936\overline{K}_T^2 \qquad (2.24.4d)$$

Example 2.24.1

A collector in Madison, WI has a slope of $60°$ and a surface azimuth angle of $0°$. For the month of March, when $\overline{K}_T = 0.50$, $\overline{H} = 12.89$ MJ/m^2, $T_a = 2$ C, $\rho_g = 0.4$, and the critical radiation level is 145 W/m^2, calculate $\overline{\phi}$ and the utilizable energy.

Solution

For the average March day (March 16) the sunset hour angle is $87.7°$ from Equation 1.6.10. Then from Equations 2.13.2 and 2.13.4, $r_{t,n} = 0.146$ and $r_{d,n} = 0.134$. For K_T of 0.50 (i.e., a day in which $H = \overline{H}$), H_d/H from Figure 2.11.2 is 0.62.

From Equation 1.8.2, $R_{b,n}$ is 1.38. Then R_n can be calculated using Equation 2.24.2:

$$R_n = \left(1 - \frac{0.134 \times 0.62}{0.146}\right) 1.38 + \frac{0.134 \times 0.62}{0.146}\left(\frac{1 + \cos 60}{2}\right)$$
$$+ 0.4\left(\frac{1 - \cos 60}{2}\right) = 1.12$$

Equation 2.19.2 is used to calculate $\bar{R}$. From Figure 2.19.1 or Appendix D, $\bar{R}_b$ is 1.42. From Figure 2.12.2 $\bar{H}_d/\bar{H}$ is 0.43 at $\bar{K}_T = 0.50$. (See Example 2.19.1 for more details.) Then

$$\bar{R} = (1 - 0.43)1.42 + 0.43\left(\frac{1 + \cos 60}{2}\right) + 0.4\left(\frac{1 - \cos 60}{2}\right) = 1.23$$

and $R_n/\bar{R} = 1.12/1.23 = 0.91$

From Equation 2.24.3 the dimensionless average critical radiation level is

$$\bar{X}_c = \frac{145 \times 3600}{0.146 \times 1.13 \times 12.89 \times 10^6} = 0.25$$

We can now get the utilizability $\bar{\phi}$ from Figure 2.24.2c or from Equations 2.24.4. With $\bar{K}_T = 0.50$, a = −0.685, b = −0.819, and c = 0.411, $\bar{\phi} = 0.68$. The month's utilizable energy is thus

$$\bar{H}_T N \bar{\phi} = \bar{H}\bar{R}N\bar{\phi} = 12.89 \times 1.23 \times 31 \times 0.68 = 334 \text{ MJ/m}^2 \quad \blacksquare$$

The $\bar{\phi}$ depend on $\bar{R}$ and R_n, which in turn depend on the division of total radiation into beam and diffuse components. As noted in Section 2.11, there are substantial uncertainties in determining this division. The correlation of $\bar{H}_d/\bar{H}$ versus $\bar{K}_T$ of Liu and Jordan (1960) was used by Klein (1978) to generate the $\bar{\phi}$ charts. The correlation of Ruth and Chant (1976), which indicates significantly higher fractions of diffuse radiation, was also used to generate $\bar{\phi}$ charts, and the results were not significantly different from those of Figure 2.24.2. A ground reflectance of 0.2 was used in generating the charts, but a value of 0.7 was also used and it made no significant difference. Consequently, even if the diffuse-to-total correlation is changed as a result of new experimental evidence, the $\bar{\phi}$ curves will remain valid. Of course, using different correlations will change the predictions of radiation on a tilted surface, which will change the performance estimates.

Utilizability can be thought of as a radiation statistic that has built into it critical radiation levels. The ϕ and $\bar{\phi}$ concepts can be applied to a variety of design problems, for heating systems, combined solar energy-heat pump systems, and many others. The concept of utilizability has been extended to apply to passively heated buildings, where the excess energy (unutilizable energy) that cannot be stored in a building structure can be estimated. The unutilizability idea can also apply to photovoltaic systems with limited storage capacity.

2.25   SUMMARY

In this chapter we have described the instruments (pyrheliometers and pyranometers) used to measure solar radiation. Radiation data are available in several forms, with the most widely available being pyranometer measurements of total (beam plus diffuse) radiation on horizontal surfaces. These data are available on an hourly basis from a limited number of stations and on a daily basis for many stations.

Solar radiation information is needed in several different forms, depending on the kinds of calculations that are to be done. These calculations fall into two major categories. First (and most detailed), we may wish to calculate on an hour-by-hour basis the long-time performance of a solar process system; for this we want hourly information of solar radiation and other meteorological measurements. Second, monthly average solar radiation is useful in estimating long-term performance of some kinds of solar processes. It is not possible to predict what solar radiation will be in the future, and recourse is made to use of past data to predict what solar processes will do.

We have presented methods (and commented on their limitations) for the estimation of solar radiation information in the desired format from the data that are available. This includes estimation of beam and diffuse radiation from total radiation, time distribution of radiation in a day, and radiation on surfaces other than horizontal. We introduced the concept of utilizability, a solar radiation statistic based on levels of radiation available above critical levels. Determination of critical radiation levels for collectors will be treated in Chapters 6 and 7, and the utilizability concepts will be the basis for most of Part III, on design of solar energy processes.

REFERENCES

Bannerot, R. B. and J. R. Howell, *Solar Energy*, **22**, 229 (1979). "Predicted Daily and Yearly Average Radiation Performance of Optimal Trapezoidal Groove Solar Energy Collectors."

Bannister, J. W., Solar Radiation Records, Division of Mechanical Engineering, Commonwealth Scientific and Industrial Research Organization, Australia (1966-1969).

Beckman, W. A., S. A. Klein, and J. A. Duffie, *Solar Heating Design by the f-Chart Method*, Wiley-Interscience, New York (1977).

Bendt, P., M. Collares-Pereira, and A. Rabl, *Solar Energy*, **27**, 1 (1981). "The Frequency Distribution of Daily Radiation Values."

Bennett, I., *Solar Energy*, **9**, 145 (1965). "Monthly Maps of Mean Daily Insolation for the United States."

Benseman, R. F. and F. W. Cook, *New Zealand J. of Sciences*, **12**, 696 (1969). "Solar Radiation in New Zealand - The Standard Year and Radiation on Inclined Slopes."

Boes, E. C., Sandia Report SAND 75-0565 (1975). "Estimating the Direct Component of Solar Radiation."

Chiam, H. F., *Solar Energy*, **26**, 503 (1981). "Planar Concentrators for Flat-Plate Collectors."

Chiam, H. F., *Solar Energy*, **29**, 65 (1982). "Stationary Reflector-Augmented Flat-Plate Collectors."

Choudhury, N. K. D., *Solar Energy*, **7**, 44 (1963). "Solar Radiation at New Delhi."

Cinquemani, V., J. R. Owenby, and R. G. Baldwin, NOAA report to Department of Energy (Nov. 1978). "Input Data for Solar Systems."

Clark, D. R., S. A. Klein, and W. A. Beckman, *Trans. ASME, J Solar Energy Engrg.*, **105**, 281 (1983). "Algorithm for Evaluating the Hourly Radiation Utilizability Function."

Collares-Pereira, M. and A. Rabl, *Solar Energy*, **22**, 155 (1979a). "The Average Distribution of Solar Radiation - Correlations between Diffuse and Hemispherical and between Daily and Hourly Insolation Values."

Collares-Pereira, M. and A. Rabl, *Solar Energy*, **23**, 235 (1979b). "Simple Procedure for Predicting Long Term Average Performance of Nonconcentrating and of Concentrating Solar Collectors."

Commission of the European Communities (CEC), *European Solar Radiation Atlas,* Vol. **1**, *Global Radiation on Horizontal Surfaces*, Vol. **2**, *Inclined Surfaces* (W. Palz, ed.), Verlag TÜV Rheinland, Koln, (1984).

Coulson, K. L., *Solar and Terrestrial Radiation*, Academic Press, New York (1975).

deJong, B., *Net Radiation Received by a Horizontal Surface at the Earth,* monograph published by Delft University Press, Netherlands (1973).

Drummond, A. J., *Archiv fur Meteorologie Geophysik Bioklimatologie*, Series B, **7**, 413 (1956). "On the Measurement of Sky Radiation."

Drummond, A. J., *J. of Applied Meteorology*, **3**, 810 (1964). "Comments on Sky Radiation Measurement and Corrections."

Erbs, D. G., S. A. Klein, and J. A. Duffie, *Solar Energy*, **28**, 293 (1982). "Estimation of the Diffuse Radiation Fraction for Hourly, Daily, and Monthly-Average Global Radiation."

Evans, D. L., T. T. Rule, and B. D. Wood, *Solar Energy*, **28**, 13 (1982). "A New Look at Long Term Collector Performance and Utilizability."

Farber, E. A. and C. A. Morrison, in *Applications of Solar Energy for Heating and Cooling of Buildings* (R. C. Jordan and B. Y. H Liu eds.) ASHRAE GRP-170, New York (1977). "Clear-Day Design Values."

Foster, N. B. and L. W. Foskett, *Bull. of the American Meteorological Society*, **34**, 212 (1953). "A Photoelectric Sunshine Recorder."

Fritz, S., *Transactions of the Conference on Use of Solar Energy*, **1**, 17, University of Arizona Press, Tucson (1958). "Transmission of Solar Energy through the Earth's Clear and Cloudy Atmosphere."

Grassie, S. L. and N. L. Sheridan, *Solar Energy*, **19**, 663 (1977). "The Use of Planar Reflectors for Increasing the Energy Yield of Flat-Plate Collectors."

Grimmer, D. P., K. G. Zinn, K. C. Herr, and B. E. Wood, Report LA-7041 of Los Alamos Scientific Lab (1978). "Augmented Solar Energy Collection Using Various Planar Reflective Surfaces: Theoretical Calculations and Experimental Results."

Hamilton, D. C. and W. R. Morgan, National Advisory Committee for Aeronautics, Technical Note 2836 (1952). "Radiant Interchange Configuration Factors."

Hay, J. E. and J. A. Davies, in *Proc. First Canadian Solar Radiation Data Workshop* (J. E. Hay and T. K. Won, eds.) Ministry of Supply and Services Canada, 59 (1980). "Calculation of the Solar Radiation Incident on an Inclined Surface."

Hay, J. E. and D. C. McKay, *Intl. J. Solar Energy*, **3**, 203 (1985). "Estimating Solar Irradiance on Inclined Surfaces: A Review and Assessment of Methodologies."

Herzog, M. E., M.S. Thesis, Trinity University (1985). "Estimation of Hourly and Monthly Average Daily Insolation on Tilted Surfaces."

Hollands, K. G. T., *Solar Energy*, **13**, 149 (1971). "A Concentrator for Thin-Film Solar Cells."

Hollands, K. G. T. and R. G. Huget, *Solar Energy*, **30**, 195 (1983). "A Probability Density Function for the Clearness Index, with Applications."

Hottel, H. C., *Solar Energy*, **18**, 129 (1976). "A Simple Model for Estimating the Transmittance of Direct Solar Radiation Through Clear Atmospheres."

Hottel, H. C. and A Whillier, in *Trans. of the Conference on Use of Solar Energy*, **2**, 74, University of Arizona Press (1958). "Evaluation of Flat-Plate Solar Collector Performance."

Hottel, H. C. and B. B. Woertz, *Trans. ASME*, **64**, 91 (1942). "Performance of Flat-Plate Solar Heat Collectors."

IGY Instruction Manual, Pergamon Press, London, Vol. V, Pt. VI, p. 426 (1958). "Radiation Instruments and Measurements."

Iqbal, M., *Solar Energy*, **22**, 87 (1979). "A Study of Canadian Diffuse and Total Solar Radiation Data - II Monthly Average Hourly Radiation."

Iqbal, M., *An Introduction to Solar Radiation*, Academic Press, Toronto (1983).

Jeys, T. H. and L. L. Vant-Hull, *Solar Energy*, **18**, 343 (1976). "The Contribution of the Solar Aureole to the Measurements of Pyrheliometers."

Klein, S. A., *Solar Energy*, **19**, 325 (1977). "Calculation of Monthly Average Insolation on Tilted Surfaces."

Klein, S. A., *Solar Energy*, **21**, 393 (1978). "Calculations of Flat-Plate Collector Utilizability."

Klein, S. A. and J. A. Duffie, in *Proc. of 1978 Annual Meeting American Section*, International Solar Energy Society, **2.2**, 672 (1978). "Estimation of Monthly Average Diffuse Radiation."

Klein, S. A. and J. C Theilacker, *Trans. ASME, J. Solar Energy Engrg.*, **103**, 29 (1981). "An Algorithm for Calculating Monthly-Average Radiation on Inclined Surfaces."

Knight, K. M., S. A. Klein, and J. A. Duffie, *Solar Energy*, **46**, 109 (1991). "A Methodology for the Synthesis of Hourly Weather Data."

Kondratyev, K. Y., *Actinometry* (translated from Russian), NASA TT F-9712 (1965); also *Radiation in the Atmosphere*, Academic Press, New York (1969).

Klucher, T. M., *Solar Energy*, **23**, 111 (1979). "Evaluating Models to Predict Insolation on Tilted Surfaces."

Latimer, J. R., in *Proc. First Canadian Solar Radiation Data Workshop* (J. E. Hay and T. K. Won, eds.) Ministry of Supply and Services Canada, 81 (1980). "Canadian Procedures for Monitoring Solar Radiation."

Liu, B. Y. H. and R. C. Jordan, *Solar Energy*, **4**(3), 1 (1960). "The Interrelationship and Characteristic Distribution of Direct, Diffuse and Total Solar Radiation."

Liu, B. Y. H. and R. C. Jordan, *ASHRAE Journal*, **3** (10), 53 (1962). "Daily Insolation on Surfaces Tilted Toward the Equator."

Liu, B. Y. H. and R. C. Jordan, *Solar Energy*, **7**, 53 (1963). "The Long-Term Average Performance of Flat-Plate Solar Energy Collectors."

Liu, B. Y. H. and R. C. Jordan, in *Application of Solar Energy for Heating and Cooling of Buildings*, ASHRAE, New York (1977). "Availability of Solar Energy for Flat-Plate Solar Heat Collectors."

Löf, G. O. G., J. A. Duffie, and C. O. Smith, Engineering Experiment Station Report 21, University of Wisconsin, Madison (July 1966a). "World Distribution of Solar Radiation."

Löf, G. O. G., J. A. Duffie, and C. O. Smith, *Solar Energy*, **10**, 27 (1966b). "World Distribution of Solar Energy."

McDaniels, D. K., D. H. Lowndes, H. Mathew, J. Reynolds, and R. Gray, *Solar Energy*, **17**, 277 (1975). "Enhanced Solar Energy Collection Using Reflector-Solar Thermal Collector Combinations."

MacDonald, T. H., *Monthly Weather Review*, **79** (8) (1951). "Some Characteristics of the Eppley Pyrheliometer."

Moon, P., *J. Franklin Institute*, **230**, 583 (1940). "Proposed Standard Solar Radiation Curves for Engineering Use."

Morse, R. N. and J. T. Czarnecki, Report E.E.6 of Engineering Section, Commonwealth Scientific and Industrial Research Organization, Melbourne, Australia (1958). "Flat-Plate Solar Absorbers: The Effect on Incident Radiation of Inclination and Orientation."

Norris, D. J., *Solar Energy*, **12**, 107 (1968). "Correlation of Solar Radiation with Clouds."

Norris, D. J., *Solar Energy*, **14**, 99 (1973). "Calibration of Pyranometers."

Norris, D. J., *Solar Energy*, **16**, 53 (1974). "Calibration of Pyranometers in Inclined and Inverted Positions."

Olseth, J. A. and A. Skartveit, *Solar Energy*, **339**, 343 (1987). "A Probability Density Model for Hourly Total and Beam Irradiance on Arbitrarily Oriented Planes."

Orgill, J. F. and K. G. T. Hollands, *Solar Energy*, **19**, 357 (1977). "Correlation Equation for Hourly Diffuse Radiation on a Horizontal Surface."

Page, J. K., *Proc. of the UN Conference on New Sources of Energy*, **4**, 378 (1964). "The Estimation of Monthly Mean Values of Daily Total Short-Wave Radiation of Vertical and Inclined Surfaces from Sunshine Records for Latitudes 40°N–40°S."

Page, J. K., *Prediction of Solar Radiation on Inclined Surfaces*, Reidel (for the Commission of the European Communities), Dordrecht, Holland (1986).

Paltridge, G. W. and D. Proctor, *Solar Energy*, **18**, 235 (1976). "Monthly Mean Solar Radiation Statistics for Australia."

Perez, R., R. Seals, P. Ineichen, R. Stewart, and D. Menicucci, *Solar Energy*, **39**, 221 (1987). "A New Simplified Version of the Perez Diffuse Irradiance Model for Tilted Surfaces."

Perez, R., R. Stewart, R. Seals, and T. Guertin, Sandia National Laboratories Contractor Report SAND88-7030 (Oct. 1988). "The Development and Verification of the Perez Diffuse Radiation Model."

Reindl, D. T., M.S. Thesis, Mechanical Engineering, U. of Wisconsin-Madison (1988). "Estimating Diffuse Radiation on Horizontal Surfaces and Total Radiation on Tilted Surfaces."

Reindl, D. T., W. A. Beckman, and J. A. Duffie, *Solar Energy*, **45**, 1 (1990a). "Diffuse Fraction Correlations."

Reindl, D. T., W. A. Beckman, and J. A. Duffie, *Solar Energy*, **45**, 9 (1990b). "Evaluation of Hourly Tilted Surface Radiation Models."

Robinson, N. (ed.), *Solar Radiation*, Elsevier, Amsterdam (1966)

Ruth, D. W. and R. E. Chant, *Solar Energy*, **18**, 153 (1976). "The Relationship of Diffuse Radiation to Total Radiation in Canada."

Saunier, G. Y., T. A. Reddy, and S. Kumar, *Solar Energy*, **38**, 169 (1987). "A Monthly Probability Distribution Function of Daily Global Irradiation Values Appropriate for Both Tropical and Temperate Locations."

Skartveit, A. and J. A. Olseth, *Solar Energy*, **36**, 333 (1986). "Modelling Sloped Irradiance at High Latitudes."

Skartveit, A. and J. A. Olseth, *Solar Energy*, **38**, 271 (1987). "A Model for the Diffuse Fraction of Hourly Global Radiation."

SOLMET Manual, U.S. National Climatic Center, Ashville, NC, Vols. 1 and 2 (1978).

Stanhill, G., *Solar Energy*, **10** (2), 69 (1966). "Diffuse Sky and Cloud Radiation in Israel."

Stewart, R., D. W. Spencer, and R. Perez, in *Advances in Solar Energy*, **2** (K. Boer and J. A. Duffie, eds.), 1 (1985). "The Measurement of Solar Radiation."

Temps, R. C. and K. L. Coulson, *Solar Energy*, **19**, 179 (1977). "Solar Radiation Incident upon Slopes of Different Orientations."

Thekaekara, M. P., *Supplement to the Proc. of 20th Annual Meeting of the Inst. for Environmental Science*, 21 (1974). "Data on Incident Solar Energy."

Thekaekara, M. P., *Solar Energy*, **28**, 309 (1976). "Solar Radiation Measurement: Techniques and Instrumentation."

Trewartha, G. T., *An Introduction to Climate*, 3rd ed., McGraw-Hill, New York (1954).

Trewartha, G. T., *The Earth's Problem Climates*, University of Wisconsin Press, Madison (1961).

Tuller, S. E., *Solar Energy*, **18**, 259 (1976). "The Relationship between Diffuse, Total and Extraterrestrial Solar Radiation."

Whillier, A., Ph.D. Thesis, MIT, Cambridge, MA (1953). "Solar Energy Collection and Its Utilization for House Heating."

Whillier, A., *Arch. Met. Geoph. Biokl.* Series B, **7**, 197 (1956). "The Determination of Hourly Values of Total Radiation from Daily Summations."

Whillier, A., *Solar Energy*, **9**, 164 (1965). "Solar Radiation Graphs."

Wiebelt, J. A. and J. B. Henderson, *Trans. ASME, J. Heat Transfer*, **101**, 101 (1979). "Selected Ordinates for Total Solar Radiation Property Evaluation from Spectral Data."

World Meteorological Organization (WMO), *Guide to Meteorological Instruments and Observing Practices*, 3rd ed., WMO and T.P.3, WMO, Geneva (1969).

World Meteorological Organization, *Solar Radiation and Radiation Balance Data (The World Network)* (Annual data compilations, 1964–1985), World Radiation Data Centre, Voeikov Main Geophysical Observatory, Leningrad, USSR.

Yellott, J. I., in *Applications of Solar Energy for Heating and Cooling of Buildings*, ASHRAE, New York (1977). "Solar Radiation Measurement."

Chapter 3

SELECTED HEAT TRANSFER TOPICS

This chapter is intended to review those aspects of heat transfer that are important in the design and analysis of solar collectors and systems. It begins with a review of radiation heat transfer, which is often given cursory treatment in standard heat transfer courses. The next sections review some convection correlations for internal flow and wind-induced flow.

The role of convection and conduction heat transfer in the performance of solar systems is obvious. Radiation heat transfer plays a role in bringing energy to the earth, but not so obvious is the significant role radiation heat transfer plays in the operation of solar collectors. In the usual engineering practice radiation heat transfer is often negligible. In a solar collector the energy flux is often two orders of magnitude smaller than in conventional heat transfer equipment, and thermal radiation is a significant mode of heat transfer.

3.1   THE ELECTROMAGNETIC SPECTRUM

Thermal radiation is electromagnetic energy that is propagated through space at the speed of light. For most solar energy applications, only thermal radiation is important. Thermal radiation is emitted by bodies by virtue of their temperature; the atoms, molecules, or electrons are raised to excited states, return spontaneously to lower energy states, and in doing so emit energy in the form of electromagnetic radiation. Because the emission results from changes in electronic, rotational, and vibrational states of atoms and molecules, the emitted radiation is usually distributed over a range of wavelengths.

The spectrum of electromagnetic radiation is divided into wavelength bands. These bands and the wavelengths representing their approximate limits are shown in Figure 3.1.1. The wavelength limits associated with the various names and the mechanism producing the radiation are not sharply defined. There is no basic distinction between these ranges of radiation other than the wavelength λ; they all travel with the speed of light C and have a frequency v such that

$$C = C_o/n = \lambda v \tag{3.1.1}$$

where C_o is the speed of light in a vacuum and n is the index of refraction.

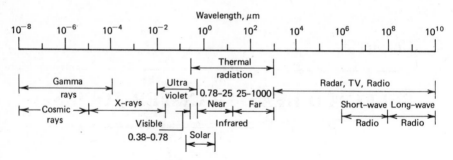

Figure 3.1.1 The spectrum of electromagnetic radiation.

The wavelengths of importance in solar energy and its applications are in the ultraviolet to near-infrared range, that is, from 0.3 to approximately 25 μm. This includes the visible spectrum, light being a particular portion of the electromagnetic spectrum to which the human eye responds. Solar radiation outside the atmosphere has most of its energy in the range of 0.3 to 3 μm, while solar energy received at the ground is substantially in the range of 0.29 to 2.5 μm as noted in Chapters 1 and 2.

3.2   PHOTON RADIATION

For some purposes in solar energy applications, the classical electromagnetic wave view of radiation does not explain the observed phenomena. In this connection, it is necessary to consider the energy of a particle or photon, which can be thought of as an "energy unit" with zero mass and zero charge. The energy of the photon is given by

$$E = h\nu \tag{3.2.1}$$

where h is Planck's constant (6.6256×10^{-34} Js). It follows that as the frequency ν increases (i.e., as the wavelength λ decreases), the photon energy increases. This fact is particularly significant where a minimum photon energy is needed to bring about a required change (for example, the creation of a hole-electron pair in a photovoltaic device). There is thus an upper limit of wavelength of radiation that can cause the change.

3.3   THE BLACKBODY - A PERFECT ABSORBER AND EMITTER

By definition, a blackbody is a perfect absorber of radiation. No matter what wavelengths or directions describe the radiation incident on a blackbody, all incident radiation will be absorbed. A blackbody is an ideal concept since all real substances will reflect some radiation.

Even though a true blackbody does not exist in nature, some materials approach a blackbody. For example, a thick layer of carbon black can absorb approximately 99% of all incident thermal radiation. This absence of reflected radiation is the reason for the name given to a blackbody. The eye would perceive a blackbody as being black. However, the eye is not a good indicator of the ability of a material to absorb radiation, since the eye is only sensitive to a small portion of the wavelength range of thermal radiation. White paints are good reflectors of visible radiation but most are good absorbers of infrared radiation.

A blackbody is also a perfect emitter of thermal radiation. In fact, the definition of a blackbody could have been put in terms of a body that emits the maximum possible radiation. A simple thought experiment can be used to show that if a body is a perfect emitter of radiation, then it must also be a perfect absorber of radiation. Suppose a small blackbody and small non-blackbody are placed in a large evacuated enclosure made from a blackbody material. If the enclosure is isolated from the surroundings, then the blackbody, the real body, and the enclosure will in time come to the same equilibrium temperature. The blackbody must, by definition, absorb all the radiation incident on it, and to maintain a constant temperature, the blackbody must also emit an equal amount of energy. The non-blackbody in the enclosure must absorb less radiation than the blackbody and will consequently emit less radiation than the blackbody. Thus a blackbody both absorbs and emits the maximum amount of radiation.

3.4 PLANCK'S LAW AND WIEN'S DISPLACEMENT LAW

Radiation in the region of the electromagnetic spectrum from approximately 0.2 to approximately 1000 μm is called thermal radiation and is emitted by all substances by virtue of their temperature. The wavelength distribution of radiation emitted by a blackbody is given by Planck's law[1] [Richtmyer and Kennard (1947)]:

$$E_{\lambda b} = \frac{2\pi h C_o^2}{\lambda^5 \left[\exp\left(h C_o / \lambda k T\right) - 1\right]} \tag{3.4.1}$$

where h is Planck's constant and k is Boltzmann's constant. The groups $2\pi h C_o^2$ and $h C_o/k$ are often called Planck's first and second radiation constants and given the symbols C_1 and C_2, respectively.[2] Recommended values are $C_1 = 3.7405 \times 10^{-16}$ m²W and $C_2 = 0.0143879$ mK.

It is also of interest to know the wavelength corresponding to the maximum intensity of blackbody radiation. By differentiating Planck's distribution and

---

[1] The symbol $E_{\lambda b}$ represents energy per unit area per unit time per unit wavelength interval at wavelength λ. The subscript b represents blackbody.

[2] Sometimes the definition of C_1 does not include the factor 2π.

equating to zero, the wavelength corresponding to the maximum of the distribution can be derived. This leads to Wien's displacement law, which can be written as

$$\lambda_{max}T = 2897.8 \ \mu mK \tag{3.4.2}$$

Planck's law and Wien's displacement law are illustrated in Figure 3.4.1, which shows spectral radiation distribution for blackbody radiation from sources at 6000, 1000, and 400 K. The shape of the distribution and the displacement of the wavelength of maximum intensity is clearly shown. Note that 6000 K represents an approximation of the surface temperature of the sun so the distribution shown for that temperature is an approximation of the distribution of solar radiation outside the earth's atmosphere. The other two temperatures are representative of those encountered in low- and high-temperature solar-heated surfaces.

The same information shown in Figure 3.4.1 has been replotted on a normalized linear scale in Figure 3.4.2. The ordinate on this figure, which ranges from zero to one, is the ratio of the spectral emissive power to the maximum value at the same temperature. This clearly shows the wavelength division between a 6000 K source and lower temperature sources at 1000 and 400 K.

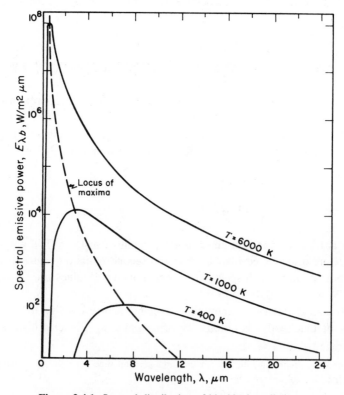

Figure 3.4.1 Spectral distribution of blackbody radiation.

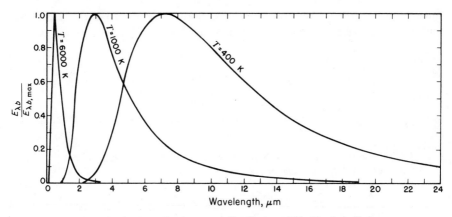

Figure 3.4.2 Normalized spectral distribution of blackbody radiation.

3.5 STEFAN-BOLTZMANN EQUATION

Planck's law gives the spectral distribution of radiation from a blackbody, but in engineering calculations the total energy is often of more interest. By integrating Planck's law over all wavelengths, the total energy emitted by a blackbody is found to be

$$E_b = \int_0^\infty E_{b\lambda} \ d\lambda = \sigma T^4 \tag{3.5.1}$$

where σ is the Stefan-Boltzmann constant and is equal to 5.6697×10^{-8} W/m²K⁴. This constant appears in essentially all radiation equations.

3.6 RADIATION TABLES

Starting with Planck's law (Equation 3.4.1) of the spectral distribution of blackbody radiation, Dunkle (1954) has presented a method for simplifying blackbody calculations. Planck's law can be written as

$$E_{\lambda b} = \frac{C_1}{\lambda^5 \left[\exp\left(C_2/\lambda T\right) - 1\right]} \tag{3.6.1}$$

Equation 3.6.1 can be integrated to give the radiation between any wavelength limits. The total emitted from zero to any wavelength λ is given by

$$E_{0-\lambda,b} = \int_0^\lambda E_{\lambda b} \ d\lambda \tag{3.6.2}$$

Substituting Equation 3.6.1 into 3.6.2 and noting that by dividing by σT^4, the integral can be made to be only a function of λT,

$$f_{0-\lambda T} = \frac{E_{0-\lambda T}}{\sigma T^4} = \int_0^{\lambda T} \frac{C_1 \, d(\lambda T)}{\sigma (\lambda T)^5 \left[\exp(C_2/\lambda T) - 1\right]} \tag{3.6.3}$$

Table 3.6.1a   Fraction of Blackbody Radiant Energy between Zero and λT for Even Increments of λT

$\lambda T, \mu\text{m K}$	$f_{0-\lambda T}$	$\lambda T, \mu\text{m K}$	$f_{0-\lambda T}$	$\lambda T, \mu\text{m K}$	$f_{0-\lambda T}$
1000	0.0003	4500	0.5643	8000	0.8562
1100	0.0009	4600	0.5793	8100	0.8601
1200	0.0021	4700	0.5937	8200	0.8639
1300	0.0043	4800	0.6075	8300	0.8676
1400	0.0077	4900	0.6209	8400	0.8711
1500	0.0128	5000	0.6337	8500	0.8745
1600	0.0197	5100	0.6461	8600	0.8778
1700	0.0285	5200	0.6579	8700	0.8810
1800	0.0393	5300	0.6693	8800	0.8841
1900	0.0521	5400	0.6803	8900	0.8871
2000	0.0667	5500	0.6909	9000	0.8899
2100	0.0830	5600	0.7010	9100	0.8927
2200	0.1009	5700	0.7107	9200	0.8954
2300	0.1200	5800	0.7201	9300	0.8980
2400	0.1402	5900	0.7291	9400	0.9005
2500	0.1613	6000	0.7378	9500	0.9030
2600	0.1831	6100	0.7461	9600	0.9054
2700	0.2053	6200	0.7541	9700	0.9076
2800	0.2279	6300	0.7618	9800	0.9099
2900	0.2506	6400	0.7692	9900	0.9120
3000	0.2732	6500	0.7763	10000	0.9141
3100	0.2958	6600	0.7831	11000	0.9318
3200	0.3181	6700	0.7897	12000	0.9450
3300	0.3401	6800	0.7961	13000	0.9550
3400	0.3617	6900	0.8022	14000	0.9628
3500	0.3829	7000	0.8080	15000	0.9689
3600	0.4036	7100	0.8137	16000	0.9737
3700	0.4238	7200	0.8191	17000	0.9776
3800	0.4434	7300	0.8244	18000	0.9807
3900	0.4624	7400	0.8295	19000	0.9833
4000	0.4829	7500	0.8343	20000	0.9855
4100	0.4987	7600	0.8390	30000	0.9952
4200	0.5160	7700	0.8436	40000	0.9978
4300	0.5327	7800	0.8479	50000	0.9988
4400	0.5488	7900	0.8521	∞	1.

Table 3.6.1b   Fraction of Blackbody Radiation Energy between Zero and λT for Even Fractional Increments

$f_{0-\lambda T}$	$\lambda T, \mu m\,K$	λT at midpoint	$f_{0-\lambda T}$	$\lambda T, \mu m\,K$	λT at midpoint
0.05	1880	1660	0.55	4410	4250
0.10	2200	2050	0.60	4740	4570
0.15	2450	2320	0.65	5130	4930
0.20	2680	2560	0.70	5590	5350
0.25	2900	2790	0.75	6150	5850
0.30	3120	3010	0.80	6860	6480
0.35	3350	3230	0.85	7850	7310
0.40	3580	3460	0.90	9380	8510
0.45	3830	3710	0.95	12500	10600
0.50	4110	3970	1.00	∞	16300

The value of this integral is the fraction of the blackbody energy between zero and λT. Sargent (1972) has calculated values for convenient intervals and the results are given in Table 3.6.1. (Note that when the upper limit of integration of Equation 3.6.3 is infinity, the value of the integral is unity.)

For use in a computer, the following polynomial approximations to Equation 3.6.3 have been given by Pivovonsky and Nagel (1961). For γ greater than or equal to 2

$$f_{0-\lambda T} = \frac{E_{0-\lambda T}}{\sigma T^4} = \frac{15}{\pi^4} \sum_{m=1,2..} \frac{e^{-m\gamma}}{m^4}\{[(m\gamma+3)m\gamma+6]m\gamma+6\} \qquad (3.6.4)$$

For γ less than 2

$$f_{0-\lambda T} = \frac{E_{0-\lambda T}}{\sigma T^4} = 1 - \frac{15}{\pi^4}\,\gamma^3\left(\frac{1}{3} - \frac{\gamma}{8} + \frac{\gamma^2}{60} - \frac{\gamma^4}{5040} + \frac{\gamma^6}{272,160} - \frac{\gamma^8}{13,305,600}\right) \qquad (3.6.5)$$

where $\gamma = C_2/\lambda T$.

Example 3.6.1

Assume that the sun is a blackbody at 5777 K. **a** What is the wavelength at which the maximum monochromatic emissive power occurs? **b** What is the energy from this source that is in the visible part of the electromagnetic spectrum (0.38 to 0.78 μm)?

Solution

a The value of λT at which the maximum monochromatic emissive power occurs is 2897.8 μm K, so the desired wavelength is 2897.8/5777, or 0.502 μm. **b** From

Table 3.6.1 the fraction of energy between zero and $\lambda T = 0.78 \times 5777 = 4506$ μmK is 56%, and the fraction of the energy between zero and $\lambda T = 0.38 \times 5777 = 2195$ μmK is 10%. The fraction of the energy in the visible is then 56% minus 10%, or 46%. These numbers are close to the values obtained from the actual distribution of energy from the sun as calculated in Example 1.3.1. ∎

3.7 RADIATION INTENSITY AND FLUX

Thus far we have considered the radiation leaving a black surface in all directions; however, it is often necessary to describe the directional characteristics of a general radiation field in space. The radiation intensity is used for this purpose and is defined as the energy passing through an imaginary plane per unit area per unit time and per unit solid angle whose central direction is perpendicular to the imaginary plane. Thus, in Figure 3.7.1, if ΔE represents the energy per unit time passing through ΔA and remaining within $\Delta \omega$, then intensity is[3]

$$I = \lim_{\substack{\Delta A \to 0 \\ \Delta \omega \to 0}} \frac{\Delta E}{\Delta A \, \Delta \omega} \tag{3.7.1}$$

The intensity I has both a magnitude and a direction and can be considered as a vector quantity. For a given imaginary plane in space, we can consider two intensity vectors that are in opposite directions. These two vectors are often distinguished by the symbol I^+ and I^-.

The radiation flux is closely related to the intensity and is defined as the energy passing through an imaginary plane per unit area per unit time and in all directions on one side of the imaginary plane. Note that the difference between intensity and flux is that the differential area for intensity is perpendicular to the direction of propagation, whereas the differential area for flux lies in a plane that forms the base of a hemisphere through which the radiation is passing.

The intensity can be used to determine the flux through any plane. Consider an elemental area ΔA on an imaginary plane covered by a hemisphere of radius r as

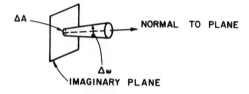

Figure 3.7.1 Schematic of radiation intensity.

---

[3] The symbol I is used for intensity when presenting basic radiation heat transfer ideas and for solar radiation integrated over an hour period when presenting solar radiation ideas. The two will seldom be used together.

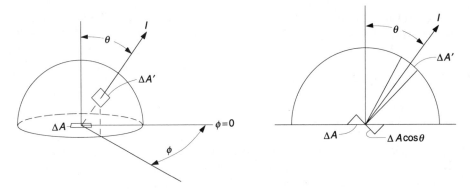

Figure 3.7.2 Schematic of radiation flux.

shown in Figure 3.7.2. The energy per unit time passing through an area $\Delta A'$ on the surface of the hemisphere from the area ΔA is equal to

$$\Delta Q = I \Delta A (\cos \theta) \frac{\Delta A'}{r^2} \tag{3.7.2}$$

Where $\Delta A'/r^2$ is the solid angle between ΔA and $\Delta A'$ and $\Delta A \cos \theta$ is the area perpendicular to the intensity vector. The energy flux per unit solid angle in the θ, ϕ direction can then be defined as

$$\Delta q = \lim_{\Delta A \to 0} \frac{\Delta Q}{\Delta A} = I \cos \theta \frac{\Delta A'}{r^2} \tag{3.7.3}$$

The radiation flux is then found by integrating over the hemisphere. The sphere incremental area can be expressed in terms of the angles θ and ϕ so that

$$q = \int_0^{2\pi} \int_0^{\pi/2} I \cos \theta \, \sin \theta \, d\theta \, d\phi \tag{3.7.4}$$

It is convenient to define $\mu = \cos \theta$ so that

$$q = \int_0^{2\pi} \int_0^1 I \mu \, d\mu \, d\phi \tag{3.7.5}$$

Two important points concerning the radiation flux must be remembered. First, the radiation flux is, in general, a function of the orientation of the chosen imaginary plane. Second, the radiation flux will have two values corresponding to each of the two possible directions of the normal to the imaginary plane. When it is necessary to emphasize which of the two possible values of the radiation flux is being considered, the superscript + or – can be used along with a definition of the positive and negative directions.

Thus far, radiation flux and intensity have been defined at a general location in space. When it is desired to find the heat transfer between surfaces in a vacuum, or at least in radiative nonparticipating media, the most useful values of radiative flux and intensity occur at the surfaces. For the special case of a surface that has intensity independent of direction, the integration of Equation 3.7.5 yields

$$q = \pi I \tag{3.7.6}$$

Surfaces that have the intensity equal to a constant are called either Lambertian or diffuse surfaces. A blackbody emits in a diffuse manner, and therefore the blackbody emissive power is related to the blackbody intensity by

$$E_b = \pi I_b \tag{3.7.7}$$

The foregoing equations were written for total radiation but apply equally well to monochromatic radiation. For example, Equation 3.7.7 could be written in terms of a particular wavelength λ:

$$E_{\lambda b} = \pi I_{\lambda b} \tag{3.7.8}$$

3.8 INFRARED RADIATION EXCHANGE BETWEEN GRAY SURFACES

The general case of infrared radiation heat transfer between many gray surfaces having different temperatures is treated in a number of textbooks [e.g., Hottel and Sarofim (1967), Siegel and Howell (1981)]. The various methods all make the same basic assumptions which, for each surface, can be summarized as follows:

1 The surface is gray. (Radiation properties are independent of wavelength.)
2 The surface is diffuse or specular-diffuse (see Section 4.3).
3 The surface temperature is uniform.
4 The incident energy over the surface is uniform.

Beckman (1971) also utilized these basic assumptions and defined a total exchange factor between pairs of surfaces of an N-surface enclosure such that the net heat transfer to a typical surface i is[4]

$$Q_i = \sum_{j=1}^{N} \varepsilon_i \varepsilon_j A_i \widehat{F}_{ij} \sigma \left(T_j^4 - T_i^4 \right) \tag{3.8.1}$$

---

[4] The emittance ε is defined by Equation 4.1.8.

The factor $\widehat{F}_{ij}$ is the total exchange factor between surfaces i and j and is found from the matrix equation

$$\left[\widehat{F}_{ij}\right] = \left[\delta_{ij} - \rho_j E_{ij}\right]^{-1}\left[E_{ij}\right]$$ (3.8.2)

where E_{ij}, the specular exchange factor, accounts for radiation going from surface i to surface j directly and by all possible specular (mirrorlike) reflections and ρ_j is the diffuse reflectance of surface j. Methods for calculating E_{ij} are given in advanced radiation texts. When the surfaces of the enclosure do not specularly reflect radiation, the specular exchange factors of Equation 3.8.2 reduce to the usual view factor (configuration factor) F_{ij}.

The majority of heat transfer problems in solar energy applications involve radiation between two surfaces. The solution of Equations 3.8.1 and 3.8.2 for diffuse surfaces with $N = 2$ is

$$Q_1 = -Q_2 = \frac{\sigma\left(T_2^4 - T_1^4\right)}{\dfrac{1 - \varepsilon_1}{\varepsilon_1 A_1} + \dfrac{1}{A_1 F_{12}} + \dfrac{1 - \varepsilon_2}{\varepsilon_2 A_2}}$$ (3.8.3)

Two special cases of Equation 3.8.3 are of particular interest. For radiation between two infinite parallel plates (i.e., as in flat-plate collectors) the areas A_1 and A_2 are equal and the view factor F_{12} is unity. Under these conditions Equation 3.8.3 becomes

$$\frac{Q}{A} = \frac{\sigma\left(T_2^4 - T_1^4\right)}{\dfrac{1}{\varepsilon_1} + \dfrac{1}{\varepsilon_2} - 1}$$ (3.8.4)

The second special case is for a small convex object (surface 1) surrounded by a large enclosure (surface 2). Under these conditions, the area ratio A_1/A_2 approaches zero, the view factor F_{12} is unity, and Equation 3.8.3 becomes

$$Q_1 = \varepsilon_1 A_1 \sigma\left(T_2^4 - T_1^4\right)$$ (3.8.5)

This result is independent of the surface properties of the large enclosure since virtually none of the radiation leaving the small object is reflected back from the large enclosure. In other words, the large enclosure absorbs all radiation from the small object and acts like a blackbody. Equation 3.8.5 also applies in the case of a flat plate radiating to the sky (i.e., a collector cover radiating to the surroundings).

3.9   SKY RADIATION

To predict the performance of solar collectors, it will be necessary to evaluate the radiation exchange between a surface and the sky. The sky can be considered as a blackbody at some equivalent sky temperature T_s so that the actual net radiation between a horizontal flat plate and the sky is given by Equation 3.8.5. The net radiation from a surface with emittance ε and temperature T to the sky at T_s is

$$Q = \varepsilon A \sigma \left(T^4 - T_s^4\right) \tag{3.9.1}$$

The equivalent blackbody sky temperature of Equation 3.9.1 accounts for the facts that the atmosphere is not at a uniform temperature and that the atmosphere radiates only in certain wavelength bands. The atmosphere is essentially transparent in the wavelength region from 8 to 14 μm, but outside of this "window" the atmosphere has absorbing bands covering much of the infrared spectrum. Several relations have been proposed to relate T_s for clear skies to measured meterological variables. Swinbank (1963) relates sky temperature to the local air temperature, Brunt (1932) relates sky temperature to the water vapor pressure, and Bliss (1961) relates sky temperature to the dew point temperature. Berdahl and Martin (1984) used extensive data from the United States to relate the effective sky temperature to the dew point temperature, dry bulb temperature, and hour from midnight t by the following equation.

$$T_s = T_a \left[0.711 + 0.0056 T_{dp} + 0.000073 T_{dp}^2 + 0.013 \cos(15t)\right]^{1/4} \tag{3.9.2}$$

where T_s and T_a are in degrees Kelvin and T_{dp} is the dew point temperature in degrees Celsius. The experimental data covered a dew point range from −20 C to 30 C. The range of the difference between sky and air temperatures is from 5 C in a hot, moist climate to 30 C in a cold, dry climate.

Clouds will tend to increase the sky temperature over that for a clear sky. It is fortunate that the sky temperature does not make much difference in evaluating collector performance. However, the sky temperature is critical in evaluating radiative cooling as a passive cooling method.

3.10   RADIATION HEAT TRANSFER COEFFICIENT

To retain the simplicity of linear equations, it is convenient to define a radiation heat transfer coefficient. The heat transfer by radiation between two arbitrary surfaces is found from Equation 3.8.3. If we define a heat transfer coefficient so that the radiation between the two surfaces is given by

$$Q = A_1 h_r \left(T_2 - T_1\right) \tag{3.10.1}$$

then it follows that

$$h_r = \frac{\sigma \left(T_2^2 + T_1^2\right)\left(T_2 + T_1\right)}{\dfrac{1 - \varepsilon_1}{\varepsilon_1} + \dfrac{1}{F_{12}} + \dfrac{(1 - \varepsilon_2)A_1}{\varepsilon_2 A_2}} \tag{3.10.2}$$

If the areas A_1 and A_2 are not equal, the numerical value of h_r depends on whether it is to be used with A_1 or with A_2.

When T_1 and T_2 are close together, the numerator of Equation 3.10.2 can be expressed as $4\sigma \bar{T}^3$ where $\bar{T}$ is the average temperature:

$$4\sigma \bar{T}^3 = \sigma \left(T_2^2 + T_1^2\right)(T_2 + T_1) \tag{3.10.3}$$

It is not difficult to estimate $\bar{T}$ without actually knowing both T_1 and T_2. Once $\bar{T}$ is estimated, the equations of radiation heat transfer are reduced to linear equations that can be easily solved along with the linear equations of conduction and convection. If more accuracy is needed, a second or third iteration may be required. Most of the radiation calculations in this book use the linearized radiation coefficient.

Example 3.10.1

The plate and cover of a flat-plate collector are large in extent, are parallel, and are spaced 0.25 mm apart. The emittance of the plate is 0.15 and its temperature is 70 C. The emittance of the glass cover is 0.88 and its temperature is 50 C. Calculate the radiation exchange between the surfaces and a radiation heat transfer coefficient for this situation.

Solution

Exact and approximate solutions are possible for this problem. The exact solution is based on Equations 3.8.4 and 3.10.1. The radiation exchange is given by Equation 3.8.4:

$$\frac{Q}{A} = \frac{\sigma(343^4 - 323^4)}{\dfrac{1}{0.15} + \dfrac{1}{0.88} - 1} = 24.6 \text{ W/m}^2$$

Then from the defining equation for the radiation coefficient (Equation 3.10.1),

$$h_r = \frac{24.6}{70 - 50} = 1.232 \text{ W/m}^2\text{C}$$

(The use of Equation 3.10.2 produces the identical result.)

We can also get an approximate solution using the average of the two plate temperatures, 60 C or 333 K, in Equation 3.10.3:

$$4\sigma \bar{T}^3 = 4(333)^3 = 1.477 \times 10^8$$

Then

$$h_r = \frac{\sigma \times 1.477 \times 10^8}{\dfrac{1}{0.15} + \dfrac{1}{0.88} - 1} = 1.231 \text{ W/m}^2\text{C}$$

This result is essentially the same as that calculated by the defining equation. ■

3.11 NATURAL CONVECTION BETWEEN FLAT PARALLEL PLATES

The rate of heat transfer between two plates inclined at some angle to the horizon is of obvious importance in the performance of flat-plate collectors. Free convection heat transfer data are usually correlated in terms of two or three dimensionless parameters: the Nusselt number Nu; the Rayleigh number Ra; and the Prandtl number Pr. Some authors correlate data in terms of the Grashof number, which is the ratio of the Rayleigh number to the Prandtl number.

The Nusselt, Rayleigh, and Prandtl numbers are given by[5]

$$Nu = \frac{hL}{k} \tag{3.11.1}$$

$$Ra = \frac{g\beta'\Delta T L^3}{\nu\alpha} \tag{3.11.2}$$

$$Pr = \frac{\nu}{\alpha} \tag{3.11.3}$$

where

h = heat transfer coefficient
L = plate spacing
k = thermal conductivity
g = gravitational constant
β' = volumetric coefficient of expansion (for an ideal gas, $\beta' = 1/T$)
ΔT = temperature difference between plates
ν = kinematic viscosity
α = thermal diffusivity

For parallel plates the Nusselt number is the ratio of a pure conduction resistance to a convection resistance [i.e., $Nu = (L/k)/(1/h)$] so that a Nusselt number of unity represents pure conduction.

Tabor (1958) examined the published results of a number of investigations and concluded that the most reliable data for use in solar collector calculations as of 1958 were contained in Report 32 published by the U.S. Home Finance Agency (1954).

In a more recent experimental study using air, Hollands et al. (1976) give the relationship between the Nusselt number and Rayleigh number for tilt angles from 0 to 75° as

$$Nu = 1 + 1.44 \left[1 - \frac{1708 \left(\sin 1.8\beta \right)^{1.6}}{Ra \cos \beta} \right] \left[1 - \frac{1708}{Ra \cos \beta} \right]^{+} + \left[\left(\frac{Ra \cos \beta}{5830} \right)^{1/3} - 1 \right]^{+} \tag{3.11.4}$$

[5] Fluid properties in the convection relationships of this chapter should be evaluated at the mean temperature.

where the meaning of the + exponent is that only positive values of the terms in the square brackets are to be used (i.e., use zero if the term is negative).

For horizontal surfaces, the results presented by Tabor compare favorably with the correlation of Equation 3.11.4. For vertical surfaces the data from Tabor approximate the 75° tilt data of Hollands et al. Actual collector performance will always differ from analysis, but a consistent set of data is necessary to predict the trends to be expected from design changes. Since a common purpose of this type of data is to evaluate collector design changes, the correlation of Hollands et al. is considered to be the most reliable.

Equation 3.11.4 is plotted in Figure 3.11.1. In addition to the Nusselt number, there is a second scale on the ordinate giving the value of the heat transfer coefficient times the plate spacing for a mean temperature of 10 C. The scale of this ordinate is not dimensionless but is mmW/m²C. For temperatures other than 10 C, a factor F_2, the ratio of the thermal conductivity of air at 10 C to that at any other temperature, has been plotted as a function of temperature in Figure 3.11.2. Thus to find hl at any temperature other than 10 C, it is only necessary to divide F_2hl as read from the chart by F_2 at the appropriate temperature.[6]

The abscissa also has an extra scale, $F_1\Delta Tl^3$. To find ΔTl^3 at temperatures other than 10 C, it is only necessary to divide $F_1\Delta Tl^3$ by F_1, where F_1 is the ratio of $1/Tv\alpha$ at the desired temperature to $1/Tv\alpha$ at 10 C. The ratio F_1 is also plotted in Figure 3.11.2.

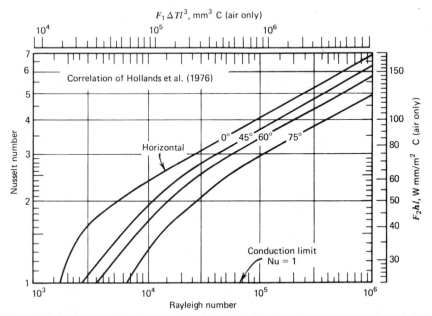

Figure 3.11.1 Nusselt Number as a function of Rayleigh number for free convection heat transfer between parallel flat plates at various slopes.

---

[6] The lowercase letter l is used as a reminder that the units are millimeters instead of meters.

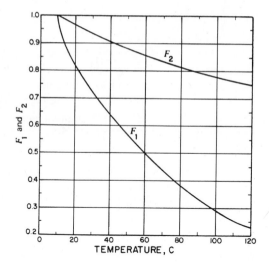

Figure 3.11.2 Air property corrections F_1 and F_2 for use with Figure 3.11.1. From Tabor (1958).

Example 3.11.1

Find the convection heat transfer coefficient between two parallel plates separated by 25 mm with a 45° tilt. The lower plate is at 70 C and the upper plate is at 50 C.

Solution

At the mean air temperature of 60 C air properties are $k = 0.029$ W/mK, $T = 333$ K so $\beta' = 1/333$, $v = 1.88 \times 10^{-5}$ m²/s, and $\alpha = 2.69 \times 10^{-5}$ m²/s. (Property data are from Appendix E.) The Rayleigh number is

$$Ra = \frac{9.81 \times 20 \times (0.025)^3}{333 \times 1.88 \times 10^{-5} \times 2.69 \times 10^{-5}} = 1.82 \times 10^4$$

From Equation 3.11.4 or Figure 3.11.1 the Nusselt number is 2.4. The heat transfer coefficient is found from

$$h = Nu\frac{k}{L} = \frac{2.4 \times 0.029}{0.025} = 2.78 \text{ W/m}^2\text{K}$$

As an alternative, the dimensional scales of Figure 3.11.1 can be used with the property corrections from Figure 3.11.2. At 60 C, $F_1 = 0.49$ and $F_2 = 0.86$. Therefore, $F_1\Delta Tl^3 = 0.49 \times 20 \times 25^3 = 1.53 \times 10^5$ K/mm³. From the 45° curve in Figure 3.11.1, $F_2hl = 59$. Finally, $h = 59/(0.86 \times 25) = 2.74$ W/m²K.

Even with the substantially reduced radiation heat transfer resulting from the low emittance in Example 3.10.1, the radiation heat transfer is about one-half of the convection heat transfer. ∎

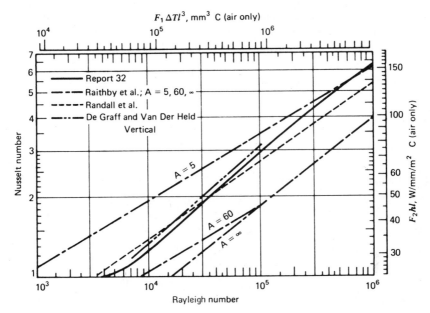

Figure 3.11.3   Nusselt number as a function of Rayleigh number for free convection heat transfer between vertical flat plates.

It is recommended that the 75° correlation of Figure 3.11.1 be used for vertical surfaces. The correlation given by Equation 3.11.1 does not cover the range from 75 to 90°, but comparisons with other correlations suggest that the 75° represents the vertical case adequately. Raithby et al. (1977) have examined vertical surface convection data from a wide range of experimental investigations. They propose a correlation that includes the influence of aspect ratio A, that is, the ratio of plate height to spacing. Their correlation is plotted in Figure 3.11.3 for aspect ratios of 5, 60, and infinity. For comparison, other correlations that do not show an aspect ratio effect are also plotted on this figure and correspond approximately to the Raithby et al. correlation with an aspect ratio of between 10 and 20.

Most of the experiments utilize a guarded hot-plate technique that measures the heat transfer only at the center of the test region. Consequently the end effects are largely excluded. However, Randall et al. (1977) used an interferometric technique that allowed determination of local heat transfer coefficients from which averages were determined; they could not find an aspect ratio effect although a range of aspect ratios from 9 to 36 was covered. The Raithby et al. (1977) correlation also includes an angular correction for angles from 70 to 110° which shows a slight increase in Nusselt number over this range of tilt angles consistent with the trends of Figure 3.11.1 [see Randall et al. (1977)].

It is unusual to find a collector sloped at angles between 75 and 90°; if they are to be that steep, they will probably be vertical. Windows and collector-storage walls are essentially always vertical. For vertical surfaces the four correlations shown in

Figure 3.11.3 (with $A \cong 15$ for the Raithby et al. result) agree within approximately 15% with the 75° correlation of Hollands et al. in Figure 3.11.1. Vertical solar collectors will have an aspect ratio on the order of 60, but at this aspect ratio the Raithby et al. result falls well below other correlations. Consequently, the 75° correlation of Figure 3.11.1 will give reasonable or conservative predictions for vertical surfaces.

3.12   CONVECTION SUPPRESSION

One of the objectives in designing solar collectors is to reduce the heat loss through the covers. This has led to studies of convection suppression by Hollands (1965), Edwards (1969), Buchberg et al. (1976), Arnold et al. (1977, 1978), Meyer et al. (1978), and others. In these studies the space between two plates, with one plate heated, is filled with a transparent or specularly reflecting honeycomb to suppress the onset of fluid motion. Without fluid motion the heat transfer between the plates is by conduction and radiation. Care must be exercised since improper design can lead to increased rather than decreased convection losses, as was first shown experimentally by Charters and Peterson (1972) and later verified by others.

For slats, as shown in Figure 3.12.1, the results of Meyer et al. (1978) can be expressed as the maximum of two numbers as

$$Nu = \max\left[1.1 C_1 C_2 Ra_L^{0.28}, 1\right] \tag{3.12.1}$$

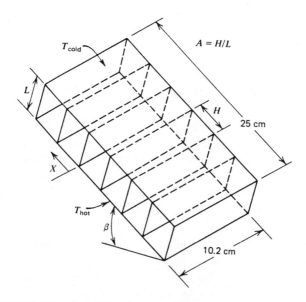

Figure 3.12.1   Slats for suppression of convection. From Meyer et al. (1978).

where C_1 and C_2 are given in Figure 3.12.2 and the subscript L indicates that the plate spacing L is the characteristic length. Note that the coefficient C_1 has a maximum near an aspect ratio of 2.

To assess the magnitude of the convection suppression with slats, it is possible to compare Equation 3.12.1 with the correlation of Randall et al. (1977) obtained from data taken on the same equipment. Although the Randall correlation uses an exponent of 0.29 on the Rayleigh number, the correlation can be slightly modified to have an exponent of 0.28. The ratio of the two correlations is then

$$\frac{Nu_{\text{slats}}}{Nu_{\text{no slats}}} = \frac{\max\left[1.1C_1C_2Ra_L^{0.28}, 1\right]}{\max\left[0.13Ra_L^{0.28}\left[\cos(\beta - 45)\right]^{0.58}, 1\right]} \qquad (3.12.2)$$

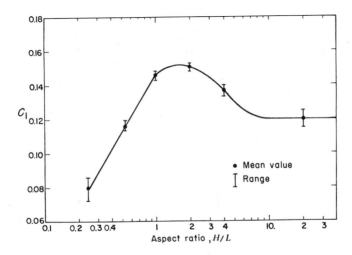

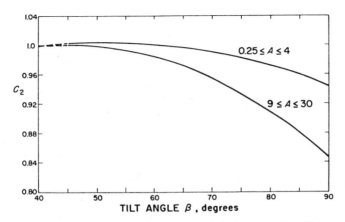

Figure 3.12.2 Coefficients C_1 and C_2 for use in Equation 3.12.1. From Meyer et al. (1978).

As long as fluid motion is not suppressed (i.e., as long as Nu is greater than 1), the ratio of the two Nusselt numbers is independent of the Rayleigh number.

At a collector angle of 45°, the addition of slats will reduce convection as long as the aspect ratio is less than approximately 0.5 (i.e., $C_1 = 0.12$, $C_2 = 1.0$). At an aspect ratio of 0.25, the slats reduce convection by one-third. At a Rayleigh number of 5800 and a tilt of 45°, fluid movement is just beginning with an aspect ratio of 0.25, and the Nusselt number is 1.0. From Randall's correlation without slats and with a Rayleigh number of 5800, the Nusselt number is 1.47, a nearly 50% reduction in convection heat transfer.

Arnold et al. (1977, 1978) experimentally investigated cores with aspect ratios between 0.125 and 0.25 but with additional partitions that produced rectangular honeycombs having horizontal aspect ratios (width-to-plate spacing) ranging from 0.25 to 6.25. The results of these experiments, using silicone oil as the working fluid to suppress thermal radiation, can be correlated within ±15% with the following equation:

$$Nu = 1 + 1.15\left[1 - \frac{Ra_1}{Ra\cos\beta}\right]^+ + 1.25\left[1 - \frac{Ra_2}{Ra\cos\beta}\right]^+ \tag{3.12.3}$$

for
$$0 \le \beta \le 60$$
$$Ra\cos\beta \le Ra_3$$
$$4 \le L/H \le 8$$
$$1 \le W/H \le 24$$

where
$$Ra_k = \frac{(a_k + b_k)^3}{a_k}$$

$$a_k = a_o + b_k/2$$

$$b_k = (k\pi + 0.85)^2$$

$$k = 1, 2, \text{ or } 3$$

$$a_o = \left[\frac{7}{1 + L/7D} + 18\left(\frac{H}{W}\right)^2\right]\left(\frac{L}{H}\right)^2$$

For vertical orientation ($\beta = 90°$), the results can be correlated by

$$Nu = 1 + \frac{10^{-4}\left(\frac{H}{L}\right)^{4.65}}{1 + 5\left(\frac{H}{W}\right)^4} Ra^{1/3} \tag{3.12.4}$$

for the same L/H and W/H limits as given for Equation 3.12.3.

These equations show little effect on heat transfer of horizontal aspect ratios beyond unity. Consequently, the results of Meyer et al. for slats should be directly comparable. At an angle of 45° and a Rayleigh number of 4×10^4 both experiments give a Nusselt number of approximately 1.7, but the slope of the data on a Nusselt-Rayleigh plot from Arnold et al. is approximately 0.48 and the slope from Meyer et al. is 0.28. Since the Rayleigh number range of the two experiments was not large, these two very different correlations give similar Nusselt numbers, but extrapolation beyond the range of test data (i.e., $Ra > 10^5$) could lead to large differences.

This discrepancy points out a problem in estimating the effect of collector design options based on heat transfer data from two different experiments. It is the nature of heat transfer work that sometimes significant differences are observed in carefully controlled experiments using different equipment or techniques. Consequently in evaluating an option such as the addition of honeycombs, heat transfer data with and without honeycombs measured in the same laboratory will probably be the most reliable.

The addition of a honeycomb in a solar collector will modify the collector's radiative characteristics. The honeycomb will certainly decrease the solar radiation reaching the absorbing plate of the collector. Hollands et al. (1978) have analyzed the solar transmittance[7] of a square-celled honeycomb and compared the results with measurements at normal incidence. For the particular polycarbonate plastic configuration tested, the honeycomb transmittance at normal incidence was 0.98. Its transmittance decreased in a nearly linear manner with incidence angle to approximately 0.90 at an angle of 70°.

The infrared radiation characteristics will also be affected in a manner largely dependent upon the honeycomb material. If the honeycomb is constructed of either an infrared transparent material or an infrared specularly reflective material, then the infrared radiative characteristics of the collector will not be significantly changed. If the honeycomb material is constructed of a material that is opaque in the infrared, then the radiative characteristics of the collector will approach that of a blackbody. As will be shown in the next two chapters, this is undesirable.

Transparent aerogels can be used to eliminate convection heat transfer. Properly made silica aerogels transmit most solar radiation. They consist of very fine silica particles and micropores that are smaller than the mean free path of air molecules; convection is supressed and conduction is less than that of still air. Thermal stability and weatherability may be problems.

3.13   VEE-CORRUGATED ENCLOSURES

Vee-corrugated absorber plates with the corrugations running horizontally have been proposed for solar collectors to improve the radiative characteristics of the absorber

---

[7] Transmittance of cover systems is discussed in Chapter 5.

plate (see Section 4.9). Also, this configuration approximates the shape of some concentrating collectors (see Chapter 7). One problem with this configuration is that the improved radiative properties are, at least in part, offset by increased convection losses. Elsherbiny et al. (1977) state that free convection losses from a vee-corrugated surface to a single plane above is as much as 50% greater than for two plane surfaces at the same temperatures and mean plate spacing.

Randall (1978) investigated vee-corrugated surfaces and correlated the data in terms of the Nusselt and Rayleigh numbers in the form

$$Nu = \max\left[(CRa^n),\ 1\right] \tag{3.13.1}$$

where the values of C and n are given in Table 3.13.1 as functions of the tilt angle β and the vee aspect ratio A', the ratio of mean plate spacing l to vee height h shown in Figure 3.13.1.

Table 3.13.1 Constants for Use in Equation 3.13.1

β	A'	C	n
0	0.75	0.060	0.41
	1	0.060	0.41
	2	0.043	0.41
45	0.75	0.075	0.36
	1	0.082	0.36
	2	0.037	0.41
60	0.75	0.162	0.30
	1	0.141	0.30
	2	0.027	0.42

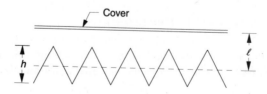

Figure 3.13.1 Section of Vee-Corrugated Absorber and Plane Cover.

3.14   HEAT TRANSFER RELATIONS FOR INTERNAL FLOW

Heat transfer coefficients for common geometries are given in many heat transfer books [e.g., McAdams (1954), Kays and Crawford (1980), and Incropera and DeWitt (1985)]. For fully developed turbulent flow inside tubes (i.e., $Re = \rho v D_h / \mu > 2200$) Kays and Crawford suggests that the Petukhov equation be used:

$$Nu = \frac{(f/8)\, Re\, Pr}{1.07 + 12.7\, \sqrt{f/8}\, \left(Pr^{2/3} - 1\right)} \left(\frac{\mu}{\mu_w}\right)^n \qquad (3.14.1)$$

where $n = 0.11$ for heating and 0.25 for cooling and the Darcy friction factor f for smooth pipes is given by

$$f = (0.79 \ln Re - 1.64)^{-2} \qquad (3.14.2)$$

For noncircular tubes the hydraulic diameter can be used for the characteristic length in the preceding two equations. The hydraulic diameter is defined as

$$D_h = \frac{4\,(\text{flow area})}{\text{wetted perimeter}} \qquad (3.14.3)$$

For short tubes with L/D greater than 1.0 and a sharp-edged entry, McAdams recommends that the Nusselt number be calculated from

$$Nu_{\text{short}} = Nu_{\text{long}} \left[1 + \left(\frac{D}{L}\right)^{0.7}\right] \qquad (3.14.4)$$

For laminar flow in tubes the thermal boundary condition is important. With fully developed hydrodynamic and thermal profiles, the Nusselt number is 3.7 for constant wall temperature and 4.4 for constant heat flux. In a solar collector the thermal condition is closely represented by a constant resistance between the flowing fluid and the constant-temperature environment.[8] If this resistance is large, the thermal boundary condition approaches constant heat flux, and if this resistance is small, the thermal boundary condition approaches constant temperature. Consequently, the theoretical performance of a solar collector should lie between the results for constant heat flux and constant temperature. Since a constant wall temperature assumption yields somewhat lower heat transfer coefficients, this is the recommended assumption for conservative design.

For short tubes the developing thermal and hydrodynamic boundary layers will result in a significant increase in the heat transfer coefficient near the entrance. Heaton et al. (1964) present local Nusselt numbers for the case of constant heat rate.

---

[8] This will become apparent in Chapter 6.

Their data is well represented by an equation of the form

$$Nu = Nu_\infty + \frac{a(Re\,Pr\,D_h/L)^m}{1 + b\,(Re\,Pr\,D_h/L)^n} \qquad (3.14.5)$$

where the constants a, b, m, and n are given in Table 3.14.1.

Goldberg (1958), as reported by Rohsenow and Choi (1961), presents average Nusselt numbers for the case of constant wall temperature. The results for Prandtl numbers of 0.7, 5, and infinity are shown in Figure 3.14.1. The data of this figure can also be represented by an equation of the form of Equation 3.14.5 but with values of a, b, m, n, and Nu_∞ given in Table 3.14.2.

Example 3.14.1

What is the heat transfer coefficient inside the tubes of a solar collector in which the tubes are 10 mm in diameter and separated by a distance 100 mm. The collector is 1.5 m wide and 3 m long, and has total flow rate of water of 0.075 kg/s. The water is at 80 C.

Table 3.14.1   Constants for Equation 3.14.5 for Calculation of Local Nu for Circular Tubes with Constant Heat Rate.

Prandtl Number	a	b	m	n
0.7	0.00398	0.0114	1.66	1.12
10	0.00236	0.00857	1.66	1.13
∞	0.00172	0.00281	1.66	1.29
	$Nu_\infty = 4.4$			

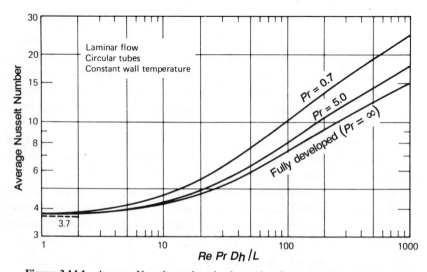

Figure 3.14.1   Average Nusselt numbers in short tubes for various Prandtl numbers.

Table 3.14.2   Constants for Equation 3.14.5 for Calculation of Average Nu for Circular Tubes with Constant Wall Temperature.

Prandtl Number	a	b	m	n
0.7	0.0791	0.0331	1.15	0.82
5	0.0534	0.0335	1.15	0.82
∞	0.0461	0.0316	1.15	0.84
		$Nu_\infty = 3.7$		

Solution

The collector has 15 tubes so that the flow rate per tube is 0.005 kg/s. The Reynolds number is

$$\frac{VD}{v} = \frac{4\dot{m}}{\pi D \mu} = \frac{4 \times 0.005}{\pi \times 0.01 \times 3.6 \times 10^{-4}} = 1800$$

which indicates laminar flow. The Prandtl number is 2.2 so that

$$Re\, Pr\, D/L = 1800 \times 2.2 \times 0.01/3 = 13$$

From Figure 3.14.1 the average Nusselt number is 4.6 so the average heat transfer coefficient is

$$h = Nu\, k/D = 4.6 \times 0.67/0.01 = 310 \text{ W/m}^2\text{C} \qquad \blacksquare$$

In the study of solar air heaters and collector-storage walls it is necessary to know the forced convection heat transfer coefficient between two flat plates. For air the following correlation can be derived from the data of Kays and Crawford (1980) for fully developed turbulent flow with one side heated and the other side insulated:

$$Nu = 0.0158\, Re^{0.8} \qquad (3.14.6)$$

where the characteristic length is the hydraulic diameter (twice the plate spacing). For flow situations in which L/D_h is 10, Kays indicates that the average Nusselt number is approximately 16% higher than that given by Equation 3.14.6. At L/D_h equal to 30, Equation 3.14.6 still underpredicts by 5%. At L/D_h equal to 100, the effect of the entrance region has largely disappeared.

Tan and Charters (1970) have experimentally studied flow of air between parallel plates with small aspect ratios for use in solar air heaters. Their results give higher heat transfer coefficients by about 10% than those given by Kays with an infinite aspect ratio.

Table 3.14.3   Constants for Equation 3.14.5 for Calculation of Local Nu **for Infinite Flat Plates, One Side Insulated and Constant Heat Flux on Other Side.**

Prandtl Number	a	b	m	n
0.7	0.00190	0.00563	1.71	1.17
10	0.00041	0.00156	2.12	1.59
∞	0.00021	0.00060	2.24	1.77

$$Nu_\infty = 5.4$$

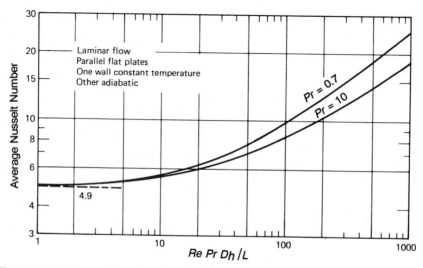

Figure 3.14.2 Average Nusselt numbers in short ducts with one side insulated and one side at constant wall temperature for various Prandtl numbers.

The local Nusselt number for laminar flow between two flat plates with one side insulated and the other subjected to a constant heat flux has been obtained by Heaton et al. (1964). The results have been correlated in the form of Equation 3.14.5 with the constants given in Table 3.14.3.

For the case of parallel plates with constant temperature on one side and insulated on the other side, Mercer et al. (1967) obtained the average Nusselt numbers shown in Figure 3.14.2. They also correlated these data into the form of Equation 3.14.7 for $0.1 < Pr < 10$:

$$Nu = 4.9 + \frac{0.0606(Re\ Pr\ D_h/L)^{1.2}}{1 + 0.0909(Re\ Pr\ D_h/L)^{0.7}Pr^{0.17}} \tag{3.14.7}$$

The results of Sparrow (1955) indicate that for $Re\ Pr\ D_h/L < 1000$, the $Pr = 10$ Nusselt numbers are essentially the same as for the case when the hydrodynamic profile is fully developed.

Example 3.14.2

a Determine the convective heat transfer coefficient for air flow in a channel 1 m wide by 2 m long. The channel thickness is 15 mm and the air flow rate is 0.03 kg/s. The average air temperature is 35 C.
b If the channel thickness is halved, what is the heat transfer coefficient?
c If the flow rate is halved, what is the heat transfer coefficient?

Solution

a At a temperature of 35 C the viscosity is 1.88×10^{-5} m^2/s and the thermal conductivity is 0.0268 W/mC. The hydraulic diameter D_h is twice the plate spacing t, and the Reynolds number can be expressed in terms of the flow rate per unit width $\dot{m}/W$. The Reynolds number is then

$$Re = \frac{\rho V D_h}{\mu} = \frac{2\rho V t W}{W\mu} = \frac{2\dot{m}}{W\mu} = \frac{2 \times 0.03}{1 \times 1.88 \times 10^{-5}} = 3200$$

so that the flow is turbulent. From Equation 3.14.6 the Nusselt number is

$$Nu = 0.0158(3200)^{0.8} = 10.1$$

and the heat transfer coefficient is $h = Nu \, k/D_h = 9$ W/m^{2}C. Since D_h/L is less than 100, 9 W/m^{2}C is probably a few percent too low.

    b If the channel thickness is halved, the Reynolds number remains the same but the heat transfer coefficient will double to 18 W/m^{2}C.

    c If the flow rate is halved, the Reynolds number will be 1600, indicating laminar flow. Equation 3.14.7 or Figure 13.14.2 should be used. The value of $Re \, Pr \, D_h/L$ is $1600 \times 0.7 \times 0.03/2 = 16.8$ so the Nusselt number is 6.0 and the heat transfer coefficient is 6.2 W/m^{2}C. ∎

3.15 WIND CONVECTION COEFFICIENTS

The heat loss from flat plates exposed to outside winds is important in the study of solar collectors. Sparrow et al. (1979) did wind tunnel studies on rectangular plates at various orientations and found the following correlation over the Reynolds number range of 2×10^4 to 9×10^4:

$$Nu = 0.86 \, Re^{1/2} Pr^{1/3} \tag{3.15.1}$$

where the characteristic length is four times the plate area divided by the plate perimeter. For laminar flow (i.e., Reynolds numbers less than 10^6, the critical Re for flow over a flat plate) over a very wide flat plate at zero angle of attack, the analysis of Pohlhausen [see Kays and Crawford (1980)] yields a coefficient for Equation 3.15.1 of 0.94.[9]

This agreement at low Reynolds numbers suggests that Equation 3.15.1 may be valid at Reynolds numbers up to 10^6 where direct experimental evidence is lacking. This extrapolation is necessary since a solar collector array 2 m by 5 m has a characteristic length of 2.9 m and Reynolds number of 9.4×10^5 in a 5 m/s wind. From Equation 3.15.1, the heat transfer coefficient under these conditions is approximately 7 W/m²C.

McAdams (1954) reports the data of Jurges for a 0.5 m² plate in which the convection coefficient is given by the dimensional equation

$$h = 5.7 + 3.8V \tag{3.15.2}$$

where V is the wind speed in m/s and h is in W/m²C. It is probable that the effects of free convection and radiation are included in this equation. For this reason Watmuff et al. (1977) report that this equation should be

$$h = 2.8 + 3.0V \tag{3.15.3}$$

For a 0.5 m² plate, Equation 3.15.1 yields a heat transfer coefficient of 16 W/m²C at a 5 m/s wind speed and a temperature of 25 C. Equation 3.15.3 yields a value of 18 W/m²C at these conditions. Thus there is agreement between the two at a characteristic length of 0.5 m. It is not reasonable to assume that Equation 3.15.3 is valid at other plate lengths.

The flow over a collector mounted on a house is not always well represented by wind tunnel tests of isolated plates. The collectors will sometimes be exposed directly to the wind and other times will be in the wake region. The roof itself will certainly influence the flow patterns. Also, nearby trees and buildings will greatly effect local flow conditions. Mitchell (1976) investigated the heat transfer from various shapes (actually animal shapes) and showed that many shapes were well represented by a sphere when the equivalent sphere diameter is the cube root of the volume. The heat transfer obtained in this manner is an average that includes stagnation regions and wake regions. A similar situation might be anticipated to occur in solar systems. Mitchell suggests that the wind tunnel results of these animal tests should be increased by approximately 15% for outdoor conditions. Thus, assuming a house to be a sphere, the Nusselt number can be expressed as

$$Nu = 0.42 \, Re^{0.6} \tag{3.15.4}$$

where the characteristic length is the cube root of the house volume.

---

[9] To be consistent with Equation 3.15.1 the characteristic length in the Pohlhausen solution must be changed to twice the plate length. This changes the familiar coefficient of 0.664 to 0.94.

When the wind speed is very low, free convection conditions may dominate. Free convection data for hot inclined flat plates facing upward are not available. However, results are available for horizontal and vertical flat plates. For hot horizontal flat plates with aspect ratios up to 7:1, Lloyd and Moran (1974) give the following equations:

$$Nu = 0.76 Ra^{1/4} \quad \text{for } 10^4 < Ra < 10^7 \qquad (3.15.5)$$

$$Nu = 0.15 Ra^{1/3} \quad \text{for } 10^7 < Ra < 3 \times 10^{10} \qquad (3.15.6)$$

where the characteristic length is four times the area divided by the perimeter. (The original reference used A/P.) For vertical plates McAdams gives

$$Nu = 0.59 \ Ra^{1/4} \text{ for } 10^4 < Ra < 10^9 \qquad (3.15.7)$$

$$Nu = 0.13 \ Ra^{1/3} \text{ for } 10^9 < Ra < 10^{12} \qquad (3.15.8)$$

where the characteristic length is the plate height.

For large Rayleigh numbers, as would be found in most solar collector systems, Equations 3.15.6 and 3.15.8 apply and the characteristic length drops out of the calculation of the heat transfer coefficient. The heat transfer coefficients from these two equations are nearly the same since the coefficient on the Rayleigh numbers differ only slightly. This means that horizontal and vertical collectors have a minimum heat transfer coefficient (i.e., under free convection conditions) of about 5 W/m²C for a 25 C temperature difference and a value of about 4 W/m²C at a temperature difference of 10 C.

From the preceding discussion it is apparent that the calculation of wind-induced heat transfer coefficients is not well established. Until additional experimental evidence becomes available, the following guidelines are recommended. When free and forced convection occur simultaneously, McAdams recommends that both values be calculated and the larger value used in calculations. Consequently, it appears that a minimum value of approximately 5 W/m²C occurs in solar collectors under still air conditions. For forced convection conditions over buildings the results of Mitchell can be expressed as

$$h_w = \frac{8.6 V^{0.6}}{L^{0.4}} \qquad (3.15.9)$$

The heat transfer coefficient (in W/m²C) for flush-mounted collectors can then be expressed as

$$h_w = \max\left[5, \frac{8.6 V^{0.6}}{L^{0.4}}\right] \qquad (3.15.10)$$

where V is wind speed in meters per second and L is the cube root of the house volume in meters. At a wind speed of 5 m/s (which is close to the world average wind speed) and a characteristic length of 8 m, Equation 3.15.10 yields a heat transfer coefficient of 10 W/m²C.

For flow of air across a single tube in an outdoor environment the equations recommended by McAdams have been modified to give[10]

$$Nu = 0.40 + 0.54\,Re^{0.52} \quad \text{for } 0.1 < Re < 1000 \tag{3.15.11}$$

$$Nu = 0.30\,Re^{0.6} \quad \text{for } 1000 < Re < 50{,}000 \tag{3.15.12}$$

3.16 HEAT TRANSFER AND PRESSURE DROP IN PACKED BEDS

In solar air heating systems the usual energy storage media is a packed bed of small rocks or crushed gravel. The heat transfer and pressure drop characteristics of these storage devices are of considerable interest and have been extensively reviewed by Shewen et al. (1978). Although many correlations were found for both heat transfer coefficients and friction factors in packed beds, none of the correlations were entirely satisfactory in predicting the measured performance of their experimental packed bed. The following relationships are based on the recommendations of Shewen et al.

The physical characteristics of pebbles vary widely between samples. Three quantities have been used to describe pebbles, the average particle diameter D, the void fraction ε, and the surface area shape factor α. The void fraction can be determined by weighing pebbles placed in a container of volume V before and after it is filled with water. The void fraction is then equal to

$$\varepsilon = \frac{m_w/\rho_w}{V} \tag{3.16.1}$$

Where m_w is the mass of water and ρ_w is the density of water. The density of the rock material is then

$$\rho_r = \frac{m}{V(1-\varepsilon)} \tag{3.16.2}$$

where m is the mass of the rocks alone. The average particle diameter is the diameter of a spherical particle having the same volume and can be calculated from

$$D = \left(\frac{6m}{\pi\rho_r N}\right)^{1/3} \tag{3.16.3}$$

where N is the number of pebbles in the sample. The surface area shape factor α is

[10] To account for outdoor conditions, the original coefficients have been increased by 25%.

the ratio of the surface area of the pebble to the surface area of the equivalent sphere and is difficult to evaluate. For smooth river gravel α appears to be independent of pebble size and approximately equal to 1.5. For crushed gravel α varies with the pebble size and decreases linearly from approximately 2.5 at very small sizes to approximately 1.5 for 50 mm diameter particles. However, large scatter is observed.

The three pebble bed parameters D, ε, and α do not fully take into account all the observed behavior of packed bed storage devices. However, exact predictions are not needed since the performance of a solar system is not a strong function of the storage unit design as long as certain criteria are met.[11] When measurements of the void fraction ε and the surface area shape factor α are available, the pressure drop relationship recommended by Shewen et al. is that of McCorquodale et al. (1978):

$$\Delta P = \frac{LG_o^2}{\rho_{air}D} \frac{(1-\varepsilon)\alpha}{\varepsilon^{3/2}}\left[4.74 + 166\frac{(1-\varepsilon)\alpha}{\varepsilon^{3/2}}\frac{\mu}{G_oD}\right] \tag{3.16.4}$$

where G_o is the mass velocity of the air (air mass flow rate divided by the bed frontal area) and L is the length of the bed in the flow direction. When measurements of α and ε are not available, Shewen et al. recommend the equation of Dunkle and Ellul (1972):

$$\Delta P = \frac{LG_o^2}{\rho_{air}D}\left(21 + 1750\frac{\mu}{G_oD}\right) \tag{3.16.5}$$

For heat transfer Shewen et al. recommend the Löf and Hawley (1948) equation.

$$h_v = 650\left(\frac{G_o}{D}\right)^{0.7} \tag{3.16.6}$$

where h_v is the volumetric heat transfer coefficient in W/m³C, G_o is the mass velocity in kg/m²s, and D is the particle diameter in meters. The relationship between volumetric heat transfer coefficient h_v and area heat transfer coefficient h is

$$h_v = 6h(1-\varepsilon)\frac{\alpha}{D} \tag{3.16.7}$$

Example 3.16.1

A pebble bed is used for energy storage in a solar heating system. Air is the working fluid and flows vertically through the bed. The bed has the following dimensions and characteristics: depth is 2.10 m; length and width are 4.0 and 3.7 m; equivalent diameter of the pebbles is 23.5 mm; and void fraction is 0.41. The superficial air velocity is 0.143 m/s. The average air temperature is 40 C. Estimate the pressure drop through the bed and the volumetric heat transfer coefficient.

---

[11] See Table 13.2.1 and Section 8.5.

Solution

Use Equation 3.16.5. From Appendix E, for air at 40 C, $\rho = 1.127$ kg/m³ and $\mu = 1.90 \times 10^{-5}$ Pa s. The mass velocity is then

$$G_o = 0.143 \times 1.127 = 0.161 \text{ kg/m}^2\text{s}$$

Using the Dunkle and Ellul correlation,

$$\Delta P = \frac{2.10 \times (0.161)^2}{1.127 \times 0.0235}\left(21 + \frac{1750 \times 1.90 \times 10^{-5}}{0.161 \times 0.0235}\right) = 61.2 \text{ Pa}$$

Use Equation 3.16.6 to estimate the volumetric heat transfer coefficient:

$$h_v = 650 \ (0.161/0.0235)^{0.7} = 2500 \text{ W/m}^3\text{C} \qquad \blacksquare$$

3.17 EFFECTIVENESS-NTU CALCULATIONS FOR HEAT EXCHANGERS

It is convenient in solar process system calculations to use the effectiveness-NTU (number of transfer units) method of calculation of heat exchanger performance. A brief discussion of the method is provided here, based on the example of a countercurrent exchanger. The working equation is the same for other heat exchanger configurations; the expressions for effectiveness vary from one configuration to another. [See Kays and London (1964).]

A schematic of an adiabatic countercurrent exchanger with inlet and outlet temperatures and capacitance rates of the hot and cold fluids is shown in Figure 3.17.1. The overall heat transfer coefficient-area product is UA. The maximum possible temperature drop of the hot fluid is from T_{hi} to T_{ci}; the heat transfer for this situation would be

$$Q_{max} = (\dot{m}C_p)_h(T_{hi} - T_{ci}) \qquad (3.17.1)$$

The maximum possible temperature rise of the cold fluid would be from T_{ci} to T_{hi}.

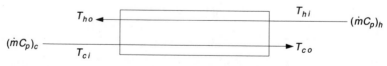

Figure 3.17.1 Schematic of an adiabatic counterflow heat exchanger showing temperatures and capacitance rates of the hot and cold fluids.

The corresponding maximum heat exchange would be

$$Q_{max} = (\dot{m}C_p)_c(T_{hi} - T_{ci}) \tag{3.17.2}$$

The maximum heat transfer that could occur in the exchanger is thus fixed by the lower of the two capacitance rates, $(\dot{m}C_p)_{min}$, and

$$Q_{max} = (\dot{m}C_p)_{min}(T_{hi} - T_{ci}) \tag{3.17.3}$$

The actual heat exchange Q is given by

$$Q = (\dot{m}C_p)_c(T_{co} - T_{ci}) = (\dot{m}C_p)_h(T_{hi} - T_{ho}) \tag{3.17.4}$$

Effectiveness ε is defined as the ratio of the actual heat exchange that occurs to the maximum possible, Q/Q_{max}, so

$$\varepsilon = \frac{Q}{Q_{max}} = \frac{(\dot{m}C_p)_h(T_{hi} - T_{ho})}{(\dot{m}C_p)_{min}(T_{hi} - T_{ci})} = \frac{(\dot{m}C_p)_c(T_{co} - T_{ci})}{(\dot{m}C_p)_{min}(T_{hi} - T_{ci})} \tag{3.17.5}$$

Since either the hot or cold fluid has the minimum capacitance rate, the effectiveness can always be expressed in terms of the temperatures only. The working equation for the heat exchanger is

$$Q = \varepsilon(\dot{m}C_p)_{min}(T_{hi} - T_{ci}) \tag{3.17.6}$$

For a counterflow exchanger, the effectiveness is given by

$$\varepsilon = \frac{1 - e^{-NTU(1 - C^*)}}{1 - C^*e^{-NTU(1 - C^*)}} \qquad \text{if } C^* \neq 1 \tag{3.17.7a}$$

or $$\varepsilon = \frac{NTU}{1 + NTU} \qquad \text{if } C^* = 1 \tag{3.17.7b}$$

where NTU is the number of transfer units,

$$NTU = (UA)/(\dot{m}C_p)_{min} \tag{3.17.8}$$

and the dimensionless capacitance rate is given by

$$C^* = (\dot{m}C_p)_{min}/(\dot{m}C_p)_{max} \tag{3.17.9}$$

Kays and London give equations and graphs for effectivenesses for many heat exchanger types.

The utility of this approach to heat exchanger calculations will be evident in Chapter 10, where the temperatures of streams entering exchangers between collectors and storage tanks and between storage tanks and loads are known.

Example 3.17.1

A heat exchanger like that in Figure 3.17.1 is located between a collector and a storage tank. The fluid on the collector side is an antifreeze, a glycol-water mixture with C_p = 3850 J/kgC. Its flow rate is 1.25 kg/s. The fluid on the tank side is water, and its flow rate is 0.864 kg/s. The UA of the heat exchanger is estimated to be 6500 W/C.

If the hot glycol from the collector enters the exchanger at 62 C and the cool water from the tank enters at 35 C, what is the heat exchange rate?

Solution

First calculate the capacitance rates on the hot (collector) and cold (tank) sides of the heat exchanger and C^*. Use the symbols C_h and C_c for the hot and cold side capacitance rates:

$$C_h = 1.25 \times 3850 = 4812 \text{ W/C}$$

$$C_c = 0.864 \times 4180 = 3610 \text{ W/C}$$

The tank (cold) side capacitance rate is the minimum of the two, and from Equation 3.17.9,

$$C^* = 3610/4812 = 0.75$$

From Equation 3.17.8,

$$NTU = \frac{UA}{C_{min}} = \frac{6500}{3610} = 1.80$$

The effectiveness is now calculated from Equation 3.17.7:

$$\varepsilon = \frac{1 - e^{-1.8(1 - 0.75)}}{1 - 0.75e^{-1.8(1 - 0.75)}} = 0.69$$

The heat transfer is now calculated from Equation 3.17.6:

$$Q = 0.70 \times 3610(62 - 35) = 67,300 \text{ W}$$

The temperatures of the fluids leaving the exchanger can also be calculated using

Equation 3.17.4. The leaving water temperature is

$$T_{co} = 35 + \frac{67,300}{4180} = 51.1 \text{ C}$$

and the leaving glycol temperature is

$$T_{ho} = 62 - \frac{67,300}{3850} = 44.6 \text{ C}$$

■

REFERENCES

Arnold, J. N., D. K. Edwards, and I. Catton, *Trans. ASME, J. Heat Transfer*, **99**, 120 (1977). "Effect of Tilt and Horizontal Aspect Ratio on Natural Convection in a Rectangular Honeycomb."

Arnold, J. N., D. K. Edwards, and P. S. Wu, ASME Paper No. 78-WA/HT-5 (1978). "Effect of Cell Size on Natural Convection in High L/D Tilted Rectangular Cells Heated and Cooled on Opposite Faces."

Beckman, W. A., *Solar Energy*, **13**, 3 (1971). "The Solution of Heat Transfer Problems on a Digital Computer."

Berdahl, P., and M. Martin, *Solar Energy*, **32**, 5 (1984). "Emissivity of Clear Skies."

Bliss, R. W., *Solar Energy*, **5**, 103 (1961). "Atmospheric Radiation Near the Surface of the Ground."

Brunt, D., *Quarterly J. Royal Meteorological Soc.*, **58**, 389 (1932). "Notes on Radiation in the Atmosphere."

Buchberg, H., I. Catton, and D. K. Edwards, *Trans. ASME, J. Heat Transfer*, **98**, 182 (1976). "Natural Convection in Enclosed Spaces: A Review of Application to Solar Energy Collection."

Charters, W. W. S. and L. J. Peterson, *Solar Energy*, **13**, 4 (1972). "Free Convection Suppression Using Honeycomb Cellular Materials."

DeGraff, J. and E. Van Der Held, *Applied Science Research*, Sec. A, **3** (1952). "The Relation Between the Heat Transfer and the Convection Phenomena in Enclosed Plane Air Layers."

Dunkle, R. V., *Trans. ASME*, **76**, 549 (1954). "Thermal Radiation Tables and Applications."

Dunkle, R. V. and W. H. J. Ellul, *Trans. of the Inst. of Engineers (Australia)*, **MC8**, 117 (1972). "Randomly Packed Particulate Bed Regenerators and Evaporative Coolers."

Edwards, D. K., *Trans. ASME, J. Heat Transfer*, **91**, 145 (1969). "Suppression of Cellular Convection by Lateral Walls."

Elsherbiny, S. M., K. G. T. Hollands, and G. D. Raithby, in *Heat Transfer in Solar Energy Systems* (J. R. Howell and T. Min, eds.), American Society of Mechanical Engineers, New York (1977). "Free Convection Across Inclined Air Layers with One Surface V-Corrugated."

Heaton, H. S., W. C. Reynolds, and W. M. Kays, *Int. J. Heat and Mass Transfer*, **7**, 763 (1964). "Heat Transfer in Annular Passages. Simultaneous Development of Velocity and Temperature Fields in Laminar Flow."

Hollands, K. G. T., *Solar Energy*, **9**, 159 (1965). "Honeycomb Devices in Flat-Plate Solar Collectors."

Hollands, K. G. T., T. E. Unny, G. D. Raithby, and L. Konicek, *Trans. ASME, J. Heat Transfer*, **98**, 189 (1976). "Free Convection Heat Transfer Across Inclined Air Layers."

Hollands, K. G. T., K. N. Marshall, and R. K. Wedel, *Solar Energy*, **21**, 231 (1978). "An Approximate Equation for Predicting the Solar Transmittance of Transparent Honeycombs."

Home Finance Agency Report No. 32, U.S. Government Printing Office, Washington, DC (1954). "The Thermal Insulating Value of Airspaces."

Hottel, H. C. and A. F. Sarofim, *Radiative Transfer*, McGraw-Hill, New York (1967).

Incropera, F. P. and D. P. DeWitt, *Fundamentals of Heat and Mass Transfer*, Wiley, New York (1985).

Kays, W. M. and M. E. Crawford, *Convective Heat and Mass Transfer,* 2nd ed., McGraw-Hill, New York (1980).

Kays, W. M. and A. L. London, *Compact Heat Exchangers*, McGraw-Hill, New York (1964).

Lloyd, J. R. and W. P. Moran, *Trans. ASME, J. Heat Transfer*, **96**, 443 (1974). "Natural Convection Adjacent to Horizontal Surface of Various Planforms."

Löf, G. O. G. and R. W. Hawley, *Ind. and Engr. Chemistry*, **40**, 1061 (1948). "Unsteady State Heat Transfer between Air and Loose Solids."

McAdams, W. H., *Heat Transmission*, 3rd ed., McGraw-Hill, New York (1954).

McCorquodale, J. A., A. A. Hannourae, and M. S. Nasser, *J. Hydraulic Research*, **16** (2), 123 (1978). "Hydraulic Conductivity of Rock Fill."

Mercer, W. E., W. M. Pearce, and J. E. Hitchcock, *Trans. ASME, J. Heat Transfer*, **89**, 251 (1967). "Laminar Forced Convection in the Entrance Region Between Parallel Flat Plates."

Meyer, B. A., M. M. El-Wakil, and J. W. Mitchell, in *Thermal Storage and Heat Transfer in Solar Energy Systems* (F. Kreith, R. Boehm, J. Mitchell, and R. Bannerot, eds.), American Society of Mechanical Engineers, New York (1978). "Natural Convection Heat Transfer in Small and Moderate Aspect Ratio Enclosures - An Application to Flat-Plate Collectors."

Mitchell, J. W., *Biophysical J.*, **16**, 561 (1976). "Heat Transfer from Spheres and Other Animal Forms."

Pivovonsky, M. and M R. Nagel, *Tables of Blackbody Radiation Properties*, Macmillan, New York (1961).

Raithby, G. D., K. G. T. Hollands, and T. R. Unny, *Trans. ASME, J. Heat Transfer*, **99**, 287 (1977). "Analysis of Heat Transfer by Natural Convection Across Vertical Fluid Layers."

Randall, K. R., Ph.D. Thesis, Mech. Engr., University of Wisconsin-Madison (1978). "An Interferometric Study of Natural Convection Heat Transfer in Flat-Plate and Vee-Corrugated Enclosures."

Randall, K. R., J. W. Mitchell, and M. M. El-Wakil, in *Heat Transfer in Solar Energy Systems* (J. R. Howell and T. Min, eds.), American Society of Mechanical Engineers, New York (1977). "Natural Convection Characteristics of Flat-Plate Collectors."

Richtmyer, F. K. and E. H. Kennard, *Introduction to Modern Physics*, 4th ed., McGraw-Hill, New York (1947).

Rohsenow, W. M. and H. Choi, *Heat Mass and Momentum Transfer*, Prentice-Hall, Englewood Cliffs, NJ (1961).

Sargent, S. L., *Bull. of the Am. Meteorological Soc.*, **53**, 360 (Apr. 1972). "A Compact Table of Blackbody Radiation Functions."

Shewen, E. C., H. F. Sullivan, K. G. T. Hollands, and A. R. Balakrishnan, Report STOR-6, Waterloo Research Institute, University of Waterloo (Aug., 1978), "A Heat Storage Subsystem for Solar Energy."

Siegel, R. and J. R. Howell, *Thermal Radiation Heat Transfer*, 2nd ed., McGraw-Hill, New York (1981).

Sparrow, E. M., NACA TN 3331 (1955). "Analysis of Laminar Forced-Convection Heat Transfer in Entrance Regions of Flat Rectangular Ducts."

Sparrow, E. M., J. W. Ramsey, and E. A. Mass, *Trans. ASME, J. Heat Transfer*, **101**, 2 (1979). "Effect of Finite Width on Heat Transfer and Fluid Flow About an Inclined Rectangular Plate."

Swinbank, W. C., *Quarterly J. Royal Meteorological Soc.*, **89**, 339 (1963). "Long-Wave Radiation from Clear Skies."

Tabor, H., *Bull. Research Council of Israel*, **6C**, 155 (1958). "Radiation Convection and Conduction Coefficients in Solar Collectors."

Tan, H. M. and W. W. S. Charters, *Solar Energy*, **13**, 121 (1970). "Experimental Investigation of Forced-Convection Heat Transfer."

Watmuff, J. H., W. W. S. Charters, and D. Proctor, *COMPLES*, No. 2, 56 (1977). "Solar and Wind Induced External Coefficients for Solar Collectors."

Chapter 4

RADIATION CHARACTERISTICS OF OPAQUE MATERIALS

This chapter begins with a detailed discussion of radiation characteristics of surfaces. For many solar energy calculations only two quantities are required, the solar absorptance and the long-wave or infrared emittance, usually referred to as just absorptance and emittance. Although values of these two quantities are often quoted, other radiation properties may be the only available information on a particular material. Since relationships exist between the various characteristics, it may be possible to calculate a desired quantity from available data. Consequently, it is necessary to understand exactly what is meant by the radiation terms found in the literature, to be familiar with the type of information available, and to know how to manipulate these data to get the desired information. The most common type of data manipulation is illustrated in the examples, and readers may wish to go directly to Section 4.5.

The names used for the radiation surface characteristics were chosen as the most descriptive of the many names found in the literature. In many cases, the names will seem to be cumbersome, but they are necessary to distinguish one characteristic from another. For example, both a monochromatic angular-hemispherical reflectance and a monochromatic hemispherical-angular reflectance will be defined. Under certain circumstances, these two quantities are identical, but in general they are different, and it is necessary to distinguish between them.

Both the name and the symbol should be aids for understanding the significance of the particular characteristic. The monochromatic directional absorptance $\alpha_\lambda(\mu,\phi)$ is the fraction of the incident energy from the direction μ,ϕ at the wavelength λ that is absorbed.[1] The directional absorptance $\alpha(\mu,\phi)$ includes all wavelengths, and the hemispherical absorptance α includes all directions as well as all wavelengths. We will also have a monochromatic hemispherical absorptance α_λ which is the fraction of the energy incident from all directions at a particular wavelength that is absorbed. Thus by careful study of the name the definition should be clear.

The middle sections of the chapter are concerned with calculation of broadband properties from spectral properties. The last part of the chapter is concerned with

---

[1] The angles θ and ϕ are shown in Figure 3.7.2; μ is equal to cos θ.

184

selective surfaces that have high absorptance in the solar energy spectrum and low emittance in the long-wave spectrum. Agnihotri and Gupta (1981) have reviewed this topic extensively, and Lampert (1990) provides an overall review of optical properties of materials for solar energy applications.

4.1 ABSORPTANCE AND EMITTANCE

The monochromatic directional absorptance is a property of a surface and is defined as the fraction of the incident radiation of wavelength λ from the direction μ, ϕ (where μ is the cosine of the polar angle and ϕ is the azimuthal angle) that is absorbed by the surface. In equation form

$$\alpha_\lambda(\mu, \phi) = \frac{I_{\lambda,a}(\mu, \phi)}{I_{\lambda,i}(\mu, \phi)} \tag{4.1.1}$$

where subscripts a and i represent absorbed and incident.

The fraction of all the radiation (over all wavelengths) from the direction μ, ϕ that is absorbed by a surface is called the directional absorptance and is defined by the following equation:

$$\alpha(\mu, \phi) = \frac{\int_0^\infty \alpha_\lambda(\mu, \phi) I_{\lambda,i}(\mu, \phi) \, d\lambda}{\int_0^\infty I_{\lambda,i} \, d\lambda}$$

$$= \frac{1}{I_i(\mu, \phi)} \int_0^\infty \alpha_\lambda(\mu, \phi) I_{\lambda,i}(\mu, \phi) \, d\lambda \tag{4.1.2}$$

Unlike the monochromatic directional absorptance, the directional absorptance is not a property of the surface since it is a function of the wavelength distribution of the incident radiation.[2]

The monochromatic directional emittance of a surface is defined as the ratio of the monochromatic intensity emitted by a surface in a particular direction to the monochromatic intensity that would be emitted by a blackbody at the same temperature:

$$\varepsilon_\lambda(\mu, \phi) = \frac{I_\lambda(\mu, \phi)}{I_{\lambda b}} \tag{4.1.3}$$

[2] Although $\alpha(\mu, \phi)$ and some other absorptances are not properties in that they depend upon the wavelength distribution of the incoming radiation, we can consider them as properties if the incoming spectral distribution is known. As the spectral distribution of solar radiation is essentially fixed, we can consider solar absorptance as a property.

The monochromatic directional emittance is a property of a surface, as is the directional emittance, defined by [3]

$$\varepsilon(\mu,\phi) = \frac{\int_0^\infty \varepsilon_\lambda(\mu,\phi) I_{\lambda b}\, d\lambda}{\int_0^\infty I_{\lambda b}\, d\lambda} = \frac{1}{I_b} \int_0^\infty \varepsilon_\lambda(\mu,\phi) I_{\lambda b}\, d\lambda \qquad (4.1.4)$$

In words, the directional emittance is defined as the ratio of the emitted total intensity in the direction μ,ϕ to the blackbody intensity. Note that $\varepsilon(\mu,\phi)$ is a property, as its definition contains the intensity $I_{\lambda b}$ which is specified when the surface temperature is known. In contrast, the definition of $\alpha(\mu,\phi)$ contains the unspecified function $I_{\lambda,i}(\mu,\phi)$ and is therefore not a property. It is important to note that these four quantities and the four to follow are all functions of surface conditions such as temperature, roughness, cleanliness, etc.

From the definitions of the directional absorptance and emittance of a surface, the corresponding hemispherical properties can be defined. The monochromatic hemispherical absorptance and emittance are obtained by integrating over the enclosing hemisphere, as was done in Section 3.7:

$$\alpha_\lambda = \frac{\int_0^{2\pi} \int_0^1 \alpha_\lambda(\mu,\phi) I_{\lambda,i}(\mu,\phi) \mu\, d\mu\, d\phi}{\int_0^{2\pi} \int_0^1 I_{\lambda,i}(\mu,\phi) \mu\, d\mu\, d\phi} \qquad (4.1.5)$$

$$\varepsilon_\lambda = \frac{\int_0^{2\pi} \int_0^1 \varepsilon_\lambda(\mu,\phi) I_{\lambda b}(\mu,\phi) \mu\, d\mu\, d\phi}{\int_0^{2\pi} \int_0^1 I_{\lambda b}(\mu,\phi) \mu\, d\mu\, d\phi} = \frac{1}{\pi} \int_0^{2\pi} \int_0^1 \varepsilon_\lambda(\mu,\phi) I_{\lambda b}(\mu,\phi) \mu\, d\mu\, d\phi$$

$$(4.1.6)$$

The monochromatic hemispherical emittance is thus a property. The monochromatic hemispherical absorptance is not a property but is a function of the incident intensity.

The hemispherical absorptance and emittance are obtained by integrating over all wavelengths and are defined by

$$\alpha = \frac{\int_0^\infty \int_0^{2\pi} \int_0^1 \alpha_\lambda(\mu,\phi) I_{\lambda,i}(\mu,\phi) \mu\, d\mu\, d\phi d\lambda}{\int_0^\infty \int_0^{2\pi} \int_0^1 I_{\lambda,i}(\mu,\phi) \mu\, d\mu\, d\phi d\lambda} \qquad (4.1.7)$$

$$\varepsilon = \frac{\int_0^\infty \int_0^{2\pi} \int_0^1 \varepsilon_\lambda(\mu,\phi) I_{\lambda b}(\mu,\phi) \mu\, d\mu d\phi d\lambda}{\int_0^\infty \int_0^{2\pi} \int_0^1 I_{\lambda b}(\mu,\phi) \mu\, d\mu\, d\phi d\lambda} = \frac{1}{E_\lambda} \int_0^\infty \varepsilon_\lambda E_{\lambda b}\, d\lambda \qquad (4.1.8)$$

---

[3] Both the numerator and denominator could be multiplied by π so that the definition of $\varepsilon(\mu,\phi)$ could have been in terms of ε_b and $\varepsilon_{\lambda b}$.

Again the absorptance (in this case the hemispherical absorptance) is a function of the incident intensity whereas the hemispherical emittance is a surface property.

If the monochromatic directional absorptance is independent of direction [i.e., $\alpha_\lambda(\mu,\phi) = \alpha_\lambda$], then Equation 4.1.7 can be simplified by integrating over the hemisphere to yield

$$\alpha = \frac{\int_0^\infty \alpha_\lambda q_{\lambda,i}\, d\lambda}{\int_0^\infty q_{\lambda,i}\, d\lambda} \tag{4.1.9}$$

where $q_{\lambda,i}$ is the incident monochromatic radiant energy. If the incident radiation in either Equation 4.1.7 or 4.1.9 is radiation from the sun, then the calculated absorptance is called the solar absorptance.

4.2 KIRCHHOFF'S LAW

A proof of Kirchhoff's law is beyond the scope of this book. [See Siegel and Howell (1972) for a complete discussion.] However, a satisfactory understanding can be obtained without a proof. Consider an evacuated isothermal enclosure at temperature T. If the enclosure is isolated from the surroundings, then the enclosure and any substance within the enclosure will be in thermodynamic equilibrium. In addition, the radiation field within the enclosure must be homogeneous and isotropic. If this were not so, we could have a directed flow of radiant energy at some location within the enclosure, but this is impossible since we could then extract work for an isolated and isothermal system.

If we now consider an arbitrary body within the enclosure, the body must absorb the same amount of energy as it emits. An energy balance on an element of the surface of the body yields

$$\alpha q = \varepsilon E_b \tag{4.2.1}$$

If we place a second body with different surface properties in the enclosure, the same energy balance must apply, and the ratio q/E_b must be constant:

$$\frac{q}{E_b} = \frac{\varepsilon_1}{\alpha_1} = \frac{\varepsilon_2}{\alpha_2} \tag{4.2.2}$$

Since this must also apply to a blackbody in which $\varepsilon = 1$, the ratio of ε to α for any body in thermal equilibrium must be equal to unity. Therefore, for conditions of thermal equilibrium

$$\varepsilon = \alpha \tag{4.2.3}$$

It must be remembered that α is not a property, and since this equation was developed for the condition of thermal equilibrium, it will not be valid if the incident radiation comes form a source at a different temperature (e.g., if the source of radiation is the sun). This distinction is very important in the performance of solar collectors.

Equation 4.2.3 is sometimes referred to as Kirchhoff's law, but his law is much more general. Within an enclosure the radiant flux is everywhere uniform and isotropic. The absorptance of a surface within the enclosure is then given by Equation 4.1.7 with $I_{\lambda,i}(\mu,\phi)$ replaced by $I_{\lambda b}$ and the emittance is given by Equation 4.1.8. Since the hemispherical absorptance and emittance are equal under conditions of thermal equilibrium, we can equate Equations 4.1.7 and 4.1.8 to obtain

$$\int_0^\infty I_{\lambda b} \int_0^{2\pi} \int_0^1 \left[\alpha_\lambda(\mu,\phi) - \varepsilon_\lambda(\mu,\phi)\right]\mu \, d\mu \, d\phi \, d\lambda = 0 \qquad (4.2.4)$$

It is mathematically possible to have this integral equal to zero without $\alpha_\lambda(\mu,\phi)$ being identical to $\varepsilon_\lambda(\mu,\phi)$, but this is a very unlikely situation in view of the very irregular behavior of $\alpha_\lambda(\mu,\phi)$ exhibited by some substances. Thus we can say

$$\varepsilon_\lambda(\mu,\phi) = \alpha_\lambda(\mu,\phi) \qquad (4.2.5)$$

This result is true for all conditions, not just thermal equilibrium, since both $\alpha_\lambda(\mu,\phi)$ and $\varepsilon_\lambda(\mu,\phi)$ are properties.[4]

If the surface does not exhibit a dependence on the azimuthal angle, then Equation 4.2.5 reduces to

$$\alpha_\lambda(\mu) = \varepsilon_\lambda(\mu) \qquad (4.2.6)$$

and if the dependence on polar angle can also be neglected, then Kirchhoff's law further reduces to

$$\alpha_\lambda = \varepsilon_\lambda \qquad (4.2.7)$$

Finally, if the surface does not exhibit a wavelength dependency, then the absorptance α is equal to the emittance ε. This is the same result obtained for any surface when in thermal equilibrium, as given by Equation 4.2.3.

---

[4] Kirchhoff's law actually applied to each component of polarization and not to the sum of the two components as implied by Equation 4.2.5.

4.3 REFLECTANCE OF SURFACES

Consider the spatial distribution of radiation reflected by a surface. When the incident radiation is in the form of a narrow "pencil" (i.e., contained within a small solid angle), two limiting distributions of the reflected radiation exist. These two cases are called specular and diffuse. Specular reflection is mirrorlike, that is, the incident polar angle is equal to the reflected polar angle and the azimuthal angles differ by 180°. On the other hand, diffuse reflection obliterates all directional characteristics of the incident radiation by distributing the radiation uniformly in all directions. In practice, the reflection from a surface is neither all specular nor all diffuse. The general case along with the two limiting situations is shown in Figure 4.3.1.

In general, the magnitude of the reflected intensity in a particular direction for a given surface is a function of the wavelength and the spatial distribution of the incident radiation. The biangular reflectance or reflection function is used to relate the intensity of reflected radiation in a particular direction by the following equation:

$$\rho_\lambda(\mu_r, \phi_r, \mu_i, \phi_i) = \lim_{\Delta\omega_i \to 0} \frac{\pi I_{\lambda,r}(\mu_r, \phi_r)}{I_{\lambda,i}\mu_i \Delta\omega_i} \tag{4.3.1}$$

The numerator is π times the intensity reflected in the direction μ_r, ϕ_r when an energy flux of amount $I_{\lambda,i}\mu_i \Delta\omega_i$ is incident on the surface from the direction μ_i, ϕ_i. The π has been included so that the numerator "looks like" an energy flux. The physical situation is shown schematically in Figure 4.3.2.

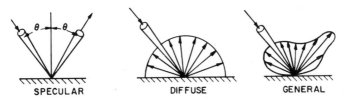

SPECULAR   DIFFUSE   GENERAL

Figure 4.3.1 Reflection from surfaces.

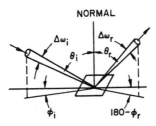

Figure 4.3.2 Coordinate system for the reflection function.

Since the energy incident in the solid angle $\Delta\omega_i$ may be reflected in all directions, the reflected intensity in the direction μ_r, ϕ_r will be of infinitesimal size compared to the incident intensity. By multiplying the incident intensity by its solid angle (which must be finite in any real experiment) and the cosine of the polar angle, we obtain the incident radiation flux which will have values on the same order of magnitude as the reflected intensity. The biangular reflectance can have numerical values between zero and infinity; its values do not lie only between zero and 1.

From an experimental point of view, it is not practical to use the scheme depicted in Figure 4.3.2 since all the radiation quantities would be extremely small. An equivalent experiment is to irradiate the surface with a nearly monodirectional flux (i.e., with a small solid angle $\Delta\omega_i$) as shown in Figure 4.3.3. The reflected energy in each direction is measured. This measured energy divided by the measurement instrument solid angle $(\Delta\omega_r)$ will be approximately equal to the reflected intensity. The incident flux will be on the same order and can be easily measured.

Two types of hemispherical reflectances exist. The angular-hemispherical reflectance is found when a narrow pencil of radiation is incident on a surface and all the reflected radiation is collected. The hemispherical-angular reflectance results from collecting reflected radiation in a particular direction when the surface is irradiated from all directions.

The monochromatic angular-hemispherical reflectance will be designated by $\rho_\lambda(\mu_i, \phi_i)$ where the subscript i indicates that the incident radiation has a specified direction. This reflectance is defined as the ratio of the monochromatic radiant energy reflected in all directions to the incident radiant flux within small solid angle $\Delta\omega_i$. The incident energy $I_{\lambda,i}\mu_i\Delta\omega_i$ that is reflected in all directions can be found using the reflection function:

$$q_{\lambda,r} = \frac{1}{\pi} \int_0^{2\pi} \int_0^1 \rho_\lambda(\mu_r, \phi_r, \mu_i, \phi_i) I_{\lambda,i}\mu_i\Delta\omega_i\,\mu_r\,d\mu_r\,d\phi_r \tag{4.3.2}$$

The monochromatic angular-hemispherical reflectance can then be expressed as

$$\rho_\lambda(\mu_i, \phi_i) = \frac{q_{\lambda,r}}{I_{\lambda,i}\mu_i\Delta\omega_i} = \frac{1}{\pi} \int_0^{2\pi} \int_0^1 \rho_\lambda(\mu_r, \phi_r, \mu_i, \phi_i)\mu_r\,d\mu_r\,d\phi_r \tag{4.3.3}$$

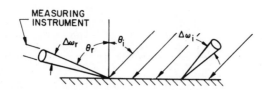

Figure 4.3.3   Schematic representation of an experiment for measuring the reflection function.

Examination of Equation 4.3.3 shows that $\rho_\lambda(\mu_i,\phi_i)$ is a property of the surface. The angular-hemispherical reflectance, $\rho(\mu_i,\phi_i)$, can be found by integrating the incident and reflected fluxes over all wavelengths, but it is not a property as it depends upon the wavelength distribution of the incoming radiation.

The monochromatic hemispherical-angular reflectance is defined as the ratio of the reflected monochromatic intensity in the direction μ_r,ϕ_r to the monochromatic energy from all directions divided by π (which then "looks like" an intensity). The incident energy can be written in terms of the incident intensity integrated over the hemisphere:

$$q_{\lambda,i} = \int_0^{2\pi} \int_0^1 I_{\lambda,i}\mu_i \, d\mu_i \, d\phi_i \tag{4.3.4}$$

and the monochromatic hemispherical-angular reflectance is then

$$\rho_\lambda(\mu_r,\phi_r) = \frac{I_{\lambda,r}(\mu_r,\phi_r)}{q_{\lambda,i}/\pi} \tag{4.3.5}$$

where the subscripts r in $\rho_\lambda(\mu_r,\phi_r)$ are used to specify the reflected radiation as being in a specified direction. In terms of the reflectance function, Equation 4.3.5 can be written as

$$\rho_\lambda(\mu_r,\phi_r) = \frac{\int_0^{2\pi}\int_0^1 \rho_\lambda(\mu_r,\phi_r,\mu_i,\phi_i) I_{\lambda,i}\mu_i \, d\mu_i \, d\phi_i}{\int_0^{2\pi}\int_0^1 I_{\lambda,i}\mu_i \, d\mu_i \, d\phi_i} \tag{4.3.6}$$

Since $\rho_\lambda(\mu_r,\phi_r)$ is dependent upon the angular distribution of the incident intensity, it is not a surface property. For the special case when the incident radiation is diffuse, the monochromatic hemispherical-angular reflectance is identical to the monochromatic angular-hemispherical reflectance. To prove the equality of $\rho_\lambda(\mu_r,\phi_r)$ and $\rho_\lambda(\mu_i,\phi_i)$ under the condition of constant $I_{\lambda,i}$, it is necessary to use the symmetry of the reflection function as given by

$$\rho_\lambda(\mu_i,\phi_i,\mu_r,\phi_r) = \rho_\lambda(\mu_r,\phi_r,\mu_i,\phi_i) \tag{4.3.7}$$

and compare Equation 4.3.6 (with $I_{\lambda,i}$ independent of incident direction) with Equation 4.3.3. The proof of Equation 4.3.7 is beyond the scope of this book [see Siegel and Howell (1981)].

The equality of $\rho_\lambda(\mu_i,\phi_i)$ and $\rho_\lambda(\mu_r,\phi_r)$ when $I_{\lambda,i}$ is uniform is of great importance since the measurement of $\rho_\lambda(\mu_r,\phi_r)$ is much easier than $\rho_\lambda(\mu_i,\phi_i)$. This is discussed in Section 4.7.

Both $\rho_\lambda(\mu_i,\phi_i)$ and $\rho_\lambda(\mu_r,\phi_r)$ can be considered on a total basis by integration over all wavelengths. For the case of the angular-hemispherical reflectance, we have

$$\rho(\mu_i,\phi_i) = \frac{\int_0^\infty q_{\lambda r}\, d\lambda}{\int_0^\infty I_{\lambda,i}\mu_i\Delta\omega_i\, d\lambda}$$

$$= \frac{1}{\pi I_i}\int_0^\infty \int_0^{2\pi}\int_0^1 \rho_\lambda(\mu_i,\phi_i,\mu_r,\phi_r)I_{\lambda,i}\mu_r\, d\mu_r\, d\phi_r\, d\lambda \qquad (4.3.8)$$

which, unlike the monochromatic angular-hemispherical reflectance, is not a property.

When a surface element is irradiated from all directions and all the reflected radiation is measured, we characterize the process by the monochromatic hemispherical reflectance, defined as

$$\rho_\lambda = \frac{q_{\lambda,r}}{q_{\lambda,i}} \qquad (4.3.9)$$

The reflected monochromatic energy can be expressed in terms of the reflection function and the incident intensity by

$$q_{\lambda,r} = \int_0^{2\pi}\int_0^1\left[\int_0^{2\pi}\int_0^1 \frac{\rho_\lambda(\mu_r,\phi_r,\mu_i,\phi_i)}{\pi}I_{\lambda,i}\mu_i\, d\mu_i\, d\phi_i\right]\mu_r\, d\mu_r\, d\phi_r \quad (4.3.10)$$

The incident energy, expressed in terms of the incident intensity, is

$$q_{\lambda,i} = \int_0^{2\pi}\int_0^1 I_{\lambda,i}\mu_i\, d\mu_i\, d\phi_i \qquad (4.3.11)$$

Division of Equation 4.3.10 by 4.3.11 yields the monochromatic hemispherical reflectance. For the special case of a diffuse surface (i.e., the reflection function is a constant), the monochromatic hemispherical reflectance is numerically equal to the reflection function and is independent of the spatial distribution of the incident intensity.

The hemispherical reflectance is found by integration of Equations 4.3.10 and 4.3.11 over all wavelengths and finding the ratio

$$\rho = \frac{q_r}{q_i} = \frac{\int_0^\infty q_{\lambda,r}\, d\lambda}{\int_0^\infty q_{\lambda,i}\, d\lambda} \qquad (4.3.12)$$

The hemispherical reflectance depends on both the angular distribution and wavelength distribution of the incident radiation.

For low-temperature applications that do not include solar radiation, a special form of the hemispherical reflectance (often the name is shortened to "reflectance") will be found to be the most useful. The special form is Equation 4.3.12, which is based on the assumption that the reflection function is independent of direction (diffuse approximation) and wavelength (gray approximation). The diffuse approximation for the hemispherical reflectance has already been discussed and was found to be equal to $\rho_\lambda(\mu_i, \phi_i, \mu_r, \phi_r)$. When the gray approximation is made in addition to the diffuse approximation, the surface reflectance becomes independent of everything except possibly the temperature of the surface, and even this is usually neglected.

4.4 RELATIONSHIPS AMONG ABSORPTANCE, EMITTANCE, AND REFLECTANCE

It is now possible to show that it is necessary to know only one property, the monochromatic angular-hemispherical reflectance, and all absorptance and emittance properties for opaque surfaces can be found.

Consider a surface located in an isothermal enclosure maintained at temperature T. The monochromatic intensity in a direction μ, ϕ from an infinitesimal area of the surface consists of emitted and reflected radiation and must be equal to $I_{\lambda b}$:

$$I_{\lambda b} = I_\lambda(\mu, \phi)\big|_{\text{emitted}} + I_\lambda(\mu, \phi)\big|_{\text{reflected}} \tag{4.4.1}$$

The emitted and reflected intensities are

$$I_\lambda(\mu, \phi)\big|_{\text{emitted}} = \varepsilon_\lambda(\mu, \lambda) I_{b\lambda} \tag{4.4.2}$$

$$I_\lambda(\mu, \phi)\big|_{\text{reflected}} = I_{\lambda b}\rho_\lambda(\mu, \lambda) \tag{4.4.3}$$

but $\rho_\lambda(\mu_r, \phi_r)$ is equal to the monochromatic angular-hemispherical reflectance, $\rho_\lambda(\mu_i, \phi_i)$, since the incident intensity is diffuse. Since $I_{\lambda b}$ can be canceled from each term, we have

$$\varepsilon_\lambda(\mu, \phi) = 1 - \rho_\lambda(\mu_i, \phi_i) \tag{4.4.4}$$

But from Kirchhoff's law

$$\varepsilon_\lambda(\mu, \phi) = \alpha_\lambda(\mu, \phi) = 1 - \rho_\lambda(\mu_i, \phi_i) \tag{4.4.5}$$

Thus the monochromatic directional emittance and the monochromatic directional absorptance can both be calculated from knowledge of the monochromatic angular-hemispherical reflectance. Also, all emittance properties (Equations 4.1.4, 4.1.6, and 4.1.8) can be found once $\rho_\lambda(\mu_i,\phi_i)$ is known. The absorptances (Equations 4.1.2, 4.1.5, and 4.1.7) can be found if the incident intensity is specified.

The relationship between the reflectance and absorptance[5] of Equation 4.4.5 can be considered as a statement of conservation of energy. The incident monochromatic energy from any direction is either reflected or absorbed. Similar arguments can be used to relate other absorptances to reflectances. For example, for an opaque surface, energy from all directions, either monochromatic or total, is either absorbed or reflected so that

$$\rho_\lambda + \alpha_\lambda = \rho_\lambda + \varepsilon_\lambda = 1 \tag{4.4.6}$$

and $\qquad \rho + \alpha = 1 \tag{4.4.7}$

4.5   BROADBAND EMITTANCE AND ABSORPTANCE

The concepts and analyses of the previous sections are greatly simplified if it is assumed that there is no directional dependence of ε or α. Figure 4.5.1 shows monochromatic emission as a function of wavelength for a blackbody and for a real surface, both at the same surface temperature. The monochromatic emittance at wavelength λ is $E_\lambda/E_{\lambda\beta}$, the ratio of the energy emitted at a wavelength to what it would be if it were a blackbody, i.e., the ratio A/B.

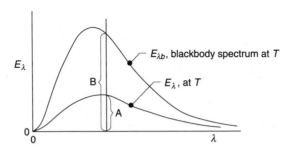

Figure 4.5.1   Monochromatic emission E_λ vs. wavelength for a black surface and a real surface, both at the same temperature.

[5] There are no generally accepted names used in the literature except for the simple "absorptance," "emittance," and "reflectance" which, for clarity, were prefixed with the name hemispherical. In the remainder of this book, the modifier hemispherical will generally be omitted since most available data are hemispherical. If it is necessary to distinguish directional quantities, then the full name will be used.

The total emittance is found by integrating over wavelengths from zero to infinity:

$$\varepsilon = \frac{\int_0^\infty \varepsilon_\lambda E_{\lambda b}\, d\lambda}{\int_0^\infty E_{\lambda b}\, d\lambda} = \frac{\int_0^\infty \varepsilon_\lambda E_{\lambda b}\, d\lambda}{\sigma T^4} \qquad (4.5.1)$$

This is the same as Equation 4.1.8. If the nature of the surface (i.e., ε_λ) and its temperature are known, the emittance ε can be determined. Since ε is not dependent on any external factors it is a property of the surface.

Monochromatic absorptance is the fraction of the incident radiation at wavelength λ that is absorbed. This is shown in Figure 4.5.2, where the incident energy spectrum $I_{\lambda,i}$ is shown as an arbitrary function of λ. The symbol α_λ is the monochromatic absorptance at λ, the ratio of C/D, or $I_{\lambda,a}/I_{\lambda,i}$. The total absorptance for this surface for the indicated incident spectrum is found by integration over wavelengths from zero to infinity:

$$\alpha = \frac{\int_0^\infty \alpha_\lambda I_{\lambda,i}\, d\lambda}{\int_0^\infty I_{\lambda,i}\, d\lambda} \qquad (4.5.2)$$

This is the same as Equation 4.1.9. In contrast to emittance, which is specified by the nature of the surface and its temperature, absorptance depends on an external factor, the spectral distribution of incident radiation. A specification of α is meaningless unless the incident radiation is described. In the context of solar energy we are usually interested in absorptance for solar radiation [as described by a terrestrial solar energy spectrum (Table 2.6.1), the extraterrestrial spectrum (Table 1.3.1), or an equivalent blackbody spectrum (described by a temperature and Table 3.6.1)]. For usual solar energy applications, the terrestrial solar spectrum of Table 2.6.1 will provide a realistic basis for computation of α, and henceforth in this book reference to absorptance without other specification of the incident radiation will mean absorptance for the terrestrial solar spectrum.

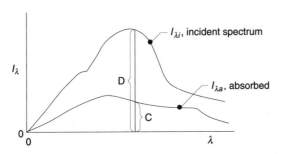

Figure 4.5.2 Monochromatic incident and absorbed energy.

4.6   CALCULATION OF EMITTANCE AND ABSORPTANCE

The data that are generally available are measurements of monochromatic reflectance ρ_λ. This is related to α_λ and ε_λ by Equation 4.4.6. With these data, we can conveniently divide the spectrum (the blackbody spectrum for emittance or the incident energy spectrum for absorptance) into segments and numerically integrate to obtain α or ε.

Consider first the calculation of emittance. As shown in Figure 4.6.1, for a segment j of the blackbody spectrum at the surface temperature T, there is a "monochromatic" emittance $\varepsilon_{\lambda,j}$ that is the ratio of the shaded area to the unshaded area. The ratio $E_\lambda/E_{\lambda b}$ at an appropriate wavelength in the segment (often its energy midpoint) is taken as characteristic of the segment. The energy increment Δf_j in the blackbody spectrum can be determined from Table 3.6.1 as the difference in $f_{0-\lambda T}$ at the wavelengths defining the segment. The contribution of the jth increment to ε is $\varepsilon_{\lambda j}\Delta f_j$.

Thus the emittance is

$$\varepsilon = \sum_{j=1}^{n} \varepsilon_j \, \Delta f_j \tag{4.6.1}$$

or in terms of reflectance

$$\varepsilon = \sum_{j=1}^{n} \left(1 - \rho_{\lambda j}\right) \Delta f_j = 1 - \sum_{j=1}^{n} \rho_{\lambda j} \, \Delta f_j \tag{4.6.2}$$

If the increments Δf_j are equal,

$$\varepsilon = \frac{1}{n} \sum_{j=1}^{n} \varepsilon_{\lambda j} = 1 - \frac{1}{n} \sum_{j=1}^{n} \rho_{\lambda j} \tag{4.6.3}$$

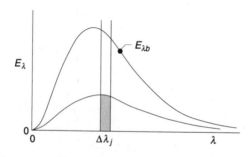

Figure 4.6.1 The jth segment in the emission spectrum for which the monochromatic emittance is $\varepsilon_{\lambda j}$.

The calculation of absorptance is similar, except that the incident radiation must be specified. In general, it will not be blackbody radiation, and other information must be available on which to base the calculation. As our interest is in absorptance for solar radiation, Table 2.6.1 provides this information for calculation of α for terrestrial applications.

The incident radiation is divided into increments, and the contributions of these increments are summed to obtain α for that incident radiation. For an increment in incident radiation Δf_j, the contribution to α is $\alpha_j \, \Delta f_j$. Summing,

$$\alpha = \sum_{j=1}^{n} \alpha_{\lambda,j} \, \Delta f_j = \sum_{j=1}^{n} \left(1 - \rho_{\lambda,j}\right)\Delta f_j = 1 - \sum_{j=1}^{n} \rho_{\lambda,j} \, \Delta f_j \qquad (4.6.4)$$

and if the increments are equal,

$$\alpha = \frac{1}{n} \sum_{j=1}^{n} \alpha_{\lambda,j} = 1 - \frac{1}{n} \sum_{j=1}^{n} \rho_{\lambda,j} \qquad (4.6.5)$$

The calculations of α and ε are illustrated in the following example, where the incident radiation is taken as the terrestrial solar spectrum of Table 2.6.1.

Example 4.6.1

Calculate the absorptance for the terrestrial solar spectrum and emittance at 177 C (450 K) of the surface having the monochromatic reflectance characteristics shown in the following figure.

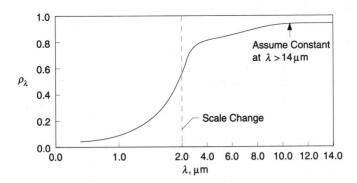

Solution

First calculate the emittance, using five equal increments of blackbody radiation from Table 3.6.1b. For each increment, $\lambda_j T$ at the midpoint is determined from the table, the midpoint wavelength λ_j for that increment is calculated from $\lambda_j T/T$, and $\rho_{\lambda,j}$ is determined at λ_j from the figure. For the first increment, which has wavelength limits of 0 and 2680/450 = 5.96 μm, the midpoint $\lambda_j T$ is 2200 μmK, λ_j is 2200/450 = 4.89 μm, and ρ_λ = 0.83. Tabulating for the five equal increments results in the following:

Increment, Δf_j	$\lambda_j T_{\text{mid}}$, μmK	$\lambda_{j,\text{mid}}$, μm	$\rho_{\lambda,j}$
0.0 - 0.2	2200	4.89	0.83
0.2 - 0.4	3120	6.93	0.87
0.4 - 0.6	4110	9.13	0.94
0.6 - 0.8	5590	12.42	0.94
0.8 - 1.0	9380	20.84	0.94
			$\Sigma = 4.52$

Using Equation 4.6.3, since all increments are equal,

$$\varepsilon = 1 - 4.52/5 = 0.10$$

Note that if 10 increments are used, the emittance is calculated to be 0.09. As there is no change of ρ_λ with λ at wavelength beyond 10 μm, smaller increments (perhaps $\Delta f_j = 0.1$) could be used for $\lambda < 10$ μm and a single large increment for $\lambda > 10$ μm.

The calculation of absorptance for terrestrial radiation is similar. Taking ten equal energy increments from Table 2.6.1, the results of these calculations are as follows:

Increment, Δf_j	λ_{mid}, μm	ρ_λ
0.0 - 0.1	0.434	0.04
0.1 - 0.2	0.517	0.05
0.2 - 0.3	0.595	0.06
0.3 - 0.4	0.670	0.06
0.4 - 0.5	0.752	0.06
0.5 - 0.6	0.845	0.07
0.6 - 0.7	0.975	0.08
0.7 - 0.8	1.101	0.10
0.8 - 0.9	1.310	0.14
0.9 - 1.0	2.049	0.55
		$\Sigma = 1.21$

And from Equation 4.6.5, $\alpha = 1 - 1.21/10 = 0.88$.   ■

4.7   MEASUREMENT OF SURFACE RADIATION PROPERTIES

In the preceding discussion many radiation surface properties have been defined. Unfortunately, in much of the literature the exact nature of the surface being reported is not clearly specified. This situation requires that caution be exercised.

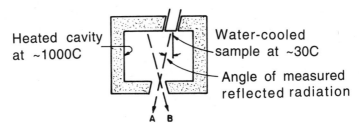

Heated cavity
at ~1000C

Water-cooled
sample at ~30C

Angle of measured
reflected radiation

A   B

Figure 4.7.1 Schematic of a hohlraum for measurement of monochromatic hemispherical - angular reflectance. Radiation A is blackbody radiation reflected from the sample. Radiation B is blackbody radiation from the cavity. The ratio A_λ/B_λ is $\rho_\lambda(\mu,\phi)$.

Many of the reflectance data reported in the literature have been measured by a method devised by Gier et al. (1954).[6] In this method a cool sample is exposed to blackbody radiation from a high-temperature source (a hohlraum), and the monochromatic radiation reflected from the surface is compared to monochromatic blackbody radiation from the cavity. The data are thus hemispherical-angular monochromatic reflectances (or angular-hemispherical monochromatic reflectances since they are equal for diffuse incident radiation). A hohlraum is shown schematically in Figure 4.7.1. In many systems the angle between the surface normal and the measured radiation is often fixed at a small value so that measurements can be made at only one angle (approximately normal). In some designs the sample can be rotated so that all angles can be measured. With measurements of this type, emittance and absorptance values can be found from the equations of Section 4.6.

Table 4.7.1 gives data on surface properties for a few common materials. The data are total hemispherical or total normal emittances at various temperatures and normal solar absorptance at room temperature. Most of these data were calculated from monochromatic data as was done in Example 4.6.1. Table 4.7.1 was compiled from *Thermophysical Properties of Matter* by Touloukian et al., Vols. 7, 8, and 9 (1970, 1972, and 1973). These three volumes are the most complete reference to radiation properties available today. In addition to total hemispherical and normal emittance, such properties as angular spectral reflectance, angular total reflectance, angular solar absorptance, and others are given in this extensive compilation.

4.8   SELECTIVE SURFACES

Solar collectors must have high absorptance for radiation in the solar energy spectrum. At the same time, they lose energy by a combination of mechanisms[7] including thermal radiation from the absorbing surface, and it is desirable to have the

[6] See Agnihotri and Gupta (1981) for a more extensive review of methods of measurement of absorptance and emittance.

[7] This will be discussed in detail in Chapter 6.

Table 4.7.1   Radiation Properties

Material	Type[a]	Emittance/Temperature, K	Absorptance[b]
Aluminum, pure	H	$\dfrac{0.102}{573}, \dfrac{0.130}{773}, \dfrac{0.113}{873}$	0.09-0.10
Aluminum, anodized	H	$\dfrac{0.842}{296}, \dfrac{0.720}{484}, \dfrac{0.669}{574}$	0.12 - 0.16
Aluminum with SiO$_2$ Coating	H	$\dfrac{0.366}{263}, \dfrac{0.384}{293}, \dfrac{0.378}{324}$	0.11
Carbon black in acrylic binder	H	$\dfrac{0.83}{278}$	0.94
Chromium	N	$\dfrac{0.290}{722}, \dfrac{0.355}{905}, \dfrac{0.435}{1072}$	0.415
Copper, polished	H	$\dfrac{0.041}{338}, \dfrac{0.036}{463}, \dfrac{0.039}{803}$	0.35
Gold	H	$\dfrac{0.025}{275}, \dfrac{0.040}{468}, \dfrac{0.048}{668}$	0.20 - 0.23
Iron	H	$\dfrac{0.071}{199}, \dfrac{0.110}{468}, \dfrac{0.175}{668}$	0.44
Lampblack in epoxy	N	$\dfrac{0.89}{298}$	0.96
Magnesium oxide	H	$\dfrac{0.73}{380}, \dfrac{0.68}{491}, \dfrac{0.53}{755}$	0.14
Nickel	H	$\dfrac{0.10}{310}, \dfrac{0.10}{468}, \dfrac{0.12}{668}$	0.36 - 0.43
Paint Parsons black	H	$\dfrac{0.981}{240}, \dfrac{0.981}{462}$	0.98
Acrylic white	H	$\dfrac{0.90}{298}$	0.26
White (ZnO)	H	$\dfrac{0.929}{295}, \dfrac{0.926}{478}, \dfrac{0.889}{646}$	0.12 - 0.18

From Duffie and Beckman (1974)

[a] H is total hemispherical emittance; N is total normal emittance.

[b] Normal solar absorptance.

long-wave emittance of the surface as low as possible to reduce losses. The temperature of this surface in most flat-plate collectors is less than 200 C (473 K), while the effective surface temperature of the sun is approximately 6000 K. Thus the wavelength range of the emitted radiation overlaps only slightly the solar spectrum. (Ninety eight percent of the extraterrestrial solar radiation is at wavelengths less than 3.0 μm, whereas less than 1% of the blackbody radiation from a 200 C surface is at wavelengths less than 3.0 μm.) Under these circumstances, it is possible to devise

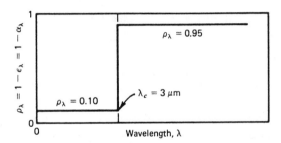

Figure 4.8.1 A hypothetical selective surface with the cutoff wavelength at 3 μm.

surfaces having high solar absorptance and low long-wave emittance, that is, selective surfaces.[8]

The concept of a selective surface is illustrated in Figure 4.8.1. This idealized surface is called a semigray surface, since it can be considered gray in the solar spectrum (i.e., at wavelengths less than approximately 3.0 μm) and also gray, but with different properties, in the infrared spectrum (i.e., at wavelengths greater than approximately 3.0 μm). For this idealized surface, the reflectance below the cutoff wavelength is very low. For an opaque surface $\alpha_\lambda = 1 - \rho_\lambda$, so in this range α_λ is very high. At wavelengths greater than λ_c the reflectance is nearly unity, and since $\varepsilon_\lambda = \alpha_\lambda = 1 - \rho_\lambda$, the emittance in this range is low.

The absorptance for solar energy and emittance for long-wave radiation are determined from the monochromatic reflectance data by integration over the appropriate spectral range. The absorptance for solar radiation, usually designated in the solar energy literature simply as α, and the emittance, usually designated simply as ε, are calculated as shown in Section 4.6. For normal operation of flat-plate solar collectors, the temperatures will be low enough that essentially all energy will be emitted at wavelengths greater than 3 μm.

Example 4.8.1

For the surface shown in Figure 4.8.1, calculate the absorptance for blackbody radiation from a source at 5777 K and the emittance at surface temperatures of 150 and 500 C.

Solution

The absorptance for blackbody radiation from a source at 5777 K is found by Equation 4.6.4 with the incident radiation $q_{\lambda,i}$ given by Planck's law, Equation 3.4.1.

For this problem, α_λ has two values, α_S in the short wavelengths below λ_c, and α_L in the long wavelengths:

$$\alpha = \alpha_S f_{0\text{-}\lambda T} + \alpha_L (1 - f_{0\text{-}\lambda T})$$

[8] Agnihotri and Gupta (1981) provide a very extensive coverage of selective surfaces.

where $f_{0\text{-}\lambda T}$ is the fraction of the incident blackbody radiation below the critical wavelength and is found from Table 3.6.1 at $\lambda T = 3 \times 5777 = 17{,}331$. Therefore the absorptance is

$$\alpha = (1 - 0.10)(0.979) + (1 - 0.95)(1 - 0.979) = 0.88$$

The emittances at 150 and 500 C are found with Equation 4.6.1. Again Table 3.6.1 is used in performing this integration. Equation 4.6.1 reduces to the following:

$$\varepsilon = \varepsilon_s f_{0\text{-}\lambda T} + \varepsilon_L(1 - f_{0\text{-}\lambda T})$$

where $f_{0\text{-}\lambda T}$ is now the fraction of the blackbody energy that is below the critical wavelength but at the surface temperature rather than the source temperature, as was used in calculating the absorptance. For a surface temperature of 150 C (423 K), λT is 1269 and $f_{0\text{-}\lambda T}$ is 0.004. The emittance at 150 C is then

$$\varepsilon_{150} = (1 - 0.10)(0.004) + (1 - 0.95)(0.996) = 0.05$$

at a surface temperature of 500 C, $f_{0\text{-}\lambda T}$ is 0.124 and the emittance at 500 C is

$$\varepsilon_{500} = (1 - 0.10)(0.124) + (1 - 0.95)(0.876) = 0.16 \qquad \blacksquare$$

In practice, the wavelength dependence of ρ_λ does not approach the ideal curve of Figure 4.8.1. Examples of ρ_λ vs. λ for several real surfaces are shown in Figures 4.8.2 and 4.8.3. Real selective surfaces do not have a well-defined critical

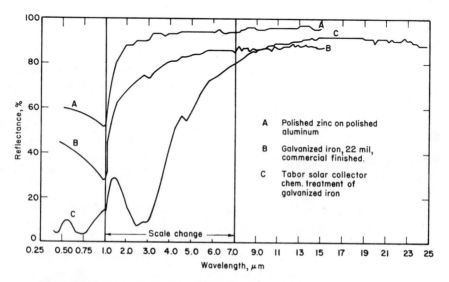

Figure 4.8.2 Spectral reflectance of several surfaces. From Edwards et al. (1960).

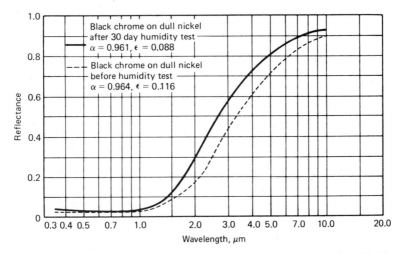

Figure 4.8.3 Spectral reflectance of black chrome on nickel before and after humidity tests. From Lin (1977).

wavelength λ_c or uniform properties in the short- and long-wavelength ranges. Values of emittance will be more sensitive to surface temperature than those of the ideal semigray surface of Figure 4.6.1. The integration procedure is the same as in Examples 4.6.1 and 4.8.1, but smaller spectral increments must be used.

Example 4.8.2

Calculate the solar absorptance and the emittance at 100 C for the surface shown in curve C of Figure 4.8.2.

Solution

The solar absorptance α should be calculated from Equation 4.6.5, with the incident radiation $q_{\lambda,i}$ having the spectral distribution of solar radiation at the collector surface. Assume that the spectral distribution of Table 2.6.1 for air mass 2 represents the distribution of solar radiation. The table that follows gives the midpoints of the spectral bands that each contain 10% of the extraterrestrial solar radiation. The monochromatic reflectances of the selective surface corresponding to these midpoint wavelengths are shown. The monochromatic absorptances are just $1 - \rho_\lambda$ and are assumed to hold over their wavelength intervals. Since the intervals are all the same, the solar absorptance is the sum of these values divided by the number of intervals as in Equation 4.6.5. The result of the calculation is $\alpha = 0.89$.

The emittance at a temperature of 100 C is found in the same manner as described in Example 4.6.1, but here 10 increments are used rather than 5 increments. The midpoint λT of each increment is found from Table 3.6.1b, the wavelength is determined from λT with $T = 373$ K, and the monochromatic reflectance is read from the curve of Figure 4.8.2. Using Equation 4.6.3, this procedure leads to an emittance of 0.15.

The details of the α and ε calculations are shown in the table.

Increment in Spectrum	Midpoint	Incident Spectrum		Emitted Spectrum		
		λ at Midpoint	$\alpha_\lambda = 1 - \rho_\lambda$	λT at Midpoint	λ at $T = 373$ K	$\varepsilon_\lambda = 1 - \rho_\lambda$
0.0 - 0.1	0.05	0.43	0.95	1880	5.0	0.43
0.1 - 0.2	0.15	0.52	0.93	2450	6.6	0.24
0.2 - 0.3	0.25	0.60	0.91	2900	7.8	0.16
0.3 - 0.4	0.35	0.67	0.96	3350	9.0	0.14
0.4 - 0.5	0.45	0.75	0.96	3830	10.3	0.11
0.5 - 0.6	0.55	0.85	0.93	4910	11.8	0.10
0.6 - 0.7	0.65	0.98	0.86	5130	13.8	0.09
0.7 - 0.8	0.75	1.10	0.78	6150	16.5	0.08
0.8 - 0.9	0.85	1.31	0.72	7850	21.3	0.10
0.9 - 1.0	0.95	2.05	0.90	12500	33.5	0.10
		Average	0.89		Average	0.15

(The reflectance of this surface for solar radiation is $1 - \alpha = 0.11$. The reflectance for incident radiation from a source at 373 C would be $1 - \varepsilon = 0.85$; if the source temperature were close to 373 K, assuming the reflectance equal to 0.85 would be a good approximation.) ∎

The potential utility of selective surfaces in solar collectors was inferred by Hottel and Woertz (1942) and noted by Gier and Dunkle (1958) and Tabor (1956, 1967) and Tabor et al. (1964). Interest in designing surfaces with a variety of ρ_λ versus λ characteristics for applications to space vehicles and to solar energy applications resulted in considerable research and compilation of data [e.g., Martin and Bell (1961), Edwards et al. (1960), Schmidt et al. (1964)]. Tabor (1967, 1977) reviewed selective surfaces and presents several methods for their preparation. Buhrman (1986) presents a review of the physics of these surfaces. Selective surfaces are in commercial use.

4.9   MECHANISMS OF SELECTIVITY

Several methods of preparing selective surfaces have been developed which depend on various mechanisms or combinations of mechanisms to achieve selectivity.

Coatings that have high absorptance for solar radiation and high transmittance for long-wave radiation can be applied to substrates with low emittance. The coating absorbs solar energy, and the substrate is the (poor) emitter of long-wave radiation. Coatings may be homogeneous or have particulate structure; their properties are then the inherent optical properties of either the coating material or of the material properties and the coating structure. Many of the coating materials used are metal oxides and the substrates are metals. Examples are copper oxide on aluminum [e.g., Hottel and Unger (1959)] and copper oxide on copper [e.g., Close (1962)]. A nickel-zinc sulfide coating can be applied to galvanized iron [Tabor (1956)].

Black chrome selective surfaces have been widely adopted in the United States. The substrate is usually nickel plating on a steel or copper base. The coatings are formed by electroplating in a bath of chromic acid and other agents.[9] In laboratory specimens, absorptances of 0.95 to 0.96 and emittances of 0.08 to 0.14 were obtained, while the average properties of samples of production run collector plates were $\alpha = 0.94$ and $\varepsilon = 0.08$ [Moore (1976)]. Reflectance properties of these surfaces are described by McDonald (1974, 1975) and others. The surfaces appear to have good durability on exposure to humid atmospheres, as shown in Figure 4.8.3. Many references are available on preparation of chrome black surfaces, for example, Benning (1976), Pettit and Sowell (1976), and Sowell and Mattox (1976). The structure and properties of black chrome coatings have been examined by Lampert and Washburn (1979). They found the wavelengths of transition from low to high reflectance to be in the 1.5 to 5 μm range, with increasing thickness of the coating shifting the transition to longer wavelengths. The coatings are aggregates of particles and voids, with particles of 0.05 to 0.30 μm diameter that are combinations of very much smaller particles of chromium and an amorphous material that is probably chromium oxide.

Selective surfaces have been in use on Israeli solar water heaters since about 1950. A base of galvanized iron is carefully cleaned and a black nickel coating is applied by immersion of the plate as the cathode in an aqueous electroplating bath containing nickel sulfate, zinc sulfate, ammonium sulfate, ammonium thiocyanate, and citric acid. Details of this process are provided by Tabor (1967).

Copper oxide on copper selective blacks are formed on carefully degreased copper plates by treating the plates for various times in hot (140 C) solutions of sodium hydroxide and sodium chlorite, as described by Close (1962). Similar propriety blackening processes have been used in the United States under the name Ebanol.

Absorptance of coatings can be enhanced by taking advantage of interference phenomena. Some coatings used on highly reflective (low-ε) substrates are semiconductors which have high absorptance in the solar energy spectrum but which have high transmittance for long-wave radiation. Many of these materials also have a high index of refraction and thus reflect incident solar energy. This reflection loss can be reduced by secondary antireflective coatings. It has been shown by Martin and Bell (1960) that three-layer coatings such as SiO_2-Al-SiO_2 on substrates such as aluminum could have absorptances for solar energy greater than 0.90 and long-wave emittances less than 0.10. The selectivity of surfaces using silicon and germanium with antireflecting coatings has been demonstrated by Seraphin (1975) and Meinel et al. (1973).

---

[9] Plating bath chemicals are available from Harshaw Chemical Co. (Chromonyx chromium process) and du Pont Company (Durimir BK black chromium process).

Vacuum sputtering processes for selective surfaces in evacuated tube collectors have been studied by Harding (1976) and Harding et al. (1976). The sputtering can be done in inert atmosphere (argon) to make metal coatings or in reactive atmospheres (argon plus 1 to 2% methane) to produce metal and metal carbide coatings. These coatings reportedly have extremely low emittances ($\varepsilon = 0.03$) but moderate absorptance ($\alpha \approx 0.8$); higher absorptances are optimum in most applications.

Sputtering processes are used in the application of cermet selective surfaces on the receivers of Luz concentrating collectors, which operate at temperatures between 300 and 400 C. [See Harats and Kearney (1989).] Four layers are deposited on the steel pipe receiving surface: an antidiffusing oxide layer to prevent diffusion of molecules of the steel substrate into the coatings, an infrared reflective layer (to provide low emittance), the cermet absorbing layer, and an antireflective oxide layer. These surfaces, which are used in vacuum jackets, have absorptance for solar radiation of 0.96 and design emittance for long-wave radiation of 0.16 at 350 C. The stability is excellent at temperatures well above 400 C. [The process used in making these surfaces is based on the work of Thornton and Lamb (1987).]

The surface structure of a metal of high reflectance can be designed to enhance its absorptance for solar radiation by grooving or pitting the surface to create cavities of dimensions near the desired cutoff wavelength of the surface. The surface acts as an array of cavity absorbers for solar radiation, thus having reduced reflectance in this part of the spectrum. The surface radiates as a flat surface in the long-wave spectrum and thus shows its usual low emittance. Desirable surface structures have been made by forming tungsten dendritic crystal in substrates by reduction of tungsten hexafluoride with hydrogen [Cuomo et al. (1976)] or by chemical vapor deposition of dendritic nickel crystals from nickel carbonyl [Grimmer et al. (1976)]. Intermetallic compounds, such as Fe_2Al_5, can be formed with highly porous structures and show some selectivity [Santala (1975)]. The degree of selectivity obtainable by this method is limited, and the emittances obtained to date have been 0.5 or more. However, roughening the substrate over which oxide (or other) coatings are applied can result in improved absorptance.

Directional selectivity can be obtained by proper arrangement of the surface on a large scale. Surfaces of deep V-grooves, large relative to all wavelengths of radiation concerned, can be arranged so that radiation from near-normal directions to the overall surface will be reflected several times in the grooves, each time absorbing a fraction of the beam. This multiple absorption gives an increase in the solar absorptance but at the same time increases the long-wavelength emittance. However, as shown by Hollands (1963), a moderately selective surface can have its effective properties substantially improved by proper configuration. For example, a surface having nominal properties of $\alpha = 0.60$ and $\varepsilon \doteq 0.05$, used in a fixed optimally oriented flat-plate collector over a year, with 55° grooves, will have an average effective α of 0.90 and an equivalent ε of 0.10. Figure 4.9.1, from Trombe et al. (1964), illustrates the multiple absorptions obtained for various angles of incidence of solar radiation on a 30° grooved surface. Figure 4.9.2, from Hollands (1963),

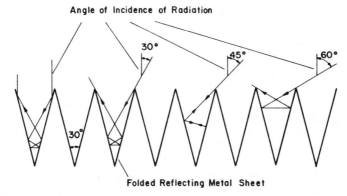

Figure 4.9.1 Absorption of radiation by successive reflections on folded metal sheets. Adapted form Trombe et al. (1964).

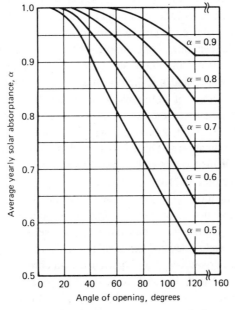

Figure 4.9.2 Average yearly solar absorptance vs. groove angle for several values of absorptance of plane surfaces. From Hollands (1963).

shows the variation of average yearly solar absorptance as a function of angle of the grooves and the absorptance of the plane surface.

The physical structure of coatings on reflective substrates will affect the reflectance of the surface. Williams et al. (1963) showed that the reflectance of coatings of lead sulfide is a function of the structure of the coating and that finely divided particulate coatings of large void fraction have a low effective refractive index and a low reflectance in the solar spectrum. [The black chrome surfaces of Lampert

Table 4.9.1   Properties of Some Selective Surfaces

Surface	α	ε	Reference
Black chrome on Ni plated steel	0.95	0.09	Mar et al. (1976), Lampert and Washburn (1979), and others
Sputtered cermet coating on steel	0.96	0.16	Harats and Kearney (1989)
"Nickel black" on galvanized steel	0.81	0.17	Tabor et al. (1964)
"Cu black" on Cu, by treating Cu with solution of NaOH and NaClO$_2$	0.89	0.17	Close (1962)
Ebanol C on Cu; commercial Cu-blackening treatment giving coatings largely CuO	0.90	0.16	Edwards et al. (1962)

and Washburn (1979) have a similar structure, in that they are particulates in voids.] This phenomenon is the basis for experimental studies of selective paints, in which binders transparent (insofar as is possible) to solar radiation are used to provide physical strength to the coatings. For example, PbS coatings of void fraction 0.8 to 0.9 on polished pure (99.99) aluminum substrates showed α of 0.8 to 0.9 and ε of 0.2 to 0.3 without a binder and $\varepsilon = 0.37$ with a silicone binder. Lin (1977) has reported studies of a range of pigments (mostly metal oxides) and binders on aluminum substrates and notes the best laboratory results obtained for an iron-manganese-copper oxide paint with a silicone binder are $\alpha = 0.92$ and $\varepsilon = 0.13$. Quality control on application to substrates is a difficult problem (thickness of a coating has a strong effect on α and ε) that remains to be solved before these selective paints become practical for applications.

A critical consideration in the use of selective surfaces is their durability. Solar collectors must be designed to operate essentially without maintenance for many years, and the coatings and substrates must retain useful properties in humid, oxidizing atmospheres and at elevated temperatures. Data from Lin (1977) and Mar et al. (1976) and from other sources, plus experience with chrome black in other kinds of applications, suggest that this surface will retain its selective properties in a satisfactory way. Years of experience with Israeli nickel black, Australian copper oxide on copper coatings, and more recently with chrome blacks have shown that these coatings can be durable.

Table 4.9.1 shows absorptance for solar radiation and emittance for long-wave radiation of surfaces that have been produced by commercial processes.

4.10   OPTIMUM PROPERTIES

In flat-plate collectors, it is generally more critical to have high absorptance than low emittance.[10] It is a characteristic of many surfaces that there is a relationship between

---

[10] This will become evident in Chapter 6.

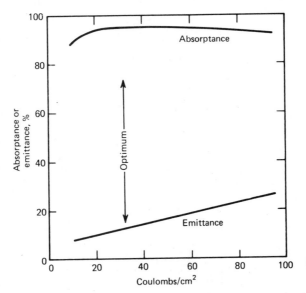

Figure 4.10.1 Variations of α and ε with the product of plating current density and time for chrome black. Adapted from Sowell and Mattox (1976).

α and ε as typified by data shown in Figure 4.10.1. In the case of the chrome black surface, the optimum plating time (coating thickness) is obvious. In the case of the PbS coating, the optimum mass per unit area is not immediately obvious. The best combination must ultimately be selected on the basis of the effects of the two properties on the annual operation of the complete solar energy system.[11] But the generalization can be made that α should be near its maximum for best performance.

4.11 ANGULAR DEPENDENCE OF SOLAR ABSORPTANCE

The angular dependence of solar absorptance of most surfaces used for solar collectors is not available. The directional absorptance of ordinary blackened surfaces (such as are used for solar collectors) for solar radiation is a function of the angle of incidence of the radiation on the surface. An example of dependence of absorptance on angle of incidence is shown in Figure 4.11.1. The limited data available suggest that selective surfaces may exhibit similar behavior [Pettit and Sowell (1976)]. A polynomial fit to the curve of Figure 4.11.1 for angles of incidence between zero and 80° is

$$\frac{\alpha}{\alpha_n} = 1 + 2.0345 \times 10^{-3}\theta - 1.990 \times 10^{-4}\theta^2$$
$$+ 5.324 \times 10^{-6}\theta^3 - 4.799 \times 10^{-8}\theta^4 \tag{4.11.1}$$

---

[11] Methods for this evaluation are in Chapter 14.

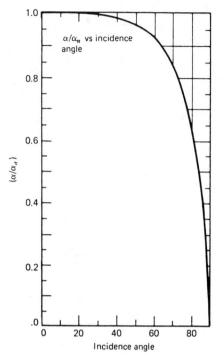

Figure 4.11.1 Ratio of solar absorptance to solar absorptance at normal incidence for a flat black surface. From Beckman et al. (1977).

4.12 ABSORPTANCE OF CAVITY RECEIVERS

Some solar energy applications require that solar radiation be absorbed in cavities rather than on flat surfaces. The effective absorptance of a cavity without a cover on its aperture, i.e., the fraction of incident radiation that is absorbed by the cavity,[12] is a function of the absorptance of the inner surfaces of the cavity and the ratio of the areas of the cavity aperture and inner surfaces. It is approximated by

$$\alpha_{\text{eff}} = \frac{\alpha_i}{\alpha_i + (1 - \alpha_i) A_a / A_i} \tag{4.12.1}$$

where α_i is the absorptance of the inner surface of the cavity, A_a is the area of the aperture of the cavity, and A_i is the area of the inner surface. As an approximation, α_i can be evaluated at the effective angle of incidence of diffuse radiation, about 60°.

---

[12] The presence of a transparent cover over the aperture of a cavity (such as a window in a room acting as a cavity receiver in passive heating) modifies Equation 4.12.1. See Section 5.10.

Example 4.12.1

A cylindrical cavity receiver has its length the same as its diameter. The aperture is in the end of the cylinder and has a diameter of two-thirds of that of the cylinder. The inner surface of the cavity has an absorptance at normal incidence of 0.60. Estimate the effective absorptance of the cavity.

Solution

Assume that the angular dependence of α_i is as shown in Figure 4.11.1. At an incident angle of $60°$, $\alpha_i/\alpha_n = 0.93$ and $\alpha_i = 0.93 \times 0.60 = 0.56$.

The ratio A_a/A_i, if d is the diameter of the cavity, is

$$\frac{A_a}{A_i} = \frac{(0.667)^2 \pi d^2/4}{(2 - 0.667^2)\pi d^2/4 + \pi d^2} = 0.098$$

The effective absorptance of the cavity is then

$$\alpha_{eff} = \frac{0.56}{0.56 + (1 - 0.56)0.098} = 0.93 \qquad ∎$$

4.13 SPECULARLY REFLECTING SURFACES

Concentrating solar collectors require the use of reflecting materials (or possibly refracting materials) to direct the beam component of solar radiation onto a receiver. This requires surfaces of high specular reflectance for radiation in the solar spectrum.

Specular surfaces are usually metals or metallic coatings on smooth substrates. Opaque substrates must be front surfaced. Examples are anodized aluminum and rhodium-plated copper. The specular reflectivity of such surfaces is a function of the quality of the substrate and the plating.

Specular surfaces can also be applied to transparent substrates, including glass or plastic. If back-surface coatings are applied, the transparency of the substrate is important, as the radiation will pass through the equivalent of twice the thickness of the substrate and twice through the front surface-air interface. (See Chapter 5 for discussion of radiation transmission through partially transparent media.) If front surface coatings are used on these substrates, the nature of the substrate, other than its smoothness and stability, is unimportant.

Specular reflectance is, in general, wavelength dependent, and in principle, monochromatic reflectances should be integrated for the particular spectral distribution of incident energy. Thus, the monochromatic specular reflectance is defined as

$$\rho_{\lambda s} = \frac{I_{\lambda s}}{I_{\lambda i}} \qquad (4.13.1)$$

where $I_{\lambda s}$ is the specularly reflected monochromatic intensity, and $I_{\lambda i}$ is the incident monochromatic intensity. Then the specular reflectance is

$$\rho_s = \frac{\int_0^\infty \rho_{\lambda s} I_{\lambda i}\, d\lambda}{\int_0^\infty I_{\lambda i}\, d\lambda} \qquad (4.13.2)$$

Typical values of specular reflectance of surfaces for solar radiation are shown in Table 4.13.1. The table includes data on front-surface and second-surface reflectors. Back-silvered glass can have excellent specular reflectance, and if the reflective coatings are adequately protected, durability is excellent. The aluminized acrylic film is one of a number of aluminized polymeric films that have been evaluated for durability in weather, and it appears to be the best of those reported by the University of Minnesota and Honeywell (1973). Many other such materials have short lifetimes (on the order of weeks or months) under practical operating conditions.

Maintenance of high specular reflectance presents practical problems. Front-surface reflectors are subject to degradation by oxidation, abrasion, dirt, and so on. Back-surface reflectors may lose reflectance because of dirt or degradation of the overlying transparent medium or degradation of the reflecting coating.

Front-surface reflectors may be covered by thin layers of protective materials to increase their durability. For example, anodized aluminum is coated with a thin stable layer of aluminum oxide deposited by electrochemical means, and silicon monoxide has been applied to front-surface aluminum films. In general, each coating reduces the initial value of specular reflectance but may result in more satisfactory levels of reflectance over long periods of time.

Table 4.13.1   Normal Specular Solar Reflectances of Surfaces

Surface	ρ
Back-silvered low-reflectance glass	0.94
Electroplated silver, new	0.96
High-purity Al, new clean	0.91
Sputtered aluminum optical reflector	0.89
Brytal processed aluminum, high purity	0.89
Back-silvered water white plate glass, new, clean	0.88
Al, SiO coating, clean	0.87
Aluminum foil, 99.5% pure	0.86
Back-aluminized 3M acrylic, new	0.86
Back-aluminized 3M acrylic[a]	0.85
Commercial Alzac process aluminum	0.85

[a] Exposed to equivalent of 1 yr solar radiation.

REFERENCES

Agnihotri, O. P. and B. K. Gupta, *Solar Selective Surfaces,* Wiley, New York (1981).

Beckman, W. A., S. A. Klein, and J. A. Duffie, *Solar Heating Design*, Wiley, New York (1977).

Benning, A. C., in *AES Coatings for Solar Collectors Symposium*, American Electroplaters' Society, Winter Park, FL, p. 57 (1976). "Black Chromium - A Solar Selective Coating."

Buhrman, R. A., in *Advances in Solar Energy,* Vol. **3** (K. Boer, ed.), American Solar Energy Society, Boulder, CO, 207 (1986). "Physics of Solar Selective Surfaces."

Close, D. J., Report E.D.7, Engineering Section, Commonwealth Scientific and Industrial Research Organization, Melbourne, Australia (1962). "Flat-Plate Solar Absorbers: The Production and Testing of a Selective Surface for Copper Absorber Plates."

Cuomo, J. J., J. M. Woodall, and T. W. DiStefano, in *AES Coatings for Solar Collectors Symposium*, Electroplaters' Society, Winter Park, FL (1976). "Dendritic Tungsten for Solar Thermal Conversion."

Duffie, J. A. and W. A. Beckman, *Solar Energy Thermal Processes*, Wiley, New York (1974).

Edwards, D. K., K. E. Nelson, R. D. Roddick, and J. T. Gier, Report No. 60-93, Dept. of Engineering, University of California at Los Angeles (Oct. 1960). "Basic Studies on the Use and Control of Solar Energy."

Edwards, D. K., J. T. Gier, K. E. Nelson, and R. Roddick, *Solar Energy*, **6**, 1 (1962). "Spectral and Directional Thermal Radiation Characteristics of Selective Surfaces for Solar Collectors."

Gier, J. T. and R. V. Dunkle, *Trans. of the Conference of the Use of Solar Energy*, Vol. **2**, P. I, p. 41, University of Arizona Press, Tucson (1958). "Selective Spectral Characteristics as an Important Factor in the Efficiency of Solar Collectors."

Gier, J. T., R. V. Dunkle, and J. T. Bevans, *J. Optical Society of America*, **44**, 558 (1954). "Measurements of Absolute Spectral Reflectivity from 1.0 to 15 Microns."

Grimmer, C. P., K. C. Herr, and W. V. McCreary, in *AES Coatings for Solar Collectors Symposium*, American Electroplaters' Society, Winter Park, FL, 79 (1976). " A Possible Selective Solar Photothermal Absorber: Ni Dendrites Formed on Al Surfaces by the CVD of $NiCO_4$."

Harats, Y. and D. Kearney, paper presented at ASME meeting, San Diego (1989). "Advances in Parabolic Trough Technology in the SEGS Plants."

Harding, G. L., *J. Vacuum Science and Technology*, **13**, 1070 (1976). "Sputtered Metal Carbide Solar-Selective Absorbing Surfaces."

Harding, G. L., D. L. McKenzie, and B. Window, *J. Vacuum Science and Technology*, **13**, 1073 (1976). "The dc Sputter Coating of Solar-Selective Surfaces onto Tubes."

Hollands, K. G. T., *Solar Energy*, **7**, 108 (1963). "Directional Selectivity, Emittance and Absorptance Properties of Vee Corrugated Specular Surfaces."

Hottel, H. C. and T. A. Unger, *Solar Energy*, **3** (3), 10 (1959). "The Properties of a Copper Oxide-Aluminum Selective Black Surface Absorber of Solar Energy."

Hottel, H. C. and B. B. Woertz, *Trans. ASME*, **14**, 91 (1942). "Performance of Flat-Plate Solar-Heat Exchangers."

Lampert, C. M. and J. Washburn, *Solar Energy Materials*, **1**, 81 (1979). "Microstructure of a Black Chrome Solar Selective Surface."

Lampert, C. M., in *Advances in Solar Energy*, Vol. **5**, American Solar Energy Society and Plenum Press,New York, 99, (1990). "Advances in Solar Optical Materials"

Lin, R. J. H., Report C00/2930-4 to ERDA (Jan. 1977). "Optimization of Coatings for Flat-Plate Solar Collectors."

Mar, H. Y. B., R. E. Peterson, and P. B. Zimmer, *Thin Solid Films*, **39**, 95 (1976). "Low Cost Coatings for Flat-Plate Solar Collectors."

Martin, D. C. and R. Bell, *Proc. of the Conference on Coatings for the Aerospace Environment*, WADD-TR-60-TB, Wright Air Development Division, Dayton, OH (November 1960). "The Use of Optical Interference to Obtain Selective Energy Absorption."

McDonald, G. E., NASA Technical Memo NASA TMX 0171596 (1974). "Solar Reflectance Properties of Black Chrome for Use as a Solar Selective Coating."

McDonald, G. E., *Solar Energy*, **17**, 119 (1975). "Spectral Reflectance Properties of Black Chrome for Use as a Solar Selective Coating."

Meinel, A. B., M. P. Meinel, C. B. McDenney, and B. O. Seraphin, paper E13 presented at Paris Meeting of International Solar Energy Society (1973). "Photothermal Conversion of Solar Energy for Large-Scale Electrical Power Production."

Moore, S. W., in *AES Coatings for Solar Collectors Symposium*, American Electroplaters' Society, Winter Park, FL (1976). "Results Obtained from Black Chrome Production Run of Steel Collectors."

Pettit, R. B. and R. P. Sowell, *J. Vacuum Science and Technology*, **13**, No. 2, 596 (1976), "Solar Absorptance and Emittance Properties."

Santala, T., in *Workshop in Solar Collectors for Heating and Cooling of Buildings*, Report, National Science Foundation RANN-75-919, p 233 (May 1975). "Intermetallic Absorption Surface Material Systems for Collector Plates."

Schmidt, R. N., K. C. Park, and E. Janssen, Tech. Doc. Report No. ML-TDR-64-250 from Honeywell Research Center to Air Force Materials Laboratory (Sept. 1964). "High Temperature Solar Absorber Coatings, Part II."

Seraphin, B. O., Reports on National Science Foundation Grant GI-36731X of the Optical Sciences Center, University of Arizona (1975).

Siegel, R. and J. R. Howell, *Thermal Radiation Heat Transfer,* 2nd ed., McGraw-Hill, New York (1981).

Sowell, R. R. and D. M. Mattox, in *AES Coatings for Solar Collectors Symposium*, American Electroplaters' Society, Winter Park, FL, 21 (1976). "Properties and Composition of Electroplated Black Chrome."

Tabor, H., *Bull. of the Research Council of Israel*, **5A** (2), 119 (Jan. 1956). "Selective Radiation."

Tabor, H., *Low Temperature Engineering Applications of Solar Energy*, ASHRAE, New York (1967). "Selective Surfaces for Solar Collectors."

Tabor, H., in *Applications of Solar Energy for Heating and Cooling of Buildings,* ASHRAE GRP-170, New York (1977). "Selective Surfaces for Solar Collectors."

Tabor, H., J. Harris, H. Weinberger, and B. Doron, *Proc. UN Conference on New Sources of Energy*, **4**, 618 (1964). "Further Studies on Selective Black Coatings."

Thornton, J. A. and J. A. Lamb, Report SERI/STR-255-3040, Solar Energy Research Institute (1987). "Evaluation of Cermet Selective Absorber Coatings Deposited by Vacuum Sputtering."

Touloukian, Y. S., et al., *Thermophysical Properties of Matter*, Plenum Data Corporation, New York, Vol. **7** (1970): *"Thermal Radiative Properties-Metallic Elements and Alloys;"* Vol. **8** (1972) *"Thermal Radiative Properties-Nonmetallic Solids;"* Vol. **9** (1973): *"Thermal Radiative Properties-Coatings."*

Trombe, F., M. Foex, and M. LePhat Vinh, *Proc. of the UN Conference on New Sources of Energy*, **4**, 625, 638 (1964). "Research on Selective Surfaces for Air Conditioning Dwellings."

University of Minnesota and Honeywell Corp. Progress Report No. 2 to the National Science Foundation, NSF/RANN/SE/GI-34871/PR/73/2 (1973). "Research Applied to Solar-Thermal Power Systems."

Williams, D. A., T. A. Lappin, and J. A. Duffie, *Trans. ASME, J. Engineering for Power*, **85A**, 213 (1963). "Selective Radiation Properties of Particulate Coatings."

Chapter 5

RADIATION TRANSMISSION THROUGH GLAZING; ABSORBED RADIATION

The transmission, reflection, and absorption of solar radiation by the various parts of a solar collector are important in determining collector performance. The transmittance, reflectance, and absorptance are functions of the incoming radiation, thickness, refractive index, and extinction coefficient of the material. Generally, the refractive index n and the extinction coefficient K of the cover material are functions of the wavelength of the radiation. However, in this chapter, all properties initially will be assumed to be independent of wavelength. This is an excellent assumption for glass, the most common solar collector cover material. Some cover materials have significant optical property variations with wavelength, and spectral dependence of properties is considered in Section 5.7. Incident solar radiation is unpolarized (or only slightly polarized). However, polarization considerations are important as radiation becomes partially polarized as it passes through collector covers.

The last sections of this chapter treat the absorption of solar radiation by collectors, collector-storage walls, and rooms on an hourly and on a monthly average basis.

Reviews of important considerations of transmission of solar radiation have been presented by Dietz (1954, 1963) and by Siegel and Howell (1981).

5.1 REFLECTION OF RADIATION

For smooth surfaces Fresnel has derived expressions for the reflection of unpolarized radiation on passing from medium 1 with a refractive index n_1 to medium 2 with refractive index n_2:

$$r_\perp = \frac{\sin^2(\theta_2 - \theta_1)}{\sin^2(\theta_2 + \theta_1)} \tag{5.1.1}$$

$$r_\parallel = \frac{\tan^2(\theta_2 - \theta_1)}{\tan^2(\theta_2 + \theta_1)} \tag{5.1.2}$$

$$r = \frac{I_r}{I_i} = \frac{1}{2}(r_\perp + r_\parallel) \tag{5.1.3}$$

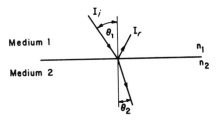

Figure 5.1.1 Angles of incidence and refraction in media with refractive indices n_1 and n_2.

where θ_1 and θ_2 are the angles of incidence and refraction, as shown in Figure 5.1.1. Equation 5.1.1 represents the perpendicular component of unpolarized radiation, $r_\perp$, and Equation 5.1.2 represents the parallel component of unpolarized radiation, $r_\parallel$. (Parallel and perpendicular refer to the plane defined by the incident beam and the surface normal.) Equation 5.1.3 then gives the reflection of unpolarized radiation as the average of the two components. The angles θ_1 and θ_2 are related to the indices of refraction by Snell's law,

$$\frac{n_1}{n_2} = \frac{\sin \theta_2}{\sin \theta_1} \tag{5.1.4}$$

Thus if the angle of incidence and refractive indices are known, Equations 5.1.1 through 5.1.4 are sufficient to calculate the reflectance of the single interface.

For radiation at normal incidence both θ_1 and θ_2 are zero, and Equations 5.1.3 and 5.1.4 can be combined to yield

$$r(0) = \frac{I_r}{I_i} = \left(\frac{n_1 - n_2}{n_1 + n_2}\right)^2 \tag{5.1.5}$$

If one medium is air (i.e., a refractive index of nearly unity), Equation 5.1.5 becomes

$$r(0) = \frac{I_r}{I_i} = \left(\frac{n - 1}{n + 1}\right)^2 \tag{5.1.6}$$

Example 5.1.1

Calculate the reflectance of one surface of glass at normal incidence and at 60°. The average index of refraction of glass for the solar spectrum is 1.526.

Solution

At normal incidence, Equation 5.1.6 can be used:

$$r(0) = \left(\frac{0.526}{2.526}\right)^2 = 0.0434$$

At an incidence angle of 60°, Equation 5.1.4 gives the refraction angle θ_2:

$$\theta_2 = \sin^{-1}\left(\frac{\sin 60}{1.526}\right) = 34.58$$

From Equation 5.1.3, the reflectance is

$$r(60) = \frac{1}{2}\left[\frac{\sin^2(-25.42)}{\sin^2(94.58)} + \frac{\tan^2(-25.42)}{\tan^2(94.58)}\right] = \frac{1}{2}(0.185 + 0.001) = 0.093$$

In solar applications, the transmission of radiation is through a slab or film of material so there are two interfaces per cover to cause reflection losses. At off-normal incidence, the radiation reflected at an interface is different for each component of polarization, so the transmitted and reflected radiation becomes partially polarized. Consequently, it is necessary to treat each component of polarization separately.

Neglecting absorption in the cover material shown in Figure 5.1.2 and considering for the moment only the perpendicular component of polarization of the incoming radiation, $(1 - r_\perp)$ of the incident beam reaches the second interface. Of this, $(1 - r_\perp)^2$ passes through the interface and $r_\perp(1 - r_\perp)$ is reflected back to the first, and so on. Summing the transmitted terms, the transmittance for the perpendicular component of polarization is

$$\tau_\perp = (1 - r_\perp)^2 \sum_{n=0}^{\infty} r_\perp^{2n} = \frac{(1 - r_\perp)^2}{1 - r_\perp^2} = \frac{1 - r_\perp}{1 + r_\perp} \tag{5.1.7}$$

Exactly the same expansion results when the parallel component of polarization is considered. The components $r_\perp$ and $r_\parallel$ are not equal (except at normal incidence),

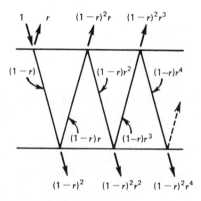

Figure 5.1.2 Transmission through one nonabsorbing cover.

and the transmittance of initially unpolarized radiation is the average transmittance of the two components,

$$\tau_r = \frac{1}{2}\left(\frac{1-r_\parallel}{1+r_\parallel} + \frac{1-r_\perp}{1+r_\perp}\right) \tag{5.1.8}$$

where the subscript r is a reminder that only reflection losses have been considered.

For a system of N covers all of the same materials, a similar analysis yields

$$\tau_{rN} = \frac{1}{2}\left[\frac{1-r_\parallel}{1+(2N-1)r_\parallel} + \frac{1-r_\perp}{1+(2N-1)r_\perp}\right] \tag{5.1.9}$$

Example 5.1.2

Calculate the transmittance of two covers of nonabsorbing glass at normal incidence and at 60°.

Solution

At normal incidence, the reflectance of one interface from Example 5.1.1 is 0.0434. From Equation 5.1.9, with both polarization components equal, the transmittance is

$$\tau_r(0) = \frac{1-0.0434}{1+3(0.0434)} = 0.85$$

Also from Example 5.1.1 but at a 60° incidence angle, the reflectances of one interface for each component of polarization are 0.185 and 0.001. From Equation 5.1.9, the transmittance is

$$\tau_r(60) = \frac{1}{2}\left[\frac{1-0.185}{1+3(0.185)} + \frac{1-0.001}{1+3(0.001)}\right] = 0.76 \qquad \blacksquare$$

The solar transmittance of *nonabsorbing* glass, having an average refractive index of 1.526 in the solar spectrum, has been calculated for all incidence angles in the same manner illustrated in Examples 5.1.1 and 5.1.2. The results for from one to four glass covers are given in Figure 5.1.3. This figure is a recalculation of the results presented by Hottel and Woertz (1942).

The index of refraction of materials that have been considered for solar collector covers are given in Table 5.1.1. The values correspond to the solar spectrum and can be used to calculate the angular dependence of reflection losses similar to Figure 5.1.3.

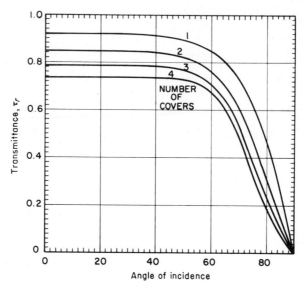

Figure 5.1.3 Transmittance of 1, 2, 3, and 4 nonabsorbing covers having an index of refraction of 1.526.

Table 5.1.1 Average Refractive Index n in Solar Spectrum of Some Cover Materials

Cover Material	Average n
Glass	1.526
Polymethyl methacrylate	1.49
Polyvinylfluoride	1.45
Polyfluorinated ethylene propylene	1.34
Polytetrafluoroethylene	1.37
Polycarbonate	1.60

5.2 ABSORPTION BY GLAZING

The absorption of radiation in a partially transparent medium is described by Bouguer's law, which is based on the assumption that the absorbed radiation is proportional to the local intensity in the medium and the distance x the radiation has traveled in the medium:

$$dI = - IK \, dx \tag{5.2.1}$$

where K is the proportionality constant, the extinction coefficient, which is assumed to be a constant in the solar spectrum. Integrating along the actual path length in the medium (i.e., from zero to $L/\cos \theta_2$) yields

$$\tau_a = \frac{I_{\text{transmitted}}}{I_{\text{incident}}} = \exp\left(-\frac{KL}{\cos \theta_2}\right) \tag{5.2.2}$$

where the subscript a is a reminder that only absorption losses have been considered. For glass, the value of K varies from approximately 4 m^{-1} for "water white" glass (which appears white when viewed on the edge) to approximately 32 m^{-1} for poor (greenish cast of edge) glass.

5.3 OPTICAL PROPERTIES OF COVER SYSTEMS

The transmittance, reflectance, and absorptance of a single cover, allowing for both reflection and absorption losses, can be determined by ray-tracing techniques similar to that used to derive Equation 5.1.7. For the perpendicular component of polarization, the transmittance $\tau_\perp$, reflectance $\rho_\perp$, and absorptance $\alpha_\perp$ of the cover are

$$\tau_\perp = \frac{\tau_a(1-r_\perp)^2}{1-(r_\perp \tau_a)^2} = \tau_a \frac{1-r_\perp}{1+r_\perp}\left(\frac{1-r_\perp^2}{1-(r_\perp \tau_a)^2}\right) \tag{5.3.1}$$

$$\rho_\perp = r_\perp + \frac{(1-r_\perp)^2 \tau_a^2 r_\perp}{1-(r_\perp \tau_a)^2} = r_\perp(1 + \tau_a \tau_\perp) \tag{5.3.2}$$

$$\alpha_\perp = (1 - \tau_a)\left(\frac{1-r_\perp}{1-r_\perp \tau_a}\right) \tag{5.3.3}$$

Similar results are found for the parallel component of polarization. For incident unpolarized radiation, the optical properties are found by the average of the two components.

The equation for the transmittance of a collector cover can be simplified by noting that the last term in Equation 5.3.1 (and its equivalent for the perpendicular component of polarization) is nearly unity, since τ_a is seldom less than 0.9 and r is on the order of 0.1 for practical collector covers. With this simplification and with Equation 5.1.8, the transmittance of a single cover becomes

$$\tau \cong \tau_a \tau_r \tag{5.3.4}$$

This is a satisfactory relationship for solar collectors with cover materials and angles of practical interest.

The absorptance of a solar collector cover can be approximated by letting the last term in Equation 5.3.3 be unity so that

$$\alpha \cong 1 - \tau_a \tag{5.3.5}$$

Although the neglected term in Equation 5.3.3 is larger than the neglected term in Equation 5.3.1, the absorptance is much smaller than the transmittances so that the overall accuracy of the two approximations is essentially the same.

The reflectance of a single cover is then found from $\rho = 1 - \alpha - \tau$, so that

$$\rho \cong \tau_a(1 - \tau_r) = \tau_a - \tau \tag{5.3.6}$$

The advantage of Equations 5.3.4 through 5.3.6 over Equations 5.3.1 through 5.3.3 is that polarization is accounted for in the approximate equations through the single term τ_r rather than by the more complicated expressions for each individual optical property. Example 5.3.1 shows a solution for transmittance by the exact equations and also by the approximate equations.

Example 5.3.1

Calculate the transmittance, reflectance, and absorptance of a single glass cover 2.3 mm thick at an angle of 60°. The extinction coefficient of the glass is 32 m^{-1}.

Solution

At an incidence angle of 60°, the optical path length is

$$\frac{KL}{\cos \theta_2} = 32 \times \frac{0.0023}{\cos 34.58} = 0.0894$$

where 34.58 is the refraction angle calculated in Example 5.1.1. The transmittance, τ_a, from Equation 5.2.2 is then

$$\tau_a = \exp(-0.0894) = 0.915$$

Using the results of Example 5.1.1 and Equation 5.3.1, the transmittance is found by averaging the transmittances for the parallel and perpendicular components of polarization,

$$\tau = \frac{0.915}{2} \left[\frac{1 - 0.185}{1 + 0.185} \left(\frac{1 - 0.185^2}{1 - (0.915 \times 0.185)^2} \right) + \frac{1 - 0.001}{1 + 0.001} \left(\frac{1 - 0.001^2}{1 - (0.915 \times 0.001)^2} \right) \right]$$

$$= 0.5(0.624 + 0.912) = 0.768$$

The reflectance is found using Equation 5.3.2 for each component of polarization and averaging:

$$\rho = 0.5[0.185(1 + 0.915 \times 0.624) + 0.001 (1 + 0.915 \times 0.912)]$$

$$= 0.5(0.291 + 0.003) = 0.147$$

In a similar manner, the absorptance is found using Equation 5.3.3:

$$\alpha = \frac{1 - 0.915}{2}\left(\frac{1 - 0.185}{1 - 0.185 \times 0.915} + \frac{1 - 0.001}{1 - 0.001 \times 0.915}\right)$$

$$= \frac{0.085}{2}(0.981 + 1.000) = 0.085$$

Alternate Solution

The approximate equations can also be used to find these properties. From Equations 5.3.4 and 5.1.8 the transmittance is

$$\tau = \frac{0.915}{2}\left(\frac{1 - 0.185}{1 + 0.185} + \frac{1 - 0.001}{1 + 0.001}\right) = 0.771$$

From Equation 5.3.5, the absorptance is

$$\alpha = 1 - 0.915 = 0.085$$

and the reflectance is then

$$\rho = 1 - 0.771 - 0.085 = 0.144$$

Note that even though the incidence angle was large and poor quality glass was used in this example so that the approximate equations tend to be less accurate, the approximate method and the exact method are essentially in agreement. ∎

Although Equations 5.3.4 through 5.3.6 were derived for a single cover, they apply equally well to identical multiple covers. The quantity τ_r should be evaluated from Equation 5.1.9 and the quantity τ_a from Equation 5.2.2 with L equal to the total cover system thickness.

Example 5.3.2

Calculate the solar transmittance at incidence angles of zero and 60° for two glass covers each 2.3 mm thick. The extinction coefficient of the glass is 16.1 m⁻¹, and the refractive index is 1.526.

Solution

For one sheet at normal incidence,

$$KL = 16.1 \times 2.3/1000 = 0.0370$$

The transmittance τ_a is

$$\tau_a(0) = \exp[-2(0.0370)] = 0.93$$

The transmittance accounting for reflection, from Example 5.1.2, is 0.85. The total transmittance is then found from Equation 5.3.4.

$$\tau(0) = \tau_r(0)\,\tau_a(0) = 0.85(0.93) = 0.79$$

From Example 5.1.1, when $\theta_1 = 60°$, $\theta_2 = 34.57°$, and

$$\tau_a(60) = \exp[-2(0.0370)/\cos 34.57] = 0.91$$

and the total transmittance (with $\tau_r = 0.76$ from Example 5.1.2) becomes

$$\tau(60) = \tau_r(60)\tau_a(60) = 0.76(0.91) = 0.69 \qquad\blacksquare$$

Figure 5.3.1 gives curves of transmittance as a function of angle of incidence for systems of one to four identical covers of three different kinds of glass. These curves were calculated from Equation 5.3.4 and have been checked by experiments [Hottel and Woertz (1942)].

In a multicover system, the ray-tracing technique used to develop Equation 5.1.7 can be used to derive the appropriate equations. Whillier (1953) has generalized the ray-tracing method to any number or type of covers, and modern radiation heat transfer calculation methods have also been applied to these complicated situations [e.g., Edwards (1977)]. If the covers are identical, the approximate method illustrated in Example 5.3.2 is recommended, although the following equations can also be used.

For a two-cover system with covers not necessarily identical, ray tracing yields the following equations for transmittance and reflectance, where subscript 1 refers to outer cover and subscript 2 to the inner cover:

$$\tau = \frac{1}{2}\left[\left(\frac{\tau_1 \tau_2}{1 - \rho_1\rho_2}\right)_\perp + \left(\frac{\tau_1 \tau_2}{1 - \rho_1\rho_2}\right)_\parallel\right] \qquad (5.3.7)$$

$$\rho = \frac{1}{2}\left[\left(\rho_1 + \frac{\tau\rho_2 \tau_1}{\tau_2}\right)_\perp + \left(\rho_1 + \frac{\tau\rho_2 \tau_1}{\tau_2}\right)_\parallel\right] \qquad (5.3.8)$$

Note that the reflectance from the cover system depends upon which cover first intercepts the radiation.

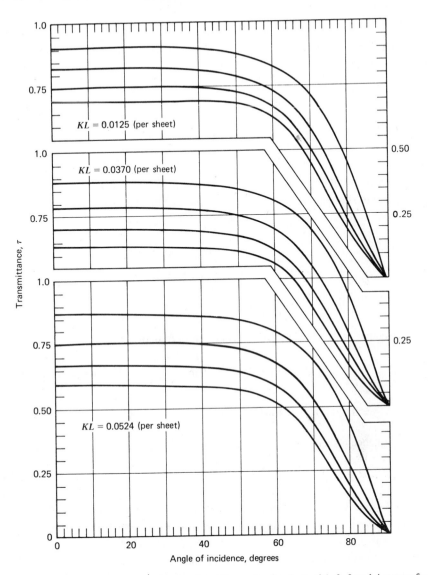

Figure 5.3.1 Transmittance (considering absorption and reflection) of 1, 2, 3 and 4 covers for three types of glass.

Example 5.3.3

Calculate the optical properties of a two-cover solar collector at an angle of 60°. The outer cover is glass with $K = 16.1$ m^{-1} and thickness of 2.3 mm. The inner cover is polyvinyl fluoride with refractive index equal to 1.45. The plastic film is thin enough so that absorption within the plastic can be neglected.

Solution

The optical properties of the glass and plastic covers alone, as calculated from Equations 5.3.1 through 5.3.3, are

glass: $\tau_{\parallel} = 0.961$ $\tau_{\perp} = 0.660$

                 $\rho_{\parallel} = 0.002$ $\rho_{\perp} = 0.304$

                 $\alpha_{\parallel} = 0.037$ $\alpha_{\perp} = 0.036$

plastic: $\tau_{\parallel} = 0.995$ $\tau_{\perp} = 0.726$

                 $\rho_{\parallel} = 0.005$ $\rho_{\perp} = 0.274$

                 $\alpha_{\parallel} = 0.$ $\alpha_{\perp} = 0.$

Equations 5.3.4 through 5.3.6 could have been used with each component of polarization to simplify the calculation of the preceeding properties.

The transmittance of the combination is found from Equation 5.3.7:

$$\tau = \frac{1}{2}\left(\frac{0.961 \times 0.995}{1 - 0.002 \times 0.005} + \frac{0.660 \times 0.726}{1 - 0.304 \times 0.274} \right)$$

$$= 0.5\,(0.956 + 0.523) = 0.739$$

The reflectance, with the glass first, is found from Equation 5.3.8:

$$\rho = \frac{1}{2}\left(0.002 + \frac{0.956 \times 0.005 \times 0.961}{0.995} + 0.304 + \frac{0.523 \times 0.274 \times 0.660}{0.726} \right)$$

$$= 0.5(0.007 + 0.434) = 0.221$$

The absorptance is then

$$\alpha = 1 - 0.221 - 0.739 = 0.040 \qquad\blacksquare$$

Equations 5.3.7 and 5.3.8 can be used to calculate the transmittance of any number of covers by repeated application. If subscript 1 refers to the properties of a cover system and subscript 2 to the properties of an additional cover placed under the stack, then these equations yield the appropriate transmittance and reflectance of the new system. The reflectance ρ_1 in Equation 5.3.7 is the reflectance of the original cover system from the bottom side. If any of the covers exhibit strong wavelength-dependent properties, integration over the wavelength spectrum is necessary (see Section 5.6).

5.4   TRANSMITTANCE FOR DIFFUSE RADIATION

The preceding analysis applies only to the beam component of solar radiation. Radiation incident on a collector also consists of scattered solar radiation from the sky and possibly reflected solar radiation from the ground. In principle, the amount of this radiation that passes through a cover system can be calculated by integrating the transmitted radiation over all angles. However, the angular distribution of this radiation is generally unknown.

For incident isotropic radiation (i.e., independent of angle), the integration can be performed. The presentation of the results can be simplified by defining an equivalent angle for beam radiation that gives the same transmittance as for diffuse radiation. For a wide range of conditions encountered in solar collector applications, this equivalent angle is essentially 60°. In other words, beam radiation incident at an angle of 60° has the same transmittance as isotropic diffuse radiation.

Circumsolar diffuse radiation can be considered as having the same angle of incidence as the beam radiation. Diffuse radiation from the horizon is usually a small contribution to the total and as an approximation can be taken as having the same angle of incidence as the isotropic diffuse radiation.

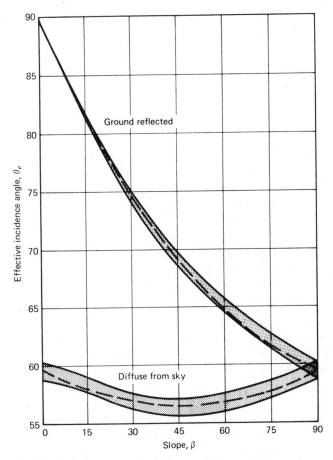

Figure 5.4.1 Effective incidence angle of isotropic diffuse radiation and isotropic ground-reflected radiation on sloped surfaces. From Brandemuehl and Beckman (1980).

Solar collectors are usually oriented so that they "see" both the sky and the ground. If the diffuse radiation from the sky and the radiation reflected from the ground are both isotropic, then the transmittance of the glazing systems can be found by integrating the beam transmittance over the appropriate incidence angles. This integration has been done by Brandemuehl and Beckman (1980); the results are presented in Figure 5.4.1 in terms of a single effective incidence angle. Thus all of the diffuse radiation can be treated as having a single equivalent angle of incidence, and all of the ground-reflected radiation can be considered as having another equivalent angle of incidence. The shaded region includes a wide range of glazings. The upper curve is for a one-cover polyflorinated ethylene propylene glazing with no internal absorption, whereas the lower curve represents a two-cover glass glazing with extinction length KL of 0.0524. All one- and two-cover systems with indices of refraction between 1.34 and 1.526 and extinction lengths less than 0.0524 lie in the shaded region.

The dashed lines shown in Figure 5.4.1 are given for ground-reflected radiation by

$$\theta_e = 90 - 0.5788\beta + 0.002693\beta^2 \tag{5.4.1}$$

and for diffuse radiation by

$$\theta_e = 59.7 - 0.1388\beta + 0.001497\beta^2 \tag{5.4.2}$$

5.5 TRANSMITTANCE-ABSORPTANCE PRODUCT

To use the analysis of the next chapter, it is necessary to evaluate the transmittance-absorptance product $(\tau\alpha)$.[1] Of the radiation passing through the cover system and incident on the plate, some is reflected back to the cover system. However, all this radiation is not lost since some of it is, in turn, reflected back to the plate.

The situation is illustrated in Figure 5.5.1, where τ is the transmittance of the cover system at the desired angle and α is the angular absorptance of the absorber plate. Of the incident energy, $\tau\alpha$ is absorbed by the absorber plate and $(1 - \alpha)\tau$ is reflected back to the cover system. The reflection from the absorber plate is assumed to be diffuse (and unpolarized) so that the fraction $(1 - \alpha)\tau$ that strikes the cover plate is diffuse radiation and $(1 - \alpha)\tau\rho_d$ is reflected back to the absorber plate. The quantity ρ_d refers to the reflectance of the cover system for diffuse radiation incident from the bottom side and can be estimated from Equation 5.3.6 as the difference between τ_a and τ at an angle of 60°.[2] If the cover system consists of two (or

---

[1] The transmittance-absorptance product $(\tau\alpha)$ should be thought of as a symbol representing a property of a cover-absorber combination rather than as a product of two properties.

[2] For single covers of the three kinds of glass of Figure 5.3.1, ρ_d at 60° is 0.15 (for KL = 0.0125), 0.12 (for KL = 0.0370), and 0.11 (for KL = 0.0524). For two covers, the corresponding values are 0.22, 0.17, and 0.14.

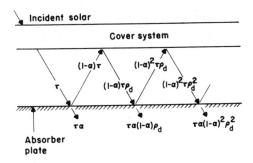

Figure 5.5.1 Absorption of solar radiation by absorber plate under a cover system.

more) covers of dissimilar materials, ρ_d will be different (slightly) from the diffuse reflectance of the incident solar radiation (see Equation 5.3.8). The multiple reflection of diffuse radiation continues so that the fraction of the incident energy ultimately absorbed is[3]

$$(\tau\alpha) = \tau\alpha \sum_{n=0}^{\infty} [(1-\alpha)\rho_d]^n = \frac{\tau\alpha}{1-(1-\alpha)\rho_d} \qquad (5.5.1)$$

Example 5.5.1

For a two-cover collector using glass with $KL = 0.0370$ per plate and an absorber plate with $\alpha = 0.90$ (independent of direction), find the transmittance-absorptance product at an angle of 50°.

Solution

From Figure 5.3.1, τ at 50° is 0.75 and τ at 60° (the effective angle of incidence of radiation reflected back to the cover) is 0.69. From Equation 5.2.2 with $\theta_2 = 34.58°$, τ_a is 0.91. From Equation 5.3.6, ρ_d is 0.91 − 0.69 = 0.22. From Equation 5.5.1

$$(\tau\alpha) = \frac{(0.75)(0.90)}{1-(1-0.90)(0.22)} = 0.69$$

Note that it is also possible to estimate $\rho_d = 1 - \tau_r$, where τ_r can be estimated from Figure 5.1.3 at 60°. For two covers, τ_r is 0.77, so ρ_d is 0.23. ■

---

[3] The absorptance α of the absorber plate for the reflected radiation should be the absorptance for diffuse radiation. Also, the reflected radiation may not all be diffuse, and it may be partially polarized. However, the resulting errors should be negligible in that the difference between $\tau\alpha$ and $(\tau\alpha)$ is small.

The value of $(\tau\alpha)$ in this example is very nearly equal to 1.01 times the product of τ times α. This is a reasonable approximation for most practical solar collectors. Thus,

$$(\tau\alpha) \cong 1.01\,\tau\alpha \tag{5.5.2}$$

can be used as an estimate of $(\tau\alpha)$ in place of Equation 5.5.1.

5.6 ANGULAR DEPENDENCE OF $(\tau\alpha)$

The dependence of absorptance and transmittance on the angle of incidence of the incident radiation has been shown in Sections 4.7 and 5.1 to 5.4. For ease in determining $(\tau\alpha)$ as a function of angle of incidence θ, Klein (1979) developed a relationship between $(\tau\alpha)/(\tau\alpha)_n$ and θ based on the angular dependence of α shown in Figure 4.11.1 and on the angular dependence of τ for glass covers with $KL = 0.04$. The result is not sensitive to KL and can be applied to all covers having a refractive index close to that of glass. Klein's curves are shown in Figure 5.6.1. The results obtained by using this figure are essentially the same as those obtained by independently finding the angular dependence of τ and α. This is illustrated in examples to follow in this chapter.

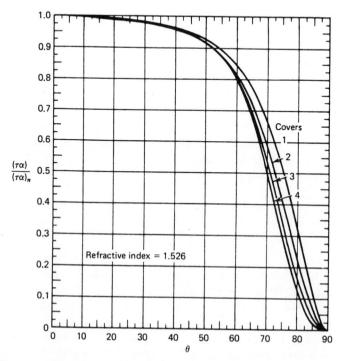

Figure 5.6.1 Typical $(\tau\alpha)/(\tau\alpha)_n$ curves for 1 to 4 covers. Adapted from Klein (1979).

5.7   SPECTRAL DEPENDENCE OF TRANSMITTANCE

Most transparent media transmit selectively; that is, transmittance is a function of wavelength of the incident radiation. Glass, the material most commonly used as a cover material in solar collectors, may absorb little of the solar energy spectrum if its Fe_2O_3 content is low. If its Fe_2O_3 content is high, it will absorb in the infrared portion of the solar spectrum. The transmittance of several glasses of varying iron content is shown in Figure 5.7.1. These show clearly that "water white" (low-iron) glass has the best transmission; glasses with high Fe_2O_3 content have a greenish appearance and are relatively poor transmitters. Note that the transmission is not a strong function of wavelength in the solar spectrum except for the "heat absorbing" glass. Glass becomes substantially opaque at wavelengths longer than approximately 3 μm and can be considered as opaque to long-wave radiation.

Some collector cover materials may have transmittances that are more wavelength dependent than low-iron glass, and it may be necessary to obtain their transmittance for monochromatic radiation and then integrate over the entire spectrum. If there is no significant angular dependence of monochromatic transmittance, the transmittance for incident radiation of a given spectral distribution is calculated by an equation analogous to Equation 4.6.4:

$$\tau = \sum_{j=1}^{n} \tau_{\lambda,j} \, \Delta f_j \tag{5.7.1}$$

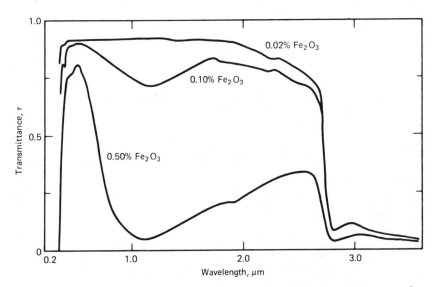

Figure 5.7.1 Spectral transmittance of 6 mm thick glass with various iron oxide contents for incident radiation at normal incidence. Data from Dietz (1954).

If there is an angular dependence of τ_λ, the total transmittance at angle θ can be written as

$$\tau(\theta) = \frac{\int_0^\infty \tau_\lambda(\theta) I_{\lambda i}(\theta)\, d\lambda}{\int_0^\infty I_{\lambda i}(\theta)\, d\lambda} \tag{5.7.2}$$

where $\tau_\lambda(\theta)$ is calculated by the equations of the preceding section, using monochromatic values of the index of refraction and absorption coefficient, and $I_{\lambda i}(\theta)$ is the incident monochromatic intensity arriving at the cover system from angle θ.

Example 5.7.1

For the glass of Figure 5.7.1, having an iron oxide content of 0.50%, estimate the transmittance at normal incidence for terrestrial solar radiation.

Solution

Use Table 2.6.1 to represent the spectral distribution of the incident radiation and Equation 5.7.1 to calculate τ. Dividing the spectrum into 10 equal increments, the increments, the wavelength at the energy midpoints of the increments, and τ_λ at the midpoints of the increments are shown in the table. The sum of the third and sixth columns is 3.65. Since 10 equal increments were chosen, the transmittance is the average of the τ_λ. Thus the transmittance of the glass at normal incidence is 0.36.

Increment, μm	λ_{mid}	$\tau_{\lambda,mid}$	Increment, μm	λ_{mid}	$\tau_{\lambda,mid}$
0.00 - 0.10	0.434	0.73	0.50 - 0.60	0.845	0.16
0.10 - 0.20	0.517	0.79	0.60 - 0.70	0.975	0.07
0.20 - 0.30	0.595	0.70	0.70 - 0.80	1.101	0.06
0.30 - 0.40	0.670	0.56	0.80 - 0.90	1.310	0.09
0.40 - 0.50	0.752	0.25	0.90 - 1.00	2.049	0.24

If the absorptance of solar radiation by an absorber plate is independent of wavelength, then Equation 5.5.1 can be used to find the transmittance-absorptance product with the transmittance as calculated from Equation 5.7.1 or 5.7.2. However, if both the solar transmittance of the cover system and the solar absorptance of the absorber plate are functions of wavelength and angle of incidence, the fraction absorbed by an absorber plate is given by

$$\tau\alpha(\theta) = \frac{\int_0^\infty \tau_\lambda(\theta) \alpha_\lambda(\theta) I_{\lambda i}(\theta)\, d\lambda}{\int_0^\infty I_{\lambda i}(\theta)\, d\lambda} \tag{5.7.3}$$

To account for multiple reflections in a manner analogous to Equation 5.5.1, it would be necessary to evaluate the spectral distribution of each reflection and integrate over all wavelengths. It is unlikely that such a calculation would ever be necessary for solar collector systems, since the error involved by directly using Equation 5.7.3 with Equation 5.5.1 would be small if α is near unity.

In a multicover system in which the covers have significant wavelength-dependent properties, the spectral distribution of the solar radiation changes as it passes through each cover. Consequently, if all covers are identical, the transmittance of individual covers increases in the direction of propagation of the incoming radiation. If the covers are not all identical, the transmittance of a particular cover may be greater or less than other similar covers in the system. Equations 5.7.1 to 5.7.3 account for this phenomenon. At any wavelength λ, the transmittance is the product of the monochromatic transmittances of the individual covers. Thus for N covers

$$\tau\alpha(\theta) = \frac{\int_0^\infty \tau_{\lambda,1}(\theta)\tau_{\lambda,2}(\theta)...\tau_{\lambda,N}(\theta)\alpha_\lambda(\theta)I_{\lambda i}(\theta)\,d\lambda}{\int_0^\infty I_{\lambda i}(\theta)\,d\lambda} \tag{5.7.4}$$

If a cover system has one cover with wavelength-independent properties (e.g., glass) and one cover with wavelength-dependent properties (e.g., some plastics), then a simplified procedure can be used. The transmittance and reflectance of each cover can be obtained separately, and the combined system transmittance and reflectance can be obtained from Equations 5.3.7 and 5.3.8.

For most plastics, the transmittance will also be significant in the infrared spectrum at $\lambda > 3$ μm. Figure 5.7.2 shows a transmittance curve for a polyvinyl fluoride ("Tedlar") film for wavelengths longer than 2.5 μm. Whillier (1963) calculated the transmittance of a similar film using Equation 5.6.2. The incident radiation, $I_{\lambda i}$, was for radiation from blackbody sources at temperatures from 0 to 200 C. He found that transmittance was 0.32 for radiation from the blackbody source at 0 C, 0.29 for the source at 100 C, and 0.32 for the source at 200 C.

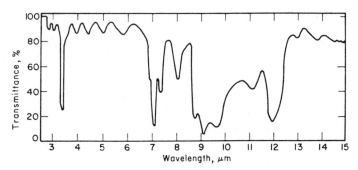

Figure 5.7.2 Infrared spectral transmittance of a polyvinyl fluoride Tedlar film. Courtesy of du Pont.

5.8   EFFECTS OF SURFACE LAYERS ON TRANSMITTANCE

If a film of low refractive index is deposited at an optical thickness of $\lambda/4$ onto a transparent slab, radiation of wavelength λ reflected from the upper and lower surfaces of the film will have a phase difference of π and will cancel. The reflectance will be decreased, and the transmittance will be increased relative to the uncoated material. This is the principal type of coating used on camera lenses, binoculars, and other expensive optical equipment.

Inexpensive and durable processes have been developed for treating glass to reduce its reflectance by the addition of films having a refractive index between that of air and the transparent medium [e.g., Thomsen (1951)]. The solar reflectance of a single pane of untreated glass is approximately 8%. Surface treatment, by dipping glass in a silica-saturated fluosilic acid solution, can reduce the reflection losses to 2%, and a double layer coating can, as shown by Mar et al. (1975), reduce reflection losses to less than 1%. Such an increase in solar transmittance can make a very significant improvement in the thermal performance of flat-plate collectors. Figure 5.8.1 shows typical monochromatic reflectance data before and after etching. Note that unlike unetched glass, it is necessary to integrate monochromatic reflectance over the solar spectrum to obtain the reflectance for solar radiation.

Experimental values for the angular dependence of solar transmission for unetched and etched glass are given in Table 5.8.1. Not only does the etched sample exhibit higher transmittance than the unetched sample at all incidence angles, but also the transmittance degrades less at high incidence angles.

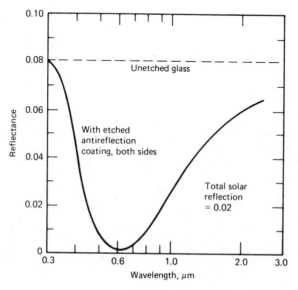

Figure 5.8.1. Monochromatic reflectance of one sheet of etched and unetched glass. From Mar et al. (1975).

Table 5.8.1 Solar Transmittance for Etched and Unetched Glass as a Function of Incidence Angle[a]

Type of Glass	Incidence Angle						
	0^0	20^0	40^0	50^0	60^0	70^0	80^0
Etched	0.941	0.947	0.945	0.938	0.916	0.808	0.562
Unetched	0.888	0.894	0.903	0.886	0.854	0.736	0.468

[a] From Mar et al. (1975).

Glass is treated by other means to decrease its emittance for use as transparent insulation for glazing applications. These treatments also change the transmittance, in many cases decreasing it substantially.

5.9 ABSORBED SOLAR RADIATION

The prediction of collector performance requires information on the solar energy absorbed by the collector absorber plate. The solar energy incident on a tilted collector can be found by the methods of Chapter 2. This incident radiation has three different spatial distributions: beam radiation, diffuse radiation, and ground-reflected radiation, and each must be treated separately. The details of the calculation depend on which diffuse sky model (Sections 2.14 to 2.16) is used. Using the isotropic diffuse concept on an hourly basis, Equation 2.15.5 can be modified to give the absorbed radiation, S, by multiplying each term by the appropriate transmittance-absorptance product:

$$S = I_b R_b (\tau\alpha)_b + I_d (\tau\alpha)_d \left(\frac{1 + \cos\beta}{2}\right) + \rho_g (I_b + I_d)(\tau\alpha)_g \left(\frac{1 - \cos\beta}{2}\right) \quad (5.9.1)$$

where $(1 + \cos\beta)/2$ and $(1 - \cos\beta)/2$ are the view factors from the collector to the sky and from the collector to the ground, respectively. The subscripts b, d, and g represent beam, diffuse, and ground. For a given collector tilt, Figure 5.4.1 gives the effective angle of incidence of the diffuse and ground-reflected radiation, and Figures 4.11.1 and 5.3.1 can be used to find the proper absorptance and transmittance values. Equation 5.5.1 or 5.5.2 can then be used to find $(\tau\alpha)_d$ and $(\tau\alpha)_g$. The angle θ for the beam radiation, which is needed in evaluating R_b, is used to find $(\tau\alpha)_b$. Alternatively, $(\tau\alpha)_n$ can be found from the properties of the cover and absorber and Figure 5.6.1 can be used at the appropriate angles of incidence for each radiation stream to determine the three transmittance-absorptance products.

The results of the preceding sections are summarized in the following example in which the solar radiation absorbed by a collector is calculated.

Example 5.9.1

For an hour 11 to 12 AM on a clear winter day, I is 1.79 MJ/m^2, I_b is 1.38 MJ/m^2, and I_d is 0.41 MJ/m^2. Ground reflectance is 0.6. For this hour, θ for the beam radiation is 17°, and R_b is 2.11. A collector with one glass cover is sloped 60° to the south. The glass has $KL = 0.0370$, and the absorptance of the plate at normal incidence, α_n, is 0.93. Using the isotropic diffuse model (Equation 5.9.1), calculate the absorbed radiation per unit area of absorber.

Solution

Two approaches to the solution are possible. The angular dependence of τ and α can be individually determined, or the the angular dependence of $(\tau\alpha)$ can be determined.

   In the first method, use Figure 4.11.1 to get angular dependence of α and Figure 5.3.1 to get angular dependence of τ. For the 60° slope, from Figure 5.4.1, the effective angle of incidence of the diffuse radiation is 57° and that of the ground-reflected radiation is 65°.

For the beam radiation, at $\theta = 17°$:
> From Figure 4.11.1, $\alpha/\alpha_n = 0.99$
> From Figure 5.3.1, $\tau = 0.88$
> $(\tau\alpha)_b = 1.01 \times 0.88 \times 0.99 \times 0.93 = 0.82$

For the (isotropic) diffuse radiation, at $\theta = 57°$:
> From Figure 4.11.1, $\alpha/\alpha_n = 0.94$
> From Figure 5.3.1, $\tau = 0.83$
> $(\tau\alpha)_d = 1.01 \times 0.83 \times 0.94 \times 0.93 = 0.73$

For the ground-reflected radiation, at $\theta = 65°$:
> From Figure 4.11.1, $\alpha/\alpha_n = 0.88$
> From Figure 5.3.1, $\tau = 0.76$
> $(\tau\alpha)_g = 1.01 \times 0.76 \times 0.88 \times 0.93 = 0.63$

Equation 5.9.1 is now used to calculate S:

$$S = 1.38 \times 2.11 \times 0.82 + 0.41 \times 0.73 \left(\frac{1 + \cos 60}{2} \right) + 1.79 \times 0.6 \times 0.63 \left(\frac{1 - \cos 60}{2} \right)$$

$$= 2.39 + 0.22 + 0.17 = 2.78 \text{ MJ/m}^2$$

   In the second method, use Figure 5.6.1 to get the angular dependence of $(\tau\alpha)/(\tau\alpha)_n$. The effective angles of incidence of the diffuse and ground-reflected radiation are $57°$ and $65°$, as before. From Figure 5.3.1, τ_n is 0.88, α_n is 0.93 (given), so

$$(\tau\alpha)_n = 1.01 \times 0.88 \times 0.93 = 0.83$$

From Figure 5.6.1 with the beam radiation at $\theta = 17°$, $(\tau\alpha)/(\tau\alpha)_n$ is 0.99 and

$$(\tau\alpha)_b = 0.83 \times 0.99 = 0.82$$

From Figure 5.6.1 with the diffuse radiation at $\theta = 57°$, $(\tau\alpha)/(\tau\alpha)_n$ is 0.87 and

$$(\tau\alpha)_d = 0.83 \times 0.87 = 0.72$$

From Figure 5.6.1 with the ground-reflected radiation at $\theta = 65°$, $(\tau\alpha)/(\tau\alpha)_n$ is 0.76. and

$$(\tau\alpha)_g = 0.83 \times 0.76 = 0.63$$

These are essentially identical to the results of the first solution. ■

The calculation of absorbed radiation using the HDKR model of diffuse radiation (Equation 2.16.7) is similar to that based on the isotropic model except that the circumsolar diffuse is treated as an increment to the beam radiation, horizon brightening is considered, and the diffuse component is correspondingly reduced. It is assumed that the angle of incidence of the circumsolar diffuse is the same as that of the beam and that the angle of incidence of the diffuse from the horizon is the same as the isotropic. The energy absorbed by the absorbing surface is given by

$$S = (I_b + I_d A_i)R_b(\tau\alpha)_b + I_d(1 - A_i)(\tau\alpha)_d\left(\frac{1 + \cos\beta}{2}\right)\left[1 + f\sin^3\left(\frac{\beta}{2}\right)\right]$$

$$+ I\rho_g(\tau\alpha)_g\left(\frac{1 - \cos\beta}{2}\right) \tag{5.9.2}$$

Example 5.9.2

Redo Example 5.9.1 using the HDKR model. For this hour I_o is 2.40 MJ/m^2.

Solution

The calculations of the transmittance-absorptance products are the same as in Example 5.9.1. The anisotropy index is calculated from Equation 2.16.3:

$$A_i = 1.38/2.40 = 0.58$$

Using Equation 2.16.6,

$$f = \sqrt{I_b/I} = \sqrt{1.38/1.79} = 0.88$$

Then using Equation 5.9.2, S can be calculated:

$$S = \left(1.38 + 0.41 \times 0.58\right) \times 2.11 \times 0.82$$

$$+ 0.41 \times (1 - 0.58)\left(\frac{1 + \cos 60}{2}\right)\left[1 + 0.88 \sin^3\left(\frac{60}{2}\right)\right] \times 0.73$$

$$+ 1.79 \times 0.6 \left(\frac{1 - \cos 60}{2}\right) \times 0.63$$

$$= 2.80 + 0.10 + 0.17 = 3.07 \text{ MJ/m}^2 \qquad \blacksquare$$

Under the clear sky conditions of these two examples, the HDKR sky model leads to substantially higher estimates of absorbed radiation, as much of the diffuse radiation is taken as circumsolar and added to the beam radiation.

In these two examples, each of the radiation streams on the collector is treated separately. At times it is convenient to define an average transmittance-absorptance as the ratio of the absorbed solar radiation, S, to the incident solar radiation, I_T. Thus,

$$S = (\tau\alpha)_{av} I_T \qquad\qquad (5.9.3)$$

This is convenient when direct measurements are available for I_T. In Example 5.9.2, S, the solar radiation absorbed by the collector for the hour 10 to 11, is 3.07 MJ/m², and I_T is 3.81 MJ/m². The average transmittance-absorptance product for this hour is then 0.80, which is slightly less than the value of $(\tau\alpha)_b$ of 0.82. When the beam fraction is high, as in these examples, $(\tau\alpha)_{av}$ is close to $(\tau\alpha)_b$. When the diffuse fraction is high, using the value of $(\tau\alpha)_d$ for $(\tau\alpha)_{av}$ may be a reasonable assumption. As will be seen in Chapter 6, useful energy gain by the collector is highest when beam radiation is high, and as an approximation when I_T data are available, the following can be assumed:

$$(\tau\alpha)_{av} \cong 0.96(\tau\alpha)_b \qquad\qquad (5.9.4)$$

(Comparisons of the measured and calculated operation of solar processes are often made, with the incident solar radiation measured on the plane of the collector, I_T. Equation 5.9.4 provides a convenient way to estimate the absorbed radiation S under these circumstances. This S is then used in the performance calculations.)

5.10   MONTHLY AVERAGE ABSORBED RADIATION

Methods for the evaluation of long-term solar system performance[4] require that the average radiation absorbed by a collector be evaluated for monthly periods. The solar transmittance and absorptance are both functions of the angle at which solar radiation is incident on the collector. Example 5.9.1 illustrated how to calculate the absorbed solar radiation for an hour. This calculation can be repeated for each hour of each day of the month, from which the monthly average absorbed solar radiation can be found. Klein (1979) calculated the monthly average absorbed solar radiation in this manner using many years of data. He defined a monthly average transmittance-absorptance product,[5] which when multiplied by the monthly average radiation incident on a collector yields the monthly average absorbed radiation $\overline{S}$:

$$\left(\overline{\tau\alpha}\right) = \frac{\overline{S}}{\overline{H}_T} = \frac{\overline{S}}{\overline{H}\ \overline{R}} \qquad (5.10.1)$$

The following methods, analogous to the hourly evaluations of S, can be used to find $\overline{S}$.

Using the isotropic diffuse assumption, Equation 2.19.1 becomes

$$\overline{S} = \overline{H}_b\overline{R}_b\left(\overline{\tau\alpha}\right)_b + \overline{H}_d\left(\overline{\tau\alpha}\right)_d\left(\frac{1 + \cos\beta}{2}\right) + \overline{H}\rho_g\left(\overline{\tau\alpha}\right)_g\left(\frac{1 - \cos\beta}{2}\right) \qquad (5.10.2)$$

For the diffuse and ground-reflected terms, $\left(\overline{\tau\alpha}\right)_d$ and $\left(\overline{\tau\alpha}\right)_g$ can be evaluated using the effective incidence angles given in Figure 5.4.1. These are functions of the properties of the cover and absorber and β, the collector slope, and so do not change with time for collectors mounted at fixed β. The hourly and monthly values are thus the same, and they can be written with or without the overbars.

For the monthly average beam radiation, Klein (1979) has worked out the monthly average (equivalent) beam incident angle $\overline{\theta}_b$ as a function of collector slope, month, latitude, and azimuth angle. These are shown in Figures 5.10.1(a-f). These values of $\overline{\theta}_b$ were evaluated using the angular distribution of $(\tau\alpha)/(\tau\alpha)_n$ shown in Figure 5.6.1.

The Klein and Theilacker equations can also be used to calculate $\overline{S}$, the product $\overline{R}\ \overline{H}$. Each of the $\overline{R}$ equations 2.20.4a and 2.20.5a includes three terms: the first is multiplied by $\left(\overline{\tau\alpha}\right)_b$, the second by $\left(\overline{\tau\alpha}\right)_d$, and the third by $\left(\overline{\tau\alpha}\right)_g$, as was shown in Equation 5.10.2. For surfaces with surface azimuth angles other than zero (or 180° in the southern hemisphere), the use of the modified Equation 2.20.5a is recommended.

Calculation of $\overline{S}$ is illustrated in the following example.

---

[4] See Chapters 20 to 22.
[5] This could be designated as $\left(\overline{\tau\alpha}\right)_{av}$, as it is a time-weighted and energy-weighted average Common usage is to designate it as $\left(\overline{\tau\alpha}\right)$.

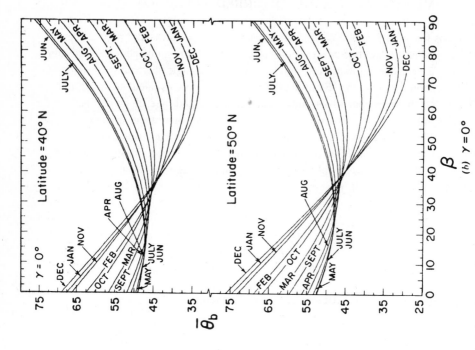

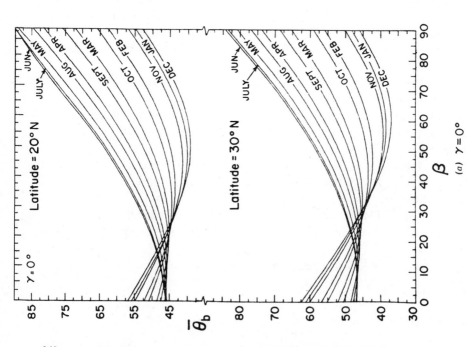

Figure 5.10.1 Monthly average beam incidence angle for various surface locations and orientations. For southern hemisphere interchange months as shown in Figure 1.8.2. From Klein (1979).

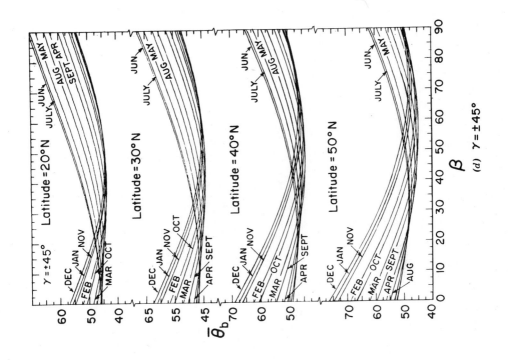

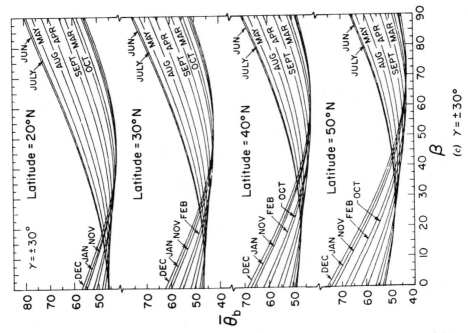

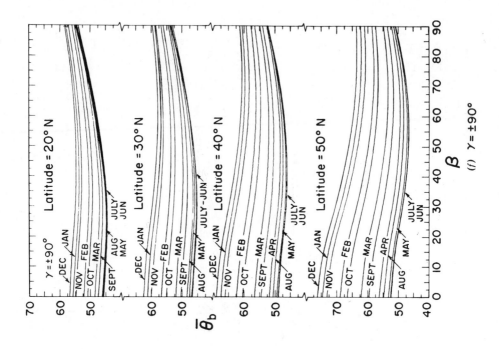

(e) $\gamma = \pm 60°$

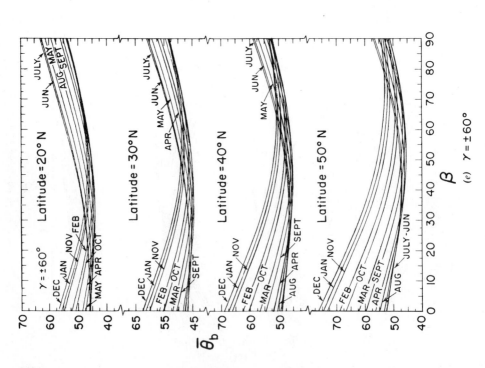

(f) $\gamma = \pm 90°$

Example 5.10.1

Estimate $\overline{S}$ for a south-facing vertical collector-storage wall at Springfield, IL, 40°N latitude. The wall consists of double glazing with a black-finished absorbing surface behind the glass. The monthly average daily radiation on a horizontal surface $\overline{H}$, in MJ/m², is shown in the table that follows. The ground reflectance is assumed to be 0.3 for all months. The angular dependence of $(\tau\alpha)$ for the two-cover glazing is as shown in Figure 5.6.1. The glass has $KL = 0.0125$ per sheet. Calculate the monthly radiation on the wall $\overline{H}_T$, the monthly absorbed radiation $\overline{S}$, and the monthly average $\overline{(\tau\alpha)}$.

Solution

The calculations are shown in detail for January, and the results for all months are shown in the table.

For these two covers, from Figure 5.3.1 at normal incidence, the transmittance is 0.83. With the absorber normal incidence absorptance of 0.90, $(\tau\alpha)_n = 1.01 \times 0.83 \times 0.90 = 0.754$. For the vertical collector the effective incidence angle of both the diffuse and the ground-reflected radiation is 59° from Figure 5.4.1. From Figure 5.6.1 at 59°, $(\tau\alpha)/(\tau\alpha)_n$ is 0.83 so that $(\tau\alpha)_d = (\tau\alpha)_g = 0.83 \times 0.754 = 0.626$. These values apply to all months.

For January, from Figure 5.10.1b, at 40° latitude and 90° slope, $\overline{\theta}_b$ is 41°. From Figure 5.6.1, $(\tau\alpha)_b/(\tau\alpha)_n = 0.96$ for the two-cover system. Thus $(\tau\alpha)_b = 0.96 \times 0.754 = 0.724$.

For January, $\overline{H}_o = 15.21$ MJ/m², so $\overline{K}_T = 6.63/15.21 = 0.436$. For the mean day of the month, from Table 1.6.1, $\delta = -20.9°$. Thus

$$\omega_s = \cos^{-1}(-\tan(-20.9)\tan 40) = 71.3°$$

Equation 2.12.1 is used to calculate the diffuse fraction. For January this gives $\overline{H}_d/\overline{H} = 0.458$.

Then $\qquad \overline{H}_d = 6.63 \times 0.458 = 3.04$ MJ/m²

and $\qquad \overline{H}_b = 3.59$ MJ/m²

Use Equation 2.19.3 to calculate $\overline{R}_b$. This results in $\overline{R}_b = 2.32$.

We can now calculate $\overline{H}_T$ with Equation 2.19.1 based on the isotropic diffuse assumption:

$$\overline{H}_T = 3.59 \times 2.32 + 3.04\left(\frac{1 + \cos 90}{2}\right) + 6.63 \times 0.3\left(\frac{1 - \cos 90}{2}\right)$$

$$= 8.33 + 1.52 + 0.99 = 10.84 \text{ MJ/m}^2$$

With the transmittance-absorptance products determined above, again using the isotropic assumption (Equation 5.10.2),

$$\overline{S} = 3.59 \times 2.32 \times 0.724 + 3.04 \times 0.626\left(\frac{1 + \cos 90}{2}\right)$$

$$+ 6.63 \times 0.3 \times 0.626\left(\frac{1 - \cos 90}{2}\right)$$

$$= 6.03 + 0.95 + 0.62 = 7.60 \text{ MJ/m}^2$$

The average transmittance-absorptance product for the month is then

$$\overline{(\tau\alpha)} = \overline{S}/\overline{H}_T = 7.60/10.84 = 0.70$$

The monthly results are as follows:

| Month | $\overline{H}$ | $\overline{K}_T$ | $\overline{H}_T$ | $\overline{(\tau\alpha)}_b$ | Absorbed Radiation, MJ/m^2 | | | | |
					Beam	Diffuse	Gr. Refl.	$\overline{S}$	$\overline{(\tau\alpha)}$
Jan.	6.63	0.44	10.84	0.72	6.03	0.95	0.62	7.60	0.70
Feb.	9.77	0.48	12.59	0.69	6.28	1.27	0.92	8.47	0.67
Mar.	12.97	0.47	11.65	0.63	4.25	1.85	1.22	7.32	0.63
Apr.	17.20	0.50	11.03	0.51	2.41	2.33	1.62	6.36	0.58
May	21.17	0.53	10.59	0.38	1.22	2.64	1.99	5.84	0.55
June	23.80	0.57	10.52	0.26	0.68	2.72	2.23	5.63	0.54
July	23.36	0.57	10.79	0.27	0.83	2.64	2.19	5.66	0.53
Aug.	20.50	0.56	11.69	0.44	2.11	2.39	1.92	6.42	0.55
Sep.	16.50	0.55	13.18	0.59	4.45	1.98	1.55	7.98	0.61
Oct.	12.13	0.54	14.23	0.67	6.72	1.49	1.14	9.35	0.66
Nov.	7.68	0.47	12.09	0.72	6.70	1.02	0.72	8.44	0.70
Dec.	5.57	0.40	9.46	0.72	5.22	0.86	0.52	6.61	0.70

The Klein and Theilacker equations could have been used for this calculation. ■

For collectors that face the equator, Klein (1976) found that $\overline{(\tau\alpha)}_b$ could be approximated by $\overline{(\tau\alpha)}$ evaluated at the incidence angle that occurs 2.5 hours from solar noon on the average day of the month. This angle can be calculated from Equation 1.6.7a (or Equation 1.6.7b for the southern hemisphere) or obtained from Figure 5.10.2. This rule leads to acceptable results for solar space heating systems for which it was derived, but inaccurate results are obtained for other types of systems. Klein also found that the value of $\overline{(\tau\alpha)}/(\tau\alpha)_n$ during the winter months is nearly constant and equal to 0.96 for a one-cover collector and suggests using this constant value for collectors tilted toward the equator with a slope approximately equal to the latitude plus 15° in heating system analysis. For two-cover collectors a constant value of 0.94 was suggested.

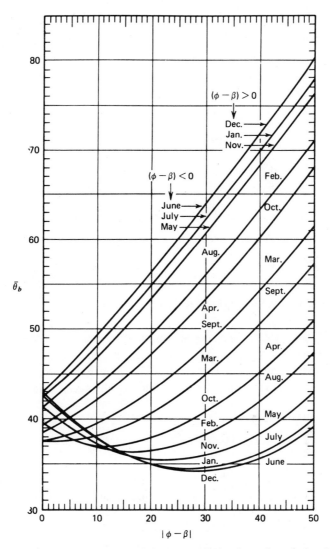

Figure 5.10.2 Monthly mean incidence angle for beam radiation for surfaces facing the equator in the northern hemisphere. For the southern hemisphere, interchange the two inequality signs. From Klein (1979).

It is useful to be able to calculate $(\overline{\tau\alpha})$ from $(\tau\alpha)_n$ and the information on $(\tau\alpha)/(\tau\alpha)_n$ in Figure 5.6.1, where $(\overline{\tau\alpha})$ is defined by Equation 5.10.1. Dividing by $(\tau\alpha)_n$, we have

$$\frac{(\overline{\tau\alpha})}{(\tau\alpha)_n} = \frac{\overline{S}}{\overline{H}_T(\tau\alpha)_n} \tag{5.10.3}$$

The appropriate equation for $\overline{S}$ is substituted in this relationship to provide a useful working equation. If the isotropic model is used, Equation 5.10.3 becomes

$$\frac{(\overline{\tau\alpha})}{(\tau\alpha)_n} = \frac{\overline{H}_b \overline{R}_b}{\overline{H}_T} \frac{(\tau\alpha)_b}{(\tau\alpha)_n} + \frac{\overline{H}_d}{\overline{H}_T} \frac{(\tau\alpha)_d}{(\tau\alpha)_n} \left(\frac{1 + \cos\beta}{2}\right)$$
$$+ \frac{\overline{H} \rho_g}{\overline{H}_T} \frac{(\tau\alpha)_g}{(\tau\alpha)_n} \left(\frac{1 - \cos\beta}{2}\right)$$

(5.10.4)

In Equation 5.10.4, the $(\overline{\tau\alpha})/(\tau\alpha)_n$ ratios are obtained from Figure 5.6.1 for the beam component at the effective angle of incidence θ_b from Figure 5.10.1 and for the diffuse and ground-reflected components at the effective angles of incidence at β from Figure 5.3.1.

The Klein and Theilacker equations can be used in a similar manner.

5.11   ABSORPTANCE OF ROOMS

Direct-gain passive solar heating depends on absorption of solar radiation in rooms or sunspaces which are cavity receivers with apertures (windows) covered with one or more glazings. Equation 4.12.1 can be modified to give the fraction of the incident solar energy that is absorbed by such a receiver for radiation incident on the glazing.

$$\tau_c \alpha_{\text{eff}} = \tau_c \frac{\alpha_i}{\alpha_i + (1 - \alpha_i)\tau_d \dfrac{A_a}{A_i}}$$

(5.11.1)

Here τ_c is the transmittance of the glazing for the incident solar radiation; τ_d is the transmittance of the glazing for isotropic diffuse solar radiation (the solar radiation reflected from the inner walls of the cavity), which is at an effective angle of incidence of about $60°$; A_a is the area of the aperture (the window); A_i is the area of the inside of the room; and α_i is the absorptance for diffuse radiation of the inner surface of the cavity. A room may have various surfaces on floor, walls, ceiling, and furnishings, and a mean value of α_i can be used.

Example 5.11.1

Calculate $\tau_c \alpha_{\text{eff}}$, the fraction of solar radiation incident on a window which is absorbed in a room that has dimensions $5 \times 4 \times 2.5$ m. The double glazed window is 1.5×3 m. The mean absorptance of the surfaces in the room is 0.45. The transmittance of the glazing for incident radiation τ_c is 0.87. The glass has $KL = 0.0125$ per glazing.

Solution

The area of the window, A_a, is $1.5 \times 3 = 4.5$ m^2. The area of the room A_i (not including the window) is $2(5 \times 4 + 5 \times 2.5 + 4 \times 2.5) - 4.5 = 80.5$ m^2. From Figure 5.3.1, the transmittance of the glazing for diffuse radiation τ_d at an effective angle of incidence of 59° is 0.74. Therefore,

$$\tau_c \alpha_{\text{eff}} = 0.87 \frac{0.45}{0.45 + (1 - 0.45)0.74(4.5/80.5)} = 0.87 \times 0.95 = 0.83$$

Thus this room will absorb 0.83 of the incident solar radiation (and 0.95 of the solar energy that is transmitted into it by the glazing). ■

Example 5.11.2

A direct-gain passive heating system is to be located in Springfield, IL ($\phi = 40°$). The receiver (the window) and the space in which solar radiation is to be absorbed have dimensions and characteristics described in the previous example (i.e., an effective absorptance of 0.95). For January, calculate $\overline{S}$, the absorbed radiation per unit area of window, if the window is not shaded. (This problem is the same as Example 5.10.1, except that the energy is absorbed in the room rather than on the black surface of an absorbing wall behind the glazing.)

Solution

For the cavity receiver the calculations are similar to those of Example 5.10.1, but the absorptance is constant and the transmittance-absorptance product is given by $\tau\alpha$. (The correction factor of 1.01 from Equation 5.5.2 is not used, as the radiation reflected back into the cavity from the cover is accounted for in the calculation of α.) From Example 5.10.1, the month's average beam radiation incidence angle is 41° and the mean incidence angle of both the diffuse and the ground-reflected radiation is 59°.

For the beam radiation, from Figure 5.3.1 at $\theta = 41°$, the transmittance is 0.82. Then

$$\left(\overline{\tau\alpha}\right)_b = 0.82 \times 0.95 = 0.78$$

For the diffuse and ground-reflected radiation, from Figure 5.3.1 at $\theta = 59°$, the transmittance is 0.74, and

$$\left(\tau\alpha\right)_d = \left(\tau\alpha\right)_g = 0.74 \times 0.95 = 0.70$$

Again assuming isotropic diffuse and using monthly average radiation calculations from Example 5.10.1,

$$\overline{S} = 3.57 \times 2.32 \times 0.78 + 3.04 \times 0.70 \left(\frac{1 + \cos 90}{2} \right)$$

$$+ 6.63 \times 0.7 \times 0.70 \left(\frac{1 - \cos 90}{2} \right)$$

$$= 6.46 + 1.06 + 1.62 = 9.14 \text{ MJ/m}^2$$

We can calculate a month's average $(\overline{\tau\alpha})$ as $\overline{S} / \overline{H}_T$. Thus for January, with $\overline{H}_T = 12.2 \text{ MJ/m}^2$,

$$(\overline{\tau\alpha}) = 9.20/12.2 = 0.75$$

So 75% of the radiation incident on this window in January is absorbed in this room. Calculations for other months are done the same way as those for January. ■

Examples 5.10.1 and 5.11.2 will be the basis of passive system performance calculations to be shown in Chapter 22. Many receivers (windows) of passive heating systems are partially shaded by overhangs or wingwalls, and these shading devices must be taken into account in estimating the incident radiation.

5.12   SUMMARY

This chapter includes several alternative methods for calculating important parameters. For purposes of calculating radiation absorbed on surfaces that face toward the equator or nearly so, it is suggested that the hourly absorbed radiation is adequately estimated by Equation 5.9.1 and the monthly absorbed radiation by Equation 5.10.2 (both based on the isotropic model). (The HDKR model for S is almost as easy to use as the isotropic, leads to less conservative estimates of S, and is a useful alternative.) For surfaces that face other than toward the equator (e.g., for calculating winter radiation on north-facing windows), anisotropic models should be used.

REFERENCES

Brandemuehl, M. J. and W. A. Beckman, *Solar Energy*, **24**, 511 (1980). "Transmission of Diffuse Radiation Through CPC and Flat-Plate Collector Glazings."

Dietz, A. G. H., in *Space Heating with Solar Energy* (R. W. Hamilton, ed.), Massachusetts Institute of Technology Press, Cambridge, MA (1954). "Diathermanous Materials and Properties of Surfaces."

Dietz, A. G. H., in *Introduction to the Utilization of Solar Energy* (A. M. Zarem and D. D. Erway, eds.), McGraw-Hill, New York (1963). "Diathermanous Materials and Properties of Surfaces."

Edwards, D. K., *Solar Energy*, **19**, 401 (1977). "Solar Absorption of Each Element in an Absorber-Coverglass Array."

Hottel, H. C. and B. B. Woertz, *Trans. ASME*, **64**, 91 (1942). "The Performance of Flat-Plate Solar-Heat Collectors."

Klein, S. A., Ph.D. Thesis, University of Wisconsin - Madison (1976). "A Design Procedure for Solar Heating Systems."

Klein, S. A., *Solar Energy*, **23**, 547 (1979). "Calculation of the Monthly-Average Transmittance-Absorptance Product."

Mar, H. Y. B., J. H. Lin, P. P. Zimmer, R. E. Peterson, and J. S. Gross, Report to ERDA under Contract No. NSF-C-957 (Sep. 1975). "Optical Coatings for Flat-Plate Solar Collectors."

Siegel, R. and J. R. Howell. *Thermal Radiation Heat Transfer* 2nd ed. McGraw-Hill, New York (1981).

Thomsen, S. M., *RCA Review*, **12**, 143 (1951). "Low-Reflection Films Produced on Glass in a Liquid Fluosilicic Acid Bath."

Whillier, A., ScD. Thesis, MIT, 1953. "Solar Energy Collection and Its Utilization for House Heating."

Whillier, A., *Solar Energy*, **7**, 148 (1963). "Plastic Covers for Solar Collectors."

Chapter 6

FLAT-PLATE COLLECTORS

A solar collector is a special kind of heat exchanger that transforms solar radiant energy into heat. A solar collector differs in several respects from more conventional heat exchangers. The latter usually accomplish a fluid-to-fluid exchange with high heat transfer rates and with radiation as an unimportant factor. In the solar collector, energy transfer is from a distant source of radiant energy to a fluid. The flux of incident radiation is, at best, approximately 1100 W/m^2 (without optical concentration), and it is variable. The wavelength range is from 0.3 to 3 μm, which is considerably shorter than that of the emitted radiation from most energy absorbing surfaces. Thus, the analysis of solar collectors presents unique problems of low and variable energy fluxes and the relatively large importance of radiation.

Flat-plate collectors can be designed for applications requiring energy delivery at moderate temperatures, up to perhaps 100 C above ambient temperature. They use both beam and diffuse solar radiation, do not require tracking of the sun, and require little maintenance. They are mechanically simpler than concentrating collectors. The major applications of these units are in solar water heating, building heating, air conditioning, and industrial process heat. Passively heated buildings can be viewed as special cases of flat-plate collectors with the room or storage wall as the absorber. Passive systems are discussed in Chapter 14.

The importance of flat-plate collectors in thermal processes is such that their thermal performance is treated in considerable detail. This is done to develop an understanding of how the component functions. In many practical cases of design calculations, the equations for collector performance are reduced to relatively simple form.

The last sections of this chapter treat testing of collectors, the use of test data, and some practical aspects of manufacture and use of these heat exchangers. Costs will be considered in chapters on applications.

6.1 DESCRIPTION OF FLAT-PLATE COLLECTORS

The important parts of a typical liquid heating flat-plate solar collector, as shown in Figure 6.1.1, are: the "black" solar energy-absorbing surface with means for

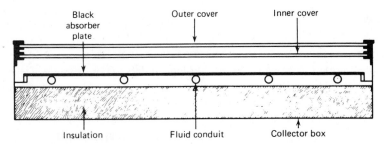

Figure 6.1.1 Cross section of a basic flat-plate solar collector.

transferring the absorbed energy to a fluid; envelopes transparent to solar radiation over the solar absorber surface that reduce convection and radiation losses to the atmosphere; and back insulation to reduce conduction losses. Figure 6.1.1 depicts a water heater, and most of the analysis of this chapter is concerned with this geometry. Air heaters are fundamentally the same except that the fluid tubes are replaced by ducts. Flat-plate collectors are almost always mounted in a stationary position (e.g., as an integral part of a wall or roof structure) with an orientation optimized for the particular location in question for the time of year in which the solar device is intended to operate.

6.2   BASIC FLAT-PLATE ENERGY BALANCE EQUATION

In steady state, the performance of a solar collector is described by an energy balance that indicates the distribution of incident solar energy into useful energy gain, thermal losses, and optical losses. The solar radiation absorbed by a collector per unit area of absorber S is equal to the difference between the incident solar radiation and the optical losses as defined by Equation 5.9.1. The thermal energy lost from the collector to the surroundings by conduction, convection, and infrared radiation can be represented as the product of a heat transfer coefficient U_L times the difference between the mean absorber plate temperature T_{pm} and the ambient temperature T_a. In steady state the useful energy output of a collector of area A_c is the difference between the absorbed solar radiation and the thermal loss:

$$Q_u = A_c \left[S - U_L \left(T_{pm} - T_a \right) \right] \tag{6.2.1}$$

The problem with this equation is that the mean absorber plate temperature is difficult to calculate or measure since it is a function of the collector design, the incident solar radiation, and the entering fluid conditions. Part of this chapter is devoted to reformulating Equation 6.2.1 so that the useful energy gain can be expressed in terms of the inlet fluid temperature and a parameter called the collector heat removal factor, which can be evaluated analytically from basic principles or measured experimentally.

Equation 6.2.1 is an energy rate equation and, in SI units, yields the useful energy gain in watts (J/s) when S is expressed in W/m^2 and U_L in W/m^2K. The most convenient time base for solar radiation is hours rather than seconds since this is the normal period for reporting of meteorological data. (For example, Table 2.5.2 gives solar radiation in J/m^2 for 1-hour time periods.) This is the time basis for S in Equation 5.9.1 since the meaning of I is hourly J/m^2. We can consider S to be an average energy rate over a 1-hour period with units of J/m^2hr, in which case the thermal loss term $U_L(T_{pm} - T_a)$ must be multiplied by 3600 s/hr to obtain numerical values of the useful energy gain in J/hr. The hour time base is not a proper use of SI units, but this interpretation is often convenient. Alternatively, we can integrate Equation 6.2.1 over a 1-hour period. Since we seldom have data over time periods less than 1 hour, this integration can be performed only by assuming that S, T_{pm}, and T_a remain constant over the hour. The resulting form of Equation 6.2.1 is unchanged except that both sides are multiplied by 3600 s/hr. To avoid including this constant in expressions for useful energy gain on an hourly basis, we could have used different symbols for rates and for hourly integrated quantities (e.g., $\dot{Q}_u$ and Q_u). However, the intended meaning is always clear from the use of either G or I in the evaluation of S, and we have found it unnecessary to use different symbols for collector useful energy gain on an instantaneous basis or an hourly integrated basis. From a calculation standpoint the 3600 must still be included since S will be known for an hour time period but the loss coefficient will be in SI units and thus on a second basis.

A measure of collector performance is the collection efficiency, defined as the ratio of the useful gain over some specified time period to the incident solar energy over the same time period:

$$\eta = \frac{\int Q_u \, dt}{A_c \int G_T \, dt} \tag{6.2.2}$$

The design of a solar energy system is concerned with obtaining minimum cost energy. Thus, it may be desirable to design a collector with an efficiency lower than is technologically possible if the cost is significantly reduced. In any event, it is necessary to be able to predict the performance of a collector, and that is the basic aim of this chapter.

6.3 TEMPERATURE DISTRIBUTIONS IN FLAT-PLATE COLLECTORS

The detailed analysis of a solar collector is a complicated problem. Fortunately, a relatively simple analysis will yield very useful results. These results show the important variables, how they are related, and how they affect the performance of a solar collector. To illustrate these basic principles, a liquid heating collector, as

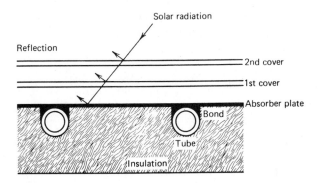

Figure 6.3.1 Sheet and tube solar collector.

shown in Figure 6.3.1, will be examined first. The analysis presented follows the basic derivation by Whillier (1953, 1977) and Hottel and Whillier (1958).

To appreciate the development that follows, it is desirable to have an understanding of the temperature distribution that exists in a solar collector constructed as shown in Figure 6.3.1. Figure 6.3.2(a) shows a region between two tubes. Some of the solar energy absorbed by the plate must be conducted along the plate to the region of the tubes. Thus the temperature midway between the tubes will be higher than the temperature in the vicinity of the tubes. The temperature above the tubes will be nearly uniform because of the presence of the tube and weld metal.

The energy transferred to the fluid will heat the fluid, causing a temperature gradient to exist in the direction of flow. Since in any region of the collector the general temperature level is governed by the local temperature level of the fluid, a situation as shown in Figure 6.3.2(b) is expected. At any location y, the general temperature distribution in the x direction is as shown in Figure 6.3.2(c) and at any location x, the temperature distribution in the y direction will look like Figure 6.3.2(d).

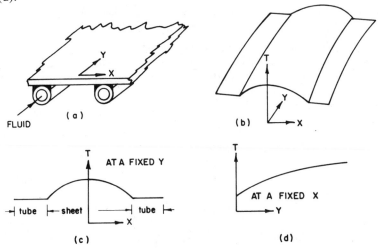

Figure 6.3.2 Temperature distribution on an absorber plate. From Duffie and Beckman (1974).

To model the situation shown in Figure 6.3.2, a number of simplifying assumptions can be made to lay the foundations without obscuring the basic physical situation. These assumptions are as follows:

1   Performance is steady state.
2   Construction is of sheet and parallel tube type.
3   The headers cover a small area of collector and can be neglected.
4   The headers provide uniform flow to tubes.
5   There is no absorption of solar energy by a cover insofar as it affects losses from the collector.
6   Heat flow through a cover is one dimensional.
7   There is a negligible temperature drop through a cover.
8   The covers are opaque to infrared radiation.
9   There is one-dimensional heat flow through back insulation.
10 The sky can be considered as a blackbody for long-wavelength radiation at an equivalent sky temperature.
11 Temperature gradients around tubes can be neglected.
12 The temperature gradients in the direction of flow and between the tubes can be treated independently.
13 Properties are independent of temperature.
14 Loss through front and back are to the same ambient temperature.
15 Dust and dirt on the collector are negligible.
16 Shading of the collector absorber plate is negligible.

In later sections of this chapter many of these assumptions will be relaxed.

6.4   COLLECTOR OVERALL HEAT LOSS COEFFICIENT

It is useful to develop the concept of an overall loss coefficient for a solar collector to simplify the mathematics. Consider the thermal network for a two-cover system shown in Figure 6.4.1. At some typical location on the plate where the temperature is T_p, solar energy of amount S is absorbed by the plate, where S is equal to the incident solar radiation reduced by optical losses as shown in Section 5.9. This absorbed energy S is distributed to thermal losses through the top and bottom and to useful energy gain. The purpose of this section is to convert the thermal network of Figure 6.4.1 to the thermal network of Figure 6.4.2.

The energy loss through the top is the result of convection and radiation between parallel plates. The steady-state energy transfer between the plate at T_p and the first cover at T_{c1} is the same as between any other two adjacent covers and is also equal to the energy lost to the surroundings from the top cover. The loss through the top per unit area is then equal to the heat transfer from the absorber plate to the first cover:

$$q_{\text{loss, top}} = h_{c,p\text{-}c1}(T_p - T_{c1}) + \frac{\sigma\left(T_p^4 - T_{c1}^4\right)}{\dfrac{1}{\varepsilon_p} + \dfrac{1}{\varepsilon_{c1}} - 1} \tag{6.4.1}$$

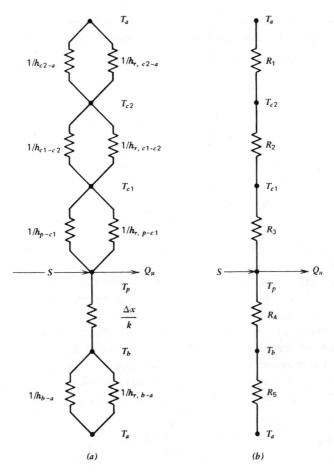

Figure 6.4.1 Thermal network for a two-cover flat-plate collector: (a) in terms of conduction, convection, and radiation resistances; (b) in terms of resistances between plates.

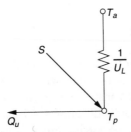

Figure 6.4.2 Equivalent thermal network for flat-plate solar collector.

where $h_{c,p-c1}$ is the convection heat transfer coefficient between two inclined parallel plates from Chapter 3. If the definition of the radiation heat transfer coefficient (Equation 3.10.1) is used, the heat loss becomes

$$q_{\text{loss, top}} = (h_{c,p-c1} + h_{r,p-c1})(T_p - T_{c1}) \tag{6.4.2}$$

where

$$h_{r,p-c1} = \frac{\sigma(T_p + T_{c1})(T_p^2 + T_{c1}^2)}{\dfrac{1}{\varepsilon_p} + \dfrac{1}{\varepsilon_{c1}} - 1} \tag{6.4.3}$$

The resistance R_3 can then be expressed as

$$R_3 = \frac{1}{h_{c,p-c1} + h_{r,p-c1}} \tag{6.4.4}$$

A similar expression can be written for R_2, the resistance between the covers. In general, we can have as many covers as desired, but the practical limit is two and most collectors use one.

The resistance from the top cover to the surroundings has the same form as Equation 6.4.4, but the convection heat transfer coefficient h_w is given in Section 3.15. The radiation resistance from the top cover accounts for radiation exchange with the sky at T_s. For convenience, we reference this resistance to the ambient temperature T_a, so that the radiation heat transfer coefficient can be written as

$$h_{r,c2-a} = \frac{\sigma\varepsilon_c(T_{c2} + T_s)(T_{c2}^2 + T_s^2)(T_{c2} - T_s)}{(T_{c2} - T_a)} \tag{6.4.5}$$

The resistance to the surroundings R_1 is then given by

$$R_1 = \frac{1}{h_w + h_{r,c2-a}} \tag{6.4.6}$$

For this two-cover system, the top loss coefficient from the collector plate to the ambient is

$$U_t = \frac{1}{R_1 + R_2 + R_3} \tag{6.4.7}$$

The procedure for solving for the top loss coefficient is necessarily an iterative process. First a guess is made of the unknown cover temperatures, from which the convective and radiative heat transfer coefficients between parallel surfaces are calculated. With these estimates, Equation 6.4.7 can be solved for the top loss coefficient. The top heat loss is the top loss coefficient times the overall temperature difference, and since the energy exchange between plates must be equal to the overall heat loss, a new set of cover temperatures can be calculated. Beginning at the absorber plate, a new temperature is calculated for the first cover. This new first cover temperature is used to find the next cover temperature, and so on. For any two adjacent plates, the new temperature of plate j can be expressed in terms of the temperature of plate i as

$$T_j = T_i - \frac{U_t(T_p - T_a)}{h_{c,i\text{-}j} + h_{r,i\text{-}j}} \qquad (6.4.8)$$

The process is repeated until the cover temperatures do not change significantly between successive iterations. The following example illustrates the process.

Example 6.4.1

Calculate the top loss coefficient for an absorber with a single glass cover having the following specifications:

Plate-to-cover spacing	25 mm
Plate emittance	0.95
Ambient air and sky temperature	10 C
Wind heat transfer coefficient	10 W/m²C
Mean plate temperature	100 C
Collector tilt	45°
Glass emittance	0.88

Solution

For this single glass cover system, Equation 6.4.7 becomes

$$U_t = \left(\frac{1}{h_{c,p\text{-}c} + h_{r,p\text{-}c}} + \frac{1}{h_w + h_{r,c\text{-}a}} \right)^{-1}$$

The convection coefficient between the plate and the cover $h_{c,p\text{-}c}$ can be found using the methods of Section 3.11. The radiation coefficient from the plate to the cover $h_{r,p\text{-}c}$ is

$$h_{r,p\text{-}c} = \frac{\sigma(T_p^2 + T_c^2)(T_p + T_c)}{\dfrac{1}{\varepsilon_p} + \dfrac{1}{\varepsilon_c} - 1}$$

The radiation coefficient for the cover to the air $h_{r,c\text{-}a}$ is

$$h_{r,c\text{-}a} = \varepsilon_c \sigma(T_c^2 + T_s^2)(T_c + T_s)$$

The equation for the cover glass temperature is based on Equation 6.4.8:

$$T_c = T_p - \frac{U_t(T_p - T_a)}{h_{c,p\text{-}c} + h_{r,p\text{-}c}}$$

The procedure is to estimate the cover temperature, from which $h_{c,p-c}$, $h_{r,p-c}$, and $h_{r,c-a}$ are calculated. With these heat transfer coefficients and h_w, the top loss coefficient is calculated. These results are then used to calculate T_c from the preceding equation. If T_c is close to the initial guess, no further calculations are necessary. Otherwise, the newly calculated T_c is used and the process is repeated.

With an assumed value of the cover temperature of 35 C, the two radiation coefficients become

$$h_{r,p-c} = 7.60 \text{ W/m}^2\text{C}$$
$$h_{r,c-a} = 5.16 \text{ W/m}^2\text{C}$$

Equation 3.11.4 is used to calculate the convection coefficient between the plate and the cover. The mean temperature between the plate and the cover is 67.5 C so the air properties are $v = 1.96 \times 10^{-5}$ m^2/s, $k = 0.0293$ W/mC, $T = 340.5$ K, and $Pr = 0.7$. The Rayleigh number is

$$Ra = \frac{9.81(100 - 35)(0.025)^3(0.7)}{340.5(1.96 \times 10^{-5})^2} = 5.33 \times 10^4$$

and from Equation 3.11.4 the Nusselt number is 3.19. The convective heat transfer coefficient is

$$h = Nu\frac{k}{L} = 3.19\frac{0.0293}{0.025} = 3.73 \text{ W/m}^2\text{C}$$

(The same result is obtained from Figures 3.11.1 and 3.11.2. From Figure 3.11.2, $F_1 = 0.46$ and $F_2 = 0.84$. The value of $F_1\Delta Tl^3$ is 4.7×10^5. From Figure 3.11.1, F_2hl is 78. Then h is 3.7 W/m^{2}C.) The first estimate of U_t is then

$$U_t = \left(\frac{1}{3.73 + 7.60} + \frac{1}{5.16 + 10.0}\right)^{-1} = 6.49 \text{ W/m}^2\text{C}$$

The cover temperature is

$$T_c = 100 - \frac{6.49(90)}{3.73 + 7.60} = 48.5 \text{ C}$$

With this new estimate of the cover temperature, the various heat transfer coefficients become

$$h_{r,p-c} = 8.03 \text{ W/m}^2 \text{ C}$$
$$h_{r,c-a} = 5.53 \text{ W/m}^2 \text{ C}$$
$$h_{c,p-c} = 3.52 \text{ W/m}^2 \text{ C}$$

and the second estimate of U_t is 6.62 W/m² C. When the cover glass temperature is calculated with this new top loss coefficient, it is found to be 48.4 C, which is essentially equal to the estimate of 48.5 C. ∎

The results of heat loss calculations for four different solar collectors, all with the same plate and ambient temperatures, are shown in Figure 6.4.3. The cover temperatures and the heat flux by convection and radiation are shown for one and two glass covers and for selective and nonselective absorber plates. Note that radiation between plates is the dominant mode of heat transfer in the absence of a selective surface. When a selective surface having an emittance of 0.10 is used, convection is the dominant heat transfer mode between the selective surface and the cover, but radiation is still the largest term between the two cover glasses in the two-cover system.

For most conditions the use of a blackbody radiation sky temperature that is not equal to the air temperature will not greatly affect the top loss coefficient or the top heat loss. For example, the top loss coefficient based on the plate-to-ambient temperature difference for condition (a) of Figure 6.4.3 is increased from 6.62 to 6.76 W/m²C when the sky temperature is reduced from 10 C to 0 C. For condition (b) the top loss coefficient is increased from 3.58 to 3.67 W/m²C.

As illustrated by Example 6.4.1, the calculation of the top loss coefficient is a tedious process. To simplify calculations of collector performance, Figures

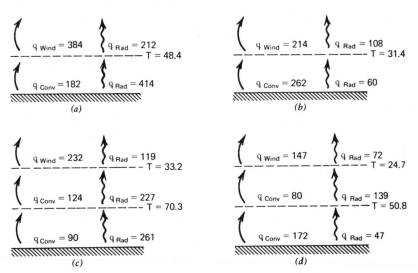

Figure 6.4.3 Cover temperature and upward heat loss for flat-plate collectors operating at 100 C with ambient and sky temperature of 10 C, plate spacing of 25 mm, tilt of 45°, and wind heat transfer coefficient of 10 W/m²C: (a) one cover, plate emittance = 0.95, U_t = 6.6 W/m²C; (b) one cover, plate emittance = 0.10, U_t = 3.6 W/m²C; (c) two covers, plate emittance = 0.95, U_t = 3.9 W/m²C; (d) two covers, plate emittance = 0.10, U_t = 2.4 W/m²C. All heat flux terms are in W/m².

6.4.4(a–f) have been prepared. These figures give the top loss coefficient for one, two, and three glass covers spaced 25 mm apart; ambient temperatures of 40, 10, and –20 C; wind heat transfer coefficients of 5, 10, and 20 W/m²C; plates having an emittance of 0.95 and 0.10; a slope of 45°; and a range of plate temperatures.

Even though the top loss coefficients of Figures 6.4.4 are for a plate spacing of 25 mm, they can be used for other plate spacings with little error as long as the spacing is greater than about 15 mm. Figure 6.4.5 illustrates the dependence of the top loss coefficient on plate spacing for selective and nonselective one- and two-cover collectors. For very small plate spacings convection is suppressed and the heat transfer mechanism through the gap is by conduction and radiation. In this range the top loss coefficient decreases rapidly as the plate spacing increases until a minimum is reached at about 10- to 15-mm plate spacing. When fluid motion first begins to contribute to the heat transfer process, the top loss coefficient increases until a maximum is reached at approximately 20 mm. Further increase in the plate spacing causes a small reduction in the top loss coefficient. Similar behavior occurs at other conditions and collector designs.

Figures 6.4.4 were prepared using a slope β of 45°. In Figure 6.4.6 the ratio of the top loss coefficient at any tilt angle to that of 45° has been plotted as a function of slope.

The graphs for U_t are convenient for hand calculations but they are difficult to use on computers. The solution to the set of equations, as was done in Example 6.4.1, is time consuming even on a computer since many thousands of solutions may be required. An empirical equation for U_t that is useful for both hand and computer calculations was developed by Klein (1979) following the basic procedure of Hottel and Woertz (1942) and Klein (1975). This relationship fits the graphs for U_t for mean plate temperatures[1] between ambient and 200 C to within ±0.3 W/m²C:

$$U_t = \left\{ \frac{N}{\dfrac{C}{T_{pm}}\left[\dfrac{(T_{pm} - T_a)}{(N+f)} \right]^e} + \frac{1}{h_w} \right\}^{-1} + \frac{\sigma(T_{pm} + T_a)\left(T_{pm}^2 + T_a^2\right)}{(\varepsilon_p + 0.00591\, N h_w)^{-1} + \dfrac{2N + f - 1 + 0.133\varepsilon_p}{\varepsilon_g} - N}$$

$$(6.4.9)$$

where N =   number of glass covers
        f = $(1 + 0.089h_w - 0.1166h_w\varepsilon_p)(1 + 0.07866N)$
        C = $520(1 - 0.000051\beta^2)$ for $0° < \beta < 70°$. For $70° < \beta < 90°$,
                  use $\beta = 70°$
        e = $0.430(1 - 100/T_{p,m})$
        β =   collector tilt (degrees)
        ε_g =   emittance of glass (0.88)
        ε_p =   emittance of plate
        T_a =   ambient temperature (K)
        T_{pm} =   mean plate temperature (K)
        h_w =   wind heat transfer coefficient (W/m²C)

[1] A method for estimating T_{pm} is given in Section 6.9.

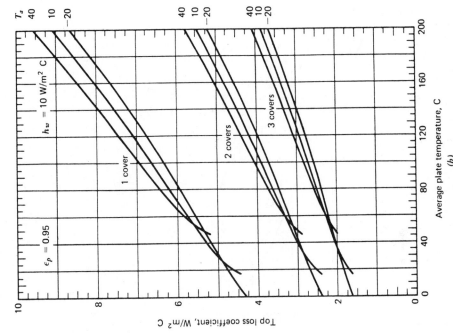

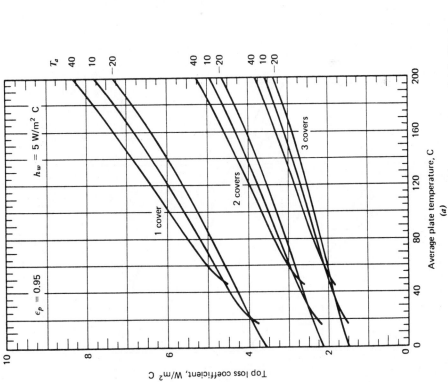

Figure 6.4.4 Top loss coefficient for slope of 45°.

261

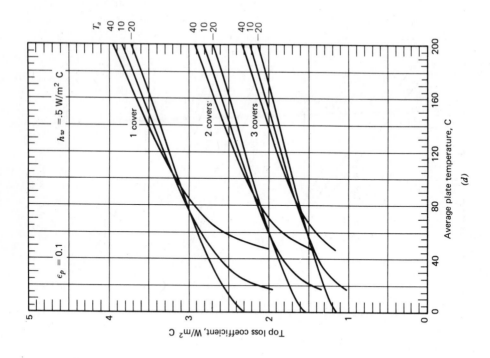

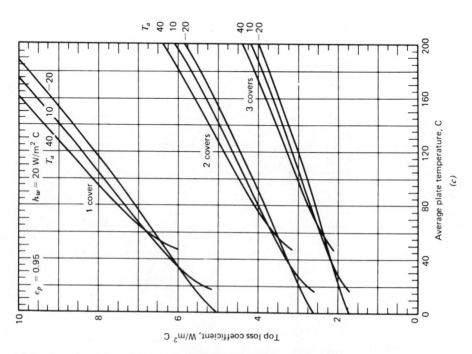

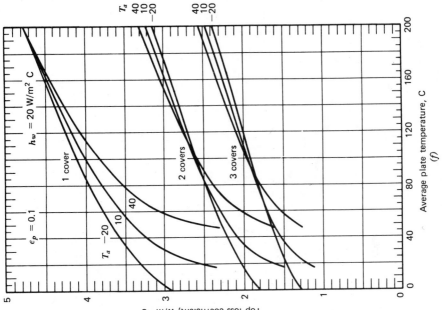

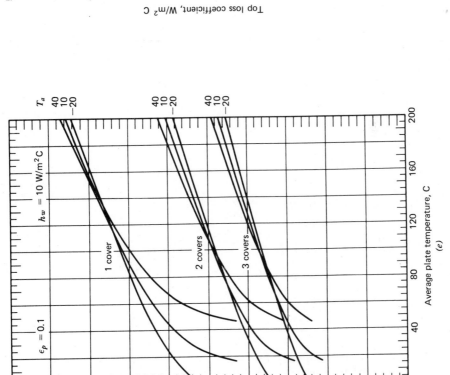

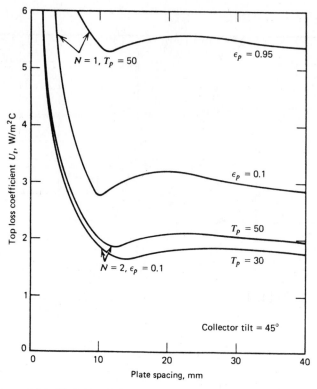

Figure 6.4.5 Typical variation of top loss coefficient with plate spacing.

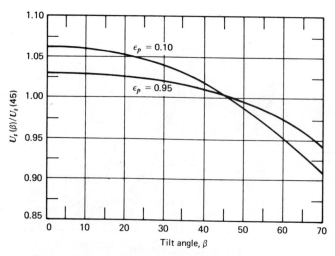

Figure 6.4.6 Dependence of top loss coefficient on slope.

Example 6.4.2

Determine the collector top loss coefficient for a single glass cover with the following specifications:

Plate-to-cover spacing	25 mm
Plate emittance	0.95
Ambient temperature	10 C
Mean plate temperature	100 C
Collector tilt	45°
Wind heat transfer coefficient	10 W/m²C

Solution

From the definitions of f, C, and e in Equation 6.4.9

$$f = [1.0 + 0.089(10) - 0.1166(10)(0.95)] \ (1 + 0.07866) = 0.846$$

$$C = 520[1 - 0.000051(45)^2] = 466$$

$$e = 0.430(1 - 100/373) = 0.315$$

From Equation 6.4.9

$$U_t = \left\{ \frac{1}{\dfrac{466}{373}\left[\dfrac{373 - 283}{1 + 0.846}\right]^{0.315} + \dfrac{1}{10}} \right\}^{-1}$$

$$+ \ \frac{5.67 \times 10^{-8}(373 + 283)(373^2 + 283^2)}{\left(0.95 + 0.00591 \times 1 \times 10\right)^{-1} + \dfrac{2 + 0.846 - 1 + 0.133 \times 0.95}{0.88} - 1}$$

$$= 2.98 + 3.65 = 6.6 \ \text{W/m}^2\text{C}$$

which is very nearly the same as found in Example 6.4.1. ■

The energy loss through the bottom of the collector is represented by two series resistors, R_4 and R_5, in Figure 6.4.1, where R_4 represents the resistance to heat flow through the insulation and R_5 represents the convection and radiation resistance to the environment. The magnitudes of R_4 and R_5 are such that it is usually possible to assume R_5 is zero and all resistance to heat flow is due to the insulation. Thus, the back loss coefficient U_b is approximately[2]

$$U_b = \frac{1}{R_4} = \frac{k}{L} \tag{6.4.10}$$

where k and L are the insulation thermal conductivity and thickness, respectively.

[2] This assumes that the back losses are to a sink at the same temperature as the front losses. This may not be the case.

For most collectors the evaluation of edge losses is complicated. However, in a well-designed system, the edge loss should be small so that it is not necessary to predict it with great accuracy. Tabor (1958) recommends edge insulation of about the same thickness as bottom insulation. The edge losses are then estimated by assuming one-dimensional sideways heat flow around the perimeter of the collector system. The losses through the edge should be referenced to the collector area. If the edge loss coefficient-area product is $(UA)_{edge}$, then the edge loss coefficient, based on the collector area A_c, is

$$U_e = \frac{(UA)_{edge}}{A_c} \qquad (6.4.11)$$

If it is assumed that all losses occur to a common sink temperature T_a, the collector overall loss coefficient U_L is the sum of the top, bottom, and edge loss coefficients:

$$U_L = U_t + U_b + U_e \qquad (6.4.12)$$

Example 6.4.3

For the collector of Example 6.4.2 with a top loss coefficient of 6.6 W/m²C, calculate the overall loss coefficient with the following additional specifications:

Back-insulation thickness	50 mm
Insulation conductivity	0.045 W/mC
Collector bank length	10 m
Collector bank width	3 m
Collector thickness	75 mm
Edge insulation thickness	25 mm

Solution

The bottom loss coefficient is found from Equation 6.4.10:

$$U_b = \frac{k}{L} = \frac{0.045}{0.050} = 0.9 \text{ W/m}^2\text{C}$$

The edge loss coefficient for the 26 m perimeter is found from Equation 6.4.11:

$$U_e = \frac{(0.045/0.025)(26)(0.075)}{30} = 0.12$$

The collector overall loss coefficient is then

$$U_L = 6.6 + 0.9 + 0.1 = 7.6 \text{ W/m}^2\text{C} \qquad \blacksquare$$

The edge loss for this 30 m² collector array is a little over 1% of the total losses. Note, however, that if this collector were 1 × 2 m, the edge losses would increase to over 5%. Thus, edge losses for well-constructed large collector arrays are usually negligible, but for small arrays or individual modules the edge losses may be significant. Also note that only the exterior perimeter was used to estimate edge losses. If the individual collectors are not packed tightly together, significant heat loss may occur from the edge of each module.

The proceeding discussion of top loss coefficients, including Equation 6.4.9, is based on covers like glass that are opaque to long-wavelength radiation. If a plastic material is used to replace one or more covers, the equation for U_t must be modified to account for some infrared radiation passing directly through the cover. For a single cover that is partially transparent to infrared radiation, the net radiant energy transfer directly between the collector plate and the sky is

$$q_{r,p\text{-}s} = \frac{\tau_c \varepsilon_p \sigma \left(T_p^4 - T_s^4\right)}{1 - \rho_p \rho_c} \tag{6.4.13}$$

where τ_c and ρ_c are the transmittance and reflectance of the cover for radiation from T_p and from T_s (assuming that the transmittance is independent of source temperature or that T_p and T_s are nearly the same), and ε_p and ρ_p are the emittance and reflectance of the plate for long-wave radiation. The top loss coefficient then becomes

$$U_t = \frac{q_{r,p\text{-}s}}{\left(T_p - T_a\right)} + \left(\frac{1}{h_{c,p\text{-}c} + h_{r,p\text{-}c}} + \frac{1}{h_w + h_{r,c\text{-}s}}\right)^{-1} \tag{6.4.14}$$

The evaluation of the radiation heat transfer coefficients in Equation 6.4.14 must take into account that the cover is partially transparent. The net radiation between the opaque plate and the partially transparent cover is given by

$$q = \frac{\sigma \varepsilon_p \varepsilon_c \left(T_p^4 - T_c^4\right)}{1 - \rho_p \rho_c} \tag{6.4.15}$$

The radiation heat transfer coefficient between the plate and cover is just the net heat transfer divided by the temperature difference:

$$h_{r,p\text{-}c} = \frac{\sigma \varepsilon_p \varepsilon_c \left(T_p + T_c\right)\left(T_p^2 + T_c^2\right)}{1 - \rho_p \rho_c} \tag{6.4.16}$$

Whillier (1977) presents top loss coefficients for collector cover systems of one glass cover over one plastic cover, two plastic covers, and one glass cover over two plastic covers.

6.5 TEMPERATURE DISTRIBUTION BETWEEN TUBES AND THE COLLECTOR EFFICIENCY FACTOR

The temperature distribution between two tubes can be derived if we temporarily assume the temperature gradient in the flow direction is negligible. Consider the sheet-tube configuration shown in Figure 6.5.1. The distance between the tubes is W, the tube diameter is D, and the sheet is thin with a thickness δ. Because the sheet material is a good conductor, the temperature gradient through the sheet is negligible. We will assume the sheet above the bond is at some local base temperature T_b. The region between the centerline separating the tubes and the tube base can then be considered as a classical fin problem.

The fin, shown in Figure 6.5.2(a), is of length $(W–D)/2$. An elemental region of width Δx and unit length in the flow direction is shown in Figure 6.5.2(b). An energy balance on this element yields

$$S\,\Delta x - U_L\Delta x\,(T - T_a) + \left(-k\delta\,\frac{dT}{dx}\right)\Big|_x - \left(-k\delta\,\frac{dT}{dx}\right)\Big|_{x+\Delta x} = 0 \qquad (6.5.1)$$

where S is the absorbed solar energy defined by Equation 5.9.1. Dividing through by Δx and finding the limit as Δx approaches zero yields

$$\frac{d^2T}{dx^2} = \frac{U_L}{k\delta}\left(T - T_a - \frac{S}{U_L}\right) \qquad (6.5.2)$$

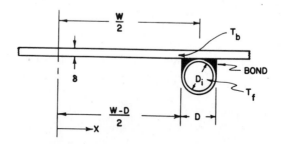

Figure 6.5.1 Sheet and tube dimensions.

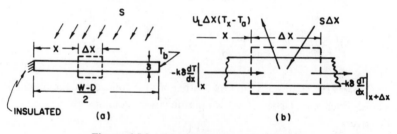

Figure 6.5.2 Energy balance on fin element.

The two boundary conditions necessary to solve this second-order differential equation are symmetry at the centerline and the known base temperature:

$$\left.\frac{dT}{dx}\right|_{x=0} = 0, \qquad T|_{x=(W-D)/2} = T_b \tag{6.5.3}$$

For convenience, we can define two variables, m and Ψ.

$$m = \sqrt{U_L/k\delta} \tag{6.5.4a}$$

$$\Psi = T - T_a - S/U_L \tag{6.5.4b}$$

and Equation 6.5.2 becomes

$$\frac{d^2\Psi}{dx^2} - m^2\Psi = 0 \tag{6.5.5}$$

which has the boundary conditions

$$\left.\frac{d\Psi}{dx}\right|_{x=0} = 0, \qquad \Psi|_{x=(W-D)/2} = T_b - T_a - \frac{S}{U_L} \tag{6.5.6}$$

The general solution is then

$$\Psi = C_1 \sinh mx + C_2 \cosh mx \tag{6.5.7}$$

The constants C_1 and C_2 can be found by substituting the boundary conditions into the general solution. The result is

$$\frac{T - T_a - S/U_L}{T_b - T_a - S/U_L} = \frac{\cosh mx}{\cosh m(W-D)/2} \tag{6.5.8}$$

The energy conducted to the region of the tube per unit of length in the flow direction can now be found by evaluating Fourier's law at the fin base:

$$q'_{fin} = -k\delta\left.\frac{dT}{dx}\right|_{x=(W-D)/2} = \frac{k\delta m}{U_L}[S - U_L(T_b - T_a)]\tanh m\frac{W-D}{2} \tag{6.5.9}$$

but $k\delta m/U_L$ is just $1/m$. Equation 6.5.9 accounts for the energy collected on only one side of a tube; for both sides, the energy collection is

$$q'_{fin} = (W-D)[S - U_L(T_b - T_a)]\frac{\tanh m(W-D)/2}{m(W-D)/2} \tag{6.5.10}$$

It is convenient to use the concept of a fin efficiency to rewrite Equation 6.5.10 as

$$q'_{\text{fin}} = (W - D)\,F\,[S - U_L(T_b - T_a)] \tag{6.5.11}$$

where

$$F = \frac{\tanh\,[m(W - D)/2]}{m(W - D)/2} \tag{6.5.12}$$

The function F is the standard fin efficiency for straight fins with rectangular profile and is plotted in Figure 6.5.3.

   The useful gain of the collector also includes the energy collected above the tube region. The energy gain for this region is

$$q'_{\text{tube}} = D\,[S - U_L(T_b - T_a)] \tag{6.5.13}$$

and the useful gain for the tube and fin per unit of length in the flow direction is the sum of Equations 6.5.11 and 6.5.13:

$$q'_u = [(W - D)F + D][S - U_L(T_b - T_a)] \tag{6.5.14}$$

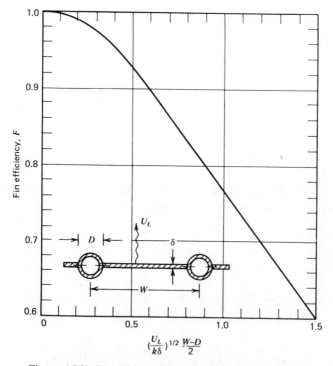

Figure 6.5.3 Fin efficiency for tube and sheet solar collectors.

Ultimately, the useful gain from Equation 6.5.14 must be transferred to the fluid. The resistance to heat flow to the fluid results from the bond and the tube-to-fluid resistance. The useful gain can be expressed in terms of the two resistances as

$$q_u' = \frac{T_b - T_f}{\dfrac{1}{h_{fi}\pi D_i} + \dfrac{1}{C_b}} \tag{6.5.15}$$

where D_i is the inside tube diameter and h_{fi} is the heat transfer coefficient between the fluid and the tube wall. The bond conductance C_b can be estimated from knowledge of the bond thermal conductivity k_b, the average bond thickness γ, and the bond width b. On a per unit length basis,

$$C_b = \frac{k_b b}{\gamma} \tag{6.5.16}$$

The bond conductance can be very important in accurately describing collector performance. Whillier and Saluja (1965) have shown by experiments that simple wiring or clamping of the tubes to the sheet results in low bond conductance and significant loss of performance. They conclude that it is necessary to have good metal-to-metal contact so that the bond conductance is greater than 30 W/mC.

We now wish to eliminate T_b from the equations and obtain an expression for the useful gain in terms of known dimensions, physical parameters, and the local fluid temperature. Solving Equation 6.5.15 for T_b, substituting it into Equation 6.5.14, and solving the result for the useful gain, we obtain

$$q_u' = WF'[S - U_L(T_f - T_a)] \tag{6.5.17}$$

where the collector efficiency factor F' is

$$F' = \frac{1/U_L}{W\left[\dfrac{1}{U_L[D + (W - D)F]} + \dfrac{1}{C_b} + \dfrac{1}{\pi D_i h_{fi}}\right]} \tag{6.5.18}$$

A physical interpretation for F' results from examining Equation 6.5.17. At a particular location, F' represents the ratio of the actual useful energy gain to the useful gain that would result if the collector absorbing surface had been at the local fluid temperature. For this and most (but not all) geometries, another interpretation for the parameter F' becomes clear when it is recognized that the denominator of Equation 6.5.18 is the heat transfer resistance from the fluid to the ambient air. This resistance will be given the symbol $1/U_o$. The numerator is the heat transfer resistance from the absorber plate to the ambient air. Thus F' is the ratio of these two heat transfer coefficients:

$$F' = \frac{U_o}{U_L} \tag{6.5.19}$$

The collector efficiency factor is essentially a constant for any collector design and fluid flow rate. The ratio of U_L to C_b, the ratio of U_L to h_{fi}, and the fin efficiency parameter F are the only variables appearing in Equation 6.5.18 that may be functions of temperature. For most collector designs F is the most important of these variables in determining F'. The factor F' is a function of U_L and h_{fi}, each of which has some temperature dependence, but it is not a strong function of temperature.

The evaluation of F' is not a difficult task. However, to illustrate the effects of various design parameters on the magnitude of F', Figure 6.5.4 has been prepared. Three values of the overall heat transfer coefficient U_L were chosen (2, 4, and 8 W/m²C) which cover the range of collector designs from a one-cover nonselective to a two cover selective. (See Figure 6.4.4 for other combinations that will yield these same overall loss coefficients.) Instead of selecting various plate materials, the curves were prepared for various values of $k\delta$, the product of the plate thermal conductivity and plate thickness. For a copper plate 1 mm thick, $k\delta$ is equal to 0.4 W/C; for a steel plate 0.1 mm thick, $k\delta$ is equal to 0.005 W/C. Thus, the probable range of $k\delta$ is from 0.005 to 0.4. The bond conductance was assumed to be very large (i.e., $1/C_b = 0$) and the tube diameter was selected as 0.01 m. Three values were chosen for the heat transfer coefficient inside the tube to cover a range from laminar flow to highly turbulent flow: 100, 300, and 1000 W/m²C. Note that increasing h_{fi} beyond 1000 W/m²C for this diameter tube does not result in significant increases in F'. As expected, the collector efficiency factor decreases with increased tube center-to-center distances and increases with increases in both material thicknesses and thermal conductivity. Increasing the overall loss coefficient decreases F' while increasing the fluid-tube heat transfer coefficient increases F'.

Example 6.5.1

Calculate the collector efficiency factor for the following specifications:

Overall loss coefficient	8.0 W/m²C
Tube spacing	150 mm
Tube diameter (inside)	10 mm
Plate thickness	0.5 mm
Plate thermal conductivity (copper)	385 W/mC
Heat transfer coefficient inside tubes	300 W/m²C
Bond resistance	0

Solution

The fin efficiency factor F, from Equations 6.5.4a and 6.5.12, is

$$m = \left(\frac{8}{385 \times 5 \times 10^{-4}}\right)^{1/2} = 6.45$$

$$F = \frac{\tanh 6.45\,[(0.15 - 0.01)/2]}{6.45\,[(0.15 - 0.01)/2]} = 0.937$$

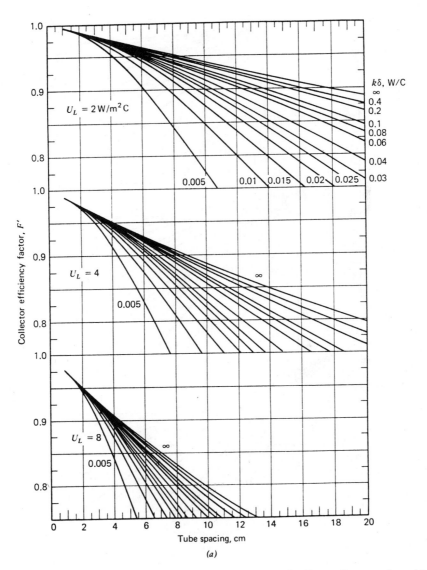

Figure 6.5.4 Collector efficiency factor F' versus tube spacing for 10 mm diameter tubes: (a) $h_{fi} = 100$ W/m²C; (b) $h_{fi} = 300$ W/m²C; (c) $h_{fi} = 1000$ W/m²C

273

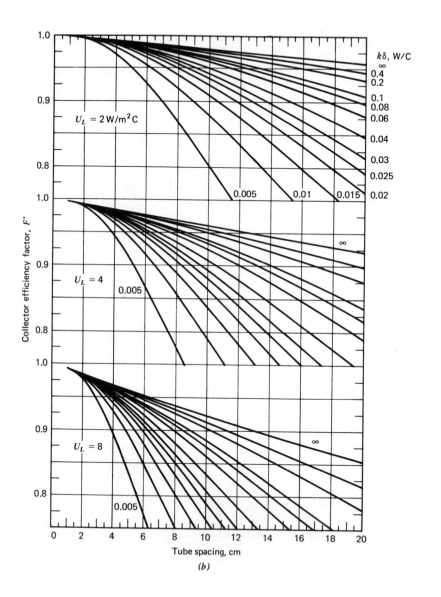

$k\delta$, W/C
∞
0.4
0.2
0.1
0.08
0.06
0.04
0.03
0.025
0.02

0.005   0.01   0.015

$U_L = 2\,\text{W/m}^2\text{C}$

$U_L = 4$

0.005

∞

$U_L = 8$

∞

0.005

Collector efficiency factor, F'

Tube spacing, cm

(b)

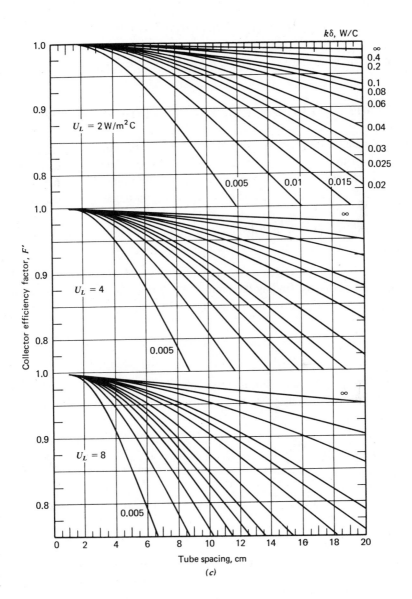

(c)

The collector efficiency factor F' is found from Equation 6.5.18:

$$F' = \frac{1/8}{0.15\left[\frac{1}{8\,[0.01 + (0.15 - 0.01)\,0.937]} + \frac{1}{\pi(0.01)\,300}\right]} = 0.841$$

The same result is obtained from Figure 6.5.4(b). ∎

6.6 TEMPERATURE DISTRIBUTION IN FLOW DIRECTION

The useful gain per unit flow length as calculated from Equation 6.5.17 is ultimately transferred to the fluid. The fluid enters the collector at temperature $T_{f,i}$ and increases in temperature until at the exit it is $T_{f,o}$. Referring to Figure 6.6.1, we can express an energy balance on the fluid flowing through a single tube of length Δy as

$$\left(\frac{\dot m}{n}\right)C_p T_f\Big|_y - \left(\frac{\dot m}{n}\right)C_p T_f\Big|_{y+\Delta y} + \Delta y\, q_u' = 0 \tag{6.6.1}$$

where $\dot m$ is the total collector flow rate and n is the number of parallel tubes. Dividing through by Δy, finding the limit as Δy approaches zero, and substituting Equation 6.5.17 for q_u', we obtain

$$\dot m C_p \frac{dT_f}{dy} - nWF'\,[S - U_L(T_f - T_a)] = 0 \tag{6.6.2}$$

If we assume that F' and U_L are independent of position,[3] then the solution for the fluid temperature at any position y (subject to the condition that the inlet fluid temperature is T_{fi}) is

$$\frac{T_f - T_a - S/U_L}{T_{fi} - T_a - S/U_L} = \exp\left(-U_L n W F' y / \dot m C_p\right) \tag{6.6.3}$$

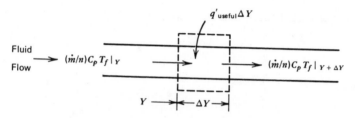

Figure 6.6.1 Energy balance on fluid element.

---

[3] Dunkle and Cooper (1975) have assumed U_L is a linear function of $T_f - T_a$.

If the collector has a length L in the flow direction, then the outlet fluid temperature T_{fo} is found by substituting L for y in Equation 6.6.3. The quantity nWL is the collector area.

$$\frac{T_{fo} - T_a - S/U_L}{T_{fi} - T_a - S/U_L} = \exp\left(-A_c U_L F'/\dot{m} C_p\right) \tag{6.6.4}$$

6.7 COLLECTOR HEAT REMOVAL FACTOR AND FLOW FACTOR

It is convenient to define a quantity that relates the actual useful energy gain of a collector to the useful gain if the whole collector surface were at the fluid inlet temperature. This quantity is called the collector heat removal factor F_R. In equation form it is

$$F_R = \frac{\dot{m} C_p (T_{fo} - T_{fi})}{A_c [S - U_L (T_{fi} - T_a)]} \tag{6.7.1}$$

The collector heat removal factor can be expressed as

$$F_R = \frac{\dot{m} C_p}{A_c U_L} \left[\frac{T_{fo} - T_{fi}}{\dfrac{S}{U_L} - (T_{fi} - T_a)} \right]$$

$$= \frac{\dot{m} C_p}{A_c U_L} \left[\frac{\left(T_{fo} - T_a - \dfrac{S}{U_L}\right) - \left(T_{fi} - T_a - \dfrac{S}{U_L}\right)}{\dfrac{S}{U_L} - (T_{fi} - T_a)} \right] \tag{6.7.2}$$

or

$$F_R = \frac{\dot{m} C_p}{A_c U_L} \left[1 - \frac{\dfrac{S}{U_L} - (T_{fo} - T_a)}{\dfrac{S}{U_L} - (T_{fi} - T_a)} \right] \tag{6.7.3}$$

which from Equation 6.6.4 can be expressed as

$$F_R = \frac{\dot{m} C_p}{A_c U_L} \left[1 - \exp\left(-\frac{A_c U_L F'}{\dot{m} C_p}\right) \right] \tag{6.7.4}$$

To present Equation 6.7.4 graphically, it is convenient to define the collector flow factor F'' as the ratio of F_R to F'. Thus

$$F'' = \frac{F_R}{F'} = \frac{\dot{m} C_p}{A_c U_L F'} \left[1 - \exp\left(-\frac{A_c U_L F'}{\dot{m} C_p}\right) \right] \tag{6.7.5}$$

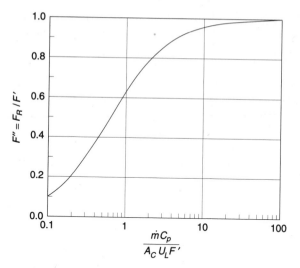

Figure 6.7.1   Collector flow factor F'' as a function of $\dot{m}C_p/A_cU_LF'$.

This collector flow factor is a function of the single variable, the dimensionless collector capacitance rate $\dot{m}C_p/A_cU_LF'$, and is shown in Figure 6.7.1.

The quantity F_R is equivalent to the effectiveness of a conventional heat exchanger, which is defined as the ratio of the actual heat transfer to the maximum possible heat transfer. The maximum possible useful energy gain (heat transfer) in a solar collector occurs when the whole collector is at the inlet fluid temperature; heat losses to the surroundings are then at a minimum. The collector heat removal factor times this maximum possible useful energy gain is equal to the actual useful energy gain Q_u:

$$Q_u = A_cF_R[S - U_L(T_i - T_a)] \tag{6.7.6}$$

This is an extremely useful equation[4] and applies to essentially all flat-plate collectors. With it, the useful energy gain is calculated as a function of the inlet fluid temperature. This is a convenient representation when analyzing solar energy systems, since the inlet fluid temperature is usually known. However, losses based on the inlet fluid temperature are too small since losses occur all along the collector from the plate and the plate has an ever increasing temperature in the flow direction. The effect of the multiplier F_R is to reduce the useful energy gain from what it would have been had the whole collector absorber plate been at the inlet fluid temperature to what actually occurs. As the mass flow rate through the collector increases, the temperature rise through the collector decreases. This causes lower losses since the average collector temperature is lower and there is a corresponding increase in the

---

[4] This is the most important equation in the book. The subscript f on the fluid inlet temperature has been dropped; whenever the meaning is not clear, it will be reintroduced.

useful energy gain. This increase is reflected by an increase in the collector heat removal factor F_R as the mass flow rate increases. Note that F_R can never exceed the collector efficiency factor F'. As the flow rate becomes very large, the temperature rise from inlet to outlet decreases toward zero but the temperature of the absorbing surface will still be higher than the fluid temperature. This temperature difference is accounted for by the collector efficiency factor F'.

Many of the equations of Sections 6.6 and 6.7 contain the ratio of the collector mass flow rate to collector area. This ratio is a convenient way to express flow rate when collector area is a design variable since increasing both in proportion will maintain a nearly constant value of F_R.

Example 6.7.1

Calculate the daily useful gain and efficiency of an array of 10 solar collector modules installed in parallel near Boulder, CO, at a slope of $60°$ and a surface azimuth of $0°$. The hourly radiation on the plane of the collector I_T, the hourly radiation absorbed by the absorber plate S, and the hourly ambient temperature T_a are given in the table. The methods of Sections 2.15, 2.16, and 5.9 can be used to find I_T and S knowing the hourly horizontal radiation, the collector orientation, and the collector optical properties. For the collector assume the overall loss coefficient U_L to be 8.0 W/m²C and the plate efficiency factor F' to be 0.841 (from Example 6.5.1). The flow rate through each 1×2 m collector panel is 0.03 kg/s and the inlet fluid temperature remains constant at 40 C.

Solution

The dimensionless collector mass flow rate is

$$\frac{\dot{m}C_p}{A_cU_LF'} = \frac{0.03 \times 4190}{2 \times 8 \times 0.841} = 9.35$$

so that the collector flow factor, from Equation 6.7.5 (or Figure 6.7.1), is

$$F'' = 9.35(1 - e^{-1/9.35}) = 0.948$$

and the heat removal factor is

$$F_R = F'F'' = 0.841 \times 0.948 = 0.797$$

The losses for the hour 10 to 11, based on an inlet temperature of 40 C, are

$$U_L(T_i - T_a) = 8(40 - 2) \times 3600 = 1.09 \text{ MJ/m}^2$$

and the useful energy gain per unit of collector area is

$$q_u = \frac{Q_u}{A_c} = 0.797(3.29 - 1.09) \times 10^6 = 1.76 \text{ MJ/m}^2$$

The collector efficiency for this hour is found from Equation 6.2.2:

$$\eta = \frac{Q_u}{I_T A_c} = \frac{q_u}{I_T} = \frac{1.76}{3.92} = 0.45$$

and the day-long collector efficiency is

$$\eta_{day} = \frac{\sum q_u}{\sum I_T} = \frac{7.57}{19.79} = 0.38$$

The daily useful energy gain of the 10 collector modules in the array is

$$Q_u = 10 \times 2 \times 7.57 \times 10^6 = 150 \text{ MJ}$$

Time	T_a C	I_T MJ/m^2	S MJ/m^2	$U_L(T_i - T_a)$ MJ/m^2	q_u MJ/m^2	η
7–8	−11	0.02	–	–	0	0
8–9	−8	0.43	0.35	1.38	0	0
9–10	−2	0.99	0.82	1.21	0	0
10–11	2	3.92	3.29	1.09	1.76	0.45
11–12	3	3.36	2.84	1.07	1.42	0.42
12–1	6	4.01	3.39	0.98	1.93	0.48
1–2	7	3.84	3.21	0.95	1.81	0.47
2–3	8	1.96	1.63	0.92	0.57	0.29
3–4	9	1.21	0.99	0.89	0.08	0.07
4–5	7	0.05	–	0.95	0	–
Sum		19.79			7.57	

A number of general observations can be made from the results of Example 6.7.1. The estimated performance is typical of a one-cover nonselective collector, although in most systems the inlet temperatures will vary throughout the day.[5] The losses are both thermal and optical, and during the early morning and late afternoon the radiation level was not sufficient to overcome the losses. The collector should not be operated during these periods.

Daily efficiency may also be based on the period while the collector is operating. The efficiency calculated in this manner is 7.57/18.39 or 41%. Reporting in this manner gives a higher value for collector efficiency. As the collector inlet temperature is reduced, these two day-long efficiencies will approach one another. Collector efficiency is a single parameter that combines collector and system characteristics and generally is not reliable for making comparisons.

---

[5] Temperature fluctuations are considered in Chapter 10.

The fluid temperature rise through the collector (calculated from $\Delta T = Q_u/\dot{m}C_p$) varies from a high of 8.5 C between 12 and 1 to a low of 2.5 C between 2 and 3. This relatively modest temperature rise is typical of liquid heating collectors. The temperature rise can be increased by reducing the flow rate, but this will reduce the useful energy gain (if the inlet fluid temperature stays the same). If the flow rate were halved and if F' remained the same (in fact, h_{fi} would decrease which would reduce F'), then F_R would decrease to 0.76 and the temperature rise during the hour 12 to 1 would be 16.2 C, which is less than twice the original temperature rise. The efficiency during this hour would be reduced from 48 to 46%.[6]

6.8   CRITICAL RADIATION LEVEL

In Chapter 2, the concept of utilizability was developed without concern for how critical radiation levels were defined. With Equation 6.7.6 established, we can now determine the critical radiation level I_{Tc} for flat-plate collectors. It is convenient to rewrite Equation 6.7.6 in the following form:

$$Q_u = A_c[F_R(\tau\alpha)I_T - F_R U_L(T_i - T_a)] \tag{6.8.1}$$

The critical radiation level is that value of I_T that makes the term in the brackets identically zero, that is, where the absorbed radiation and loss terms are equal:

$$I_{Tc} = \frac{F_R U_L(T_i - T_a)}{F_R(\tau\alpha)} \tag{6.8.2}$$

It is convenient to retain F_R in the equation for reasons that will be clear in later sections. The collector output can now be written in terms of the critical radiation level:

$$Q_u = A_c F_R(\tau\alpha)(I_T - I_{Tc}) \tag{6.8.3}$$

The equations for Q_u indicate that for the collector to produce useful output, i.e., for $Q_u > 0$, the absorbed radiation must exceed the thermal losses, and I_T must be greater than I_{Tc}. In Equations 6.7.6 and 6.8.3, only positive values of the terms in parentheses are considered. This implies that there is a controller on the collector that shuts off the flow of fluid when the value in the parentheses is not positive.

---

[6] It will be seen in later chapters that when a *system* with a thermally stratified tank is considered, a reduction in flow rate may lead to reduced T_{fi} and thus to increased Q_u even though F_R is decreased.

6.9   MEAN FLUID AND PLATE TEMPERATURES

To evaluate collector performance, it is necessary to know the overall loss coefficient and the internal fluid heat transfer coefficients. However, both U_L and h_{fi} are to some degree functions of temperature. The mean fluid temperature can be found by integrating Equation 6.6.3 from zero to L:

$$T_{fm} = \frac{1}{L} \int_0^L T_f(y)\, dy \qquad\qquad (6.9.1)$$

Performing this integration and substituting F_R from Equation 6.7.4 and Q_u from Equation 6.7.6, the mean fluid temperature was shown by Klein et al. (1974) to be

$$T_{fm} = T_{fi} + \frac{Q_u/A_c}{F_R U_L}\left(1 - F''\right) \qquad\qquad (6.9.2)$$

This is the proper temperature for evaluating fluid properties.

The mean plate temperature will always be greater than the mean fluid temperature due to the heat transfer resistance between the absorbing surface and the fluid. This temperature difference is usually small for liquid systems but may be significant for air systems.

The mean plate temperature can be used to calculate the useful gain of a collector

$$Q_u = A_c[S - U_L(T_{pm} - T_a)] \qquad\qquad (6.9.3)$$

If we equate Equations 6.9.3 and 6.7.6 and solve for the mean plate temperature, we have

$$T_{pm} = T_{fi} + \frac{Q_u/A_c}{F_R U_L}\left(1 - F_R\right) \qquad\qquad (6.9.4)$$

Equation 6.9.4 can be solved in an iterative manner with Equation 6.4.9. First an estimate of the mean plate temperature is made from which U_L is calculated. With approximate values of F_R, F'', and Q_u, a new mean plate temperature is obtained from Equation 6.9.4 and used to find a new value for the top loss coefficient. The new value of U_L is used to refine F_R and F'', and the process is repeated. With a reasonable initial guess, a second iteration is seldom necessary. A reasonable first guess for T_p for liquid heating collectors operated at typical flow rates of 0.01 to 0.02 kg/m^{2}s is $T_{fi} + 10$ C. For air heaters a reasonable first estimate is $T_{fi} + 20$ C.

Example 6.9.1

Find the mean fluid and plate temperatures for the hour 11 to 12 of Example 6.7.1.

Solution

With U_L equal to 8.0 W/m²C, $F'' = 0.948$, $F_R = 0.797$, and $q_u = 1.42$ MJ/m².
We have from Equation 6.9.2

$$T_{fm} = 40 + \frac{\left(1.42 \times 10^6\right)/3600}{8 \times 0.797}(1 - 0.948) = 43 \text{ C}$$

The mean plate temperature is found from Equation 6.9.4:

$$T_{pm} = 40 + \frac{\left(1.42 \times 10^6\right)/3600}{8 \times 0.797}(1 - 0.797) = 53 \text{ C} \qquad ■$$

In this example, U_L was assumed to be independent of temperature. If the
temperature dependence of U_L is considered, the iterative process would have been
necessary.

6.10   EFFECTIVE TRANSMITTANCE-ABSORPTANCE PRODUCT

In Section 5.5, the product of cover transmittance times plate solar absorptance was
discussed. In Section 6.4 the expressions for U_L were derived assuming that the
glazing did not absorb solar radiation. To maintain the simplicity of Equation 6.7.6
and account for the reduced thermal losses due to absorption of solar radiation by the
glass, an effective transmittance-absorptance product $(\tau\alpha)_e$ will be introduced. It
will be shown that $(\tau\alpha)_e$ is slightly greater than $(\tau\alpha)$.

All of the solar radiation that is absorbed by a cover system is not lost, since
this absorbed energy tends to increase the cover temperature and consequently reduce
the thermal losses from the plate. Consider the thermal network of a single-cover
system shown in Figure 6.10.1. Solar energy absorbed by the cover is $I_T(1 - \tau_a)$,

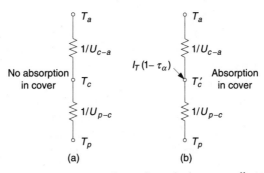

Figure 6.10.1   Thermal network for top losses for a single-cover collector with and without
absorptance in the cover.

where τ_a is the transmittance considering only absorption from Equation 5.2.2 and I_T is the incident radiation. The loss for (a), without absorption, is $U_{p\text{-}c}(T_p - T_c)$, and the loss for (b), with absorption, is $U_{p\text{-}c}(T_p - T_c')$. Here we have assumed that the small amount of absorption in the cover and consequent increased cover temperature do not change the values of $U_{p\text{-}c}$ and $U_{c\text{-}a}$. The difference D in the two loss terms is

$$D = U_{p\text{-}c}\left[(T_p - T_c) - (T_p - T_c')\right] \tag{6.10.1}$$

The temperature difference $T_p - T_c$ can be expressed as

$$T_p - T_c = \frac{U_t(T_p - T_a)}{U_{p\text{-}c}} \tag{6.10.2}$$

where U_t is the top loss coefficient and is equal to $U_{p\text{-}c}U_{c\text{-}a}/(U_{p\text{-}c} + U_{c\text{-}a})$. The temperature difference $T_p - T_c'$ can be expressed as

$$T_p - T_c' = \frac{U_{c\text{-}a}(T_p - T_a) - I_T(1 - \tau_a)}{U_{p\text{-}c} + U_{c\text{-}a}} \tag{6.10.3}$$

Therefore

$$D = U_t(T_p - T_a) - \frac{U_{p\text{-}c}\, U_{c\text{-}a}(T_p - T_a)}{U_{p\text{-}c} + U_{c\text{-}a}} + \frac{I_T U_{p\text{-}c}(1 - \tau_a)}{U_{p\text{-}c} + U_{c\text{-}a}} \tag{6.10.4}$$

or

$$D = \frac{I_T U_t(1 - \tau_a)}{U_{c\text{-}a}} \tag{6.10.5}$$

The quantity D represents the reduction in collector losses due to absorption in the cover but can be considered an additional input in the collector equation. The useful gain of a collector is then

$$q_u = F_R\left[S + \frac{I_T U_t(1 - \tau_a)}{U_{c\text{-}a}} - U_L(T_i - T_a)\right] \tag{6.10.6}$$

In general the quantity I_T has three components, beam, diffuse, and ground-reflected radiation. Each of these terms is multiplied by a separate value of $(\tau\alpha)$ to determine S [i.e., $(\tau\alpha)_b$, $(\tau\alpha)_d$, or $(\tau\alpha)_g$ as shown in Section 5.8]. We can divide the radiation absorbed in the cover into these same three components. By defining the quantity $(\tau\alpha) + (1 - \tau_a)\, U_t/U_{c\text{-}a}$ as the effective transmittance-absorptance product for each of the three components, the simplicity of Equation 6.7.6 can be maintained. For this one-cover system

$$(\tau\alpha)_e = (\tau\alpha) + (1 - \tau_a)\frac{U_t}{U_{c\text{-}a}} \tag{6.10.7}$$

When evaluating S, the appropriate value of $(\tau\alpha)_e$ should be used in place of $(\tau\alpha)$. As noted below, $(\tau\alpha)_e$ is on the order of 1% greater than $(\tau\alpha)$ for a typical single cover collector with normal glass. For a collector with low-iron (water-white) glass, $(\tau\alpha)_e$ and $(\tau\alpha)$ are nearly identical.

A general analysis for a cover system of n identical plates yields

$$(\tau\alpha)_e = (\tau\alpha) + (1 - \tau_a) \sum_{i=1}^{n} a_i \tau^{i-1} \qquad (6.10.8)$$

where a_i is the ratio of the top loss coefficient to the loss coefficient from the ith cover to the surroundings and τ_a is the transmittance of a single cover from Equation 5.2.2. This equation was derived assuming that the transmittance to the ith cover could be approximated by the transmittance of a single cover raised to the $i - 1$ power.

For a cover system composed of different materials (e.g., a combination of glass and plastic) the effective transmittance-absorptance product is

$$(\tau\alpha)_e = (\tau\alpha) + (1 - \tau_{a,1})a_1 + (1 - \tau_{a,2})a_2\tau_1 + (1 - \tau_{a,3})a_3\tau_2 + \cdots \quad (6.10.9)$$

where τ_i is the transmittance of the cover system above the $i + 1$ cover and $\tau_{a,i}$ is the transmittance due to absorption for the ith cover. The angular dependence of $(\tau\alpha)_e$ can be evaluated using the proper angular dependency of $(\tau\alpha)$, $\tau\alpha$, and τ_a.

The values of a_i actually depend upon the plate temperature, ambient temperature, plate emittance, and wind speed. Table 6.10.1 gives values of a_i for one, two, and three covers and for plate emittances of 0.95, 0.50, and 0.10. These values were calculated using a wind heat transfer coefficient of 24 W/m²C, a plate temperature of 100 C, and an ambient air and sky temperature of 10 C. The dependence of a_i on wind speed may be significant. However, lower wind heat transfer coefficients will increase the a_i values leading to slightly higher useful energy gains. Since the total amount absorbed by the glass is small, relatively large errors in a_i will not cause a significant error in the calculation of Q_u.

Table 6.10.1   Constants for Use in Equations 6.10.8 and 6.10.9

Covers	a_i	$\varepsilon_p = 0.95$	$\varepsilon_p = 0.50$	$\varepsilon_p = 0.10$
1	a_1	0.27	0.21	0.13
2	a_1	0.15	0.12	0.09
	a_2	0.62	0.53	0.40
3	a_1	0.14	0.08	0.06
	a_2	0.45	0.40	0.31
	a_3	0.75	0.67	0.53

Although the value of $(\tau\alpha)_e$ can be calculated from Equation 6.10.9 with some precision, $(\tau\alpha)_e$ is seldom more than 1 to 2% greater than $(\tau\alpha)$. For a one-cover nonselective collector, a_1 is 0.27. If the cover absorbs 4% of the incident radiation, i.e., the cover is ordinary glass with KL of about 0.03, then $(\tau\alpha)_e$ is 1% greater than $(\tau\alpha)$. For a one-cover selective collector with this glass, the difference is 0.5%. For a one-cover selective collector with low-iron glass with KL about 0.01, $(\tau\alpha)_e$ is about 0.1% greater than $(\tau\alpha)$. For a two-cover nonselective system $(\tau\alpha)_e$ is almost 2% greater than $(\tau\alpha)$. As discussed in Section 5.5, $(\tau\alpha)$ is approximately 1% greater than the product of τ and α. Since surface radiation properties are seldom known to within 1% , the effective transmittance-absorptance product can be approximated for collectors with ordinary glass by

$$(\tau\alpha)_e \cong 1.02\tau\alpha \qquad\qquad (6.10.10a)$$

and for collectors with covers with negligible absorption by

$$(\tau\alpha)_e \cong 1.01\tau\alpha \qquad\qquad (6.10.10b)$$

6.11   EFFECTS OF DUST AND SHADING

The effects of dust and shading are difficult to generalize. Data reported by Dietz (1963) show that at the angles of incidence of interest (0 to 50°) the maximum reduction of transmittance of covers due to dirt was 2.7%. From long-term experiments on collectors in the Boston area, Hottel and Woertz (1942) found that collector performance decreased approximately 1% due to dirty glass. In a rainless 30-day experiment in India, Garg (1974) found that dust reduced the transmittance by an average of 8% for glass tilted at 45°. To account for dust in temperate climates, it is suggested that radiation absorbed by the plate be reduced by 1%; in dry and dusty climates, absorbed radiation can be reduced by 2%.

Shading effects can also be significant. Whenever the angle of incidence is off normal, some of the structure will intercept solar radiation. Some of this radiation will be reflected to the absorbing plate if the sidewalls are of a high-reflectance material. Hottel and Woertz, based on experiments with two-cover collectors, recommend that the radiation absorbed by the plate be reduced by 3% to account for shading effects if the net (unobstructed) glass area is used in all calculations. The net glass area accounts for the blockage by the supports for the glass. Most modern collectors use one cover, and module areas are larger, both of which reduce shading effects. A reduction of S of 1% may be a more appropriate correction for these collectors.

Example 6.11.1

In Example 6.7.1, the effects of dust, shading, and absorption by the cover were all neglected. Reevaluate the daily performance taking these quantities into account. The single glass cover has $KL = 0.037$, and the plate has a flat black (nonselective) absorbing surface.

Solution

This glass ($KL = 0.037$) absorbs approximately 4% of the incident radiation and, according to Table 6.10.1 and Equation 6.10.8, 27% of this is not lost. Thus $(\tau\alpha)_e$ is 1.01 times $(\tau\alpha)$. The effects of dust and shading each reduce the absorbed radiation by 1%. The net effect is to decrease S by 2%. The table that follows gives new values for S and the hourly energy gains with $F_R = 0.8$ and $U_L = 8.0$ W/m²C. The daily efficiency is reduced from 38 to 36%.

Time	I_T MJ/m²	S MJ/m²	T C	q_u MJ/m²	η
7–8	0.02	–	–11	–	–
8–9	0.43	0.34	–8	0	0
9–10	0.99	0.79	–2	0	0
10–11	3.92	3.16	2	1.65	0.42
11–12	3.36	2.73	3	1.33	0.40
12–1	4.01	3.25	6	1.82	0.45
1–2	3.84	3.08	7	1.70	0.44
2–3	1.96	1.56	8	0.51	0.26
3–4	1.21	0.95	9	0.05	0.04
4–5	0.05	–	7		–
	19.79			7.06	

$$\eta_{day} = \frac{7.06}{19.79} = 0.36$$

6.12 HEAT CAPACITY EFFECTS IN FLAT-PLATE COLLECTORS

The operation of most solar energy systems is inherently transient; there is no such thing as steady-state operation when one considers the transient nature of the driving forces. This observation has led to numerical studies by Klein et al. (1974), Wijeysundera (1978), and others on the effects of collector heat capacity on collector performance. The effects can be regarded in two distinct parts. One part is due to the heating of the collector from its early morning low temperature to its final operating temperature in the afternoon. The second part is due to intermittent behavior during the day whenever the driving forces such as solar radiation and wind change rapidly.

Klein et al. (1974) showed that the daily morning heating of the collector results in a loss that can be significant but is negligible for many situations. For example, the radiation on the collector of Example 6.11.1 before 10 AM was 1.44 MJ/m². The calculated losses exceeded this value during this time period because these calculated losses assumed that the fluid entering the collector was at 40 C. In reality, no fluid would be circulating and the absorbed solar energy would heat the collector without reducing the useful energy gain.

The amount of preheating that will occur in a given collector can be estimated by solving the transient energy balance equations for the various parts of the collector. Even though these equations can be developed to almost any desired degree of accuracy, the driving forces such as solar radiation, wind speed, and ambient temperature are usually known only at hour intervals. This means that any predicted transient behavior between the hourly intervals can only be approximate, even with detailed analysis. Consequently, a simplified analysis is warranted to determine if more detailed analysis is desirable.

To illustrate the method, consider a single-cover collector. Assume the absorber plate, the water in the tubes, and one-half of the back insulation are all at the same temperature. Also assume that the cover is at a uniform temperature that is different from the plate temperature. An energy balance on the absorber plate, water, and the part of the back insulation yields

$$(mC)_p \frac{dT_p}{dt} = A_c[S - U_{p\text{-}c}(T_p - T_c)] \tag{6.12.1}$$

The subscripts c and p represent cover and plate; $U_{p\text{-}c}$ is the loss coefficient from the plate to the cover and t is time. An energy balance on the cover yields

$$(mC)_c \frac{dT_c}{dt} = A_c[U_{p\text{-}c}(T_p - T_c) + U_{c\text{-}a}(T_a - T_c)] \tag{6.12.2}$$

where $U_{c\text{-}a}$ is the loss coefficient from the cover to the ambient air and T_a is the ambient temperature. It is possible to solve these two equations simultaneously; however, a great simplification occurs if we assume $(T_c - T_a)/(T_p - T_a)$ remains constant at its steady-state value. In other words, we must assume that the following relationship holds:[7]

$$U_{c\text{-}a}(T_c - T_a) = U_L(T_p - T_a) \tag{6.12.3}$$

Differentiating Equation 6.12.3, assuming T_a is a constant, we have

$$\frac{dT_c}{dt} = \frac{U_L}{U_{c\text{-}a}} \frac{dT_p}{dt} \tag{6.12.4}$$

Adding Equation 6.12.1 to 6.12.2 and using Equation 6.12.4, we obtain the following differential equation for the plate temperature:

$$\left[(mC)_p + \frac{U_L}{U_{c\text{-}a}}(mC)_c\right] \frac{dT_p}{dt} = A_c[S - U_L(T_p - T_a)] \tag{6.12.5}$$

---

[7] The back and edge losses are assumed to be small.

The term in the square brackets on the left-hand side represents an effective heat capacity of the collector and is written as $(mC)_e$. By the same reasoning, the effective heat capacity of a collector with n covers would be

$$(mC)_e = (mC)_p + \sum_{i=1}^{n} a_i (mC)_{c,i} \tag{6.12.6}$$

where a_i is the ratio of the overall loss coefficient to the loss coefficient from the cover in question to the surroundings. This is the same quantity as presented in Table 6.10.1.

If we assume that S and T_a remain constant for some period t, the solution to Equation 6.12.5 is

$$\frac{S - U_L(T_p - T_a)}{S - U_L(T_{p,\text{initial}} - T_a)} = \exp\left(-\frac{A_c U_L t}{(mC)_e}\right) \tag{6.12.7}$$

The simplification introduced through the use of Equation 6.12.3 is significant in that the problem of determining heat capacity effects has been reduced to solving one differential equation. The error introduced by this simplification is difficult to assess for all conditions without solving the set of differential equations. Wijeysundera (1978) compared this one-node approximation against a two-node solution and experimental data and found good agreement for single-cover collectors. For two- and three-cover collectors the predicted fractional temperature rise was less than 15% in error.

The collector plate temperature T_p can be evaluated at the end of each time period by knowing S, U_L, T_a, and the collector plate temperature at the beginning of the time period. Repeated application of Equation 6.12.7 for each hour before the collector actually operates serves to estimate the collector temperature as a function of time. An estimate of the reduction in useful gain can then be obtained by multiplying the collector effective heat capacity by the temperature rise necessary to bring the collector to its initial operating temperature.

A similar loss occurs due to collector heat capacity whenever the final average collector temperature in the afternoon exceeds the initial average temperature. This loss can be easily estimated by multiplying collector effective heat capacity times this temperature difference.

Finally, Klein et al. (1974) showed that the effects of intermittent sunshine, wind speed, and ambient air temperature were always negligible for normal collector construction.

Example 6.12.1

For the collector described in Example 6.11.1, estimate the reduction in useful energy gain due to heat capacity effects. The plate and tubes are copper. The collector has the following specifications:

Plate thickness	0.5 mm
Tube inside diameter.	10.0 mm
Tube spacing	150.0 mm
Glass cover thickness	3.5 mm
Back insulation thickness	50.0 mm

The collector materials have the following properties:

	C, kJ/kgC	ρ, kg/m^3
Copper	0.48	8800
Glass	0.80	2500
Insulation	0.80	50

Solution

Since the collector operates with a constant inlet temperature, only the early morning heating will influence the useful gain. The collector heat capacity includes the glass, plate, tubes, water in tubes, and insulation. The heat capacity of the glass is

$$0.0035 \text{ m} \times 2500 \text{ kg/m}^3 \times 0.8 \text{ kJ/kg C} = 7 \text{ kJ/m}^2\text{C}$$

For the plate, tubes, water in tubes, and insulation, the heat capacities are 2, 1, 2, and 2 kJ/m^{2}C respectively. The insulation exposed to the ambient remains near ambient temperature so that the effective insulation heat capacitance is one-half of its actual value. The effective collector capacity is thus $2 + 1 + 2 + 1 + 0.27 \times 7 = 8$ kJ/m^{2}C. From Equation 6.12.7, the collector temperature at the end of the period from 8 to 9, assuming that the initial collector temperature is equal to the ambient temperature, is

$$T_p^+ = T_a + \frac{S}{U_L}\left[1 - \exp\left(-\frac{A_c U_L t}{(mC)_e}\right)\right]$$

$$= -8 + \frac{0.34 \times 10^6/3600}{8}\left[1 - \exp\left(-\frac{8 \times 3600}{8000}\right)\right] = 3 \text{ C}$$

For the second hour period, the initial temperature is 3 C and the temperature at 10:00 AM becomes

$$T_p^+ = T_a + \frac{S}{U_L} - \left[\frac{S}{U_L} - (T_i - T_a)\right]\exp\left(-\frac{A U_L t}{(mC)_e}\right)$$

$$= -2 + \frac{0.79 \times 10^6/3600}{8} - \left[\frac{0.79 \times 10^6/3600}{8} - (3 + 2)\right]$$

$$\times \exp\left(-\frac{8 \times 3600}{8000}\right) = 25 \text{ C}$$

By 10:00 AM, the collector has been heated to within 15 C of its operating temperature at 40 C. The reduction in useful gain is the energy required to heat the collector the last 15 C, or 120 kJ/m^2. Thus the useful energy gain from 10 to 11 should be reduced from 1.65 to 1.53 MJ/m^2. Note that this collector responds quickly to the various changes as the exponential term in the preceding calculation was small. [The collector "time constant" is $(mC)_e/A_c U_L$, which is approximately 15 minutes. The time constant with liquid flowing is on the order of 1 to 2 minutes, as shown in Section 6.17.]   ∎

6.13 LIQUID HEATER PLATE GEOMETRIES

In the preceding sections, we have considered only one basic collector design: a sheet and tube solar water heater with parallel tubes on the back of the plate. There are many different designs of flat-plate collectors, but fortunately it is not necessary to develop a completely new analysis for each situation. Hottel and Whillier (1958), Whillier (1977), and Bliss (1959) have shown that the generalized relationships developed for the tube and sheet case apply to most collector designs. It is necessary to derive the appropriate form of the collector efficiency factor F', and Equations 6.7.5 and 6.7.6 then can be used to predict the thermal performance. Under some circumstances, the loss coefficient U_L will have to be modified slightly. In this and the next section the analyses of the basic design are applied to other configurations.

Figure 6.13.1 shows seven different liquid heater designs. The first three have parallel tubes (risers) thermally fastened to a plate and connected at the top and bottom by headers to admit and remove the liquid. The first of these is the basic design discussed in the previous sections; the important equations for F, F', and U_L are shown. The design shown in (b) is the same, except that the tubes are mounted on top of the plate rather than under it. A design shown in (c) has the tubes centered in the plane of the plate forming an integral part of the plate structure.

In types (d), (e), and (f), long narrow flat absorbers are mounted inside evacuated glass tubes.[8] Convective heat losses from collectors are usually significant, but they can be reduced or eliminated by evacuating the space between the absorber and the cover. As the pressure is reduced to a moderately low level, convection ceases but conduction through the gas remains constant until the mean free path of the molecules is on the order of the characteristic spacing. There are flat-plate collectors built with seals at the edges, posts to support the cover, and evacuated spaces above (and below) the absorber plate However, most practical designs have been based on evacuated tubes, which provide the structural strength to withstand the pressure differences.

---

[8] The first evacuated tubular collector was built by Speyer (1965). The design shown in Figure 6.13.1(e) is similar to Speyer's.

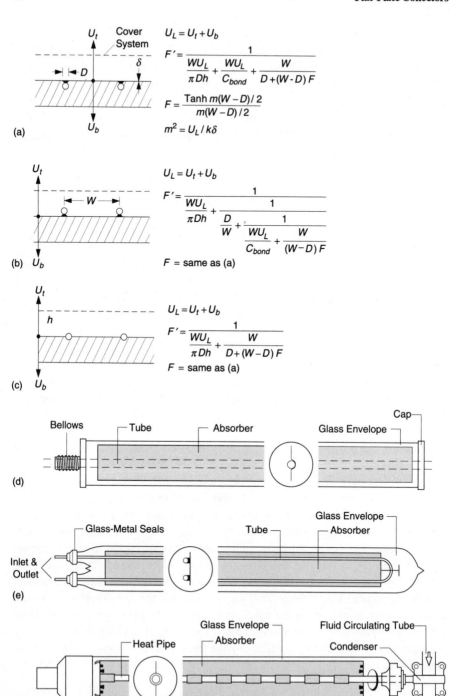

Panel (a):
$$U_L = U_t + U_b$$

$$F' = \cfrac{1}{\cfrac{WU_L}{\pi Dh} + \cfrac{WU_L}{C_{bond}} + \cfrac{W}{D+(W-D)F}}$$

$$F = \frac{\text{Tanh } m(W-D)/2}{m(W-D)/2}$$

$$m^2 = U_L/k\delta$$

Panel (b):
$$U_L = U_t + U_b$$

$$F' = \cfrac{1}{\cfrac{WU_L}{\pi Dh} + \cfrac{1}{\cfrac{D}{W} + \cfrac{1}{\cfrac{WU_L}{C_{bond}} + \cfrac{W}{(W-D)F}}}}$$

F = same as (a)

Panel (c):
$$U_L = U_t + U_b$$

$$F' = \cfrac{1}{\cfrac{WU_L}{\pi Dh} + \cfrac{W}{D+(W-D)F}}$$

F = same as (a)

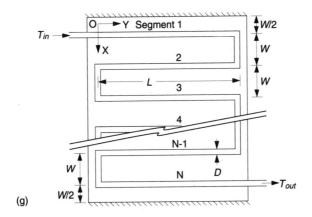

Figure 6.13.1(a-g) Liquid heater designs and collector efficiency factor equations.

The collector type shown in (d) is constructed with a single fin and tube with glass-to-metal seals at both ends; with this configuration, bellows are used to accommodate differential expansion of the glass envelope and the metal fin and tube. This configuration is like type a with a single riser tube.

Liquid flow in the type (e) collector is "down and back," with a U-tube joining the two conduits. Two glass-to-metal seals are provided at the same end of the tube. The two flow conduits down and back are in close proximity; with this arrangement, it is necessary that the thermal resistance between the two conduits be high, i.e., that the two streams be thermally decoupled. If the resistance were zero, the two conduits would be at the same temperature, and collection would be zero. Ortabasi and Buehl (1975) constructed such a collector with the plate split into two parts to avoid the coupling problem. If coupling is not significant, the analysis of type (e) collectors is basically the same as for the types (a), (b), or (c) even though the inlet and outlet manifolds are at the same end of the tubes.

Heat pipes can be used to extract energy from evacuated collectors, as shown in Figure 6.13.1(f). In the arrangement shown, the portion of the heat pipe in contact with the fin is the boiler portion. The condenser is a short section in good thermal contact with the pipe or duct through which the fluid to be heated is pumped; the condenser is shown fastened with a conducting clamp to the pipe carrying the fluid to be heated. These designs have the advantage that they have only one seal, at one end of the tube, and the fin and heat pipe are free to expand inside of the evacuated space. In contrast to most other flat-plate collectors, the temperature gradients along the length of the heat pipe will be small, but there will be gradients along the header from one heat pipe connector to the next.

There are important differences between flat-plate collectors with flat covers like types (a) to (c) and cylindrical covers like types (d) to (f). The fin width (or diameter) must be less than the tube diameter, so the absorbing surface will have projected areas less than that of the tube. The tubes are usually closely packed. The planes of the absorbers may be different from the plane of the tube arrays. The angular dependence of solar transmittance will be different from that of a flat cover.[9] Manifold designs vary widely, and manifold losses may be important. And, metal-to-glass seals must be provided.

Other important collector geometries exist for which F' and F_R cannot easily be expressed in a simple form. The risers in (a), (b), and (c) are all parallel tubes; an alternative design is the serpentine tube arrangement shown in Figure 6.13.1(g). If a thermal break is made midway between the tubes, e.g., by cutting through the absorber plate, then the collector can be analyzed as a conventional collector. If a break is not provided, reduced performance can be expected and a more complicated analysis is necessary.

Abdel-Khalik (1976) analytically solved the case of a single bend, and Zhang and Lavan (1985) obtained solutions for three and four bends. For a single bend, Zhang and Lavan show that the solution for F_R is given by Equation 6.13.1 in terms of three dimensionless parameters, F_1, F_2, and F_3. (The parameters F_4, F_5, and F_6 are functions of F_2 only.)

$$\frac{F_R}{F_1} = F_3 \left\{ \frac{2F_5F_3}{F_6 \exp\left[-F_3\left(1 - F_2^2\right)^{1/2}\right] + F_5} - F_5 \right\} \tag{6.13.1}$$

The parameters F_1 through F_6 are given by

$$F_1 = \frac{\kappa}{U_L W} \frac{\kappa R(1 + \gamma)^2 - 1 - \gamma - \kappa R}{\left[\kappa R(1 + \gamma) - 1\right]^2 - (\kappa R)^2} \tag{6.13.2a}$$

$$F_2 = \frac{1}{\kappa R(1 + \gamma)^2 - 1 - \gamma - \kappa R} \tag{6.13.2b}$$

$$F_3 = \frac{\dot{m}C_p}{F_1 U_L A_c} \tag{6.13.2c}$$

$$F_4 = \left(\frac{1 - F_2^2}{F_2^2}\right)^{1/2} \tag{6.13.2d}$$

$$F_5 = 1/F_2 + F_4 - 1 \tag{6.13.2e}$$

$$F_6 = 1 - 1/F_2 + F_4 \tag{6.13.2f}$$

and $$\kappa = \frac{\left(k\delta U_L\right)^{1/2}}{\sinh\left[(W - D)\left(U_L/k\delta\right)^{1/2}\right]} \tag{6.13.2g}$$

---

[9] See Section 6.19 for further information on angular dependence of $(\tau\alpha)$ of evacuated tube collectors.

$$\gamma = -2 \cosh\left[(W - D)(U_L/k\delta)^{1/2}\right] - \frac{DU_L}{\kappa} \qquad (6.13.2h)$$

$$R = \frac{1}{C_b} + \frac{1}{\pi D_i h_{fi}} \qquad (6.13.2i)$$

Zhang and Lavan point out that Equation 6.13.1 is valid for any number of bends if the group $\dot{m}C_p/F_1 U_L A_c$ is greater than about 1.0. For smaller values of this group, their paper should be consulted.

Example 6.13.1

Determine the heat removal factor for a collector with serpentine tube having the following specifications [see Figure 6.13.1(g)]:

Length of one serpentine segment L	1.2 m
Distance between tubes W	0.1 m
Number of segments N	6
Plate thickness δ	1.5 mm
Tube outside diameter D	7.5 mm
Tube inside diameter D_i	6.5 mm
Plate thermal conductivity k	211 W/mC
Overall loss coefficient U_L	5 W/m²C
Fluid mass flow rate $\dot{m}$	0.014 kg/s
Fluid specific heat C_p	3352 J/kgC
Fluid-to-tube heat transfer coefficient h_{fi}	1500 W/m²C
Bond conductance C_b	∞

Solution

From Equation 6.13.2(g):

$$\kappa = \frac{(211 \times 0.0015 \times 5)^{1/2}}{\sinh\left[(0.1 - 0.0075)\left(\dfrac{5}{211 \times 0.0015}\right)^{1/2}\right]} = 3.346 \text{ W/mC}$$

From Equation 6.13.2(h):

$$\gamma = -2 \cosh\left[(0.1 - 0.0075)\left(\frac{5}{211 \times 0.0015}\right)^{1/2}\right] - \frac{0.0075 \times 5}{3.346} = -2.148$$

From Equation 6.13.2(i):

$$R = \frac{1}{\pi \times 0.0065 \times 1500} = 0.0326 \text{ mC/W}$$

Then $kR = 3.346 \times 0.0326 = 0.1091$. From Equation 6.13.2(a):

$$F_1 = \frac{3.346}{5 \times 0.1} \frac{0.1091(1 - 2.148)^2 - 1 + 2.148 - 0.1091}{[0.1091(1 - 2.148) - 1]^2 - (0.1091)^2} = 6.310$$

From Equation 6.13.2(b):

$$F_2 = \frac{1}{1.1827} = 0.846$$

The collector area is $NWL = 6 \times 0.1 \times 1.2 = 0.72$ m^2. From Equation 6.13.2(c) the dimensionless capacitance rate is

$$F_3 = \frac{\dot{m}C_p}{F_1 U_L A_c} = \frac{0.014 \times 3352}{6.309 \times 5 \times 0.72} = 2.066$$

From Equations 6.13.2(d) to 6.13.2(f), $F_4 = 0.631$, $F_5 = 0.814$, and $F_6 = 0.449$. Finally from Equation 6.13.1,

$$F_R/F_1 = 0.148 \quad \text{and} \quad F_R = 0.148 \times 6.309 = 0.93 \qquad \blacksquare$$

6.14   AIR HEATERS

Figure 6.14.1 shows six designs for air heating collectors. Also on this figure are equations for the collector efficiency factors that have been derived for these geometries. For (e) and (f), the Löf overlapped glass plates and the matrix air heater, the analyses to date have not put the results in a generalized form. For these two situations, it is necessary to resort to numerical techniques for analysis. Selcuk (1971) has analyzed the overlapped glass plate system. Chiou et al. (1965), Hamid and Beckman (1971), and Neeper (1979) have studied the matrix-type air heaters. Hollands and Shewen (1981) have analyzed the effects of flow passage geometry on h_{fi} and F'.

To illustrate the procedure for deriving F' and U_L for an air heater, we derive the equation for type (a) of Figure 6.14.1. Although type (b) is the more common design for an air heater, type (a) is somewhat more complicated to analyze. Also, type (a) is similar to a collector-storage wall in a passive heating system. A schematic of the collector and thermal network are shown in Figure 6.14.2. The derivation follows that suggested by Jones (1979).

At some location along the flow direction the absorbed solar energy heats up the plate to a temperature T_p. Energy is transferred from the plate to the ambient air at T_a through the back loss coefficient U_b, to the fluid at T_f through the convection heat transfer coefficient h_2, and to the bottom of the cover glass through the

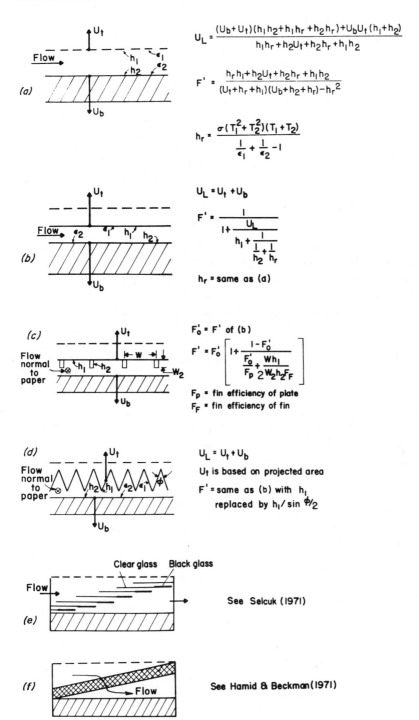

$$U_L = \frac{(U_b + U_t)(h_1 h_2 + h_1 h_r + h_2 h_r) + U_b U_t (h_1 + h_2)}{h_1 h_r + h_2 U_t + h_2 h_r + h_1 h_2}$$

$$F' = \frac{h_r h_1 + h_2 U_t + h_2 h_r + h_1 h_2}{(U_t + h_r + h_1)(U_b + h_2 + h_r) - h_r^2}$$

$$h_r = \frac{\sigma (T_1^2 + T_2^2)(T_1 + T_2)}{\frac{1}{\epsilon_1} + \frac{1}{\epsilon_2} - 1}$$

$$U_L = U_t + U_b$$

$$F' = \frac{1}{1 + \dfrac{U_L}{h_1 + \dfrac{1}{\dfrac{1}{h_2} + \dfrac{1}{h_r}}}}$$

h_r = same as (a)

$F_o' = F'$ of (b)

$$F' = F_o' \left[1 + \frac{1 - F_o'}{\dfrac{F_o'}{F_p} + \dfrac{W h_1}{2 W_2 h_2 F_F}} \right]$$

F_p = fin efficiency of plate
F_F = fin efficiency of fin

$$U_L = U_t + U_b$$

U_t is based on projected area

F' = same as (b) with h_1
    replaced by $h_1 / \sin \phi/2$

See Selcuk (1971)

See Hamid & Beckman (1971)

Figure 6.14.1(a-f) Air heater designs and efficiency factors.

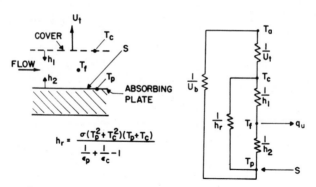

Figure 6.14.2 Type (a) solar air heater and thermal network.

linearized radiation heat transfer coefficient h_r. Energy is transferred to the cover glass from the fluid through the heat transfer coefficient h_1. Energy is lost to the ambient air though the combined convection and radiation coefficient U_t. Note that U_t can account for multiple covers.

Energy balances on the cover, the plate, and the fluid yield the following equations:

$$U_t(T_a - T_c) + h_r(T_p - T_c) + h_1(T_f - T_c) = 0 \qquad (6.14.1)$$

$$S + U_b(T_a - T_p) + h_2(T_f - T_p) + h_r(T_c - T_p) = 0 \qquad (6.14.2)$$

$$h_1(T_c - T_f) + h_2(T_p - T_f) = q_u \qquad (6.14.3)$$

These three equations are solved so that the useful gain is expressed as a function of U_t, h_1, h_2, h_r, T_f, and T_a. In other words, T_p and T_c must be eliminated. The algebra is somewhat tedious and only a few of the intermediate steps are given. Solving the first two equations for $T_p - T_f$ and $T_c - T_f$,

$$T_p - T_f = \frac{S(U_t + h_r + h_1) - (T_f - T_a)(U_t h_r + U_t U_b + U_b h_r + U_b h_1)}{(U_t + h_r + h_1)(U_b + h_2 + h_r) - h_2{}^2}$$
$$(6.14.4)$$

and
$$T_c - T_f = \frac{h_r S - (T_f - T_a)(U_t h_2 + U_t U_b + U_t h_r + U_b h_r)}{(U_t + h_r + h_1)(U_b + h_2 + h_r) - h_r{}^2} \qquad (6.14.5)$$

Substituting these into the equation for q_u and rearranging, we obtain

$$q_u = F'[S - U_L(T_f - T_a)] \qquad (6.14.6)$$

where
$$F' = \frac{h_r h_1 + U_t h_2 + h_2 h_r + h_1 h_2}{(U_t + h_r + h_1)(U_b + h_2 + h_r) - h_r{}^2} \qquad (6.14.7)$$

and
$$U_L = \frac{(U_b + U_t)(h_r h_1 + h_2 h_r + h_1 h_2) + U_b U_t(h_1 + h_2)}{h_r h_1 + U_t h_2 + h_2 h_r + h_1 h_2} \tag{6.14.8}$$

Note that U_L for this collector is not just the top loss coefficient in the absence of back losses but also accounts for heat transfer between the absorbing surface and the bottom of the cover. Whenever the heat removal fluid is in contact with a transparent cover, U_L will be modified in a similar fashion.

The equations for type (b) air heaters are derived in a similar manner but the working fluid does not contact the cover system. For simplicity, back losses are assumed to occur from the absorber plate temperature. The following example shows calculation of the performance of a type (b) air heater.

Example 6.14.1

Calculate the outlet temperature and efficiency of a single-cover type (b) air heater of Figure 6.14.1 at a 45° slope when the radiation incident on the collector is 900 W/m². The plate-to-cover spacing is 20 mm and the air channel depth is 10 mm. The collector is 1 m wide by 4 m long. The absorber plate is selective with an emittance of 0.1 and the effective transmittance-absorptance product is 0.82. The inlet air temperature is 60 C, the ambient air temperature is 10 C, and the mass flow rate of the air is 0.056 kg/s. The wind heat transfer coefficient is 10 W/m²C and the sum of the back and edge loss coefficients is 1.0 W/m²C (see Example 6.4.3).

Solution

From Figure 6.4.4(e) with an assumed average plate temperature of 70 C, the top loss coefficient is 3.3 W/m²C, and with the back and edge loss coefficient of 1.0 W/m²C, the overall loss coefficient is 4.3 W/m²C. The radiation coefficient between the two air duct surfaces is estimated by assuming a mean radiant temperature equal to the mean fluid temperature. With an estimated mean fluid temperature of 70 C, we have, from Equations 3.10.2 and 3.10.3,

$$h_r = \frac{4\sigma \bar{T}^3}{(1/\varepsilon_1) + (1/\varepsilon_2) - 1} = \frac{4 \times 5.67 \times 10^{-8} \times 343^3}{(2/0.95) - 1} = 8.3 \text{ W/m}^2 \text{ C}$$

The heat transfer coefficients between the air and two duct walls will be assumed to be equal. The characteristic length is the hydraulic diameter, which for flat plates is twice the plate spacing. The Reynolds number, at an assumed average fluid temperature of 70 C, is

$$Re = \frac{\rho V D_h}{\mu} = \frac{\dot{m} D_h}{A_f \mu} = \frac{0.056(2 \times 0.01)}{(0.01 \times 1)2.04 \times 10^{-5}} = 5490$$

The length-to-diameter ratio is

$$\frac{L}{D_h} = \frac{4}{2 \times 0.01} = 200$$

Since Re is greater than 2100 and L/D_h is large, the flow is turbulent and fully developed. From Equation 3.14.6

$$Nu = 0.0158(5490)^{0.8} = 15.5$$

The heat transfer coefficients inside of the duct, h_1 and h_2, are then

$$h = 15.5 \frac{k}{D_h} = \frac{15.5 \times 0.029}{2 \times 0.01} = 22 \text{ W/m}^2 \text{ C}$$

From Figure 6.14.1b, with $h_1 = h_2$,

$$F' = \left[1 + \frac{U_L}{h + [(1/h) + (1/h_r)]^{-1}}\right]^{-1} = 0.87$$

The dimensionless capacitance rate is

$$\frac{\dot{m}C_p}{A_c U_L F'} = \frac{0.056 \times 1009}{4 \times 4.3 \times 0.87} = 3.78$$

From Equation 6.7.5 or Figure 6.7.1,

$$F'' = 3.78[1 - e^{-1/3.78}] = 0.88$$

or $$F_R = F''F' = 0.88 \times 0.87 = 0.77$$

The useful gain, from Equation 6.7.6, is

$$Q_u = 4 \times 0.77[900 \times 0.82 - 4.3(60 - 10)] = 1610 \text{ W}$$

The outlet temperature is

$$T_o = T_i + \frac{Q_u}{\dot{m}C_p} = 60 + \frac{1610}{0.056 \times 1009} = 89 \text{ C}$$

It is now necessary to check the assumed mean fluid and absorber plate temperatures.

The mean plate temperature is found from Equation 6.9.4,

$$T_{pm} = 60 + \frac{1610/4}{4.3 \times 0.77}(1 - 0.77) = 88 \text{ C}$$

and the mean fluid temperature is found from Equation 6.9.2,

$$T_{fm} = 60 + \frac{1610/4}{4.3 \times 0.77}(1 - 0.88) = 74 \text{ C}$$

Since the initial guess of the plate and fluid temperatures was 70 C another iteration is necessary. With a new assumed average plate temperature of 88 C, U_t is 3.4 W/m²C and U_L is 4.4 W/m²C. The radiation heat transfer coefficient between the two duct surfaces is 8.7 W/m²C (with $\overline{T}$ assumed to be the same as T_{fm}, or 348 K), the Reynolds number is 5400, and the heat transfer coefficient is 23 W/m²C. The parameters F', F'', and F_R are unchanged so that the useful energy gain remains at 1610 W. Note that even though the first iteration used an estimate of the plate temperature that was 18 C in error, only minor changes resulted from the second iteration.

The efficiency is

$$\eta = \frac{Q_u}{A_c G_T} = \frac{1610}{4 \times 900} = 0.45, \text{ or } 45\%$$

■

6.15 MEASUREMENTS OF COLLECTOR PERFORMANCE

The first detailed experimental study of the performance of flat-plate collectors, by Hottel and Woertz (1942), was based on energy balance measurements on an array of collectors on an experimental solar-heated building. The analysis was basically similar to that of the previous sections, but with performance calculations based on mean plate temperature rather than on inlet temperature and F_R. They developed a correlation for thermal losses which was a forerunner of Equation 6.4.9. Their experimental data were for time periods of many days, and calculated and measured performance agreed within approximately 13% before effects of dust and shading were taken into account.

Tabor (1958) modified the Hottel and Woertz loss calculation by use of new correlations for convection transfer between parallel planes and values of emittance of glass lower than those used by Hottel and Woertz. These modifications permitted estimation of loss coefficients for collectors with selective surfaces. He found equilibrium (no fluid flow) temperatures from experiment and theory for a particular collector to be 172 and 176 C, indicating satisfactory agreement. He also recalculated Hottel and Woertz results using his modified heat loss coefficients and found calculated and measured losses for two sets of conditions to be 326 versus 329 W/m^2C and 264 versus 262 W/m^2C, again indicating excellent agreement. Moore et al. (1974) made extensive comparisons of the performance of a flat-plate liquid heating collector with results predicted by use of the original Hottel and Woertz method. The operating conditions were similar to those of Hottel and Woertz, and agreement was good.

As shown by these examples and by many other measurements, there is substantial experimental evidence that the energy balance calculation methods developed in this chapter are very satisfactory representations of the performance of most flat-plate collectors.

6.16   COLLECTOR CHARACTERIZATIONS

Based on the theory outlined in the previous sections and the laboratory measurements that support the theory, several methods for characterizing collectors can be noted. These characterizations, or models, have various numbers of parameters and are thus of varying complexity, and they serve different purposes. At one extreme very detailed models include all of the design features of the collector (plate thickness, tube spacing, number of covers and cover material, back and edge insulation dimensions, etc). At the other extreme is a model that includes only two parameters, one that shows how the collector absorbs radiation and the other how it loses heat. (The simplest model would be a one-parameter model, a single efficiency; this is essentially useless as the efficiency is dependent on the collector operating temperature and ambient temperature, and in most collector applications this changes with time.)

For collector design (i.e., the specification of the details of the design such as plate thickness, tube spacing, etc.) and for detailed understanding of how collectors function, detailed models are appropriate. The most complete steady-state model includes all of the design parameters entering the terms in Equation 6.7.6, i.e., those that determine F_R, U_L, and $(\tau\alpha)$. To this can be added an analysis of transient behavior, which can be a single lumped capacitance analysis (like that of Section 6.12) or a multinode analysis. For most flat-plate collectors, the single-capacitance model is adequate, and for many it may not be necessary to consider thermal capacitance at all.

For use in simulations of thermal processes, less complex models are usually adequate. Several of these have been used [e.g., in the simulation program TRNSYS (1990) and by Wirsum (1988)]; in these models, one, two, or all of the three major terms in Equation 6.7.6[10] may be considered as variables: an angular-dependent $(\tau\alpha)_{av}$, a temperature-dependent U_L, and a temperature-dependent F_R.

---

[10] S in Equation 6.7.6 can be written as $G_T(\tau\alpha)_{av}$.

The definition of instantaneous efficiency, combined with Equation 6.7.6, provides the basis for simulation models:

$$\eta_i = \frac{Q_u}{A_c G_T} = \frac{F_R[G_T(\tau\alpha)_{av} - U_L(T_i - T_a)]}{G_T} \tag{6.16.1}$$

(Note that this equation is usually written without the subscript "av.") If most of the radiation is beam radiation that is nearly normal to the collector, and if F_R and U_L do not vary greatly in the range of operation of the collector, $F_R(\tau\alpha)_n$ and $F_R U_L$ are two parameters that describe how the collector works. $F_R(\tau\alpha)_n$ is an indication of how energy is absorbed and $F_R U_L$ is an indication of how energy is lost. These two parameters constitute the simplest practical collector model.

A third parameter describes the effects of the angle of incidence of the radiation. An **incidence angle modifier** can be defined as the ratio $(\tau\alpha)/(\tau\alpha)_n$; it is a function of θ_b. As shown in the next section, the incidence angle modifier can be approximately represented in terms of a single coefficient b_o, the third parameter in this simple collector model.

We now have a three-parameter model $[F_R(\tau\alpha)_n, F_R U_L,$ and $b_o]$ for a flat-plate collector which takes into account the major phenomena associated with collector operation. This model will be utilized in Part III of this book on design of systems.

Other models can be written which account for temperature dependence of U_L and F_R and for complex angular relationships for $(\tau\alpha)$. Thus there is a range of collector thermal performance models that vary in their level of detail. The selection of a model depends on the purposes at hand. If the collector operates over restricted ranges of conditions, which is true of many space and water heating applications, then the simplest model is quite adequate. If operating conditions are highly variable, more complex models may be needed.

6.17 COLLECTOR TESTS: EFFICIENCY, INCIDENCE ANGLE MODIFIER, AND TIME CONSTANT

This section is concerned with how collector tests are done and how test data are presented in useful ways. In the next section, test data are presented for typical flat-plate collectors of the types shown in Sections 6.13 and 6.14. The following two sections discuss methods for manipulating test data to get them in forms that are needed for designing solar energy systems.

In the mid-1970s many new collector designs appeared on the commercial market. Needs developed for standard tests to provide collector operating data. Information was needed on how a collector absorbs energy, how it loses heat, the effects of angle of incidence of solar radiation, and the significant heat capacity effects. In response to this need in the USA, the National Bureau of Standards [Hill and Kusuda (1974)] devised a test procedure [see also Hill and Streed (1976, 1977) and Hill et al. (1979)][11] which has been modified by ASHRAE (1977). The ASHRAE 93-77 standard procedure is the basis for this section. At the end of the section, references are provided to more detailed discussions of test procedures, to other test procedures, to analyses of test results, and to procedures that expand the capabilities of the tests to additional kinds of collectors and to high temperature operation.

Collector thermal performance tests can be considered in three parts. The first is determination of instantaneous efficiency with beam radiation nearly normal to the absorber surface. The second is determination of effects of angle of incidence of the solar radiation. The third is determination of collector time constant, a measure of effective heat capacity.

The basic method of measuring collector performance is to expose the operating collector to solar radiation and measure the fluid inlet and outlet temperatures and the fluid flow rate. The useful gain is

$$Q_u = \dot{m}C_p(T_o - T_i) \tag{6.17.1}$$

In addition, radiation on the collector, ambient temperature, and wind speed are also recorded. Thus two types of information are available: data on the thermal output and data on the conditions producing that thermal performance. These data permit the characterization of a collector by parameters that indicate how the collector absorbs energy and how it loses energy to the surroundings.

Equation 6.16.1, which describes the thermal performance of a collector operating under steady conditions, can be written without the subscript on $(\tau\alpha)$:

$$Q_u = A_c F_R[G_T(\tau\alpha) - U_L(T_i - T_a)] \tag{6.17.2}$$

Here $(\tau\alpha)$ is a transmittance-absorptance product that is weighted according to the proportions of beam, diffuse, and ground-reflected radiation on the collector, as discussed in Section 5.8; it is customary to omit the subscript. $(\tau\alpha)$ is determined under test conditions which are generally similar to conditions under which the collectors will provide most of their useful output, i.e., when radiation is high and most of the incident radiation is beam radiation.

Equations 6.17.1 and 6.17.2 can be used to define an instantaneous efficiency:

$$\eta_i = \frac{Q_u}{A_c G_T} = F_R(\tau\alpha) - \frac{F_R U_L(T_i - T_a)}{G_T} \tag{6.17.3}$$

and $$\eta_i = \frac{\dot{m}C_p(T_o - T_i)}{A_c G_T} \tag{6.17.4}$$

These equations are the basis of the standard test methods outlined in this section.

---

[11] European test practices are discussed in detail by Gillett and Moon (1985).

Other equations are also used. European practice is to base collector test results on $T_{f,\mathrm{av}}$ the arithmetic average of the fluid inlet and outlet temperatures.[12] Thus

$$\eta_i = F_{av}(\tau\alpha)_n - \frac{F_{av}U_L(T_{f,av} - T_a)}{G_T} \tag{6.17.5}$$

ASHRAE 93-77 sets forth three standard test procedures for liquid heaters and one for air heaters. A schematic of one of these, a closed-loop system for liquid heating collectors, is shown in Figure 6.17.1. Although details differ, the essential features of all of the procedures can be summarized as follows:

1   Means are provided to feed the collector with fluid at a controlled inlet temperature; tests are made over a range of inlet temperatures.
2   Solar radiation is measured by a pyranometer on the plane of the collector.
3   Means of measuring flow rate, inlet and outlet fluid temperatures, and ambient conditions are provided.
4   Means are provided for measurements of pressure and pressure drop across the collector.

The ASHRAE method for air collectors includes the essential features of those for liquid heaters, with the addition of detailed specifications of conditions relating to air flow, air mixing, air temperature measurements, and pressure drop measurements.

Measurements may be made either outdoors or indoors. Indoor tests are made using a solar simulator, that is, a source producing radiant energy that has spectral distribution, intensity, uniformity in intensity, and direction closely resembling that of solar radiation. Means must also be provided to move air to produce wind. [See Vernon and Simon (1974), Simon (1976), and Gillett (1980).] There are not many test facilities of this kind available, the results are not always comparable to those of outdoor tests, and most collector tests are done outdoors. Gillett (1980) has compared the results of outdoor tests with those of mixed indoor and outdoor tests and notes that diffuse fraction and long-wave radiation exchange (which are not the same indoors and outdoors) can affect the relative results of the tests.

The general test procedure is to operate the collector in the test facility under nearly steady conditions, measure the data to determine Q_u from Equation 6.17.1, and measure G_T, T_i, and T_a which are needed for analysis based on Equation 6.17.3. Of necessity, this means outdoor tests are done in the midday hours on clear days when the beam radiation is high and usually with the beam radiation nearly normal to the collector. Thus the transmittance-absorptance product for these test conditions is approximately the normal incidence value and is written as $(\tau\alpha)_n$.

Tests are made with a range of inlet temperature conditions. To minimize effects of heat capacity of collectors, tests are usually made in nearly symmetrical pairs, one before and one after solar noon, with results of the pairs averaged. Instantaneous efficiencies are determined from $\eta_i = \dot{m}C_p(T_o - T_i)/A_cG_T$ for the

---

[12] Methods for converting between F_R and F_{av} are given in Section 6.19.

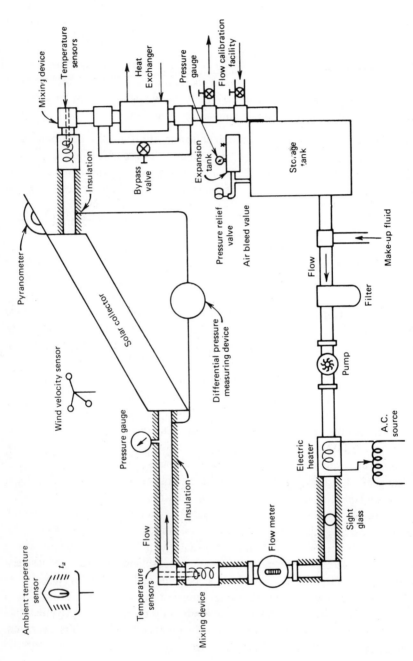

Figure 6.17.1 Closed-loop test setup for liquid heating flat-plate collectors. From ASHRAE Standard 93-77 (1977), with permission.

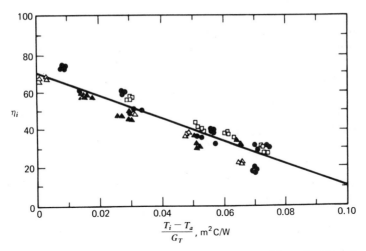

Figure 6.17.2 Experimental collector efficiency data measured for a liquid heating flat-plate collector with one cover and a selective absorber. Sixteen points are shown for each of five test sites. The curve represents the theoretical characteristic derived from points calculated for the test conditions. Adapted from Streed et al. (1979).

averaged pairs, and are plotted as a function of $(T_i - T_a)/G_T$. A sample plot of data taken at five test sites under conditions meeting ASHRAE 93-77 specifications, from Streed et al. (1979), is shown in Figure 6.17.2.

If U_L, F_R, and $(\tau\alpha)_n$ were all constant, the plots of η_i versus $(T_i - T_a)/G_T$ would be straight lines with intercept $F_R(\tau\alpha)_n$ and slope $-F_R U_L$. However, they are not, and the data scatter. From Section 6.4 it is clear that U_L is a function of temperatures and wind speed, with decreasing dependence as the number of covers increases. Also, F_R is a weak function of temperature. And some variations of the relative proportions of beam, diffuse, and ground-reflected components of solar radiation will occur. Thus scatter in the data are to be expected, because of temperature dependence, wind effects, and angle of incidence variations. In spite of these difficulties, long-time performance estimates of many solar heating systems, collectors can be characterized by the intercept and slope [i.e., by $F_R(\tau\alpha)_n$ and $F_R U_L$].

Example 6.17.1

A water heating collector with an aperture area of 4.10 m² is tested by the ASHRAE method, with beam radiation nearly normal to the plane of the collector. The following information comes from the test:

Q_u	G_T	T_i	T_a
9.05 MJ/hr	864 W/m²	18.2 C	10.0 C
1.98	894	84.1	10.0

What are $F_R(\tau\alpha)_n$ and $F_R U_L$ for this collector, based on aperture area?

Solution

For the first data set,

$$\eta_i = \frac{9.05 \times 1000}{864 \times 3.6 \times 4.10} = 0.71$$

and $\qquad \dfrac{T_i - T_a}{G_T} = \dfrac{18.2 - 10.0}{864} = 0.0095 \ \text{m}^2\text{C/W}$

For the second data set, $\eta_i = 0.15$ and $(T_i - T_a)/G_T = 0.083 \ \text{m}^2\text{C/W}$. These two points can be plotted. The slope is

$$\frac{0.71 - 0.15}{0.0095 - 0.083} = -7.62 \ \text{W/m}^2\text{C}$$

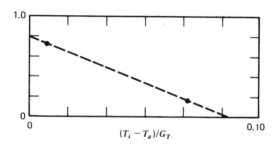

Then $F_R U_L = -\text{slope} = 7.62 \ \text{W/m}^2\text{C}$. The intercept of the line of the η_i axis is 0.78, which is $F_R(\tau\alpha)_n$. (In practice, tests produce multiple data points and a least squares fit would be used to find the best constants.) ∎

The collector area appears in the denominator of the definition of η_i. Various areas have been used: gross area, glass area, aperture area, unshaded absorber plate area, etc. Gross collector area is defined as the total area occupied by a collector module, that is, the total area of a collector array divided by the number of modules in the array. Aperture area is defined as the unobstructed cover area or the total cover area less the area of cover supports. The unshaded absorber plate area is sometimes used. ASHRAE 93-77 uses gross collector area, a satisfactory choice. It does not matter which area is used as long as it is specified clearly so that the same area basis can be used in subsequent design calculations based on the results of the collector tests.

In Example 6.17.1, if the collector has a gross area of 4.37 m^2, $F_R(\tau\alpha)_n$ and $F_R U_L$ based on the gross area would be the numbers based on the aperture area multiplied by 4.10/4.37.

Additional methods of treating test data may be encountered. For example, Cooper and Dunkle (1981) assume a linear temperature dependence of the overall loss coefficient and the formulation of the instantaneous collector efficiency as a second-order fit:

$$U_L = a + b(T_i - T_a) \tag{6.17.6}$$

$$\eta_i = F_R(\tau\alpha) - a\frac{(T_i - T_a)}{G_T} - b\left[\frac{(T_i - T_a)}{G_T}\right]^2 \tag{6.17.7}$$

This formulation considers the temperature dependence of U_L, and a and b are constants at a particular wind speed; wind speed variations can dominate the temperature effect for one-cover collectors. This model[13] has additional parameters if wind speed dependence is considered.

The second important aspect of collector testing is the determination of effects of angle of incidence of the incident radiation. An **incidence angle modifier** $K_{\tau\alpha}$ can be introduced into Equation 6.17.2. The dependence of $(\tau\alpha)$ on the angle of incidence of radiation on the collector varies from one collector to another, and the standard test methods include experimental estimation of this effect. Note that the requirement of a clear test day means that the experimental value of $(\tau\alpha)$ will be essentially the same as $(\tau\alpha)_b$. The incidence angle modifier is written as

$$K_{\tau\alpha} = \frac{(\tau\alpha)}{(\tau\alpha)_n} \tag{6.17.8}$$

Then $$Q_u = A_c F_R[G_T K_{\tau\alpha}(\tau\alpha)_n - U_L(T_i - T_a)] \tag{6.17.9}$$

A general expression has been suggested by Souka and Safwat (1966) for angular dependence of $K_{\tau\alpha}$ for collectors with flat covers as

$$K_{\tau\alpha} = 1 + b_o\left(\frac{1}{\cos\theta} - 1\right) \tag{6.17.10}$$

where b_o is a constant called the **incidence angle modifier coefficient**. (Note that this equation follows the ASHRAE 93-77 convention, and b_o is generally a negative number.) Figure 6.17.3 shows an incidence angle modifier, $K_{\tau\alpha}$, plotted as a function of both θ and $(1/\cos\theta - 1)$. At incidence angles less than 60°, Equation 6.17.10 is a useful approximation to account for angle-of-incidence effects. At larger angles of incidence, the linear relationship no longer applies, but most of the useful energy absorbed in a collector system will be at times when θ is less than 60°.

Strictly speaking, the incident radiation should be considered as consisting of beam, diffuse, and ground-reflected radiation, with the three components treated

---

[13] Cooper and Dunkle also write this model in terms of average fluid temperature, $T_{f,av}$. This will be noted in the next section.

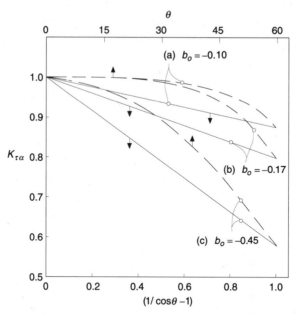

Figure 6.17.3 Incidence angle modifier as a function of θ and $(1/\cos\theta - 1)$ for a collector with a flat cover: (a) a one-glass-cover collector, (b) a two-glass-cover collector, and (c) a one-glass-cover collector with a nylon honeycomb. In each case, $\alpha = 0.9$. Adapted from ASHRAE 93-77 (1977).

separately. The beam component would be treated as outlined here, and the diffuse and ground-reflected radiation would be considered to be at the appropriate fixed angles of incidence (from Figure 5.4.1). However, collector tests are made under clear sky conditions when the fraction of beam is high so the data are indicative of the effects of the angle of incidence of beam radiation.

ASHRAE 93-77 recommends that experimental determination of $K_{\tau\alpha}$ be done by positioning a collector in indoor tests so that θ is 0, 30, 45, and 60°. For outdoor tests it is recommended that pairs of tests symmetrical about solar noon be done early and late in the day, when angles of incidence of beam radiation are approximately 30, 45, and 60°.

Evacuated tubular collectors of the types shown in Figure 6.13.1(d–f) have covers that are optically nonsymmetric. As recommended by McIntire and Read (1983), biaxial incidence angle modifiers are used. Figure 6.17.4 shows the two planes of the modifiers, the transverse and the longitudinal planes. An overall incidence angle modifier is taken as the product of the two:

$$K_{\tau\alpha} = (K_{\tau\alpha})_t \, (K_{\tau\alpha})_l \qquad (6.17.11)$$

The third aspect of collector testing is the determination of the heat capacity of a collector in terms of a **time constant**. The time constant is defined as the time required for a fluid leaving a collector to change through $1/e$ (i.e., 0.632) of the total

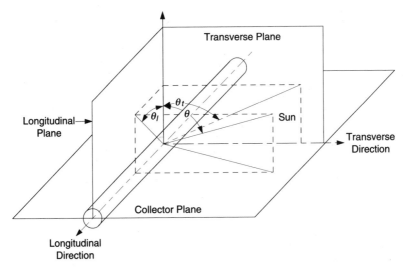

Figure 6.17.4.   The planes of the evacuated tube incidence angle modifiers.   From Theunissen and Beckman (1985).

change from its initial to its ultimate steady value after a step change in incident radiation or inlet fluid temperature. The ASHRAE standard test procedure outlines two procedures for estimating the time constant. The first is to operate a collector at nearly steady conditions with inlet fluid temperature controlled at or very near ambient temperature. The solar radiation is abruptly shut off by shading or repositioning the collector and the decrease in outlet temperature (with the pump running) is noted as a function of time. The time t at which the equality for Equation 6.17.12 is reached is the time constant of the collector:

$$\frac{T_{o,t} - T_i}{T_{o,\text{init}} - T_i} = \frac{1}{e} = 0.368 \tag{6.17.12}$$

where $T_{o,t}$ is the outlet temperature at time t and $T_{o,\text{init}}$ is the outlet temperature when the solar radiation is interrupted. A typical time-temperature history is shown in Figure 6.17.5.

An alternative method for measuring a collector time constant is to test a collector not exposed to radiation (i.e., at night, indoors, or shaded) and impose a step change in the temperature of the inlet fluid from a value well above ambient (e.g., 30 C) to a value very near ambient. Equation 6.17.12 also applies to this method (which may not give results identical with the first method).

There are other test procedures and analyses which produce more detailed information about collectors. Proctor (1984) provides a detailed discussion of evaluation of instruments used in tests and the sensitivity of test results to instrumental errors. He shows that uncertainties in radiation measurements dominate the uncertainties. He develops collector efficiency equations based on a mean fluid

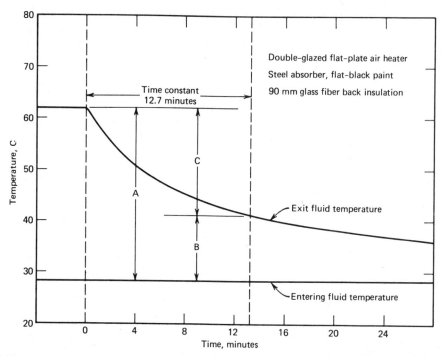

Figure 6.17.5 A time-temperature plot for a flat-plate air heater showing temperature drop on sudden reduction of the solar radiation on the collector to zero. The time constant is the time for the temperature to drop to 1/e of the total potential drop, that is, for B/A to reach 0.368. Adapted from Hill et al. (1979).

temperature $T_{f,m}$ that includes nonlinearities in ΔT, and separately accounts for ambient and sky temperatures. The result is a six-constant model with various combinations of the constants applicable to various types of collectors.

Gillett and Moon (1985) review European practices and standards for collector testing. Proctor and James (1985) note that there can be as much as ±10% variations in the measured values of $F_R(\tau\alpha)_n$ in production runs of collectors with selective surfaces.

6.18   TEST DATA

For purposes of illustrating the kinds of data that are available and the differences that exist among collectors, test results for several flat-plate collectors are shown. [See Hill et al. (1978) and the *Solar Products Specification Guide* (1983).] These data are based on the ASHRAE 93-77 standard procedures. Figure 6.18.1 shows test points and correlations for a double-glazed liquid heater. The points and solid curve are based on the aperture area of the collector, that is, the unobstructed glass area. The dashed curve shows the correlation based on the gross area of the collector. In some

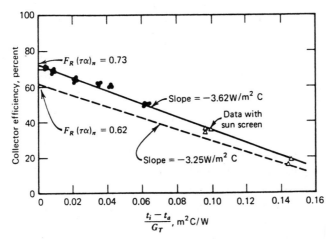

Figure 6.18.1 Test points for a liquid heater, based on collector aperture area. Also shown is the correlation based on gross collector area. The collector is double glazed with antireflective coatings on three glass surfaces and has a black chrome selective surface. Gross collector area is 1.66 m^2 and aperture area is 1.40 m^2. Adapted from Hill et al. (1979).

of the tests the intensity of incident radiation G_T was reduced by a shading screen to obtain points over a range of values of $(T_i - T_a)/G_T$.

Figure 6.18.2 shows a set of curves for liquid heating collectors based on net collector area. There are obvious differences in these collector characteristics. The selection of one or another of these collectors would depend on the operating conditions, i.e., on the range of values of $(T_i - T_a)/G_T$ that are expected in an

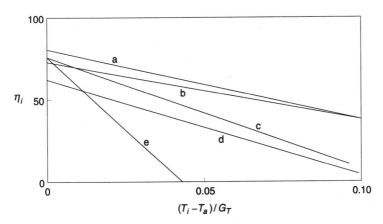

Figure 6.18.2 Characteristics of five flat-plate liquid heaters: (a) One-cover, selective black chrome absorber; (b) Two-cover, selective black chrome absorber; (c) One-cover, flat-black absorber; (d) two polycarbonate covers, polymeric flat black absorber with close spaced tubes; (e) unglazed flat-black absorber. Details of plate thickness, tube spacing, etc., vary among these collectors. The slope of (e) will be very sensitive to wind speed. Data are from Solar Products Specification Guide (1983).

application, and on costs. The fact that one collector may be more efficient than another in part or all of its range of operating conditions does not mean that it necessarily is a better collector in an economic context; as will be pointed out later, collectors must be evaluated in terms of their performance in systems and ultimately in terms of costs.

Hill et al. have measured the effects of angle of incidence of radiation on the collector of Figure 6.18.1; these are shown in Figure 6.18.3.

Figure 6.18.4 shows characteristic curves for two air heating flat-plate collectors, one of them operating at two different air flow rates. These curves are based on gross collector area. It is characteristic of most air heaters that they are operated at relatively low capacitance rates, have low heat transfer coefficients between absorber surface and fluid, and show relatively low values of F_R. The low flow rates are used to minimize pressure drop in the collectors. The collector represented by (a) is operated at higher than normal air flow rates. (Also, low air flow rates through pebble bed storage units used with air heaters in space heating applications result in high thermal stratification, which has operating advantages. See Chapters 9 and 13.)

Test data for flat-plate collectors in evacuated tubes are not as neatly categorized as those for collectors with flat covers, as there is more variability in their design and they may operate in higher temperature ranges where temperature dependence of losses is more important. Figure 6.18.5 shows η_i vs. $T_i - T_a)/G_T$ data for an evacuated tube collector with flat absorbers and close-packed tubes. There is a spread of the data and a dependence on radiation level that is not evident in ordinary flat-plate collectors.

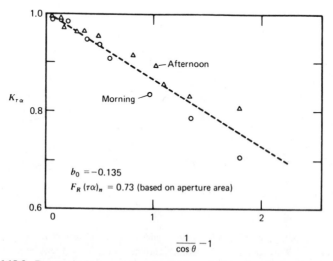

Figure 6.18.3 Data on incidence angle modifier for a double-glazed water heating collector. The conditions of operation were: collector tilted 25° to south; ambient temperature 34 C; wind speed 4.5 m/s; and insolation 230-830 W/m² of which an estimated 20% was diffuse. Adapted from Hill et al. (1979).

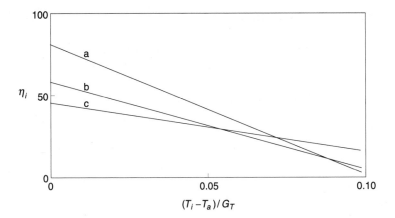

Figure 6.18.4 Characteristic curves for two air heaters: (a) one-cover, selective absorber, operated at 25 l/m²s; (b) same collector, operated at 10 l/m²s; (c) one-cover selective black chrome absorber, operated at 20 l/m²s. Data are from Solar Products Specification Guide (1983).

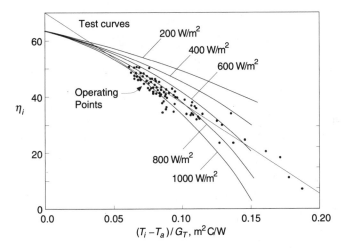

Figure 6.18.5 Test curves, operating points, and straight-line fit to the operating points for a Sanyo evacuated tube collector of the type shown in Figure 6.13.1(d), with tubes close packed, flat absorbers, and tubes oriented on horizontal axis. Adapted from IEA (1986).

As noted, the presence of cylindrical covers dictates that the optical properties of these collectors will not be symmetrical. For collectors with tubes on the sloped axis, if the slope is approximately the same as the latitude, the component of the angle of incidence in the longitudinal plane will vary with declination, but not greatly, and the major angle-of-incidence effects will be in the transverse plane.

6.19   THERMAL TEST DATA CONVERSION

Data from collector tests are usually shown in the form indicated by Figures 6.18.2 and 6.18.4 in the United States. This is a format useful in evaluating flat-plate collectors for many applications. Practice in some other countries is to plot η_i vs. $(T_{av} - T_a)/G_T$. Test data for air systems may be plotted as η_i vs. $(T_o - T_a)/G_T$. In this section, methods are presented for converting test results from one format to another.

If test data are plotted as η_i vs. $(T_{av} - T_a)/G_T$, where T_{av} is the arithmetic average of inlet and outlet fluid temperatures, the equation of the straight line is

$$\eta_i = F_{av}(\tau\alpha)_n - F_{av}U_L \frac{(T_{av} - T_a)}{G_T} \tag{6.19.1}$$

where F_{av} is approximately the same as F'; it would be equal to F' if the temperature rise through the collector were linear with distance. If the flow rate of the fluid is known, the intercept $F_{av}(\tau\alpha)_n$ and slope $-F_{av}U_L$ of the curve of Equation 6.19.1 are related to $F_R U_L$ and $F_R(\tau\alpha)_n$ by the following [from Beckman et al. (1977)]:

$$F_R(\tau\alpha)_n = F_{av}(\tau\alpha)_n \left[1 + \frac{A_c F_{av}U_L}{2\dot{m}C_p}\right]^{-1} \tag{6.19.2}$$

$$F_R U_L = F_{av}U_L \left[1 + \frac{A_c F_{av}U_L}{2\dot{m}C_p}\right]^{-1} \tag{6.19.3}$$

For air heater test data that are presented as plots of η_i vs. $(T_o - T_a)/G_T$, the intercept $F_o(\tau\alpha)_n$ and slope $-F_o U_L$ of these curves can be converted to $F_R U_L$ and $F_R(\tau\alpha)_n$ by the following:

$$F_R(\tau\alpha)_n = F_o(\tau\alpha)_n \left[1 + \frac{A_c F_o U_L}{\dot{m}C_p}\right]^{-1} \tag{6.19.4}$$

$$F_R U_L = F_o U_L \left[1 + \frac{A_c F_o U_L}{\dot{m}C_p}\right]^{-1} \tag{6.19.5}$$

Example 6.19.1

What are $F_R U_L$ and $F_R(\tau\alpha)_n$ for the two-cover air heater having $F_o U_L = 3.70$ W/m^{2}C and $F_o(\tau\alpha)_n = 0.64$? The volumetric flow rate per unit area is 10.1 l/m^{2}s.

Solution

For air at 20 C, $C_p = 1006$ J/kg and $\rho = 1.204$ kg/m^3 or 0.001204 kg/l. Then

$$\frac{A_c F_o U_L}{\dot{m}C_p} = \frac{3.7}{10.1 \times 0.001204 \times 1006} = 0.302$$

Then from Equations 6.19.2 and 6.19.3,

$$F_R U_L = 3.7(1 + 0.302)^{-1} = 2.84 \text{ W/m}^2\text{C}$$

$$F_R(\tau\alpha)_n = 0.64(1 + 0.302)^{-1} = 0.49$$ ∎

Figure 6.19.1 shows linear test results for an air heating collector plotted in all three of the common formats. The three curves have a common intercept at $\eta_i = 0$. For air heaters the differences in the three methods of plotting are substantial; for liquid heaters operating at normal flow rates, the differences among the three are less.

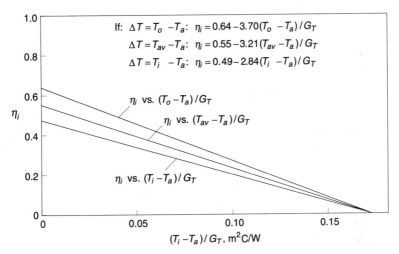

Figure 6.19.1 Efficiency curves for an air heating flat-plate collector plotted as functions of T_i, T_{av}, and T_o.

6.20 FLOW RATE CORRECTIONS TO $F_R(\tau\alpha)_n$ AND $F_R U_L$

Ideally, test data should be measured at flow rates corresponding to those to be used in applications. If a collector is to be used at a flow rate other than that of the test conditions, an approximate analytical correction to $F_R(\tau\alpha)_n$ and $F_R U_L$ can be obtained from the ratios of values of F_R determined by use of Equation 6.7.4 or 6.7.5. Assume that the only effect of changing flow rate is to change the temperature gradient in the flow direction, and that changes in F' due to changes of h_{fi} with flow rate are small (reasonable assumptions for liquid heating collectors operating at normal flow rates). The ratio r by which $F_R U_L$ and $F_R(\tau\alpha)_n$ are to be corrected is then given by

$$r = \frac{F_R U_L|_{use}}{F_R U_L|_{test}} = \frac{F_R(\tau\alpha)_n|_{use}}{F_R(\tau\alpha)_n|_{test}} \tag{6.20.1}$$

$$r = \frac{\dfrac{\dot{m}C_p}{A_c F' U_L}\left[1 - \exp\left(-A_c F' U_L/\dot{m}C_p\right)\right]\Big|_{use}}{\dfrac{\dot{m}C_p}{A_c F' U_L}\left[1 - \exp\left(-A_c F' U_L/\dot{m}C_p\right)\right]\Big|_{test}} \tag{6.20.2}$$

and

$$r = \frac{\dfrac{\dot{m}C_p}{A_c}\left[1 - \exp\left(-A_c F' U_L/\dot{m}C_p\right)\right]\Big|_{use}}{F_R U_L|_{test}} \tag{6.20.3}$$

To use these equations, it is necessary to know or estimate $F' U_L$. For the test conditions, it can be calculated from $F_R U_L$. Rearranging Equation 6.7.4 and solving for $F' U_L$ yields

$$F' U_L = -\frac{\dot{m}C_p}{A_c} \ln\left(1 - \frac{F_R U_L A_c}{\dot{m}C_p}\right) \tag{6.20.4}$$

For liquid collectors, $F' U_L$ calculated for the test conditions is approximately equal to $F' U_L$ at use conditions and can be used in both numerator and denominator of Equation 6.20.2.

Example 6.20.1

The water heating collector of Example 6.17.1 is to be used at a flow rate of 0.020 kg/s rather than at the 0.040 kg/s at which the test data were obtained. Estimate the effect of reducing the flow rate on $F_R(\tau\alpha)_n$ and $F_R U_L$.

Solution

First, calculate $F' U_L$ for the test conditions from Equation 6.20.4:

$$F' U_L = -\frac{0.040 \times 4187}{4.10} \ln\left(1 - \frac{7.62 \times 4.10}{0.040 \times 4187}\right) = 8.43 \text{ W/m}^2\text{C}$$

Then r is obtained with Equation 6.20.2. For use conditions,

$$\frac{\dot{m}C_p}{A_c F' U_L} = \frac{0.020 \times 4187}{4.10 \times 8.43} = 2.42$$

For test conditions

$$\frac{\dot{m}C_p}{A_c F' U_L} = \frac{0.040 \times 4187}{4.10 \times 8.43} = 4.85$$

So

$$r = \frac{2.42\left(1 - e^{-1/2.42}\right)}{4.85\left(1 - e^{-1/4.85}\right)} = 0.91$$

Then at the reduced flow rate

$$F_R(\tau\alpha)_n = 0.78 \times 0.91 = 0.71$$

$$F_R U_L = 7.62 \times 0.91 = 6.90 \text{ W/m}^2\text{C} \qquad \blacksquare$$

The procedure used in Example 6.20.1 should not be used for air heating collectors, or for liquid heating collectors where there is a strong dependence of h_{fi} and thus F' on flow rate. The defining equation for F' (Equation 6.5.18 or its equivalent from Figure 6.13.1 or 6.14.1) must be used at both use and test flow rates to estimate the effect of flow rate on F'. The collector tests do not provide adequate information with which to do this calculation, and theory must be used to estimate F' at use and test conditions.

Example 6.20.2

Test evaluations of the air heater of Example 6.14.1 give $F_R(\tau\alpha)_n = 0.63$ and $F_R U_L = 3.20$ W/m²C when the flow rate is 0.056 kg/s through the 4.00 m² collector. The conditions of the tests and of anticipated use (other than flow rate) are as outlined in Example 6.14.2. What will be $F_R(\tau\alpha)_n$ and $F_R U_L$ if the flow rate through the collector is reduced to 0.028 kg/s?

Solution

The effects of flow rate on F' need to be checked for this situation, since the convection coefficient from duct to air will change significantly. This requires calculation of the convection and radiation coefficients h_{fi} and h_r as was done in Example 6.14.2 and recalculation for the reduced flow rate. The details of that example are not repeated here. It will be assumed that h_r and U_L are not greatly changed from the test to the use conditions, so $h_r = 8.1$ W/m²C and $U_L = 4.3$ W/m²C.

When the flow rate is reduced, the Reynolds number becomes

$$Re = \frac{0.028(2 \times 0.01)}{(0.01 \times 1)2.04 \times 10^{-5}} = 2740$$

This is still in the turbulent range, so

$$Nu = 0.0158(2740)^{0.8} = 8.9$$

and $\qquad h_{fi} = \dfrac{8.9 \times 0.029}{2 \times 0.01} = 12.9 \text{ W/m}^2\text{C}$

The value of F' can now be calculated at the reduced flow rate in the same way it was calculated in Example 6.12.2 for the test conditions:

$$F' = \left[1 + \frac{4.3}{12.9 + \left(\dfrac{1}{12.9} + \dfrac{1}{8.1}\right)^{-1}}\right]^{-1} = 0.81$$

With the test condition calculation of $F' = 0.89$ and the use conditions $F' = 0.81$, we can go to Equation 6.20.2. For test conditions,

$$\frac{\dot{m}C_p}{A_c F' U_L} = 3.69$$

For use conditions,

$$\frac{\dot{m}C_p}{A_c F' U_L} = \frac{0.028 \times 1009}{4.00 \times 0.81 \times 4.3} = 2.03$$

So $\qquad r = \dfrac{2.03\left(1 - e^{-1/2.03}\right)}{3.69\left(1 - e^{-1/3.69}\right)} = 0.90$

Then for use conditions,

$$F_R(\tau\alpha)_n = 0.90 \times 0.63 = 0.57$$

and $\qquad F_R U_L = 0.90 \times 3.20 = 2.89 \text{ W/m}^2\text{C}$ ∎

6.21   FLOW DISTRIBUTION IN COLLECTORS

Performance calculations of collectors are based on an implicit assumption of uniform flow distribution in all of the risers in a single or multiple collector units. If flow is not uniform, the parts of the collectors with low flow through risers will have lower F_R than parts with higher flow rates. Thus the design of both headers and risers is important in obtaining good collector performance. This problem has been studied analytically and experimentally by Dunkle and Davey (1970). It is of particular significance in large forced-circulation systems; natural circulation systems tend to be self-correcting and the problem is not as critical.

Based on the assumptions that flow is turbulent in headers and laminar in risers (assumptions logical for Australian and many other water heaters), the analysis by Dunkle and Davey shows pressure drop along the headers for the common situation of water entering the bottom header at one side of the collector and leaving the top header at the other side. An example of calculated pressure distributions in top and bottom headers is shown in Figure 6.21.1. The implications of these pressure distributions are obvious; the pressure drops from bottom to top are greater at the ends than the center portion, leading to high flows in the end risers and low flows in the center risers.

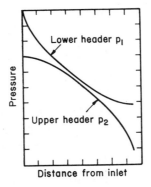

Figure 6.21.1 Pressure distribution in headers of an isothermal absorber bank. Adapted from Dunkle and Davey (1970).

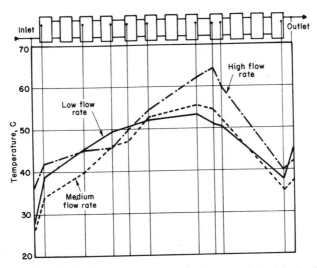

Figure 6.21.2 Experimental temperature measurements on plates in a bank of collectors connected in parallel. Adapted from Dunkle and Davey (1970).

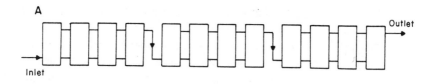

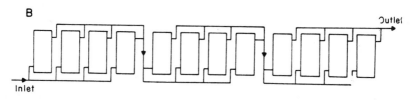

Figure 6.21.3 Examples of alternative methods of connecting arrays in (a) series-parallel and (b) parallel-series arrangements, as recommended by Dunkle and Davey (1970)

This situation is found experimentally. Temperatures of absorber plates are measures of how effectively energy is removed, and thus differences among temperatures measured at the same relative location on individual collectors in banks is a measure of the lack of uniformity of flow in risers. Figure 6.21.2, from Dunkle and Davey, shows measured temperatures for a bank of 12 collectors connected in parallel. The data show temperature differentials of 22 C from center to ends. Connecting the units in a series-parallel or multiple-parallel arrangements such as shown on Figure 6.21.3 results in more uniform flow distribution and temperatures.

Manufacturers of collectors have worked out recommended practices for piping collectors to avoid maldistribution of flow and minimize pressure drops, and the designer of a system should refer to those recommendations. Otherwise, standard references on design of piping networks should be consulted.

6.22 *IN SITU* COLLECTOR PERFORMANCE

There is no fundamental reason why a collector would not operate as well installed in an application as it does on a test stand. However, there are some practical considerations that can influence measured performance on site. Differences between predicted and measured performance may arise from several sources.

1   Flow of fluid through the collector may not be uniform through all parts of the collector array. Parts of a collector array receiving reduced fluid flow will have lower F_R and poorer performance, resulting in degradation of array performance.

2   Flow rates may not be those at which collectors were tested. F_R is a function of flow rate for both liquid and air heaters (particularly so for air heating collectors), and changes in flow rate can make significant differences in collector performance.

3   Leaks in air collectors may also introduce differences between predicted and measured performance. Collectors are usually operated at slightly less than atmospheric pressure, resulting in leakage of cool ambient air in and reduced collector outlet temperatures. [See Close and Yusoff (1978).]

4   Edge and back losses may be different in tests and applications. Edge losses may be reduced in large arrays resulting in smaller U_L than is obtained from measurements on a single module.

5.   Duct and pipe losses may be more significant in applications than in tests, as runs may be longer and ducts and pipes may not be as well insulated.

6.23 PRACTICAL CONSIDERATIONS FOR FLAT-PLATE COLLECTORS

In this chapter we have discussed the thermal performance of collectors in tests and installed in systems. There are many other practical consideration in the design, manufacture, shipment, installation, and long-term use of flat-plate collectors. In this section we briefly illustrate some of these considerations and show two commercial designs (one liquid and one air). The industry has advanced to the point that manufacturers have developed installation and service manuals; see these manuals for more details. The discussion here is based largely on the concept of factory-manufactured, modular collectors that are assembled in large arrays on a job; the same practical considerations hold for site-built collectors. Discussions of many aspects of installation and operation of large collector arrays are included in IEA (1985).

Equilibrium temperatures, encountered under conditions of high radiation with no fluid flowing through the collector, are substantially higher than ordinary operating temperatures, and collectors must be designed to withstand these temperatures. It is inevitable that at some time power failure, control problems, servicing, summer shut-down, or other causes will lead to no-flow conditions. (They may be encountered first during installation of the collectors.) Maximum plate temperatures can be estimated by evaluating the fluid temperature in Equation 6.7.6 with Q_u equal to zero. The fluid and plate temperatures are the same for the no-flow condition. The equilibrium temperatures of other parts of the collector can be estimated from the ratios of thermal resistances between those parts and ambient to that of the plate to ambient. These maximum equilibrium temperatures place constraints on the materials, which must retain their important properties during and after exposure to these temperatures, and on mechanical design to accommodate thermal expansion.

Extremely low temperatures must also be considered. This is particularly important for liquid heating collectors where freeze protection must be provided. This may be done by use of antifreeze fluids or by arranging the system so that the collector will drain during periods when it is not operating. Corrosion is an important consideration with collectors containing antifreeze solutions. Methods of freeze protection are in part system considerations and will be treated in later chapters.

Materials of wide variety are used in collectors, including structural materials (such as metal used in boxes), glazing (which is usually glass), insulation, caulking materials, etc. Skoda and Masters (1977) have compiled a survey of practical experience with a range of materials and illustrate many problems that can arise if materials are not carefully selected for withstanding weather, temperature extremes, temperature cycling, and compatibility. A series of articles in *Solar Age* (1977) deals with fabrication and corrosion prevention problems in use of copper, steel, and aluminum absorber plates.

Covers and absorbers are particularly critical; their properties determine $(\tau\alpha)$ and ε_p and thus strongly affect thermal performance. Degradation of these properties can seriously affect long-term performance, and materials should be selected that have stable properties. Condensation of moisture sometimes occurs under covers or between covers, and this imparts an increased reflectance to covers; energy is required to evaporate the condensate on start-up and collector performance is diminished, although the effect on long-term performance may be small. Some collectors use sealed spaces under the top cover, and some use breathing tubes containing desiccants to dry the air in the space under the top cover. Covers and supports should be designed so that dirt will not get under the cover. Materials used in the collector should not contain volatiles that can be evaporated during periods of high collector temperatures; these volatiles condense on the underside of covers and reduce transmittance.

Mechanical design affects thermal performance. It is important in that structural strength of collectors must be adequate to withstand handling and installation and the conditions to be expected over the lifetime of the units. Collectors must be water tight or provided with drains to avoid rain damage and in high latitudes should be mounted so as to allow snow to slide off. Wind loads may be high and tie-downs must be adequate to withstand these loads. Some collectors may be designed to provide structural strength or serve as the water-tight envelope of a building; these functions may impose additional requirements on collector design and manufacture.

Hail damage to collectors has been a matter of concern. Löf (1980) reports the effects on collectors of a hail storm in Colorado in 1979. Hail averaging 2 to 3 cm in diameter fell for a period of several minutes, followed by a lull and then 1 or 2 minutes of hail with sizes in the 3 to 10 cm range. Eleven solar heating systems were in the path of the storms; the slopes were 32 to 56°, and most of the collector panels were about 1×2 m in size. Of the 1010 collector modules on these systems, 11 suffered broken glass. In the same storm, a greenhouse had about a third of its roof glass broken by the hail. Löf concludes that "the risk of hail damage to collectors glazed with 3 mm tempered glass mounted on slopes greater than 30° is negligibly small."

Installation costs can be an important item of cost of solar collectors. These are largely determined by three factors: handling, tie-down, and manifolding. Moving and positioning of collector modules may be done by hand if modules are light enough or may be done by machine for heavy modules or inaccessible mountings. Tie-downs must be adequate to withstand wind loads; tie-down and weather-proofing should be accomplished with a minimum of labor. Manifolding, connecting the inlets and outlets of many collectors in an array to obtain the desired flow distribution, can be a very time-consuming and expensive operation, and module design should be such as to minimize the labor and materials needed for this operation.

Safety must also be considered in two contexts. First, it must be possible to handle and install modules with a minimum of hazards. Second, materials used in collectors must be capable of withstanding the maximum equilibrium temperatures without hazard of fire or without evolution of toxic or flammable gases.

All of these considerations are important and must be viewed in the light of the many years of low-maintenance service that should be expected of a well-designed collector. In later chapters on economic calculations it will be noted that lifetimes of decades are used in economic calculations, and collectors in most applications should be designed to last 10 to 30 years with minimum degradation in thermal performance or mechanical properties.

Here we illustrate two collectors that have been commercially manufactured. These collectors are selected as typical of good design. Figure 6.23.1 shows a cross-

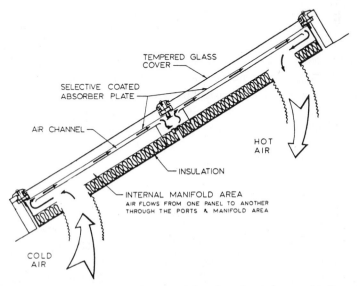

Figure 6.23.1   Cross-sectional view of two modules of a solar air heater. Air flows from one module to another through side ports and manifold area. Courtesy Solaron Corp.

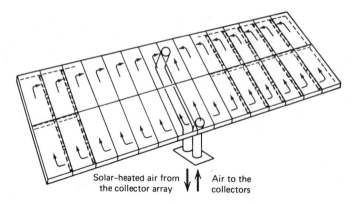

Figure 6.23.2 Arrangement of collector array for a solar air heater. Courtesy Solaron Corp.

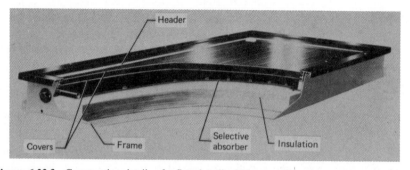

Figure 6.23.3 Construction details of a flat-plate liquid heater. Courtesy of Lennox Industries.

sectional view of two modules of an air heating collector. Manifolds are built into the structure of the collectors, and matching openings are provided on the sides of the manifold to interconnect units placed side by side in an array. Figure 6.23.2 shows a schematic of a typical collector array assembly; one pair of openings in the back of the array will serve as entrance and exit ports for multiple modules.

Figure 6.23.3 shows details of a liquid heating collector. The absorber plate is steel with bright nickel plating and a chrome black absorbing surface.

6.24   SUMMARY

In this chapter, we have gone through the theory of flat-plate collectors, shown performance measurements on them, outlined methods of testing these units, and shown standard methods of presenting the thermal characteristics. The thermal performance data taken on a test stand provides valuable information that will be used later in process design calculations. Performance of a collector in an application will depend on the design of the particular installation and also on the skill with which the equipment is assembled and installed.

REFERENCES

Abdel-Khalik, S. I., *Solar Energy*, **18**, 59 (1976). "Heat Removal Factor for a Flat-Plate Solar Collector with a Serpentine Tube."

ASHRAE Standard 93-77, *Methods of Testing to Determine the Thermal Performance of Solar Collectors*, American Society of Heating, Refrigeration, and Air Conditioning Engineers, New York (1977).

Beckman, W. A., S. A. Klein, and J. A. Duffie, *Solar Heating Design by the f-Chart Method*, Wiley, New York (1977).

Bliss, R. W., *Solar Energy*, **3** (4), 55 (1959). "The Derivations of Several 'Plate Efficiency Factors' Useful in the Design of Flat-Plate Solar-Heat Collectors."

Chiou, J. P., M. M. El-Wakil, and J. A. Duffie, *Solar Energy*, **9**, 73 (1965). "A Slit-and-Expanded Aluminum-Foil Matrix Solar Collector."

Close, D. J. and M. B. Yusoff, *Solar Energy*, **20**, 459 (1978). "Effects of Air Leaks on Solar Air Collector Behavior."

Cooper, P. I. and R. V. Dunkle, *Solar Energy*, **26**, 133 (1981). "A Non-Linear Flat-Plate Collector Model."

Dietz, A. G. H., in *Introduction to the Utilization of Solar Energy* (A. M. Zarem and D. D. Erway, eds.), McGraw-Hill, New York (1963). "Diathermanous Materials and Properties of Materials."

Dufffie, J. A., and W. A. Beckman, *Solar Energy Thermal Processes*, Wiley-Interscience, New York (1974).

Dunkle, R. V. and P. I. Cooper, paper presented at the Los Angeles Meeting of the International Solar Energy Society (1975). "A Proposed Method for the Evaluation of Performance Parameters of Flat-Plate Solar Collectors."

Dunkle, R. V. and E. T. Davey, paper presented at Melbourne International Solar Energy Society Conference (1970). "Flow Distribution in Absorber Banks."

Garg, H. P., *Solar Energy*, **15**, 299 (1974). "Effect of Dirt on Transparent Covers in Flat-Plate Solar Energy Collectors."

Gillett, W. B., *Solar Energy*, **25**, 543 (1980). "The Equivalence of Outdoor and Mixed Indoor/Outdoor Solar Collector Testing."

Gillett, W. B. and J. E. Moon, *Solar Collectors. Test Methods and Design Guidelines.* Reidel (for the Commission of the European Communities), Dordrecht (1985).

Hamid, Y. H. and W. A. Beckman, *Trans. ASME, J. Engrg for Power*, **93**, 221 (1971). "Performance of Air-Cooled, Radiatively Heated Screen Matrices."

Hill, J. E. and T. Kusuda, National Bureau of Standards Interim Report NBSIR 74-635 to NSF/ERDA (1974). "Proposed Standard Method of Testing for Rating Solar Collectors Based on Thermal Performance."

Hill, J. E. and E. R. Streed, *Solar Energy*, **18**, 421 (1976). "Method of Testing for Rating Solar Collectors Based on Thermal Performance."

Hill, J. E. and E. R. Streed, in *Application of Solar Energy of Heating and Cooling of Buildings* (R. Jordan and B. Y. Liu, eds.), ASHRAE GRP 170 (1977). "Testing and Rating of Solar Collectors."

Hill, J. E., J. P. Jenkins, and D. E. Jones, *ASHRAE Trans.*, **84**, 2 (1978). "Testing of Solar Collectors According to ASHRAE Standard 93-77."

Hill, J. E., J. P. Jenkins, and D. E. Jones, NBS Building Science Series 117, U.S. Dept. of Commerce (1979). "Experimental Verification of A Standard Test Procedure for Solar Collectors."

Hollands, K. G. T. and E. C. Shewen, *Trans. ASME, J. Solar Energy Engrg.*, **103**, 323 (1981). "Optimization of Flow Passage Geometry for Air Heating Plate-Type Solar Collectors."

Hottel, H. C. and A. Whillier, *Trans. of the Conference on the Use of Solar Energy*, Vol. **2**, P. I, 74, University of Arizona Press (1958). "Evaluation of Flat-Plate Collector Performance."

Hottel, H. C. and B. B. Woertz, *Trans. ASME*, **64**, 91 (1942). "Performance of Flat-Plate Solar-Heat Collectors."

International Energy Agency (IEA), *Design and Performance of Large Solar Thermal Collector Arrays*, Proc. of the IEA Workshop, San Diego, CA (1985).

International Energy Agency (IEA), *Results from Eleven Evacuated Collector Installations*, Report IEA-SHAC-TIV-4 (1986).

Jones, D. E., personal communication (1979).

Klein, S. A., *Solar Energy*, **17**, 79 (1975). "Calculation of Flat-Plate Loss Coefficients."

Klein, S. A., personal communication (1979).

Klein, S. A., J. A. Duffie, and W. A. Beckman, *Trans. ASME., J. Engrg. for Power*, **96A**, 109 (1974). "Transient Considerations of Flat-Plate Solar Collectors."

Löf, G. O. G., *Solar Energy*, **25**, 555 (1980). "Hail Resistance of Solar Collectors with Tempered Glasss Covers."

McIntire, W. R., and K. A. Read, *Solar Energy*, **31**, 405 (1983). "Orientational Relationships for Optically Non-Symmetric Solar Collectors."

Moore, S. W., J. D. Balcomb, and J. C. Hedstrom, paper presented at Ft. Collins ISES meeting (August 1974). "Design and Testing of a Structurally Integrated Steel Solar Collector Unit Based on Expanded Flat Metal Plates."

Neeper, D. A., *Proc. ISES Congress, Atlanta*, Vol **1**, Pergamon Press, Elmsford, NY, 298(1979). "Analysis of Matrix Air Heaters."

Ortabasi, U. and W. M. Buehl, *Proc. of ISES Conference, Los Angeles* paper 32/11, 222 (1975), "Analysis and Performance of an Evacuated Tubular Collector."

Proctor, D., *Solar Energy*, **32**, 377 (1984). "A Generalized Method for Testing All Classes of Solar Collectors. I. Attainable Accuracy. II. Evaluation of Collector Thermal Constants. III. Linearized Efficiency Equations."

Proctor, D. and S. R. James, *Solar Energy*, **35**, 387 (1985). "Analysis of the Sanyo Evacuated Tube Solar Collector Test Results from the IEA Task III Programme."

Selcuk, K., *Solar Energy*, **13**, 165 (1971). "Thermal and Economic Analysis of the Overlapped-Glass Plate Solar-Air Heaters."

Simon, F. F., *Solar Energy*, **18**, 451 (1976). "Flat-Plate Solar Collector Performance Evaluation with a Solar Simulation as a Basis for Collector Selection and Performance Prediction."

Skoda, L. F. and L.W. Masters, Report NBSIR 77-1314 of the National Bureau of Standards (1977). " Solar Energy Systems - Survey of Materials Performance."

Solar Age, **2**, 5 (1977), a series of articles including: Butt, S. H. and J. W. Popplewell, "Absorbers: Questions and Answers"; Lyman, W. S. and P. Anderson, "Absorbers: Copper"; Kruger, P., "Absorbers: Steel"; Byrne, S. C., "Absorbers: Aluminum."

Solar Products Specification Guide, SolarVision, Harrisville, NH (1983).

Souka, A. F. and H. H. Safwat, *Solar Energy*, **10**, 170 (1966). "Optimum Orientations for the Double Exposure Flat-Plate Collector and Its Reflectors."

Speyer, E., *Trans. ASME., J. Engrg. for Power*, **86**, 270 (1965). "Solar Energy Collection with Evacuated Tubes."

Streed, E. R., J. E. Hill, W. C. Thomas, A. G. Dawson, and B. D. Wood, *Solar Energy*, **22**, 235 (1979). "Results and Analysis of a Round Robin Test Program for Liquid-Heating Flat-Plate Solar Collectors."

Tabor, H., *Bull. of the Research Council of Israel*, **6C**, 155 (1958). "Radiation, Convection and Conduction Coefficients in Solar Collectors."

Theunissen, P-H. and W. A. Beckman, *Solar Energy*, **35**, 311 (1985). "Solar Transmittance Characteristics of Evacuated Tubular Collectors with Diffuse Back Reflectors."

TRNSYS Users Manual, Version 13, EES Report 38, University of Wisconsin Engineering Experiment Station (1990). [The first public version was 7 (1976).]

Vernon, R. W. and F. F. Simon, paper presented at Ft. Collins ISES meeting, August 1974, and NASA TMX-71602. "Flat-Plate Collector Performance Determined Experimentally with a Solar Simulator."

Whillier, A., Sc.D. Thesis, MIT (1953). "Solar Energy Collection and Its Utilization for House Heating."

Whillier, A., in *Applications of Solar Energy for Heating and Cooling of Buildings*, ASHRAE, New York, (1977). "Prediction of Performance of Solar Collectors."

Whillier, A., and Saluja, G., *Solar Energy*, **9**, 21 (1965). "Effects of Materials and of Construction Details on the Thermal Performance of Solar Water Heaters."

Wijeysundera, N. E., *Solar Energy*, **21**, 517 (1978). "Comparison of Transient Heat Transfer Models for Flat-Plate Collectors."

Wirsum, M. C., M.S. Thesis, University of Wisconsin-Madison (1988). "Simulation Study of a Large Solar Air Heating System."

Zhang, H-F. and Lavan, Z., *Solar Energy*, **34**, 175 (1985). "Thermal Performance of a Serpentine Absorber Plate."

Chapter 7

CONCENTRATING COLLECTORS

For many applications it is desirable to deliver energy at temperatures higher than those possible with flat-plate collectors. Energy delivery temperatures can be increased by decreasing the area from which heat losses occur. This is done by interposing an optical device between the source of radiation and the energy-absorbing surface. The small absorber will have smaller heat losses compared to a flat-plate collector at the same absorber temperature. In this chapter we discuss two related approaches: the use of nonimaging concentrators and imaging concentrators.

Many designs have been set forth for concentrating collectors. Concentrators can be reflectors or refractors, can be cylindrical or surfaces of revolution, and can be continuous or segmented. Receivers can be convex, flat, or concave and can be covered or uncovered. Many modes of tracking are possible. Concentration ratios (the ratios of collector aperture area to absorber area, which are approximately the factors by which radiation flux on the energy-absorbing surface is increased) can vary over several orders of magnitude. With this wide range of designs, it is difficult to develop general analyses applicable to all concentrators. Thus concentrators are treated in two groups: nonimaging collectors with low concentration ratio and linear imaging collectors with intermediate concentration ratios. We also note some basic considerations of three-dimensional concentrators that can operate at the high end of the concentration ratio scale.

Concentrators can have concentration ratios from low values less than unity to high values of the order of 10^5. Increasing ratios mean increasing temperatures at which energy can be delivered and increasing requirements for precision in optical quality and positioning of the optical system. Thus the cost of delivered energy from a concentrating collector is a function of the temperature at which it is available. At the highest range of concentration and correspondingly highest precision of optics, concentrating collectors are termed **solar furnaces**; these are laboratory tools for studying properties of materials at high temperatures and other high temperature processes. Laszlo (1965) and the *Proceedings of the 1957 Solar Furnace Symposium* include extensive discussions of solar furnaces. The main concerns in this chapter are with energy delivery systems operating at low or intermediate concentrations.

From an engineering point of view, concentrating collectors present problems in addition to those of flat-plate collectors. They must (except at the very low end of the

330

concentration ratio scale) be oriented to "track" the sun so that beam radiation will be directed onto the absorbing surface. However, the designer has available a range of configurations that allow new sets of design parameters to be manipulated. There are also new requirements for maintenance, particularly to retain the quality of optical systems for long periods of time in the presence of dirt, weather, and oxidizing or other corrosive atmospheric components. The combination of operating problems and collector cost has restricted the utility of concentrating collectors. New materials and better engineering of systems have now led to large-scale applications, as will be noted in Chapter 17.

To avoid confusion of terminology, the word **collector** will be applied to the total system including the receiver and the concentrator. The **receiver** is that element of the system where the radiation is absorbed and converted to some other energy form; it includes the **absorber**, its associated covers, and insulation. The **concentrator**, or **optical system**, is the part of the collector that directs radiation onto the receiver. The **aperture** of the concentrator is the opening through which the solar radiation enters the concentrator.

The first four sections in this chapter deal with general information on optical principles and heat transfer that are important in concentrating collectors. The next is concerned with arrays of cylindrical absorbers over diffuse reflectors. The following three treat the performance of compound parabolic concentrator (CPC) collectors. The balance of the chapter is concerned with imaging collectors and discusses linear collectors of types being used in experimental industrial process heat and pumping applications and multiple heliostat "central receiver" collectors.

7.1 COLLECTOR CONFIGURATIONS

Many concentrator types are possible for increasing the flux of radiation on receivers. They can be reflectors or refractors. They can be cylindrical to focus on a "line" or circular to focus on a "point." Receivers can be concave, flat, or convex. Examples of six configurations are shown in Figure 7.1.1.

The first two are arrays of evacuated tubes with cylindrical absorbers spaced apart, with back reflectors to direct radiation on the area between the tubes to the absorbers. The first uses a flat diffuse back reflector and the second cusp-shaped specular reflectors. The configuration shown in Figure 7.1.1(c) has a plane receiver with plane reflectors at the edges to reflect additional radiation onto the receiver. The concentration ratios of this type are low, with a maximum value of less than 4. Some of the diffuse component of radiation incident on the reflectors would be absorbed at the receiver. These collectors can be viewed as flat-plate collectors with augmented radiation. Analyses of these concentrators have been presented by Hollands (1971), Selcuk (1979), and others. Figure 7.1.1(d) shows a reflector of parabolic section, which could be a cylindrical surface (with a tubular receiver) or a surface of revolution (with a spherical or hemispherical receiver). Cylindrical collectors of this type have been studied in some detail and are being applied.

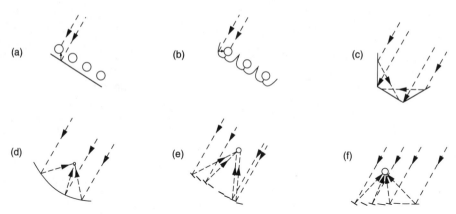

Figure 7.1.1 Possible concentrating collector configurations: (a) tubular absorbers with diffuse back reflector; (b) tubular absorbers with specular cusp reflectors; (c) plane receiver with plane reflectors; (d) parabolic concentrator; (e) Fresnel reflector; (f) array of heliostats with central receiver.

The continuous parabolic reflector can be replaced by a Fresnel reflector, a set of flat reflectors on a moving array as shown in Figure 7.1.1(e), or by its refracting equivalent. The facets of the reflector can also be individually mounted and adjusted in position as shown in Figure 7.1.1(f). Large arrays of heliostats of this type, with receivers mounted on a tower, are the basis of designs of "central receiver" collectors.

For the concentrators shown on Figure 7.1.1(c)-(f), single-sided flat receivers may be used (if the receiver is not "inside" of the reflector). Cylindrical, hemispherical, or other convex shapes may also be possible, and cavity receivers may also be used.

In general, concentrators with receivers much smaller than the aperture are effective only on beam radiation. It is evident also that the angle of incidence of the beam radiation on the concentrator is important and that sun tracking will be required for these collectors. A variety of orienting mechanisms has been designed to move focusing collectors so that the incident beam radiation will be reflected to the receiver. The motions required to accomplish tracking vary with the design of the optical system, and a particular resultant motion may be accomplished by more than one system of component motions.

Linear (cylindrical) optical systems will focus beam radiation to the receiver if the sun is in the central plane of the concentrator (the plane including the focal axis and the vertex line of the reflector). These collectors can be rotated about a single axis of rotation, which may be north-south, east-west, or inclined and parallel to the earth's axis (in which case the rate of rotation is 15°/h). There are significant differences in both quantity of incident beam radiation, its time dependence, and the image quality obtained with these three modes of orientation.

Reflectors that are surfaces of revolution (circular concentrators) generally must be oriented so that the axis and sun are in line and thus must be able to move about two axes. These axes may be horizontal and vertical, or one axis of rotation may be inclined so that it is parallel to the earth's axis of rotation (i.e., a polar axis) and the other perpendicular to it. The angle of incidence of beam radiation on a moving plane is indicated for the most probable modes of orientation of that plane by the equations in Section 1.7.

Orientation systems can provide continuous or nearly continuous adjustments, with movement of the collector to compensate for the changing position of the sun. For some low-concentration linear collectors it is possible to adjust their position intermittently, with weekly, monthly, or seasonal changes possible for some designs. Continuous orientation systems may be based on manual or mechanized operation. Manual systems depend on the observations of operators and their skill at making the necessary corrections and may be adequate for some purposes if concentration ratios are not too high and if labor costs are not prohibitive; they have been suggested for use in areas of very low labor cost.

Mechanized orienting systems can be sun-seeking systems or programmed systems. Sun-seeking systems use detectors to determine system misalignment and through controls make the necessary corrections to realign the assembly. Programmed systems, on the other hand, cause the collector to be moved in a predetermined manner (e.g., 15°/hr about a polar axis) and may need only occasional checking to assure alignment. It may also be advantageous to use a combination of these tracking methods, for example, by superimposing small corrections by a sun-seeking mechanism on a programmed "rough positioning" system. Any mechanized system must have the capability of adjusting the position of the collector from end-of-day position to that for operation early the next day, adjusting for intermittent clouds, and adjusting to a stowed position where it can best withstand very high winds without damage.

7.2   CONCENTRATION RATIO

The most common definition of concentration ratio, and that used here, is an **area concentration ratio**,[1] the ratio of the area of aperture to the area of the receiver. (A **flux concentration ratio** is defined as the ratio of the average energy flux on the receiver to that on the aperture, but generally there are substantial variations in energy flux over the surface of a receiver. A **local flux concentration ratio** can also be defined as the ratio of the flux at any point on the receiver to that on the aperture, which will vary across the receiver.)

The area concentration ratio is

$$C = \frac{A_a}{A_r} \tag{7.2.1}$$

This ratio has an upper limit that depends on whether the concentration is a three-

---

[1] Usually termed simply concentration ratio.

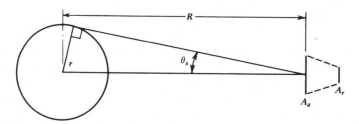

Figure 7.2.1 Schematic of sun at T_s at distance R from a concentrator with aperture area A_a and receiver area A_r. Adapted from Rabl (1976a).

dimensional (circular) concentrator such as a paraboloid or a two-dimensional (linear) concentrator such as a cylindrical parabolic concentrator. The following development of the maximum concentration ratio, from Rabl (1976a), is based on the second law of thermodynamics applied to radiative heat exchange between the sun and the receiver. Consider the circular concentrator with aperture area A_a and receiver area A_r viewing the sun of radius r at distance R, as shown in Figure 7.2.1. The half-angle subtended by the sun is θ_s. (The receiver is shown beyond the aperture for clarity; the argument is the same if it is on the same side of the aperture as the sun.)

If the concentrator is perfect, the radiation from the sun on the aperture (and thus also on the receiver) is the fraction of the radiation emitted by the sun which is intercepted by the aperture. Although the sun is not a blackbody, for purposes of an approximate analysis it can be assumed to be a blackbody at T_s:

$$Q_{s \to r} = A_a \frac{r^2}{R^2} \sigma T_s^4 \tag{7.2.2}$$

A perfect receiver (i.e., blackbody) radiates energy equal to $A_r T_r^4$, and a fraction of this, $E_{r\text{-}s}$, reaches the sun:[2]

$$Q_{r \to s} = A_r \sigma T_r^4 E_{r\text{-}s} \tag{7.2.3}$$

When T_r and T_s are the same, the second law of thermodynamics requires the $Q_{s \to r}$ be equal to $Q_{r \to s}$. So from Equations 7.2.2 and 7.2.3

$$\frac{A_a}{A_r} = \frac{R^2}{r^2} E_{r\text{-}s} \tag{7.2.4}$$

and since the maximum value of $E_{r\text{-}s}$ is unity, the maximum concentration ratio for circular concentrators is

$$\left(\frac{A_a}{A_r}\right)_{circular,max} = \frac{R^2}{r^2} = \frac{1}{\sin^2 \theta_s} \tag{7.2.5}$$

[2] $E_{r\text{-}s}$ is an exchange factor as used in Equation 3.8.2.

A similar development for linear concentrators leads to

$$\left(\frac{A_a}{A_r}\right)_{linear,max} = \frac{1}{\sin\theta_s} \tag{7.2.6}$$

Thus with $\theta_s = 0.27°$, the maximum possible concentration ratio for circular concentrators is 45,000, and for linear concentrators the maximum is 212.

The higher the temperature at which energy is to be delivered, the higher must be the concentration ratio and the more precise must be the optics of both the concentrator and the orientation system. Figure 7.2.2 shows practical ranges of concentration ratios and types of optical systems needed to deliver energy at various temperatures.

Concentrators can be divided into two categories: nonimaging and imaging. Nonimaging concentrators, as the name implies, do not produce clearly defined

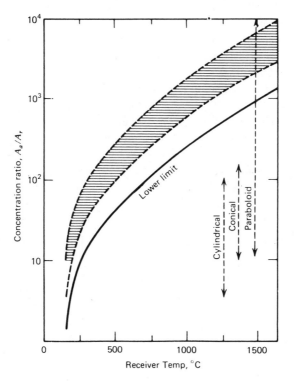

Figure 7.2.2 Relationships between concentration ratio and temperature of receiver operation. The "lower limit" curve represents concentration ratios at which the thermal losses will equal the absorbed energy; higher ratios will then result in useful gain. The shaded range corresponds to collection efficiencies of 40 to 60% and represents a probable range of operation. Also shown are approximate ranges in which several types of reflectors might be used. (This figure is not to be used for design. It is based on an assumed set of conditions determining the absorbed radiation and the thermal losses and on reasonable design practice at various temperatures. The positions of these curves would shift under conditions other than those assumed.) From Duffie and Löf (1962).

images of the sun on the absorber but rather distribute radiation from all parts of the solar disc onto all parts of the absorber. The concentration ratios of linear nonimaging collectors are in the low range and are generally below 10. Imaging concentrators, in contrast, are analogous to camera lenses in that they form images (usually of very low quality by ordinary optical standards) on the absorber.

7.3   THERMAL PERFORMANCE OF CONCENTRATING COLLECTORS

Calculation of the performance of concentrating collectors follows the same general outlines as for flat-plate collectors. The absorbed radiation per unit area of aperture S must be estimated from the radiation and the optical characteristics of the concentrator and receiver. Estimation of S is discussed in sections to follow. Thermal losses from the receiver must be estimated, usually in terms of a loss coefficient U_L which is based on the area of the receiver. In principle, temperature gradients on the receiver can be accounted for by a flow factor F_R to allow the use of inlet fluid temperatures in energy balance calculations. This section is concerned with the estimation of U_L and F_R.

The methods for calculating thermal losses from receivers are not as easily summarized as in the case of flat-plate collectors. The shapes and designs are widely variable, the temperatures are higher, the edge effects are more significant, conduction terms may be quite high, and the problems may be compounded by non-uniformity of radiation flux on receivers which can result in substantial temperature gradients across the energy-absorbing surfaces. It is difficult to present a single general method of estimating thermal losses, and ultimately each receiver geometry must be analyzed as a special case.

The nature of the thermal losses for receivers of concentrating collectors is the same as for flat-plate exchangers. Receivers may have covers transparent to solar radiation. If so, the outward losses from the absorber by convection and radiation to the atmosphere are correspondingly modified and equations similar to those of Chapter 6 can be used to estimate their magnitude. As with flat-plate systems, the losses can be estimated as being independent of the intensity of incident radiation (although this may not be strictly true if a transparent cover absorbs appreciable solar radiation). In any event, an effective transmittance-absorptance product can also be defined for focusing systems. Furthermore, with focusing systems the radiation flux at the receiver is generally such that only cover materials with very low absorptance for solar radiation can be used without thermal damage to the cover. Conduction losses occur through the supporting structure and through insulation on parts of the receiver that are not irradiated.

The generalized thermal analysis of a concentrating collector is similar to that of a flat-plate collector. It is necessary to derive appropriate expressions for the collector efficiency factor F', the loss coefficient U_L, and the collector heat removal factor F_R. With F_R and U_L known, the collector useful gain can be calculated from an expression that is similar to that for a flat-plate collector.

As an example of calculation of thermal loss coefficient U_L, consider an uncovered cylindrical absorbing tube that might be used as a receiver with a linear concentrator. Assume that there are no temperature gradients around the receiver tube. The loss coefficient considering convection and radiation from the surface and conduction through the support structure is

$$U_L = h_w + h_r + U_{cond} \tag{7.3.1}$$

The linearized radiation coefficient can be calculated from

$$h_r = 4\sigma\varepsilon\overline{T}^3 \tag{7.3.2}$$

where $\overline{T}$ is the mean temperature for radiation and ε is the emittance of the absorbing surface. If a single value of h_r is not acceptable due to large temperature gradients in the flow direction, the collector can be considered as divided into two or more segments each with constant h_r. The estimation of h_w for cylinders is noted in Section 3.15. Estimation of conductive losses must be based on knowledge of the details of construction or on measurements on a particular collector.

Linear concentrators may be fitted with cylindrical absorbers surrounded by tubular covers. For this geometry, the outside convective coefficient h_w is calculated with Equation 3.15.12 and the radiation coefficient $h_{r,c-a}$ from Equation 7.3.2. The radiation heat transfer coefficient from receiver (absorber) to cover is calculated with Equation 3.10.2. If the annulus is evacuated so that the convection coefficient $h_{c,r-c}$ is negligible,[3] U_L based on the absorber area A_r is

$$U_L = \left[\frac{A_r}{(h_w + h_{r,c-a})A_c} + \frac{1}{h_{r,r-c}}\right]^{-1} \tag{7.3.3}$$

The procedure is to estimate T_c (which will under these conditions be much closer to T_a than T_r), calculate U_L, and check T_c by an energy balance on the cover. If solar radiation absorbed by the cover is negligible, this balance is

$$A_c(h_{r,c-a} + h_w)(T_c - T_a) = A_r h_{r,r-c}(T_r - T_c) \tag{7.3.4}$$

Solving for the cover temperature,

$$T_c = \frac{A_r h_{r,r-c} T_r + A_c(h_{r,c-a} + h_w)T_a}{A_r h_{r,r-c} + A_c(h_{r,c-a} + h_w)} \tag{7.3.5}$$

which provides a value of T_c for a second iteration if needed.

It may be necessary to account for absorption of radiation by the cover or for convection transfer from absorber to cover. If so, appropriate terms must be added to the right side of Equation 7.3.4. The principles are identical with those shown in Chapter 6 for flat-plate collectors.

---

[3] See Tabor (1955) for correlations for $h_{c,r-c}$ if the annulus is not evacuated.

Example 7.3.1

Calculate the overall loss coefficient for a 60 mm cylindrical receiver at 200 C. The absorber surface has an emittance of 0.91. The absorber is covered by a glass tubular cover 90 mm in diameter and the space between the two is evacuated. The wind speed is 5 m/s and the sky and air temperatures are 10 C.

Solution

Assume the cover temperature is 50 C. To estimate the wind heat transfer coefficient, it is necessary to find the Reynolds number for an air temperature that is the average of the cover and ambient temperatures, or 30 C:

$$Re = \frac{\rho VD}{\mu} = \frac{1.16 \times 5 \times 0.090}{1.86 \times 10^{-5}} = 28,100$$

The Nusselt number is then found from Equation 3.15.12:

$$Nu = 0.30(28,100)^{0.6} = 140$$

$$h_w = 140 \times \frac{0.0265}{0.090} = 41 \text{ W/m}^2\text{C}$$

The radiation coefficient from the cover tube to ambient is

$$h_{r,c-a} = 0.88 \times 5.67 \times 10^{-8} \times 4 \times 303^3 = 5.55 \text{ W/m}^2\text{C}$$

The radiation heat transfer coefficient between the receiver tube and cover tubes is found from Equation 3.10.2:

$$h_{r,r-c} = \frac{5.67 \times 10^{-8} \left(473^2 + 323^2\right)(473 + 323)}{\dfrac{1 - 0.91}{0.91} + 1 + \dfrac{1 - 0.88}{0.88} \times \dfrac{0.06}{0.09}} = 12.45 \text{ W/m}^2\text{C}$$

There is no convection transfer from receiver tube to cover through the evacuated space, so the overall loss coefficient U_L based on the absorber area is calculated from Equation 7.3.3:

$$U_L = \left[\frac{0.06}{(41 + 5.5)\,0.09} + \frac{1}{12.45} \right]^{-1} = 10.6 \text{ W/m}^2\text{C}$$

With this first estimate of U_L, it is necessary to check the assumed cover temperature of 50 C. Using Equation 7.3.5,

$$T_c = \frac{12.45 \times 200 + \dfrac{0.09}{0.06}(5.5 + 41)10}{12.45 + \dfrac{0.09}{0.06}(5.5 + 41)} = 39 \text{ C}$$

A second iteration with the cover temperature equal to 39 C does not significantly change the value of U_L. Therefore $U_L = 10.6$ W/m²C. ∎

Next consider the factors which account for temperature gradients on the receiver. The development is analogous to that for flat-plate collectors, but the different geometries require modified procedures. Again treating linear concentrating systems with cylindrical receivers, the heat transfer resistance from the outer surface of the receiving tube to the fluid in the tube should include the tube wall as the heat flux in a concentrating system may be high. The overall heat transfer coefficient (based on the outside tube diameter) from the surroundings to the fluid is

$$U_o = \left[\frac{1}{U_L} + \frac{D_o}{h_{fi}D_i} + \frac{D_o \ln(D_o/D_i)}{2k} \right]^{-1} \tag{7.3.6}$$

where D_i and D_o are the inside and outside tube diameters, h_{fi} is the heat transfer coefficient inside the tube, and k is the thermal conductivity of the tube.

The useful energy gain per unit of collector length q'_u expressed in terms of the local receiver temperature T_r and the absorbed solar radiation per unit of aperture area S, is

$$q'_u = \frac{A_a S}{L} - \frac{A_r U_L}{L}(T_r - T_a) \tag{7.3.7}$$

where A_a is the unshaded area of the concentrator aperture and A_r is the area of the receiver ($\pi D_o L$ for the cylindrical absorber). In terms of the energy transfer to the fluid at local fluid temperature T_f,

$$q'_u = \frac{(A_r/L)(T_r - T_f)}{\dfrac{D_o}{h_{fi}D_i} + \left(\dfrac{D_o}{2k} \ln\dfrac{D_o}{D_i} \right)} \tag{7.3.8}$$

If T_r is eliminated from Equations 7.3.7 and 7.3.8, we have

$$q'_u = F' \frac{A_a}{L} \left[S - \frac{A_r}{A_a} U_L(T_f - T_a) \right] \tag{7.3.9}$$

where the collector efficiency factor F' is

$$F' = \frac{1/U_L}{\dfrac{1}{U_L} + \dfrac{D_o}{h_{fi}D_i} + \left(\dfrac{D_o}{2k} \ln\dfrac{D_o}{D_i} \right)} \tag{7.3.10}$$

or $\qquad F' = U_o/U_L \tag{7.3.11}$

The form of Equations 7.3.9 to 7.3.11 is identical to that of Equations 6.5.16 to

6.5.19. If the same procedure is followed as was used to derive Equation 6.7.6, the following equation results:

$$Q_u = F_R A_a \left[S - \frac{A_r}{A_a} U_L (T_i - T_a) \right] \tag{7.3.12}$$

In a manner analogous to that for a flat-plate collector, the collector flow factor F'' is

$$F'' = \frac{F_R}{F'} = \frac{\dot{m} C_p}{A_r U_L F'} \left[1 - \exp\left(-\frac{A_r U_L F'}{\dot{m} C_p} \right) \right] \tag{7.3.13}$$

The differences between covered and uncovered receivers are in the calculations of S and U_L.

If a receiver of this type serves as a boiler, F' is the same as is given by Equation 7.3.10, but F_R is then identically equal to F' as there is no temperature gradient in the flow direction. If a part of the receiver serves as a boiler and other parts as fluid heaters, the two or three segments of the receiver must be treated separately.

Example 7.3.2

A cylindrical parabolic concentrator with width 2.5 m and length 10 m has an absorbed radiation per unit area of aperture of 430 W/m^2. The receiver is a cylinder painted flat black and surrounded by an evacuated glass cylindrical envelope. The absorber has a diameter of 60 mm, and the transparent envelope has a diameter of 90 mm. The collector is designed to heat a fluid entering the absorber at 200 C at a flow rate of 0.0537 kg/s. The fluid has C_p = 3.26 kJ/kgC. The heat transfer coefficient inside the tube is 300 W/m^{2}C and the overall loss coefficient is 10.6 W/m^{2}C (from Example 7.3.1). The tube is made of stainless steel (k = 16 W/mC) with a wall thickness of 5 mm. If the ambient temperature is 25 C, calculate the useful gain and exit fluid temperature.

Solution

The solution is based on Equation 7.3.12. The area of the receiver is

$$A_r = \pi D L = \pi \times 0.06 \times 10 = 1.88 \text{ m}^2$$

Taking into account shading of the central part of the collector by the receiver,

$$A_a = (2.5 - 0.09)\, 10 = 24.1 \text{ m}^2$$

To calculate F_R, we first calculate F' for this situation from Equation 7.3.10:

$$F' = \frac{1/10.6}{\dfrac{1}{10.6} + \dfrac{0.06}{300 \times 0.05} + \dfrac{0.06 \ln(0.06/0.05)}{2 \times 16}} = 0.96$$

Then F_R from Equation 7.3.13 is calculated as follows:

$$\frac{\dot{m}C_p}{A_r U_L F'} = \frac{0.0537 \times 3260}{1.88 \times 10.6 \times 0.96} = 9.15$$

$$F'' = 9.15\left(1 - e^{-1/9.15}\right) = 0.95$$

$$F_R = F'' \times F' = 0.95 \times 0.96 = 0.91$$

The useful gain is

$$Q_u = 24.1 \times 0.91\left[430 - \frac{1.88 \times 10.6}{24.1}(200 - 25)\right] = 6260 \text{ W}$$

The exit fluid temperature is

$$T_o = T_i + \frac{Q_u}{\dot{m}C_p} = 200 + \frac{6260}{0.0537 \times 3260} = 236 \text{ C} \qquad ■$$

7.4   OPTICAL PERFORMANCE OF CONCENTRATING COLLECTORS

Concentrating collectors have optical properties that vary substantially with the geometry of the device. The following general concepts can be applied to all concentrators, although the ways in which they are applied vary with configuration. An equation for S, the absorbed radiation per unit area of unshaded aperture, can be written as

$$S = I_b \rho (\gamma \tau \alpha)_n K_{\gamma\tau\alpha} \qquad (7.4.1)$$

The terms in this equation have implications that are different from those for flat-plate collectors, and the treatment of them depends on collector geometry.

The effective incident radiation measured on the plane of the aperture I_b includes only beam radiation for all concentrators except those of very low concentration ratio (i.e., perhaps 10 or below). For systems of low concentration ratio, part of the diffuse radiation will be reflected to the receiver, with the amount depending on the acceptance angle of the concentrator.

Generally ρ is the specular reflectance of the concentrator. For diffuse reflectors used with cylindrical absorbers, it will be the diffuse reflectance. If the concentrator is a refractor, it will be the transmittance of the refractor.

The next three factors, γ, τ, and α, are functions of the angle of incidence of radiation on the aperture. The effects of angle of incidence on these characteristics may be considered individually, or they may, as implied by Equation 7.4.1, be combined in an incidence angle modifier.

The **intercept factor** γ is defined as the fraction of the reflected radiation that is incident on the absorbing surface of the receiver. This concept is particularly useful in describing imaging concentrators. An example of an image formed in the focal plane of a linear concentrator is shown in Figure 7.4.1; if a receiver extends from A to B, the intercept factor will be

$$\gamma = \frac{\int_{A}^{B} I(y)\, dy}{\int_{-\infty}^{\infty} I(y)\, dy} \tag{7.4.2}$$

The objective in using concentrating systems is to reduce heat losses from the absorber by reducing its area. Most imaging collectors are built with receivers large enough to intercept a large fraction of the reflected radiation but not large enough to intercept the low-intensity fringes of the images. Values of γ greater than 0.9 are common.

Here τ is the transmittance of any cover system on the receiver. Its calculation may be difficult as the angles of incidence of the radiation from the reflector on the cover may be uncertain. α is the absorptance of the absorber for the reflected (and transmitted) solar radiation, and may be difficult to determine for the same reason as that of τ.

An incidence angle modifier $K_{\gamma\tau\alpha}$ can be used to account for deviations from the normal of the angle of incidence of the radiation on the aperture. For cylindrical systems, biaxial incidence modifiers are required, with separate treatment for the longitudinal and transverse planes. For circular systems that are fully symmetric, a single modifier will be adequate. If the system is not fully symmetric,[4] biaxial or

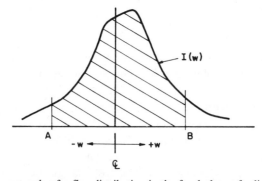

Figure 7.4.1   An example of a flux distribution in the focal plane of a linear imaging concentrator. The receiver extends from A to B.

[4] The central receiver power systems described in Section 17.5 are examples of unsymmetrical circular systems.

possibly more complex modifiers will be needed. As written in Equation 7.4.1, $K_{\gamma\tau\alpha}$ includes the effects of angle of incidence on the intercept factor; in general, as the angle of incidence of radiation increases, the size of the image will increase and the fraction of the reflected radiation intercepted by the receiver will decrease. With cylindrical (linear) systems there may be end effects if the collector is not long compared to its focal length; this effect can be included in $K_{\gamma\tau\alpha}$.

These optical terms will be discussed for two classes of concentrators, nonimaging and imaging, in the following sections.

7.5 CYLINDRICAL ABSORBER ARRAYS

In Chapter 6, we discussed collectors with selective flat absorbers placed in evacuated tubes and noted that low loss coefficients can be obtained. Another type of evacuated tubular collector is the Dewar type in which the vacuum is maintained between two concentric glass tubes; a cylindrical absorbing surface is on the outside of the inner of the two tubes or on a cylindrical fin inside of the inner tube. Two important configurations of these tubes are shown in Figure 7.5.1(a) and (b).

In type (a), the selective absorbing surface is on the outside of the inner glass tube of the vacuum jacket, in the evacuated space. A third tube, the delivery tube, is used to move fluid into (or out of) the end of the Dewar and thus move the fluid past the solar-heated inner tube of the jacket. The working fluid fills the space inside of the Dewar. The delivery tube should be designed to reduce thermal coupling between the in and out streams. Figure 7.5.1(b) shows a design with a fin-and-tube absorber inserted in the Dewar. The fin is rolled into a cylindrical form to provide a "spring" fit when inserted. The absorbing surface is on the fin. In this design, the quantity of fluid in the Dewar is very much less, and breakage of a tube does not result in loss of fluid and failure of an array.

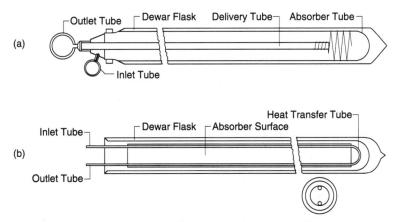

Figure 7.5.1 Two types of evacuated Dewar tubes with cylindrical absorbers: (a) Dewar with delivery tube: (b) Dewar with inserted fin and tube. Adapted from IEA (1986).

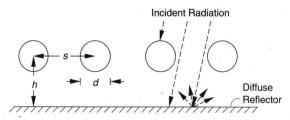

Figure 7.5.2   Section of a collector with an array of cylindrical absorbers and a diffuse back reflector.

Figure 7.5.2 shows a cross section of an array of cylindrical absorbers spaced approximately a diameter apart and a diameter above a diffuse back reflector. Some of the incident radiation is absorbed directly by the cylinders. Some that is incident on the back reflector is reflected to the cylinders, and some is reflected back to the sky and lost. The optical properties of these reflectors and tube arrays are nonsymmetrical (as are those of flat absorbers with cylindrical covers) and biaxial incidence angle modifiers are used. Theunissen and Beckman (1985) have computed the optical properties of arrays of tubes for incident beam, diffuse, and reflected radiation, for arrays sloped toward the south with tube axes north-south, and with the diffuse reflectance of the back reflector equal to 0.8.

Values of $F_R(\gamma\tau\alpha)_n$ for these collectors with flat diffuse back reflectors are typically 0.65 to 0.70. Transverse plane incidence angle modifiers for a collector of this type are shown in Figure 7.5.3. Data are from Chow et al. (1984), who also did ray trace studies to derive theoretical incidence angle modifier (IAM) curves; they found that the measured curves agreed with the theoretical within the experimental errors. Mather (1980) notes similar experimental results.

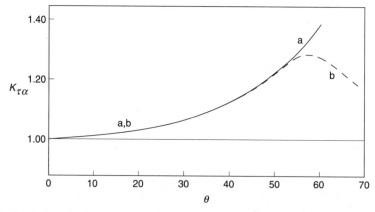

Figure 7.5.3   Measured transverse incidence angle modifiers for 30 mm diameter cover tubes and 22 mm absorber tubes over white diffuse back reflectors. Curve (a) tube-to-tube spacing s is 60 mm, and height h is 40 mm. Curve (b) tube-to-tube spacing is 45 mm and height is 32 mm. Data are from Chow et al. (1984).

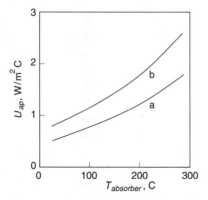

Figure 7.5.4 Calculated loss coefficients for evacuated tubular collectors of the type shown in Figure 7.5.1(a) considering (a) radiation transfer only across the evacuated space, and (b) adding an estimate of 50 percent to allow for manifold losses. U values are based on aperture area of an array with tubes spaced a tube diameter apart. Based on Rabl (1985).

Loss coefficients for these collectors are more temperature dependent than those for flat-plate collectors. Radiation, the major mechanism for heat transfer across the evacuated space, is dependent on the difference of the fourth power of the temperatures, and the collectors in many applications run at higher temperatures than flat-plate collectors. Rabl (1985) shows calculations of overall loss coefficients for the tubes based on the absorber area; for typical spacing of tubes with a tube diameter space between them, the results based on aperture area U_{ap} are shown as curve (a) in Figure 7.5.4. Rabl suggests that losses in manifolds can be as much as 50% of those from the tubes. Curve (b) reflects this increase and can be considered as an overall loss coefficient U_L from an array of tubes and its associated manifold.

7.6 OPTICAL CHARACTERISTICS OF NONIMAGING CONCENTRATORS

It is possible to construct concentrating collectors that can function seasonally or annually with minimum requirements for tracking (with its attendant mechanical complications). These nonimaging concentrators have the capability of reflecting to the receiver all of the incident radiation on the aperture over ranges of incidence angles within wide limits. The limits define the **acceptance angle** of the concentrator. As all radiation incident within the acceptance angle is reflected to the receiver, the diffuse radiation within these angles is also useful input to the collector.

Most of this section is devoted to **compound parabolic concentrators** (designated CPCs).[5] These concentrators had their origins in instruments for detection of Cherenkov radiation in high-energy physics experiments, a development

---

[5] The term CPC is applied to many nonimaging concentrators even though their geometry differs from parabolic.

noted by Hinterberger and Winston (1966). An independent and parallel development occurred in the USSR [Baranov and Melnikov (1966)]. Their potential as concentrators for solar energy collectors was pointed out by Winston (1974), and they have been the basis of detailed study since then by Welford and Winston (1978), Rabl (1976a,b), and many others.

The basic concept of the compound parabolic concentrator is shown in Figure 7.6.1. These concentrators are potentially most useful as linear or trough-type concentrators (although the analysis has also been done for three-dimensional concentrators), and the following is based on the two-dimensional CPC. Each side of the CPC is a parabola; the focus and axis of only the right-hand parabola are indicated. Each parabola extends until its surface is parallel with the CPC axis. The angle between the axis of the CPC and the line connecting the focus of one of the parabolas with the opposite edge of the aperture is the acceptance half-angle θ_c. If the reflector is perfect, any radiation entering the aperture at angles between $\pm\theta_c$ will be reflected to a receiver at the base of the concentrator by specularly reflecting parabolic reflectors.

Concentrators of the type of Figure 7.6.1 have area concentration ratios which are functions of the acceptance half-angle θ_c. For an ideal two-dimensional system the relationship is[6]

$$C_i = \frac{1}{\sin \theta_c} \qquad (7.6.1)$$

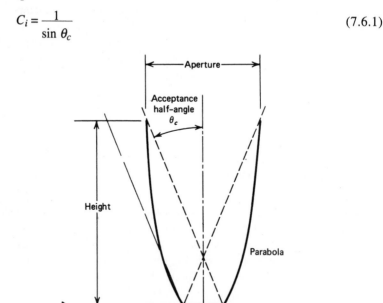

Figure 7.6.1 Cross section of a symmetrical nontruncated CPC.

[6] This can be shown by the same arguments that led to Equations 7.2.6.

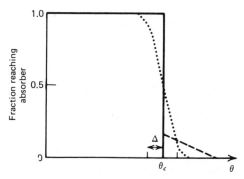

Figure 7.6.2 Fraction of radiation incident on the aperture of a CPC at angle θ which reaches the absorber surface if $\rho = 1$. θ_c is the acceptance half-angle and Δ is an angular surface error. Full CPC with no surface errors, ———; truncated CPC with no surface errors, ------; full CPC with surface errors, ·······. Adapted from Rabl (1976b).

An ideal CPC in this context is one which has parabolas with no errors. Thus an ideal CPC with an acceptance half-angle of 23.5° will have C_i of 2.51, and one with an acceptance half-angle of 11.75° will have C_i of 4.91. Figure 7.6.2 shows the fraction of radiation incident on the aperture at angle θ which reaches the absorber as a function of θ. For the ideal CPC, the fraction is unity out to θ_c and zero beyond. For a real CPC with surface errors, some radiation incident at angles less the θ_c does not reach the absorber, and some at angles greater than θ_c does reach it.

At the upper end points of the parabolas in a CPC, the surfaces are parallel to the central plane of symmetry of the concentrator. The upper ends of the reflectors thus contribute little to the radiation reaching the absorber, and the CPC can be truncated to reduce its height from h to h' with a resulting saving in reflector area and little sacrifice in performance. A truncated CPC is shown in Figure 7.6.3. The dashed plot in Figure 7.6.2 shows the spread of the image for the truncated concentrator. Limited truncation affects the acceptance angle very little, but it does change the height-to-aperture ratio, the concentration ratio, and the average number of

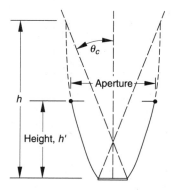

Figure 7.6.3 A CPC truncated so its height-aperture ratio is about one-half of the full CPC.

reflections undergone by radiation before it reaches the absorber surface. The effects of truncation are shown for otherwise ideal CPCs in Figure 7.6.4 to 7.6.6. Figure 7.6.4 shows the height-to-aperture ratio (see Figure 7.6.1) and Figure 7.6.5 shows the ratio of reflector area to aperture area. Figure 7.6.6 shows the average number of reflections undergone by radiation entering the aperture before it reaches the absorber. If the truncation is such that the average number of reflections is below the $(N)_{min}$ curve, that average number is at least $(1 - 1/C)$. [Rabl (1976a) gives equations for all of these quantities.]

The use of these plots can be illustrated as follows. An ideal full CPC has an acceptance half-angle θ_c of $12°$. From Figure 7.6.4 the height-to-aperture ratio is 2.8 and the concentration ratio is 4.8. From Figure 7.6.5, the area of reflector required is 5.6 times the aperture. The average number of reflections undergone by radiation before reaching the absorber surface is 0.97 from Figure 7.6.6. If this CPC is truncated so that its height-to-aperture ratio is 1.4, from Figure 7.6.4 the concentration ratio will drop to 4.2. Then from Figure 7.6.5 the reflector area-aperture area ratio is 3.0 and from Figure 7.6.6 the average number of reflections will be at least $1 - 1/4.2 = 0.76$.

CPCs with flat receivers should have a gap between the receiver and the reflector to prevent the reflector from acting as a fin conducting heat from the receiver. The gap results in a loss of reflector area and a corresponding loss in performance and should be kept small [Rabl (1985)].

The preceding discussion has been based on flat receivers occupying the plane between the two foci (Figure 7.6.3). Other receiver shapes can be used; Winston and Hinterberger (1975) showed that a CPC can be developed with aperture l which will concentrate incident radiation with incidence angles between $\pm\theta_c$ onto any

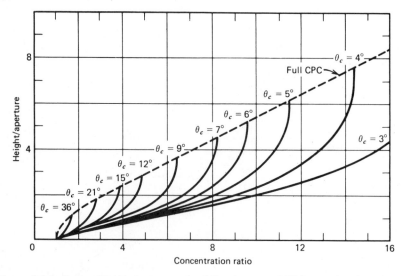

Figure 7.6.4 Ratio of height to aperture for full and truncated CPCs as a function of C and θ_c. Adapted from Rabl (1976b).

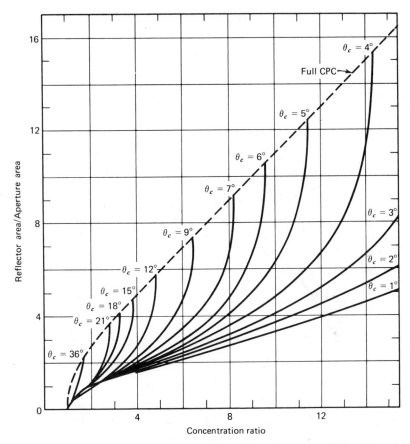

Figure 7.6.5 Ratio of reflector area to aperture area for full and truncated CPCs. Adapted from Rabl (1976b).

convex absorber with circumference $l \sin \theta_c$. The method of generation of the shape of the CPC is illustrated by Figure 7.6.7, which shows a special case of interest, a cylindrical absorber. Parts AB and AC of the reflector are convolutes of parts AF and AG of the absorber. The requirement for the rest of the reflector is that at any point P the normal to the reflector NP must bisect the angle between the tangent to the absorber PT and the line QP, which is at an angle θ_c to the axis of the CPC. This CPC is used with evacuated tubular receivers. They can be truncated in the same way as other CPCs. An example is shown in Figure 7.6.8.

This method can be used to generate a reflector for any convex receiver shape. Thus a set of CPC-type concentrators (not necessarily parabolas) can be evolved that permit a range of choices of receiver shape. CPCs can be used in series; the receiver for a primary concentrator can be the aperture of a secondary concentrator. The concentrators need not be symmetrical.

Tubular absorbers are often used with CPC reflectors. O'Gallagher et al. (1980) have shown that the reflector shape leading to maximum absorption of radiation by cylindrical absorbers is an involute, shown in Figure 7.6.9.

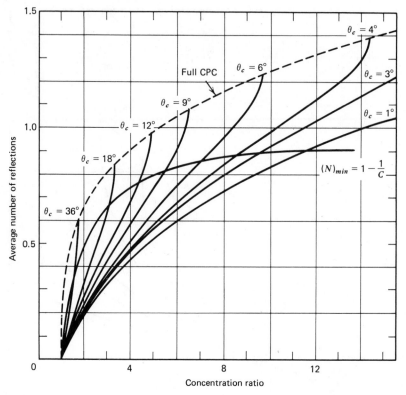

Figure 7.6.6 The average number of reflections undergone by radiation within the acceptance angle reaching the absorber surface of full and truncated CPCs. Adapted from Rabl (1976b).

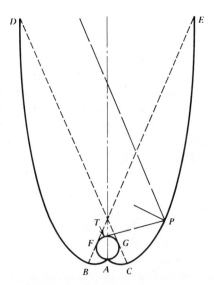

Figure 7.6.7 A CPC for a tubular receiver. Adapted from Rabl (1978).

Figure 7.6.8 An array of truncated CPC reflectors with evacuated tubular receivers with a glass cover over the array. Courtesy of Energy Design Corporation.

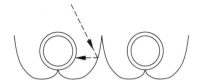

Figure 7.6.9 Involute reflector for use with cylindrical absorber. Adapted from O'Gallagher et al. (1980).

7.7 ORIENTATION AND ABSORBED ENERGY FOR CPC COLLECTORS

The advantages of CPCs are that they can function without continuous tracking and still achieve some concentration. However, they must be properly oriented to maximize the absorbed radiation when output from the collector is needed. To determine the operation at any time (which is a step to finding the best orientation), it is necessary to calculate the absorbed radiation. A particularly critical question is whether or not the beam radiation will be absorbed.

A logical orientation for such a collector is along a horizontal east-west axis, sloped toward the equator, and more or less adjustable about that axis. The CPC is arranged so that the pseudo incidence angle[7] of beam radiation (the projection of the angle of incidence on the north-south vertical plane) lies within the limits $\pm\theta_c$ during the times when output is needed from the collector. In practice, compromises are necessary between frequency of movement of the collector and concentration ratio, with high ratios associated with small acceptance angles and relatively frequent positioning.

To estimate the radiation absorbed by the receiver of a CPC, it is necessary to determine if the angle of incidence of the beam radiation is within the acceptance angle

---

[7] This angle is referred to in architectual literature as the solar profile angle, and by Hollands (1971) as the east-west vertical (EWV) angle.

$2\theta_c$, and then estimate the contributions of the beam and diffuse radiation, plus the ground-reflected radiation if it is within the acceptance angle. The absorbed radiation can be estimated as

$$S= A_a \left(G_{b,\text{CPC}} \tau_{c,b} \tau_{\text{CPC},b} \alpha_b + G_{d,\text{CPC}} \tau_{c,d} \tau_{\text{CPC},d} \alpha_d + G_{g,\text{CPC}} \tau_{c,g} \tau_{\text{CPC},g} \alpha_g \right) \tag{7.7.1a}$$

$$G_{b,\text{CPC}} = F G_{bn} \cos \theta \tag{7.7.1b}$$

$$G_{d,\text{CPC}} = \begin{cases} \dfrac{G_d}{C} & \text{if } (\beta + \theta_c) < 90^\circ \\[2mm] \dfrac{G_d}{2}\left(\dfrac{1}{C} + \cos \beta\right) & \text{if } (\beta + \theta_c) > 90^\circ \end{cases} \tag{7.7.1c}$$

$$G_{g,\text{CPC}} = \begin{cases} 0 & \text{if } (\beta + \theta_c) < 90^\circ \\[2mm] \dfrac{G_g}{2}\left(\dfrac{1}{C} - \cos \beta\right) & \text{if } (\beta + \theta_c) > 90^\circ \end{cases} \tag{7.7.1d}$$

The first term in Equation 7.7.1a is the beam contribution to S, the second is the contribution of the diffuse, and the third is the contribution of the ground-reflected radiation. In the first term, $G_{b,\text{CPC}}$ is the beam radiation on the aperture that is within the acceptance angle, τ_{cb} is the transmittance for beam radiation of any cover which may be placed over the concentrator array, and α_b is the absorptance of the receiver for the beam radiation. The factor $\tau_{\text{CPC},b}$ is a "transmittance" of the CPC which accounts for reflection losses and is a function of the average number of reflections. The factors in the terms for diffuse and ground-reflected radiation are analogous to those for the beam radiation.

The ground-reflected radiation is only effective if $(\beta + \theta_c) > 90^\circ$, i.e., if the receiver "sees" the ground. The angles are shown in Figure 7.7.1. Equations 7.7.1c and 7.7.1d account for the contribution or lack thereof of the ground-reflected radiation. Here the diffuse radiation is assumed to be isotropic.[8]

Figure 7.7.1 shows the acceptance angle of a CPC on a vertical north-south plane for a CPC oriented east-west. Two angles, $(\beta - \theta_c)$ and $(\beta + \theta_c)$, are the angles from the vertical in this plane to the two limits describing the acceptance angle. Mitchell (1979) has shown that the following condition must be met in order for the beam radiation to be effective:

$$(\beta - \theta_c) \le \tan^{-1}(\tan \theta_z \cos \gamma_s) \le (\beta + \theta_c) \tag{7.7.2}$$

It is convenient to introduce the **control function** F in Equation 7.7.1b which is 1 if the criterion of Equation 7.7.2 is met and 0 otherwise. If beam radiation is incident on the aperture within the acceptance angle, $F = 1$ and the beam radiation term will be included in the calculation of S.

---

[8] If the diffuse has a significant circumsolar contribution, the method described in this section will underestimate the effective diffuse radiation.

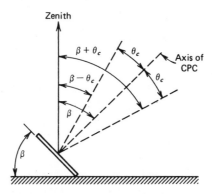

Figure 7.7.1 Projection on a north-south plane of CPC acceptance angles and slope for a CPC on an east-west axis.

A CPC collector will probably have a transparent cover over the array of reflectors. This serves both to protect the reflecting and absorbing surfaces and to reduce thermal losses from the absorber. Then beam and diffuse radiation effectively entering the CPC are reduced by the transmittance of the cover τ_c.

Only part of the incident diffuse radiation effectively enters the CPC, and that part is a function of the acceptance angle. A relationship between the mean angle of incidence of effective diffuse radiation and the acceptance half-angle θ_c is shown in Figure 7.7.2. This is based on the assumption that the diffuse radiation is isotropic. The relationship depends on the nature of the cover system, and the figure shows a

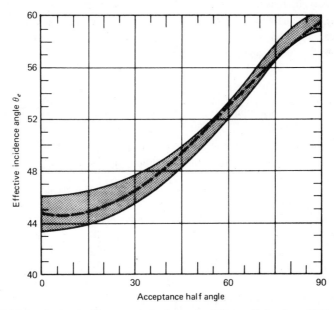

Figure 7.7.2 Equivalent incidence angle for isotropic diffuse radiation for a CPC as a function of acceptance half-angle. From Brandemuehl and Beckman (1980).

band of solutions including one and two covers, refractive indices from 1.34 to 1.526, and KL per cover up to 0.0524. An equation for the equivalent angle of incidence θ_e (the dashed line) is

$$\theta_e = 44.86 - 0.0716\theta_c + 0.00512\theta_c^2 - 0.00002798\theta_c^3 \qquad (7.7.3)$$

Thus for a CPC with $\theta_c = 20°$, the mean angle of incidence of effective diffuse radiation is $45°$ and the transmittance of the cover for this radiation is that of beam radiation at $45°$.

The terms $\tau_{CPC,b}$, $\tau_{CPC,d}$, and $\tau_{CPC,g}$ in Equation 7.7.1a are transmittances of the CPC that account for the specular reflectance of the concentrator and the average number of reflections. The terms are usually treated as the same, and an appropriate value to be applied to all three of them is estimated from the number of reflections n_r from Figure 7.6.6 and the reflectance ρ by the equation

$$\tau_{CPC} = \rho^{n_r} \qquad (7.7.4)$$

Here it is assumed that end effects are negligible or that the ends are highly reflective.

The absorptances α_b, α_d, and α_g of the receiver surface depend on the angles of incidence of the radiation on the surface. The angles of incidence vary depending on the angle of incidence of the radiation on the aperture of the CPC, the number of reflections undergone by the radiation, and the shape of the reflecting and receiving surfaces. If the average number of reflections is small, as it will be for large acceptance angles, then as a first approximation the absorptance can be taken as that corresponding to the angle of incidence of the effective radiation on the aperture of the CPC; however, this will generally result in overestimation of α. Due to lack of design information, there may be no alternative.

Example 7.7.1

A CPC collector array is mounted on a horizontal east-west axis and oriented at a slope of $25°$ from the horizontal. The latitude is $35°N$. The acceptance angle $2\theta_c$ is $24°$. At 10 AM on August 1, the beam normal radiation G_{bn} is 805 W/m². Estimate the effective contribution of incident beam radiation on the CPC.

Solution

The first step is to see if the criterion of Equation 7.7.2 is met, that is, whether F is 0 or 1. For this date, $n = 213$ and $\delta = 17.9°$. From Equation 1.6.4, $\cos \theta_z = 0.851$ and $\theta_z = 31.6°$. From Equations 1.6.6, $\omega_{ew} = 73°$, $\omega = -30°$, $C_1 = 1$, $C_2 = 1$, $C_3 = -1$, and $\gamma_s' = \gamma_s$, so

$$\gamma_s = \sin^{-1}\left[\frac{\cos 17.9 \sin(-30)}{\sin 31.6}\right] = -65.2°$$

For this collector, $(\beta - \theta_c) = 25 - 12 = 13°$, and $(\beta + \theta_c) = 25 + 12 = 37°$. Then $\tan^{-1}[\tan 31.6 \cos(-65.2)] = 14.5°$. Since $14.5°$ lies between $13°$ and $37°$, F is 1. The angle of incidence of the beam radiation on the aperture can be obtained with Equation 1.6.7a:

$$\cos\theta = \cos(35 - 25)\cos 17.9 \cos 30 + \sin(35 - 25)\sin 17.9 = 0.865$$

The effective beam radiation incident on the plane of the CPC is

$$G_{b,\text{CPC}} = 1 \times 805 \times 0.865 = 700 \text{ W/m}^2 \qquad \blacksquare$$

Example 7.7.2

At the location and time of Example 7.7.1 the diffuse radiation on a horizontal surface is 320 W/m². The CPC of Example 7.7.1 is truncated and has a concentration ratio of 4.5. The average number of reflections is 0.75 (from Figure 7.6.6) and the reflectance of the CPC reflector is 0.88. Estimate the effective diffuse radiation on the aperture. Also, estimate the total radiation reaching the absorber if the CPC array has a glass cover with $KL = 0.0125$.

Solution

For these circumstances $(\beta + \theta_c) = 37°$, so there is no contribution of ground-reflected radiation. Equation 7.7.1c is used to estimate the effective diffuse radiation:

$$G_{d,\text{CPC}} = 320/4.5 = 71 \text{ W/m}^2$$

The angle of incidence of the beam radiation on the cover is $\cos^{-1}(0.865)$, or $30°$. The transmittance of the cover for this radiation, from Figure 5.3.1, is 0.90. The mean angle of incidence for the diffuse radiation from Figure 7.7.2 is $45°$. From Figure 5.3.1 the transmittance for this radiation is 0.88. The radiation reaching the absorber per unit area of aperture is then the contribution of the beam and diffuse components

$$(0.70 \times 0.90 + 0.071 \times 0.88)0.88^{0.75} = 630 \text{ W/m}^2 \qquad \blacksquare$$

Example 7.7.3

What are the useful incident diffuse and ground-reflected radiation terms $G_{d,\text{CPC}}$ and $G_{g,\text{CPC}}$ for a full CPC collector with an acceptance half-angle of $18°$ sloped at $80°$ to south at latitude $35°$? The horizontal total radiation is 530 W/m², the diffuse is 175 W/m², and the ground reflectance is 0.7.

Solution

In this case the ground-reflected radiation must be considered, as $(\beta + \theta_c) > 90°$. Equations 7.7.1c and 7.7.1d are used with C replaced by $1/\sin \theta_c$ for this full CPC:

$$G_{d, \text{CPC}} = \frac{175}{2} (\sin 18 + \cos 80) = 42.2 \text{ W/m}^2$$

$$G_{g,\text{CPC}} = \frac{530 \times 0.7}{2} (\sin 18 - \cos 80) = 25.1 \text{ W/m}^2 \qquad ■$$

There are a variety of CPC geometries that require special treatment. For example, a common form of CPC uses evacuated tubular receivers with cylindrical absorbing surfaces, interposing an additional cover that must be considered. Evacuated tubes with flat receivers can also be used with cusp reflectors similar to those shown in Figure 7.6.9. With the procedures outlined in this section, it is possible to estimate the energy absorbed by many CPC collectors. In the next section, this is combined with the analysis of thermal performance of Section 7.2 to show how collector performance can be estimated for any set of operating conditions.

7.8   PERFORMANCE OF CPC COLLECTORS

The basic equation summarizing the performance of a CPC collector is Equation 7.3.12, where S is calculated by Equations 7.7.1. The remaining question is the calculation of the appropriate thermal loss coefficient U_L. Losses from CPCs will include losses from the receivers (which are calculated by methods that are basically the same as for other collectors) plus losses from manifolds, as shown in Figure 7.5.5.

Rabl (1976b) presents a discussion of calculation of loss coefficients for a CPC collector geometry using a flat absorber. Figure 7.8.1 shows his estimations of overall loss coefficients per unit area of absorber for use in Equation 7.3.12. The coefficients are functions of concentration ratio; the method used in their estimation assumed fixed conduction losses independent of concentration ratio, and both radiation and convection from plate to cover are to some degree functions of the concentration ratio. Estimates are shown for two plate emittances and two plate temperatures.

With estimates of U_L from Figure 7.8.1, from measurements, or from calculations, and with a knowledge of the meteorological conditions and inlet fluid temperature, the procedure for calculating collector output is the same as for a flat-plate collector. Then F_R is calculated from F', U_L, and the fluid flow rate $\dot{m}$, and F'' is calculated by Equation 7.3.13; Q_u is calculated with Equation 7.3.12, and the average plate temperature is calculated as the basis for a check of the assumed T_p (and U_L).

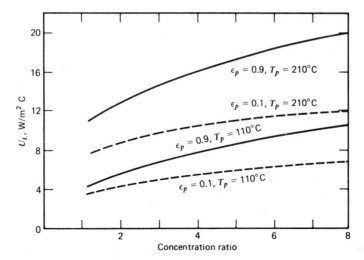

Figure 7.8.1 Estimated loss coefficients for CPC collectors with flat absorbers. Data are from Rabl (1976b).

Example 7.8.1

The CPC collector of the examples in Section 7.7 has a flat receiver with $\varepsilon_p = 0.10$. Under the conditions of its use, it is expected that F' will be 0.87. The inlet fluid temperature is 130 C and the ambient temperature is 28 C. The collector array of 10 m² total aperture area has 10 concentrators each with an aperture width of 0.30 m, with flow in parallel through the 10 receivers. The total flow rate is 0.135 kg/s and the fluid heat capacity is 3560 J/kg C. Estimate the useful gain from the collector when the absorbed radiation S is 590 W/m². Base U_L on Rabl's estimates as indicated in Figure 7.8.1.

Solution

Assume the absorber temperature T_p to be 140 C. From Figure 7.8.1 the overall loss coefficient U_L is approximately 6.1 W/m²C for this CPC with a concentration ratio of 4.5. The flow rate per unit area of aperture is $\dot{m}/A_a = 0.135/10 = 0.0135$ kg/m²s. With these, F_R can be calculated from Equation 7.3.13:

$$F_R = \frac{0.0135 \times 3560}{6.1}\left[1 - \exp\left(-\frac{6.1 \times 0.87}{0.0135 \times 3560}\right)\right] = 0.82$$

The total useful gain can be estimated from Equation 7.3.12:

$$Q_u = 10 \times 0.82\left[590 - \frac{6.1(130 - 28)}{4.5}\right] = 3.7 \text{ kW}$$

The temperature rise through the collector is

$$\Delta T = \frac{3700}{0.135 \times 3560} = 7.7 \text{ C}$$

Equation 6.8.4 could be used to estimate the mean receiver temperature of 134 C, and a new U_L estimated based on the revised mean absorber temperature T_p. It will be close to the 140 C assumed, and it is not possible to improve significantly on the estimate already made. ∎

Rabl et al. (1980) have measured the performance of several experimental CPC collectors. A collector with a cylindrical absorber, a concentration ratio of 5.2, an acceptance half-angle of 6.5°, and a flat cover over the array had $F'(\tau\alpha)_n = 0.68$ and $F'U_L = 0.185$ W/m²C. Operating data from some CPC arrays with evacuated tubular receivers have been reported to be similar to that shown in Figure 6.18.5, which indicate a more complex dependence of efficiency on incident radiation and operating temperature than is indicated by the two-parameter model.

7.9   LINEAR IMAGING CONCENTRATORS - GEOMETRY

Linear concentrators with parabolic cross section have been studied extensively both analytically and experimentally, and have been proposed and used for applications requiring intermediate concentration ratios and temperatures in the range of 100 to 500 C. Figure 7.9.1 shows a collector of this type which is part of a power generation system in California. The receiver used with this concentrator is cylindrical and is enclosed in an evacuated tubular cover; flat receivers have also been used with reflectors of this type.

To understand how these collectors operate, it is necessary to describe the optical properties of the concentrators and the images (the distribution of solar radiation flux across the focus) they produce. This section treats the geometry of reflectors and the width of the images they produce. The next section is concerned with the distribution of flux in images from perfect reflectors. Section 7.11 then treats images produced by imperfect reflectors. The final section in this series is on energy balances on these collectors.

For collectors of this type, the absorbed radiation per unit area of unshaded aperture is given by Equation 7.4.1. In order to evaluate S, it is necessary to know the characteristics of the images produced by reflectors. Theoretical images, i.e., those produced by perfect concentrators that are perfectly aligned, depend on concentrator geometry. Cross sections of a linear parabolic concentrator are shown in Figures 7.9.2 and 7.9.3. The equation of the parabola, in terms of the coordinate system shown, is

$$y^2 = 4fx \tag{7.9.1}$$

Figure 7.9.1 Collector with linear parabolic concentrator used in a power generation installation. Photo courtesy Luz Corporation.

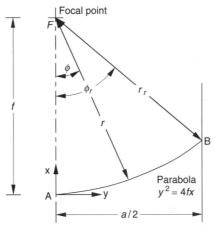

Figure 7.9.2 Section of a linear parabolic concentrator showing major dimensions and the x,y,z coordinates.

The aperture is a, and the focal length (the distance from the focal point to the vertex) is f.

The radiation beam shown in Figure 7.9.3 is incident on the reflector at point B at the rim where the **mirror radius** is a maximum at r_r. The angle ϕ_r is the rim angle, described by AFB, and is given by

$$\phi_r = \tan^{-1}\left[\frac{8(f/a)}{16(f/a)^2 - 1}\right] = \sin^{-1}\left(\frac{a}{2r_r}\right) \tag{7.9.2}$$

For convenience, ϕ_r is plotted as a function of f/a in Figure 7.9.4.

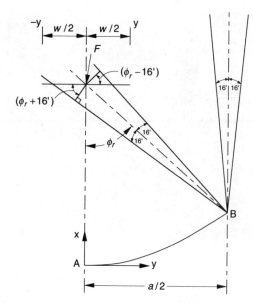

Figure 7.9.3 Image dimensions for a linear concentrator.

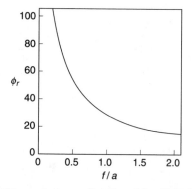

Figure 7.9.4 Rim angle ϕ_r as a function of focal length-aperture ratio.

For any point of the parabolic reflector the **local mirror radius** is

$$r = \frac{2f}{1 + \cos \phi} \tag{7.9.3}$$

An incident beam of solar radiation is a cone with an angular width of 0.53°
(i.e., a half-angle θ_s of 0.267°, or 16'). For present purposes, assume that the
concentrator is symmetrical and that the beam radiation is normal to the aperture.
Thus the beam radiation is incident on the concentrator in a direction parallel to the
central plane of the parabola (the x-z plane described by the axis and focus of the
parabola). (The effects of alignment errors will be noted in Section 7.11.)

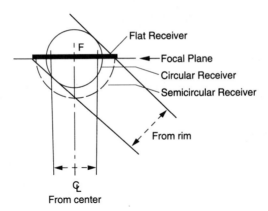

Figure 7.9.5 Schematic of reflected radiation from center and rim of a (half) parabolic reflector, with minimum plane, circular, and semicircular receivers to intercept all of the reflected radiation from a full parabola.

Figure 7.9.5 shows schematically how the reflected radiation from the rim of the parabola determines the width of the focal zone. The width of the solar image in the focal plane increases with increasing rim angle. The minimum sizes of flat, circular, and semicircular receivers centered at the focal point to intercept all of the reflected radiation are shown. It is clear from this diagram that the angle of incidence of radiation on the surface of any of these receiver shapes is variable.

For specular parabolic reflectors of perfect shape and alignment, the size of the receiver to intercept all of the solar image can be calculated. The diameter D of a cylindrical receiver is

$$D = 2r_r \sin 0.267 = \frac{a \ \sin 0.267}{\sin \phi_r} \qquad (7.9.4)$$

For a flat receiver in the focal plane of the parabola (the y-z plane through F, as shown in Figure 7.9.3) the width W is

$$W = \frac{2r_r \sin 0.267}{\cos(\phi_r + 0.267)} = \frac{a \ \sin 0.267}{\sin \phi_r \cos(\phi_r + 0.267)} \qquad (7.9.5)$$

Note that W is also the the diameter of a semicircular receiver.

For a flat receiver, as ϕ varies from zero to ϕ_r, r increases from f to r_r and the theoretical image size in the focal plane increases from D (evaluated with r_r equal to f) to W. The focal length is a determining factor in image size, and the aperture is the determining factor in total energy; thus the image brightness or energy flux concentration at the receiver of a focusing system will be a function of the ratio a/f.

7.10   IMAGES FORMED BY PERFECT LINEAR CONCENTRATORS

We turn now to a more detailed consideration of the theoretical images produced by perfectly oriented cylindrical parabolic reflectors.[9] Only images formed on planes perpendicular to the axis of the parabola will be considered; these examples provide the basis for an appreciation of the important factors in the operation of concentrators.

The radiation incident on a differential element of area of a reflector can be thought of as a cone having an apex angle of 32', or a half-angle of 16'. The reflected radiation from the element will be a similar cone and will have the same apex angle if the reflector is perfect. The intersection of this cone with the receiver surface determines the image size and shape for that element, and the total image is the sum of the images for all of the elements of the reflector.

Consider a flat receiver perpendicular to the axis of a perfect parabola at its focal point, with beam radiation normal to the aperture. The intersection of the focal plane and a cone of reflected radiation from an element is an ellipse, with minor axis of $2r$ sin 16' and major axis of $(y_1 - y_2)$, where y_1 is r sin $16'/\cos(\phi - 16')$ and y_2 is r sin $16'/\cos(\phi + 16')$. The total image is the sum of the ellipses for all of the elements of the reflector.

Evans (1977) has determined the distribution of energy flux in the integrated images on the focal plane for three models of the sun, including a nonuniform solar disk with an intensity distribution suggested by Lose (1957) based on data of Abetti (1938). The nonuniform disk model takes into account the fact that the sun radiates more from its central portion than it does from the limb, or edge. Figures 7.10.1(a), 7.10.2, and 7.10.3, from Evans, show the distribution of energy across images for several perfect concentrators for the nonuniform solar disk.

The local concentration ratio $C_l = I(y)/I_{b,\mathrm{ap}}$ is the ratio of intensity at any position y in the image to the intensity on the aperture of the concentrator. Local concentration ratios are shown in Figure 7.10.1a. The abscissa is the distance from the image center, expressed in dimensionless form as y/f. A reflectance of 1 was assumed in these calculations. Distributions are shown for five rim angles of the reflector. As rim angle increases, the local concentration ratio increases, as does the size of the image. (If a uniform solar disk is assumed, the distributions show lower concentration ratios in the center of the image and higher ratios in the outer portions of the images than for the nonuniform model.)

The distributions of Figure 7.10.1(a) can be integrated from zero to y/f to show the intercept factor γ, the fraction of the total image that is intercepted by the flat receiver in the focal plane. Intercept factors are shown in Figure 7.10.1(b).

---

[9] This and the following sections are based on numerical integrations of images formed by parabolic cylinder reflectors on focal planes and draws on the work of Evans (1977). An alternative approach extensively developed by Rabl (1985) is based on the concept that images can be described as Gaussian in character. Many concentrators use cylindrical receivers, which require modifications of the analyses. Ray-trace methods are used in the analysis and design of these collectors.

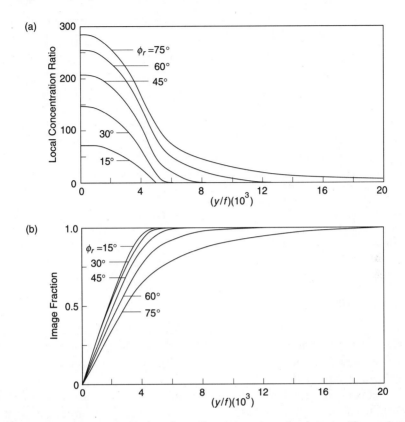

Figure 7.10.1 (a) Image distributions for perfect concentrators for the nonuniform solar disk. From Evans (1977). (b) Intercept factors for images from perfect concentrators, obtained by integrating areas under curves of (a).

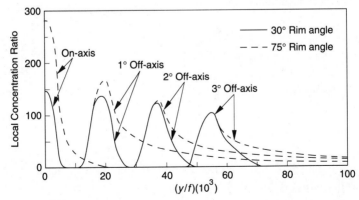

Figure 7.10.2 Image distributions for the nonuniform solar disk for several pointing errors in the *x-y* plane. From Evans (1977).

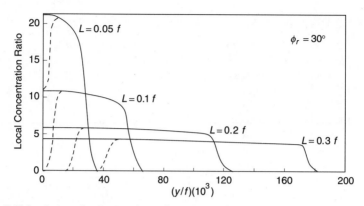

Figure 7.10.3 Image distributions for 30° rim angle reflectors for several displacements of the plane of the image from the focal plane. The dashed curve shows the effect of shading of the reflector by the absorber. From Evans (1977).

Example 7.10.1

A perfect cylindrical parabolic reflector has an aperture of 7.00 m and a focal length of 6.53 m. It is long in the z direction so that end effects are negligible. The incident beam radiation normal to the aperture is 805 W/m². The specular reflectance of the concentrator is 0.85.

What is the image width? What is the radiation intensity at the center of the image for the nonuniform solar disk? What is the intensity of the image at a point in the focal plane 26 mm from the axis, and how much of the total solar energy is incident on a symmetrical receiver 52 mm wide?

Solution

The f/a ratio of this concentrator is $6.53/7.00 = 0.933$. From Equation 7.9.2 the rim angle for a concentrator of these dimensions is

$$\phi_r = \tan^{-1}\left[\frac{8 \times 0.933}{16 \times 0.933^2 - 1}\right] = 30°$$

From Equation 7.9.5 the image width is

$$W = \frac{7.00 \sin 0.267}{\sin 30 \cos(30 + 0.267)} = 0.076 \text{ m}$$

From Figure 7.10.1(a), for the nonuniform solar disk, the local concentration ratio C_l at the axis is 149. Thus the flux in the image at the axis is

$$I(y = 0) = I_{b,\text{ap}}\rho C_l = 805 \times 0.85 \times 149 = 102 \text{ kW/m}^2$$

At a distance 0.026 m from the axis, $(y/f)10^3 = 0.026 \times 10^3/6.53 = 4.0$. Again from Figure 7.10.1(a) at $\phi_r = 30°$, $C_l = 69$ and

$$I(y = 0.026) = 805 \times 0.85 \times 69 = 47 \text{ kW/m}^2$$

From Figure 7.10.1(b), at $(y/f)10^3 = 4.0$, the intercept factor is 0.92, so the energy per unit length on a receiver of width of 52 mm, neglecting shading by the receiver, is

$$7.00 \times 805 \times 0.85 \times 0.92 = 4.41 \text{ kW/m} \qquad ∎$$

To this point, orientation of the collector has been considered to be perfect, i.e., with the beam radiation normal to the aperture. Two kinds of variation from this situation arise. The first is pointing errors in the x-y plane. The second is variation of the angle of incidence in the x-z plane that result from the mode of tracking.

Calculations similar to those resulting in Figure 7.10.1(a) have been done by Evans for perfect reflectors with alignment errors in the x-y plane. The results are shown in Figure 7.10.2 for two rim angles and three alignment errors. Figure 7.10.3 shows the effects of displacement L of the image plane from the focal plane for a perfect reflector with a rim angle of 30° with the radiation normal to the aperture. The dashed lines show the effect of shading of the reflector by the receiver for receiver widths of $0.05f$, $0.1f$, $0.2f$, and $0.3f$. It is clear that significant displacements of a receiver from the focal plane (or gross reflector distortions that lead to variation in focal length) lead to major changes in the nature of the image. The effect increases as rim angle increases.

Linear parabolic concentrators may be oriented in several ways to track the beam radiation. In general the beam radiation will be in the principal plane (the x-z plane) of the collector as described by the focus and vertex of the parabola but will not be normal to the aperture. Rabl (1985) shows that under these circumstances a parabolic trough collector can be analyzed by projections in the x-y plane. Common modes of orientation are adjustment about horizontal axes, aligned either east-west or north-south; rotation about a polar axis is occasionally used. In these situations the angle of incidence θ of beam radiation on the aperture of the collector, if it is parallel to the central plane of the reflector, is described by Equations 1.7.2a, 1.7.3a or 1.7.5a.

As the angle of incidence in the x-z plane increases, the apparent half-angle subtended by the sun (the projection on the x-y plane of the 0.267° half-angle) increases as $1/\cos \theta$. The effect on image width W is obtained from Equation 7.9.5 by substituting $0.267/\cos \theta$ for 0.267.

$$W = \frac{2 \, r_r \sin (0.267/\cos \theta)}{\cos (\phi + 0.267/\cos \theta)} \qquad (7.10.1)$$

The effect of θ on image width is best illustrated by W/W_0, the ratio of image width at incidence angle θ to the width at $\theta = 0$.

$$\frac{W}{W_0} = \frac{\sin (0.267/\cos \theta)}{\sin 0.267} \frac{\cos (\phi + 0.267)}{\cos (\phi + 0.267/\cos \theta)} \approx \frac{1}{\cos \theta} \qquad (7.10.2)$$

This ratio does not change significantly with θ until the rim angle exceeds about $80°$. Until θ becomes large, the second fraction varies little with θ and the dominant effect is that of the first fraction. Also, the sine of a small angle is nearly equal to the angle, so W/W_0 is very nearly equal to $1/\cos \theta$.

The result of this image widening is that a linear parabolic concentrator oriented on a horizontal east-west axis would have an image on the receiver very much enlarged in the early and late hours of a day and of minimum size at noon. A collector oriented on a north-south horizontal axis would have an image size that would be less variable through the day. The effects of θ on image size are in addition to the effect of reduction of energy in the image (for a given I_{bn}) by the factor $\cos \theta$.

Example 7.10.2

The concentrator of Example 7.10.1 is oriented with its z axis horizontal in the E-W direction, and it is rotated continuously to minimize the angle of incidence of the beam radiation. Assume that there is no pointing error in the x-y plane. At 3 PM on June 11, what is the image width? What is the image intensity in the focal plane at the axis? What will the image intensity be at a point 0.026 m from the axis? What is the fraction of the specularly reflected radiation that is intercepted by a symmetrical receiver in the focal plane that is 0.052 m wide?

Solution

On June 11 at 3 PM, $\delta = 23.1°$ and $\omega = 45°$. From Equation 1.7.2a

$$\cos \theta = (1 - \cos^2 23.1 \sin^2 45)^{1/2} = 0.760$$

$$\theta = 40.6°$$

From Example 7.10.1 $\phi_r = 30°$ and $f = 6.53$ m, so from Equation 7.9.3

$$r_r = \frac{2 \times 6.53}{1 + \cos 30} = 7.00 \text{ m}$$

The image width is found from Equation 7.10.1

$$W = \frac{2 \times 7.00 \sin(0.267/0.760)}{\cos(30 + 0.267/0.760)} = 0.099 \text{ m}$$

(The same information can be obtained from Equation 7.10.2, where W_0 is the width calculated in Example 7.10.1. The width $W = 0.076/\cos 40.6 = 0.099$ m.)

The intensity at the center will be reduced by two effects, the reduced radiation incident on the aperture resulting from off-normal solar radiation and the image spread due to the apparent increase in the sun's half-angle θ_s. From Example 7.10.1 the intensity at the center is 102 kW/m^2. For the off-normal condition of this example,

$$I(y = 0, \theta = 40.6) = I(y = 0, \theta = 0)(\cos 40.6)\frac{W_0}{W}$$

$$= 102 \times 0.759 \times \frac{0.076}{0.099} = 59 \text{ kW/m}^2$$

The area under a C_l vs. y/f curve is fixed, so at a given y/f as the image widens the local intensity diminishes. In Figure 7.10.1(a) the ordinate becomes $C_l/\cos \theta$ and the abscissa becomes $(y/f) \times 10^3 \times \cos \theta$. The value of $(y/f) \times 10^3$ of 4.0 of Example 7.10.1 is equivalent to a value of $4.0 \cos 40.6 = 3.0$ for this example. From Figure 7.10.1(a), at an abscissa of 3.0, the ordinate is $C_l/\cos \theta = 105$, so the local concentration ratio C_l is 80. The intensity is

$$I(y = 0.026, \theta = 40.6) = 805(\cos 40.5)\,0.85 \times 80 = 42 \text{ kW/m}^2$$

From Figure 7.10.1(b) at an abscissa of 3.0, the intercept factor is 0.78 and the energy per unit length on a receiving surface of width 0.052 m, neglecting receiver shading, is

$$7.00 \times 805\,(\cos 40.6)0.85 \times 0.78 = 2.8 \text{ kw/m}$$ ■

Comparing the results of the two examples (at the same I_{bn}), when the angle of incidence in the x-z plane went from 0° to 40.6°, the image width increased from 76 to 99 mm, the intensity at the center of the image dropped from 102 to 59 kW/m^2, the intensity at $y = 26$ mm decreased from 47 to 42 kW/m^2, and the energy per unit length incident on the receiving surface 52 mm wide dropped from 4.4 to 2.8 kW/m.

7.11 IMAGES FROM IMPERFECT LINEAR CONCENTRATORS

The distributions shown in Figures 7.10.1 to 7.10.3 are for perfect parabolic cylinders. If a reflector has small, two-dimensional surface slope errors that are normally distributed, the images in the focal plane created by these reflectors for perfect alignment will be as shown in Figure 7.11.1. Distributions for reflectors with rim angles of 30° and 75° are shown for various values of the standard deviation of the normally distributed slope errors. Imperfect reflectors will, as is intuitively obvious, produce larger images than the theoretical.

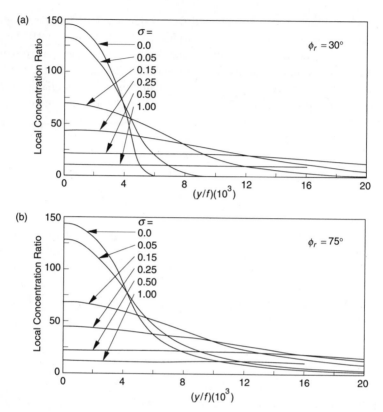

Figure 7.11.1 (a) Image distributions for imperfect reflectors for 30° rim angle for various standard deviations of normally distributed reflector slope errors. (b) Image distributions for imperfect reflectors for 75° rim angle for various standard deviations of normally distributed reflector slope errors. From Evans (1977).

A second method of accounting for imperfections in the shape of a parabola is to consider the reflected beam as having an angular width of (0.53 + δ) degrees, where δ is a **dispersion angle**, as shown in Figure 7.11.2. Here δ is a measure of the limits of angular errors of the reflector surface. With δ, equations can be written for the size of images produced on cylinders or planes at the fucus. The diameter of a cylindrical receiver that would intercept all of the image would be

$$D = 2r_r \sin\left(0.267 + \delta/2\right) = \frac{a \sin\left(0.267 + \delta/2\right)}{\sin \phi_r} \tag{7.11.1}$$

The image width on the focal plane, if the incident radiation were normal to the aperture, would be

$$W = \frac{2r_r \sin\left(0.267 + \delta/2\right)}{\cos\left(\phi_r + 0.267 + \delta/2\right)} = \frac{a \sin\left(0.267 + \delta/2\right)}{\sin \phi_r \cos\left(\phi_r + 0.267 + \delta/2\right)} \tag{7.11.2}$$

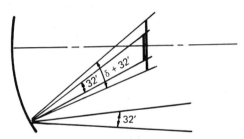

Figure 7.11.2 Schematic of a portion of a concentrator with a dispersion angle δ added to the $0.53°$ solar intercept angle.

It may be possible to estimate δ from either flux distribution measurements, by ray trace methods if the distribution of angular errors is known, or by knowledge of the distribution of angular errors to be expected from the process used in manufacturing a reflector. An assumption of normally distributed angular errors may be closer to reality.

Equations 7.11.1 and 7.11.2 give the diameter and width of images of imperfect parabolic troughs that can be described by the dispersion angle δ. (Equations 7.9.4 and 7.9.5 give the same information for perfect reflectors.) The ratio $(a/W)_{\gamma=1}$ is defined as the ratio of the area of the aperture to the total area of the receiver when the receiver is just large enough to intercept all of the specularly reflected radiation (i.e., when $\gamma = 1$). From $(a/W)_{\gamma=1}$ we can get C_{max}, the maximum area concentration ratio that leads to interception of the total image. For a concentrator producing an image with well-defined boundaries and without pointing errors or mispositioning of the receiver, for a flat receiver of width W with a shadow W wide, the maximum area concentration ratio for $\gamma = 1$ is $(a/W)_{\gamma=1} - 1$. Then

$$C_{max} = \frac{\sin\phi_r \cos(\phi_r + 0.267 + \delta/2)}{\sin(0.267 + \delta/2)} - 1 \qquad (7.11.3)$$

7.12 RAY-TRACE METHODS FOR EVALUATING CONCENTRATORS

Analysis of concentrating collectors (and other optical systems) is commonly done by ray-trace methods. This is the process of following the paths of a large number of rays of incident radiation through the optical system (the set of refracting and reflecting surfaces) to determine the distribution of the processed rays on the surface of interest. Thus for a concentrating collector, the ray trace starts with the assembly of rays of incident radiation on the aperture and determines the distribution and intensity of those rays on the receiver. This subject is treated in detail in many books on optics; a useful summary in the solar energy context is provided by Welford and Winston (1981). Spencer and Murty (1962) present a useful general treatment of the

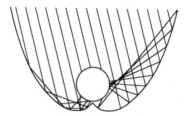

Figure 7.12.1   A ray trace diagram for a CPC reflector.

method, and Bendt et al. (1979) show application of ray tracing to linear parabolic concentrators.

Ray tracing in solar collector systems is done with vectors. For a reflecting surface, the direction and point of intersection of an incident ray with the reflecting surface are determined. The normal to the surface is also determined from its shape, and the direction of the reflected ray then follows from the principle that the angle of reflection equals the angle of incidence. This is done in an appropriate set of coordinates. Refracting surfaces are similarly treated using Snell's law. Figure 7.12.1 shows an example of a ray trace analysis of a CPC; the sizes and shapes of receivers can be shown on the diagram to determine the distribution and angles of incidence of radiation on those receivers.

7.13   INCIDENCE ANGLE MODIFIERS AND ENERGY BALANCES

In practice, images often do not have well-defined boundaries, and it usually is best to use a receiver that will intercept less than all of the specularly reflected radiation. A trade-off between increasing thermal losses with increasing area and increasing optical losses with decreasing area is necessary to optimize long-term collector performance. This optimization problem has been studied by Löf et al. (1962) and Löf and Duffie (1963), with a result that for a wide range of conditions the optimum size receiver will intercept 90 to 95% of the possible radiation. Thus an optical loss, often in a range of 5 to 10%, will be incurred in this type of collector. This has been expressed in terms of the intercept factor γ, the fraction of the specularly reflected radiation which is intercepted by the receiver. Figure 7.10.1(b) shows intercept factors as a function of receiver width in the focal plane for symmetrical collectors with perfect concentrators.

Errors in concentrator contours, tracking (pointing) errors, and errors in displacement of receivers from the focus all lead to enlarged or shifted images and consequently affect γ. These errors can also affect the transmittance of a cover system and the absorptance of a receiver. In addition, a spread of the image at the receiver of a linear concentrator depends on the angle of incidence of beam radiation θ as indicated in Section 7.10. These effects can be represented by biaxial incidence angle modifiers, in the transverse (x-y) and longitudinal (x-z) planes.

In the transverse plane, γ will drop off sharply as the transverse component of the angle of incidence increases. This effect can be determined from information in curves like those of Figure 7.10.2.

In the longitudinal plane, the two major effects will be that of image spread as a function of θ (as noted in Section 7.10), and variation with θ of the reflectance, transmittance, and absorptance. It is not practical to generalize these effects, and experimental data are most useful to determine their magnitude. Figure 7.13.1, from Gaul and Rabl (1980), shows an example of measured incidence angle modifiers as a function of θ. In addition, if the trough is not long in extent, there will be end effects as the image from the end of the trough is formed beyond the end of the receiver. Rabl (1985) gives an example of an end effect correction for receivers that are the same length l as the reflector and placed symmetrically over it:

$$K(\theta) = 1 - \frac{f}{l}\left(1 + \frac{a^2}{48f^2}\right)\tan\theta \qquad (7.13.1)$$

As a first approximation the overall incidence angle modifier can be taken as the product of the transverse and longitudinal components. These factors are in general not simply expressed as analytic functions, and it is not now possible to write general equations for overall effects of the angles of incidence in the longitudinal and transverse planes.

In order to estimate the useful output of the collector, it is necessary to estimate F_R and U_L. Methods for calculating F_R are basically the same as for flat-plate collectors, except that fin and bond conductance terms generally will not appear in F'; these considerations are outlined in Section 7.3.

Values of U_L may be difficult to determine. [For examples of calculation methods for uncovered cylinders and cylinders with cylindrical covers, see Section 7.3 and Tabor (1955 and 1958).] An additional complication arises in that conduction losses through supporting structures are highly variable with design and are

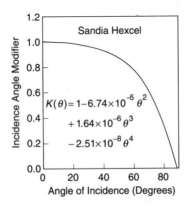

Figure 7.13.1 Incidence angle modifier as a function of angle of incidence in the x-z plane for a Hexcel linear parabolic collector. Adapted from Gaul and Rabl (1980).

Table 7.13.1   Receiver Design and Collector Characteristics[a]

Type		Description	C	a, m	f, m
A.	Hexcel	Steel pipe, black chrome coated. Back side insulated. On absorbing side, cover is glass semicylinder. Space not evacuated.	67	2.6	0.914
B.	Solar Kinetics	Steel tube, black chrome plated, inside a borosilicate glass tube, not evacuated.	41	1.3	0.267
C.	Suntec Systems	Two parallel steel pipes, black chrome plated, with down- and back-flow path. Back insulated, front covered with flat glass.	35	3.5	3.05

[a] From Leonard (1978).

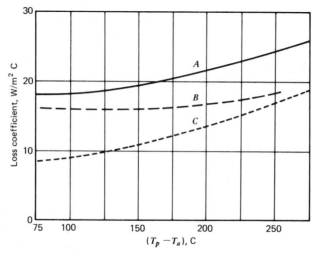

Figure 7.13.2 Loss coefficients per unit of area of receiver as a function of the difference between average receiver temperature and ambient temperature for three receivers described in Table 7.13.1. Data from Leonard (1978).

dependent on temperature. Some experimental data are available on loss coefficients for several receiver designs. Table 7.13.1 indicates a brief description of three receivers, and Figure 7.13.2 shows measured loss coefficients as a function of the difference between receiver temperature and ambient temperature. These loss coefficients are based on receiver area.

Example 7.13.1

A linear parabolic concentrator with $a = 2.50$ m and $f = 1.00$ m is continuously adjusted about a horizontal east-west axis. It is to be fitted with a liquid-heating receiver very similar to Type C in Table 7.13.1. The unit is 10.5 m long. A strip of the reflector 0.21 m wide is shaded by the receiver. The receiver is designed to be just large enough to intercept all of the specularly reflected beam radiation when the incident beam radiation is normal to the aperture, and under those conditions the distribution of radiation in the focal plane can be approximated as shown by the solid line on the figure below. The loss coefficient per unit area of receiver is found from curve C in Figure 7.13.2.

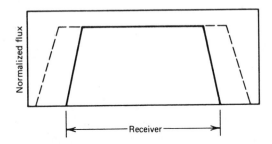

The normal beam radiation G_{bn} at noon on 20 April is 975 W/m², and at 3 PM it is 810 W/m². Here $(\tau\alpha)$ for the receiver is estimated at 0.78 with radiation normal to the aperture, and ρ is 0.86. The inlet fluid temperature is 170 C, the ambient temperature is 25 C and F_R is estimated to be 0.85.

What will be the output of the collector at noon and at 3 PM?

Solution

The basic equation to be used for the estimation of useful gain is Equation 7.3.9. The first step is to estimate an optical efficiency, then S, and then Q_u. At noon, the radiation will be normal to the aperture, and according to design specifications $\gamma = 1$. A fraction of the reflector of 0.21/2.50, or 0.084, is shaded by the receiver, so (1 − 0.084) or 0.916 of the reflector is effective. The product $\rho\gamma\tau\alpha$ is

$$\rho\gamma\tau\alpha = 0.86 \times 1.00 \times 0.78 \times 0.916 = 0.61$$

Based on the area of the unshaded aperture,

$$S = 975 \times 0.61 = 600 \text{ W/m}^2$$

At an estimated mean receiver surface temperature of 200 C, U_L from Figure 7.13.2 is 13.7 W/m²C. Then from Equation 7.3.9, at a concentration ratio A_a/A_r of (2.5 − 0.21)/0.21= 10.9,

$$Q_u = 0.85 \times 10.5(2.5 - 0.21)\left[600 - \frac{13.7}{10.9}(170 - 25)\right] = 8.5 \text{ kW}$$

At 3 PM the angle of incidence of beam radiation on the aperture is 43.9° (from Equation 1.7.2a). This leads to significant image spread and changes in γ, τ, and α. The dashed line in the figure shows how the image will spread, with its width (1/cos 43.9) greater than at noon. An approximate integration between the limits of the receiver dimensions indicates $\gamma = 0.80$. From Figures 4.11.1 and 5.3.1, α and τ will be reduced by approximately 3% each, so an approximation to $(\tau\alpha)$ is $0.78(0.97)^2 = 0.73$. Then $\rho\gamma\tau\alpha = 0.86 \times 0.80 \times 0.73 \times 0.916 = 0.46$.

At 3 PM, based on the unshaded aperture,

$$S = 810 \times 0.46 \cos 43.9 = 268 \text{ W/m}^2$$

The useful gain is

$$Q_u = 0.85 \times 10.5(2.5 - 0.21)\left[268 - \frac{13.7}{10.9}(170 - 25)\right] = 1.75 \text{kW} \quad \blacksquare$$

In this example, some shortcuts have been taken. The mean angles of incidence of the radiation on the receiver are not known when γ is significantly less than unity, and only simple corrections accounting for incidence angles on the plane of the receiver have been made. The integration to find γ is approximate. More detailed knowledge of the optical characteristics of reflector orienting systems and receivers would have to be combined with ray-trace techniques to improve on these calculations.

An experimental study of energy balances on a linear parabolic collector was reported by Löf et al. (1962). While it is a special case, and while the concentrator f/a was smaller than optimum, it provides an illustration of the importance of several factors affecting performance of collectors of this type. The collector consisted of a parabolic cylinder reflector of aperture 1.89 m, length 3.66 m, and focal length 0.305 m, with bare tubular receivers of three sizes coated with nonselective black paint having an absorptance of 0.95. The collector was mounted so as to rotate on a polar axis at 15°/h. It was operated over a range of temperatures from near ambient to approximately 180 C with each of the three receivers.

The intercept factors for various receiver sizes were determined from measurements of the flux distribution at many locations on the focal tube. These distributions were averaged, and the resulting mean distribution is shown in Figure

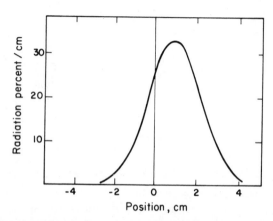

Figure 7.13.3 Experimental mean flux distribution for a parabolic cylinder reflector. From Löf et al. (1962).

7.13.3. This distribution is similar to a Gaussian distribution and is displaced from the position of the theoretical focus. The intercept factors that resulted from this distribution are shown in Figure 7.13.4.

The results of many energy balance measurements are summarized in Figure 7.13.5, which shows the distribution of incident beam solar energy (during operation at steady state in clear weather) into useful gain and various losses for a tube diameter of 60 mm. The relative magnitudes of the losses are evident.

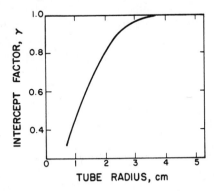

Figure 7.13.4 Intercept factors for tubes centered at position 0 of the reflector of Figure 7.13.3. From Löf et al. (1962).

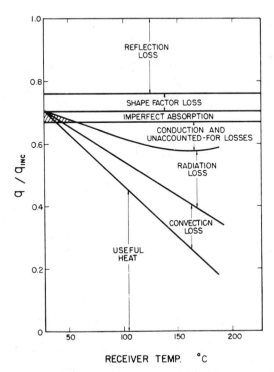

Figure 7.13.5 Distribution of incident energy for the 1.89 m reflector with 60 mm diameter receiver as a function of receiver temperature. From Löf et al. (1962).

From these figures it is possible to estimate the effects of design changes. For example, for this collector, the use of selective surface of emittance 0.2 would reduce the radiation loss by 79% of the values shown. However, radiation loss is not the dominant loss [a generalization made by Edwards and Nelson (1961)]. The most obvious initial improvements for this exchange would be in reduction of optical losses by using surfaces of high reflectance and by intercept factor improvements.

In this study, the reflector and receiver tubes were supported by plates at each end; these result in heat loss by conduction from the tubes. These losses were estimated from temperature measurements along the supporting plates. Although not shown in the figures, they were estimated at 3, 6, and 10% of the incident clear sky radiation for receiver-surface temperatures of 100, 135, and 175 C, respectively, for the conditions of these experiments.

7.14   PARABOLOIDAL CONCENTRATORS

In the previous sections methods were outlined for calculation of absorbed radiation for collectors with linear parabolic concentrators. A similar analysis can be done for collectors with three-dimensional parabolic reflectors, that is, concentrators that are surfaces of revolution. Rabl (1976a and 1985) summarizes important optical aspects of these collectors. In a section through the axis of the paraboloid these collectors are represented by Figures 7.9.2 and 7.9.3, and the rim angle ϕ_r and mirror radius r are analogous to those for the linear concentrator. Dispersion also occurs in paraboloidal concentrators, and equations analogous to Equation 7.11.2 can be written for collectors without tracking errors. For spherical receivers (allowing for minimum shading by the receiver)

$$C_{\max} = \frac{\sin^2 \phi_r}{4 \sin^2(0.267 + \delta/2)} - 1 \qquad (7.14.1)$$

For flat receivers

$$C_{\max} = \frac{\sin^2 \phi_r \cos^2(\phi_r + 0.267 + \delta/2)}{4 \sin^2(0.267 + \delta/2)} - 1 \qquad (7.14.2)$$

$C_{\max}$ is again defined as the maximum that can be obtained, based on interception of all of the specularly reflected radiation which is within the cone of angular width $(0.53 + \delta)$.

Cavity receivers may be used with paraboloidal concentrators to increase absorptance and to reduce convective losses from the absorbing surface. The equations for optical properties of systems with flat receivers will also apply to cavity receivers if appropriate absorptances of the cavities are used.

Absorbed energy for a paraboloidal collector depends on properties in the same way as linear parabolic collectors. However, at the higher concentration ratios achieved by these collectors, any absorption of solar radiation in the cover material would lead to heating of the cover, and as a result covers are usually not used. In the absence of a cover, the absorbed energy for the unshaded aperture area of the collector is then

$$S = I_{b,a} \rho (\gamma\alpha)_n K_{\gamma\alpha} \tag{7.14.3}$$

where the factors have the same meaning as in Equation 7.4.1. The incidence angle modifier may be circularly symmetrical; if the optical system is not symmetrical, biaxial or other modifiers may have to be used. The modifier accounts for pointing errors in the alignment of the collector.

Calculation of thermal losses is very much a function of receiver geometry and may be complicated by the existence of temperature gradients on the surface of a receiver. For examples of these estimations, see reports of Martin Marietta, Boeing, and McDonnell-Douglas on the receiver subsystems of central receiver solar thermal power systems.

7.15 CENTRAL RECEIVER COLLECTORS

The central receiver or "power tower" concept for generation of electrical energy from solar energy is based on the use of very large concentrating collectors. The optical system consists of a field of a large number of heliostats, each reflecting beam radiation onto a central receiver. The result is a Fresnel-type concentrator, a parabolic reflector broken up into small segments, as shown in Figure 7.1.1(f). Several additional optical phenomena must be taken into account. Shading and blocking can occur (shading of incident beam radiation from a heliostat by another heliostat and blocking of reflected radiation from a heliostat by another which prevents that radiation from reaching the receiver). As a result of these considerations, the heliostats are spaced apart, and only a fraction of the ground area ψ is covered by mirrors. A ψ of about 0.3 to 0.5 has been suggested as a practical value.

The maximum concentration ratio for a three-dimensional concentrator system with radiation incident at an angle θ_i on the plane of the heliostat array ($\theta_i = \theta_z$ for a horizontal array), for a rim angle of ϕ_r, and a dispersion angle of δ, if all reflected beam radiation is to be intercepted by a spherical receiver, is

$$C_{max} = \frac{\psi \sin^2 \phi_r}{4 \sin^2(0.267 + \delta/2)} - 1 \tag{7.15.1}$$

For a flat receiver, the concentration ratio is

$$C_{max} = \psi \left[\frac{\sin\phi_r \cos(\phi_r + 0.267 + \delta/2)}{\sin(0.267 + \delta/2)} \right]^2 - 1 \tag{7.15.2}$$

Figure 7.15.1 Solar One, a central receiver collector system. Photo courtesy of Sandia Laboratories.

As with linear concentrators, the optimum performance may be obtained with intercept factors less than unity. The designer of these collectors has additional considerations to take into account. The heliostat field need not be symmetrical, the ground cover ψ does not have to be uniform, and the heliostat array is not necessarily all in one plane. These collectors operate with high concentration ratios and at relatively high receiver temperatures.

There has been considerable work done on the design and performance of collectors of this type in the 1970s, and a large body of literature is available. For example, see the University of Houston-Sandia Proceedings of ERDA Workshops on Central Receiver Systems (1977). Also, many papers have appeared in journals, and as a starting point the reader could go to Hildebrandt and Vant-Hull (1977) or Vant-Hull (1977). The *Proceedings of the IEA Workshop on Design and Performance of Large Solar Thermal Collector Arrays* (1984) includes performance reports on several central receiver collectors and components thereof. The *ASME Transactions, Journal of Solar Energy Engineering*, Volume 106, Number 1 (1984) was devoted entirely to papers on central receiver systems. Radosevich and Skinrood (1989) review experience with Solar One, a power-generating facility based on a central receiver collector. A photograph of this facility is shown in Figure 7.15.1.

7.16   PRACTICAL CONSIDERATIONS

Many of the practical considerations for flat-plate collectors also apply to concentrators; the necessity to move the collectors and the higher temperature ranges over which they operate impose additional requirements on focusing collectors. For more information on these topics,[10] see Radosevich and Skinrood (1989) for a report on Solar One and Harats and Kearney (1989) for a description of experience with the large-scale Luz linear concentrator systems used in power generation in California.

---

[10] See Chapter 17 for more information on these two systems.

For acceptable performance, all of the terms in the optical efficiency $\rho\gamma\tau\alpha$ must be maintained at acceptably high levels over the years of the system operating life. To maintain high specular reflectance, mechanized methods of cleaning reflectors have been developed for heliostats for central receiver systems and for reflectors of linear concentrators, and plant layout has to be arranged to allow access to the reflectors by the cleaning machinery. Protection of the coatings on the back-silvered glass reflectors is necessary to reduce corrosion of the reflecting surfaces to very low values.

The intercept factor is a function of quality of construction and also of collector orientation. Control systems of precision appropriate to the quality of the optics of the collector are needed. Each of the individual heliostats are controlled in Solar One. The Luz systems have a controller and tracking drive for each solar collector assembly (235 m^2 area each for LS-2 and 545 m^2 area each for LS-3). There are obvious advantages to using fewer controllers and drives, each controlling larger areas, but stiffness of the structures becomes a limiting factor.

Receivers of concentrators operate at elevated temperatures, and absorbing surfaces must be capable of prolonged exposure to the surrounding atmospheres at elevated temperature while maintaining acceptable absorptance and emittance.

Tracking collectors are by nature not rigidly mounted on buildings or other structures and are subject to wind loads that dictate structural requirements. Concentrators are typically designed to operate in winds up to 15 to 20 m/s; above that speed the reflectors are usually moved to a stowed position in which they are capable of withstanding the forces imposed by the maximum anticipated winds. Wind forces on outer collectors in very large arrays are higher than those in interior rows, and the structures and tracking drives must be designed to cope with these higher forces.

Concentrating collectors operate at temperatures that can range from 150 C to 800 C and cycle on and off at least once each operating day. This means that receivers, fluid circulation pumps, flexible connections, piping runs, heat exchangers, and other equipment in the hot fluid loops will cycle through substantial temperature excursions at least once a day.

Concentrating collectors must, as is obvious, be carefully designed and constructed if they are to provide reliable energy delivery over periods of many years.

REFERENCES

Abetti, G., *The Sun*, D. Van Nostrand Co., Princeton NJ (1938)

Baranov, V. K. and G. K. Melnikov, *Soviet J. of Optical Technology*, **33**, 408 (1966).

Bendt, P., A. Rabl, H. W. Gaul, and K. A. Reed, Report SERI/TR-34-092 of the Solar Energy Research Institute, Golden, CO (1979). "Optical Analysis and Optimization of Line Focus Solar Collectors."

Brandemuehl, M. J. and W. A. Beckman, *Solar Energy*, **24**, 511 (1980). "Transmission of Diffuse Radiation through CPC and Flat-Plate Collector Glazings."

Chow, S. P., G. L. Harding, B. Window, and K. J. Cathro, *Solar Energy*, **32**, 251 (1984). "Effects of Collector Components on the Collection Efficiency of Tubular Evacuated Collectors with Diffuse Reflectors."

Duffie, J. A. and G. O. G. Löf, Paper 207 III.7/5, World Power Conference, Melbourne (1962). "Focusing Solar Collectors for Power Generation."

Edwards, D. E. and K. E. Nelson, Paper 61-WA-158, New York ASME Meeting (1961). "Radiation Characteristics in the Optimization of Solar-Heat Power Conversion Systems."

Evans, D. L., *Solar Energy*, **19**, 379 (1977). "On the Performance of Cylindrical Parabolic Solar Concentrators with Flat Absorbers."

Gaul, H., and A. Rabl, *Trans. ASME, J. Solar Energy Engrg.*, **102**, 16 (1980). "Incidence-Angle Modifier and Average Optical Efficiency of Parabolic Trough Collectors."

Harats, Y. and D. Kearney, Paper presented at ASME meeting, San Diego, CA (1989). "Advances in Parabolic Trough Technology in the SEGS Plants."

Hildebrandt, A. F. and L. L. Vant-Hull, *Science*, **198**, 1139 (1977). "Power with Heliostats."

Hinterberger, H. and R. Winston, *Review of Scientific Instruments*, **37**, 1094 (1966).

Hollands, K. G. T., *Solar Energy*, **13**, 149 (1971). "A Concentrator for Thin-Film Solar Cells."

International Energy Agency (IEA), *Proc. IEA Workshop on Design and Performance of Large Solar Thermal Collector Arrays* (C.A. Bankston, ed.), San Diego, CA (1984).

International Energy Agency (IEA), *Results from Eleven Evacuated Collector Installations*, Report IEA-SHAC-TIV-4 (1986).

Laszlo, T. S., *Image Furnace Techniques*, Interscience, New York (1965).

Leonard, J. A., Paper presented at Solar Thermal Concentrating Collector Technology Symposium, Denver, June (1978). "Linear Concentrating Solar Collectors - Current Technology and Applications."

Löf, G. O. G. and J. A. Duffie, *Trans ASME, J. Engrg. for Power*, **85A**, 221 (1963). "Optimization of Focusing Solar-Collector Design."

Löf, G. O. G., D. A. Fester, and J. A. Duffie, *Trans. ASME, J. Engrg. for Power*, **84A**, 24 (1962). "Energy Balance on a Parabolic Cylinder Solar Reflector."

Lose, P. D., *Solar Energy*, **1**, No.2,3, 1 (1957). "The Flux Distribution through the Focal Spot of a Solar Furnace."

Mather, G. R. Jr., *Trans. ASME, J. Solar Energy Engrg.* **102**, 294 (1980). "ASHRAE 93-77 Instantaneous and All-Day Tests of the Sunpak Evacuated Tube Collector."

Mitchell, J.C., Personal communication (1979).

O'Gallagher, J. J., A. Rabl, and R. Winston, *Solar Energy*, **24**, 323 (1980). "Absorption Enhancement in Solar Collectors by Multiple Reflections."

Proceedings of the 1957 Solar Furnace Symposium, Solar Energy, **1**(2, 3), (1957).

Rabl, A., *Solar Energy*, **18**, 93 (1976a). "Comparison of Solar Concentrators."

Rabl, A., *Solar Energy*, **18**, 497 (1976b). "Optical and Thermal Properties of Compound Parabolic Concentrators."

Rabl A., Paper presented at Solar Thermal Concentrating Collector Technology Symposium, Denver, June (1978). "Optical and Thermal Analysis of Concentrators."

Rabl, A. *Active Solar Collectors and their Applications.* Oxford University Press, New York and Oxford (1985).

Rabl, A., J. J. O'Gallagher, and R. Winston, *Solar Energy,* **25** 335 (1980). "Design and Test of Non-evacuated Solar Collectors with Compound Parabolic Concentrators."

Radosevich, L. G. and A. C. Skinrood, *Trans ASME, J. Solar Energy Engrg.,* **111**, 144 (1989). "The Power Production Operation of Solar One, the 10MWe Solar Thermal Central Receiver Power Plant."

Reports of Boeing Engineering and Construction Co. (Contract No. EY-76-C-03-1111), Martin Marietta Corp. (Contract EY-77-C-03-1110), and McDonnell Douglas Astronautics Co. (Contract EY-76-C-03-1108), to the U.S. Department of Energy (1977) on "Central Receiver Solar Thermal Power System."

Selcuk, M. K., *Solar Energy,* **22**, 413 (1979), "Analysis, Development and Testing of a Fixed Tilt Solar Collector Employing Reversible Vee-trough Reflectors and Vacuum Tube Receivers."

Spencer, G. H. and M. V. R. K. Murty, *J. Optical Soc. of America,* **52**, 672 (1962). "General Ray-Tracing Procedure."

Tabor, H., *Bull. of the Research Council of Israel,* **5C**, 5 (1955). "Solar Energy Collector Design."

Tabor, H., *Solar Energy,* **2**(1), 3 (1958). "Solar Energy Research: Program in the New Desert Research Institute in Beersheba."

Theunissen, P-H., and W. A. Beckman, *Solar Energy,* **35**, 311 (1985). "Solar Transmittance Characteristics of Evacuated Tubular Collectors with Diffuse Back Reflectors."

University of Houston-Sandia Laboratories, *Proc. of the ERDA Solar Workshop on Methods for Optical Analysis of Central Receiver Systems* (1977). Available from National Technical Information Services, U.S. Dept. of Commerce.

Vant-Hull, L. L., *Optical Engineering,* **16**, 497 (1977). "An Educated Ray-Trace Approach to Solar Tower Optics."

Welford, W. T. and R. Winston, *The Optics of Nonimaging Concentrators.* Academic Press, New York (1978).

Welford, W. T.and R. Winston, in *Solar Energy Handbook* (J. Kreider and F. Kreith, eds.), McGraw-Hill, New York (1981). "Principles of Optics Applied to Solar Energy Concentrators."

Winston, R., *Solar Energy,* **16**, 89, (1974). "Solar Concentrations of Novel Design."

Winston, R. and H. Hinterberger, *Solar Energy,* **17**, 255 (1975). "Principles of Cylindrical Concentrators for Solar Energy."

Chapter 8

ENERGY STORAGE

Solar energy is a time-dependent energy resource. Energy needs for a very wide variety of applications are also time dependent but in a different fashion than the solar energy supply. Consequently, the storage of energy or other product of a solar process is necessary if solar energy is to meet substantial portions of these energy needs.

Energy (or product) storage must be considered in the light of a solar process system, the major components of which are the solar collector, storage units, conversion devices (such as air conditioners or engines), loads, auxiliary (supplemental) energy supplies, and control systems. The performance of each of these components is related to that of the others. The dependence of the collector performance on temperature makes the whole system performance sensitive to temperature. For example, in a solar-thermal power system, a thermal energy storage system which is characterized by high drop in temperature between input and output will lead to unnecessarily high collector temperature and/or low heat engine inlet temperature, both of which lead to poor system performance.

In passive solar heating, collector and storage components are integrated into the building structure. The performance of storage media in passive heating systems is interdependent with the absorption of energy, and we reserve discussion of part of this aspect of solar energy storage for Chapter 14.

The optimum capacity of an energy storage system depends on the expected time dependence of solar radiation availability, the nature of loads to be expected on the process, the degree of reliability needed for the process, the manner in which auxiliary energy is supplied, and an economic analysis that determines how much of the annual load should be carried by solar and how much by the auxiliary energy source.

In this chapter we set forth the principles of several energy storage methods and show how their capacities and rates of energy input and output can be calculated. In the example problems, as in the collector examples, we arbitrarily assume temperatures or energy quantities. In reality these must be found by simultaneous solutions of the equations representing all of the system components. These matters are taken up in Chapter 10; we will see there that the differential equations for storage are the key equations for most systems, with time the independent variable and storage (and other) temperatures the dependent variables.

8.1 PROCESS LOADS AND SOLAR COLLECTOR OUTPUTS

Consider an idealized solar process in which the time dependence of the load L and gain from the collector Q_u are shown in Figure 8.1.1(a). During part of the time, available energy exceeds the load on the process, and at other times it is less. A storage subsystem can be added to store the excess collector output and return it when needed. Figure 8.1.1(b) shows the energy stored as a function of time. Energy storage is clearly important in determining system output. If there were no storage, the useful solar gain would be reduced on the first and third days by the amount of energy added to storage on those days. This would represent a major drop in solar contribution.

In most applications it is not practical to meet all of the loads L on a process from solar energy over long periods of time, and an auxiliary energy source must be used. The total load L is met by a combination of solar energy L_S (which in practice will be somewhat less than Q_u because of losses) and L_A (the auxiliary energy supplied).

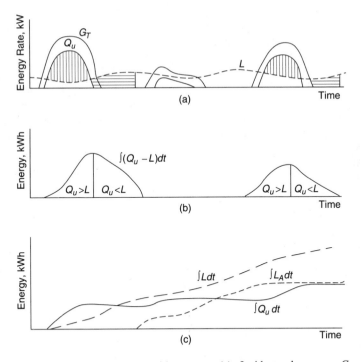

Figure 8.1.1 A solar energy process with storage. (a) Incident solar energy G_T, collector useful gain Q_u, and loads $\dot{L}$, as functions of time for a 3-day period. Vertical shaded areas show times of excess energy to be added to storage. Horizontal shaded areas show energy withdrawn from storage to meet loads. Dotted areas show energy supplied to load from collector during collector operation. (b) Energy added to or removed from storage, taking time $t = 0$ as a base. (c) Integrated values of the useful gain from collector, the load, and the auxiliary energy for the same 3-day period. In this example solar energy collected is slightly more than half the integrated load.

It is also useful to show the integrated values of the major parameters Q_u (i.e., approximately L_S), L, and L_A. Examples of these are shown in Figure 8.1.1(c). A major objective of system performance analysis is a determination of long-term values of L_A, the amount of energy that must be purchased; this is needed to assess the cost of delivering energy or product from the solar energy process and to estimate the fraction of total energy or product needs met from solar and auxiliary energy sources. In practice, these integrations must be done over long periods (typically a year), and both collector area and storage capacity are variables to be considered.

8.2 ENERGY STORAGE IN SOLAR PROCESS SYSTEMS

Energy storage may be in the form of sensible heat of a liquid or solid medium, as heat of fusion in chemical systems, or as chemical energy of products in a reversible chemical reaction. Mechanical energy can be converted to potential energy and stored in elevated fluids. Products of solar processes other than energy may be stored; for example, distilled water from a solar still may be stored in tanks until needed, and electrical energy can be stored as chemical energy in batteries.

The choice of storage media depends on the nature of the process. For water heating, energy storage as sensible heat of stored water is logical. If air heating collectors are used, storage in sensible or latent heat effects in particulate storage units is indicated, such as sensible heat in a pebble bed heat exchanger. In passive heating, storage is provided as sensible heat in building elements. If photovoltaic or photochemical processes are used, storage is logically in the form of chemical energy.

The major characteristics of a thermal energy storage system are (a) its capacity per unit volume; (b) the temperature range over which it operates, that is, the temperature at which heat is added to and removed from the system; (c) the means of addition or removal of heat and the temperature differences associated therewith; (d) temperature stratification in the storage unit; (e) the power requirements for addition or removal of heat; (f) the containers, tanks, or other structural elements associated with the storage system; (g) the means of controlling thermal losses from the storage system; and (h) its cost.

Of particular significance in any storage system are those factors affecting the operation of the solar collector. The useful gain from a collector decreases as its average plate temperature increases. A relationship between the average collector temperature and the temperature at which heat is delivered to the load can be written as

$$T(\text{collector}) - T(\text{delivery}) = \quad \Delta T \text{ (transport from collector to storage)}$$
$$+ \Delta T \text{ (into storage)}$$
$$+ \Delta T \text{ (storage loss)}$$
$$+ \Delta T \text{ (out of storage)}$$
$$+ \Delta T \text{ (transport from storage to application)}$$
$$+ \Delta T \text{ (into application)}$$

Thus, the temperature of the collector, which determines the useful gain for the collector, is higher than the temperature at which the heat is finally used by the sum of

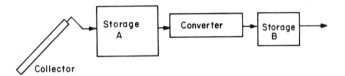

Figure 8.2.1   Schematic of alternative storage location at *A* or *B*.

a series of temperature difference driving forces. An objective of system design, and particularly of storage unit design, is to minimize or eliminate these temperature drops within economic constraints.

A solar process designer may have alternatives in locating the energy storage component. As an example, consider a process in which a heat engine converts solar energy into electrical energy. In such a system storage can be provided as thermal storage between the solar collector and the engine, as mechanical energy storage between the engine and generator, or as chemical storage in a battery between the generator and the end application. Solar cooling with an absorption air conditioner provides another example. Thermal energy from the collector can be stored to be used when needed by the air conditioner, or alternatively, the "cooling" produced by the air conditioner can be stored in a low-temperature (below ambient) thermal storage unit. These are illustrated schematically in Figure 8.2.1.

These two alternatives are not equivalent in capacity, costs, or effects on overall system design and performance. The storage capacity required of a storage unit in position *B* is less than that required in position *A* by (approximately) the efficiency of the intervening converter. Thus if the conversion process is operating at 25% efficiency, the capacity of storage at *B* must be approximately 25% of the capacity of *A*. Thermal energy storage at *A* has the advantage that the converter can be designed to operate at a more nearly constant rate, leading to better conversion efficiency and higher use factor on the converter; it can lower converter capacity requirements by removing the need for operation at peak capacities corresponding to direct solar input. Storage at *B* may be energy of a different form than that at *A*. The choice between energy storage at *A* or at *B* may have very different effects on the operating temperature of the solar collector, collector size, and ultimately on cost. These arguments may be substantially modified by requirements for use of auxiliary energy.

8.3   WATER STORAGE

For many solar systems water is the ideal material in which to store useable heat. Energy is added to and removed from this type of storage unit by transport of the storage medium itself, thus eliminating the temperature drop between transport fluid and storage medium. The typical systems in which water tanks are used can be represented by the water heating system shown in Figure 8.3.1. A forced circulation system is shown, but it could be natural circulation. Implicit in the following discussion is the idea that flow rates into and out of the tanks, to collector and load, can be determined.

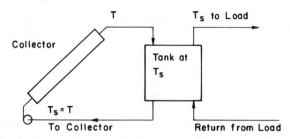

Figure 8.3.1 A typical system using water tank storage, with water circulation through collector to add energy and through the load to remove energy.

The energy storage capacity of a water (or other liquid) storage unit at uniform temperature (i.e., fully mixed, or unstratified) operating over a finite temperature difference is given by

$$Q_s = (mC_p)_s \Delta T_s \tag{8.3.1}$$

where Q_s is the total heat capacity for a cycle operating through the temperature range ΔT_s and m is the mass of water in the unit. The temperature range over which such a unit can operate is limited at the lower extreme for most applications by the requirements of the process. The upper limit may be determined by the process, the vapor pressure of the liquid, or the collector heat loss.

For a nonstratified tank, as shown in Figure 8.3.2, an energy balance on the tank yields

$$(mC_p)_s \frac{dT_s}{dt} = Q_u - L_S - (UA)_s (T_s - T_a') \tag{8.3.2}$$

where Q_u and L_S are rates of addition or removal of energy from the collector and to the load and T_a' is the ambient temperature for the tank (which may not be the same as that for a collector supplying energy to the tank).

Equation 8.3.1 is to be integrated over time to determine the long-term performance of the storage unit and the solar process. There are many possible integration methods. Using simple Euler integration (i.e., rewriting in finite difference form and solving for the tank temperature at the end of a time increment),

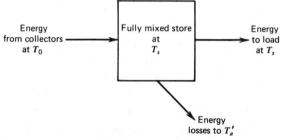

Figure 8.3.2 Unstratified storage of mass m operating at time-dependent temperature T_s in ambient temperature T_a'.

$$T_s^+ = T_s + \frac{\Delta t}{(mC_p)_s}\left[Q_u - L_S - (UA)_s(T_s - T_a')\right]$$ (8.3.3)

With this approximation, the temperature at the end of an hour (or other time period) is calculated from that at the beginning of the hour with the known inputs and outputs, assuming that the loss term can be assumed constant throughout the hour. (The most common time period used in these calculations is an hour, as radiation data are usually available on an hourly basis.)

Example 8.3.1 illustrates how the energy balances on a storage tank can be used to predict its temperature as a function of time. In this example, loads and input from the collector are given as functions of time, and the dependence of Q_u on storage tank temperature is not shown. Examples in later chapters will illustrate this dependence.

Example 8.3.1

A fully mixed water tank storage containing 1500 kg of water has a loss coefficient-area product of 11.1 W/C. The tank starts a particular 12-hour period at 45 C and is in a room at a constant temperature of 20 C. Energy Q_u is added to the tank from a solar collector, and energy L_S is extracted from the tank to meet a load. The first column of the table in the solution indicates the time at the end of an hourly period; the second and third columns indicate values of Q_u and L_S for those hours. Using Euler integration, calculate the temperature of the tank through the 24-hour period.

Solution

The energy balance on the tank in finite difference form is Equation 8.3.3. Inserting the appropriate constants, with a time increment of 1 hour, yields

$$T_s^+ = T_s + \frac{1}{1500 \times 4190}\left[Q_u - L_S - 11.1 \times 3600(T_s - 20)\right]$$

This equation is applied each hour. The results are shown in the table that follows, where column 4 is the old temperature and column 5 is the new temperature for the end of that hour calculated from the equation.

Hour	Q_u, MJ	L_s, MJ	T_s, C	T_s^+, C
1	0	12	45	42.9
2	0	12	42.9	40.9
3	0	11	40.9	39.0
4	0	11	39.0	37.1
5	0	13	37.1	35.0
6	0	14	35.0	32.6
7	0	18	32.6	29.7
8	0	21	29.7	26.3
9	21	20	26.3	26.4
10	41	20	26.4	29.7
11	60	18	29.7	36.3
12	75	16	36.3	45.6

In this example, the input Q_u and the output L_S have been given. The general situation is that these are calculated, most often as functions of T_s; this will be illustrated in examples to follow.

8.4   STRATIFICATION IN STORAGE TANKS

Water tanks may operate with significant degrees of stratification, that is, with the top of the tank hotter than the bottom. Many stratified tank models have been developed; they fall into either of two categories. In the first, the **multinode** approach, a tank is modeled as divided into N nodes (sections), with energy balances written for each section of the tank; the result is a set of N differential equations that can be solved for the temperatures of the N nodes as functions of time. In the second, the **plug flow** approach, segments of liquid at various temperatures are assumed to move through the tank in plug flow, and the models are essentially bookkeeping methods to keep track of the size, temperature, and position of the segments. Each of these approaches has many variations, and the selection of a model depends on the use to which it will be put. Examples of each approach are discussed. None of them lend themselves to hand calculations.

The degree of stratification in a real tank will depend on the design of the tank, the size, location, and design of the inlets and outlets, and flow rates of the entering and leaving streams. It is possible to design tanks with low inlet and outlet velocities that will be highly stratified [see Gari and Loehrke (1982) or van Koppen et al. (1979)]. The effects of stratification on solar process performance can be bracketed by calculating performance with fully mixed tanks and with highly stratified tanks.

To formulate the equations for a multinode tank, it is necessary to make assumptions about how the water entering the tank is distributed to the various nodes. For example, for the five-node tank shown in Figure 8.4.1, water from the collector enters at a temperature T_{co} which lies between $T_{s,2}$ and $T_{s,3}$. It can be assumed

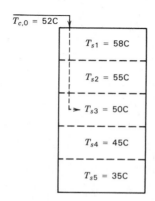

$\overline{T_{c,0}} = 52C$

$T_{s1} = 58C$

$T_{s2} = 55C$

$T_{s3} = 50C$

$T_{s4} = 45C$

$T_{s5} = 35C$

Figure 8.4.1   A hypothetical five-node tank with $T_{s,2} > T_{co} > T_{s,3}$. Water can be considered to enter at node 3 or to be distributed among nodes 1, 2, and 3.

Figure 8.4.2 Dyed water entering a stratified tank through a low-velocity manifold. From Gari and Loehrke (1979).

that it all finds its way down inside the tank to node 3, where its density nearly matches that of the water in the tank. (That this can be done is clearly illustrated in Figure 8.4.2, which shows dyed water entering a tank from a carefully designed inlet manifold.) Alternatively, it can be assumed that the incoming water distributes itself in some way to nodes 1, 2, and 3. In the following discussion, a model is developed that can represent a high degree of stratification; it is assumed that the water in Figure 8.4.1 finds its way into node 3.

Stratification is difficult to evaluate without considering the end use. If the load can use energy at the same efficiency without regard to its temperature level (that is, thermodynamic availability), then maximum stratification would provide the lowest possible temperature near the bottom of the tank and this would maximize collector output. On the other hand, if the quality of the energy to the load is important, then minimizing the destruction of available energy may be the proper criteria for defining maximum stratification (although all parts of the system should be considered simultaneously in such an analysis). The following analysis is intended to provide a limiting case in which the bottom of the tank is maintained at a minimum temperature, but other criteria could be used.

For a three-node tank, as shown in Figure 8.4.3, the flow to the collector always leaves from the bottom, node 3, and the flow to the load always leaves from the top, node 1. The flow returning from the collector will return to the node that is closest to but less than the collector outlet temperature. Suppose the three node temperatures are 45, 35, and 25 C, with, of course, the hottest at the top. Return water from the collector lower than 35 C will go to node 3 and between 35 and 45 C would go to node 2.

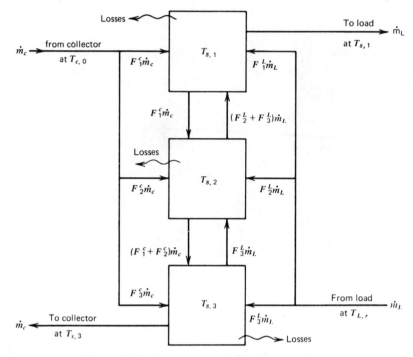

Figure 8.4.3 Three-node stratified liquid storage tank.

A collector control function F_i^c can be defined to determine which node receives water from the collector:

$$
F_i^c = \begin{cases}
1 & \text{if } i = 1 \text{ and } T_{co} > T_{s,i} \\
1 & \text{if } T_{s,i-1} \geq T_{co} > T_{s,i} \\
0 & \text{if } i = 0 \text{ or if } i = N+1 \\
0 & \text{otherwise}
\end{cases}
\tag{8.4.1}
$$

Note that if the collector is operating, then one and only one control function can be nonzero. Also, a fictitious temperature $T_{s,0}$ of the nonexistent node zero is assumed to be a large number. The three branches of the collector return water are controlled by this collector return control function, as shown in Figure 8.4.3. For the three-node tank of the previous paragraph with return water between 45 and 35 C, $F_1^c = 0$, $F_2^c = 1$, $F_3^c = 0$.

The liquid returning from the load can be controlled in a similar manner with a load return control function F_i^L.

$$
F_i^L = \begin{cases}
1 & \text{if } i = N \text{ and } T_{Lr} < T_{s,N} \\
1 & \text{if } T_{s,i-1} \geq T_{Lr} > T_{s,i} \\
0 & \text{if } i = 0 \text{ or if } i = N+1 \\
0 & \text{otherwise}
\end{cases}
\tag{8.4.2}
$$

The net flow between nodes can be either up or down depending upon the magnitudes of the collector and load flow rates and the values of the two control functions at any particular instant. It is convenient to define a mixed-flow rate that represents the net flow into node i from node $i - 1$, excluding the effects of flow, if any, directly into the node from the load:

$$\dot{m}_{m,1} = 0 \tag{8.4.3a}$$

$$\dot{m}_{m,i} = \dot{m}_c \sum_{j=1}^{i-1} F_j^c - \dot{m}_L \sum_{j=i+1}^{N} F_j^L \tag{8.4.3b}$$

$$\dot{m}_{m,N+1} = 0 \tag{8.4.3c}$$

With these control functions, an energy balance on node i can be expressed as

$$m_i \frac{dT_{s,i}}{dt} = \left(\frac{UA}{C_p}\right)_i \left(T_a' - T_{s,i}\right) + F_i^c \dot{m}_c (T_{co} - T_{s,i}) + F_i^L \dot{m}_L (T_{L,r} - T_{s,i})$$

$$+ \begin{cases} \dot{m}_{m,i}(T_{s,i-1} - T_{s,i}) & \text{if } \dot{m}_{m,i} > 0 \\ \dot{m}_{m,i+1}(T_{s,i} - T_{s,i+1}) & \text{if } \dot{m}_{m,i+1} < 0 \end{cases} \tag{8.4.4}$$

where a term has been added to account for losses from node i to an environment at T_a'.

With a large number of nodes, the tank model given by Equation 8.4.4 represents a high level of stratification that may not be achievable in actual experiments. There is very little experimental evidence to support the use of this model to represent high stratification, but the model is based on first principles.

As a practical matter, many tanks show some degree of stratification, and it is suggested that three or four nodes may represent a reasonable compromise between conservative design (represented by systems with one-node tanks) and the limiting situation of carefully maintained high degrees of stratification.

Two other factors may be significant. First, stratified tanks will have some tendency to destratify over time due to diffusion and wall conduction. This has been experimentally studied by Lavan and Thompson (1977). Second, some tanks have sources of energy in addition to that in fluids pumped into or out of the tank. If, for example, an auxiliary heater coil were present in one of the nodes, an additional term could be added to Equation 8.4.4 to account for its effect.

Numerical integration of Equations 8.3.1 or 8.4.4 can be accomplished by several techniques that are discussed in texts on numerical methods. The explicit Euler, the implicit Crank-Nicolson, predictor-corrector, and Runge-Kutta methods are the most common. Because of the complicated nature of the tank equations when coupled to a load and a collector, particularly with the stratified tank model, a computer must be used to obtain solutions.

An alternative approach suggested by Kuhn et al. (1980) for modeling a stratified storage tank is based on the assumption of plug flow of the fluid up or down in the tank. [The resulting model is not amenable to the step-by-step solutions of a differential equation as in Example 8.3.1 and is useful only in programs such as TRNSYS (1990).] Increments of volume (segments) of fluid from either the collector (or other heat source) or load enter the tank at an appropriate location, shifting the position of all existing segments in the tank between the inlet and the return. The size of a segment depends on the flow rate and the time increment used in the computation.

The point at which heated fluid is considered to enter the tank can be at a fixed position or at a variable position. If the fluid is considered to enter at a fixed position, it may be necessary to combine nodes above or below the entry to avoid temperature inversions. If the inlet is considered to be at a variable position (in a manner

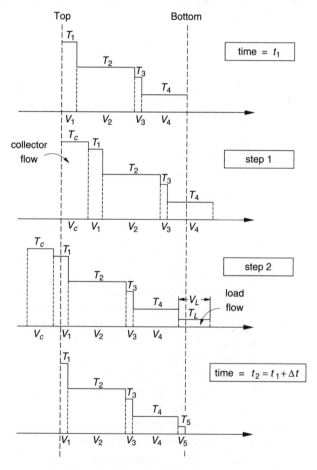

Figure 8.4.4 An example of the plug flow algebraic tank model. Adapted from Kuhn et al. (1980).

analogous to that shown in Figure 8.4.1), the position is selected so as to avoid temperature inversions.

Figure 8.4.4 illustrates the concept of the plug flow tank model. In this example, the tank is initially divided into four segments of volume V_i (represented on the figure by segments of the horizontal axis) and temperatures T_i without temperature inversions. In one time step, the collector delivers a volume of liquid, V_c, equal to $\rho/(\dot{m}_c \Delta t)$ at a temperature T_c. Assuming that T_c is greater than T_L, a new segment is added at the top of the tank and the existing profile is shifted down. At the same time, a volume of fluid V_L returns from the load, with V_L equal to $\rho/(\dot{m}_L \Delta t)$ at a temperature of T_L. (These steps are shown sequentially, although they occur simultaneously.) If T_L is less than T_4, then a segment is added at the bottom of the tank and the whole profile is shifted upward. The net shift of the profile in the tank is volume $V_c - V_L$, or $\rho/(\dot{m}_c - \dot{m}_L)\Delta t$. The segments and/or fraction of segments whose positions fall outside the bounds of the tank are returned to the collector and/or load. The average temperature of the fluid delivered to the load for the example of Figure 8.4.4. is

$$T_D = \frac{V_c T_c + (V_L - V_c) T_L}{V_L} \tag{8.4.5}$$

and the average heat source return temperature T_R is equal to T_L.

The plug flow model, in contrast to the differential equations of the type of Equation 8.4.5, is algebraic. Its use in simulations is primarily a matter of book-keeping. This model can represent a somewhat higher degree of stratification than that of multinode differential equation models. [See TRNSYS (1990) or Kuhn et al. (1980).]

8.5 PACKED BED STORAGE

A packed bed (pebble bed or rock pile) storage unit uses the heat capacity of a bed of loosely packed particulate material to store energy. A fluid, usually air, is circulated through the bed to add or remove energy. A variety of solids may be used, rock being the most widely used material.

A packed bed storage unit is shown in Figure 8.5.1. Essential features include a container, a screen to support the bed, support for the screen, and inlet and outlet ducts. In operation, flow is maintained through the bed in one direction during addition of heat (usually downward) and in the opposite direction during removal of heat. Note that heat cannot be added and removed at the same time; this is in contrast to water storage systems where simultaneous addition to and removal from storage is possible.

Well-designed packed beds using rocks have several characteristics that are desirable for solar energy applications: the heat transfer coefficient between the air and solid is high, which promotes thermal stratification; the costs of the storage

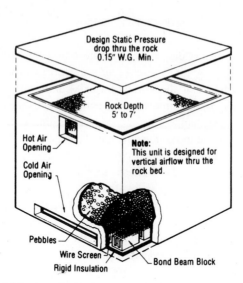

Figure 8.5.1 A packed bed storage unit. Courtesy of Solaron Corp.

material and container are low; the conductivity of the bed is low when there is no air flow; and the pressure drop through the bed can be low.

A major advantage of a packed bed storage unit is its high degree of stratification. This can be visualized by consideration of a hypothetical situation of a bed initially at a fixed temperature, which has air blown into it at a higher fixed temperature. The temperature profiles in the bed during heating are shown in Figure 8.5.2. The high heat transfer coefficient-area product between the air and pebbles means that high-temperature air entering the bed quickly loses its energy to the pebbles. The pebbles near the entrance are heated but the temperature of the pebbles near the exit remains unchanged and the exit air temperature remains very close to the initial bed temperature. As time progresses, a temperature front passes through the bed. By hour 5 the front reaches the end of the bed and the exit air temperature begins to rise.

When the bed is fully charged, its temperature is uniform. Reversing the flow with a new reduced inlet temperature results in a constant outlet temperature at the original inlet temperature for 5 hours and then a steadily decreasing temperature until the bed is fully discharged.

If the heat transfer coefficient between air and pebbles were infinitely large, the temperature front during charging or discharging would be square. The finite heat transfer coefficient produces a "smeared" front that becomes less distinct as time progresses.

A packed bed in a solar heating system does not normally operate with constant inlet temperature. During the day the variable solar radiation, ambient temperature, collector inlet temperature, load requirements, and other time-dependent conditions

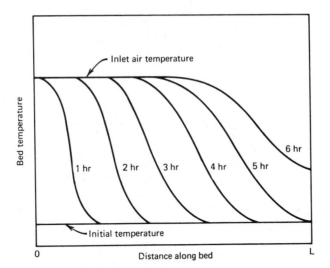

Figure 8.5.2 Temperature distributions in a pebble bed while charging with inlet air at constant temperature.

result in a variable collector outlet temperature. A set of typical measured temperature profiles in an operating bed is shown in Figure 8.5.3. This bed was heated during the day with air from collectors flowing down, and energy was removed during the evening and night by air at temperatures near 20 C flowing upward.

Many studies are available on the heating and cooling of packed beds. The first analytical study was by Schumann (1929), and the following equations describing a packed bed are often referred to as the Schumann model. The basic assumptions leading to this model are one-dimensional plug flow; no axial conduction or dispersion; constant properties; no mass transfer; no heat loss to the environment; and no temperature gradients within the solid particles. The differential equations for the fluid and bed temperatures are

$$(\rho C_p)_f \varepsilon \frac{\partial T_f}{\partial t} = -\frac{(\dot{m} C_p)_f}{A} \frac{\partial T_f}{\partial x} + h_v(T_b - T_f) \tag{8.5.1}$$

$$(\rho C_p)_b (1 - \varepsilon) \frac{\partial T_b}{\partial t} = h_v(T_f - T_b) \tag{8.5.2}$$

where ε is the bed void fraction, h_v is the volumetric heat transfer coefficient between the bed and the fluid (i.e., the usual area heat transfer coefficient times the bed particulate surface area per unit bed volume), and other terms have their usual meanings. A correlation relating h_v to the bed characteristics and to the fluid flow conditions is given in Section 3.16.

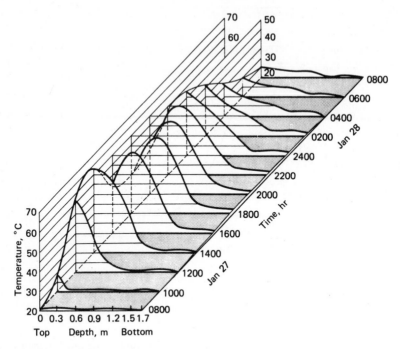

Figure 8.5.3 Temperature profiles in the Colorado State University House II pebble bed during charging and discharging. From Karaki et al. (1977).

For an air-based system, the first term of Equation 8.5.1 can be neglected and the equations can be written as

$$\frac{\partial T_f}{\partial(X/L)} = NTU\,(T_b - T_f) \tag{8.5.3}$$

$$\frac{\partial T_b}{\partial \theta} = NTU\,(T_f - T_b) \tag{8.5.4}$$

$$NTU = \frac{h_v A L}{(\dot{m}C_p)_f} \tag{8.5.5}$$

and the dimensionless time is

$$\theta = \frac{t(\dot{m}C_p)_f}{(\rho C_p)_f(1 - \varepsilon)A L} \tag{8.5.6}$$

where A is bed cross-sectional area and L is bed length. Analytical solutions to these equations exist for a step change in inlet conditions and for cyclic operation.

For the long-term study of solar energy systems, these analytical solutions are not useful and numerical techniques must be employed. Kuhn et al. (1978) investigated a large number of finite difference schemes to numerically solve Equations 8.5.3 and 8.5.4 and concluded that the "complicated effectiveness-NTU" method of Hughes (1975) was best suited for solar system simulation. The following development follows the simpler "effectiveness-NTU" method of Hughes, which is the same as the method proposed by Mumma and Marvin (1976). For practical designs the two Hughes models give essentially the same results, and the simpler one is recommended.

Over a length of bed ΔX as shown in Figure 8.5.4, the bed temperature can be considered to be uniform. (The more complicated Hughes equations assumed the bed temperature to have a linear variation with distance.) The air temperature has an exponential profile and the air temperature leaving bed element i is found from

$$\frac{T_{f,i+1} - T_{b,i}}{T_{f,i} - T_{b,i}} = e^{-NTU\left(\Delta X/L\right)} \tag{8.5.7}$$

This equation is exactly analogous to a heat exchanger operating as an evaporator. The energy removed from the air and transferred to the bed in length ΔX is then

$$\left(\dot{m}C_p\right)_f\left(T_{f,i} - T_{f,i+1}\right) = \left(\dot{m}C_p\right)_f\left(T_{f,i} - T_{b,i}\right)\left(1 - e^{-NTU/N}\right) \tag{8.5.8}$$

where $N = L/\Delta X$.

With Equation 8.5.8, an energy balance on the rock within region ΔX can then be expressed as

$$\frac{dT_{b,i}}{d\theta} = \eta N(T_{f,i} - T_{b,i}) \tag{8.5.9}$$

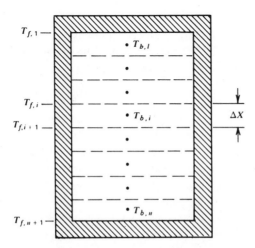

Figure 8.5.4 Packed bed divided into N segments.

where η is a constant and equal to $1 - e^{-NTU/N}$. Equation 8.5.9 represents N ordinary differential equations for the N bed temperatures. Fluid temperatures are found from Equation 8.5.8. An extension to Equation 8.5.9 permits energy loss to an environment at T_a' to be included. Then

$$\frac{dT_{b,i}}{d\theta} = \eta N \left(T_{f,i} - T_{b,i}\right) + \frac{(U\Delta A)_i}{(\dot{m}C_p)_f} \left(T_a' - T_{b,i}\right) \qquad (8.5.10)$$

Where $(U\Delta A)_i$ is the loss area-loss coefficient product for node i.

Hughes suggests that a Crank-Nicolson approach be used to solve Equation 8.5.10. The time derivative is replaced by $(T_{b,i}^+ - T_{b,i})/\Delta\theta$ and the bed temperatures on the right-hand side of Equation 8.5.10 are replaced by $(T_{b,i}^+ + T_{b,i})/2$. With all bed temperatures known, the process starts at node 1 so that the inlet fluid temperature is known. A new bed temperature is calculated from Equation 8.5.10 and an outlet fluid temperature from Equation 8.5.8. This new fluid temperature becomes the inlet fluid temperature for node 2.

The repetitive solution of the Schumann model, even in the form of Equation 8.5.10, is time consuming for year-long solar process calculations. This observation led Hughes et al. (1976) to investigate an infinite NTU model. When the complete Schumann equations are solved for various values of NTU, the long-term performance of a solar air heating system with NTU equal to 25 is virtually the same as that with NTU equal to infinity.

For infinite NTU, Equations 8.5.3 and 8.5.4 can be combined into a single partial differential equation since the bed and fluid temperatures are everywhere equal. With the addition of a container heat loss term the result is

$$\frac{dT}{d\theta} = -L\frac{\partial T}{\partial x} + \frac{(UA)_b}{(\dot{m}C_p)_f} \left(T_a' - T\right) \qquad (8.5.11)$$

This is a single partial differential equation instead of the two coupled partial differential equations of the Schumann model. The single equation represents both the bed and air temperatures since the infinite NTU model assumes the two are identical.

Figure 8.5.5 shows the ratio of the predicted long-term fraction of a heating load met by solar energy for a system with a bed having a finite NTU to the solar fraction with an infinite NTU bed. At NTU greater than 25 the ratio is unity, but at values as low as 10 the ratio is above 0.95. Consequently, the infinite NTU model can be used for long-term performance predictions even if NTU is as low as 10. Short term predictions of bed temperature profiles should be based on the full Schumann equations unless the bed NTU is large.

The Schumann model is based on the assumption that the temperature gradients within the particles of the packed bed are not significant; this can be relaxed by defining a corrected value of NTU. Jefferson (1972) has shown that temperature

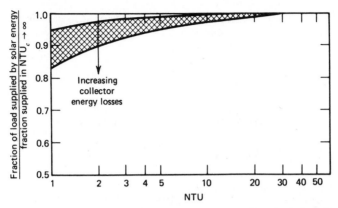

Figure 8.5.5 Long-term performance on a pebble bed as a function of NTU. The band represents different values of collector and bed characteristics for solar heating systems. From Hughes et al. (1976).

gradients within the rocks can be accommodated by defining a corrected NTU:

$$NTU_c = \frac{NTU}{1 + Bi/5} \qquad (8.5.12)$$

where Bi is the Biot number defined as hR/k, R is an equivalent spherical rock radius, k is the rock conductivity, and h is the fluid-to-rock heat transfer coefficient. The NTU_c can be used in any one of the equations of this section instead of NTU to include the effects of temperature gradients in the rock. If the Biot number is less than 0.1, the temperature gradients within the pebbles can be neglected.

Example 8.5.1

A pebble bed has the following characteristics: length in flow direction is 1.80 m; cross-sectional area, 14.8 m²; air velocity, 0.053 m/s; equivalent diameter of pebbles, 12.5 mm; void fraction, 0.47; density of pebble material, 1350 kg/m³; specific heat of pebbles, 0.90 kJ/kgC; thermal conductivity of pebble material, 0.85 W/mC; surface area of pebbles per unit volume, 255 m²/m³. For this pebble bed, calculate the Biot number and NTU. Will there be significant temperature gradients in the pebbles? Can the infinite NTU model be used to calculate the performance of this storage unit?

Solution

The temperature of the air is unknown. If we use a low temperature to evaluate properties (e.g., 20 C), the Biot number and NTU criteria will be more severe than if a high temperature were used. The volumetric heat transfer coefficient from fluid to pebbles is estimated from the Löf and Hawley equation of Section 3.16:

$$h_v = 650 \, (G/D)^{0.7}$$

$$G = 0.053 \times 1.204 = 0.0638 \text{ kg/m}^2\text{s}$$

so $$h_v = 650(0.0638/0.0125)^{0.7} = 2030 \text{ W/m}^3\text{C}$$

and $$h = 2030/255 = 8.0 \text{ W/m}^2\text{C}$$

$$Bi = \frac{hR}{k} = \frac{8.0 \times 0.0125}{2 \times 0.85} = 0.059$$

Since the Biot number is less than 0.1, NTU_c is nearly the same as NTU. From the definition of NTU, Equation 8.4.5,

$$NTU = h_v AL/(\dot{m}C_p)_f$$

$$= \frac{2030 \times 14.8 \times 1.80}{0.053 \times 14.8 \times 1.204 \times 1010} = 57$$

Thus the number of transfer units is much larger than 10, and the infinite NTU model is appropriate.   ■

8.6   STORAGE WALLS

In passive heating systems, storage of thermal energy is provided in the walls and roofs of the buildings. A case of particular interest is the collector-storage wall, which is arranged so that solar radiation transmitted through glazing is absorbed on one side of the wall. The temperature of the wall increases as energy is absorbed, and time-dependent temperature gradients are established in the wall. Energy is lost through the glazing and is transferred from the room side of the wall to the room by radiation and convection. Some of these walls may be vented, i.e., have openings in the top and bottom through which air can circulate from and to the room by natural convection, providing an additional mechanism for transfer of energy to the room. Figure 8.6.1 shows a section of such a wall.

A solid storage wall [e.g., a wall of concrete as used by Trombe et al. (1977)] can be considered as a set of nodes connected together by a thermal network, each with a temperature and capacitance. The network used in TRNSYS (1990), which is similar to that used by Balcomb et al. (1977), Ohanessian and Charters (1978), and others, is shown in Figure 8.6.2. The wall is shown divided into n nodes across its

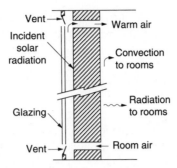

Figure 8.6.1 Section of a storage wall with glazing and energy-absorbing surface on one surface.

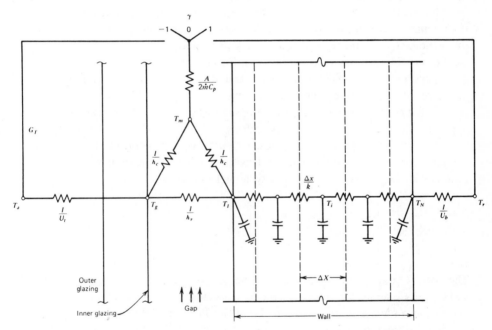

Figure 8.6.2 A thermal circuit diagram for a collector storage wall with double glazing. From TRNSYS (1990).

thickness, with the surface nodes having half of the mass of the interior nodes. A loss coefficient U_t is based on the inner glazing temperature, as it is in Figure 6.14.1(a). Heat is transferred by radiation across the gap and by convection between air flowing in the gap and the absorbing surface and the inner glazing. (The glazing will at times be colder than the air, and at those times the convective transfer will be from air to glazing.) If there is no air flow, heat transfer will occur across the gap by convection in the enclosed gap space and by radiation.

Energy balances are written for each node, resulting in a set of ordinary differential equations with terms that represent its time-dependent temperature and energy flows to all adjacent nodes. The energy balance for the ith node in the wall is

$$\frac{dT_i}{dt} = \frac{k}{\rho C}(T_{i-1} + T_{i+1} - 2T_i) \tag{8.6.1}$$

The set of n equations like Equation 8.6.1 are simultaneously solved for the time-dependent temperatures at each of the nodes, and from this the energy stored in the wall (relative to a base temperature T_{room}) can be calculated. The balance for node N will be in terms of T_N, the surface temperature of the wall on the room side, and the rate of heat transfer into the room. The balance for node 1 will include heat transfer across the gap by convection and radiation.

If there is flow through vents and to the room, the energy added to the room by this mechanism will be $(\dot{m}C_p)_a(T_o - T_r)$. Air may circulate by either natural or forced convection, and in either case it is necessary to estimate flow rates and convection coefficients. Flow by natural convection is difficult to estimate. Trombe et al. (1977) have published measurements that indicate that most of the pressure drop is through the vents. Analyses have been done for laminar flow [e.g., see Fender and Dunn (1978)], but have not included pressure drops through vents. It is probable that flow will be turbulent due to entrance and exit effects.

A solution of Bernoulli's equation can be used to estimate the mean velocity in the gap, based on the assumption that density and air temperature in the gap vary linearly with height. The average velocity through the gap is

$$\overline{V} = \left[\frac{2gh}{\left[C_1\left(\frac{A_g}{A_v}\right)^2 + C_2\right]}\frac{T_m - T_r}{T_m}\right]^{1/2} \tag{8.6.2}$$

The term $[C_1(A_g/A_v)^2 + C_2]$ represents the pressure drop in the gap and vents, and C_1 and C_2 are dimensionless empirical constants. The wall height is h and T_m is the mean air temperature in the gap. From data in Trombe et al. (1977), values of C_1 and C_2 have been determined by Utzinger (1979) to be 8.0 and 2.0, respectively.

The heat transfer coefficient between air in the gap and the wall and glazing depends on whether the vents are open. If the vents are closed, the methods of Sections 3.8 and 3.11 for vertical collectors can be used. If the vents are open and there is flow through the gap, Equation 3.14.6 can be used for flow in the turbulent region and Equation 3.14.7 can be used for the laminar region. The loss coefficient U_t can be estimated by the standard methods of Chapter 6. Heat transfer from the room side of the wall to the room is calculated by conventional methods.

This analysis for collector-storage walls can be simplified to apply to walls that are not vented, or that are not glazed.

8.7   SEASONAL STORAGE

Large-scale solar heating systems that supply energy to district heating systems for building and water heating require large-scale storage facilities. These are in most cases in the ground, as there is no other feasible way to gain the necessary capacity. The objective of very large scale storage is to store summer energy for winter use. Thus the time spans over which losses from storage occur is of the order of a year, and losses occur in summer months when they would not occur in systems with storage capacities of the order of days. The volume of a storage unit increases (roughly) as the cube of the characteristic dimension, and its area for heat loss increases as the square, so increasing the size reduces the loss-to-capacity ratio. Figure 8.7.1 shows how annual solar fraction can vary with storage capacity (on a log scale) for building heating. There are two "knees" in the curve; the first is where sufficient energy is stored in a day to carry the loads through the night, and the second is where energy is stored in the summer to carry loads in the next winter. Storage capacity per unit collector area must be two to three orders of magnitude larger for seasonal storage than for overnight storage. These factors have led to consideration of very large scale thermal energy storage.

Bankston (1988) provides a review of seasonal storage solar heating systems, much of which is concerned with design and performance of the storage units. Several ground storage methods have been explored and/or used, including water-filled tanks, pits, and caverns underground; storage in the ground itself; and storage in aquifers.

Large underground water storage tanks or caverns, or water filled pits in the ground that are roofed over, are the same in their principles of operation as the tanks discussed in Sections 8.3 and 8.4. Energy is added to or removed from the store by pumping water into or out of the storage unit. Their larger capacity makes stratification more likely. The major difference will be in the mechanisms for heat loss and possible thermal coupling with the ground. Units that have been built and operated vary widely in design and characteristics, and the calculation of performance

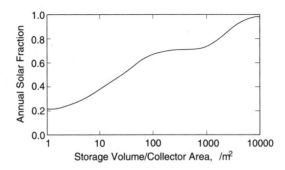

Figure 8.7.1   Variation of annual fraction of heating loads carried by solar energy in a space heating application with storage capacity per unit collector area. Adapted from Braun (1980).

of each must be based on the particular circumstances of its design. A brief description follows of one such system, at Lyckebo, Sweden, with note of the important factors in its operation.

The Lyckebo system [see, e.g., Brunstrom and Larsson (1986)] is a cavern of 10^5 m^3 capacity, cut out of bedrock using standard mining methods, of cylindrical shape, with a central column of rock left to support the overhead rock. The cavern is about 30 m high and its top is about 30 m below ground level. It is water filled, and inlet and outlet pipes can be moved up and down to inject and remove water from controlled levels. The water is highly stratified with top to bottom temperatures of about 80 to 30 C. Figure 8.7.2 shows temperature profiles in the store at various dates in the second year of operation; the high degree of stratification is evident. No thermal insulation is used, and there is a degree of coupling with surrounding rock which adds some effective capacity to the system. Losses occur to a semi-infinite solid and can be estimated by standard methods. Observed losses from this system are higher than those calculated; this is attributed to small but significant thermal circulation of water through the tunnel used in cavern construction and back through fissures in the rock. It takes several years of cycling through the annual weather variations for a storage system of this size to reach a "steady periodic" operation. In the second year of its operation, while it was still in a "warm-up" stage, 74% of the energy added to the store was recovered.

Ground (i.e., the soil, sand, rocks, clay, etc.) can be used for thermal energy storage. Means must be provided to add energy to and remove it from the medium. This is done by pumping heat transfer fluids through pipe arrays in the ground. The pipes may be U-tubes inserted in wells that are spaced at appropriate intervals in the storage field or they may be horizontal pipes buried in trenches. The rates of charging and discharging are limited by the area of the pipe arrays and the rates of heat transfer through the ground surrounding the pipes. If the storage medium is

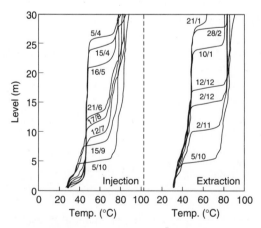

Figure 8.7.2 Temperature profiles in the Lyckebo store during the second year of operation. From Brunstrom and Larsson (1986).

porous, energy transport may occur by evaporation and condensation and by movement of water through the medium, and a complete analysis of such a store must include consideration of both heat and mass transfer. These storage systems are usually not insulated, although insulation may be provided at the ground surface. These storage units usually operate over smaller temperature ranges than the water stores noted previously.

A borehole storage system at Lulea, Sweden, provides an example of ground storage. [See Bankston (1988).] It consists of 120 boreholes drilled to a depth of 65 m into bedrock which lies under 2 to 6 m of soil. The holes are on a square pattern 4 m apart. The total volume of rock in the active store is about 10^5 m^3. Each of the holes has inserted in it a tube that delivers the fluid to the bottom of the hole; the fluid then passes up through the annulus in contact with the rock. Flow patterns are so arranged as to have the central part of the store the hottest, achieving a degree of stratification. This store, which is heated to 80 C by wastewater from an industrial process rather than solar energy, shows maximum and minimum temperatures in the center of the array of 60 and 40 C; the far-field rock temperature is 5 C. About 60% of the energy that is added to the store is usefully extracted to provide heat for buildings.

Aquifer storage is closely related to ground storage, except that the primary storage medium is water which flows at low rates through the ground. Water is pumped out of and into the ground to heat it and extract energy from it. Water flow also provides a mechanism for heat exchange with the ground itself. As a practical matter, aquifers cannot be insulated. Only aquifers that have low natural flow rates through the storage field can be used. A further limitation may be in chemical reactions of heated water with the ground materials. Aquifers, as with ground storage, operate over smaller temperature ranges than water stores.

8.8 PHASE CHANGE ENERGY STORAGE

Materials that undergo a change of phase in a suitable temperature range may be useful for energy storage if several criteria can be satisfied. The phase change must be accompanied by a high latent heat effect, and it must be reversible over a very large number of cycles without serious degradation. The phase change must occur with limited supercooling or superheating, and means must be available to contain the material and transfer heat into it and out of it. Finally, the cost of the material and its containers must be reasonable. If these criteria can be met, phase change energy storage systems can operate over small temperature ranges, have relatively low volume and mass, and have high storage capacities (relative to energy storage in specific-heat-type systems).

The storage capacity of a phase change material heated from T_1 to T_2, if it undergoes a phase transition at T^*, is the sum of the sensible heat change of the solid (the lower temperature phase) from T_1 to T^*, the latent heat at T^*, and the sensible heat of the liquid (the melt, or higher temperature phase) from T^* to T_2:

$$Q_s = m\left[C_{so}(T^* - T_1) + \lambda + C_{li}(T_2 - T^*)\right] \qquad (8.8.1)$$

where m is the mass of material, C_{so} and C_{li} are the heat capacities of the solid and liquid phases, and λ is the latent heat of phase transition. $Na_2SO_4 \cdot 10H_2O$ has C_{so} of approximately 1950 J/kg C, λ of 2.43×10^5 J/kg at 34 C, and C_{li} of about 3550 J/kg C; on heating 1 kg of this medium from 25 to 50 C the energy stored would be $1950(34 - 25) + 2.43 \times 10^5 + 3350(50 - 34)$, or 0.315 MJ.

$Na_2SO_4 \cdot 10H_2O$ (Glauber's salt) was the earliest phase change storage material to be studied experimentally for house heating applications [Telkes (1955)]. On heating at approximately 34 C it gives a solution and solid Na_2SO_4:

$$Na_2SO_4 \cdot 10H_2O + energy \leftrightarrow Na_2SO_4 + 10H_2O$$

Energy storage is accomplished by the reaction proceeding from left to right on addition of heat. The total energy added depends on the temperature range over which the material is heated since it will include sensible heat to heat the salt to the transition temperature, heat of fusion to cause the phase change, and sensible heat to heat the Na_2SO_4 and solution to the final temperature. Energy extraction from storage is the reverse procedure, with the reaction proceeding from right to left and the thermal effects reversed.

Practical difficulties have been encountered with this material. It has been found that performance degrades on repeated cycling, with the thermal capacity of the system reduced. As shown in Figure 8.8.1, $Na_2SO_4 \cdot 10H_2O$ has an incongruent melting point, and as its temperature increases beyond the melting point, it separates into a liquid (solution) phase and solid Na_2SO_4. Since the density of the salt is higher than the density of the solution, phase separation occurs. Many studies have been done to develop means to avoid phase separation: through the use of containers of

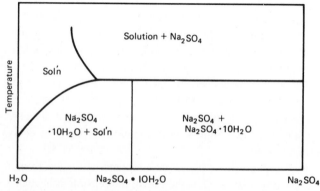

Figure 8.8.1 Phase diagram for sodium sulfate decahydrate, with incongruent melting point.

thin cross section, through the use of gels, thickeners, and other agents, or by mechanically agitating the melt mix.

Johnson (1977) has applied phase change storage to a passively heated building, where operation over small temperature ranges is needed. The melting point of $Na_2SO_4 \cdot 10H_2O$ is too high for this application; the addition of 9% NaCl to the Glauber's salt reduced the melting point to about 23 C. In addition, borax, a thickener, and surfactants were added. Laboratory tests showed no noticeable phase separation after 500 cycles. This phase change material is designed to be used in thin layers in ceiling tiles in passively heated buildings.

Other possibilities for phase change storage media for temperatures in ranges of interest in comfort heating include, for example, $Na_2SO_4 \cdot 12H_2O$ and $Fe(NO_3)_2 \cdot 6H_2O$ (which has a congruent melting point). Paraffin waxes have also been studied for this use; they can be obtained with various melting points and as mixtures with ranges of melting points. Eutectic mixtures have also been studied [see Kauffman and Grunfest (1973)], such as $CaCl_2 \cdot MgCl_2 \cdot H_2O$, urea$\cdot NH_4NO_3$, and others. Phase change materials studied for application at higher temperatures include $AlSO_4 \cdot 10H_2O$ (melting point 112 C), $MgCl_2 \cdot 6H_2O$ (melting point 115 C), and $NaNO_3 + NaOH$ (melting point 245 C).

A further consideration with phase change materials lies in the possibility of supercooling on energy recovery. If the material supercools, the latent heat of fusion may not be recovered, or it may be recovered at a temperature significantly below the melting point. This question has been approached from three standpoints: by selection of materials that do not have a strong tendency to supercool, by addition of nucleating agents, and by ultrasonic means of nucleation. These considerations are reviewed by Belton and Ajami (1973), who note that the viscosity at the melting point of a material is a major factor in determining the glass-forming ability of a melt and thus its tendency to supercool.

Carlsson et al. (1978) have carried out an extensive experimental study of the use of $CaCl_2 \cdot 6H_2O$, including consideration of heat transfer through the storage medium, supercooling, temperature stratification, and the use of $SrCl_2 \cdot 6H_2O$ to avoid the formation of a stable tetrahydrate that interferes with hydration and dehydration reactions.

Phase change storage media are usually contained in small containers or shallow trays to provide a large heat transfer area. The general principles of packed bed storage units outlined in Section 8.5 will apply to these units. The heat transfer fluid is usually air, although water has been used. Two additional phenomena must be considered. First, the latent heat must be taken into account; it is in effect a high specific heat over a very small temperature range. Second, the thermal resistance to heat transfer within the material is variable with the degree of solidification and whether heating or cooling of the material is occurring. [Heat transfer in situations of this type have been studied, for example, by Hodgins and Hoffman (1955), Murray and Landis (1959), and Smith et al. (1980).] As heat is extracted from a phase change material, crystallization will occur at the walls and then progressively inward into the material; at the end of the crystallization, heat must be transferred across layers of solid to the container walls. As a solidified material is heated, melting occurs first at the walls and then inward toward the center of the container. These effects can be minimized by design of the containers to give very short path lengths for internal heat transfer, and the following development assumes the internal gradients to be small.

Morrison and Abdel-Khalik (1978) developed a model applicable to phase change materials in such containers, where the length in flow direction is L, the cross-sectional area of the material is A, and the wetted perimeter is P. The heat transfer fluid passes through the storage unit in the x direction at the rate $\dot{m}$ and with inlet temperature T_{fi} as shown in Figure 8.8.2.

A model can be based on three assumptions: (a) during flow, axial conduction in the fluid is negligible; (b) the Biot number is low enough that temperature gradients normal to the flow can be neglected; and (c) heat losses from the bed are negligible. An energy balance on the material gives

$$\frac{\partial u}{\partial t} = \frac{k}{\rho}\frac{\partial^2 T}{\partial x^2} + \frac{UP}{\rho A}(T_f - T) \tag{8.8.2}$$

where u, T, K, and ρ are the specific internal energy, temperature, thermal conductivity, and density of the phase change material, T_f and U are the circulating fluid temperature and overall heat transfer coefficient between the fluid and phase change material, and t is time.

An energy balance on the fluid is

$$\frac{\partial T_f}{\partial t} + \frac{\dot{m}}{\rho_f A_f}\frac{\partial T_f}{\partial x} = \frac{UP}{\rho_f A_f C_f}(T - T_f) \tag{8.8.3}$$

where ρ_f, A_f, and C_f are the density, flow area, and specific heat of the fluid.

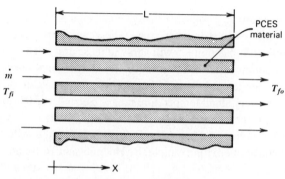

Figure 8.8.2   Schematic of a phase change storage unit. From Morrison and Abdel-Khalik (1978).

The specific internal energy is related to temperature T, liquid fraction X, and the specific heats of the liquid and solid phases, C_{li} and C_{so}, by

$$u = C_{so}(T - T_{ref}) \qquad\qquad\qquad \text{if } T < T^*$$
$$u = C_{so}(T^* - T_{ref}) + \chi\lambda \qquad\qquad \text{if } T = T^*$$
$$u = C_{so}(T^* - T_{ref}) + \lambda + C_{li}(T - T^*) \qquad \text{if } T > T^* \qquad (8.8.4)$$

where T^* is the melting temperature and T_{ref} is the reference temperature at which the internal energy is taken as zero.

The equation and boundary conditions for phase change energy storage can be simplified for particular cases [see Morrison and Abdel-Khalik (1978) for details]. It has been shown that axial conduction during flow is negligible, and if the fluid capacitance is small, Equations 8.8.2 and 8.8.3 become

$$\frac{\partial u}{\partial t} = \frac{UP}{\rho A}(T_f - T) \tag{8.8.5}$$

$$\frac{\partial T_f}{\partial x} = \frac{UP}{(\dot{m}C_p)_f}(T - T_f) \tag{8.8.6}$$

These two equations can be rewritten in terms of NTU

$$\frac{1}{C}\frac{\partial u}{\partial \theta} = NTU\,(T_f - T) \tag{8.8.7}$$

$$\frac{\partial T_f}{\partial (x/L)} = NTU\,(T - T_f) \tag{8.8.8}$$

where $\theta = t(\dot{m}C_p)_f / \rho A L \overline{C}$ and $NTU = UPL/(\dot{m}C_p)_f$.

Equations 8.8.7 and 8.8.8 are essentially the same as Equations 8.5.2 and 8.5.3 for air and rock beds. Morrison and Abdel-Khalik have developed these models, explored the effects of assumptions made in their development and used them in simulation studies of solar heating systems with ideal phase change energy storage that has no superheating or supercooling and complete reversibility (see Section 13.10). They showed that the infinite NTU model is a good approximation for practical systems.

Jurinak and Abdel-Khalik (1979) have shown that for building heating operations where the melting point is between an optimum value (usually 5 to 10 C above room temperature) and 50 C, an empirical equation for equivalent heat capacity can be written as

$$(mC_p)_e = m\left[\frac{18.3}{(T^*)^2}\lambda + \left(1 - \frac{18.3}{(T^*)^2}\right)C_{so} + \frac{18.3}{(T^*)^2}C_{li}\right] \tag{8.8.9}$$

where all temperatures are in degrees Celsius. This empirical equation has utility in design of house-heating systems with phase change storage. Here λ is in kJ/kg.

Other practical considerations include corrosion, reactions of the phase change material with containers or other side reactions, vapor pressure, and toxicity. Cost is a major factor in many applications and rules out all but the least expensive materials.

Comparisons of sensible heat storage and phase change storage can only be made in the context of systems in particular applications. The system analysis methods outlined in following chapters can be used to evaluate the relative merits of the several possible methods. In doing so, it should be kept in mind that the formulations for sensible heat storage are reliable (in the sense that temperature changes of sensible heat media are very predictably related to quantities of heat added to or removed), whereas those for phase change materials may be more uncertain because of possible (and unpredictable) superheating, supercooling, and lack of complete reversibility of the phase change.

8.9 CHEMICAL ENERGY STORAGE

The previous section dealt with storage of thermal energy in materials undergoing phase transitions. In this section we consider chemical reactions that might be used to store energy. None of these possibilities have yet been used in practical applications, and both technical and economic questions are yet to be answered for all of the possibilities. A review of chemical storage is presented by Offenhartz (1976).

An ideal thermochemical reaction for energy storage is an endothermic reaction in which the reaction products are easily separable and do not undergo further reactions. For example, decomposition reactions of the type $AB + \text{heat} \rightarrow A + B$ are candidates if the reaction can be reversed to permit recovery of the stored energy. The products A and B can be stored separately, and thermal losses from the storage units are restricted to sensible heat effects, which are usually small compared to heats of reaction. Unfortunately, there are not very obvious candidates of this type useful for low-temperature solar applications. Reactions in which water vapor is a product have the difficulty that the heat of condensation is usually lost. Reactions in which products like atomic chlorine are produced lose energy on formation of the dimer Cl_2.

Thermal decomposition of metal oxides for energy storage has been considered by Simmons (1976). These reactions may have the advantage that the oxygen evolved can be used for other purposes or discarded and oxygen from the atmosphere used in the reverse reactions. Two examples are the decomposition of potassium oxide,

$$4KO_2 \leftrightarrow 2K_2O + 3O_2$$

which occurs over a temperature range of 300 to 800 C with a heat of decomposition of 2.1 MJ/kg, and lead oxide,

$$2PbO_2 \leftrightarrow 2PbO + O_2$$

which occurs over a temperature range of 300 to 350 C with a heat of decomposition of 0.26 MJ/kg. There are many practical problems yet to be faced in the use of these reactions.

Energy storage by thermal decomposition of $Ca(OH)_2$ has been extensively studied by Fujii et al. (1985, 1989). The reaction is $Ca(OH)_2 \leftrightarrow CaO + H_2O$. The forward reaction will proceed at temperatures above about 450 C; the rates of reaction can be enhanced by the addition of zinc or aluminum. The product CaO is stored in the absence of water. The reverse exothermic reaction proceeds easily; in experiments reported in Fujii et al. (1985) energy was recovered at temperatures in the 100 to 200 C range, and temperature rises to 300 C were mentioned.

Another reaction of possible interest in higher temperature ranges is the dehydration of mixtures of $MgO/Mg(OH)_2$ [see Bauerle et al. (1976), who also considered $CaO/Ca(OH)_2$]. The storage reaction proceeds at temperatures of 350 to 550 C. The reverse reactions occur at lower temperatures on the addition of water or steam. The system sulfuric acid-water has also been proposed [e.g., by Huxtable and Poole (1976)], with condensed water and acid stored until needed. There are obvious practical problems with these systems.

An example of a photochemical decomposition reaction is the decomposition of nitrosyl chloride, which can be written as

$$NOCl + Photons \rightarrow NO + Cl$$

The atomic chlorine produced forms chlorine gas, Cl_2, with the release of a substantial part of the energy added to the NOCl in decomposition. Thus the overall reaction is:

$$2NOCl + photons \rightarrow 2NO + Cl_2$$

The reverse reaction can be carried out to recover part of the energy of the photons entering the reaction [see Marcus and Wohlers (1960, 1961a,b, 1964)].

Processes that produce electrical energy may have storage provided as chemical energy in electrical storage batteries or their equivalent. Several types of battery systems can be considered for these applications, including lead-acid, nickel-iron, and nickel-cadmium batteries. For low discharge rates and moderate charge rates the efficiencies of these systems range from 60 to 80% (ratio of watt-hour output to watt-hour input). It is also possible to electrolyze water with solar-generated electrical energy, to store oxygen and hydrogen, and to recombine in a fuel cell to regain electrical energy [see Bacon (1964)]. These storage systems are characterized by the relatively high cost per kilowatt-hour of storage capacity and can now be considered for low-capacity special applications such as auxiliary power supply for space vehicles, isolated telephone repeater power supplies, instrument power supplies, and so on.

REFERENCES

Bacon, F. T., *Proc. of the UN Conference on New Sources of Energy*, **1**, 174 (1964). "Energy Storage Based on Electrolyzers and Hydrogen-Oxygen Fuel Cells."

Balcomb, J. D., J. C. Hedstrom, and R. D. McFarland, *Solar Energy*, **19**, 277 (1977). "Simulation Analysis of Passive Solar Heated Buildings – Preliminary Results."

Bankston, C. A., in *Advances in Solar Energy* **4**, (K. W. Boer, ed.), 352 (1988). "The Status and Potential of Central Solar Heating Plants with Seasonal Storage: An International Report."

Bauerle, G., D. Chung, G. Ervin, J. Guon, and T. Springer, *Proc. of the ISES Meeting*, Winnipeg, **8**, 192 (1976). "Storage of Solar Energy by Inorganic Oxide Hydrides."

Belton, G. and F. Ajami, Report NSF/RANN/SE/G127979 TR/73/4 to the National Science Foundation by the University of Pennsylvania National Center for Energy Management and Power (1973). "Thermochemistry of Salt Hydrates."

Braun, J., M.S. Thesis, University of Wisconsin-Madison (1980). "Seasonal Storage of Energy in Solar Heating."

Brunstrom, C. and M. Larsson, *Proc. ASES Meeting,* American Solar Energy Soc., Boulder, CO (1986). "The Lyckebo Rock Cavern Seasonal Storage Plant: Performance and Economy."

Carlsson, B., H. Stymne, and G. Wettermark, Document D12:1978, Swedish Council for Building Research (1978). "Storage of Low-Temperature Heat in Salt-Hydrate Melts - Calcium Chloride Hexahydrate."

Fender, D. A. and J. R. Dunn, *Proc. ASME Winter Annual Meeting*, 78-WA/Sol-11 (1978). "A Theoretical Analysis of Solar Collector/Storage Panels."

Fujii, I., K. Tsuchiya, M. Higano, and J. Yamada, *Solar Energy*, **34**, 367 (1985). "Studies of an Energy Storage System by Use of the Reversible Chemical Reaction: $CaO + H_2O \leftrightarrow Ca(OH)_2$."

Fujii, I., K. Tsuchiya, Y. Shikakura, and M. S. Murthy, *Trans. ASME, J. Solar Energy Engrg.*, **111**, 245 (1989). "Consideration on Thermal Decomposition of Calcium Hydroxide Pellets for Energy Storage."

Gari, H. N. and R. I. Loehrke, *Trans. ASME., J. Fluids Engrg.*, **104**, 475 (1982). "A Controlled Bouyant Jet for Enhancing Stratification in a Liquid Storage Tank."

Hodgins, J. W. and T. W. Hoffman, *Canadian J. of Technology*, **33**, 293 (1955). "The Storage and Transfer of Low Potential Heat."

Hughes, P. J, M.S. Thesis in Mechanical Engineering, University of Wisconsin - Madison (1975). "The Design and Predicted Performance of Arlington House."

Hughes, P. J., S. A. Klein, and D. J. Close, *Trans. ASME, J. Heat Transfer*, **98**, 336 (1976). "Packed bed Thermal Storage Models for Solar Air Heating and Cooling Systems."

Huxtable, D. D. and D. R. Poole, *Proc. of the ISES Meeting*, Winnipeg (1976). "Thermal Energy Storage by the Sulfuric Acid-Water System."

Jeffreson, C. P., *American Institute of Chemical Engineers J.*, **18**(2), 409 (1972). "Prediction of Breakthrough Curves in Packed Beds."

Johnson, T. E., *Solar Energy*, **19**, 669 (1977). "Lightweight Thermal Storage for Solar Heated Buildings."

Jurinak, J. J. and S. I. Abdel-Khalik, paper for meeting of ISES, Atlanta (1979). "On the Performance of Solar Heating Systems Utilizing Phase Change Energy Storage."

Karaki, S., P. R. Armstrong, and T. N. Bechtel, Report COO-2868-3 from Colorado State University to the U.S. Department of Energy (1977). "Evaluation of a Residential Solar Air Heating and Nocturnal Cooling System."

Kauffman, K. and I. Gruntfest, Report NCEMP-20 of the University of Pennsylvania National Center for Energy Management and Power, to the National Science Foundation (1973). "Congruently Melting Materials for Thermal Energy Storage."

Kuhn, J. K., G. F. Von Fuchs, A. W. Warren, and A. P. Zob, Report of Boeing Computer Services Company, to the U. S, Dept. of Energy (1980). "Developing and Upgrading of Solar System Thermal Energy Storage Simulation Models."

Lavan, Z. and T. Thompson, *Solar Energy*, **19**, 519 (1977). "Experimental Study of Thermally Stratified Hot Water Storage Tanks."

Marcus, R. J. and H. C. Wohlers, *Proc. of the UN Conference on NEw Sources of Energy*, **1** (1964). "Chemical Conversion and Storage of Concentrated Solar Energy." See also *Solar Energy*, **5**, 121 (1961); **5**, 44 (1961); **4** (2), 1 (1960).

Morrison, D. J. and S. I. Abdel-Khalik, *Solar Energy*, **20**, 57 (1978). "Effects of Phase-Change Energy Storage on the Performance of Air-Based and Liquid-Based Solar Heating Systems."

Mumma, S. D. and W. C. Marvin, ASME paper 76-HT-73 (1976). "A Method of Simulating The Performance of a Pebble Bed Thermal Energy Storage and Recovery System."

Murray, W. D. and F. Landis, *Trans. ASME, J. Heat Transfer*, **81C**, 107 (1959). "Numerical and Machine Solutions of Transient Heat Conduction Problems Involving Melting or Freezing."

Offenhartz, P. O., *Proc. of the ISES Meeting*, Winnipeg, **8**, 48 (1976). "Chemical Methods of Storing Thermal Energy."

Ohanessian, P. and W. W. S. Charters, *Solar Energy*, **20**, 275 (1978). "Thermal Simulation of a Passive Solar House using a Trombe-Michel Wall Structure."

Schumann, T. E. W., *J. Franklin Inst.*, **208**, 405 (1929). "Heat Transfer: A Liquid Flowing Through a Porous Prism."

Simmons, J. A., *Proc. of the ISES Meeting*, Winnipeg, **8**, 219 (1976). "Reversible Oxidation of Metal Oxides for Thermal Energy Storage."

Smith, R. N., T. E. Ebersole, and F. P. Griffin, *Trans. ASME, J. Solar Energy Engrg.*, **102**, 112 (1980). "Heat Exchanger Performance in Latent Heat Thermal Energy Storage."

Telkes, M., *Solar Energy Research*, University of Wisconsin Press, Madison (1955). "Solar Heat Storage."

TRNSYS, A Transient Simulation Program, User's Manual, Version 13, Engineering Experiment Station Report 38, University of Wisconsin-Madison (1990). [The first published version was 7 (1976).]

Trombe, F., J. F. Robert, M. Cabanot, and B. Sesolis, *Solar Age*, **2**, 13 (1977). "Concrete Walls to Collect and Hold Heat."

Utzinger, D. M., M.S. Thesis, University of Wisconsin-Madison (1979). "Analysis of Building Components Related to Direct Solar Heating of Buildings."

van Koppen, C. W. J., J. P. S. Thomas, and W. B. Veltkamp, paper in *Sun II, Proc. of ISES Biennial Meeting*, Atlanta, GA, **2**, 576 (1979). "The Actual Benefits of Thermally Stratified Storage in Small and Medium Sized Solar Systems."

Chapter 9

SOLAR PROCESS LOADS

The supply of solar energy from a collector is time dependent. In general, loads to be met by an energy system are also time dependent. Energy storage, as outlined in the previous chapter, provides a buffer between these two time-dependent functions. In Chapters 6 and 7 we discussed calculation of collector output. In this chapter, we outline briefly the common methods for calculating loads and provide references which will provide much more detailed information.

The term loads, as used here, refers to the time-dependent energy needs which are to be met by an energy system.[1] The system will generally be a combination of a solar system and another system, the auxiliary, or backup, system. We designate the total load as L, that part that is supplied by solar energy as L_S, and that by auxiliary as L_A. These loads will usually be integrated over some time period, such as an hour or a month. (Load rates in this chapter are identified by $\dot{L}$, $\dot{L}_S$, or $\dot{L}_A$; in later chapters where the distinction between rates and integrated loads is clear, the dot will not be used.) There may be times when all of the loads are met by solar energy and times when all are met by auxiliary energy.

The objective of solar process performance calculations can be (a) to determine the detailed hour-by-hour supply of solar and auxiliary energy to meet the loads, and/or (b) how much of a long-term load (usually for a year) is met by solar energy and thus how much auxiliary energy must be purchased.

The hour-by-hour load patterns, together with the solar process characteristics, determine the dynamics of a system; the simulation methods outlined in Chapters 10 and 19 are hourly calculations that provide information on how much auxiliary energy must be supplied and when it must be supplied. The monthly loads that an energy system must meet contain no information on process dynamics. However, monthly loads are needed in the design procedures described in Part III which provide estimates of annual solar energy and auxiliary energy.

In this chapter, we show some samples of time-dependent loads for several applications and outline methods for estimating monthly loads. This presentation is brief, and the reader is referred to sources such as ASHRAE publications and Mitchell (1983) for details.

---

[1] Loads in ASHRAE specifically refer to the rate of energy transfer. Here we use loads for both rates and integrated quantities. The meaning is always clear from the context.

9.1 EXAMPLES OF HOURLY LOADS

To determine solar process dynamics, load dynamics must be known. There are no general methods for predicting the time dependence of loads; detailed information on heat requirements (energy and temperature), hot water requirements, cooling needs, etc., must be available. Lacking data, it may be possible to assume typical standard load distributions. An illustration of each of these situations is noted.

The possibility of application of a combination of a solar energy system and a waste heat recovery system to supply industrial process heat to a meat packing operation was considered by Arny (1982). The loads to be met were for hot water at 85 C for sanitary purposes during production operations and at 60 to 71 C during nonproduction hours. Water enters the plant at 12.8 C. The load profile (the variation of energy, temperature, and mass flow rate requirements with time, in this example for 2-hour intervals) for a production day are shown in Figure 9.1.1. Other profiles apply to nonproduction days, i.e., weekends and holidays. This profile is unique to the particular industrial operation studied; it is not possible to develop a general profile.

In contrast, a commonly used domestic hot water load pattern is shown in Figure 9.1.2 [Mutch (1974)]. Hot water use is concentrated in the morning and

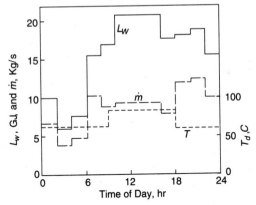

Figure 9.1.1. Process hot water sensible heat load L_w, delivery temperature T_d, and water mass flow rate $\dot{m}$, (kg/s), as a function of time for 2-hour intervals, for a production day for a meat processing operation. Adapted from Arny (1982).

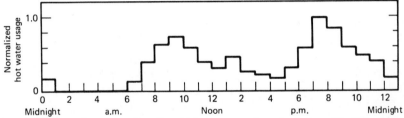

Figure 9.1.2 A normalized profile of hourly hot water use for a domestic application. Adapted from Mutch (1974).

evening hours; to supply these loads, stored solar energy must be used. Minor changes in time dependence of loads do not have major effect on annual performance of solar domestic hot water systems. However, variations such as those caused by closure of commercial buildings on weekends can have significant impact on design and performance.

If a solar process system is to meet only a small part of the total loads, i.e., if the instantaneous (hourly) solar energy delivered is always smaller than the instantaneous load, then the hourly load distribution has no effect on system performance. Also, energy storage is not needed and, if provided, will not increase the net solar contribution to the load.

9.2   HOT WATER LOADS

The loads to be met by water heating systems can be considered to include three parts. First, the sensible heat requirements of the water must be met. If water flowing at a rate of $\dot{m}$ is to be heated from a supply temperature T_s to a delivery temperature T_d, the required rate of addition of sensible heat will be

$$\dot{L}_w = \dot{m}C_p(T_d - T_s) \qquad (9.2.1)$$

Second, losses from the distribution system must be met. These can be estimated by conventional methods. If continuous recirculation is used (as in large buildings where "instant" hot water is required at locations far from the water heater), the losses from the piping may be of the same magnitude as the sensible heat requirements.

Third, losses from storage tanks may be significant. The rate of tank losses $\dot{L}_t$ is estimated from the tank loss coefficient-area product $(UA)_t$ and the temperature difference between the water in the tank and the ambient temperature surrounding the tank:

$$\dot{L}_t = (UA)_t(T_t - T_a') \qquad (9.2.2)$$

However, estimation of the loss coefficient based on the thickness and thermal conductivity of insulation will lead to underestimation of tank losses if significant effects of supports, piping, and other heat leaks are ignored. Measured values of tank loss coefficients are usually two to four times those calculated from insulation thickness and conductivity.

In the case of water heaters located in heated buildings, losses from the hot water tanks are uncontrolled heat gains for the building.

9.3 SPACE HEATING LOADS; DEGREE DAYS; BALANCE TEMPERATURE

A detailed discussion of the calculation of heating loads is beyond the scope of this book, and for such a discussion the reader is referred to the ASHRAE *Handbook – Fundamentals* (1989), the ASHRAE *Cooling and Heating Load Calculation Manual* (1979), Mitchell (1983), or similar publications. Here we briefly outline the degree-day method. Other methods can be used at the discretion of the designer.

The degree-day method of estimating loads is based on the principle that the energy loss from a building is proportional to the difference in temperature between indoors and outdoors. If energy (from a source such as a furnace, space heater, or solar heating system) is added to a building at a rate $\dot{L}$, if the rate of heat generation in the building due to occupants, lights, appliances, incoming solar radiation, etc., is $\dot{g}$, and if the rate of energy transfer from the building is proportional to the difference between the house temperature and the ambient temperature, then a steady-state energy balance on the house yields

$$\dot{L} + \dot{g} = (UA)_h(T_h - T_a) \tag{9.3.1}$$

There is an ambient temperature, called the balance point temperature T_b, at which the internal energy generation offsets the losses:

$$\dot{g} = (UA)_h(T_h - T_b) \tag{9.3.2}$$

or $\quad\quad T_b = T_h - \dot{g}/(UA)_h \tag{9.3.3}$

If Equation 9.3.3 is solved for the house temperature and this is substituted into Equation 9.3.1, the instantaneous load on the building can be expressed in terms of the balance temperature as

$$\dot{L} = (UA)_h(T_b - T_a) \tag{9.3.4}$$

A month's heating load on a building is obtained by integrating Equation 9.3.4:

$$L = \int_{mo} (UA)_h(T_b - T_a)^+ dt \tag{9.3.5}$$

where the superscript + indicates that only positive values of $T_b - T_a$ contribute to the heating load. If the integral is approximated by a summation over the month and $(UA)_h$ and T_b are assumed to be constant, then the monthly heating load is given by

$$L = (UA)_h DD \tag{9.3.6}$$

where DD is the number of degree-days. The number of degree-days in a day is approximated by the difference between the balance temperature T_b and the day's average temperature T_{av}, defined by the U.S. Weather Service as $(T_{max} + T_{min})/2$, with only positive values of the difference considered.[2] A month's degree-days are the sum of the daily values:

$$DD = \sum_{mo} (T_b - T_{av})^+ \tag{9.3.7}$$

Note that if $(UA)_h$ is in W/C and DD is in C-days, a factor of 3600×24 must be used to obtain the monthly load in joules.

Although the balance temperature varies throughout the day, the traditional base (balance) temperature T_b is 18.3 C (65 F), which was based on average room temperatures at 24 C and internal energy generation rates and construction practices typical of residences that were built prior to 1940 in the United States. Most data are available to this base (e.g., those in Appendix G). In recent years, buildings have been better insulated and are kept at lower temperatures, and rates of internal energy generation are higher, all of which contribute to a lower base temperature. Thus, degree-day information is needed to other base temperatures. Tables of degree-days to several base temperatures are available [e.g., Balcomb et al. (1984) or Butso and Hatch]. A method for calculating degree-days to any base temperature from monthly average ambient temperature was developed by Thom (1954, 1966) and modified by Erbs et al. (1983).

The monthly average degree-days can be approximated by

$$DD = \sigma_m \, n^{3/2} \left[\frac{h}{2} + \frac{\ln[\cosh(1.698h)]}{3.396} + 0.2041 \right] \tag{9.3.8}$$

where n is the number of days in the month and h is defined as

$$h = \frac{T_b - \overline{T}_a}{\sigma_m \sqrt{n}} \tag{9.3.9}$$

The standard deviation of the monthly average ambient temperature is σ_m, which is generally not available from weather summaries. It can be approximated by the following with essentially no loss in accuracy:

$$\sigma_m = 1.45 - 0.0290\overline{T}_a + 0.0664\,\sigma_{yr} \tag{9.3.10}$$

where σ_{yr} is the standard deviation of the monthly average ambient temperature from the annual average ambient temperature.

When the base temperature is within 6 C of the monthly average ambient temperature, the monthly degree-days can by found from the simple formula

$$DD = n\left(T_b - \overline{T}_a\right) \tag{9.3.11}$$

The degree-days tabulated in Appendix G are calculated using Equation 9.3.8.

---

[2] The use of T_{av} in the definition of degree-days instead of using the actual average temperature does not lead to significant errors.

Example 9.3.1

Calculate the 12 monthly heating degree-days to a base temperature of 18.3 C for Madison, Wisconsin, and compare them to the values listed in Appendix G.

Solution

All of the winter months, October through April, can use the simple formula given by Equation 9.3.11. The calculations will be shown in detail for the month of September, which has an average ambient temperature of 15 C, using Equations 9.3.8 through 9.3.10. The annual average temperature for Madison is 7.1 C. The standard deviation of the monthly temperatures is found from

$$\sigma_{yr} = \sqrt{\frac{\sum_{i=1}^{12} (\overline{T}_{a,i} - 7.1)^2}{12}} = 10.4$$

Then σ_m is found from Equation 9.3.10 and is equal to 1.71, and h is equal to 0.353 from Equation 9.3.9. The September degree-days are found from Equation 9.3.8:

$$DD = 1.71 \times 30^{1.5} \left[\frac{0.353}{2} + \frac{\ln\left[\cosh\left(1.698 \times 0.353\right)\right]}{3.396} + 0.2041 \right] = 121$$

which is within 25 degree-days of the measured value of 96. The measured and calculated degree-days for each month are given in the table. The differences between the calculated values in this example and those in Appendix G are due to the fact that T_a is not known to better than 0.5 C.

Month	$\overline{T}_a$	Meas. DD	Calc. DD
Jan	-8	830	815
Feb	-7	696	708
Mar	-1	599	599
Apr	7	328	341
May	13	165	178
Jun	19	40	44
Jul	21	8	23
Aug	20	22	33
Sep	15	96	121
Oct	10	263	263
Nov	2	505	490
Dec	-6	742	753
Yr	7.1	4294	4368

Some buildings (particularly passively heated buildings described in Chapter 14) may have insulation added at night. If night insulation is used, the building loss coefficient varies with time, and the loads can be approximated by integrating Equation 9.3.5 in two parts, day and night.

Thermostat settings may also vary through the day, with the most common shift being night set-back. Changing thermostat settings does not result in an immediate change in room temperature for real buildings that have finite thermal capacitance and heating systems with finite capacity. Typically a reduction in thermostat setting results in a slow reduction of room temperature to the new set-point, and a rise in thermostat setting results in a faster rise back to the high set-point brought about by heating systems which normally have excess capacity. An estimate of the upper limit of reduction of heating loads with night set-back can be obtained by time-weighted degree-days to appropriate base temperatures. See Mitchell (1983) for a discussion of this problem.

For existing structures where fuel consumption records are available, $(UA)_h$ may be calculated from

$$(UA)_h = \frac{N_F H_F \eta}{DD} \tag{9.3.12}$$

The numerator is the energy delivered from fuel to the building; N_F is the number of fuel units used, H_F is the appropriate heating value of the fuel, and η is the efficiency with which it is burned.

9.4   BUILDING LOSS COEFFICIENTS

For new or proposed structures there are standard methods of estimating building UA. The total heat loss is the sum of the losses through walls, windows, doors, floor, and roof, plus infiltration loss. For details of these calculation methods, see Mitchell (1983), ASHRAE (1989), or other standard works on heating loads.

Loss coefficients of windows and walls are of particular importance in passive solar heating. As window areas (i.e., solar aperture) increase, replacing walls which have lower overall loss coefficients, total loads on the building rise. The increases can be modified by use of night insulation. Typical coefficients for several window and wall types are shown in Table 9.4.1.

Important advances have been made in transparent insulation, e.g., by manipulating the spectral dependence of transmittance and emittance of glazings to achieve properties that are desirable for particular applications. Where reduction of building heating loads is the major concern, it is desirable to use glazings with low long-wave emittance to reduce outward heat losses through glazings. Products are now in use with ε reduced from the usual values of about 0.88 to as low as 0.10; the use of one or more such coatings on double glazed windows (usually on the inner surfaces) can reduce the loss coefficient U to a half or less of its value without the

Table 9.4.1   Overall Loss Coefficients U for Walls and Windows with and without Movable Insulation

Wall, Insulation	U, W/m^{2}C
Stud wall, 14 cm fiberglass insulation between studs, plaster board interior, 2.5 cm (1 in.) thick foam insulating board.	0.45
Masonry wall, 19 cm thick concrete blocks with insulation in voids	1.70

Window, Insulation	U, W/m^{2}C
Single glazed, no insulation	6.0
Double glazed, 12 mm gap, no insulation	3.2
Double glazed with low emittance coating on one interior surface, no insulation	2.1
Single glazed, with tight fitting shade or drape	2.6
Double glazed, with tight fitting shade or drape	1.2
Double glazed with 20 mm of foam insulation, tight fitting, with air gap between glazing and foam	0.9
Double glazed, with foam insulation as above but 100 mm thick	0.3

coating. Windows are also constructed with low-ε polymeric films between glazings; these provide additional reduction in convective transport and reduced radiation exchange. All of these coatings and films have reduced transmittance for the solar spectrum; for passively heated buildings in which the solar gain is a direct function of window transmittance the use of these low-loss glazing systems will depend on the balance between reduced solar gains and reduced losses.

Transparent insulation can be applied to the outside of opaque building walls, making the walls into passive collectors.[3] The application reduces losses and at times results in net gains into the building. For these applications, consideration has been given to multiple parallel covers, honeycombs, layers of transparent beads, or aerogels. Properties and applications are discussed in proceedings of a series of international workshops on Transparent Insulation Technology (1988, 1989).

Where cooling loads are the major concern, control of solar heat gains can be accomplished by controlling spectral dependence of transmittance for the solar spectrum. Coated glass is available which transmits much of the visible part of the spectrum but not the infrared part. These coated glasses also have low emittances to reduce heat transfer between building interior and ambient.

See Lampert (1989) for detailed discussion of these and related matters and an extensive bibliography; the *ASHRAE Handbook - Fundamentals* is a useful source of information on heat loss coefficients and transmittances of various glazing.

---

[3] Section 14.6 discusses collector-storage walls for passive heating.

The UA of a building is the sum of the products of the areas and U values for all of the areas of the building envelope.

It is also necessary, for satisfactory estimation of heating loads, to include effects of infiltration of outside air into the building. In old residential buildings there are typically 1 to 3 air change per hour due to infiltration; modern buildings have 0.3 to 0.6 changes per hour; very tight buildings may have infiltration rates of 0.1 change per hour. The sensible heat to bring ambient air up to room temperature (plus that to heat any ventilation air) must be added to the losses through the building envelope in estimating the total heating loads.

Measurements of heating loads for buildings of supposedly identical construction show wide variations (often by factors of 2 or 3) in measured energy use. This may be due to differences in the care with which insulation and vapor barriers are installed, to differences in the ways in which the occupants use the buildings, or to both.

9.5   BUILDING ENERGY STORAGE CAPACITY

Building thermal capacitance is the effective heat capacity of a structure per unit change of interior temperature. It is important in a building that undergoes significant temperature changes. (Additional thermal storage that may be provided in media such as water tanks or pebble beds.) Building thermal capacitance is of particular significance in passive heating or in hybrid systems where storage is provided by the building itself.

A procedure for calculating the effective thermal capacitance is to sum the products of mass, heat capacity, and temperature changes of all of the elements of a building:[4]

$$C = \sum (mC_p)_i \, \Delta T_i'$$
(9.5.1)

Table 9.5.1   Effective Building Thermal Capacity for Typical Construction per Unit Floor Area[a]

Description		Thermal Capacity MJ/m^{2}K
Light	Standard frame construction, 13 mm gyproc walls and ceilings, carpet over over wooden floor.	0.060
Medium	As above, but 51 mm gyproc walls and 25 mm gyproc ceiling	0.153
Heavy	Interior wall finish of 102 mm brick, 13 mm, gyproc ceiling, carpet over wooden floor	0.415
Very heavy	Commercial office building with 305 mm concrete floor	0.810

[a] From Barakat and Sander (1982)

[4] This procedure is analogous to that for calculating the effective thermal capacitance of collectors, as shown in Section 6.12.

where $\Delta T_i'$ is the expected change in temperature of the ith element per degree of change of interior temperature. This is difficult to calculate exactly as the temperature gradients in walls, floors furniture, etc., depend on the rates of heat transfer into and out of those elements. However, tables of effective thermal capacity have been worked out for typical building constructions. Table 9.5.1 shows recommended values of Barakat and Sander (1982).

9.6 COOLING LOADS

Air conditioning loads arise from energy that flows into a building through its envelope, solar gains through windows, infiltration, and ventilation which brings in outside air that needs to be cooled and/or dried, plus heat and moisture that are generated within the building. Building thermal capacitance is often more important in summer cooling than in winter heating, as summer ambient temperatures may fluctuate around room temperature with resulting heat flow into and out of the structure. Building capacitance attenuates the peak loads and may substantially reduce cooling loads over time if ambient temperatures drop below room temperature.

Cooling degree-days have been used to calculate cooling loads in a manner similar to the technique described in Section 9.3 for heating loads. The procedure is generally less reliable. Erbs et al. (1983) also show how to estimate cooling degree-days from knowledge of the average ambient temperature in a manner similar to that described in Section 9.3 for heating degree-days.

If energy storage tanks (for water heating and/or for solar space heating) are located inside of the spaces to be cooled, losses from the tanks will add to the cooling loads.

See Mitchell (1983) and ASHRAE (1979 or 1989) for details of calculation of cooling loads on an hourly or monthly basis.

9.7 SWIMMING POOL HEATING LOADS

Swimming pools have been economically heated by solar energy for many years. The heat loss from outdoor, in-ground pools is by radiation to the sky and both convection and evaporation to the air. Many studies have shown that conduction into the ground is a minor heat loss mechanism.

In some pool systems the heat losses are made up from a solar collector system, possibly uncovered. In others, the heat losses are reduced by placing a plastic cover on the pool surface whenever it is not being used. The cover lets solar radiation pass into the pool and greatly reduces evaporation losses. Some pool covers are made of "bubble" plastic sheet, similar to the common packing material, to act as an insulator and reduce convection losses as well as evaporation losses. Some pool systems use both collectors and covers. If plastic collectors or covers are used, they must be made of UV stabilized materials or they may last only a few months.

To properly size an active collector system for a pool or to estimate the temperature rise that will result from a pool cover, it is necessary to estimate the pool heat losses. The best way to estimate the monthly heat loss from a pool is to use measured fuel data from the pool if it already exists or use data from a nearby similar pool. Often these data are not available and it is necessary to estimate the losses by calculations. These calculations are very uncertain since the wind conditions in the vicinity of the pool are seldom known with any degree of certainty.

To be useful, pools operate in a narrow temperature range, typically between 24 and 32 C. Since the pool has a large mass, its temperature does not change quickly. In fact, monthly energy balances are sufficient to estimate either the pool temperature in the absence of auxiliary energy or the auxiliary energy necessary to maintain a desired temperature.

The radiation heat loss from an outdoor pool is to the sky. Equation 3.9.2 provides estimates of the clear sky temperature for hourly periods. For a whole month, the cosine term in the equation can be neglected and the monthly average dew point temperature can be used to estimate the clear sky temperature. For dew point temperatures of 10, 20, and 30 C, the ratio $\overline{T}_s/\overline{T}_a$ is 0.938, 0.961, and 0.986, respectively. The sky is not always clear. The monthly average sky temperature can be estimated as a weighted average of the air and clear sky temperature. Wei et al. (1979) propose that cloud cover be used as the weighting factor. In the absence of cloud cover data, $\overline{K}_T$ may be a reasonable alternative. Wei et al. measured the outgoing radiation from covered (with clear plastic) and uncovered pools and concluded that the pool surface acts like a blackbody for radiation losses, even with bubble-type covers, at the pool temperature. Thus, Equation 3.9.1 can be used with a pool emittance of 1.0 to estimate the radiation heat loss.

Wei et al. also propose the following dimensional equations for evaporation heat loss from an uncovered pool and for convection losses from covered or uncovered pools in W/m^2:

$$q_e = P_a\left(35V + 43(T_p - T_a)^{1/3}\right)(\omega_p - \omega_a) \qquad (9.7.1)$$

where V is the wind speed in m/s in the vicinity of the pool, P_a is the ambient air pressure in kPa, ω_p is the saturation humidity ratio at the temperature of the pool, and ω_a is the humidity ratio of the ambient air above the pool. Convection losses are expressed in terms of the evaporation losses as

$$q_c = q_e \times 0.0006 \frac{T_p - T_a}{\omega_p - \omega_a} \qquad (9.7.2)$$

The wind speed V is almost always unknown in the vicinity of the pool. Wei et al. suggest that one-fifth to one-tenth of the measured wind speed at the local weather station be used to account for the usual shelter around a pool.

All of the solar radiation incident on a pool is not absorbed. For an uncovered pool with an average depth of 2 m, Francey and Golding (1981) estimate that the pool surface transmittance is about 92% and the pool absorptance is on the order of 60%. However, Wei et al. estimate that the pool absorptance is on the order of 90%. When covers are present, additional reflection losses occur, reducing the transmittance to 70 to 80%. In light of the uncertainties associated with the heat loss mechanisms, further refinement of these optical properties is not warranted.

Example 9.7.1

A 72 m^2 pool is located in Atlanta, Georgia. Estimate the energy required to maintain the pool at 25 C in April. The pool has a cover installed for 12 hours each night.

Solution

From Appendix G, the monthly average ambient temperature and horizontal radiation in April are 16 C (289 K) and 19.1 MJ/m^2 ($\overline{K_T} = 0.53$). For the uncovered pool, the absorbed solar radiation is the pool area times the incident solar radiation times the surface transmittance times the pool absorptance times the number of days in April. Here we use the lower estimate of pool absorptance of 0.6:

$$Q_s = 72 \times 19.1 \times 10^6 \times 0.92 \times 0.6 \times 31 = 24 \text{ GJ}$$

If the monthly average relative humidity is 70%, then from a psychrometric chart, the monthly average dew point and wet bulb temperatures are 10 and 12 C. Also, the monthly average humidity ratio is 0.008 and the saturation humidity ratio for the 25 C pool is 0.020. The clear sky temperature, from Equation 3.9.2, neglecting the cosine term, is $0.938T_a$ or 271 K. Since cloud cover data are not available, the monthly average sky temperature is estimated to be $(1 - 0.53) \times 271 + 0.53 \times 289 = 281$ K. The calculations will begin assuming the pool is uncovered at all times.

The radiation heat losses are found from Equation 3.9.1:

$$Q_r = 72 \times 5.67 \times 10^{-8} \left(298^4 - 281^4\right) \times 2.68 \times 10^6 = 18 \text{ GJ}$$

where 2.68×10^6 is the number of seconds in April. Evaporation losses are found from Equation 9.7.1:

$$Q_e = 72 \times 101.3\left[35 \times 3 + 43(298 - 289)^{1/3}\right](0.020 - 0.008) \times 2.68 \times 10^6 = 46 \text{ GJ}$$

The convection losses are found from Equation 9.7.2:

$$Q_c = 46 \times 0.006 \left(\frac{298 - 289}{0.020 - 0.008}\right) = 21 \text{ GJ}$$

If the pool is uncovered for the whole month, the monthly load is

$$L = 18 + 46 + 21 - 24 = 61 \text{ GJ}$$

If the pool is covered one-half of the time with a film cover, only the evaporation losses will be affected and the monthly load is

$$L = 18 + 0.5 \times 46 + 21 - 24 = 38 \text{ GJ}$$

If the cover is of the bubble type, the evaporation losses will be eliminated and convection losses will be reasonably small. Then the load is

$$L = 18 + 0.5 \times (46 + 21) - 24 = 28 \text{ GJ} \qquad \blacksquare$$

REFERENCES

Arny, M. D., M.S. Thesis, University of Wisconsin-Madison (1982). "Thermal Analysis for the Design of an Industrial Process Water Heating System Utilizing Waste Heat Recovery and Solar Energy."

ASHRAE *Cooling and Heating Load Calculation Manual*, GRP 158, American Society of Heating, Refrigerating and Air-Conditioning Engineers, New York. (1979).

ASHRAE *Handbook - Fundamentals,* American Society of Heating, Refrigerating and Air-Conditioning Engineers, Atlanta, GA. (1989).

Balcomb, J. D., R. W, Jones, R. D. McFarland, and W. O. Wray, *Passive Solar Heating Analysis - A Design Manual*, ASHRAE, Atlanta (1984).

Barakat, S. A. and D. M. Sander, Building Research Note No. 184, Division of Building Research, National Research Council of Canada (1982).

Butso, K. D. and W. L. Hatch, Document 4.11, U.S. Dept. of Commerce. "Selective Guide to Climatic Data Sources."

Erbs, D. G., S. A. Klein, and W. A. Beckman, *ASHRAE J.*, p. 60 (June 1983). "Estimation of Degree-Days and Ambient Temperature Bin Data from Monthly Average Temperatures."

Francey, J. L. and P. Golding, *Solar Energy*, **26**,259 (1981). "The Optical Characteristics of Swimming Pool Covers Used for Direct Solar Heating".

Mitchell, J. W., *Energy Engineering*, Wiley, New York (1983).

Mutch, J. J., RAND Report R 1498 (1974). "Residential Water Heating, Fuel Consumption, Economics and Public Policy."

Thom, H. C. S., *Monthly Weather Review*, **82**, 5 (1954). "Normal Degree-Days Below Any Base."

Thom, H. C. S., *Monthly Weather Review*, **94**, 7 (1966). "Normal Degree Days Above Any Base by the Universal Truncation Coefficient."

Transparent Insulation Technology for Solar Energy Conversion, Proc. of the Third International Workshop (1989), and *The 2nd International Workshop on Transparent Insulation in Solar Energy Conversion for Buildings and other Applications* (1988), available from the The Franklin Co., 192 Franklin Road, Birmingham B30 2HE, UK.

Wei, J., H. Sigworth Jr., C. Lederer, and A. H. Rosenfeld, Lawrence Berkeley Laboratory report 9388 (July 1979). "A Computer Program for Evaluating Swimming Pool Heat Conservation."

Chapter 10

SYSTEM THERMAL CALCULATIONS

In Chapters 6 through 9 we have developed mathematical models for two of the key components in solar energy systems, i.e., collectors and storage units. In this chapter we show how other components can be modeled and how the component models can be combined into system models. With information on the magnitude and time distribution of the system loads and the weather, it is possible to simultaneously solve the set of equations to estimate the thermal performance of a solar process over any time period. These estimates (simulations) are usually done numerically with computers and provide information on the expected dynamic behavior of the system and long-term integrated performance.

The collector performance is a function of the temperature of the fluid entering the collector. This temperature, neglecting (for the moment) heat losses from the connecting pipes, is the same as the temperature in the exit portion of the storage unit. The outlet temperature from the collector becomes the inlet temperature to the storage unit. In these equations, time is the independent variable and the solution is in the form of temperature as a function of time. Meteorological data (radiation, temperature, and possibly other variables such as wind speed and humidity) are forcing functions that are applied hourly (or at other time steps) to obtain numerical solutions through time.

Once the temperatures are known, energy rates can be determined. It is then possible to integrate the energy quantities over time to develop information such as that in Figure 8.1.1(c), and thus assess the annual thermal performance of a system. This *simulation* approach can be used to estimate, for any process application, the amount of energy delivered from the solar collector to meet a load and the amount of auxiliary energy required. The simulation also can indicate whether the temperature variations for a particular system design are reasonable, for example, whether a collector temperature would rise above the boiling point of the liquid being heated.

In this chapter we provide a brief review of collector and storage models and then show how models of controls, heat exchangers, and pipe and duct losses in collector circuits can be formulated. Methods are given for calculating the effects of partial shading of collectors and the output of collector arrays where sections of the arrays are at different orientations. The need for meteorological data and information on the energy loads a system is to supply is noted. In brief, this chapter treats system models (the combination of individual component models) and their solution.

427

10.1   COMPONENT MODELS

Chapters 6, 7, and 8 present developments of collector and storage models, performance, measurements, and data. For flat-plate collectors, Equation 6.7.6 is appropriate, and for focusing collectors, Equation 7.3.12 or its equivalent can be used. The rate of useful gain from a flat-plate collector can be written as

$$Q_u = A_c F_R [S - U_L (T_i - T_a)]^+ \tag{10.1.1}$$

where the + sign implies the presence of a controller and that only positive values of the term in the brackets should be used. Operation of a forced-circulation collector will not be carried out when $Q_u < 0$ (or when $Q_u < Q_{min}$, where Q_{min} is a minimum level of energy gain to justify pumping the fluid through the system). In real systems, this is accomplished by comparing the temperature of the fluid leaving the collector (i.e., in the top header) with the temperature of the fluid in the exit portion of the storage tank and running the pump only when the difference in temperatures is positive and energy can be collected.

The rate of useful gain is also given by

$$Q_u = \dot{m} C_p (T_o - T_i) \tag{10.1.2}$$

where $\dot{m}$ is the output of the pump circulating fluid through the collector.

If the storage unit is a fully mixed sensible heat unit, its performance is given by Equation 8.3.2.

$$Q_u - L_S - (UA)_s (T_s - T_a') = (\dot{m} C_p)_s \frac{dT_s}{dt} \tag{10.1.3}$$

The equivalent equations for stratified water tank storage systems, pebble bed exchangers, or heat of fusion systems are used in lieu of Equation 10.1.3, as appropriate. These equations are the basic equations to be solved in the analysis of systems such as a simple solar water heater with collector, pump, and controller and storage tank. The rate of energy removal to meet all or part of a load is L_S and is time-dependent; S and T_a are also time dependent.

The performance models discussed so far have been based on the fundamental equations describing the behavior of the equipment. Models may also be expressed as empirical or stochastic representations of operating data from particular items of equipment. An example is the model[1] of a LiBr-H_2O absorption cooler, which relates cooling capacity to temperatures of the fluid streams entering the machine. These empirical relations may be in the form of equations, graphs, or tabular data. In whatever form the models are expressed, they must represent component performance over the range of operating conditions to be encountered in the solar operation.

[1] See Chapter 15.

There is very often a heat exchanger between the collector and the storage tank, the **collector heat exchanger**, when antifreeze is used in collectors. If piping or ducting to and from a collector is not well insulated, the losses from the piping or ducting may have to be taken into account. Each of these can be accounted for by modifications of Equation 10.1.1, as noted in the following sections.

10.2 COLLECTOR HEAT EXCHANGER FACTOR

Collectors are often used in combination with a heat exchanger between collector and storage allowing the use of antifreeze solutions in the collector loop. A common circuit of this type is shown in Figure 10.2.1.

A useful analytical combination of the equations for the collector and the heat exchanger has been derived independently by deWinter (1975) and by Brinkworth (1975). In this development the collector equation and the heat exchanger equation will be combined into a single expression that has the same form as the collector equation alone. The combination of a collector and a heat exchanger performs exactly like a collector alone, but with a reduced value of F_R. The useful gain of the collector is represented by Equations 10.1.1 and 10.1.2 written in terms of T_{ci} and T_{co}. The heat exchanger performance is expressed in terms of effectiveness [see Kays and London (1964)] by Equation 3.17.6:

$$Q_{HX} = \varepsilon (\dot{m} C_p)_{min} (T_{co} - T_i) \tag{10.2.1}$$

where $(\dot{m} C_p)_{min}$ is the smaller of the fluid capacitance rates (flow rate $\dot{m}$ times fluid heat capacity C_p) on the collector side $(\dot{m} C_p)_c$ and tank side $(\dot{m} C_p)_t$ of the heat exchanger, T_{co} is the outlet fluid temperature from the collector, and T_i is the inlet water temperature to the heat exchanger (very nearly the temperature in the bottom of the tank). For a counterflow heat exchanger, a common configuration, the effectiveness ε is given by Equation 3.17.7.

Combining Equations 10.1.1, 10.1.2, and 10.2.1,

$$Q_u = A_c F_R'[S - U_L(T_i - T_a)]^+ \tag{10.2.2}$$

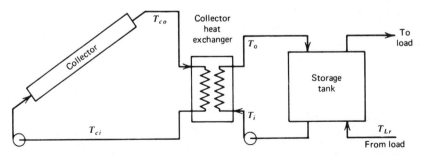

Figure 10.2.1 Schematic of liquid system with a collector heat exchanger between collector and tank.

where the modified collector heat removal factor F_R' accounts for the presence of the heat exchanger and is given by

$$\frac{F_R'}{F_R} = \left[1 + \left(\frac{A_c F_R U_L}{(\dot{m}C_p)_c}\right)\left(\frac{(\dot{m}C_p)_c}{\varepsilon(\dot{m}C_p)_{min}} - 1\right)\right]^{-1} \qquad (10.2.3)$$

The factor F_R'/F_R is an indication of the penalty in collector performance incurred because the heat exchanger causes the collector to operate at higher temperatures than it otherwise would. Another way of looking at the penalty is to consider the ratio F_R'/F_R as the fractional increase in collector area required for the system with the heat exchanger to give the same energy output as without the heat exchanger. Equation 10.2.3 now represents the performance of a subsystem including the collector and the heat exchanger (and, implicitly, the controller and pumps) and is of the same form as that for the collector only. The ratio F_R'/F_R can be calculated from Equation 10.2.3. It can also be determined from Figure 10.2.2, which shows F_R'/F_R as a function of $\varepsilon(\dot{m}C_p)_{min}/(\dot{m}C_p)_c$ and $(\dot{m}C_p)_c/A_c F_R U_L$.

Example 10.2.1

A collector to be used in a solar heating system like that of Figure 10.2.1 heats antifreeze; heat is transferred to water through a collector heat exchanger. The collector $F_R U_L$ is 3.75 W/m²C. Flow rates through both sides of the heat exchanger are 0.0139 kg/s per square meter of collector. The fluid on the collector side is a glycol solution having $C_p = 3350$ J/kgC. The effectiveness of the heat exchanger is 0.7. What is F_R'/F_R?

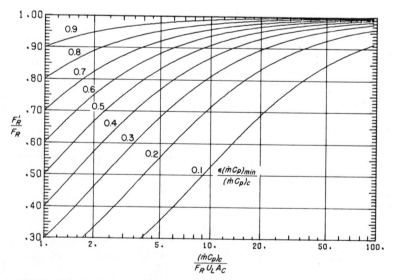

Figure 10.2.2 Collector heat exchanger correction factor as a function of $\varepsilon(\dot{m}C_p)_{min}/(\dot{m}C_p)_c$ and $(\dot{m}C_p)_c/F_R U_L A_c$. Adapted from Beckman et al. (1977).

Solution

The capacitance rate per unit area on the water side of the exchanger is 0.0139×4190 $= 58.2$ W/m²C. On the glycol side it is $0.0139 \times 3350 = 46.6$ W/m²C. Thus the minimum $(\dot{m}C_p)$ product is that of the glycol, $\varepsilon(\dot{m}C_p)_{min}/(\dot{m}C_p)_c = \varepsilon = 0.7$, and

$$\frac{(\dot{m}C_p)_c}{A_c F_R U_L} = \frac{0.0139 \times 3350}{3.75} = 12.4$$

From Figure 10.2.2 or Equation 10.2.4 $F'_R/F_R = 0.97$.   ■

10.3   DUCT AND PIPE LOSS FACTORS

The energy lost from ducts and pipes leading to and returning from the collector in a solar energy system can be significant. Beckman (1978) has shown that the combination of pipes or ducts plus the solar collector is equivalent in thermal performance to a solar collector with different values of U_L and $(\tau\alpha)$. (For simplicity in terminology, the term duct will be used, but the same analysis holds for pipes. Losses from ducts are more likely to be a problem than those from pipes.)

Consider the fluid temperature distribution shown in Figure 10.3.1. Fluid enters the portion of the duct from which losses occur[2] at temperature T_i. Due to heat losses to the ambient at temperature T_a, the fluid is reduced in temperature by an amount ΔT_i before it enters the solar collector. The fluid passes through the

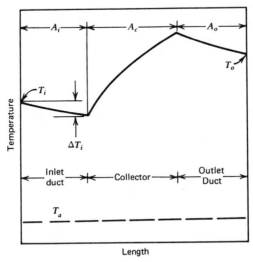

Figure 10.3.1 Temperature distribution through a duct-collector system. From Beckman (1978).

---

[2] In a house heating system, duct and pipe losses inside the heated space are not net losses but are uncontrolled gains. These gains may not be desirable during warm weather.

collector and is heated to the collector outlet temperature. This temperature is then reduced to T_o as the fluid loses heat to the ambient while passing through the outlet ducts.

From energy balance considerations the useful energy gain of this collector-duct combination is equal to

$$Q_u = (\dot{m}C_p)_c(T_o - T_i) \qquad (10.3.1)$$

This energy gain can also be related to the energy gain of the collector minus the duct losses by the following rate equation:

$$Q_u = A_c F_R\left[G_T(\tau\alpha) - U_L(T_i - \Delta T_i - T_a)\right] - \text{losses} \qquad (10.3.2)$$

The duct losses are equal to the integrated losses over the inlet and outlet ducts and are given by

$$\text{Losses} = U_d \int (T - T_a)\, dA \qquad (10.3.3)$$

where U_d is the loss coefficient from the duct. It is possible to integrate Equation 10.3.3, but in any well-designed system the losses from ducts must be small and the integral can be approximated to an adequate degree of accuracy in terms of the inlet and outlet temperatures:

$$\text{Losses} = U_d A_i(T_i - T_a) + U_d A_o(T_o - T_a) \qquad (10.3.4)$$

where A_i and A_o are the areas for heat loss of the inlet and outlet ducts. Upon rearranging Equations 10.3.1 and 10.3.4, the losses can be expressed in terms of the useful energy gain and the inlet fluid temperature as

$$\text{Losses} = U_d(A_i + A_o)(T_i - T_a) + \frac{U_d A_o Q_u}{(\dot{m}C_p)_c} \qquad (10.3.5)$$

The decrease in temperature, ΔT_i, due to heat losses on the inlet side of the collector can be approximated by

$$\Delta T_i = \frac{U_d A_i(T_i - T_a)}{(\dot{m}C_p)_c} \qquad (10.3.6)$$

Substituting Equations 10.3.5 and 10.3.6 into 10.3.2 and rearranging, the useful energy gain of the collector and duct system can be expressed as

$$Q_u = \frac{A_c F_R\left[G_T(\tau\alpha) - U_L\left(1 - \dfrac{U_d A_i}{(\dot{m}C_p)_c} + \dfrac{U_d(A_i + A_o)}{A_c F_R U_L}\right)(T_i - T_a)\right]}{1 + \dfrac{U_d A_o}{(\dot{m}C_p)_c}} \qquad (10.3.7)$$

Equation 10.3.7 can be made to look like the usual collector equation by defining modified values of $(\tau\alpha)$ and U_L so that

$$Q_u = A_c F_R \left[G_T (\tau\alpha)' - U_L'(T_i - T_a) \right] \tag{10.3.8}$$

where

$$\frac{(\tau\alpha)'}{(\tau\alpha)} = \frac{1}{1 + \dfrac{U_d A_o}{(\dot{m}C_p)_c}} \tag{10.3.9}$$

and

$$\frac{U_L'}{U_L} = \frac{1 - \dfrac{U_d A_i}{(\dot{m}C_p)_c} + \dfrac{U_d(A_i + A_o)}{A_c F_R U_L}}{1 + \dfrac{U_d A_o}{(\dot{m}C_p)_c}} \tag{10.3.10}$$

Example 10.3.1

Compare the performance equations of an air collector system with ducts insulated and not insulated. The collector has an area of 50 m², an $F_R U_L$ product of 3.0 W/m²C and $F_R(\tau\alpha)$ of 0.60. The mass flow rate-specific heat product of the air through the collector is 500 W/C. The area of the inlet duct is 10 m², as is the area of the outlet duct. Insulation 33 mm thick is available with a conductivity of 0.033 W/mC. The heat transfer coefficient outside the duct is 10 W/m²C, and the coefficient inside the duct is assumed to be large, leading to a loss coefficient U_d for the insulated duct of 1.0 W/m²C and for the uninsulated duct of 10 W/m²C.

Solution

For the insulated duct, the modified transmittance-absorptance product is found using Equation 10.3.9:

$$\frac{(\tau\alpha)'}{(\tau\alpha)} = \frac{1}{1 + \dfrac{1.0 \times 10}{500}} = 0.98$$

and

$$F_R(\tau\alpha)' = 0.60 \times 0.98 = 0.59$$

The modified loss coefficient is obtained with Equation 10.3.10:

$$\frac{U_L'}{U_L} = \frac{1 - \dfrac{1.0 \times 10}{500} + \dfrac{1.0(10 + 10)}{50 \times 3.0}}{1 + \dfrac{1.0 \times 10}{500}} = 1.09$$

and

$$F_R U'_L = 3.27 \text{ W/m}^2\text{C}$$

For the system with insulated ducts the collector performance equation becomes

$$Q_u = 50[0.59G_T - 3.27(T_i - T_a)]$$

If the ducts are not insulated, U_d is 10 W/m^{2}C, $F_R(\tau\alpha)'$ is 0.50, and $F_R U'_L$ is 5.3 W/m^{2}C. The performance equation is then

$$Q_u = 50[0.50G_T - 5.3(T_i - T_a)]$$

The addition of insulation has a very substantial effect on the useful gain to be expected from the system.   ■

In a liquid system when both pipe losses and a heat exchanger are present and when the heat exchanger is near the tank, the collector characteristics should first be modified to account for pipe losses and these modified collector characteristics used in Equation 10.2.3 to account for the heat exchanger.

Close and Yusoff (1978) have developed an analysis of the effects of leakage of air into air heating collectors. Their results can also be expressed in terms of modified collector characteristics.

10.4   CONTROLS

Two types of control schemes are commonly used on solar collectors on building scale applications: on-off and proportional. With an on-off controller, a decision is made to turn the circulating pumps on or off depending on whether or not useful output is available from the collectors. With a proportional controller, the pump speed is varied in an attempt to maintain a specified temperature level at the collector outlet. Both strategies have advantages and disadvantages, largely depending on the ultimate use of the collected energy.

The most common control scheme requires two temperature sensors, one in the bottom of the storage unit and one on the absorber plate at the exit of a collector (or on the pipe near the plate). Assume the collector has low heat capacity. When fluid is flowing, the collector transducer senses the exit fluid temperature. When the fluid is not flowing, the mean plate temperature T_p is measured. A controller receives this temperature and the temperature at the bottom of the storage unit. This storage temperature will be called T_i; when the pump turns on, the temperature at the bottom of storage will equal the inlet fluid temperature if the connecting pipes are lossless. Whenever the plate temperature at no-flow conditions exceeds T_i by a specific amount ΔT_{on}, the pump is turned on.

When the pump is on and the measured temperature difference falls below a specified amount ΔT_{off}, the controller turns the pump off. Care must be exercised when choosing both ΔT_{on} and ΔT_{off} to avoid having the pump cycle on and off.

When the collector pump is off, the useful output is zero and the absorber plate reaches an equilibrium temperature given by

$$[S - U_L(T_p - T_a)] = 0 \qquad (10.4.1)$$

The value of S when T_p is equal to $T_i + \Delta T_{on}$ is

$$S_{on} = U_L\left(T_i + \Delta T_{on} - T_a\right) \qquad (10.4.2)$$

When the pump does turn on, the useful gain is

$$Q_u = A_c F_R[S_{on} - U_L(T_i - T_a)] \qquad (10.4.3)$$

which when Equation 10.4.2 is substituted for S_{on} becomes

$$Q_u = A_c F_R U_L \Delta T_{on} \qquad (10.4.4)$$

The outlet temperature under these conditions is found from

$$Q_u = (\dot{m} C_p)(T_o - T_i) \qquad (10.4.5)$$

The temperature difference $(T_o - T_i)$ is the temperature measured by the controller after flow begins. Consequently, the turn-off criterion must satisfy the following inequality or the system will be unstable (until the solar radiation increases to levels significantly higher than that corresponding to S_{on}):

$$\Delta T_{off} \le \frac{A_c F_R U_L}{\dot{m} C_p} \Delta T_{on} \qquad (10.4.6)$$

Example 10.4.1

For a water collector with $A_c = 2$ m², $F_R U_L = 3$ W/m²C, and $\dot{m} = 0.030$ kg/s, what is the ratio of turn-on criterion to turn-off criterion?

Solution

From Equation 10.4.6

$$\frac{\Delta T_{on}}{\Delta T_{off}} \ge \frac{0.030 \times 4190}{2 \times 3} = 21 \qquad \blacksquare$$

From the preceding example, the turn-on criterion must be significantly greater than the turn-off criterion. Another way of looking at this situation is to assume a 5 C turn-on setting. Then the controller of the example will have to be sensitive to temperature differences of 0.25 C or oscillations may result if the radiation stays near S_{on}. This is a small temperature difference to detect with inexpensive controllers. Raising the turn-on setting to 20 C or more will not significantly reduce the useful energy collection. In fact, the pump should not be operated until the value of the useful energy collected exceeds the cost of pumping.

Even if the criterion of Equation 10.4.6 is satisfied, cycling may occur, particularly in the morning. The fluid in the pipes or ducts between the storage unit and collector will be colder than the temperature at the bottom of storage. Consequently, when the pump first turns on, cold fluid will enter the collector resulting in lower temperatures detected by the outlet sensor than expected. The controller may turn the pump off until the fluid in the collector is heated to the proper temperature for the pump to again turn on. Other than wear and tear on the pump and motor, this is an efficient way to heat the fluid in the collector and inlet ducts to the proper temperature.

Proportional controllers have been used to maintain either the collector outlet temperature or the temperature rise through the collector at or near a predetermined value. In such a control system the temperature sensors are used to control the pump speed. Although higher outlet temperatures can be obtained with a proportional controller than with an on-off controller, the useful energy collected by an on-off system (with a fully mixed tank and with both turn-off and turn-on criteria equal to zero) will be greater than that by a system with a proportional controller. This becomes clear when the basic collector equation is examined, as the highest value of Q_u is obtained when F_R is a maximum.[3] With a proportional controller F_R is reduced when the flow rate is reduced.

There is a growing body of literature on control of solar energy systems. For example, Kovarik and Lesse (1976) have considered optimization of flow rates through collectors. Winn and Hull (1978) have further studied this concept by calculation of an F_R which leads to a maximum difference between energy collected and energy required for pumping fluid through the collector. Winn (1982) provides a broad review of control technologies in solar energy systems.

10.5   COLLECTOR ARRAYS;   SERIES CONNECTIONS

Collector modules in arrays may be connected in series, in parallel, or in combinations. The performance of collectors in arrays is dependent on how they are arranged, i.e., on the flow rate through risers and on the inlet temperatures to individual modules.

---

[3] There are other considerations that affect the choice of flow rates. There may be advantages to low collector flow rates if they lead to improved stratification in storage tanks. The resulting reduction in T_i may more than compensate for the reduced F_R. See Section 12.5.

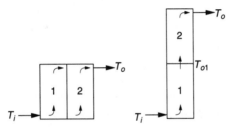

Figure 10.5.1 Collector modules in parallel and series. Flow divides in the parallel set; full flow goes through each module in the series set.

Figure 10.5.1 shows arrays of two modules (or two groups of modules) with parallel and series connections. Assume for the moment that the modules are identical. If the flow rate to the arrays are the same, the velocity through the risers of the series-connected array will be twice that in the parallel array. If this difference in velocity does not make an appreciable difference in F' through its effect on h_{fi}, then there will be no appreciable difference in performance between the two arrays as the terms in Equation 6.7.6 will be identical. If there is a significant difference in F' and thus in F_R or if the design of the modules is not the same (e.g., with the first having one cover and the second two covers), the performance of the arrays will not be the same. In the series arrangement, the performance of the second (and subsequent) module will not be the same as the first as its inlet temperature will be the outlet temperature of the first.

For collector modules in series, an analysis can be developed that results in an equation identical in form to that of Equation 6.7.6. It is not necessary that the modules be identical; the constraint is that each can be described by its set of two parameters, $F_R(\tau\alpha)_n$ and $F_R U_L$. The useful output of the combination is

$$Q_{u,1} + Q_{u,2} = A_1 F_{R1}[(\tau\alpha)_1 I_T - U_{L1}(T_i - T_a)]$$
$$+ A_2 F_{R2}[(\tau\alpha)_2 I_T - U_{L2}(T_{o,1} - T_a)] \tag{10.5.1}$$

where T_i is the inlet fluid temperature to the pair of collectors and $T_{o,1}$ is the inlet temperature to the second collector, which is found from the outlet of the first collector:

$$T_{o,1} = T_i + \frac{Q_{u,1}}{\dot{m}C_p} \tag{10.5.2}$$

The values of $F_R(\tau\alpha)$ and $F_R U_L$ for each collector must be the values corresponding to the actual fluid flow rate through the pair. By eliminating $T_{o,1}$ from these two equations the useful output of the combination can be expressed as

$$Q_{u,1+2} = [A_1 F_{R1}(\tau\alpha)_1(1 - K) + A_2 F_{R2}(\tau\alpha)_2]I_T$$
$$- [A_1 F_{R1} U_{L1}(1 - K) + A_2 F_{R2} U_{L2}](T_i - T_a) \tag{10.5.3}$$

where K is given by

$$K = \frac{A_2 F_{R2} U_{L2}}{\dot{m} C_p}$$
(10.5.4)

The form of Equation 10.5.3 suggests that the combination of the two collectors can be considered as a single collector with the following characteristics:

$$A_c = A_1 + A_2$$
(10.5.5)

$$F_R(\tau\alpha) = \frac{A_1 F_{R1}(\tau\alpha)_1(1 - K) + A_2 F_{R2}(\tau\alpha)_2}{A_c}$$
(10.5.6)

$$F_R U_L = \frac{A_1 F_{R1} U_{L1}(1 - K) + A_2 F_{R2} U_{L2}}{A_c}$$
(10.5.7)

If three or more collectors are placed in series, then these equations can be used for the first two collectors to define a new equivalent first collector. The equations are applied again with this equivalent first collector and the third collector becoming the second collector. The process can be repeated for as many collectors as desired.

If the two collectors are identical, then Equations 10.5.6 and 10.5.7 reduce to the following:

$$F_R(\tau\alpha) = F_{R1}(\tau\alpha)_1\left(1 - \frac{K}{2}\right)$$
(10.5.8)

$$F_R U_L = F_{R1} U_{L1}\left(1 - \frac{K}{2}\right)$$
(10.5.9)

For N identical collectors in series, Oonk et al. (1979) have shown that repeated applications of Equations 10.5.6 and 10.5.7 yield

$$F_R(\tau\alpha) = F_{R1}(\tau\alpha)_1\left[\frac{1 - (1 - K)^N}{NK}\right]$$
(10.5.10)

$$F_R U_L = F_{R1} U_{L1}\left[\frac{1 - (1 - K)^N}{NK}\right]$$
(10.5.11)

Example 10.5.1

Calculate $F_R(\tau\alpha)$ and $F_R U_L$ for the combination of two air heating collectors connected in series with a flow rate of 0.056 kg/s. The characteristics of a single air heater are $F_R(\tau\alpha) = 0.67$ and $F_R U_L = 3.6$ W/m²C at a flow rate of 0.056 kg/s. Each collector is 1.00 m wide by 2.00 m long. C_p is 1008 J/kgC.

Solution

From Equation 10.5.4,

$$K = \frac{3.63 \times 2.00}{0.056 \times 1008} = 0.129$$

From Equations 10.5.8 and 10.5.9,

$$F_R(\tau\alpha) = 0.67 \left(1 - \frac{0.129}{2}\right) = 0.63$$

$$F_R U_L = 3.6 \left(1 - \frac{0.129}{2}\right) = 3.4 \text{ W/m}^2\text{C}$$ ∎

10.6 PERFORMANCE OF PARTIALLY SHADED COLLECTORS

Two circumstances can lead to nonuniform radiation on collectors. In arrays, shading of the bottom of a collector row from beam radiation by the row in front may occur so that beam and diffuse are incident on the top portion and diffuse only on the bottom portion. Also, collectors with specular planar reflectors may have different levels of radiation on upper and lower portions. It is usually adequate to use an average level of radiation over the whole collector area to calculate its output. However, an analysis similar to that of the previous section can be written for situations where flow is in series through the parts of the collector at two radiation levels; this indicates whether or not there are significant changes in performance and if changing flow direction will result in improved performance.

The collector of Figure 10.6.1 is in two zones: 1, on which the incident radiation is $I_{T,1}$, and 2, on which the incident radiation is $I_{T,2}$. The fraction of the total area A in zone 1 is A_1, and that in zone 2 is A_2. The temperature of the fluid at the outlet of the first zone is $T_{o,1}$. The values of F_R and U_L should be essentially the same for both zones. If the angles of incidence of the radiation on the two zones

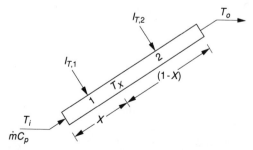

Figure 10.6.1 Schematic of a collector with zone 1 irradiated at $I_{T,1}$ and zone 2 irradiated at $I_{T,2}$.

are different (which will probably be the case if one zone is shaded from beam radiation), the $(\tau\alpha)$ for the two zones will be different. The useful gains are

$$Q_{u,1} = A_1[F_R(\tau\alpha)_1 I_{T,1} - F_R U_L(T_i - T_a)] \tag{10.6.1}$$

$$Q_{u,2} = A_2[F_R(\tau\alpha)_2 I_{T,2} - F_R U_L(T_{o,1} - T_a)] \tag{10.6.2}$$

and $\qquad T_{o,1} = T_i + Q_{u,1}/(\dot{m}C_p) \tag{10.6.3}$

By eliminating $T_{o,1}$ from Equations 10.6.2 and 10.6.3 and adding the two gains, a single equation can be written in the same general form as Equation 10.5.3:

$$Q_u = A_1 F_R(\tau\alpha)_1(1 - K)I_{T,1} + A_2 F_R(\tau\alpha)_2 I_{T,2}$$

$$- F_R U_L[A_1(1 - K) + A_2](T_i - T_a) \tag{10.6.4}$$

where $\qquad K = \dfrac{A_2 F_R U_L}{\dot{m}C_p} \tag{10.6.5}$

In contrast to the series-connected different collectors of Section 10.5 where the areas of the two collectors are fixed, the areas A_1 and A_2 in Equation 10.6.4 will be a function of time where the differences in $I_{T,1}$ and $I_{T,2}$ are caused by shading.

Example 10.6.1

A collector is situated so that it is partially shaded from beam radiation. One zone that has 22% of the area has an incident radiation of 220 W/m². The other zone that has 78% of the area has an incident radiation of 1050 W/m². Water is pumped at a rate of 0.0060 kg/m²s through the collector and goes through the two zones in series. The characteristics of the collector at this flow rate are $F_R = 0.96$ and $F_R U_L = 6.20$ W/m²C. Temperature $T_i = 48$ C and $T_a = 9$ C. As defined by Equation 5.8.2, $(\tau\alpha)_{av}$ is 0.86 for the zone with $I_T = 1050$ W/m² and 0.73 for the zone with $I_T = 220$ W/m².

Calculate the output from this collector **a** with flow from the low-intensity to the high-intensity zone, and **b** with flow from the high-intensity to the low-intensity zone.

Solution

a   For flow from the low- to high-intensity zone

$$K = 0.78 \times 6.20/(0.0060 \times 4190) = 0.192$$

$$Q_u/A_c = 0.96[0.22 \times 0.73(1 - 0.192)220 + 0.78 \times 0.86 \times 1050]$$

$$- 6.20[0.22(1 - 0.192) + 0.78](48 - 9) = 472 \text{ W/m}^2$$

b   For flow from the high- to low-intensity zone,

$$K = 0.22 \times 6.20/(0.0060 \times 4190) = 0.054$$

$$Q_u/A_c = 0.96\big[0.78 \times 0.86(1 - 0.054)1050 + 0.22 \times 0.73 \times 220\big]$$
$$- 6.20[0.78(1 - 0.054) + 0.22](48 - 9) = 442 \text{ W/m}^2$$

In this situation there is a 7% advantage to pumping the fluid through the collector in the correct direction.

These results can be checked by calculating the outputs of the two zones independently and summing them. For case **b**, per square meter of total area,

$$Q_{u,1}/A_c = 0.78(0.96 \times 0.86 \times 1050 - 6.20(48 - 9)) = 488 \text{ W/m}^2$$

$$T_{o,1} = 48 + \frac{488}{0.006 \times 4180} = 67.5 \text{ C}$$

$$Q_{u,2}/A_c = 0.22\big[0.96 \times 0.73 \times 220 - 6.20(67.5 - 9)\big] = -46 \text{ W/m}^2$$

$$\Sigma Q_u/A_c = 488 - 46 = 442 \text{ W/m}^2$$

In this example there is a net loss from the shaded part of the collector. However, the useful gain from the whole collector is positive.   ■

For shading of a collector module or set of modules that is part of a larger array piped or ducted in parallel, the gain is simply the sum of the gains from the shaded and unshaded zones.

10.7   SERIES ARRAYS WITH SECTIONS WITH DIFFERENT ORIENTATIONS

Collector arrays are occasionally constructed in sections, with the sections having different orientations (i.e., with different azimuth angles and/or slopes). These arrangements may be imposed by the locations and structure available to mount the collector. If these collector sections are series connected, the analysis of Section 10.6 can be applied to calculate the output.

Consider an array of two sections, 1 and 2, connected in series. Areas of the two sections A_1 and A_2 will be fixed. Flow is in the direction of 1 to 2. The F_R and U_L values will be the same for both sections, assuming that $\dot{m}/A_c$ is the same and neglecting effects of orientation and collector temperature on U_L. The values of $(\tau\alpha)$ for the two sections will be different and will depend on orientation. Equation 10.6.4 applies to this array, as shown in the following example.

Example 10.7.1

A collector array is in two sections of equal area. The slopes of both sections are 55°; section 1 has $\gamma = -45°$ and section 2 has $\gamma = 45°$. The latitude is 40°. The collector has one glass cover with $K_L = 0.0125$ and α_n is 0.93. F_R is 0.933 and U_L is 4.50 W/m²C. Water is pumped in series through sections 1 and 2 at a rate of 0.01 kg/m²s.

For the hour 2 to 3 on May 15, $I = 2.83$ MJ/m². The ambient temperature is 10 C. The fluid inlet temperature is 45 C and ground reflectance is estimated to be 0.25. Estimate the output per square meter of total array area for this hour.

Solution

For this hour, $I_o = 3.72$ MJ/m², $k_T = 0.76$, and $I_d/I = 0.177$: $\delta = 18.8°$ and $\omega = 37.5°$.

For both sections, $I_d = 0.177 \times 2.83 = 0.50$ MJ/m² and $I_b = 2.33$ MJ/m². For diffuse, $\theta_e = 57°$ (from Figure 5.4.1),

$$\tau_d = 0.85 \text{ and } (\alpha/\alpha_n)_d = 0.94$$

$$1.01\,\tau\alpha = 1.01 \times 0.85 \times 0.93 \times 0.94 = 0.751$$

For ground reflected, $\theta_e = 66°$,

$$\tau_g = 0.77 \text{ and } (\alpha/\alpha_n)_g = 0.88$$

$$1.01\,\tau\alpha = 1.01 \times 0.77 \times 0.93 \times 0.88 = 0.636$$

For Section 1:
For beam radiation, from Equation 1.6.2,

$$\cos\theta = 0.277 \text{ and } \theta = 74°$$

$$\tau_b = 0.65 \text{ and } (\alpha/\alpha_n)_b = 0.79$$

$$1.01\,\tau\alpha = 1.01 \times 0.65 \times 0.93 \times 0.79 = 0.482$$

From Equation 1.6.5, $\cos\theta_z = 0.782$,

$$R_{b,1} = 0.277/0.782 = 0.355$$

Then $I_{T,1} = 2.33 \times 0.355 + 0.50(1 + \cos 55)/2 + 2.83 \times 0.25(1 - \cos 55)/2$

$$= 0.827 + 0.393 + 0.151 = 1.37 \text{ MJ/m}^2$$

and $S_1 = 0.827 \times 0.482 + 0.393 \times 0.751 + 0.151 \times 0.636 = 0.790$ MJ/m²

So $(\tau\alpha)_{av,1} = 0.790/1.37 = 0.577$

For Section 2:
For beam radiation,

$$\cos\theta = 0.942 \text{ and } \theta = 19.6°$$

$$\tau_b = 0.91, \text{ and } (\alpha/\alpha_n)_b = 1.00$$

$$1.01 \ \tau\alpha = 1.01 \times 0.91 \times 0.93 = 0.855$$

$$R_{b,2} = 0.942/0.782 = 1.18$$

Then $I_{T,2} = 2.33 \times 1.18 + 0.393 + 0.151 = 3.30 \text{ MJ/m}^2$

and $S_2 = 2.33 \times 1.18 \times 0.855 + 0.393 \times 0.751 + 0.151 \times 0.636$

$$= 2.75 \text{ MJ/m}^2$$

So $(\tau\alpha)_{av,2} = 2.75/3.30 = 0.833$

We now use Equations 10.6.4 and 10.6.5 to calculate the performance of the two sections in series:

$$K = 0.933 \times 4.5/(0.01 \times 4190) = 0.100$$

$$Q_u/A_c = 0.933\big\{[0.5 \times 0.577(1 - 0.100)1.37 + 0.5 \times 0.833 \times 3.30]10^6$$

$$-4.50 \times 3600[0.5(1 - 0.100) + 0.5](45 - 10)\big\}$$

$$= 1.11 \text{ MJ/m}^2 \qquad ∎$$

Note that for the hour 9 to 10 for this example, which corresponds to a reversal of flow direction for the hour 2 to 3, the output of the array would be 1.02 MJ/m², or about 8% less than for the conditions in the example. These calculations show how much difference flow direction makes and thus assists in determining control strategy. (Note that there may be conditions under which the section with the lower irradiation should be bypassed. This can be determined by calculating the output of the first section. If the output is negative, it should be bypassed. If its output is higher than that of the combined sections, the second section should be bypassed.)

If sections of a collector array are connected in parallel, those sections perform as separate collectors and the total output is the sum of the outputs of the sections.

10.8   USE OF MODIFIED COLLECTOR EQUATIONS

The equations that have been developed in this chapter up to this point are all modifications of Equation 6.7.6, the basic collector performance model. The result is a series of equations, all of the same form, that are combinations of the collector model plus other components: a controller (by the superscript +), a collector heat exchanger (by F_R'), pipe and duct losses (by $F_R(\tau\alpha)'$ and $F_R U_L'$), and various combinations of series connections (by the factor K). For many purposes, it is convenient to use these combined component models and consider the entire collector subsystem (collector, control, piping, and heat exchanger) as a modified collector.[4]

Combinations of these models are often encountered, and it is the purpose of this section to illustrate the logical way to account for the combinations. The example that follows includes many of the modifications noted in previous sections. The logic illustrated can be extended to cover additional modifications.

Example 10.8.1

An active solar water heating system is shown schematically in the diagram that follows. The collector is located at a distance from the tank, and the collector heat exchanger is next to the tank.

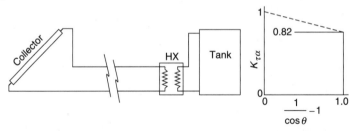

There are three modules in the collector array, connected in parallel. The array area A_c is 5.90 m². Operating data from the manufacturer indicate that $F_{av}(\tau\alpha)_n$ is 0.77 and $F_{av}U_L$ is 3.88 W/m²C when the flow rate of water is 0.020 kg/m²s. Data on the incidence angle modifier are shown in the diagram.

The pipe runs between the collector and the heat exchanger are each 12.2 m long. The pipes are 25.5 mm in outer diameter and are insulated with 30.0 mm of foam insulation with $k = 0.0342$ W/mC. It can be assumed that most of the resistance to heat loss from the pipes is in the insulation and that losses occur to ambient temperature over essentially all of the length of the runs.

A solution of 40% propylene glycol in water (an antifreeze solution) is pumped through the collector loop at a rate of 0.110 kg/s, and water from the storage tank is pumped through the heat exchanger at 0.122 kg/s. The manufacturer of the heat exchanger specifies that its UA is 260 W/C.

---

[4] Other modifications have been made. Zarmi (1982) has developed a method for accounting for the transient response of collector subsystems in systems where there are quantities of fluid in the collector and piping that are significant compared to that in a storage tank.

a   For this system, write an equation in the form of Equation 10.1.1 for Q_u in terms of G_T, T_a, T_i, and θ_b.

b   For an hour when G_T is 1010 W/m², T_a is 13 C, T_i is 34 C, and θ_b is 35°, what is Q_u?

Solution

There is a logical order in which to account for the several modifications that are to be made in the collector equation to convert it to the subsystem equation. Start with the collector parameters, convert them to functions of F_R, correct for capacitance rate, add the pipe losses, and then include the effect of the heat exchanger. In other words, do the calculations in the order in which the components occur, starting with the known collector characteristics and working toward the tank.

The collector data are applicable to the parallel configuration, and no correction is required to account for series connections. The effects of the incidence angle modifier can be entered into the equation at any point; for convenience in writing the equations it will be done last. At the end of each of the five steps that follow, modified values of $F_R(\tau\alpha)_n$ and $F_R U_L$ to that point will be shown. These in turn are inputs to the succeeding step.

a   First, convert the collector parameters to $F_R(\tau\alpha)_n$ and $F_R U_L$ using Equations 6.19.2 and 6.19.3. The capacitance rate $\dot{m}C_p$ through the collector under test conditions, with $C_p = 4180$ J/kgK, is $0.118 \times 4180 = 493$ W/C.

$$F_R(\tau\alpha)_n = \frac{0.77}{1 + \dfrac{5.90 \times 3.88}{2 \times 493}} = \frac{0.77}{1.0232} = 0.753$$

$$F_R U_L = \frac{3.88}{1.0232} = 3.79 \text{ W/m}^2\text{C}$$

Second, correct $F_R(\tau\alpha)_n$ and $F_R U_L$ to the capacitance rate to be used in the operation. The test capacitance rate is 493 W/C. The use capacitance rate, with $C_p = 3750$ J/kgK, is $0.110 \times 3750 = 413$ W/C.

With Equation 6.20.4, calculate $F'U_L$ from the test conditions:

$$F'U_L = -\frac{493}{5.90} \ln\left(1 - \frac{3.79 \times 5.90}{493}\right) = 3.88 \text{ W/m}^2\text{C}$$

(Note that if the temperature rise through the collector is linear, F' and F_{av} will be the same. In this case, with high flow rates through the collector, they are essentially equal.) Then from Equation 6.20.2

$$r = \frac{\dfrac{413}{3.88 \times 5.90}\left(1 - e^{-5.90 \times 3.88/413}\right)}{\dfrac{493}{3.88 \times 5.90}\left(1 - e^{-5.90 \times 3.88/493}\right)} = 0.995$$

The change in capacitance rate thus produces a very small change in collector parameters, which under use conditions are

$$F_R(\tau\alpha)_n = 0.995 \times 0.753 = 0.749$$

$$F_R U_L = 3.79 \times 0.995 = 3.77 \text{ W/m}^2\text{C}$$

Third, the pipe losses are accounted for using Equations 10.3.9 and 10.3.10. If the significant resistance to heat loss is in the insulation, the loss coefficient for the pipes to and from the collector, based on the outside area of the insulation, will be

$$U_{out} = \frac{2k}{D_o \ln(D_o/D_i)} = \frac{2 \times 0.0342}{0.0855 \times \ln(0.0855/0.0255)} = 0.661 \text{ W/m}^2\text{C}$$

Then $\qquad (UA)_i = (UA)_o = 0.661 \times \pi \times 0.0855 \times 12.2 = 2.17 \text{ W/C}$

$$\frac{(\tau\alpha)'}{(\tau\alpha)} = \frac{1}{1 + 2.17/413} = 0.995$$

$$\frac{U_L'}{U_L} = \frac{1 - \dfrac{2.17}{413} + \dfrac{2 \times 2.17}{5.90 \times 3.77}}{1 + 2.17/413} = 1.184$$

The collector parameters with the additional modification for pipe losses are

$$F_R(\tau\alpha)_n' = 0.995 \times 0.749 = 0.745$$

$$F_R U_L' = 1.184 \times 3.77 = 4.47 \text{ W/m}^2\text{C}$$

Fourth, the parameters are modified to account for the heat exchanger. The effectiveness of the heat exchanger is estimated by the method of Section 3.17. The capacitance rate on the tank side of the heat exchanger is $0.122 \times 4180 = 510$ W/C, so the capacitance rate on the collector side (413 W/C) is the minimum. Thus C^* is $413/510 = 0.810$.

From Equation 3.17.8,

$$NTU = 260/413 = 0.630$$

Equation 3.17.7 gives the heat exchanger effectiveness:

$$\varepsilon = \frac{1 - e^{-0.630(1 - 0.810)}}{1 - 0.810e^{-0.630(1 - 0.810)}} = 0.40$$

Using Equation 10.2.3 (or Figure 10.2.2) to get F_R'/F_R

$$\frac{F_R'}{F_R} = \left[1 + \frac{5.90 \times 4.47}{413}\left(\frac{1}{0.4} - 1\right)\right]^{-1} = 0.913$$

Thus the parameters with modification for the heat exchanger become

$$F_R'(\tau\alpha)_n = 0.745 \times 0.913 = 0.680$$

$$F_R'U_L' = 4.47 \times 0.913 = 4.08 \text{ W/m}^2\text{C}$$

Fifth, the incidence angle modifier is needed. The slope b_o of the $K_{\tau\alpha}$ curve is –0.18, so

$$K_{\tau\alpha} = 1 - 0.18\left(\frac{1}{\cos\theta_b} - 1\right)$$

Using Equation 10.1.1,

$$Q_u = 5.90\left[0.749\left[1 - 0.18\left(\frac{1}{\cos\theta_b} - 1\right)\right]G_T - 3.77(T_i - T_a)\right]$$

The desired working equation is thus

$$Q_u = 4.42\left[1 - 0.18\left(\frac{1}{\cos\theta_b} - 1\right)\right]G_T - 22.3(T_i - T_a)$$

b We can now enter the operating conditions in this equation to get the useful gain from the subsystem:

$$Q_u = 4.42\left[1 - 0.18\left(\frac{1}{\cos 35} - 1\right)\right]1010 - 22.3(34 - 13) = 3820 \text{ W} \qquad \blacksquare$$

In this example, the corrections for collector capacitance rate and the differences between the parameters based on F_{av} and F_R were small. These corrections can be very much larger if the capacitance rates in the collector are small, as is the case for some water heating systems in which high degrees of stratification are obtained in tanks by use of flow rates an order of magnitude smaller than those noted here. This will be discussed in Chapter 12.

10.9 SYSTEM MODELS

The previous section dealt with the collector equation and modifications thereof. The next step is the inclusion of models of other system components, including equations for storage, loads, and whatever other components are included in the physical system. System models are assemblies of appropriate component models. The net effect of the assembly is a set of coupled algebraic and differential equations with time as the independent variable. Inputs to these equations include meteorological data as forcing functions that operate on the collector, and possibly also on the load, depending on the application.

These equations can be manipulated and combined algebraically or they can be solved simultaneously without formal combinations. Each procedure has some advantages in solar process simulation. If the equations are all linear (and if there are not too many to manipulate), the algebraic equations can be solved and substituted into the differential equations, which can then be solved by standard methods [e.g., Hamming (1962)]. If the algebraic equations are nonlinear or if there is a large number of them coupled so that they are difficult to solve, it may be advantageous to leave them separated and solve the set of combined algebraic and differential equations numerically. Also, leaving the component equations uncombined makes it possible to extract information about the performance of components in the system that is not readily available from solution of combined equations.

As an example of a simple system that yields a single differential equation, consider a solar water heater with an unstratified storage unit supplying a load at a fixed flow rate and returning water back to the tank at a constant temperature T_{Lr}. This is illustrated in Figure 10.2.1. Equation 10.1.1 for the collector can be combined with Equation 10.1.3 for the tank to give

$$(mC_p)_s \frac{dT_s}{dt} = A_c F_R[S - U_L(T_s - T_a)]^+ - (UA)_s(T_s - T_a')$$

$$- \varepsilon_L(\dot{m}C_p)_{\min}(T_s - T_{Lr}) \tag{10.9.1}$$

Once the collector parameters, the storage size and loss coefficient, the magnitude of the load, and the meteorological data are specified, the storage tank temperature can be calculated as a function of time. Also, gain from the collector, losses from storage, and energy to load can be determined for any desired period of time by integration of the appropriate rate quantities.

Various methods are available to numerically integrate equations like Equation 10.9.1.[5] For example, simple Euler integration can be used, the same technique that was used in Example 8.3.1. Using simple Euler integration, we express the temperature derivative dT_s/dt as $(T_s^+ - T_s)/\Delta t$ and obtain an expression for the change in storage tank temperature for the time period in terms of known quantities. Equation 10.9.1 then becomes

$$T_s^+ = T_s + \frac{\Delta t}{(mC_p)_s} \{A_c F_R[S - U_L(T_s - T_a)]^+ - (UA)_s(T_s - T_a')$$

$$- \varepsilon_L(\dot{m}C_p)_{\min}(T_s - T_{Lr})\} \tag{10.9.2}$$

[5] Equation 10.9.1 can be integrated analytically by making various simplifying assumptions.

Integration schemes must be used with care to insure that they are stable for the desired time step and that reasonably accurate solutions are being attained. When performing hand calculations, both stability and accuracy can be problems. However, most computer facilities have subroutines that will solve systems of differential equations to a specified accuracy and automatically take care of stability problems.

In order to insure stability in Equation 10.9.2, the time step Δt must be small enough so that the coefficient on T_s is positive. The critical time step occurs when the coefficient on T_s is zero. In order to minimize truncation errors on a digital computer, use about one-sixth of the critical time step. However, weather data are usually known in hour increments so hour time steps are often used if stability is maintained.

Example 10.9.1

The performance of the collector of Example 6.7.1 was based upon a constant water supply temperature of 40 C to the collector. Assume the collector area is 4 m², F_R is 0.80 and U_L is 8.0 W/m²C. The collector is connected to a water storage tank containing 150 kg of water initially at 40 C. The storage tank loss coefficient-area product is 1.70 W/C and the tank is located in a room at 25 C. Assume water is withdrawn to meet a load at a constant rate of 10 kg/hr and is replenished from the mains at a temperature of 15 C. Calculate the performance of this system for the period from 7 AM to 5 PM using the collector and meteorological data from Example 6.11.1. Check the energy balance on the tank.

Solution

Equation 10.9.2 is used to calculate hourly temperatures of the tank. For this problem a time step of 1 hour is sufficient to guarantee stability. For this problem, the equations will be left uncombined in order that the individual terms in the energy balance can be seen.

$$T_s^+ = T_s + \frac{1}{150 \times 4190} \left\{ 4 \times 0.80 \left[S - 8.0 \times 3.6 \times 10^3 (T_s - T_a) \right]^+ \right.$$

$$\left. - 1.70 \times 3.6 \times (T_f - 25) - 10 \times 4190 (T_f - 15) \right\}$$

The first term in the bracket is the useful output of the collector and can have only positive values. The second term is the thermal loss from the tank. The third term is the energy delivered to the load. Simplifying the individual terms, this equation becomes

$$T_s^+ = T_s + 1.50 \{ 3.20[S - 0.0288(T_s - T_a)]^+$$

$$-0.00612(T_s - 25) - 0.04(T_s - 15) \}$$

where S is in MJ/m^2. The data in the first four columns are from Example 6.11.1. The new tank temperature calculated for the end of the hour is T_s^+. The three terms in the bracket of the equation are Q_u, load met by solar, and loss, assuming T_s to be fixed for the hour at its initial value. The change in internal energy of the water should be equal to $\Sigma Q_u - \Sigma L_S - \Sigma$ Losses.

$$150 \times 4190\ (53.0 - 40.0) = (23.43 - 1.41 - 13.87) \times 10^6$$

$$8.17\ \text{MJ} \cong 8.15\ \text{MJ}$$

Hour	I_T, MJ/m^2	S, MJ/m^2	T_a, C	T_s^+, C	Q_u, MJ	Loss, MJ	Load, MJ
Start				40.0			
7-8	0.02	–	−11	38.2	0	0.09	1.05
8-9	0.43	0.34	−8	36.5	0	0.08	0.97
9-10	0.99	0.79	−2	35.0	0	0.07	0.90
10-11	3.92	3.16	2	44.8	7.07	0.06	0.84
11-12	3.36	2.73	3	50.4	4.88	0.12	1.25
12-1	4.01	3.25	6	57.8	6.31	0.16	1.48
1-2	3.84	3.08	7	62.9	5.17	0.20	1.79
2-3	1.96	1.56	8	59.3	0	0.23	2.01
3-4	1.21	0.95	9	56.0	0	0.21	1.86
4-5	0.05	–	7	53.0	0	0.19	1.72
Total	19.79				23.43	1.41	13.87

This is a satisfactory check. (The calculations shown in the table were carried out to 0.01 MJ to facilitate checking the energy balance. The result is certainly no better than ±0.1 MJ.) The day's efficiency is

$$\eta_{\text{day}} = \frac{23.4}{19.8 \times 4} = 0.30, \text{ or } 30\%$$

∎

The preceding example was simple enough that hand calculation was possible to simulate a few hours of real time. Most problems in solar simulation are not so easy, and we are usually interested in more than just a few hours of simulated data. In general, it is necessary to use a computer and simulate over a season or a year to obtain useful solutions.

10.10   SOLAR FRACTION

System thermal analyses (and measurements) result in long-term system performance, which is then used in economic analyses. It is convenient in the economic analyses to express the solar energy contribution to the total load in terms of the fractional reduction in the amount of energy that must be purchased. Let the purchased energy with a zero area solar energy system (i.e., a fuel only system) be L_0, the auxiliary (purchased) energy for the system with solar energy be L_A, and the solar energy delivered be L_S. For the ith month, the fractional reduction of purchased energy when a solar energy system is used is

$$f_i = \frac{L_{0,i} - L_{A,i}}{L_{0,i}} = \frac{L_{S,i}}{L_{0,i}} \tag{10.10.1}$$

The same concept applies on an annual basis, with energy quantities integrated over the year.

$$\mathcal{F} = \frac{L_0 - L_A}{L_0} = \frac{L_S}{L_0} = \frac{\displaystyle\sum_{yr} f_i L_{0,i}}{\displaystyle\sum_{yr} L_{0,i}} \tag{10.10.2}$$

For many solar processes the energy delivered to meet loads does not depend on collector area. For other solar processes the total amount of energy that is supplied to meet a load is a function of collector area, i.e., the total energy delivered varies with the size of the solar energy system.

For processes such as active building heating systems the total energy supplied to meet the load L ($L = L_S + L_A$) is very close to L_0 regardless of the size of the solar heating system. In other words, the energy supplied from the solar system plus the auxiliary energy sources is essentially independent of A_c. For these systems the denominator of the solar fraction $\mathcal{F}$ is well defined and $\mathcal{F}$ is an obvious indicator of the reduction in energy purchases.

For some processes, the load L is a function of the size of the solar energy system. (For a passively heated building that has large south facing windows to admit solar radiation, the losses through the building envelope will vary with the receiver (i.e., window) area if the overall loss coefficient for the window is different than that of the walls.) With the total load a function of solar energy size, $L \neq L_0$ and the fraction L_S/L is not itself an index of reduction of energy purchase. In order to use the economic analysis methods to be presented in Chapter 11, the solar fraction must be defined as in Equation 10.10.2, i.e., as L_S/L_0.

For most active processes the definition of $\mathcal{F}$ is widely accepted and is unambiguous. For passive processes usage of $\mathcal{F}$ varies and care must be exercised to understand what a particular author means by the term.

10.11 SUMMARY

The concepts in this chapter are all concerned with techniques for combining component models (i.e., equations) into system models and obtaining solutions for the system models. The solutions of these sets of equation over time, using meteorological and load data as forcing functions, provides valuable information about the dynamics and the integrated performance of solar processes. These *simulations* can be viewed as numerical experiments in which the processes are "operated" in computers quickly and inexpensively. They are the applications of theory and, in combination with physical experiments on components and systems, provide the basis for much of our existing knowledge of solar applications.

Chapters in Part II, on applications, include many studies of effects of design parameters on seasonal or annual performance of systems, and much of this information is derived from simulations. Simulations, their validity, their utility, their use in design of processes, and computer programs for doing them will be noted in Chapter 19.

REFERENCES

Beckman, W. A., *Solar Energy*, **21**, 531 (1978). "Duct and Pipe Losses in Solar Energy Systems."

Beckman, W. A., S. A. Klein, and J. A. Duffie, *Solar Heating Design by the f-Chart Method*, Wiley, New York (1977).

Brinkworth, B. *Solar Energy*, **17**, 331 (1975). "Selection of Design Parameters for Closed-Circuit Forced-Circulation Solar Heating Systems."

Close, D. J. and M. B. Yusoff, *Solar Energy*, **20**, 459 (1978). "The Effects of Air Leaks on Solar Air Collector Behavior."

deWinter, F., *Solar Energy*, **17**, 335 (1975). "Heat Exchanger Penalties in Double-Loop Solar Water Heating Systems."

Hamming, R. W., *Numerical Methods for Scientists and Engineers*, McGraw-Hill, New York (1962).

Kays, W. M. and A. L. London, *Compact Heat Exchangers*, McGraw-Hill, New York (1964).

Kovarik, M. and P. F. Lesse, *Solar Energy*, **18**, 431 (1976). "Optimal Control of Flow in Low Temperature Solar Heat Collectors."

Oonk, R., D. E. Jones, and B. E. Cole-Appel, *Solar Energy*, **23**, 535 (1979). "Calculation of Performance of N Collectors in Series from Test Data on a Single Collector."

Winn, C. B. and D. E. Hull, *Proc. of the 1978 Annual Meeting of the American Section of ISES*, Denver, **2**(2), 493 (1978). "Optimal Controllers of the Second Kind."

Winn, C. B., in *Advances in Solar Energy*, (K. Boer and J. A. Duffie, eds.) Am. Solar Energy Soc. and Plenum Press, New York **1**, 209 (1982). "Controls in Solar Energy Systems."

Zarmi, Y., *Solar Energy*, **29**, 3 (1982). "Transition Time Effects in Collector Systems Coupled to a Well Mixed Storage."

Chapter 11

SOLAR PROCESS ECONOMICS

In the first 10 chapters, we have discussed in some detail the thermal performance of components and systems and showed how the long-term thermal performance can be estimated in terms of the design parameters of the components. We also want to be able to assess the value of a solar process in economic terms. Given the performance, we need methods for making economic evaluations.

Solar processes are generally characterized by high first cost and low operating costs. Thus the basic economic problem is one of comparing an initial known investment with estimated future operating costs. Most solar energy processes require an auxiliary (i.e., conventional) energy source so that the system includes both solar and conventional equipment and the annual loads are met by a combination of the sources. In essence, solar energy equipment is bought today to reduce tomorrow's fuel bill.

The cost of any energy delivery process includes all of the items of hardware and labor that are involved in installing the equipment plus the operating expenses. Factors which may need to be taken into account include interest on money borrowed, property and income taxes, resale of equipment, maintenance, insurance, fuel, and other operating expenses. The objective of the economic analysis can be viewed as the determination of the least cost method of meeting the energy need, considering both solar and nonsolar alternatives. For solar energy processes, the problem is to determine the size of the solar energy system that gives the lowest cost combination of solar and auxiliary energy.

In this chapter we note several ways of doing economic evaluations, with emphasis on the life cycle savings method. This method takes into account the time value of money and allows detailed consideration of the complete range of costs. It is introduced by an outline of cost considerations, note of economic figures of merit (design criteria), and comments on design variables which are important in determining system economics. For additional discussion of economic analyses, see Riggs (1968), De Garmo and Canada (1973), Ruegg (1975), and White et al. (1977).

Section 11.8 of the chapter describes the P_1, P_2 method of doing life cycle savings analyses. This is a quick and convenient way of carrying out the computations described in detail in earlier sections and is the method used in economic analyses of particular processes in following chapters.

11.1   COSTS OF SOLAR PROCESS SYSTEMS

Investments in buying and installing solar energy equipment are important factors in solar process economics. These include the delivered price of equipment such as collectors, storage unit, pumps and blowers, controls, pipes and ducts, heat exchangers, and all other equipment associated with the solar installation. Costs of installing this equipment must also be considered, as these can match or exceed the purchase price. Also to be included are costs of structures to support collectors and other alterations made necessary by the solar energy equipment. Under some circumstances credits may be taken for the solar process if its installation results in reduction of costs; for example, a collector may serve as part of the weatherproof envelope of a building, eliminating the need for some of the conventional siding or roofing.

Installed costs of solar equipment can be shown as the sum of two terms, one proportional to collector area and the other independent of collector area:

$$C_S = C_A A_C + C_E \tag{11.1.1}$$

where C_S = total cost of installed solar energy equipment ($)
C_A = total area-dependent costs ($/m^2)
A_C = collector area (m^2)
C_E = total cost of equipment which is independent of collector area ($)

The area-dependent costs C_A include such items as the purchase and installation of the collector and a portion of storage costs. The area-independent costs C_E include such items as controls and bringing construction or erection equipment to the site, which do not depend on collector area.

Operating costs are associated with a solar process. These continuing costs include cost of auxiliary energy; energy costs for operation of pumps and blowers (often termed parasitic energy, which should be minimized by careful design); extra real estate taxes imposed on the basis of additional assessed value of a building or facility; interest charges on any funds borrowed to purchase the equipment; and others.

There may be income tax implications in the purchase of solar equipment. In the United States, interest paid on a mortgage[1] for its purchase and extra property tax on an increased assessment due to solar equipment are both deductible from income for tax purposes if the owner itemizes his deductions. States may allow similar deductions. The income tax reduction associated with these payments depends on the tax bracket of the owner and serves to reduce the cost of the solar process.[2]

---

[1] In this chapter we use the term "mortgage," commonly applied to funds borrowed for building projects, for any loan for purchase of solar energy equipment.
[2] This chapter is written with U.S. tax law in mind. For other countries, treatment of tax implications will have to be modified.

Equipment purchased by businesses have other tax implications. Income-producing property and equipment may be depreciated, resulting in reduced taxable income and thus reduced income tax. But the value of fuel saved by the use of solar equipment is effectively reduced because a business already deducts the cost of fuel from its income for tax purposes. If the equipment is for purposes other than building heating or air conditioning, there may be investment tax credits available in the first year. Further, the equipment may have salvage or resale value which may result in a capital gains tax. Finally, federal and state governments may allow special tax credits to encourage the use of solar energy. (Federal and state tax laws relating to solar energy are often changed, and current law should be used in any economic analysis.)

In equation form, the annual costs for both solar and nonsolar systems to meet an energy need can be expressed as

$$\text{Yearly cost} = \text{Fuel expense} + \text{Mortgage payment}$$
$$+ \text{Maintenance and insurance} + \text{Parasitic energy cost}$$
$$+ \text{Property taxes} - \text{Income tax savings} \qquad (11.1.2)$$

Fuel expense is for energy purchase for auxiliary or for the conventional (nonsolar) system. The mortgage payment includes interest and principal payment on funds borrowed to install the system. Maintenance and insurance are recurring costs to keep a system in operating condition and protected against fire or other losses. Parasitic energy costs are for blowing air or pumping liquids and other electrical or mechanical energy uses in a system. Property taxes are levied on many installations. Income tax savings for a non-income-producing system (such as a home heating system) can be expressed as

$$\binom{\text{Income tax}}{\text{savings}} = \binom{\text{Effective}}{\text{tax rate}} \times (\text{Interest payment} + \text{Property tax}) \quad (11.1.3)$$

If the system is an income-producing installation,

$$\binom{\text{Income tax}}{\text{savings}} = \binom{\text{Effective}}{\text{tax rate}} \times \begin{cases} \text{Interest payments} \\ + \text{Property tax} \\ + \text{Fuel expense} \\ + \text{Maintenance and insurance} \\ + \text{Parasitic energy costs} \\ + \text{Depreciation} \end{cases} \quad (11.1.4)$$

State income taxes are deductible from income for federal tax purposes. Where federal taxes are not deductible for state tax purposes, the effective tax rate is given by

$$\binom{\text{Effective}}{\text{Tax rate}} = \binom{\text{Federal}}{\text{tax rate}} + \binom{\text{State}}{\text{tax rate}} - \binom{\text{Federal}}{\text{tax rate}} \times \binom{\text{State}}{\text{tax rate}} \qquad (11.1.5)$$

The concept of **solar savings**, as outlined by Beckman et al. (1977), is a useful one. Solar savings are the difference between the cost of a conventional system and a solar system. (Savings can be negative; they are then losses.) In equation form it is simply

$$\text{Solar savings} = \begin{pmatrix} \text{Costs of} \\ \text{conventional} \\ \text{energy} \end{pmatrix} - \begin{pmatrix} \text{Costs of} \\ \text{solar} \\ \text{energy} \end{pmatrix} \qquad (11.1.6)$$

In this equation it is not necessary to evaluate costs that are common to both the solar and the nonsolar system. For example, the auxiliary furnace and much of the ductwork or plumbing in a solar heating system are often the same as would be installed in a nonsolar system. With the savings concept, it is only necessary to estimate the incremental cost of installing a solar system. If the furnaces or other equipment in the two systems are different, the difference in their costs can be included as an increment or decrement to the cost of installing a solar system. Solar savings can be written as

$$
\begin{aligned}
\text{Solar savings} = \text{Fuel savings} & \\
& - \text{Incremental mortgage payment} \\
& - \text{Incremental insurance and maintenance} \\
& - \text{Incremental parasitic energy cost} \\
& - \text{Incremental property tax} \\
& + \text{Income tax savings} \qquad (11.1.7)
\end{aligned}
$$

The significance of the terms is the same as for Equation 11.1.2, except that here they refer to the increments in the various costs, that is, the differences between the costs for the solar energy system compared to a nonsolar system. Equations analogous to 11.1.3 and 11.1.4 can be written for the last term in Equation 11.1.7. They are, for a non-income-producing system,

$$\begin{pmatrix} \text{Income tax} \\ \text{savings} \end{pmatrix} = \begin{pmatrix} \text{Effective} \\ \text{tax rate} \end{pmatrix} \times \left[\begin{pmatrix} \text{Incremental} \\ \text{interest} \\ \text{payment} \end{pmatrix} + \begin{pmatrix} \text{Incremental} \\ \text{property} \\ \text{tax} \end{pmatrix} \right] \qquad (11.1.8)$$

and for an income-producing system

$$\begin{pmatrix} \text{Income tax} \\ \text{savings} \end{pmatrix} = \begin{pmatrix} \text{Effective} \\ \text{tax rate} \end{pmatrix} \times \begin{pmatrix} \text{Incremental interest payment} \\ + \text{Incremental property tax} \\ + \text{Incremental maintenance and insurance} \\ + \text{Incremental parasitic energy cost} \\ + \text{Incremental depreciation} \\ - \text{Value of fuel saved} \end{pmatrix} \qquad (11.1.9)$$

Fuel saved is a negative tax deduction since a business already deducts fuel expenses; the value of fuel saved is effectively taxable income.

11.2 DESIGN VARIABLES

The economic problem in solar process design is to find the lowest cost system. In principle, the problem is a multivariable one, with all of the components in the system and the system configuration having some effect on the thermal performance and thus on cost. The design of the load system (the building, the industrial process using energy, or other load) must also be considered in the search for optimum design. Barley (1978) and Balcomb (1986) have investigated the economic trade-off between energy conservation and solar heating. In practice, the problem often resolves to a simpler one of determining the size of a solar energy system for a known load, with storage capacity and other parameters fixed in relationship to collector area. Given a load that is some function of time through a year, a type of collector, and a system configuration, the primary design variable is collector area. System performance is much more sensitive to collector area than to any other variable.

Three examples of the dependence of annual thermal performance on collector area are shown in Figure 11.2.1 for a solar heating operation. Curve A is for a system with a two-cover, selective-surface collector, while B is a for a one-cover, nonselective collector. Curve C is for double the storage capacity and the same collector as for curve A. Figure 11.2.2 shows the dependence of annual thermal performance on storage capacity for 60 m² of type A collectors.

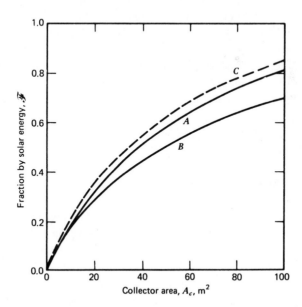

Figure 11.2.1 An example of annual fraction of heating loads carried by an active solar heating system for a building in Madison, WI. Curve A is for a system with a two-cover, selective-surface collector. B is for one with a one-cover, nonselective collector. C is the same system as A but with twice the storage capacity.

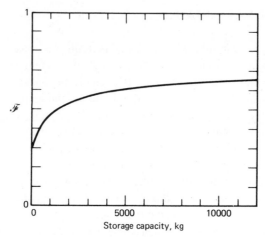

Figure 11.2.2 Annual solar fraction for the collector of type A of Figure 11.2.1 for a collector area of 60 m^2 as a function of storage capacity.

For this active solar heating example, relative sensitivity of annual performance to collector area and relative insensitivity to the differences in these two collectors are apparent. The solar contribution is relatively insensitive to storage capacity within the range shown in Figure 11.2.2 once a critical storage capacity (about 3000 kg in this example) is exceeded. Similar results are found for solar heating for a wide range of locations. It is difficult to generalize this experience with active solar heating systems to other applications. The general procedure for determining which variables are most critical to thermal performance is to do a sensitivity analysis such as that of Close (1967). The economic analysis for solar heating to meet a given load is simplified by the fact that collector area is the primary design parameter; economic analyses for other solar processes must take into account the possibility that other design variables might be of comparable importance.

11.3   ECONOMIC FIGURES OF MERIT

Several economic criteria have been proposed and used for evaluating and optimizing solar energy systems, and there is no universal agreement on which should be used. This section outlines some of the possible figures of merit; two of them are discussed in more detail in following sections, and one of these, maximum life cycle savings, is applied in later chapters on solar energy applications.

Least cost solar energy is a reasonable figure of merit for systems in which solar energy is the only energy resource. The system yielding least cost can be defined as that showing minimum owning and operating cost over the life of the system, considering solar energy only. However, the optimum design of a combined solar plus auxiliary energy system based on minimum total cost of delivering energy will generally be different from that based on least cost solar energy, and the use of least cost solar energy as a criterion is not recommended for systems using solar in combination with other energy sources.

Life cycle cost (LCC) is the sum of all the costs associated with an energy delivery system over its lifetime or over a selected period of analysis, in today's dollars, and takes into account the time value of money. The basic idea of life cycle costs is that anticipated future costs are brought back to present cost (discounted) by calculating how much would have to be invested at a market discount rate[3] to have the funds available when they will be needed. A life cycle cost analysis includes inflation when estimating future expenses. This method can include only major cost items or as many details as may be significant. **Life cycle savings (LCS) (net present worth)** is defined as the difference between the life cycle costs of a conventional fuel-only system and the life cycle cost of the solar plus auxiliary energy system. Life cycle savings analysis is outlined in Sections 11.6 and 11.8 and is applied in later chapters to solar processes.

A special case of life cycle savings is based on cash flow, that is, the sum of the items on the right-hand side of Equation 11.1.2. The major items of cash flow are principal payment, interest, taxes, and insurance, referred to as **PITI**; this is compared to reduction in fuel costs. Cash flow may be an important consideration in residential solar heating applications where the willingness of lending institutions to provide mortgage funds may be dependent on a borrower's ability to meet periodic obligations. The main impacts of solar heating will be increased mortgage payments and decreased fuel costs.

Annualized life cycle cost (ALCC) is the average yearly outflow of money (cash flow). The actual flow varies with year, but the sum over the period of an economic analysis can be converted to a series of equal payments in today's dollars that are equivalent to the varying series. The same ideas apply to an **annualized life cycle savings (ALCS)**.

Payback time is defined in many ways. Below are listed several which may be encountered; these are illustrated in Figure 11.3.1.

A   The time needed for the yearly cash flow to become positive.
B   The time needed for the cumulative fuel savings to equal the total initial investment, that is, how long it takes to get an investment back by savings in fuel. The common way to calculate this payback time is without discounting the fuel savings. It can also be calculated using discounted fuel savings.
C   The time needed for the cumulative savings to reach zero.
D   The time needed for the cumulative savings to equal the down payment on the solar energy system.
E   The time needed for the cumulative solar savings to equal the remaining debt principal on the solar energy system.

The most common definition of payback time is **B**. Each of these payback times may have significance in view of economic objectives of various solar process

---

[3] Market discount rate is the rate of return on the best alternative investment. See Section 11.4.

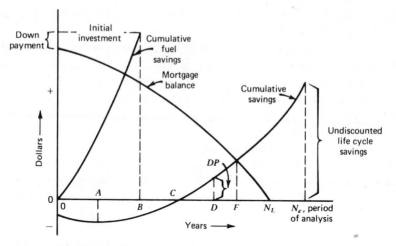

Figure 11.3.1 Changes in mortgage balance, cumulative fuel savings, and cumulative savings as a function of time through the period of a life cycle cost analysis.

users. Calculation of payback periods can be done including only major items or including many details. Care must be used in interpreting reported payback periods that the definition of the period and the items included in it are fully understood. Calculation of payback times is discussed in Section 11.8.

    Return on investment (ROI) is the market discount rate which results in zero life cycle savings, that is, the discount rate that makes the present worth of solar and nonsolar alternatives equal. This is illustrated in Figure 11.3.2, which shows an example of life cycle savings as a function of market discount rate.

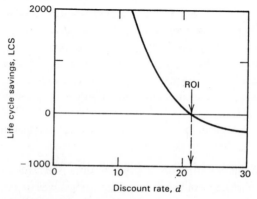

Figure 11.3.2 An example of life cycle savings as a function of market discount rate for a solar energy application.

11.4   DISCOUNTING AND INFLATION

The most complete approach to solar process economics is to use life cycle cost methods that take into account all future expenses. This method provides a means of comparison of future costs with today's costs. This is done by discounting all anticipated costs to the common basis of present worth (or present value), that is, what would have to be invested today at the best alternative investment rate to have the funds available in the future to meet all of the anticipated expenses.

Conceptually, in a life cycle cost analysis, all anticipated costs are tabulated and discounted to a present worth; the life cycle cost is the sum of all of the present worths. As a practical matter, the calculations can be simplified. For example, the cash flow (net payment) for each year can be calculated and the life cycle cost found by discounting each annual cash flow to its present value and finding the sum of these discounted cash flows. When the present values of all future costs have been determined for each of the alternative systems under consideration, including solar and nonsolar options, the system that yields the lowest life cycle cost is selected as the most cost effective. Life cycle costing requires that all costs be projected into the future; the results obtained from analyses of this type usually depend very much on predictions of future costs.

The reason that cash flow must be discounted lies in the "time value of money." The relationship of determining the present worth of one dollar needed N periods (usually years) in the future, with a market discount rate of d (fraction per time period), is

$$PW = \frac{1}{(1+d)^N} \qquad (11.4.1)$$

Thus an expense that is anticipated to be $1.00 in 5 years is equivalent to an obligation of $0.681 today at a market discount rate of 8%. To have $1000 available in 5 years, it would be necessary to make an investment of $681 today at an annual rate of return of 8%.

Many recurring costs can be assumed to inflate (or deflate) at a fixed percentage each period. Thus an expense of $1.00, when inflated at a rate i per time period, will be $1 + i$ at the end of one time period, $(1 + i)^2$ at the end of an additional time period, etc. If a cost A is considered to be incurred as of the end of the first time period (e.g., a fuel bill is to be paid at the end of the month or year), that recurring cost at the Nth period is

$$C_N = A(1 + i)^{N-1} \qquad (11.4.2)$$

Thus a cost which will be $1.00 at the end of the first period and inflates at 6% per year will at the end of five periods be $(1 + 0.06)^4$, or $1.26.

The progression of a series of payments which are expected to inflate at a rate i is shown in Figure 11.4.1. The first payment in the series is A (payable at the end

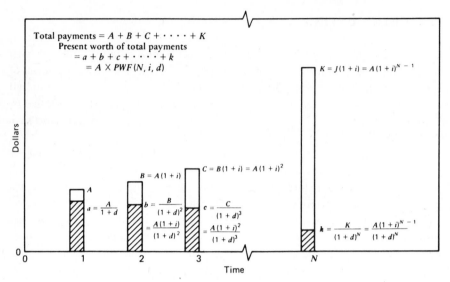

Figure 11.4.1 Present worth of a geometric series of inflating payments. Each payment is made at the end of a period. The bars show both the payment and the present worth of the payment.

of the first period), the second is B, and the Nth is K. The payments are made at the ends of the periods. The shaded portions of the bars show the present worth of the anticipated payment. At the Nth period the cost is $A(1 + i)^{N-1}$, and the present worth of the Nth cost is

$$PW_N = \frac{A(1 + i)^{N-1}}{(1 + d)^N} \tag{11.4.3}$$

This equation is useful for calculating the present worth of any one payment of a series of inflating payments. It is also useful for calculating the present worth of any one-time costs that are anticipated during the period of the analysis and for which the cost at the end of the first period is known.

Example 11.4.1

a A fuel cost is expected to inflate at the rate of 7% per year, and for the first year is $400 (payable at the end of the year). The market discount rate is 10% per year. What is the present worth of the payment to be made at the end of the third year? **b** It is expected that the blower in an air heater system will need to be replaced at the end of 10 years. The cost in the first year (payable at the end of the year) would be $300. Under the same assumption of inflation and discount rates, what is the present worth of replacing the blower?

Solution

a Using Equation 11.4.3, the present worth of the third-year fuel payment is

$$PW_3 = 400(1.07)^2/(1.10)^3 = \$344$$

b The present worth of the blower replacement after 10 years is

$$PW_{10} = 300 \ (1.07)^9/(1.10)^{10} = \$213$$ ∎

The analyses and examples in this chapter are all based on the premise that the costs are known and payable at the end of the first time period. If a cost A' is known as of the beginning of the first time period, Equation 11.4.2 becomes

$$C_N = A'(1 + i)^N \tag{11.4.4}$$

and Equation 11.4.3 becomes

$$PW_N = A'\left(\frac{1+i}{1+d}\right)^N \tag{11.4.5}$$

11.5 PRESENT WORTH FACTOR

In the previous section we dealt with calculating the present worth of a single future payment. If an obligation recurs each year and inflates at a rate i per period, a present worth factor, PWF, of the series of N such payments can be found by summing Equation 11.4.3 over N periods:

$$PWF(N, i, d) = \sum_{j=1}^{N} \frac{(1 + i)^{j-1}}{(1 + d)^j} = \begin{cases} \frac{1}{(d-i)}\left[1 - \left(\frac{1+i}{1+d}\right)^N\right] & \text{if } i \neq d \\[2mm] N/(1+i) & \text{if } i = d \end{cases}$$

$$\tag{11.5.1}$$

If the function of Equation 11.5.1 is multiplied by the first of a series of payments that are made at the end of the periods, the result is the sum of N such payments discounted to the present with a discount rate of d. This is represented by the sum of the cross-hatched bars in Figure 11.4.1. This function is tabulated in Appendix F for N from 5 to 30 in increments of 5, for values of i from 1 to 12% per period, and for values of d from 0 to 20% per period.

Example 11.5.1

What is the present worth of a series of 20 yearly payments, the first of which is $500, which are expected to inflate at the rate of 8% per year if the market discount rate is 10% per year?

Solution

From Equation 11.5.1,

$$PW = 500 \frac{1}{0.10 - 0.08} \left[1 - \left(\frac{1.08}{1.10}\right)^{20} \right] = 500 \times 15.35892 = \$7679$$

or, from Appendix F, the present worth factor for $N = 20$, $i = 0.08$, and $d = 0.10$ is 15.359, and the present worth of the series is $15.359 \times 500 = \$7680$. ■

   The present worth factor defined in Equation 11.5.1 can be used to find the periodic loan payment on a fixed-rate mortgage which involves a series of N_L equal payments over the lifetime of the loan. Since all mortgage payments are equal, we have a series of payments with an inflation rate of zero. The discount rate in Equation 11.5.1 becomes the mortgage interest rate. Thus the periodic loan payment[4] is

$$\text{Periodic payment} = \frac{M}{PWF(N_L, 0, m)} \qquad (11.5.2)$$

where m is the mortgage interest rate, N_L is the period of the mortgage, and M is the mortgage principal.

Example 11.5.2

What is the annual payment and yearly interest charge if an $11,000 solar installation is to be financed by a 10% down payment with the balance borrowed at an annual interest rate of 9% for 20 years? The payments are to be made at the end of the year. The market discount rate is 8%. What is the present worth of the series of interest payments?

Solution

The present worth factor is used in Equation 11.5.2 to calculate the annual payment. The present worth of the sum of all payments is the mortgage, or $0.9 \times 11,000 = \$9900$. The yearly payment is

$$\frac{9900}{PWF(20, 0, 0.09)} = \frac{9900}{9.129} = \$1084.46$$

---

[4] In this and following sections the examples are based on periods of a year. The identical principles and methods hold for monthly (or other) periods. Interest, inflation, and discount rates must correspond to the time period used.

Year	Mortgage Payment	Remaining Principal	Interest Payment	PW of Interest Payment
1	1084	9707	891	825
2	1084	9496	874	749
3	1084	9266	855	678
4	1084	9015	834	613
–	–	–	–	–
–	–	–	–	–
20	1084	0	90	19

Total present worth of interest payments = $6730

The interest charge varies with time, since the mortgage payment includes a principal payment and interest. In this example, the interest for the first year is $0.09 \times 9900 =$ 891. The payment is $1084.46, so the principal is reduced by $193.46 (i.e., $1084.46 - 891$) to $9706.54. The second year's interest is $0.09 \times 9706.54 =$ $873.59 and the principal is reduced by $210.87 to $9495.67. This progression of payments, remaining principal, interest payment, and the present worth of the interest payment is shown in the table (to the nearest dollar).   ■

The total present worth of all of the mortgage interest payments can be calculated as in Example 11.5.2, but it is tedious. Present worth factors can be used to obtain this total by the following equation:

$$PW_{int} = M\left[\frac{PWF(N_{min}, 0, d)}{PWF(N_L, 0, m)} + PWF(N_{min}, m, d)\left(m - \frac{1}{PWF(N_L, 0, m)}\right)\right]$$

(11.5.3)

where M is the initial mortgage principal, N_L is the term of mortgage, N_e is the term of economic analysis, N_{min} is the lesser of N_L or N_e, m is the mortgage interest rate, and other terms are as previously defined. Note that it is not necessary that the term of the economic analysis coincide with the term of the mortgage.

Example 11.5.3

Calculate the total present worth of all of the mortgage interest payments in Example 11.5.2 over the term of the mortgage.

Solution

Here $N_L = N_e = 20$, $m = 0.09$, and $d = 0.08$. The PWF values can be calculated from Equation 11.5.1 or they can be obtained from Appendix F. Using Equation 11.5.3,

$$PW_{int} = 9900\left[\frac{9.818}{9.129} + 20.242\left(0.09 - \frac{1}{9.129}\right)\right] = \$6731$$

This is in agreement with the value obtained by summing the individual present worths in Example 11.5.2. ∎

11.6   LIFE CYCLE SAVINGS METHOD

The previous two sections dealt with discounting of future costs. Now these ideas are applied in a series of examples to illustrate the principles and steps in life cycle savings analysis. The first two examples are for fuel payments for a conventional process. The next is a calculation of solar savings, the difference in present worth of a solar plus fuel system and the fuel-only system. The last is a calculation of optimum system design based on solar savings calculations for several collector areas.[5]

These examples are intended to illustrate the method, and the particular costs used are not intended to have significance. These costs vary widely from one location to another, with time, as state and federal legislation is enacted which impacts the costs of solar equipment, and as international developments occur which affect the prices of energy.

Example 11.6.1

For a nonsolar process using fuel only, calculate the present worth of the fuel cost over 20 years if the first year's cost is $1255 (i.e., 125.5 GJ at $10.00/GJ). The market discount rate is 8% per year and the fuel cost inflation rate is 10% per year.

Solution

A tabulation of the yearly progression of fuel costs and their present worth follows. Each year's fuel cost is the previous year's cost multiplied by $(1 + i)$. Each item in the PW column is calculated from the corresponding item in the Fuel Cost column using Equation 11.4.1. The same result is obtained by multiplying $PWF(20, 0.10, 0.08)$ by the first year's fuel cost.

Year	Fuel Cost	PW of Fuel Cost
1	1255	1162
2	1381	1183
3	1519	1206
.	.	.
.	.	.
20	7675	1647
	Total present worth of fuel cost = $27,822	

∎

---

[5] Section 11.8 shows an additional and more convenient way of approaching these same calculations.

The next example also shows a calculation of the present worth of a series of fuel costs, but in this case the costs of fuel are expected to inflate at rates dependent on time. It illustrates the general principle that costs need not be expected to change at fixed rates. The variation can be anything from nil to completely irregular; the basic ideas of life cycle cost hold whether payments are regular or otherwise.

Example 11.6.2

For a nonsolar process, using fuel only, calculate the present worth of fuel cost over 8 years if the first year's cost is $1200, it inflates at 10% per year for 3 years, and then it inflates at 6% per year. The market discount rate is 8% per year.

Solution

A tabulation of the yearly progression of fuel costs and their present worth follows. As in the previous example, each item in the Present Worth column is calculated from the corresponding item in the Fuel Cost column using Equation 11.4.1. In this example, the sum of the expected payments is 43% more than the present worth of the payments. Note that the present worths increase during the time when i is greater than d, and decrease when i is less than d.

Year	Fuel Cost	PW of Fuel Cost
1	1200	1111
2	1320	1132
3	1452	1153
4	1597	1174
5	1693	1152
6	1795	1131
7	1902	1110
8	2016	1089
	12,975	9052

Alternate Solution

This example can also be solved using present worth factors by considering the series to be in two sets, one at each of the inflation rates. Part of the present worth is that of the first set of four payments, the first of which is $1200 and which inflate at 10%. The other part is the present worth of the next four, which inflate at 10% per year. From Equation 11.5.1,

$$PWF(4,0.10,0.08) = \frac{1}{0.08 - 0.10}\left[1 - \left(\frac{1.10}{1.08}\right)^4\right] = 3.808$$

The present worth of this first set is thus $1200 \times 3.808 = \$4569$.
The second set starts at the beginning of the fifth year, when i is 6% per year.

The initial payment in this set is A', the fifth-year payment. To find A', the initial payment is inflated three times by 1.10 and once by 1.06. Thus

$$A' = 1200(1.10)^3(1.06) = \$1693$$

The *PWF* for this part of the series as of the beginning of year 5 is

$$PWF(4,0.06,0.08) = \frac{1}{0.08 - 0.06}\left[1 - \left(\frac{1.06}{1.08}\right)^4\right] = 3.602$$

The second set in the series is then discounted to the present by

$$PW = \frac{3.602 \times 1693}{(1.08)^4} = \$4482$$

The sum of the present worths of the two sets is $4569 + 4482 = \$9051$. ■

With either Equation 11.4.1 or 11.5.1 it is possible to discount any future cost or series of costs to a present worth. In the same way, future savings (negative costs) can be discounted to a present worth. In the following examples we apply these methods to systems using combined solar and auxiliary (conventional) sources.

Example 11.6.3

A combined solar and fuel system to meet the same energy need as in Example 11.6.1 is to be considered. The proposed collector and associated equipment will supply energy so as to reduce fuel purchase by 56%, will cost $11,000, and will be 90% financed over 20 years at an interest rate of 9%. The first year's fuel cost for a system without solar would be $1255. Fuel costs are expected to rise at 10% per year. It is expected that the equipment will have a resale value at the end of 20 years of 40% of the original cost.

In the first year, extra insurance, maintenance, and parasitic energy costs are estimated to be $110. Extra property tax is estimated to be $220. These are expected to rise at a general inflation rate of 6% per year. Extra property taxes and interest on mortgage are deductible from income for tax purposes; the effective income tax rate is expected to be 45% through the period of the analysis.

What is the present worth of solar savings for this process over a 20-year period if the market discount rate is 8%?

Solution

The table that follows shows the incremental yearly costs and savings. Year 1 includes the estimates of the first year's costs as outlined in the problem statement. The annual payment on the $9900 mortgage is calculated as 9900/*PWF*(20, 0, 0.09), or $1084.46. The entries in the tax savings column are calculated by Equation

11.1.3; for example, in years 1 and 2,

Interest in **year 1** = 0.09(9900) = $891.00
Principal payment = 1084.46 – 891.00 = $193.46
Principal balance = 9900 – 193.46 = $9706.54
Tax savings = 0.45(891.00 + 220) = $500.00
Interest in **year 2** = 0.09(9706.54) = $873.59
Principal payment = 1084.46 – 873.59 = $210.87
Principal balance = 9706.54 – 210.87 = $9495.67
Tax savings = 0.45(873.59 + 220(1.06)) = $498.06

Solar savings for each year are the sums of the items in columns 2 through 6. Each year's solar savings is brought to a present worth using a market discount rate of 8%. The down payment is $1100 and is entered in the table in year 0, i.e., now; it is a negative present worth in solar savings. The resale value of $4400 in year 20 is shown as a second entry in year 20 and is positive as it contributes to savings. The sum of the last column, $4203, is the total present worth of the gains from the solar energy system compared to the fuel-only system, and is termed "life cycle solar savings" (*LCS*), or simply "solar savings."

Year	Fuel Savings	Extra Mort. Pt.	Extra Ins., Maint., Energy	Extra Prop. Tax	Income Tax Savings	Solar Savings	PW of Solar Saving
0						−1100	−1100
1	703	−1084	−110	−220	500	−211	−195
2	773	−1084	−117	−233	498	−163	−140
3	850	−1084	−124	−247	496	−109	−87
4	935	−1084	−131	−262	493	−49	−36
5	1029	−1084	−139	−278	490	18	12
--	--	--	--	--	--	--	--
--	--	--	--	--	--	--	--
20	4298	−1084	−333	−666	340	2555	548
20						4400	944

Total present worth of solar savings = $4203

■

Another approach to this problem is to do a life cycle cost analysis of both the solar and the nonsolar systems. The difference in the present worths of the two systems (that is, the difference in the life cycle costs) is the life cycle solar savings. More information is needed to do the analyses separately if they are to be complete, as equipment common to both systems should be included. In the calculation of Example 11.6.3 common costs do not influence life cycle solar savings.

In the previous examples, collector area (i.e., system cost) and annual fraction of loads met by solar were given. In Example 11.6.4, the relationship between solar fraction $\mathcal{F}$ and the area from thermal performance calculations is given, as are C_A and C_E, and the life cycle solar savings are calculated for various sizes of solar energy systems to find the collector size (combination of solar and auxiliary) which provides the highest savings.

Example 11.6.4

A thermal analysis of the process of Example 11.6.3 indicates a relationship between collector area and solar fraction $\mathcal{F}$ as indicated in the first two columns of the table that follows. Area dependent costs are $200/m², and fixed cost is $1000. All economic parameters are as in Example 11.6.3. What is the optimum collector area which shows the maximum life cycle solar savings relative to the fuel-only system?

Solution

The costs of the solar energy system are calculated by Equation 11.1.1, with $C_A =$ $200/m² and C_E = $1000. The third column indicates the total system cost. Calculations of the kind shown in Example 11.6.3 give column 4, life cycle solar savings, as a function of collector area. A calculation is made for a very small collector area, where the cost of the system is essentially C_E, to establish the "zero area" solar savings. The solar savings of column 4 are plotted in the figure. The maximum savings are realized at a collector area of about 39 m², and positive savings are realized over an area range of approximately 3 to 110 m².

Area m²	Solar Fraction $\mathcal{F}$	Installed Cost	Solar Savings
0.01	0	1,000	−1,036
25	0.37	6,000	4,088
39	0.49	8,800	4,531
50	0.56	11,000	4,203
75	0.71	21,000	3,204
150	0.92	31,000	−6,468

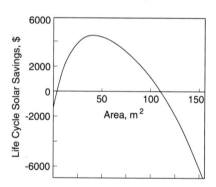

These examples show only regularly varying costs. All anticipated costs can be included, whether they are recurring, regularly varying, or however they may be incurred. The method, in essence, is to construct a table such as that in Example 11.6.3 and sum the last column. Nonrecurring items such as expected replacement covers in years hence, changes in income tax bracket, or other increments can be included in the present worth calculation, as were the down payment and resale value in the examples. Example 11.6.3 shows expenses of types associated with non-income-producing applications. Income-producing applications will require that additional terms be taken into account, for example, depreciation.

The assumptions made in these examples are not to be construed as representative or typical of situations that may be encountered. They are intended only to illustrate the methods of calculation. Consideration of the results of these examples, and those in following sections, indicates that the end results, the solar savings, are very sensitive[6] to the assumptions made in the calculations.

11.7 EVALUATION OF OTHER ECONOMIC INDICATORS

The principles of the preceding sections can be used to find payback times. Payback time **B** (Section 11.3) is the time needed for the cumulative fuel savings to equal the total initial investment in the system. Consider first the case where fuel savings are not discounted. The fuel savings in the jth year are given by $\mathcal{F}LC_{F1}(1 + i_F)^{j-1}$, where $\mathcal{F}L$ is the energy saved, C_{F1} is the first period's unit energy cost delivered from fuel,[7] and i_F is the fuel cost inflation rate. Summing these over the payback time N_p and equating to the initial investment as given by Equation 11.1.1,

$$\sum_{j=1}^{N_p} \mathcal{F}LC_{F1}(1 + i_F)^{j-1} = C_S \tag{11.7.1}$$

Summing the geometric series

$$\frac{\mathcal{F}LC_{F1}\left[(1 + i_F)^{N_p} - 1\right]}{i_F} = C_S \tag{11.7.2}$$

This can be solved for N_p, the payback period:

$$N_p = \frac{\ln\left[\dfrac{C_S i_F}{\mathcal{F}LC_{F1}} + 1\right]}{\ln(1 + i_f)} \tag{11.7.3}$$

The present worth factors of Appendix F can also be used to find this payback period. The sum of the fuel savings is the first year's saving, $\mathcal{F}LC_{F1}$, times the *PWF* at zero discount rate:

$$\mathcal{F}LC_{F1} \times PWF(N_p, i_F, 0) = C_S \tag{11.7.4}$$

By interpolating from the tables in Appendix F, the value of N_p can be found for which the *PWF* is $C_S/\mathcal{F}LC_{F1}$.

---

[6] Calculation of the sensitivity of solar savings to economic assumptions will be discussed in Section 11.9.

[7] The energy delivered from a unit of fuel is the product of the fuel lower heating value times the furnace efficiency.

Example   11.7.1

What is the undiscounted payback time **B** for an $11,000 investment in solar energy equipment which meets 56% of an annual load of 156 GJ? The first-year fuel cost is $8.00/GJ and is expected to inflate at 10% per year.

Solution

Using Equation 11.7.3,

$$N_p = \frac{\ln\left[\dfrac{11{,}000 \times 0.10}{0.56 \times 156 \times 8} + 1\right]}{\ln(1.10)} = 9.9 \text{ years} \qquad \blacksquare$$

By similar procedures it is possible to equate discounted fuel costs to initial investment. If fuel costs are discounted, an equation like 11.7.4 can be written

$$PWF(N_p, i_F, d) = \frac{C_S}{\mathcal{F}LC_{F1}} \tag{11.7.5}$$

and the appropriate value of N_p can be found (where $i_F \neq d$) from

$$N_p = \frac{\ln\left[\dfrac{C_S(i_F - d)}{\mathcal{F}LC_{F1}} + 1\right]}{\ln\left(\dfrac{1 + i_F}{1 + d}\right)} \tag{11.7.6}$$

and if $i_F = d$,

$$N_p = \frac{C_S(1 + i_F)}{\mathcal{F}LC_{F1}} \tag{11.7.7}$$

Example   11.7.2

Repeat Example 11.7.1 but discount future fuel costs at a rate of 8% per year.

Solution

Using Equation 11.7.6,

$$N_p = \frac{\ln\left[\dfrac{11{,}000(0.10 - 0.08)}{0.56 \times 156 \times 8} + 1\right]}{\ln\left(\dfrac{1 + 0.10}{1 + 0.08}\right)} = 14.9 \text{ years} \qquad \blacksquare$$

Other payback times are defined in Section 11.3 in terms of cumulative savings. For these it is necessary to use Equation 11.1.7, including in the equation whatever terms are significant, calculating the solar savings each year, and finding the year in which the cumulative savings meet whatever criteria is established for the particular payback time desired. Table 11.7.1 shows information from Example 11.6.3 and illustrates the several payback periods defined in Section 11.3.

A   The solar savings become positive (and the cumulative solar savings reach a minimum) by year 5.
B   The undiscounted cumulative fuel savings exceed the total initial investment in year 10. This is in agreement with Example 11.7.1. (The time for the cumulative discounted fuel savings to reach the initial investment cannot be determined from this table but is obtained as shown in Example 11.7.2.)
C   The cumulative solar savings reach zero during year 10.
D   The cumulative solar savings exceed the remaining debt principal (mortgage balance) by the end of year 15.
E   The cumulative solar savings exceeded the down payment of $1100 by year 12.

Table 11.7.1   Mortgage Balance Fuel Savings, Cumulative Fuel Savings, Solar Savings, and Cumulative Solar Savings from Example 11.6.4

Year	Mortgage Balance	Fuel Savings	Cumulative Fuel Savings	Solar Savings	Cumulative Solar Savings
				−1,100	−1,100
1	9,707	703	703	−211	−1,311
2	9,496	773	1,476	−163	−1,474
3	9,266	850	2,326	−109	−1,583
4	9,105	935	3,261	−49	−1,632
5	8,742	1,029	4,290	18	−1,614
6	8,445	1,132	5,422	94	−1,520
7	8,120	1,245	6,667	175	−1,345
8	7,766	1,370	8,037	268	−1,077
9	7,381	1,570	9,607	369	−708
10	6,961	1,657	11,264	481	−227
11	6,503	1,823	13,087	607	380
12	6,004	2,005	15,092	745	1,125
13	5,459	2,206	17,298	900	2,025
14	4,866	2,426	19,724	1,070	3,095
15	4,220	2,669	22,393	1,260	4,355
16	3,513	2,936	25,329	1,469	5,824
17	2,747	3,229	28,558	1,701	7,525
18	1,910	3,552	32,110	1,958	9,483
19	997	3,907	36,017	2,241	11,724
20	−	4,298	40,315	2,550	14,274
	Resale value			4,400	
	Undiscounted savings including resale value			$18,674	

Annual cash flow (or savings) vary with years (as shown in the next-to-last column of the tabulation of results in Example 11.6.3). These costs can be "annualized" or "levelized" by determining the equal payments that are equivalent" (in present dollars) to the varying series. The annualized life cycle cost, $ALCC$, or savings, $ALCS$, are determined from

$$ALCC = \frac{LCC}{PWF(N_e, 0, d)} \qquad (11.7.8)$$

$$ALCS = \frac{LCS}{PWF(N_e, 0, d)} \qquad (11.7.9)$$

Thus the annualized life cycle savings in Example 11.6.3 are

$$\frac{4203}{PWF\ (20,\ 0,\ 0.08)} = \$428 \text{ per year}$$

The series of variable savings is equivalent to annual savings of $428 in today's dollars. (An annualized cost per unit of delivered solar energy can also be calculated. It is necessary to divide annualized life cycle cost by the total annual load.)

The return on investment ROI of a solar process may be found by determining the market discount rate d_o which corresponds to zero life cycle solar savings. This can be done by trial and error, by plotting LCS versus d to find d_o at $LCS = 0$. For the 39 m^2 optimum area of Example 11.6.4, variation of d results in a plot of LCS versus d as shown in Figure 11.3.2. The return on investment under these circumstances is about 21%. (Note that maximizing the return on investment does not lead to the same area as maximizing solar savings.)

11.8   THE P_1, P_2 METHOD

It is possible to view the calculations of Example 11.6.3 in a different way, by obtaining the present worth of each of the columns and summing these (with appropriate signs) to get the present worth of the solar savings. The life cycle costs of insurance, maintenance and parasitic power, property taxes, and mortgage payments and the life cycle fuel savings are determined with the appropriate present worth factors. The life cycle benefit of tax savings can be determined by multiplying the present worth factor for property taxes by the effective income tax rate. This view of the calculation is shown in Example 11.8.1. In this example, we use several new symbols:

MS_1 = Miscellaneous costs (maintenance, insurance, parasitic power)
        payable at the end of the first period.
PT_1 = Property tax payable at the end of the first period
$\bar{t}$ = Effective federal-state income tax bracket from Equation 11.1.5

Example 11.8.1

Redo Example 11.6.3, by obtaining the life cycle costs of each of the columns in the table in that example and summing them to get the solar savings.

Solution

The present worth of the fuel savings is given by

$$FLC_{F1} \times PWF(N_e, i_F, d) = 703 \times 22.1687 = \$15585$$

The present worth of the series of mortgage payments is

$$-M \times PWF(N_L, 0, d) = -1084 \times 9.8181 = -\$10643$$

The present worth of the miscellaneous costs is given by

$$-MS_1 \times PWF(N_e, i, d) = -110 \times 15.5957 = -\$1716$$

The present worth of the extra property tax is

$$-PT_1 \times PWF(N_e, i, d) = -220 \times 15.5957 = -\$3431$$

The present worth of the income tax savings on the interest paid on the mortgage is $\bar{t}$ times the present worth of the series of interest payments. From Equation 11.5.3,

$$0.45 \times 9900\left[\frac{9.818}{9.129} + 20.242\left(0.09 - \frac{1}{9.129}\right)\right] = \$3028$$

The present worth of the income tax savings due to the property taxes is $\bar{t}$ times the present worth of the extra property tax:

$$\bar{t} \times \left[PT_1 \times PWF(N_e, i, d)\right] = 0.45 \times 3431 = \$1544$$

The down payment is \$1100. The present worth of the estimated resale value is $4400/(1.08)^{20} = \$944$. Thus the solar savings are

$$15,585 - 1716 - 10,643 - 3431 + 3028 + 1544 - 1100 + 944 = \$4211$$

Within round-off errors, this is the same result as that of Example 11.6.3. ■

An examination of the terms in the savings calculation of Example 11.8.1 suggests that a general formulation of solar savings can be developed. Two facts are apparent. First, there is one term that is directly proportional to the first year's fuel savings. Second, the remainder of the terms are all related directly to the initial investment in the system (or to the mortgage, which in turn is a fraction of the initial investment). Using these facts, Brandemuehl and Beckman (1979) have shown how the present worth factors in terms like those in Example 11.8.1 can be combined to a simple formulation for life cycle solar savings:[8]

$$LCS = P_1 C_{F1} L \mathcal{F} - P_2(C_A A_c + C_E) \tag{11.8.1}$$

where P_1 is the ratio of the life cycle fuel cost savings to the first-year fuel cost savings and P_2 is the ratio of the life cycle expenditures incurred because of the additional capital investment to the initial investment. Other terms are as defined previously.

Any costs that are proportional to the first-year fuel cost can be included in the analysis by appropriate determination of P_1, and any costs that are proportional to the investment can be included in P_2. Thus the full range of costs noted in the examples of Section 11.6 can be included as needed. The ratio P_1 is given by

$$P_1 = (1 - C\bar{t})PWF(N_e, i_F, d) \tag{11.8.2}$$

where C is a flag indicating income producing or non–income producing (1 or 0, respectively), i_F is the fuel inflation rate, and d is the discount rate. The ratio P_2 is calculated by

$$P_2 = D + (1 - D)\frac{PWF(N_{\min}, 0, d)}{PWF(N_L, 0, m)}$$

$$- \bar{t}(1 - D)\left[PWF(N_{\min}, m, d)\left(m - \frac{1}{PWF(N_L, 0, m)}\right)\right.$$

$$\left. + \frac{PWF(N_{\min}, 0, d)}{PWF(N_L, 0, m)}\right]$$

$$+ M_s(1 - \bar{t}) \times PWF(N_e, i, d) + tV(1 - \bar{t}) \times PWF(N_e, i, d)$$

$$- \frac{C\bar{t}}{N_D} PWF(N'_{\min}, 0, d) - \frac{R_v}{(1 + d)^{N_e}}(1 - C\bar{t}) \tag{11.8.3}$$

where

m	=	Annual mortgage interest rate
i	=	General inflation rate
N_e	=	Period of economic analysis

[8] Equation 11.8.1 is written with the implicit assumption that the loads are independent of the size of the solar energy system. To calculate life cycle savings where there are significant differences between loads with solar and loads without solar, the first term of the equation can be written in terms of the difference in auxiliary energy required, of $P_1 C_{F1} (L_0 - L_A)$.

$$
\begin{aligned}
N_L &= && \text{Term of loan} \\
N_{min} &= && \text{Years over which mortgage payments contribute to the analysis} \\
& && \text{(usually the minimum of } N_e \text{ or } N_L) \\
N'_{min} &= && \text{Years over which depreciation contributes to the analysis} \\
& && \text{(usually the minimum of } N_e \text{ or } N_D) \\
N_D &= && \text{Depreciation lifetime in years} \\
t &= && \text{Property tax rate based on assessed value} \\
\bar{t} &= && \text{Effective income tax rate (from Equation 11.1.5)} \\
D &= && \text{Ratio of down payment to initial investment} \\
M_s &= && \text{Ratio of first year miscellaneous costs (parasitic power,} \\
& && \text{insurance and maintenance) to initial investment} \\
V &= && \text{Ratio of assessed valuation of the solar energy system in first} \\
& && \text{year to the initial investment in the system} \\
R_v &= && \text{Ratio of resale value at end of period of analysis to initial} \\
& && \text{investment}
\end{aligned}
$$

In this equation the first term on the right represents the down payment. All other terms represent life cycle costs of payments or series of payments, are in proportion to the initial investment, and are as follows: the second term represents the life cycle cost of the mortgage principal and interest; the third, income tax deductions of the interest; the fourth, miscellaneous costs such as parasitic power, insurance, and maintenance; the fifth, net property tax costs (tax paid less income saved); the sixth, straight line depreciation tax deduction;[9] and the seventh, present worth of resale value at the end of the period of the economic analysis. Terms may be added to or deleted from P_2 as appropriate.

The contributions of loan payments to the analysis depends on N_L and N_e. If $N_L \leq N_e$, all N_L payments will contribute. If $N_L \geq N_e$, only N_e payments would be made during the period of the analysis. Accounting for loan payments past N_e depends on the rationale for choosing N_e. If N_e is a period over which the discounted cash flow is calculated without regard for costs outside of the period, then $N_{min} = N_e$. If N_e is the expected operating life of the system and all payments are expected to continue as scheduled, then $N_{min} = N_L$. If N_e is chosen as the time to anticipated sale of the facility, the remaining loan principal at N_e would be repaid at that time, and the life cycle mortgage cost would consist of the present worth of N_e loan payments plus the principal balance in year N_e. The principal balance would then be deducted from resale value.

[9] Straight line depreciation is assumed in Equation 11.8.3. For $N'_{min} > N_n$, the sixth term for double declining balance or sum of digits [from Barley and Winn (1978)] may be written as

$$
DB = C_{\bar{t}} + \frac{2C_{\bar{t}}}{N_D} \left[PWF\left((N_D - 1), \frac{-2}{N_D}, d\right) - \frac{PWF\left((N_D - 1), \frac{-2}{N_D}, 0\right)}{(1 + d)^{N_D}} \right]
$$

$$
SOD = \frac{2\bar{t}}{N_D (N_D + 1)} \left[PWF(N_D, 0, d) + \frac{N_D - 1 - PWF((N_D - 1), 0, d)}{d} \right]
$$

Similar arguments can be made about the period over which depreciation deductions contribute to an analysis if the facility is part of a business (i.e., is income producing). The contributions will depend on the relationship of N_D and N_e.

The equations of P_1 and P_2 include only present worth factors and ratios of payments to initial investments in the system. They do not include collector area or solar fraction. As P_1 and P_2 are independent of A_c and $\mathscr{F}$, systems in which the primary design variable is A_c can be optimized by use of Equation 11.8.1. At the optimum, the derivative of the savings with respect to collector area is zero:

$$\frac{\partial LCS}{\partial A_c} = 0 = P_1 C_{F1} L \frac{\partial \mathscr{F}}{\partial A_c} - P_2 C_A \tag{11.8.4}$$

Rearranging, the maximum savings are realized when the relationship between collector area and solar load fraction satisfies

$$\frac{\partial \mathscr{F}}{\partial A_c} = \frac{P_2 C_A}{P_1 C_{F1} L} \tag{11.8.5}$$

The relationship of the optimum area to the annual thermal performance curve is shown in Figure 11.8.1. The optimum area occurs where the slope of the curve $\mathscr{F}$ versus A_c is $P_2 C_A / P_1 C_{F1} L$.

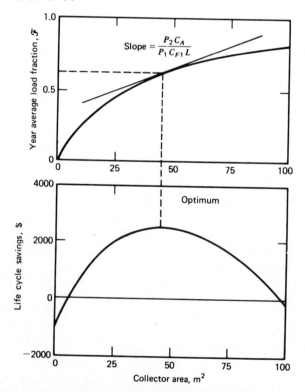

Figure 11.8.1 Optimum collector area determination from the slope of the $\mathscr{F}$ vs A_c thermal performance curve. From Brandemuehl and Beckman (1979).

Example 11.8.2

Redo Example 11.6.4 using the P_1, P_2 method.

Solution

The installation is not an income-producing one so $C = 0$. The ratio P_1 is calculated from Equation 11.8.2:

$$P_1 = PWF(20, 0.10, 0.08) = 22.169$$

The ratio P_2 is calculated from Equation 11.8.3.

$$P_2 = 0.1 + 0.9\frac{PWF(20, 0, 0.08)}{PWF(20, 0, 0.09)}$$

$$- 0.9 \times 0.45\left[PWF(20, 0.09, 0.08)\left(0.09 - \frac{1}{PWF(20, 0, 0.09)}\right)\right.$$

$$\left. + \frac{PWF(20, 0, 0.08)}{PWF(20, 0, 0.09)}\right]$$

$$+ 0.01 \times PWF(20, 0.06, 0.08)$$

$$+ 0.02 \times 0.55 \times 1.0 \times PWF(20, 0.09, 0.08) - \frac{0.4}{(1.08)^{20}}$$

and $\qquad P_2 = 0.1 + 0.968 - 0.275 + 0.156 + 0.172 - 0.086 = 1.035$

From Equation 11.8.5, with $C_A = \$200/m^2$ and $C_{F1}L = \$1255$,

$$\frac{\partial \mathcal{F}}{\partial A_c} = \frac{1.035 \times 200}{22.169 \times 1255} = 0.00744$$

A plot of the $\mathcal{F}$ versus A_c data from Example 11.6.4 is shown in the figure. The optimum collector area, where the slope is 0.00744, is about 40 m².

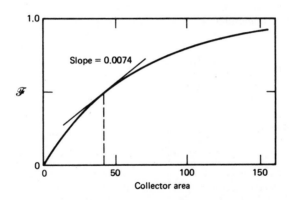

The P_1, P_2 method is not limited to regularly varying costs. The requirements are for P_1 that the fuel expenses be proportioned to the first year unit energy cost, and for P_2 that the owning costs be proportioned to the initial investment. For the irregularly varying fuel costs of Example 11.6.2 the value of P_1 can be found from the ratio of the life cycle fuel cost to the first-year fuel cost, that is, $P_1 = 10,516/1200 = 8.763$. This value of P_1 can be used with any other first-year fuel cost that has the same inflation rate schedule as Example 11.6.2. Also P_1 and P_2 can be obtained from a single detailed calculation of life cycle savings such as that in Example 11.6.3, and can then be applied to all collector areas and solar fractions. If there are highly irregular costs, this may be the easiest way to determine P_1 and P_2.

The P_1, P_2 method is quick, convenient, and extremely useful. It is used in developing economic evaluations of specific applications in later chapters.

11.9   UNCERTAINTIES IN ECONOMIC ANALYSES

Many assumptions and uncertainties are involved in the use of the economic analysis methods presented in this chapter. The analyst must make estimates of many economic parameters with varying degrees of uncertainty. In particular, the projection of future energy costs is difficult in view of unsettled international energy affairs. Thus it is desirable to determine the effects of uncertainties on the calculated values of life cycle savings and optimum system design.

For a given set of conditions, the change in life cycle savings resulting from a change in a particular parameter, Δx_j, can be approximated by

$$\Delta LCS = \frac{\partial LCS}{\partial x_j} \Delta x_j = \frac{\partial}{\partial x_j}[P_1 C_{F1} L \, \mathcal{F} - P_2(C_A A_C + C_E)] \qquad (11.9.1)$$

When there are uncertainties in more than one variable the maximum possible uncertainty is given by

$$\Delta LCS_{max} = \sum_{j=1}^{n} \left|\frac{\partial LCS}{\partial x_j}\right| \Delta x_j \qquad (11.9.2)$$

A "most probable" uncertainty in savings can be written as

$$\Delta LCS_{prob} = \left[\sum_{j=1}^{n} \left(\frac{\partial LCS}{\partial x_j} \Delta x_j\right)^2\right]^{1/2} \qquad (11.9.3)$$

From Equation 11.8.1,

$$\frac{\partial LCS}{\partial x_j} = \frac{\partial(P_1 C_{F1} L \, \mathcal{F})}{\partial x_j} - \frac{\partial P_2(C_A A_C + C_E)}{\partial x_j} \qquad (11.9.4)$$

The partial derivatives of P_1 and P_2 (from Equation 11.8.2 and 11.8.3) for selected variables are as follows:

For the fuel inflation rate

$$\frac{\partial P_1}{\partial i_F} = (1 - C\bar{t})\frac{\partial PWF(N_e, i_F, d)}{\partial i_F} \tag{11.9.5}$$

For the general inflation rate

$$\frac{\partial P_2}{\partial i} = [(1 - C\bar{t})M_s + (1 - \bar{t})tV]\frac{\partial PWF(N_e, i, d)}{\partial i} \tag{11.9.6}$$

For effective income tax bracket

$$\frac{\partial P_1}{\partial \bar{t}} = -C \times PWF(N_e, i_F, d) \tag{11.9.7}$$

$$\frac{\partial P_2}{\partial \bar{t}} = (D - 1)\left[PWF(N_{\min}, m, d)\left(m - \frac{1}{PWF(N_L, 0, m)}\right)\right.$$
$$+ \frac{PWF(N_{\min}, 0, d)}{PWF(N_L, 0, m)}\right]$$
$$- (tV)PWF(N_e, i, d) - C\left[M_s PWF(N_e, i, d) + \frac{PWF(N'_{\min}, 0, d)}{N_D}\right] \tag{11.9.8}$$

For property tax rate

$$\frac{\partial P_2}{\partial t} = V(1 - \bar{t})PWF(N_e, i, d) \tag{11.9.9}$$

For resale value:

$$\frac{\partial P_2}{\partial R_v} = \frac{1 - C\bar{t}}{(1 + d)^{N_e}} \tag{11.9.10}$$

The partial derivative of the life cycle savings with respect to the fraction by solar is

$$\frac{\partial LCS}{\partial \mathcal{F}} = P_1 C_{F1} L \tag{11.9.11}$$

A complete set of the partial derivatives is provided by Brandemuehl and Beckman (1979).

It is also necessary to know the partial derivatives of the present worth functions. From Equation 11.5.1, when $i \neq d$,

$$\frac{\partial}{\partial a} PWF(a, b, c) = -\frac{1}{c-b}\left(\frac{1+b}{1+c}\right)^a \ln\left(\frac{1+b}{1+c}\right) \tag{11.9.12}$$

$$\frac{\partial}{\partial b} PWF(a, b, c) = \frac{1}{c-b}\left[PWF(a, b, c) - \frac{a}{1+b}\left(\frac{1+b}{1+c}\right)^a\right] \tag{11.9.13}$$

$$\frac{\partial}{\partial c} PWF(a, b, c) = \frac{1}{c-b}\left[\frac{a}{1+c}\left(\frac{1+b}{1+c}\right)^a - PWF(a, b, c)\right] \tag{11.9.14}$$

If $i = d$ (i.e., if $b = c$),

$$\frac{\partial}{\partial a} PWF(a, b, c) = \frac{a}{b+1} = \frac{a}{c+1} \tag{11.9.15}$$

$$\frac{\partial}{\partial b} PWF(a, b, c) = \frac{-a}{(b+1)^2} \quad \text{and} \quad \frac{\partial}{\partial c} PWF(a, b, c) = \frac{-a}{(c+1)^2} \tag{11.9.16}$$

In the next example, the effect of uncertainty of one variable is illustrated. To estimate the effects of uncertainties of more than one variable, the same procedure is used as for one variable in determining the appropriate terms in Equations 11.9.2 and 11.9.3.

Example 11.9.1

In Example 11.8.2, the fuel inflation rate was taken as 10% per year. For a 50 m² collector area, what are the life cycle savings? What is the uncertainty in life cycle savings if the fuel inflation rate is uncertain to ±2%? From the data of Example 11.6.4, the 50 m² collector will provide 0.56 of the annual loads. The first year's fuel cost is $1255 and the installed cost is $11,000. $N_e = 20$ years, and $d = 0.08$. From Example 11.8.2, $P_1 = 22.169$ and $P_2 = 1.035$.

Solution

The life cycle savings are, from Equation 11.8.1,

$$LCS = 22.619 \times 1255 \times 0.56 - 1.035 \times 11000 = \$4195$$

The effect of fuel inflation rate is only on P_1. From Equation 11.9.5, with $C = 0$,

$$\frac{\partial P_1}{\partial i_F} = \frac{\partial PWF(N_e, i_F, d)}{\partial i_F}$$

and from Equation 11.9.13,

$$\frac{\partial}{\partial i_F} PWF(N_e, i_F, d) = \frac{1}{0.08 - 0.10}\left[PWF(20, 0.10, 0.08) - \frac{20}{1.10}\left(\frac{1.10}{1.08}\right)^{20}\right]$$

$$= 204$$

The uncertainty in LCS is is obtained from Equation 11.9.1, 11.9.2, or 11.9.3 (which all give the same result when uncertainty in only one variable is considered) and 11.9.4:

$$\Delta LCS = \frac{\partial LCS}{\partial i_F}\,\Delta i_F = C_{F1}L\,\mathcal{F}\frac{\partial P_1}{\partial i_F}\Delta i_F$$

$$= 1255 \times 0.56 \times 204 \times 0.02 = \$2867$$

The uncertainty in LCS due to the uncertainty of 2% in the fuel inflation rate is over half of the projected savings. (An increase in i_F results in an increase in LCS.) ■

The effect of any one variable on the life cycle savings is largely determined by the values of the many other variables. It is not easy to generalize, and a quantitative analysis is usually required.

11.10   SUMMARY

In this chapter we have outlined the kinds of investments and operating costs that may be expected with a solar process and indicated that in many circumstances collector area can be considered the primary design variable once a system configuration and collector type are established.

A variety of economic figures of merit have been proposed and used including payback times, cash flow, and life cycle savings. The life cycle costing method is the most inclusive and takes into account any level of detail the user wishes to include, including the time value of money. The use of the life cycle savings method will be illustrated for specific solar processes in the later chapters. The P_1, P_2 method of calculating life cycle savings is particularly useful. The results of these calculations are very dependent on values assumed for N_e, i_F, and d. If $i_F > d$ (an abnormal situation that has existed in recent years), the choice of a sufficiently high value of

N_e will make the life cycle savings of solar (or other capital-intensive fuel-saving technologies) appear positive. Clearly, the selection of values for the critical economic parameters in the analysis will have great influence on the results obtained.

REFERENCES

Balcomb, J. D., *Passive Solar J.*, **3**, 221 (1986). "Conservation and Solar Guidelines."

Barley, C. D., in Procedings of Denver Meeting of American Sect., Int'l Solar Energy Soc., **2**(1), 163 (1978). "Optimization of Space Heating Loads."

Barley, C. D. and C. B. Winn, *Solar Energy*, **21**, 279 (1978). "Optimal Sizing of Solar Collectors by the Method of Relative Areas."

Beckman, W. A., S. A. Klein, and J. A. Duffie, *Solar Heating Design by the f-Chart Method*, Wiley-Interscience, New York (1977).

Brandemuehl, M. J. and W. A. Beckman, *Solar Energy*, **23**, 1 (1979). "Economic Evaluation and Optimization of Solar Heating Systems."

Close, D. J., *Solar Energy*, **11**, 112 (1967). "A Design Approach for Solar Processes."

De Garmo, E. P. and J. R. Canada, *Engineering Economy*, McMillan, New York (1973).

Riggs, J. L., *Economic Decision Models*, McGraw-Hill, New York (1968).

Ruegg, R. T., Report No. NBSIR 75-712, U.S. Dept. of Commerce, (July 1975). "Solar Heating and Cooling in Buildings: Methods of Economic Evaluation."

White, J. A., M. H. Agee, and K. E. Case, *Principles of Engineering Economic Analysis*, Wiley, New York (1977).

Part II

APPLICATIONS

Part I was concerned with the fundamental concepts and laws that determine the operation of solar process components and systems, the formulation of working equations for collectors and other components, and economics. With this background established, we proceed to discussions of applications of solar energy to meet various energy needs. There is substantial emphasis on applications to buildings for heating, hot water, and cooling, by active and passive processes.

Much of Part II is qualitative and descriptive, although most of the chapters include results of quantitative performance calculations and/or measurements and economic evaluations. The objectives are to illustrate what applications have been made, show some of those that may be possible, and cast these applications in a quantitative framework where possible. Practical problems arising with the applications are noted.

Chapter 12 deals with active and passive water heating systems and covers many aspects of the most widespread of all modern solar processes. The extension of water heating practice to larger systems for active space heating is described in Chapter 13. Chapter 14 treats thermal aspects of passive heating processes, i.e., the engineering part of the combined engineering and architectural problem of designing buildings to admit and use solar radiation. Chapter 15 concerns cooling processes operated by solar radiation.

The next three chapters are on applications that are usually made on a larger scale. The first is Chapter 16, on industrial process heat, applications that are interesting because so much energy is used in industry at temperatures that are compatible with collector operation but which for practical and economic reasons have not been widely made. Chapter 17, on thermal processes for conversion of solar to mechanical energy, treats experiments and practical applications that have been made in small numbers; it includes a discussion of the Luz peak power plants in California that have an aggregate electrical generating capacity of several hundred megawatts. Chapter 18, on solar ponds and evaporative processes, includes information on a peaking plant in Israel that is based on a solar pond, and also traditional processes of evaporation and drying for water removal and distillation for water recovery. This chapter includes some theory and calculation methods for solar stills.

Chapter 12

SOLAR WATER HEATING: ACTIVE AND PASSIVE

This is the first of a set of chapters on thermal energy applications. In this chapter, we treat the use of solar heating for domestic or institutional hot water supplies. Descriptions of systems, components, and important design considerations are outlined. Considerations important in designing water heating systems are also basic to solar heating and cooling systems, applications that are covered in succeeding chapters. Many of the principles noted here for service hot water for buildings also apply to industrial process heat, discussed in Chapter 16.

In recent years considerable knowledge has been developed about low-flow hot water systems; these use collector flow rates a fifth or so of those that have been commonly used in forced-circulation systems. The F_R of collectors operated this way is lower than that in conventional operation, but higher stratification in storage tanks can be obtained, resulting in lower collector inlet temperatures and improved performance. These developments are of sufficient importance that they are discussed in a separate section.

12.1 WATER HEATING SYSTEMS

The basic elements in solar water heaters can be arranged in several system configurations. The most common of these are shown in Figure 12.1.1. Auxiliary energy is shown added in three different ways; these are interchangeable among the four methods of transferring heat from the collector to the tank.

A passive water heater (i.e., a natural circulation system) is shown in Figure 12.1.1(a). The tank is located above the collector, and water circulates by natural convection whenever solar energy in the collector adds energy to the water in the collector leg and so establishes a density difference. Auxiliary energy is shown added to the water in the tank near the top to maintain a hot water supply.

Figure 12.1.1(b) shows an example of a forced-circulation system. A pump is required; it is usually controlled by a differential thermostat turning on the pump when the temperature at the top header is higher than the temperature of the water in the bottom of the tank by a sufficient margin to assure control stability (as outlined in Section 10.4). A check valve is needed to prevent reverse circulation and resultant nighttime thermal losses from the collector. Auxiliary energy is shown added to the water in the pipe leaving the tank to the load.

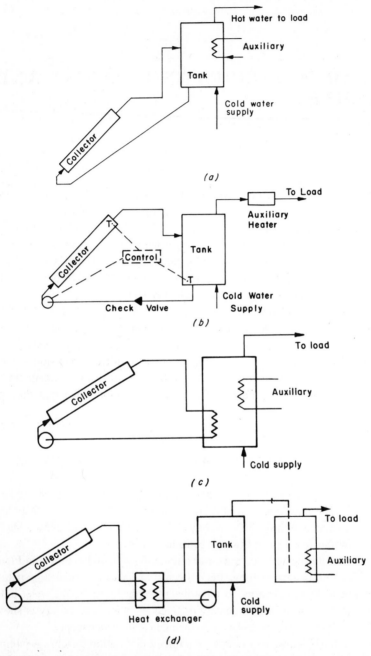

Figure 12.1.1 Schematic of common configurations of water heaters. (a) A natural circulation system. (b) One-tank forced-circulation system. (c) System with antifreeze loop and internal heat exchanger. (d) System with antifreeze loop and external heat exchanger. Auxiliary is shown added in the tank, in a line heater, or in a second tank; any of these auxiliary methods can be used with any of the collector-tank arrangements.

In climates where freezing temperatures occur, these designs are modified. Examples of systems using nonfreezing fluids in the collector are shown in Figure 12.1.1(c) and (d). The collector heat exchangers can be either internal or external to the tank. Auxiliary energy is shown added to the water in the storage tank in Figures 12.1.1(c) by a heat exchanger in the tank. The auxiliary energy supply can also be provided by a standard electric, oil, or gas water heater with storage capacity of its own; this is the two-tank system shown in Figure 12.1.1(d). Any of these systems may be fitted with tempering valves that mix cold supply water with heated water to put an upper limit on the temperature of the hot water going to the distribution system. Other equipment not shown can include surge tanks and pressure relief valves.

Solar water heaters are manufactured in Australia, Israel, Japan, Greece, the United States, and elsewhere. They were common in Florida and California early in this century, disappeared when inexpensive natural gas became available, and have again been installed as the costs of gas and other fuels have risen. Solar water heating has the advantage that heating loads are usually uniform through the year, which leads to high use factors on solar heating equipment. Figure 12.1.2 shows domestic and institutional water heaters. The Griffith and Perth systems utilize natural convection, and the others use forced convection.

The collectors in use in many water heating systems are similar to that shown in Figure 6.23.3, with parallel riser tubes 0.10 to 0.15 m apart. Plate materials may be copper or steel. Other plate designs are also used. For example, some are manufactured of two spot-welded, seam-welded, or roll-bonded plates of stainless or ordinary steel, copper, or aluminum. The fluid passages between the plates are formed by hydraulic expansion after welding. Serpentine tubes are also used and may become more common in the microflow systems to be discussed in Section 12.5. The absorber plates are mounted in a metal box, with 50 to 100 mm of insulation behind the plate and one or two glass covers over the plates. The dimensions of a typical collector module made in the United States are approximately 1×2 m; the Australian and Israeli heater modules are typically 0.6×1.2 m, $1.2. \times 1.2$ m, or 1×2 m. The thermal performance characteristics of these and other collectors can be determined by equations given in Chapter 6.

It is advantageous to maintain stratification in the storage tanks, and the location and design of tank connections is important, as noted in Chapter 8. The schematics in Figure 12.1.1 show approximate locations of connections in typical use. Close (1962) measured tank temperatures at various levels in an experimental natural circulation water heating system operated for a day with no hot water removal from the tank. These data are shown in Figure 12.1.3; this degree of stratification is characteristic of natural circulation systems. Tanks without baffles or carefully designed diffusers on inlets and outlets will stratify to some degree in forced-circulation systems; at the present time most water heater tanks are not so equipped. Storage tanks should be well insulated, and good practice is to use 0.2 m or more of mineral wool or glass wool insulation on sides, top, and bottom. Piping connections to a tank should also be well insulated.

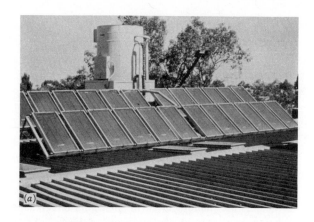

Figure 12.1.2 Solar water heaters. (a) Natural circulation system on hostel, Griffith, Australia. Photo courtesy of CSIRO. (b) Domestic system, with coupled tank and collector, Perth, Australia. (c) Hospital service hot water system, Madison, WI. Photo courtesy of Affiliated Engineers, Inc. (d) Collectors for a domestic system, Madison, WI.

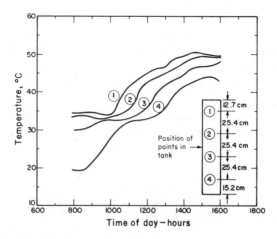

Figure 12.1.3 Temperature distribution in a vertical cylindrical water tank used in a thermal circulation water heater. From Close (1962).

12.2   FREEZING AND BOILING

Solar collectors and associated piping must be designed to avoid damage from freezing or boiling. Low ambient temperatures during periods of no solar radiation can result in plate temperatures below 0 C. If no energy is withdrawn from a system or if a circulating pump should not operate during times of high radiation, the plate temperature may exceed 100 C.

Freeze protection can be provided by draining the water from the collectors, use of nonfreezing solutions, or warming of the water in the collectors. Five approaches have been developed to protect collectors against damage by freezing.

First, antifreeze solutions can be used in the collector loop with a heat exchanger between the collector and the storage tank. As shown in Figures 12.1.1(c) and (d), the heat exchanger can be external to the tank, or it can be a coil within the tank relying on natural circulation of the water in the tank for heat transfer. For either arrangement, the performance of the collector-heat exchanger combination can be treated by the F_R' method outlined in Section 10.2. A typical overall heat transfer coefficient for a coil in a tank is 600 W/m^{2}C.

Ethylene glycol-water and propylene glycol-water solutions are common antifreeze liquids. Their physical properties are included in Appendix E. Ethylene glycol is toxic, as are some commonly used corrosion inhibitors, and many plumbing codes require the use of two metal interfaces between the toxic fluid and the potable water supply. This can be accomplished either by the use of two heat exchangers in series or more commonly by double-walled heat exchangers that can be either internal coils in the tank or external to the tank.

Second, air can be used as the heat transfer fluid in the collector-heat exchanger loop of Figure 12.1.1(d). Air heating collectors have lower $F_R(\tau\alpha)$ and $F_R U_L$ than liquid heating collectors. However, no toxic fluids are involved, no second heat exchanger interface is needed, leakage is not critical, and boiling is not a problem.

The third method of freeze protection is to circulate warm water from the tank through the collector to keep it from freezing. Thermal losses from the system are significantly increased, and an additional control mode must be provided. This method can only be considered in climates where freezes are infrequent. In emergencies, when pump power is lost, the collector and piping subject to freezing temperatures must be drained.

The fourth method is based on draining water from the collectors when they are not operating. Draining systems must be arranged so that collectors and piping exposed to freezing temperatures are completely emptied, and the collectors must be vented. These systems are of two types, **drain-back** systems in which the water drains back into the tank or a sump that is not exposed to freezing temperatures, or **drain-out** systems, which drain out of the system to waste.

The fifth method is to design the collector plate and piping so that it will withstand occasional freezing. Designs have been proposed using butyl rubber risers and headers that can expand if water freezes in them.

High collector temperatures may also be a problem, as no-load (equilibrium) temperatures of good collectors can be well above the boiling point of water under conditions of no fluid circulation, high radiation, and high ambient temperature. These conditions can be expected to occur, for example, when occupants of a residence are away from home in the summer. Several factors may mitigate this problem. First, antifreeze solutions used in collector loops have elevated boiling points (see Appendix E); 50% ethylene glycol and propylene glycol solutions in water have boiling points at atmospheric pressure of 112 and 108 C, respectively. Second, many systems are operated at pressures of several atmospheres, which further raises the boiling point; the boiling points of the 50% glycol solutions at 2 atm absolute are about 145 and 140 C, respectively. Third, collector loss coefficients rise as plate temperatures rise, as shown in Figure 6.4.4, which tends to hold down no-load temperatures. Practical systems should include pressure relief valves and vent tanks to relieve excess pressure and contain any antifreeze solution that is vented.

Example 12.2.1

A collector with $(\tau\alpha)$ equal to 0.78 has a temperature dependence of overall loss coefficient as shown on the following plot. The fluid being heated is a 50% solution of propylene glycol in water. The glycol loop is pressurized to a limit of 4.0 atm. On a summer day the radiation on the collector is 1.15 kW/m^2 and ambient temperature is 38 C. Will the solution in the collector boil if the circulating pump does not operate?

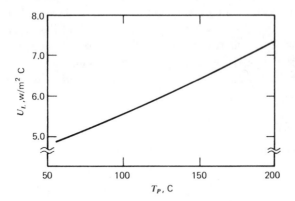

Solution

With no flow, absorbed energy equals losses:

$$G_T(\tau\alpha) = U_L(T_{p,m} - T_a) = 1150 \times 0.78 = 900 \text{ W/m}^2$$

so the plate temperature is

$$T_{p,m} = T_a + 900/U_L$$

Since the loss coefficient is a function of plate temperature, it is necessary to simultaneously solve the energy balance equation and the equation for U_L as a function of T_p. Assume for a first estimate that $U_L = 6.5$ W/m²C. Then

$$T_{p,m} = 38 + 900/6.5 = 176 \text{ C}$$

A second estimate of 6.8 W/m²C yields

$$T_{p,m} = 38 + 900/6.8 = 170 \text{ C}$$

The loss coefficient at this temperature is close to the second estimate, so $T_p = 170$ C is a reasonable estimate of the temperature of the plate and the temperature of the fluid under no-flow conditions. The boiling point of a 50% propylene glycol solution at 4 atm is approximately 145 C and the solution would boil in the collector under these circumstances. ■

12.3   AUXILIARY ENERGY

The degree of reliability desired of a solar process to meet a particular load can be provided by a combination of properly sized collector and storage units and an auxiliary energy source. In a few unique areas where there seldom are clouds of significant duration, it may be practical to provide all of the loads with the solar systems. However, in most climates auxiliary energy is needed to provide high reliability and avoid gross overdesign of the solar system.

Auxiliary energy can be provided in any of the three ways shown in Figure 12.3.1.

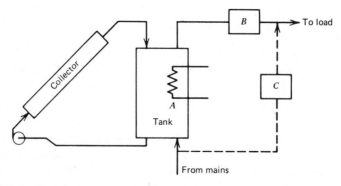

Figure 12.3.1   Schematic of alternative location for auxiliary energy supply to a one-tank forced-circulation solar water heater: *A*, in tank; *B*, in line to load; *C*, in a bypass around the tank.

A   Energy can be supplied to the water in the tank, at location A. Auxiliary energy is controlled by a thermostat in the top part of the tank which keeps the temperature in the top portion at or above a minimum setpoint. This is the simplest and probably the least expensive method. However, it has the disadvantage that auxiliary energy will usually increase the temperature of the water in the bottom of the tank and thus the collector inlet temperature, resulting in reduction of solar gain.

B   Auxiliary energy can be supplied to the water leaving the tank, thus "topping off" the solar energy with auxiliary energy. This requires a heater separate from the solar tank. This heater may be a simple line heater, or it may be a conventional water heater with storage capacity of its own. Auxiliary energy is controlled to maintain the outlet temperature from the auxiliary heater at a desired level. This method has the advantage of using the maximum possible solar energy from the tank without driving up the collector temperature, but additional heat loss will occur from the auxiliary heater if it has storage capacity.

C   Auxiliary energy can be added directly to the incoming supply water by bypassing the tank when auxiliary energy is needed. This is a simple method but has the disadvantage that collected solar energy in the tank which does not increase the water temperature above the set temperature may be lost from the tank.

The relative merits of these three methods have been examined by Gutierrez et al. (1974). By simulating a month's operation of the three systems having forced circulation, fixed collector size, and a partially stratified (three-node) storage tank under a range of conditions of ambient temperature, time of day, and magnitude of loads, the relative portion of the loads carried by solar and by auxiliary energy were estimated. It was shown for the system and climate data used, that when the setpoint temperature was not greatly above ambient temperatures, the method of adding auxiliary energy was not critical (with method **B** showing minor advantages). However, when the minimum hot water temperature was raised to setpoints usually used in domestic hot water systems, method **B** showed significant advantages over method **A**, which in turn was better than method **C**.

The major reasons for the differences in performance on changing the method of adding auxiliary energy concern the temperature at which the collector operates. Adding auxiliary energy to the top of the tank (method **A**) can result in a higher mean collector temperature, poorer collector performance, and higher requirements for auxiliary energy (depending on the degree of stratification in the tank). Method **C**, which bypasses the tank when its top section is not hot enough, results in failure to use some collected solar energy. Method **B**, with a modulated auxiliary heater (or its equivalent in an on-off auxiliary heater) maximizes the solar collector output and minimizes collector losses by operating at the lowest mean collector temperature of any of the methods. The magnitude of the differences may depend on the degree of stratification in the tank and the insulation on the tank. Losses from the auxiliary heater must be maintained at very low levels, or method **B** may not be best (see Section 12.4).

12.4   FORCED-CIRCULATION SYSTEMS

Most domestic solar water heaters in the United States (and other climates where freezing temperatures are frequently encountered) are forced-circulation systems. Typically, two or three collector modules of dimensions about 1×2 m are used. Until about 1980, flow rates were commonly around 0.015 kg/m²s. Recently, lower rates have been used (e.g., in Sweden, rates of 0.002 to 0.006 kg/m²s are used). The collectors are typically one-cover units with black chrome selective surfaces, although some use flat black absorbers. Heat exchangers are common, including jackets on tanks, external shell-and tube-exchangers, or coils immersed in the tank; propylene or ethylene glycol is used in collector loops. Tanks are usually similar to standard hot water tanks and range in size for residential applications from 200 to 400 liters. Auxiliary heaters may be immersed in the upper part of the tanks, or the solar heaters may serve as preheaters for standard water heaters. There are many variations on these designs.

In this section we review comparative simulation studies by Buckles and Klein (1980) of the long-term performance of domestic-scale systems of several configurations and designs. Daily loads were kept fixed at 300 liters of water, heated from 10 to 50 C, for an annual heating load of 18.3 GJ. The simulations were based on a typical domestic hot water load pattern shown in Figure 9.1.2 [Mutch (1974)]. Hot water use is concentrated in morning and evening hours; to supply large fractions of these loads, water heated by solar energy one day must be stored for use the following day. (Minor changes in time dependence of loads do not have a major effect in long-term performance of domestic hot water systems. However, variations of larger magnitude such as are caused by weekend closures of commercial buildings can have a significant impact on water heater performance and design.)

The collectors were well-designed single-cover selective surface collectors with $F_R(\tau\alpha)_n = 0.84$ and $F_R U_L = 4.67$ W/m²C. Collector areas were taken as two, three, or four modules each of area 1.44 m². The tanks in all cases had loss coefficients of 1.67 W/m²C. Single-tank systems were modeled as three-node tanks with auxiliary energy supplied to the top node by immersion heaters. Two-tank systems were modeled with the same total volume as the single tanks, with one-third of the volume in the auxiliary tank and the preheat tank and the remaining two-thirds treated as a two-node tank. The collector heat exchangers inside the tank were assumed to have an effectiveness of 0.5 (which is probably an overly optimistic assumption in view of measurements that have been made of performance of several systems). The direct systems were modeled without heat exchangers, with circulation from the bottom of the tank to collector and return to the top of the tank.

An index of performance of these systems is the yearly quantity of solar energy delivered to heat the water. This quantity is often calculated without consideration of thermal losses from the tanks. However, the total energy required by a conventional heater would include such losses. The data that follow are comparisons of solar system performance relative to a conventional system in which the total load is the energy required to heat the water, 18.3 GJ for a year, plus estimated losses from a conventional tank of 3.6 GJ, for a total load L_0 of 21.6 GJ. The fraction by solar, $\mathcal{F}$, is then the ratio of the solar energy delivered to the total conventional load, or $(1 - L_A/L_0)$ and is a measure of the reduction in annual consumption of energy relative to the conventional system.

Under the assumptions made in this study, that is, with a reasonable degree of stratification in the solar tank and with auxiliary energy supplied to the top third of a single tank or in the second tank of a two-tank system, there is very little difference between one- and two-tank systems. In some cases one-tank systems are slightly better because of less area for heat loss from the tanks. The two-tank direct systems without tempering valves have essentially the same performance as those with tempering valves. During times when the system is oversized (i.e., during very good weather) and the water delivered from storage is hot enough to require tempering with cold water, the higher thermal losses negate any advantages of delivery of less heated water from the tank. At times when the system is undersized, the tempering valve does not function. Direct systems without the heat exchanger performed about 5% better than those with the heat exchanger.

An additional set of simulations was done in which the average daily hot water load was kept at 300 liters but which had a 15-fold variation from the largest daily draw to the smallest daily draw in each week. The percentage of the week's hot water load drawn in each of the days of the week (starting with Sunday) was 5.7, 42.9, 2.8, 2.8, 14.4, 2.8, and 28.6%. For the one-tank, external heat exchanger system with three collector modules, the fraction of the load met by solar energy was 0.64 for the regularly recurring load and 0.58 for the irregular load. This simulation study suggests that the effectiveness of the collector heat exchanger is an important aspect of design and that the choice of system configuration otherwise makes less difference in the annual solar output than does the kind of variation in the day-to-day loads which might be expected with a domestic system.

Sizing of domestic forced-circulation systems can be done by "rule of thumb" or by thermal and economic analyses and design procedures.[1] Typical home hot water usage in the United States is 75 liters per day per person, and typical collector areas per person are about 1.5 m^2 for systems which deliver approximately 0.5 to 0.8 of the annual loads by solar energy. However, these figures vary with quality of the collector and the climate, and there are wide individual variations from average water use.

Systems for institutional, commercial, and office buildings are almost all forced-circulation systems. The magnitude and time dependence of loads in these buildings may be easier to predict than that for residences. The loads may go to zero on weekends and holidays in some office and commercial buildings.

---

[1] See Chapters 20 and 21 for details on design procedures.

12.5   LOW-FLOW PUMPED SYSTEMS

The collector flow rates noted in Section 12.4, i.e., in the range of 0.01 to 0.02 kg/m²s, lead to high values of F_R. However, in direct systems (without a collector heat exchanger) they also lead to relatively high fluid velocities in piping and subsequent mixing (or partial mixing) in tanks. Van Koppen et al. (1979) suggested the advantages of low-flow and stratified tanks, and recent work has confirmed that it can be advantageous to use reduced fluid flow rates in the collector loops, accept a lower F_R, and gain the advantage of increased stratification with resulting reduced collector inlet temperature. The result can be a net improvement in system performance. Lower flow rates result in greater temperature rise across collectors, and if tanks are perfectly stratified, the temperature differences from top to bottom will increase as the flow rate decreases. As pointed out by Hollands and Lightstone (1989) in a very useful review, the use of lower flow rates can have the additional advantages of reduction in both the first cost and operating cost of the systems through use of smaller pipes and pumps and reduction of operating costs for pump operation. Flow rates used in Swedish flat-plate collectors have typically been in the range of 0.002 to 0.007 kg/m²s. [Dalenbäck (1990).]

Figure 12.5.1, from simulation studies by Wuestling et al. (1985), illustrates for a specific example the effects of collector flow rate per unit area on system performance (expressed as solar fraction $\mathcal{F}$) for a fully mixed tank and for a highly stratified tank. The potential advantage of the low flow system are evident; the maximum performance for the stratified tank ($\mathcal{F}= 0.66$) is a third greater than that for the fully mixed tank ($\mathcal{F}= 0.48$). This level of improvement is not realized in practice, as real tanks are in general neither fully mixed nor fully stratified. As Hollands and Lightstone point out, the degree of improvement depends in part on load patterns, as loads that draw the tank down completely by the beginning of collection in the

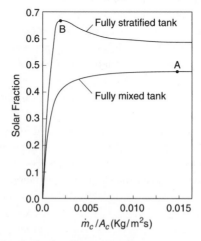

Figure 12.5.1   An example of solar fraction as a function of collector flow rate for a very highly stratified tank and for a fully mixed tank. Adapted from Wuestling et al. (1985).

morning will result in less improvement in performance than if the tank is hot in the morning. However, the gains that have been reported in experiments [e.g., Carvalho et al. (1988)] are significant.

There are design and operating problems that must be faced in low flow systems. Reduction of flow rate in standard collectors may result in poor flow distribution and performance penalties due to hot areas. Collector redesign may be needed; serpentine tubes rather than parallel risers can be used to avoid this problem. Care must be taken to see that the flow rate does not go below the peak point on the upper curve of Figure 12.5.1, as performance drops off very rapidly. Tank design should be such that stratification is maintained at the flow rates used in both collector and load loops.

Hollands et al. (1986) described a system designed for low flow. Small-diameter (4.8 mm) serpentine tubes are used on the collectors, which are connected in series. A small gear pump is used to pump the glycol heat exchanger fluid through the collector loop, and a collector heat exchanger is used with natural convection on the tank side. Collector flow rates are about one-sixth of that used in conventional systems. The heat exchanger has an effectiveness of about 0.8, and for much of the day in experimental operation the inlet temperature to the collector was close to the mains water supply temperature. This design has significant cost advantages over conventional systems. The equipment costs are reduced through use of small tubing and pump. Plumbing costs are reduced by simplified connections. Operating costs are reduced through reduced parasitic power requirements for pump operation.

12.6   NATURAL CIRCULATION SYSTEMS

Circulation in passive solar water heaters such as that shown in Figure 12.1.1(a) occurs when the collector warms up enough to establish a density difference between the leg including the collector and the leg including the tank and the feed line from tank to collector. The density difference is a function of temperature difference, and the flow rate is then a function of the useful gain of the collector which produces the temperature difference. Under these circumstances, these systems are self-adjusting, with increasing gain leading to increasing flow rates through the collector. Norton and Probert (1986) have reviewed the theory of operation of these systems.

Two approaches have been taken to the analysis of systems of this type. The first is to model the system in detail, taking into account the dimensions of the system, the temperatures in various parts thereof, the temperature-dependent densities, the pressure differences caused by the density differences, and the resulting flow rates. The solutions of the equations are iterative, as the temperatures and flow rates are interdependent. An early analysis of this type was done by Close (1962). A more recent analysis has been described by Morrison and Braun (1985) and is a standard subroutine in the solar process simulation program TRNSYS (1990).

In a very much simpler approach, Löf and Close (1967) and Cooper (1973) observed that under wide ranges of conditions the increase in temperature of water flowing through many collectors in natural circulation systems (particularly those of Australian design) is approximately 10 C. Close (1962) compared computed and experimental inlet and outlet temperatures. His results, some of which are shown in Figure 12.6.1, confirm the suggestion that temperature increases of approximately 10 C are representative of these systems if they are well designed and without serious flow restrictions. Gupta and Garg (1968) also show inlet and outlet water temperatures for two collectors that suggest nearly constant temperature rise across the collectors.

If it is assumed that there is a constant temperature increase of water flowing through the collector, it is possible to calculate the flow rate which will produce this temperature difference at the estimated collector gain. The basic collector equations are

$$Q_u = A_c F_R[S - U_L(T_i - T_a)] \tag{12.6.1}$$

and $\qquad Q_u = \dot{m}C_p(T_o - T_i) = \dot{m}C_p\Delta T_f \tag{12.6.2}$

Solving for the flow rate,

$$\dot{m} = \frac{A_c F_R[S - U_L(T_i - T_a)]}{C_p \Delta T_f} \tag{12.6.3}$$

This equation can be solved for $\dot{m}$ if it is assumed that F' is independent of flow rate.

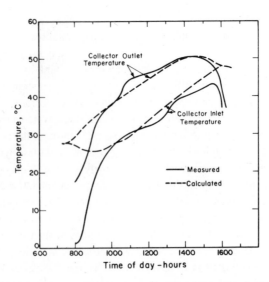

Figure 12.6.1 Collector inlet and outlet water temperatures for a natural circulation water heater. From Close (1962).

Substituting Equation 6.7.4 for F_R into Equation 12.6.3 and rearranging gives

$$\dot{m} = \frac{U_L F' A_c}{C_p \ln\left[1 - \dfrac{U_L(T_o - T_i)}{S - U_L(T_i - T_a)}\right]} \qquad (12.6.4)$$

Example 12.6.1

A natural circulation water heater operates with a nearly constant increase in water temperature of 10 C. The collector has an area of 4 m², an overall loss coefficient of 4.2 W/m²C, and an F' of 0.91. If the water inlet temperature is 30 C, the ambient temperature is 15 C, and the radiation absorbed by the collector plate is 780 W/m², what is the useful gain from the collector?

Solution

All of the information needed to estimate $\dot{m}$ using Equation 12.6.4 is available:

$$\dot{m} = \frac{-4.2 \times 0.91 \times 4}{4190 \ln\left[1 - \dfrac{4.2 \times 10}{780 - 4.2(30 - 15)}\right]}$$

$$= 0.060 \text{ kg/s}$$

$$Q_u = 0.060 \times 4190 \times 10 = 2550 \text{ W}$$

We have assumed F' is fixed. If necessary, a first estimate can be made of F', a first iteration of $\dot{m}$ is obtained as shown, and the assumed value of F' is checked using the calculated $\dot{m}$ to see if a second iteration is needed. If the fluid temperature rise were 20 C, the calculation shows $Q_u = 2450$ W, with the reduction due to lower F_R. The useful gain is only moderately sensitive to ΔT_f, except at low radiation levels or with collectors with high loss coefficients. ■

Collector operation with fluid temperature rises of approximately 10 C implies for practical systems that water circulates through the collector several times per day. Tabor (1969) suggested an alternative design in which the resistance to flow is higher so that flow rates would be lower and water in the tank would make about one pass through the collector in a day. He calculated that the daily efficiency of a "one-pass" high ΔT_f system will be about the same as a system using several passes per day and lower ΔT_f. This is dependent on maintaining good stratification in the tank. Gordon and Zarmi (1981) have analyzed single- and multipass systems and also concluded that under the approximations and assumptions they made (which included the assumption of no daytime load) the two types of systems give approximately the same daily collected energy. In these systems the effects of reduced F_R are offset by reduced mean fluid inlet temperature (in the same way as in the microflow forced-circulation systems noted in Section 12.5). Many Israeli water heaters are designed to operate this way.

In Australia, where natural circulation systems are widely used, sizing of systems is based on experience [CSIRO (1964)]. They are designed to produce water at 65 C and at daily average usage of 45 liters per person per day. If an all-solar system is to be used (for example, in Darwin, which is characterized by almost continuous good solar weather and expensive conventional fuels), a storage capacity of 2.5 times the daily requirement is suggested. For a family of four in Darwin, a collector area of about 4 m^2 is suggested. If an auxiliary energy source is to be used (e.g., in Melbourne, where radiation is more intermittent and conventional energy is less expensive), the recommended tank size is approximately 1.5 times the daily requirement. The majority of Australian solar water heaters are natural circulation systems, as freezing is not a problem in most of the country.

Year-long performance of thermosyphon water heaters has been measured by Morrison and Sapsford (1983). Two types of systems were measured: close-coupled systems, such as that shown in Figure 12.1.2(b), and systems with separate tank and collector, such as that shown in Figure 12.1.1(a). Auxiliary energy was supplied electrically, either on a continuous basis or off-peak. Loads were imposed that are typical of Australian hot water consumption. The close-coupled system showed slightly better performance than the systems with separate collector and tank when off-peak electric auxiliary was supplied. Systems operated with off-peak auxiliary delivered more solar energy than those with continuous auxiliary, as auxiliary caused less elevation of collector temperature. However, there were times in poor weather when the water output temperature dropped below control points in systems supplied with off-peak auxiliary. In contrast to pumped systems, the solar contribution of thermosyphon systems is better when the major loads were imposed in the morning; the difference is attributed to lower collector operating temperature in the morning and higher circulation rate produced by the introduction of cold water into the tank before the main solar input period.

12.7 INTEGRAL COLLECTOR STORAGE SYSTEMS

Water heaters that combine the functions of collector and storage are manufactured in Japan [Tanishita (1970)], the United States, and elsewhere. The design of these integral collector storage (ICS) systems vary. The basic configuration is that of a tank or set of interconnected tanks, with energy-absorbing surfaces on their exposed surfaces, enclosed in an insulated box with a transparent cover on the top to admit solar radiation. A cross-sectional schematic of a system of this type is shown in Figure 12.7.1. In the morning, water is put in the tubes, which in this configuration are black plastic cylinders about 0.2 m in diameter. Through the day, the water is

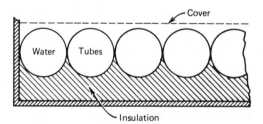

Figure 12.7.1 Schematic section of solar water heaters combining collector and storage.

heated by absorbed solar energy. The top cover and back insulation reduce energy losses from the cylinders. Analysis of the performance of this type of collector can be done by solving the differential equation describing the energy balance on the tank, as shown in the following example.

Example 12.7.1

A water heater that combines collector and storage is constructed of 150 mm black tubes placed side by side in a box, insulated on its bottom to the extent that bottom losses are small, and covered on the top by a single glass cover. A section of the heater is as shown in Figure 12.7.1. The heater is filled with water at 18 C at 6 AM. The heater is sloped toward the equator with $\beta = \phi = 40°$. Meteorological data for a 24-hour period are shown in the second and third columns of the table. Assume that no water is drawn off at any time during the 24-hour period.

Hour Ending	I_T MJ/m^2	T_a C	T_w^+ C	Hour Ending	I_T MJ/m^2	T_a C	T_w^+ C
7 am.	0.05	7	17.5	7 pm.	0	15	43.4
8	0.32	9	17.7	8	0	14	42.0
9	1.09	14	19.3	9	0	12	40.5
10	1.23	16	21.2	10	0	10	38.9
11	2.36	16	25.0	11	0	9	37.4
12	3.95	17	31.3	12	0	9	36.0
1 pm.	3.90	20	37.4	1 am.	0	9	34.7
2	3.52	20	42.5	2	0	8	33.3
3	2.55	21	45.7	3	0	6	32.0
4	1.38	22	46.9	4	0	7	30.7
5	0.46	21	46.4	5	0	7	29.5
6	0.04	16	44.9	6	0	7	28.4

Estimate the temperature history of the water through the 24-hour period.

Solution

A set of assumptions must be made to estimate the temperature history of this heater. A reasonable set is as follows:

> Absorptance = 0.95
> Emittance of black plastic = 0.95
> Solar transmittance of cover = 0.89
> Heat capacities of cover and other structure are negligible compared to that of the water.
> Water and tubes are at uniform temperature.
> Wind speed is such that h_w is 10 W/m^{2}C.
> $U_t = U_L$ = constant through the period.

First, estimate U_L. If the plate were flat, U_t would be approximately 5.5 W/m^{2}C, estimated from Figure 6.4.4(b). The area of the plate is $0.5\pi D/D$ larger than the cover, or

$$A_p = 1.57\, A_c$$

A first order estimate of U_t is

$$U_t = 5.5\left(1 + \frac{0.57}{2}\right) = 7.0 \text{ W/m}^2\text{C}$$

a value used without change in the calculation below.

The basic equation used is Equation 6.12.7. This can be written

$$T_w^+ = T_a + \frac{S}{U_L} - \left[\frac{S}{U_L} - (T_w - T_a)\right]\exp\left[-\frac{A_c U_L t}{(mC_p)_c}\right]$$

For each hour, with I_T in MJ/m^2,

$$\frac{S}{U_l} = \frac{1.02 \times 0.89 \times 0.95 I_T \times 10^6}{7.0 \times 3600} = 34.2 I_T$$

$$\frac{A_c U_L t}{(mC_p)_c} = \frac{0.15 \times 1 \times 7.0 \times 3600}{(\pi/4)(0.15)^2 \times 1 \times 1000 \times 4190} = 0.0511$$

and $e^{-0.0511} = 0.95$

A working equation is thus

$$T_w^+ = T_a + 34.2 I_T - [34.2 I_T - T_w + T_a]0.95$$

The temperatures at the end of each hour, T^+, are shown in the last column in the table. (These heaters are suitable for use where hot water is wanted at the end of the collection period. Thermal losses from the stored hot water are sufficiently high at night or during prolonged cloudy periods that their use for continuous hot water is usually impractical.)   ∎

Other types of ICS water heaters have been built. Some use a single energy-absorbing storage cylinder with an optical concentrator to increase absorbed radiation; these typically have concentration ratios of about unity. Vaxman and Sokolov (1985) describe experiments with a heater in which a flat tank is divided into two parts, a thin absorber channel under the absorbing surface, and a deeper storage zone behind the absorber channel and separated from it by an insulated partition. Water circulates from the storage zone through the channel by convection. Zollner et al. (1985) have worked out an analysis of ICS water heaters and developed a method for estimating long-term performance of these systems.

Water heaters combining collector and storage are passive systems in that they require no mechanical energy for operation. They are similar in some ways to collector-storage walls used in passive space heating (see Chapter 14).

12.8   RETROFIT WATER HEATERS

It is generally difficult to add solar equipment to buildings not designed for them. Of all solar energy systems, water heaters are the easiest to retrofit. Collector areas of residential water heating systems are manageable, and the addition of a separate solar storage tank can result in a system configuration such as shown in Figure 12.1.1(d), where the existing heater is the second tank. The solar heating system then provides preheated water to the conventional system and integration into the existing system is easy. Or a solar tank equipped with an auxiliary heater can replace the existing heater. The major problems are likely to be collector mounting and space for pipe or duct runs. As these runs are likely to be longer than in buildings designed for solar applications, they must be carefully insulated to avoid excessive losses. The methods of Section 10.3 can be used to estimate the effects of pipe or duct losses.

It is also possible to adapt existing tanks for solar operation, as shown by Czarnecki and Read (1978), by use of special fittings that allow the needed extra connections to be made to the tanks for circulation through the collector. Single openings in a tank can serve as dual openings by use of fittings having concentric center tubes; the center tube and annulus provide two connections that can be separated by lengthening the tube. Czarnecki and Read found that the measured thermal performance of such a system was poorer than that of a similar one-tank forced-circulation system using an outlet to the collector at the bottom of the tank and a return from the collector located halfway up the tank. Thus, reduced installation cost resulted in reduced solar output from the system. It is difficult to generalize on the merits of these various configurations, but as long as collector costs dominate the costs of the system, it is advisable to consider the use of the other components with good performance characteristics which will produce the most useful gain from the collectors.

12.9   WATER HEATING IN SPACE HEATING AND COOLING SYSTEMS

The following chapters treat space heating and cooling, and most systems for these applications will include provision for water heating. The water heating subsystems in these systems often include heat exchanger to extract heat either from the fluid circulated through the collector or from the main storage tank. A "solar preheat" tank to store solar-heated water and a conventional water heater (auxiliary) may be used, as shown in Figure 12.9.1, or a single-tank system including the auxiliary source can be used. The collectors for heating and cooling are usually sized for these functions and are oversized for water heating during periods when heating or cooling loads are low. Even in northern U.S. climates, annual residential hot water loads are typically one-fourth to one-fifth of annual heating loads, and significant contributions to total annual loads can be obtained by well-designed hot water facilities.

12.10   TESTING OF SOLAR WATER HEATERS

Standard methods have been devised for testing water heating systems of the types shown in Figure 12.1.1. ASHRAE-95 is an indoor test procedure in which a water heater is subjected to a defined sequence of operating conditions that are presumed to be typical of those that a solar water heater would undergo. The test conditions are specified by a rating agency, which is usually the Solar Rating and Certification Committee (SRCC) (1983) in the United States.

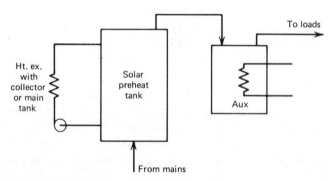

Figure 12.9.1   A two-tank hot water subsystem for solar heating/cooling systems.

Table 12.10.1   SRCC Irradiation, Incidence Angle, and Draw Profile During ASHRAE-95 Test

Time	Irradiation (W/m^2)	Incidence Angle	Draw (kg/hr)
8–9	1134	60	125
9–10	1692	45	0
10–11	2052	30	0
11–12	2376	15	0
12–1	2520	0	125
1–2	2376	15	0
2–3	2052	30	0
3–4	1692	45	0
4–5	1134	60	125

Incident radiation is simulated by the output of high-intensity lamps. The hourly irradiation, loads, and incidence angles specified by the SRCC are shown in Table 12.10.1. Ambient temperature and the temperature of the water supply are maintained at 22 C, and hot water is delivered at 50 C. The test procedure is as follows. The heater is set up according to the manufacturer's specifications in the simulator. It is then subjected to the "weather" and load profiles shown in the table. If it has an auxiliary heater (which is usually electric), the electrical energy supplied is subtracted from the load with the difference taken as the delivered solar energy. If the heater has no auxiliary heater, the delivered solar energy is determined from flow rate and temperature measurements at 10-minute intervals on water to the load. The test procedure is repeated until the delivered solar energy is within 3% of that of the previous day.

This test procedure provides comparative information on the performance of solar water heaters under the conditions of the tests. It does not indicate what comparative performance would be in various real climates and under other load conditions. Also, it does not permit calculation of the system parameters that would be the basis of evaluation of performance in any climate or under any load. Data on collector parameters [$F_R(\tau\alpha)_n$, $F_R U_L$, and b_o] tank loss coefficient $(UA)_s$, effects of controls, etc., would be needed to generalize on the results of these tests.

12.11   ECONOMICS OF SOLAR WATER HEATING

Most solar water heating installations are designed for use through the year, which results in a high use factor on the equipment and makes solar water heating more competitive with conventional methods than solar applications that are seasonal. A solar collector used in a water heater will deliver more energy per unit area per year than will a comparable installation used seasonally, such as winter space heating or harvest time crop drying.

The economic principles outlined in Chapter 11 apply directly to solar water heating. The installed costs include the purchase and installation costs of interest and principal payments, parasitic power for fluid circulation, insurance, maintenance, taxes, and any other costs which may be significant. The following example shows the application of life cycle savings analysis by the P_1, P_2 method to a solar water heater installation.

Example 12.11.1

A solar water heater is proposed to be installed on a residence in Madison. The annual load is estimated to be 22.2 GJ. The collector areas and tank capacities available are such that it is not practical to consider collector area and tank size as continuous variables, and a set of discrete system designs must be considered. The following table shows collector areas (for two, three, or four collector modules), tank capacities, installed costs, and fractions of annual hot water loads carried by solar energy.

Design	Collector Area, m^2	Tank Capacity, kg	System Cost, $	Fraction by Solar, F
A	3.66	310	1700	0.41
B	5.49	310	2100	0.55
C	5.49	454	2400	0.56
D	7.32	454	2800	0.67

The water heaters would be purchased by a 20% down payment with the balance to be paid over a 10-year period with interest at 9% per year. The present cost of electrical energy (the auxiliary and alternative energy supply) is $14.00/GJ, and it is expected to inflate at 8.5% per year. Insurance, maintenance, and parasitic power costs in the first year are expected to be 1% of the system cost. The real estate tax increment in the first year will be 1.5% of the system costs. The insurance, maintenance, parasitic power, and real estate taxes are expected to rise at a general inflation rate of 6% per year.

The owner's effective income tax bracket is 0.45. The market discount rate is to be 8% per year. If the analysis is done over 15 years and the system is assumed to have no appreciable resale value at that time, which system (if any) should be bought? For the best system, how long would it take to recover the system cost in savings on the purchase of electricity?

Solution

The analysis is done by the P_1, P_2 method of Chapter 11, with life cycle savings in each of the four cases calculated as shown in Example 11.8.2. Values of P_1 and P_2 are common to all four designs and are calculated by Equations 11.8.2 and 11.8.3.

$$P_1 = PWF(15, 0.085, 0.08) = 14.348$$

$$P_2 = 0.20 + 0.80 \frac{\text{PWF}(10, 0, 0.08)}{\text{PWF}(10, 0, 0.09)} - 0.80(0.45)$$

$$\times \left\{ \text{PWF}(10, 0.09, 0.08) \left[0.09 - \frac{1}{\text{PWF}(10, 0.09, 0.08)} \right] \right\}$$

$$+ \frac{\text{PWF}(10, 0, 0.08)}{\text{PWF}(10, 0, 0.09)}$$

$$+ 0.01\text{PWF}(15, 0.06, 0.08) + 0.015(1 - 0.45)\text{PWF}(15, 0.06, 0.08)$$

$$= 1.112$$

Then savings are calculated from Equation 11.8.1

$$LCS = 14.348 \times 14 \times 22.2 \times \mathcal{F} - 1.112 \times \text{Cost}$$

The savings for the four designs, under these assumptions of costs, are: A, –$62; B, $118; C, –$171; D, –$127.

Design B, the only one showing positive savings, is the one which should be purchased. However, the others are close, and small changes in the cost assumptions could shift all of them either way.

The time to recover the installed costs in undiscounted electricity savings for design B can be determined from Equation 11.7.3:

$$N_p = \frac{\ln \left[\dfrac{2100 \times 0.085}{0.55 \times 22.2 \times 14} + 1 \right]}{\ln(1 + 0.085)} = 8.8 \text{ years} \qquad \blacksquare$$

The competitive position of solar water heating is very much dependent on the costs of alternative energy and how rapidly it is expected to rise. It may also depend on the potential resale value of the equipment at the end of the period of economic analysis. Figure 12.11.1 was generated for design B, with economic assumptions the same as in Example 12.11.1. Fuel cost inflation rate is plotted against first-year fuel cost for combinations of C_{F1} and i_F which result in zero life cycle savings ("breakeven" combinations). Any combination of C_{F1} and its inflation rate which lies above and to the right of the line results in positive solar savings and economically feasible solar water heating.

The 0% curve is based on the assumption of zero resale value at the end of the period of the economic analysis. Two additional lines are shown that are based on the assumptions that the system is very well built and will have a resale value equal to its first cost (the curve marked 100%), or at its first cost inflated at the general inflation rate (the curve marked 240%). If the present cost of energy from fuel is

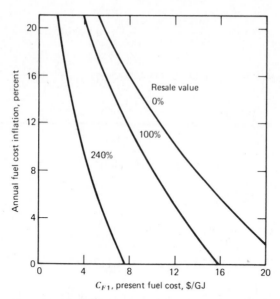

Figure 12.11.1 Examples of relations between fuel cost and its inflation rate for break even solar water heating system operation. The three curves are for three assumptions of resale value at the end of the period of the economic analyses. Areas above curves represent combinations leading to economically feasible solar water heating under the assumptions of Example 12.11.1 in Madison, WI.

$10/GJ and it is expected to increase at 10%/year, this system would show positive life cycle savings if its resale value will be 100% but not if it will have no resale value. Resale value results in significantly improved competitive positions for solar water heating for short periods of economic analysis. (Note, however, that this is a specific example. A new set of curves would have to be generated for new sets of economic parameters and locations. In particular, tax credits that effectively reduce first cost and other changes in installed cost of the water heaters may substantially alter the economic feasibility.)

12.12 SWIMMING POOL HEATING

Heating of swimming pools to prolong swimming seasons represents a significant consumption of energy in some areas and a substantial expense to pool owners and operators. The temperatures required for pools are low, usually not much above average ambient temperatures. For these reasons, solar heating of pools has become an application of increasing interest. Two general approaches have been taken: the first is based on covering the pool in such a way to use it as the collector, and the second is based on the use of separate collector systems.

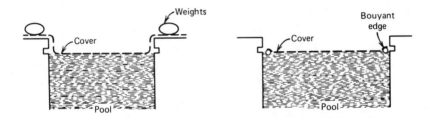

Figure 12.12.1 Schematic section of plastic pool covers: (a) held on pool edges by weights; (b) with floating edge.

Open pools lose heat by conduction to the ground, by convection to the air, and by evaporation,[2] as outlined by Löf and Löf (1977) and analyzed in some detail by Govaer and Zarmi (1981).

The simplest and least expensive method for heating pools may be to cover them when they are not in use with a transparent plastic cover which floats on the surface of the water. The cover may fit within the edges of the pool, or it may overlap the edges and be held down on the area around the pool by weights placed on the plastic to keep it in position. Section of pools with such covers are shown in Figure 12.12.1. As long as the top of the cover is dry, evaporation losses are effectively eliminated. The plastic cover, if largely transparent to solar radiation, admits solar energy which is absorbed in the pool. The cover also serves to keep dirt out of the pool. It is difficult to quantify the performance of such a system, as times of use, winds, humidity, etc., are highly variable. However, experiments in Denver have indicated that through the summer months an uncovered pool will have an average temperature below ambient, while a pool covered except for periods of occasional use will have average temperatures of 5 to 10 C above ambient temperature.

Either single-layer or double-layer covers are used. Single layers store in less volume when not on the pool and have higher transmittance. Double-layer covers float without addition of other buoyant materials but have lower transmittance for solar radiation and do not store in as compact a manner as single-layer covers.

Separate collectors are also used for heating pools. They operate by forced circulation with water inlet temperatures near ambient. Many pool heating collectors have no covers and a minimum of insulation. They may be made of metal or plastic and may be installed as a roof material for a patio or pool cabana. Swimming pool heaters have been described by Andrassy (1964) and others; deWinter (1974) has written a monograph on design and construction of pool heaters of this type, and an analysis of pool-heating collectors without covers has been presented by Farber (1978).

[2] Estimation of pool losses is discussed in Section 9.7. Section 18.5 discusses solar evaporators, an analogous situation.

12.13 SUMMARY

Water heating is a practical application of solar energy in many parts of the world. Natural circulation systems are widely used in climates where freezing does not occur. Forced-circulation systems using drain-down or nonfreezing fluids in collector-heat exchanger loops are used in climates where freezing is a problem. Auxiliary energy is used in essentially all systems where high reliability is wanted, and care must be taken to use auxiliary in such a way that it does not drive up collector temperatures.

REFERENCES

Andrassy, S., *Proc. of the UN Conference on New Sources of Energy*, **5**, 20 (1964). "Solar Water Heaters."

Buckles, W. E. and S. A. Klein, *Solar Energy*, **25**, 417 (1980). "Analysis of Solar Domestic Hot Water Heaters."

Carvalho, M. J., M. Collares-Pereira, F. M. Cunha, and C Vitorino, *Solar Energy*, **41**, 33 (1988). "An Experimental Comparison of Operating Strategies for Solar Water Systems."

Close, D. J., *Solar Energy*, **6**, (1), 33 (1962). "The Performance of Solar Water Heaters with Natural Circulation."

Cooper, P. I., personal communication (1973).

CSIRO, Div. of Mech. Engr. Circular No. 2, Commonwealth Scientific and Industrial Research Organization, (CSIRO), Melbourne, Australia (1964). "Solar Water Heaters, Principles of Design, Construction and Installation."

Czarnecki, J. T. and W. R. W. Read, *Solar Energy*, **20**, 75 (1978). "Advances in Solar Water Heating for Domestic Use in Australia."

Dalenbäck, J-O., personal communication (1990).

deWinter, F., *How to Design and Build a Solar Energy Swimming Pool Heater*, Copper Development Association, New York (1974).

Farber, J., *Proc. of the 1978 Annual Meeting of the American Section of the ISES*, Denver, 2 (1), 235 (1978). "Analysis of Low Temperature Plastic Collectors."

Gordon, J. M. and Y. Zarmi, *Solar Energy*, **27**, 441 (1981). "Thermosyphon Systems: Single vs. Multi-Pass."

Govaer, D. and Y. Zarmi, *Solar Energy*, **27**, 529 (1981). "Analytical Evaluation of Direct Solar Heating of Swimming Pools."

Gupta, C. L. and H. P. Garg, *Solar Energy*, **12**, 163 (1968). "System Design in Solar Water Heaters with Natural Circulation."

Gutierrez, G., F. Hincapie, J. A. Duffie, and W. A. Beckman, *Solar Energy*, **15**, 287 (1974). "Simulation of Forced Circulation Water Heaters; Effects of Auxiliary Energy Supply, Load Type and Storage Capacity."

Hollands. K. G. T. and M. F. Lightstone, *Solar Energy*, **43**, 97 (1989). "A Review of Low-Flow, Stratified-Tank Solar Water Heating Systems."

Hollands, K. G. T., D. A. Richmond, and D. R. Mandelstam, in *Proceedings, Intersol '85*, (the Biennial Congress of ISES), **1**, Pergamon Press, New York, 544 (1986). "Re-engineering Domestic Hot Water Systems for Low Flow."

Löf, G. O. G. and D. J. Close, paper in *Low Temperature Engineering Application of Solar Energy*, ASHRAE, New York (1967). "Solar Water Heaters."

Löf, G. O. G. and L. G. A. Löf, paper in *Proc. of the 1977 Annual Meeting of the American Section of the ISES*, Orlando, Vol. 1, 31-1 (1977). "Performance of a Solar Swimming Pool Heater-Transparent Cover Type."

Morrison, G. L. and J. E. Braun, *Solar Energy*, **34**, 389 (1985). "System Modeling and Operation Characteristics of Thermosyphon Solar Water Heaters."

Morrison, G. L. and C. M. Sapsford, *Solar Energy*, **30**, 341 (1983). "Long Term Performance of Thermosyphon Solar Water Heaters."

Mutch, J. J., RAND Report R 1498 (1974). "Residential Water Heating, Fuel Consumption, Economics and Public Policy."

Norton, B. and S. D. Probert, *Advances in Solar Energy*, **3**, (K. Boer, ed.) Plenum Press, New York, 125 (1986). "Thermosyphon Solar Energy Water Heaters."

SRCC, Standard 200-82, Solar Rating and Certification Corporation, Washington, DC, (1983). "Test Methods and Minimum Standards for Certifying Solar Water Systems."

Tabor, H., *Bulletin, Cooperation Mediterranee pour L'Energie Solaire* (COMPLES), No. 17, 33 (1969). "A Note on the Thermosyphon Solar Hot Water Heater."

Tanishita, I., paper presented at Melbourne International Solar Energy Society Conference (1970). "Present Situation of Commercial Solar Water Heaters in Japan."

TRNSYS Users Manual, Version 13.1, University of Wisconsin Solar Energy Laboratory (1990). [First public version was 7.0 (1976).]

van Koppen, C. W. J., J. P. S. Thomas, and W. B. Veltkamp, *Sun II, Proc. of ISES Biennial Meeting*, Atlanta, GA, **2**, 576 Pergamon Press, New York (1979). "The Actual Benefits of Thermally Stratified Storage in Small and Medium Sized Solar Systems."

Vaxman, B. and M. Sokolov, *Solar Energy*, **34**, 447 (1985). "Experiments with an Integral Compact Solar Water Heater."

Wuestling, M. D., S. A. Klein, and J. A. Duffie, *Trans. ASME, J. Solar Energy Engrg.*, **107**, 215 (1985). Promising Control Alternatives for Solar Water Heating Systems."

Zollner, A., S. A. Klein, and W. A. Beckman, *Trans. ASME, J Solar Energy Engrg.*, **107**, 265 (1985). "A Performance Prediction Methodology for Integral Collection-Storage Solar Domestic Hot Water Systems."

Chapter 13

BUILDING HEATING - ACTIVE

Heat for comfort in buildings can be provided from solar energy by systems that are similar in many respects to the water heater systems described in Chapter 12. The two most common heat transfer fluids are water (or water and antifreeze mixtures) and air, and systems based on each of these are described in this chapter. The basic components are the collector, storage unit, and load (i.e., the house or building to be heated). In temperate climates, an auxiliary energy source must be provided and the design problem is in part the determination of an optimum combination of solar energy and auxiliary (i.e., conventional) energy. Systems for space heating are very similar to those for water heating, and the same considerations of combination with an auxiliary source, boiling and freezing, controls, etc., apply to both applications.

In this chapter we deal with active solar heating systems which use collectors to heat a fluid, storage units to store solar energy until needed, and distribution equipment to provide the solar energy to the heated spaces in a controlled manner. In combination with conventional heating equipment solar heating provides the same levels of comfort, temperature stability, and reliability as conventional systems. A building so heated is often referred to as a "solar house."

The term solar house is also applied to buildings that include as integral parts of the building elements that admit, absorb, store, and release solar energy and thus reduce the needs for auxiliary energy for comfort heating. Architectural design can be used to maximize solar gains in the winter (and minimize them in the summer) to reduce heating (and cooling) loads that must be met by other means. Elements in the building (floors, walls) may be constructed to have high heat capacity to store thermal energy and reduce temperature variations. Moveable insulation may be used to control losses (and gains) from windows or other architectural elements in the building. These concepts will be treated in Chapter 14 on passive heating. There is no substitute for good energy-conserving architectural design which (as far as other constraints allow) maximize the solar gains in the building itself. In this chapter we are concerned with meeting the energy needs for heating (and hot water) that are not eliminated by careful design.

Since 1970 there has been a surge of interest and activity in solar heating, and many thousands of active systems have been designed, installed, and operated. Patterns in the configuration of many air and liquid systems have emerged. These "standard" configurations, which are used with many variations, are discussed here in some detail. In addition, other system types are noted. Operating data are shown for two systems that are illustrative of the long-term thermal performance data needed to evaluate the economics of solar heating.

The design of solar heating systems is a three-part problem. The first is to determine the thermal performance, and the second is to carry out an economic analysis. This book is concerned with these two. The third is to assure that the resulting building is aesthetically pleasing and functional; other works treat these aspects of the design process. Thermal analyses can be done by simulation methods that were touched on in Chapter 10 and are discussed in more detail in Chapter 19 or by design methods that are to be covered in Chapters 20 and 21. Economic analyses are done by the P_1, P_2 method.

13.1   HISTORICAL NOTES

A gift by Godfrey L. Cabot to the Massachusetts Institute of Technology for solar energy studies in 1938 marked the beginning of modern research in solar heating. The MIT program resulted in development of methods of calculating collector performance (see Chapter 6) which, with some modification, are standard methods in use today. It also resulted in the successive development of a series of five solar-heated structures. MIT House IV was a residence equipped with a carefully engineered and instrumented system based on solar water heaters and water storage. The solar heating system was designed to carry approximately two-thirds of the total winter heating loads in the Boston area, and its performance was carefully monitored. In the two years of its experimental operation it supplied 48 and 57% of the combined heating and hot water loads, the weather being substantially better in the second year than the first [Engebretson (1964)]. This series of experiments has provided much of the background information on liquid systems that has been used in development of these systems in the three decades since.

Telkes and Raymond (1949) described a solar house at Dover, Massachusetts, that utilized vertical south-facing air heater collectors and energy storage in the heat of fusion of sodium sulfate decahydrate. This system was designed to carry the total heating load, having theoretical capacity in the storage system to carry the design heating loads for 5 days.

Bliss (1955) constructed and measured the performance of a fully solar heated house in the Arizona desert, using a matrix, through-flow air heater and a rock bed energy storage unit. He noted that the system as built did not represent an economic optimum, and a smaller system using some auxiliary energy would have resulted in lower cost.

Löf designed an air heating system using overlapped glass-plate collectors and a pebble bed for energy storage and, using these concepts, built a residence near Denver in which he and his family have lived since 1959. The performance of this system during the first years of its operation was studied and reported by Löf et al. (1963, 1964) and was again measured in 1976 to 1978. The data are particularly significant in that they are the only data available on a system that has been in operation over a time span of 18 years when essentially no maintenance work was done on it. It was found that the performance in the second measurement period was about 78% of the original; the drop was attributed to breakage of glass inside of the collectors, a problem that more modern collectors do not have. Experience with this system provides evidence that well-designed and well-built air systems can operate reliably over many years with very little maintenance.

Close et al. (1968) described a heating system used for partial heating of a laboratory building in Australia that was operated for many years. This system used a 56-m^2 vee-groove air heater and a pebble bed storage unit. Air flow through the collectors was modulated to obtain a fixed 55 C air outlet temperature.

Since 1970, many varied experimental systems have been built, and the performance of a few of them has been measured and reported. These experiments have led to commercial production of both liquid and air systems, and these have been sold and installed by the thousands.

13.2 SOLAR HEATING SYSTEMS

Figure 13.2.1 is a schematic of a basic solar heating system using air as the heat transfer fluid, with a pebble bed storage unit and auxiliary furnace. The various modes of operation are achieved by appropriate damper positioning. Most air systems are arranged so that it is not practical to combine modes by both adding energy to and removing energy from storage at the same time. The use of auxiliary energy can be combined with energy supplied to the building from collector or storage if that supply is inadequate to meet the loads. In this system configuration it is possible to bypass the collector and storage unit when auxiliary alone is being used to provide heat.

A more detailed schematic of an air system is shown in Figure 13.2.2. Blowers, controls, means of obtaining service hot water, and more details of ducting

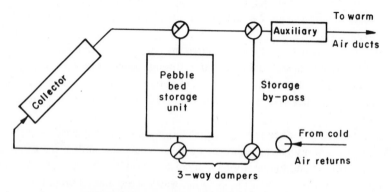

Figure 13.2.1 Schematic of basic hot air system.

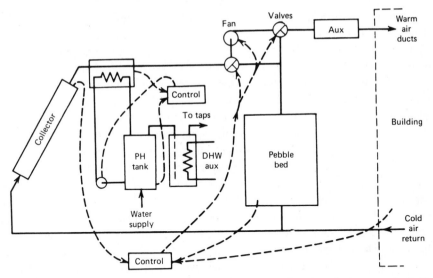

Figure 13.2.2   Detailed schematic of a solar heating system using air as the heat transfer fluid.

are shown. The hot water subsystem is the same as that shown in Figure 12.8.1. Auxiliary energy for space heating is added to "top off" that available from the solar energy system. This is the air system configuration on which the design procedure of Chapter 20 is based. Table 13.2.1 shows typical design parameters for heating systems of this type.

Air systems have a number of advantages compared to those using liquid heat transfer media. Problems of freezing and boiling in the collectors are eliminated and corrosion problems are reduced. The high degree of stratification possible in the pebble bed leads to lower collector inlet fluid temperatures. The working fluid is air, and warm air heating systems are in common use. Control equipment is readily

Table 13.2.1   Typical Design Parameters for Solar Air Heating Systems

Collector air flow rate	5 to 20 liters/m²s
Collector slope	$(\phi + 15^\circ) \pm 15^\circ$
Collector surface azimuth angle	$0^\circ \pm 15^\circ$
Storage capacity	0.15 to 0.35 m³ pebbles/m²
Pebble size (graded to uniform size)	0.01 to 0.03 m
Bed length, flow direction	1.25 to 2.5 m
Pressure drops:	
Pebble bed	55 Pa minimum
Collectors	50 to 200 Pa
Ductwork	10 Pa
Maximum entry velocity of air	
into pebble bed (at 55 Pa pressure	
drop in bed)	4 m/s
Water preheat tank capacity	1.5 × conventional heater

available that can be applied to these systems. Disadvantages include the possibility of relatively high fluid pumping costs (if the system is not carefully designed to minimize pressure drops), relatively large volumes of storage, and the difficulty of adding solar air conditioning to the systems. Air systems are relatively difficult to seal, and leakage of solar-heated air from collectors and ductwork can represent a significant energy loss from the system. Air collectors are operated at lower fluid capacitance rates and thus with lower values of F_R than are liquid heating collectors.

Air heating collectors in space heating systems are usually operated at fixed air flow rates, with outlet temperatures variable through the day. It is possible to operate them at fixed outlet temperature by varying the flow rate. This results in reduced F_R and thus reduced collector performance when flow rates are low. It leads to a reduction in required storage volume by changing the shape of the temperature profiles in storage beds from those of Figure 8.5.3 to those of Figure 8.5.2. The uniform bed temperature means that sufficiently warm air can be delivered from storage which can improve comfort conditions.

Figure 13.2.3 is a schematic of a basic water heating system with water tank storage and auxiliary energy source. This system allows independent control of the solar collector-storage loop on the one hand, and storage-auxiliary-load loop on the other, as solar-heated water can be added to storage at the same time that hot water is removed from storage to meet building loads. In the system illustrated, a bypass around the storage tank is provided to avoid heating the storage tank with auxiliary energy.

More details of a liquid-based system are shown in Figure 13.2.4. A **collector heat exchanger** is shown between the collector and storage tank, allowing the use of antifreeze solutions in the collector. Relief valves are shown for dumping excess energy should the collector run at excessive temperatures. A **load heat exchanger** is shown to transfer energy from the tank to the heated spaces. Means of extracting energy for service hot water are indicated. Auxiliary energy for heating is added so as to "top off" that available from the solar energy system. (This system is the basis of the liquid system design method discussed in Chapter 20.) Typical design parameter values for systems of this type are shown in Table 13.2.2.

The load heat exchanger, the exchanger for transferring solar heat from the storage tank to the air in the building, must be adequately designed to avoid excessive

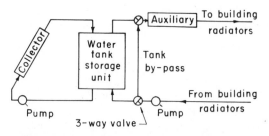

Figure 13.2.3   Schematic of basic hot water system.

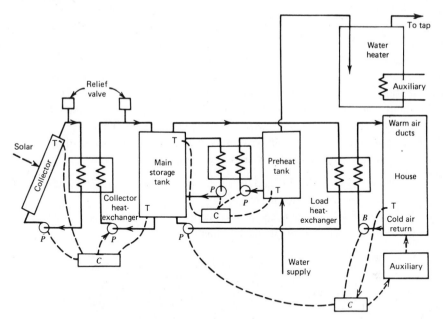

Figure 13.2.4 Detailed schematic of a liquid-based solar heating system: P, pump; C, controller; T, temperature sensor; B, blower.

Table 13.2.2 Typical Design Parameter Ranges for Liquid Solar Heating Systems

Collector flow rate	0.010 to 0.020 kg/m^{2}s
Collector slope	$(\phi + 15^\circ) \pm 15^\circ$
Collector azimuth	$0^\circ \pm 15^\circ$
Collector heat exchanger	$F'_R/F_R > 0.9$
Storage capacity	50 to 100 liters/m^2
Load heat exchanger	$1 < \varepsilon_L C_{min}/(UA)_h < 5$
Water preheat tank capacity	$1.5 \times$ conventional heater

temperature drop and corresponding increase in tank and collector temperatures. The parameter describing the exchanger is $\varepsilon_L C_{min}/(UA)_h$, where ε_L is the effectiveness of the exchanger, C_{min} is the lower of the two fluid capacitance rates (mass flow rate times specific heat of the fluid) in the heat exchanger (usually that of the air), and $(UA)_h$ is the building overall energy loss coefficient-area product.

An alternative to the use of antifreeze loops to transfer heat from the collector is a drain-back (or drain-down) system. This method of freeze protection was used in MIT House IV.

Advantages of liquid heating systems include high collector F_R, smaller storage volume, and relatively easy adaptation to supply of energy to absorption air conditioners.

Use of water also involves problems. Freezing of collectors must be avoided, as outlined in Chapter 12 for water heaters. Solar heating systems using liquids will operate at lower water temperatures than conventional hydronic systems and will require more baseboard heater area (or other heat transfer surface) to transfer heat into the building. In spring and fall, solar heaters will operate at excessively high temperatures, and means must be provided to vent energy to avoid pressure build-up and boiling. Care must also be exercised to avoid corrosion problems.

It is useful to consider solar systems as having four basic modes of operation, depending on the conditions that exist in the system at a particular time:

A   If solar energy is available and heat is not needed in the building, energy gain from the collector is added to storage.

B   If solar energy is available and heat is needed in the building, energy gain from the collector is used to supply the building need.

Figure 13.2.5 Three solar heated buildings: (a) The Denver House; (b) MIT House IV (Courtesy of H.C. Hottel); (c) CSU House II. (Courtesy of Colorado State University)

C   If solar energy is not available, heat is needed in the building, and the storage unit has stored energy in it, the stored energy is used to supply the building need.

D   If solar energy is not available, heat is needed in the building, and the storage unit has been depleted, auxiliary energy is used to supply the building need.

Note that there is a fifth situation that will exist in practical systems. The storage unit is fully heated, there are no loads to be met, and the collector is absorbing radiation. Under these circumstances, there is no way to use or store the collected energy and this energy must be discarded. This can happen through operation of pressure relief valves or other energy dumping mechanisms, or if the flow of fluid is turned off, the collector temperature will rise until the absorbed energy is dissipated by thermal losses.

Additional operational modes may also be provided, for example, to provide service hot water. It is possible with many systems to combine modes, that is, to operate in more than one mode at a time. Many systems do not allow direct heating from collector to building but transfer heat from collector to storage whenever possible and from storage to load whenever needed.

To illustrate the system configurations, modes of operation, and designs of air and water systems, the following sections describe the design and performance of residential-scale installations. These systems were carefully engineered, and their performance has been measured and the results documented. The buildings and systems are the Colorado State University (CSU) House III drain-back liquid system and the CSU House II air system. Three houses with varying architectural treatment, CSU House II, MIT House IV, and the Löf Denver residence, are shown in Figure 13.2.5.

13.3   CSU HOUSE III FLAT-PLATE LIQUID SYSTEM

The Colorado State University solar houses at Ft. Collins, CO., are residential-scale buildings that are used as offices and laboratories. The thermal loads on the buildings, for heating and air conditioning, are representative of those of conventional, well-built three-bedroom residences. The buildings serve as "test beds" for development of solar heating and air conditioning systems. Each system that is installed on one of the three houses is operated for at least a heating (and/or cooling) season, and its performance is carefully monitored. In this and the following section the results of two of these operations are described. The systems described here may be more carefully designed, built, and operated than most domestic systems, and their performance may be somewhat better than can be expected of many commercial systems. However, their performance represents achievable goals for commercial systems.

House III is a two-story frame building with a floor area of about 130 m^2 per floor. The estimated $(UA)_h$ of the building is 420 W/C, including an infiltration rate

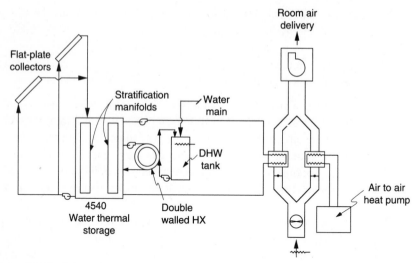

Figure 13.3.1 Schematic of the CSU House III heating system. From Karaki et al. (1984).

of one air change per hour. The 48.7 m² of collectors are mounted on a south-facing roof at a slope of 45°. Thermal energy storage is in a 4500-liter horizontal, insulated water tank on the lower level. The water heating subsystem includes a double-walled heat exchanger between the main storage tank and the 300-liter hot water tank. Auxiliary energy for building heating is supplied in parallel with the solar heat by an air-to-air heat pump (which as far as the solar operation is concerned is the equivalent of a furnace). A schematic of the system is shown in Figure 13.3.1. The building, the system, and its operation are described by Karaki et al. (1984). (The system also included an absorption air conditioner for summer cooling; this will be noted in Chapter 15.)

The collectors are Revere double-glazed, flat-plate collectors with tempered glass covers, copper absorber plates, and black chrome selective surfaces; $F_R(\tau\alpha)_n$ is 0.71 and $F_R U_L$ is 4.2 W/m²C. When the collector circulation pump is not operating, the collectors drain back into the main storage tank. Piping and manifolds are designed to be large enough so that the water in the collectors will drain completely on shutdown.

The storage tank is fitted with stratification manifolds on the inlet pipes from both collector and load. These are shown schematically in Figure 8.4.2. Water enters the inlet chamber horizontally, and momentum is dissipated radially in the chamber. The water then flows up or down in the perforated pipe to the region where its density matches that of the water in the tank. Temperatures at several levels in the tank and collector inlet and outlet temperatures are shown in Figure 13.3.2 for two flow rates. At both rates, the temperatures in the top and bottom of the tank are close to the collector inlet and outlet temperatures, indicating that the stratification obtained is close to the maximum possible.

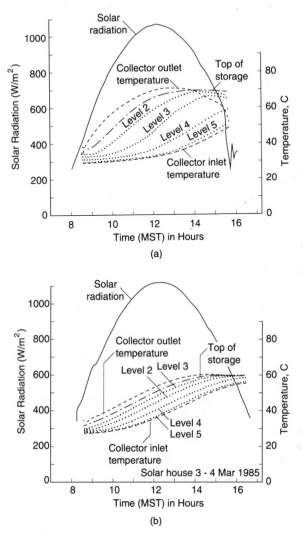

Figure 13.3.2 Temperatures in and out of the collector and at several levels in the storage tank in CSU House III at collector flow rates of: (a) 0.0034 kg/m²s and (b) 0.0086 kg/m²s. From Karaki et al. (1984).

The performance of the building heating system for a season is shown in Figure 13.3.3. During this time the solar heating system met 80% of the space heating load on the building. (The solar contribution in March was reduced when the tank was drained and refilled for experiments on effects of collector flow rate.) The performance of the system for water heating is shown in Figure 13.3.4 for the same months; the fraction of the water heating loads carried by solar was 0.36. (The March contribution was again reduced by experimentation. Summer operation of the

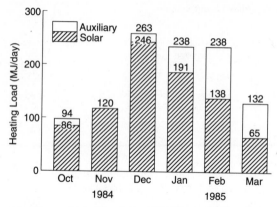

Figure 13.3.3 Monthly performance of the CSU House III flat-plate space heating system. March performance was reduced by experiments. From Karaki et al. (1984).

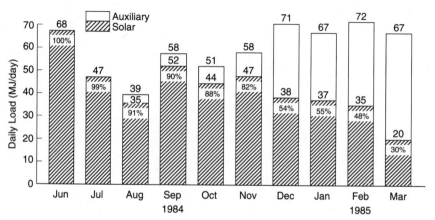

Figure 13.3.4 Monthly water heating performance of the CSU House III system. Adapted from Karaki et al. (1984).

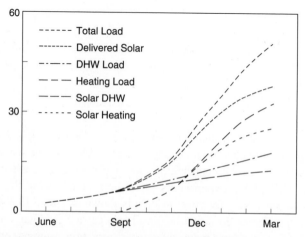

Figure 13.3.5 Measured cumulative performance of the CSU House III system.

water heating system, when the tank was at higher temperatures for air conditioning, resulted in solar contributions of 90 to 100%.)

A cumulative performance plot for the heating season showing total loads and solar contributions to meeting the loads is shown in Figure 13.3.5.

13.4 CSU HOUSE II AIR SYSTEM

An experiment on House II used an air system with conventional air heating collectors and a nearly cubic pebble bed energy store. The system configuration is shown in Figure 13.4.1 and is essentially the same as is shown in Figure 13.2.2. The house is nearly identical in appearance to CSU House III as shown in Figure 13.2.5. The system as it was installed and operated in 1977 to 1978 is described, and its performance is summarized by Ward et al. (1977) and Karaki et al. (1977, 1978). (Since 1978, the building has been used for other system experiments.)

The building serves as an office and laboratory and has 130 m^2 floor area on the first floor and an equal area on a basement level. The collectors in the experiment

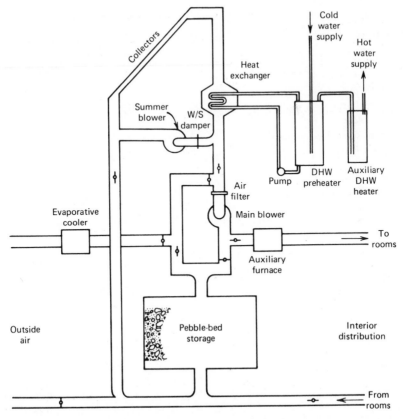

Figure 13.4.1 Schematic of the system in CSU House II. From Karaki et al. (1977).

were site built and were similar in design to air collectors available on the commercial market. They occupied an area of 68.4 m^2 on the south roof and had 64.1 m^2 net collector area. The storage unit contained 10.2 m^3 of pebbles 20 to 40 mm in diameter. The auxiliary energy source was a gas-fired duct furnace. Domestic water preheat was obtained by an air-to-water heat exchanger; the preheat tank capacity was 300 liters. The system configuration required only one blower for the heating operation; it had a second blower for providing heated air to the hot water subsystem in the summer.

Controls on the collector loop during the heating season were based on temperature measurements in the top of the collectors and the bottom of the storage unit; when the difference of these two exceeded 11 C, the blower was turned on, and it remained on until the difference dropped to 0.6 C. The solar-heated air went to the building when heat was needed; otherwise it went to storage. A thermostat in the building sensed when the rooms needed heat and positioned the dampers to obtain heat from the collector (if available) or storage. If room temperature dropped 1.1 C below the control point, a second-stage contact was made and the auxiliary gas duct heater was turned on.

Table 13.4.1   Thermal Performance Data for Solar House II[a]

	1976		1977			
	Nov 24 days	Dec 10 days	Jan 26 days	Feb 20 days	Mar 27 days	Apr 18 days
Collector (MJ/m^2 day)						
Total solar insolation	13.5	15.8	15.1	15.6	17.7	14.2
Solar insolation						
while collecting	11.2	13.8	12.9	13.6	15.6	12.2
Heat collected	3.1	4.9	4.6	4.7	5.0	3.7
Efficiency, %						
Based on total solar insolation	23	31	30	30	28	26
Based on solar insolation						
while collecting	28	36	36	35	32	30
Space and DW heating (MJ/day)						
Total energy required						
Space	311.6	323.2	432.9	319.9	319.0	192.3
DHW	40.6	52.3	73.0	82.9	75.4	62.8
Total	352.2	375.5	505.9	402.8	394.4	255.1
Solar contribution						
Space	198.7	257.5	231.4	239.6	245.8	166.1
DHW	21.3	30.8	63.5	75.4	67.4	51.8
Total	220.0	288.5	294.9	315.0	313.2	217.9
Solar fraction						
Space	0.64	0.80	0.53	0.75	0.77	0.86
DHW	0.52	0.59	0.87	0.91	0.89	0.82
Total	0.62	0.77	0.58	0.78	0.79	0.85

[a] From Karaki et al. (1977).

The house and its heating system were extensively instrumented, and data for most of the winter of 1976 to 77 (except for some days when instrumentation was not operable) are shown in Table 13.4.1. For this heating season, the solar heating system carried 72% of the heating and domestic hot water (DHW) loads. The parasitic power for operation of the solar energy system was 7.4% of the delivered solar energy for space and water heating, resulting in a "coefficient of performance" (COP), the ratio of solar energy delivered to electrical energy to operate the system, of 13.5 for the heating season.

13.5   A HEATING SYSTEM PARAMETRIC STUDY

In the preceding sections we discussed the design and measured performance of two solar heating systems. Of necessity, each of these systems was of fixed design and was operated and monitored in the particular weather sequence experienced during the operation. Simulations and design procedures,[1] on the other hand, offer the opportunity to determine the effects of changes in system configuration and design parameters and changes in climate on annual performance. In this section, computed performance is used to illustrate the effects of collector area and storage capacity on monthly and annual solar fractions. The meteorological data used are hourly data for a Madison design year, as defined by Klein et al. (1976).

The system in this simulation is similar to that in Figure 13.2.2 and is shown in Figure 13.5.1. It is to provide space heating and hot water for a well-insulated residence of 150 m^2 floor area. Internal heat generation, infiltration, and capacitance

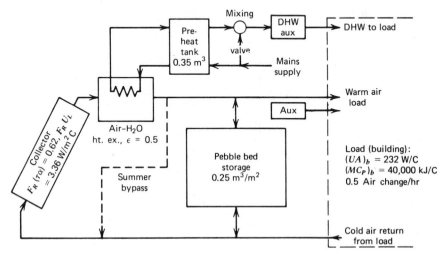

Figure 13.5.1   Schematic of the air system simulated in the Madison simulation example. Details of blowers, pumps, etc., are not shown.

---

[1] As described in Chapters 19, 20, and 21.

of the building are considered. The hot water demand is 300 liters per day, assumed distributed through the day according to the load profile shown in Figure 9.1.2, with water entering at 11 C and supplied to the building at 55 C.

The collectors are air heaters of design similar to that in Figure 6.14.1(b), with $F_R(\tau\alpha) = 0.62$. The $F_R U_L$ is assumed fixed at 3.38 W/m²C, and air flow rate to collectors is 0.0125 kg/m²s. Collector areas of 10, 30, 55, and 105 m² were used in the simulations. The collector slope is 58° and surface azimuth angle is zero. Ground reflectance is assumed to be 0.2 at all times.

The pebble bed is sized so that its volume is at a fixed ratio to the collector area of 0.25 m³/m². The bulk density of the pebble bed is 1600 kg/m³, and the specific heat of the pebbles is 840 J/kgC. The loss coefficient of the pebble bed is 0.5 W/m²C, and losses from the bed are to a 20 C environment. To show the effects of relative pebble bed size, simulations were done at 25, 50, and 200% of the nominal size for collector areas of 30 and 55 m².

The auxiliary heater has a maximum capacity of 13.9 kW. The heating system is controlled by a two-stage thermostat which is designed to keep the building near 20 C. When heat is needed, solar heat is called first. If that will not hold the temperature, the solar heating system is turned off and the auxiliary is turned on. These control modes are indicated in Figure 13.5.2. In summer (June, July, and August), the pebble bed is bypassed so that heat is supplied only to the hot water preheat tank, and the collector fan operates when the preheat tank is below its control temperature and the collector can deliver energy. The volume of the water preheat tank is 0.35 m³, and its loss coefficient is 0.5 W/m²C.

Room temp, °C	Direction of change	Solar heating system	Auxiliary heating system
21.5	↓	Off	Off
	↑	Turns off[1]	Off
20.0	↓	Turns on	Off
	↑	On	Off
19.9	↓	On	Off
	↑	Turns on[2]	Turns off[1]
18.5	↓	Turns off	Turns on
	↑	Off	On

[1] Turns off if on
[2] Turns on if off

Figure 13.5.2 The control logic used in the Madison simulations. In addition to the functions shown, it is a requirement that the solar energy system must be able to supply air at a temperature above 30 C. If it cannot, it is off regardless of the room temperature.

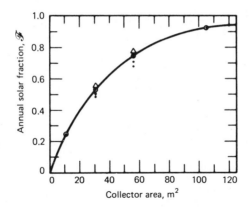

Figure 13.5.3 Solar fraction $\mathcal{F}$ as a function of collector area for the Madison simulations. The circled points and curves are for the normal storage size of 0.25 m³ of storage per m² of collector. The triangles show the effect of increasing storage size to 0.50 m³/m², and the dots show the effects of decreasing it to 0.125 m³/m² and 0.063 m³/m².

Figure 13.5.3 shows the annual solar fraction $\mathcal{F}$ as a function of the collector area for the normal storage volume-collector area ratio, and also shows points for larger and smaller storage capacities for the 30 and 55 m² collector areas. This is a typical $\mathcal{F}$ versus A_c curve for house heating, in that the slope is substantially higher at small collector areas than at large collector areas. The large collectors are oversized a greater part of the year than are the small collectors. Thus an increment in collector area produces more useful energy at small collector areas than at large areas. The effects of change of storage size are small for the 30 m² collector, as for much of the year the bottom of the bed is at a higher average temperature, resulting in increased air temperature to the collector.

Figure 13.5.4(a) shows the monthly fraction of loads carried by solar for the four collector areas for the normal storage volume-collector area ratio. All of these systems will carry most of the summer hot water loads. (There are a few times during the design year when auxiliary heat is used to supply hot water during prolonged cloudy periods, and storage capacity limits solar fractions to less than unity.) The largest system meets 95% or more of the loads through 9 months, while the smaller systems meet relatively small fractions of the monthly loads in midwinter.

Figure 13.5.4(b) shows essentially the same information but is in terms of the monthly efficiencies of the system for the four areas. System efficiency is defined here as the total useful solar energy delivered to the building for the month divided by the total energy incident on the collector in the month. Annual efficiencies η_a are also indicated for the four collector areas. It is clear that the efficiencies of this collector can vary over a very wide range, depending on the relative magnitudes of loads and insolation. (Variations of instantaneous efficiencies are greater than those shown for months.) The larger collectors are oversized for longer periods of the year and operate at higher temperatures and lower efficiencies.

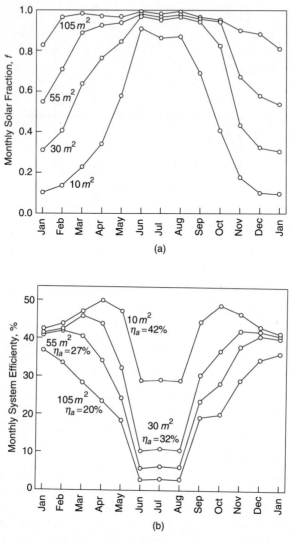

Figure 13.5.4 Data for four collector areas from the Madison simulations: (a) Monthly fraction of loads carried by solar energy. (b) Monthly system efficiencies.

Figure 13.5.5 indicates the cumulative energy quantities, starting from zero at the beginning of the heating season, for the 55 m^2 system with normal storage capacity. It indicates the cumulative total loads and the heating and hot water loads met by solar through the year.

The cumulative energy quantities are summarized in Table 13.5.1. These are used in the discussion of economics of solar heating (see Section 13.10).

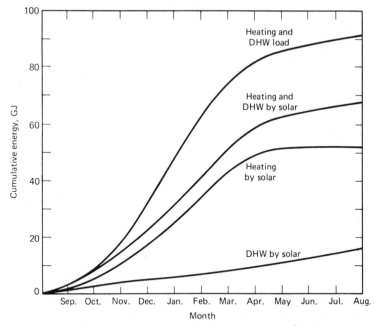

Figure 13.5.5 Cumulative loads and solar contributions for the 55 m^2 collector area and 0.25 m^3 of storage per m^2 of collector, for the Madison simulations.

Table 13.5.1   Annual Energy Quantities for the Madison Simulations

Collector Area	Storage Volume	Solar Energy to Space Heating	Solar Energy to Hot Water	Total Heating, Hot Water Loads[a]	$\mathcal{F}$
10 m^2	2.50 m^3	11.1 GJ	10.7 GJ	86.9 GJ	0.25
30	7.50	32.7	14.5	89.3	0.53
55	13.75	51.7	15.9	91.2	0.74
105	26.25	68.3	16.9	92.7	0.92

[a] The loads are to some degree a function of system size, as larger systems keep the building at slightly higher mean temperatures.

13.6   SOLAR ENERGY-HEAT PUMP SYSTEMS

Heat pumps use mechanical energy to transfer thermal energy from a source at a lower temperature to a sink at a higher temperature. Electrically driven heat pump heating systems have attracted wide interest as an alternative to electric resistance heating or expensive fuels. They have two advantages: a COP (ratio of heating capacity to electrical input) greater than unity for heating, which saves on purchase of energy, and usefulness for air conditioning in the summer. Heat pumps may use air or water sources for energy, and dual-source machines are under development that can use either.

Yanagimachi (1958, 1964) and Bliss (1964) built and operated heating and cooling systems that use uncovered collectors as daytime collectors and nighttime radiators, "hot" and "cold" water storage tanks to supply heating or cooling to the buildings, and heat pumps to assure maintenance of adequate temperature differences between the two tanks. The Yanagimachi system was applied to several houses in the Tokyo area, and the Bliss system was used on a laboratory in Tucson, Arizona.

Heat pumps have been studied by Jordan and Threlkeld (1954) and by an AEIC-EEI Heat Pump Committee (1956). An office building in Albuquerque was heated and cooled by a collector-heat pump system [Bridgers et al. (1957a,b)]. More recent systems have been built on residential buildings [e.g., Converse (1976), Kuharich (1976), and Terrell (1979)].

A schematic of an air-to-air heat pump is shown in Figure 13.6.1, operating in the heating mode. The most common types in small sizes are air-to-air units. For a discussion of the design and operation of heat pumps, see, for example, ASHRAE (1976). Typical operation characteristics of a residential-scale heat pump are shown in Figure 13.6.2. As ambient air temperature (the evaporator fluid inlet temperature) increases, the COP increases as does the capacity. As the air temperature drops, frost can form on the evaporator coils, which adds heat transfer resistance and blocks air flow. Brief operation of the heat pump in the cooling mode removes the frost (a defrost cycle) and causes the irregularity shown in the capacity and COP curves. Figure 13.6.2 also shows a typical building heating requirement curve, which crosses the capacity curve at the balance point (BP). At ambient temperatures below this point a heat pump will have inadequate capacity to heat the building, and the difference must be supplied by a supplemental source (often an electric resistance heater). At ambient temperatures above the balance point, the heat pump has excess capacity.

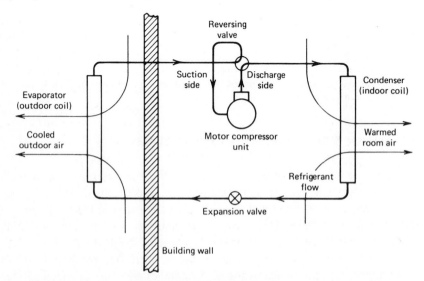

Figure 13.6.1 Schematic of a reversible heat pump system shown operating as a space heater.

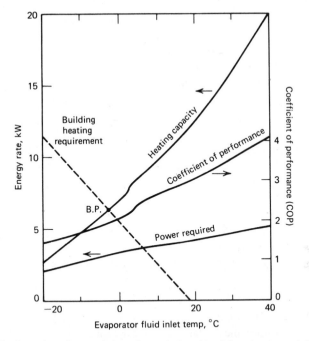

Figure 13.6.2 Operating characteristics of a typical residential scale air-to air-heat pump as a function of ambient air temperatures for delivery of energy to a building at 20 C.

Several configurations of solar energy-heat pump systems are possible. Schematics of the energy flows of the most important of these are shown in Figure 13.6.3. Part (a) shows a **parallel** configuration, with the heat pump serving as an independent auxiliary source for the solar energy system, and (b) shows a **series** configuration with the heat pump evaporator supplied with energy from the solar energy system. It indicates that energy from the collector can be supplied directly to the building if the tank is hot enough. It also shows that such a system can be a **dual-source** configuration where the heat pump evaporator is supplied with energy from either the solar-heated tank or from another source (usually ambient air).

The performance of collectors is best at low temperatures, and the performance of heat pumps is best at high evaporator temperatures. This combination leads to consideration of series systems in which the evaporator of the heat pump is supplied with energy from the solar system. These systems are arranged and controlled so that solar heat can be added to the building directly from the storage unit when the storage temperature is high enough. Systems of this type typically use liquid collectors and water tank and a water-to-air heat pump.

In a dual-source system, the heat pump evaporator is supplied with energy from either the storage unit (often a liquid system) or from ambient air (or water source if another is available). Solar energy can be supplied directly to the building. Controls can be arranged to select the source leading to best heat pump COP, that is, the higher

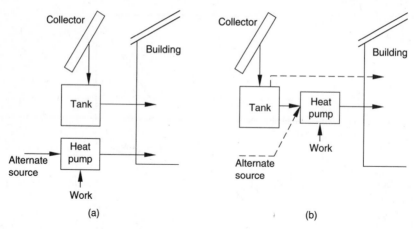

Figure 13.6.3 (a) Schematic of energy flows in a parallel system. (b) Schematic of energy flow in series and dual source solar energy–heat pump systems. In either case, an additional and independent auxiliary source (such as electric resistance heating) may have to be provided.

of the two source temperatures (although other control strategies may lead to better long-term system performance). An alternative design would use an air solar heating system and an air-to-air heat pump. The evaporator could be supplied either from ambient air or solar-heated air.

A parallel system could include either an air-based or a liquid-based solar energy system and an air-to-air heat pump. From the solar process point of view, the operation of this system is the same as a conventional solar heating system, but with the heat pump supplying the auxiliary energy.[2]

An integrated overall energy balance for a heat pump-solar energy heating system over a long time period includes Q_S, the energy supplied by the collector and tank; Q_{air}, the energy extracted from the source (air or water) by the heat pump; W_E, the electric energy used to operate the heat pump; and Q_A, the auxiliary energy needed to assure meeting total heating load L. In equation form

$$Q_S + Q_{air} + W_E + Q_A = L \tag{13.6.1}$$

A useful index of system performance of these systems is the ratio of "nonpurchased" energy to the total load, $\mathcal{F}'$:

$$\mathcal{F}' = \frac{Q_S + Q_{air}}{L} \tag{13.6.2}$$

This fraction is analogous to the solar fraction $\mathcal{F}$ defined for conventional solar heating systems.

---

[2] Design of parallel heat pump systems is discussed in Section 20.7.

Simulation studies have been used to compare these systems. Marvin and Mumma (1976) studied air systems of four configurations. Karman et al. (1976) considered two air systems and a dual-source system, and included simulations, in Madison and Albuquerque climates. Hatheway and Converse (1981) simulated several system configurations and compared their thermal performance and economics in the Vermont climate and concluded that the best combination of solar and heat pump is in a parallel configuration. Mitchell et al. (1978) have computed the performance of series, dual-source, and parallel systems and a standard liquid solar heating system of the type of Figure 13.6.3(b) using Madison meteorological data. In this study, the heat pump was modeled using the published performance data of a typical recent model commercial heat pump. The simulations were done using TRNSYS. Figure 13.6.4 summarizes the results for Madison, showing $\mathcal{F}'$ as a function of collector area for a typical residential-type building. Figure 13.6.5 shows comparative information for a collector area of 30 m^2 for each of the three combined systems and the conventional solar and heat pump only systems.

The results of these simulations show that with the same collector the parallel system is substantially better than the series system and slightly better than the dual-source system in all collector sizes, in that it delivers a greater fraction of the loads from "nonpurchased" sources. This arises because the heat pumps in the series and

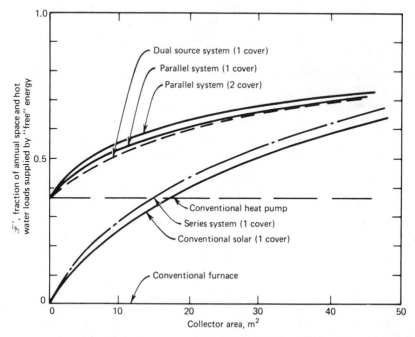

Figure 13.6.4 The fraction of energy from nonpurchased sources, $\mathcal{F}'$, for series, parallel, dual-source, heat-pump-only, and standard solar energy systems as a function of collector area, based on simulations of the systems on a residential building in the Madison climate. From Freeman et al. (1979).

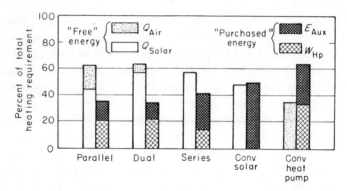

Figure 13.6.5 Sources of energy used for heating under the same circumstances at those of Figure 13.6.4. Collector areas are 30 m^2 for the solar energy systems. From Freeman et al. (1979).

dual-source systems must operate to deliver all solar energy stored below 20 C. The extra electrical energy required to deliver this energy more than compensated for the combined advantages of higher collector efficiency and higher heat pump COP. The temperature of storage in the series system (and in the dual-source system) in these simulations is rarely high enough to permit direct solar heating.

The simulations described by Mitchell et al. were done with the same collector for each of the system configurations. The series collector will run cooler for part of the time than will the collector for the parallel or dual-source systems. Thus there is the possibility that less expensive collectors will be adequate for the series system. An economic comparison must be made which takes into account the collector operating temperature ranges for each of the systems.

The combined height of the left bars of Figure 13.6.5 for each of the systems is $\mathcal{F}'$. The systems are arranged from left to right in order of decreasing $\mathcal{F}'$ or increasing purchased energy. The data for the series and dual-source heat pump systems show higher solar energy contribution, as is to be expected with reduced collector temperature, but they also show higher purchased energy. The yearly computed COP for the heat pumps for the parallel, dual-source, series, and heat pump only systems were 2.0, 2.5, 2.8, and 2.1, indicating that the series system heat pump with the solar source has the highest mean evaporator temperature and the highest annual COP.

Increasing the COP of a heat pump in a series system configuration does not lead to substantial reduction of collector area. If a load of a specified size is to be met, the energy comes from either the solar-heated tank or the (electrical) input to the heat pump. Assuming that the electrical input is not a large fraction of the total, doubling or halving it will have little impact on the amount of energy to be delivered from the solar-heated tank. (Thus if an annual load is 100 GJ and 10 GJ of electrical energy is required to operate the heat pump, doubling of the annual COP will (approximately) halve the electrical energy requirements and add about 5% to the energy that must come from the solar energy system.)

Mitchell et al. have explored system design variables including storage size, heat pump characteristics, and reduced minimum tank temperatures (i.e., antifreeze solutions in the storage tank) and found no significant changes in the results of the simulations. Marvin and Mumma (1976) also concluded that parallel operation led to best thermal performance. Many experimental systems in the series configuration have been installed, and based on economic considerations, there may be circumstances for which series systems would be better than parallel systems.

Heat pumps are capital intensive, as are solar energy systems. Consideration of the economics of solar energy-heat pump systems indicates that is may be difficult to justify the investment in a heat pump to improve solar energy system performance unless the cost of the heat pump can be justified on the basis of its use for air conditioning or unless the use of the heat pump would reduce collector costs to a small fraction of their present levels.

13.7   PHASE CHANGE STORAGE SYSTEMS

Phase change energy storage materials, methods, and problems were reviewed in Chapter 8, and a model developed by Morrison and Abdel-Khalik for a phase change energy storage unit was presented. This section reviews the results of the use of this model in simulation studies of solar heating systems to evaluate the possible impact of successful development of a phase change storage unit on heating system performance.

A number of buildings have been built using phase change storage. Telkes and Raymond (1949) described a solar house constructed at Dover, MA, that utilized vertical south-facing air heating collectors and energy storage in Glauber's salt, $Na_2SO_4 \cdot 10H_2O$, contained in 5-gal drums. The system was designed to have 5 days storage capacity and to carry the total heating load by solar energy. No data are available on its operation, which was terminated after a few years. Solar One, an experimental house at the University of Delaware [Boer (1973), Telkes (1975)], used phase change storage units for energy storage at two temperature levels in a solar energy-heat pump system. Baer (1973) reported the use of Glauber's salt in drums in a storage wall. MIT House V uses phase change materials in thin layers in the ceiling of the passively heated structure [Johnson (1977)].

Morrison and Abdel-Khalik (1978), in a simulation study of effects of use of phase change storage on solar heating system performance, postulated an idealized phase change operation (free of superheating, supercooling, and property degradation), developed models for phase change storage units for incorporation in simulations of liquid and air heating systems, and made comparative simulation studies of several systems. Their infinite NTU model of the storage unit is outlined in Section 8.8. The phase change material properties used were those of Glauber's salt and P116 wax; the properties of materials are indicated in Table 13.7.1. Meteorological data for Madison, WI and Albuquerque, NM were used.

Table 13.7.1   Properties of Phase Change Storage
Media and Rock[a]

Property	Paraffin Wax	Na$_2$SO$_4$·10H$_2$O	Rock
C_{ps}, J/kg C	2890	1920	840
C_{pl}, J/kg C	b	3260	
k_s, W/m C	0.138	0.514	0.125
k_l, W/m C	b	c	
T^*, C	46.7	32	
λ, J/kg	2.09 × 10^5	2.51 × 10^5	
ρ_s, kg/m^3	786	1460	1600
ρ_l, kg/m^3		1330	

[a] From Morrison and Abdel-Khalik (1978).
[b] Assumed equal to value for the solid phase.
[c] Assumed to be 0.475 W/m C, the value for liquid Na$_2$HPO$_4$·12H$_2$O

Typical results of this study are shown in Figure 13.7.1(a) for air systems in Albuquerque. The air systems are the same configuration as that of Figure 13.2.2, using either the pebble bed or the phase change storage unit. The model of the phase change storage allows for flow reversal and temperature stratification in the bed in a manner similar to that of the pebble bed. The figure shows annual fraction of heating loads carried by solar energy for a residential building for collector area of 50 m^2 as a function of storage volume. For the idealized phase change units, these data show that the same thermal performance is obtained with smaller volumes of the phase change materials than with pebble beds or water tanks and that the annual performance of systems with any of the three storage media is nearly the same as long as the volume of the storage media is above the "knee" of the curve. Similar results were obtained at other collector areas and with collectors having two different heat loss coefficients.

Figure 13.7.1(b) shows results of similar computations for liquid systems in Madison. The basic system configuration in these simulations was that shown in Figure 13.2.4, with storage in either water in the tank or in a phase change unit. These simulation results indicate that the system with the idealized Na$_2$SO$_4$•10H$_2$O store is better than that with the wax storage medium. The same general conclusions can be reached as for air systems, i.e., the annual performance of heating systems using an idealized phase change storage is nearly the same as that with sensible heat storage, but the storage volumes are significantly smaller.

In all of these systems, the phase change material operates in sensible heat modes as a solid part of the time, as a liquid part of the time, and in a combined phase change and sensible heat mode part of the time. Systems carrying larger fractions of annual loads by solar tend to operate as liquid phase sensible heat stores larger fractions of the time.

These results, based on ideal phase change storage characteristics, represent upper bounds to the performance of heating systems using Glauber's salt or P116

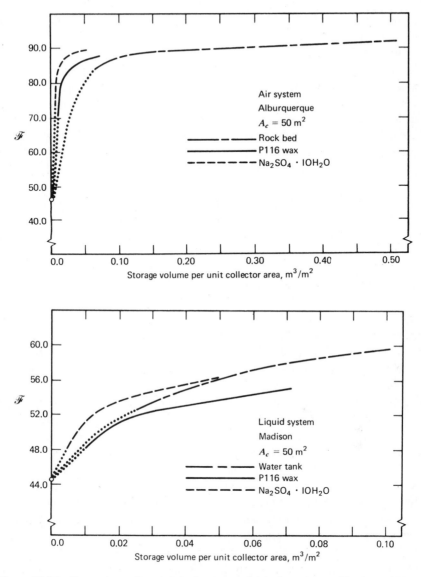

Figure 13.7.1 Comparison of annual performance of (a) air systems with phase-change or pebble-bed storage, in Albuquerque, and (b) liquid systems with phase-change or water-tank storage, in Madison. From Morrison and Abdel-Khalik (1978).

wax. Any change from the ideal can be expected to result in a greater proportion of the operation being in sensible heat modes and a reduction in effective storage capacity. This will lead to a small reduction in annual system output if the storage capacity is very large. If the storage volume is at or near the knee in the curve, deviations from idealized phase change behavior may cause large reductions in annual system output.

Jurinak and Abdel-Khalik (1978) have used the same phase change storage model and idealized behavior and explored the possibilities of improvements in system performance on changing the melting point, latent heat, and melting point range (of mixes of wax-type materials). They found that "fine-tuning" of the properties of the materials did not lead to significant increases in long-term system performance. In another paper (1979) they developed the concept of an effective heat capacity of phase change material, as indicated in Section 8.8. This concept permits use of quick design procedures[3] to estimate the performance of systems with phase change storage.

Successful application of phase change storage in solar heating depends on availability of materials that can cycle thousands of times without significant degradation and which can be packaged or handled without corrosion. The economic feasibility will depend on what the phase change material costs, the expense of packaging it, and the value of the space it saves in the building.

13.8   SEASONAL ENERGY STORAGE SYSTEMS

Early studies of the possibility of storing summer solar energy for use in the winter centered on systems for single residential-scale buildings. One of the first was that of Speyer (1959), who concluded that increasing storage capacity to high levels results in reduction of collector area requirements but that the cost of seasonal storage was not justified in terms of increased annual output of the system. Similar conclusions have been reached by others, and most solar heating studies have concerned storage capacities equivalent to approximately 1 to 3 days design heating loads of the building. A more optimistic view of long-term storage and a method of estimating heat losses from tanks to ground are provided by Hooper and Attwater (1977).[4]

Another approach to seasonal storage is to consider larger systems with storage capacity orders of magnitude more than those for single residential-scale buildings. In these systems, which are designed to provide heat for communities (i.e., district heating systems), the losses from storage are reduced by the fact that the area for heat loss goes up roughly as the square of the linear dimension of the store, and its capacity goes up roughly as the cube of the linear dimension. Considerable work has been done on these systems, largely in Europe and particularly in Sweden; the prospects for diurnal heating systems (with "overnight" storage) in northern latitudes with short winter days do not appear promising. As pointed out by Bankston (1988) in a review of central solar heating plants, such systems should have low storage losses because of the large storage size, should benefit from reduced costs of large-scale installations, would allow supply of solar energy to buildings with limited solar access, and operation and maintenance of the plants would be the responsibility of

---

[3] See Chapters 20 to 22.

[4] A simplified design method for systems with long-term storage is shown in Example 21.3.1.

paid staff rather than the individual building occupants. Most of the energy collection would occur when ambient temperatures are highest, so the collectors in these systems should perform better than those in diurnal storage systems.

The system configurations for seasonal storage are basically the same as those for short-term (overnight) storage, and most of them have been liquid systems like that of Figures 13.2.3 and 13.2.4. Some have been combined solar energy–heat pump systems with the heat pumps in series with the tanks. The differences between the two types of systems are in the relative and the absolute sizes of the collectors and the storage systems. The capacity of seasonal storage must be large enough to store most or all of the energy over and above the summer loads that is collected. Thus the ratio of storage to collector area is much higher for seasonal storage systems.

The plot of solar fraction versus water storage capacity shown in Figure 11.2.2 covers the usual range of storage capacities considered in building heating. Figure 13.8.1, from Braun (1980), is based on a similar analysis but shows four orders of magnitude variation in water tank storage capacity. The simulations were done using Madison meteorological data. System parameters (other than storage size) were like those of systems with ordinary storage capacity (except that the tank is very heavily insulated). Three collector areas are shown, which with liquid storage of 75 l/m^2 would deliver 39, 63, and 85% of the annual loads by solar energy. The 50 m^2 collector would supply 21% of the load with no storage. To approach 100% heating by solar, the tank would be of the same order of magnitude in size as the heated spaces. (This study was done with a simple model for heat losses from storage; more refined models may show different magnitudes of the effects, but the trends will be the same.)

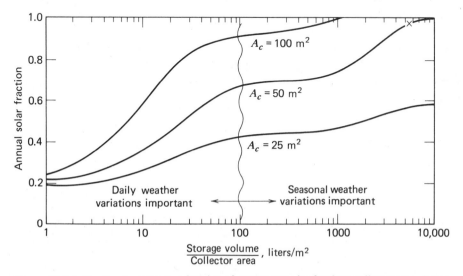

Figure 13.8.1 Fraction by solar as a function of storage capacity for three collector areas for an example based on Madison meteorological data. Adapted from Braun (1980).

The solar fraction increases sharply as storage is added until approximately 30 l/m^2 is reached. At this point the storage capacity is adequate to smooth out much of the diurnal solar variation. As more capacity is added, as more than a day's storage is provided, a gradual increase in solar fraction is noted. As the storage capacity increases by two orders of magnitude, a second "knee" in the curve appears as the capacity becomes adequate for seasonal storage, that is, of energy collected in the summer for winter use. Similar relationships with two "knees" in the curves would be observed for community-scale systems.

As the solar fraction approaches unity in a system with seasonal storage (and, for that matter, in any solar energy system), year-to-year variation in weather must be considered. The 50 m^2 system shown at point X in Figure 13.8.1 was simulated for 21 years, with the results shown in Figure 13.8.2. Useful gain from the collector Q_u, losses from storage Q_{st}, load L, and energy supplied by solar L_S are shown as integrated quantities for each year. The difference between L and L_S must be supplied from an auxiliary source. In four of these years very little auxiliary energy was required; and in other years substantial amounts were required. Again, similar considerations hold for community-scale systems. Dalenbäck and Jilar (1987) have concluded that community-scale seasonal storage systems will probably be most economical at solar fractions of about 0.7.

Figure 13.8.3 shows the solar fraction as a function of collector area for a "normal" storage capacity, for very large storage capacity, and for no storage. For large storage, annual output is limited by collector area and is nearly proportional to collector area until the solar fraction approaches unity. The difference between the two curves at constant area is the possible improvement in annual system performance ($\Delta \mathcal{F}$) that can be achieved by addition of large storage capacity. At constant solar fraction, the difference in the two curves (ΔA_c) represents the increment in area that can be saved by going to very large storage capacity. If no storage is provided, there is very little gain on increasing collector area beyond 50 m^2. (The examples shown in these figures were calculated assuming no building capacitance. At low storage capacity, including building capacitance has the effect of adding some storage.[5])

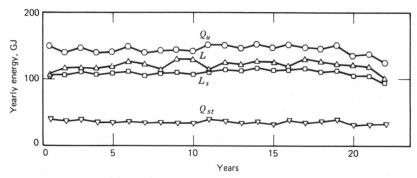

Figure 13.8.2 Year-to-year variation of performance of the 50 m^2 system with a storage capacity of 5500 l/m^2. From Braun (1980).

[5] See Chapter 22.

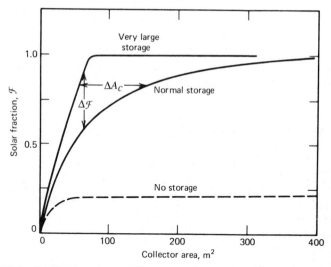

Figure 13.8.3 Fraction by solar vs collector area for systems with very large storage, standard storage of 75 l/m^2, and no storage. Adapted from Braun (1980).

Table 13.8.1   Characteristics of Seasonal Storage Solar Heating Systems[a]

	Finland	France	Netherlands	Sweden	Sweden
Location	Kerava	Aulnay-Sous-Boi	Groningen	Ingelstad	Lyckebo
	60°N	49°N	53°N	57°N	60°N
Year Built	1983	1983	1984	1979	1983
Load	44 Residences	225 Residences	96 Residences	52 Residences	550 Residences
	1980 GJ	8900 GJ	4300 GJ	3380 GJ	25200 GJ
Max, Min Delivery Temp	60, 45 C	50, 25 C	43, ? C	80, 50 C	70, 55 C
Collectors	Flat-Plate	Unglazed Flat-Plate	Evacuated	Evacuated	Parabolic
Area	1100 m^2	1275 m^2	2400 m^2	1300 m^2	4300 m^2
Storage	Pit & Duct/Rock	Aquifer	Duct/Clay	Tank	Rock Cavern
Volume	11,000 m^3	85,000 m^3	23,000 m^3	5000 m^3	100,000 m
Temp Range	10 - 70 C	4 - 14 C	30 - 60 C	40 - 95 C	40 - 90 C
Capacity	900 GJ/yr	2500 GJ/yr	2340 GJ/yr	1080 GJ/yr	19800
Annual η			0.68		0.74
Heat Pump	Electric	Electric	–	–	–
Capacity	240 kW elec.	660 kW elec.			
Ann. COP	2.7	3.9			
Solar Contribution	0.50	0.66	0.66	0.50	[b]
Heat Pump		0.24			
Other Aux.		0.10	0.34	0.50	

[a] Data are from Bankston (1988).

[b] Part of the collectors have been installed; the balance of energy added to the storage is electrical.

Methods for seasonal storage for community-scale applications were outlined in Section 8.7. Reviews of characteristics of 26 seasonal storage projects and experiments that have been built and operated are provided in publications of the International Energy Agency (IEA) [e.g., Dalenbäck (1989)] and by Bankston (1988). Characteristics of five of these are summarized in Table 13.8.1.

13.9   SOLAR AND OFF-PEAK ELECTRIC SYSTEMS

Auxiliary energy can be supplied to solar heating systems from supplies stored on site (oil or LPG) or from utilities (gas or electricity). On-site storage poses no unique problems for the distributors of oil or LPG. Supply of auxiliary energy by utilities, in contrast, could cause significant peaking problems for electric utilities if a large number of solar-heated buildings all call for energy during bad weather. A potential solution to this problem is to supply electric auxiliary energy during off-peak periods and store the energy in the building. (The same possible solutions apply to systems that are not solar, i.e., are all electric.) Hughes et al. (1977) have studied off-peak electric options in air solar heating systems; the results of that study are summarized here.

The three systems were considered. The first is a conventional air solar heating system such as that shown in Figures 13.2.2 and 13.4.1; it requires delivery of auxiliary energy whenever solar energy cannot meet the load. The second, shown in Figure 13.9.1(a), uses an electric furnace to supply energy to a separate pebble bed storage unit. The auxiliary storage subsystem acts as any other auxiliary if it is controlled so that at the end of each off-peak period there is sufficient energy in the auxiliary store to meet the maximum anticipated heating loads on the building until the next off-peak period. The solar heating system should not be affected by the nature of the auxiliary energy supply if the combined operation is such that the collectors operate at the low input temperatures characteristic of air systems.

The third system, shown in Figure 13.9.1(b), combines the storage of solar energy and auxiliary energy in a single unit. During off-peak periods, electrical energy is added to the pebble bed to assure having adequate energy stored in the bed to last until the next off-peak period. Solar energy is also added to the bed as it is available from the collectors. The pebble bed will operate at significantly higher temperatures with this system than with the system with separate storage units, which leads to higher collector temperatures and reduced collector performance. Stratification in the bed helps to diminish the elevation of collector temperature, but the effect is significant for flat-plate collectors. Compared to the two-store system, this system is mechanically simpler but suffers in performance.

The extent of this performance penalty depends on collector characteristics. Hughes et al. showed that the penalties for evacuated tube collectors with U_L of about 0.8 W/m^{2}C were substantially less than for flat-plate collectors with U_L of 4.0 W/m^{2}C; the collectors with low U_L are not nearly as sensitive to temperature as those with U_L representative of those of flat-plate air heaters.

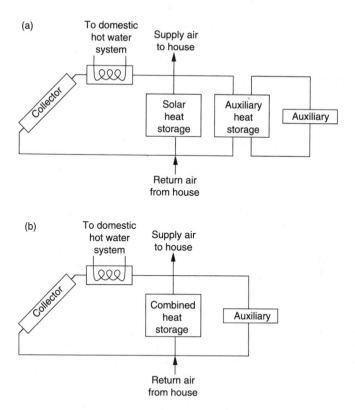

Figure 13.9.1 Schematics of solar-off-peak air systems with (a) two storage units and (b) a single store. From Hughes et al. (1977).

The off-peak addition of energy to a pebble bed can be controlled by comparing the average bed temperature to a set minimum average temperature which will assure that adequate heat supply is available until the next off-peak period. The minimum average temperature may be fixed through the year, or it may be varied to account for variable anticipated loads throughout the year. Collector performance will be influenced by the energy in storage, and there is an advantage in changing the minimum average temperature from month to month. This advantage is a few percent with low-loss collectors but is substantially larger with flat-plate collectors.

13.10   SOLAR HEATING ECONOMICS

The first major economic studies of solar heating by Tybout and Löf (1970) and Löf and Tybout (1973) were based on constant annual savings without anticipation of changing future costs. They devised a thermal model for a liquid-based solar heating system to estimate annual thermal performance (based on one year's meteorological data) and developed a set of cost assumptions to calculate costs of delivered solar energy for houses of two sizes in eight U.S. locations of differing climate types. Several system design parameters were studied in addition to collector area to establish the range of optimum values, including collector slope, number of covers, and heat storage capacity per unit collector area. Their results (which are in general agreement with the conclusions of others) can be summarized as follows:

A   The optimum tilt is in the range of the latitude plus $10°$ to the latitude plus $20°$, and variation of $10°$ either way outside of this range, that is, from latitude to latitude plus $30°$, has relatively little effect on the cost of delivered energy for heating.

B   The best number of (ordinary) glass covers with nonselective absorbers was found to be two for all locations except those in the warmest and least severe climates, that is, Miami and Phoenix, where one cover produces less expensive energy from the solar heating system.

C   The best storage tank capacity per unit collector area was indicated to be in the range of 50 to 75 l/m². Increasing the size of storage several-fold for fixed collector size had relatively small effect on the cost of delivered solar energy or on the fraction of total heating loads carried by solar.

The costs assumed by Löf and Tybout (approximately \$20 and \$40/m² of collector, installed) are unrealistically low in the light of 1990s costs (\$250 to \$700/m² of factory-built collector installed on new residential buildings). However, their conclusions on collector orientation and storage capacity have been confirmed by others. The availability of durable selective surfaces has led to wider applicability of one-cover collectors than they predicted.

The solar process economic considerations noted in Chapter 11 apply directly to solar heating. The costs of installed solar heating equipment (the first costs) include purchase and installation of all collectors, storage units, pumps, blowers, controls, ductwork, piping, heat exchangers, etc. Operating costs include costs of auxiliary energy, parasitic power, insurance, maintenance, taxes, etc. In the following discussions, as in Chapter 11, first costs of solar heating systems are taken to be the incremental costs, that is, the difference in cost between the solar heating system and a conventional heating system.

The basic method used in calculating the economics of solar heating is the P_1, P_2 life cycle savings method described in Section 11.9. In this analysis there is a large number of economic parameters that must be determined. This discussion shows how some of these parameters affect the economic viability of active solar heating and is introduced by a brief review of their nature and significance.

A primary economic consideration is the first cost of solar heating system. In the 1980s, the installed costs of many systems are in the range of \$250 to \$700/m² of collector, with fixed costs of the order of \$1000 to \$3000. (An important component in these costs which must be borne by the purchaser is that of marketing, which can vary widely with the size of the system and the level of sales effort that is involved in getting it to the purchaser.) Retrofit systems tend to be more expensive, particularly on large flat-roof buildings where new structures must be provided for collector support. Some residential-scale systems cost less if integrated into a building or constructed by the owner. If a system is well built, it can have significant salvage or resale value, and if its major components are as durable as the basic building, its value may appreciate.

Tax laws can have important impact on the first cost of solar heating (and other solar processes). In the United States, tax incentives were enacted by Congress and by the legislatures of several states; these had the effect of reducing the first cost of systems.   As of 1989 most of these incentives have expired and have not been renewed.

The costs of energy delivered from fuel at the time of installation C_{F1} are a second important economic consideration. They are widely variable with time, location, type of fuel, and efficiency of the fuel burning system. In 1978 in the United States, some regulated natural gas prices were at the low end of the scale with costs of delivered energy from \$2 to \$3/GJ. Typical costs of energy delivered from oil were \$5 to \$8/GJ. LPG energy costs were higher than oil. In the years since then, prices have risen and then fallen, driven primarily by the cost of imported crude oil. Some of the most expensive energy is electricity, which when used in resistance heating in some metropolitan areas costs more than \$20/GJ. Thus in 1990 energy costs to consumers vary over an order of magnitude.

In the decade 1970 to 1980, most energy costs inflated at higher rates than the general inflation rate, but since then energy costs have not risen with general inflation. Future fuel energy costs will be dependent on gas and oil discoveries and on technological developments related to fuel conversion and use. They will also be dependent on international political developments, as many of the world's industrialized nations are energy importers. Life cycle cost calculations require that projections be made of future energy costs. These projections are uncertain and interject corresponding uncertainties into the life cycle cost analysis.

A third category of economic factors relates to costs of operation of systems. Insurance, maintenance, and property taxes will probably increase with the general inflation rate. Parasitic power in a well-designed system should be small in relation to other costs. (A good system will have a coefficient of performance, the ratio of solar energy delivered to parasitic energy used to drive pumps, controls, etc., of 15 to 30.)

A fourth category of economic considerations relates to the costs of money and time periods over which analyses are made. Interest rates and terms of initial mortgages or building improvement loans both affect the life cycle costs of the capital-intensive solar energy systems. Time periods for economic analysis are sometimes selected as the period of time over which it is expected the building will be occupied or owned, over the expected lifetime of the equipment, or over the term of the loan used to purchase the equipment. In general, longer periods of analysis will tend to improve life cycle solar savings, although costs or savings far in the future have relatively small impact on life cycle costs when discounted to present value. Life cycle costs are sensitive to discount rates, with lower discount rates generally improving life cycle savings for solar or other energy-conserving and capital-intensive measures. Market discount rates are often taken as approximately 2% more than the general inflation rate for individuals, approximately 4% more for established businesses, and may be as high as 20% higher than general inflation rates for fast-growing industries. Most industries have developed their own discount rates.

Examples of the solar savings to a homeowner is shown in Figure 13.10.1. The configuration is the air system described in Section 13.5 on a well-insulated building in Madison. The economic parameters assumed are: term of analysis is 20 years; term of mortgage is 20 years; mortgage interest rate is 9.5% per year; down payment is 20%; market discount rate is 9% per year; general inflation rate is 7% per year; real estate taxes in first year are 2% of system first cost; insurance, maintenance, and parasitic power in first year are 2% of system first cost; income tax bracket through the period of analysis is 45%. The system is assumed to be durable and well built, having a resale value of 100% of its first cost (i.e., it does not decline in value or appreciate as the building is expected to appreciate).

First-year fuel costs are taken as $4, $8, and $12/GJ, and these are assumed to inflate at rates of 10 and 15% per year. This covers a range of present costs and possible inflation rates. Figure 13.10.1 is based on C_A = $300/m^2 and C_E = $1000, which is near the bottom of the range of 1989 costs of commercially installed systems on new residential buildings without tax credits.

Several trends are clear from this figure. Higher present fuel costs and higher fuel cost inflation rates lead to more favorable savings. As fuel costs or inflation rates

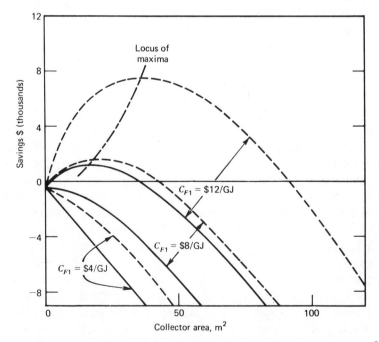

Figure 13.10.1 Life cycle solar savings for the Madison example, with C_A= $300/m^2 and C_E = $1000, for present fuel costs of $4, $8, and $12 /GJ, inflating at 0.10 (solid lines) and 0.15 (dashed lines) per year.

rise, the optimum collector area increases. The approximate locus of the maxima is shown. As the optimum collector area increases, the range of areas over which positive savings can be expected will also increase. In all cases the optima are broad and the selection of area is not critical. For example, from Figure 13.10.1 at $C_{F1}=$ \$12/GJ and $i_F = 10\%$ per year, the optimum area is about 15 m^2, any area between 10 and 25 m^2 results in essentially the same savings, and any area between 2 and 35 m^2 results in positive savings. A reduction in first cost increases life cycle savings (not shown on this figure).

Given a full set of economic parameters (other than fuel costs) and the thermal performance of a system, it is possible to determine combinations of present fuel costs and fuel cost inflation rate that will make solar heating just competitive with the alternative, i.e., when savings = \$0. At a fuel cost inflation rate of 0.10 per year, a C_{F1} of about \$6/GJ is a "breakeven" fuel cost. If fuel cost is above this level, solar savings are positive.

The results shown in the figure are for a specific system, location, and (most importantly) a set of economic assumptions. Many economic parameters have not been varied in the estimates that are shown in this figure. As was noted in Chapter 11, the explicit effect of any one variable on the life cycle savings is influenced by the values of many other variables. Thus "rule of thumb" sensitivity trends are not easily recognized, and quantitative sensitivity analyses such as that outlined in Section 11.10 are required. The numbers used in this illustration are subject to argument and are not intended to show economic feasibility (or lack thereof) of solar heating. However, the trends are shown, and the methods described in Chapter 11 should be applied to individual cases to determine the economics of a particular application.

13.11 ARCHITECTURAL CONSIDERATIONS

Active solar heating poses challenges to architectural design of buildings. Many approaches to these challenges have been devised; Shurcliff (1978) and Szokolay (1980) have complied information on a variety of solar-heated buildings. In this section we outline in general terms some of the major architectural considerations to be taken into account in solar building design. While not addressed specifically here, it is implicit throughout this discussion that any building design should be energy conserving, as solar energy and the fuels with which it competes will be expensive.

Economic studies of active solar heating indicated that the optimum fractions of total annual loads range from zero to over three-fourths. For some locations the architect must design into the building collector areas in the range up to approximately one-half of the floor area of the house (depending on the collector, the climate, and the degree of insulation in the building). A basic problem faced by architects and engineers is to integrate the collectors into the building design in such a way that thermal performance is satisfactory and the structure is aesthetically satisfying.

Collectors should be oriented with the slope and azimuth angle within or close to the ranges noted in Tables 13.2.1 and 13.2.2. Vertical collectors may be useful at high latitudes to answer problems of integration of collectors into the building and avoiding snow accumulation. Space must be provided in the structure for energy storage units, piping and ducts, controls, auxiliaries, and all associated equipment. Anderson et al. (1955) have addressed these and related questions on solar house architecture. Similar considerations will apply to institutional buildings.

The collector may be a part of the envelope of the building (as in MIT House IV) or separate (as in the Denver Solar House). The orientation of the collector is substantially fixed, and if it is part of the envelope of the house, the collector will probably become an important or dominant architectural feature of the structure. The collector may serve as part of the weatherproof enclosure and thus allow a reduction in cost of roofing or siding; such a reduction is a credit which reduces the cost of the collector. Separate collectors, on the other hand, can permit greater flexibility in house design and allow buildings that are more conventional (contemporary or traditional) in appearance.

Figure 13.2.5 shows three solar-heated residential buildings, including MIT House IV, the Denver House, and the House II Laboratory building at Colorado State University. Figure 13.11.1 shows three additional solar-heated buildings, illustrating additional architectural approaches.

Storage is usually not a major architectural problem, other than recognizing the need for appropriate space and access to it within the structure. The volume of storage per unit area of collector depends on the system used. Pebble bed storage units will usually occupy a volume of roughly 0.15 to 0.35 m^3/m^2 of collector and water tanks about 0.050 to 0.10 m^3/m^2. The volume of phase change storage systems would be less than that of water systems. The most common location for storage in buildings is in basements.

Providing solar heat to larger buildings, such as apartment buildings, presents a special set of problems. It may be necessary to consider vertical mounting for collectors, which may cause a significant reduction in their performance. This possibility has been studied by Lorsch and Niyogi (1971). Otherwise, collectors may be mounted like awnings, with improvements in performance but with increased cost of installation. If the building geometry is such that the roof area is adequate, banks of collectors can be mounted on the roof with appropriate piping or ducts leading to the space to be heated.

Most solar heating studies to date have been concerned primarily with new buildings designed to include solar heating systems. Adding solar heating systems to existing buildings presents more formidable tasks. Consideration should be given to the problems of designing buildings that can accommodate the addition of solar heating after construction.

REFERENCES

AEIC-EEI Heat Pump Committee, Edison Electric Inst. Bull., 77 (1956). "Possibilities of a Combination Solar-Heat Pump Unit." (based on a report by G.O.G. Löf).

Figure 13.11.1 Three examples of solar heated buildings: (a) and (b) residences, Madison, WI; (c) Wayside, Portage, WI.

Anderson, L. B., H. C. Hottel, and A. Whillier, *Solar Energy Research*, University of Wisconsin Press, Madison (1955), p.47. "Solar Heating Design Problems."

ASHRAE *Systems Handbook*, American Society of Heating, Refrigeration, and Air Conditioning Engineers, New York (1976).

Baer, S., *Proc. of Solar Heating and Cooling for Buildings Workshop*, (R. Allen, ed.), University of Maryland. p 186 (March, 1973). "The Drum Wall."

Bankston, C., *Advances in Solar Energy*, (K. Boer, ed.) **4**, 352 (1988). American Solar Energy Society and Plenum Press. "The Status and Potential of Central Solar Heating Plants with Seasonal Storage: An International Report."

Bliss, R. W., *Air Conditioning, Heating and Ventilating*, **52** (10), 92 (October 1955). "Design and Performance of the Nation's Only Fully Solar-Heated House." See also *Proc. of the World Symposium on Applied Solar Energy*, 151, SRI, Menlo Park, CA (1956).

Bliss, R. W., *Proc. of the UN Conferences on New Sources of Energy*, **5**, 148 (1964). "The Performance of an Experimental System Using Solar Energy for Heating, and Night Radiation for Cooling a Building."

Boer, K., paper presented at the ISES Congress, Paris (1973). "A Combined Solar Thermal Electrical House System."

Braun, J., M.S. Thesis, University of Wisconsin-Madison (1980). "Seasonal Storage of Energy in Solar Heating."

Bridgers, F. H., D. D. Paxton, and R. W. Haines, paper 57-SA-26 presented at ASME meeting (June 1957a). "Solar Heat for a Building."

Bridgers, F. H., D. D. Paxton, and R. W. Haines, *Heating, Piping, and Air Conditioning*, **29**, 165 (1957b). "Performance of a Solar Heated Office Building."

Close, D. J., R. V. Dunkle, and K. A. Robeson, *Mechanical and Chemical Engineering Trans., Inst. Engrs, Australia*, **MC4,** 45 (1968). "Design and Performance of Thermal Storage Air Conditioner System."

Converse, A. O., *Proc. Joint Conference of the American Section of ISES and SES of Canada*, Winnipeg, **3**, 277 (1976). "Solar Heating in Northern New England."

Dalenbäck, J-O., and T. Jilar, personal communication (1987).

Dalenbäck, J-O., *The Status of CSHPPS*, a report of International Energy Agency Solar Heating and Cooling Task VII (1989).

Engebretson, C. D. and N. G. Ashar, paper 60-WA-88 presented at the New York ASME Meeting (1960). "Progress in Space Heating with Solar Energy."

Engebretson, C. D., *Proc. of the UN Conference on New Sources of Energy*, **5**, 159 (1964). "The Use of Solar Energy for Space Heating-M.I.T.: Solar House IV."

Freeman, T. L., J. W. Mitchell, and T. E. Audit, *Solar Energy*, **22**, 125 (1979). "Performance of Combined Solar-Heat Pump Systems."

Hatheway, F. M., and A. O. Converse, *Solar Energy*, **27**, 561 (1981). "Economic Comparison of Solar Assisted Heat Pumps."

Hooper, F. C. and C. R. Attwater, in *Heat Transfer in Solar Energy Systems*, American Society of Mechanical Engineers, New York (1977). "Design Method for Heat Loss Calculation for In-Ground Heat Storage Tanks."

Hughes, P. J., W. A. Beckman, and J. A. Duffie, *Solar Energy*, **19**, 317 (1977). "Simulation Study of Several Solar Heating Systems with Off-Peak Auxiliary."

Johnson, T. E., *Solar Energy*, **18**, 669 (1977). "Lightweight Thermal Storage for Solar Heated Buildings."

Jordan, R. C. and J. L. Threlkeld, *Heating, Piping and Air Conditioning*, **28**, 122 (1954). "Design and Economics of Solar Energy Heat Pump Systems."

Jurinak, J. J. and S. I. Abdel-Khalik, *Solar Energy*, **21**, 377 (1978). "Properties Optimization for Phase-Change Energy Storage in Air-based Solar Heating Systems."

Jurinak, J. J. and S. I. Abdel-Khalik, *Solar Energy*, **22**, 355 (1979). "Sizing Phase-Change Energy Storage Units for Air-based Solar Heating Systems.

Karaki, S., W. S. Duff, and G. O. G. Löf, Colorado State University report COO-2868-4 to U.S. Department of Energy (1978). "A Performance Comparison Between Air and Liquid Residential Solar Heating Systems."

Karaki, S., P. R. Armstrong, and T. N. Bechtel, Colorado State University report COO-2868-3 to U.S. Department of Energy (1977). "Evaluation of a Residential Solar Air Heating and Nocturnal Cooling System."

Karaki, S., T. E. Brisbane, and G. O. G. Löf, Colorado State University Report SAN 11927-15 to U.S. Dept. of Energy (1984). "Performance of the Solar Heating and Cooling System for CSU House III - Summer Season 1983 and Winter Season 1983-84."

Karman, V. D., T. L. Freeman, and J. W. Mitchell, *Proc. Joint Conf. of the American Section of ISES and SES of Canada*, Winnipeg, **3**, 325 (1976). "Simulation Study of Solar Heat Pump Systems."

Klein, S. A., W. A. Beckman, and J. A. Duffie, *Solar Energy*, **18**, 113 (1976). "A Design Procedure for Solar Heating Systems."

Kuharich, R. R., *Proc. Joint Conf. of the American Section of ISES and SES of Canada*, Winnipeg, **3**, 378 (1976). "Operational Analysis of a Solar Optimized Heat Pump."

Löf, G. O. G., M. M. El-Wakil, and J. P. Chiou, *Trans. ASHRAE*, 77 (October 1963). "Residential Heating with Solar Heated Air-the Colorado Solar House."

Löf, G. O. G., M. M. El-Wakil, and J. P. Chiou, *Proc. of the UN Conference on New Sources of Energy*, **5**, 185 (1964). "Design and Performance of Domestic Heating System Employing Solar Heated Air-The Colorado House."

Löf, G. O. G. and R. A. Tybout, *Solar Energy*, **14**, 253 (1973). "Cost of House Heating with Solar Energy."

Lorsch, J. G. and B. Niyogi, Report NSF/RANN/SE/GI 27976/TR72/18, to NSF from University of Pennsylvania (August 1971). "Influence of Azimuthal Orientation on Collectible Energy in Vertical Solar Collector Building Walls"

Marvin, W. C. and S. A. Mumma, *Proc. of the Joint Conference of the American Section of the ISES and the SES of Canada*, Winnipeg, **3**, 321 (1976). "Optimum Combination of Solar Energy and The Heat Pump for Residential Heating."

Mitchell, J. W., T. L. Freeman, and W. A. Beckman, *Solar Age*, **3** (7), 20 (1978). "Heat Pumps."

Morrison, D. J. and S. I. Abdel-Khalik, *Solar Energy*, **20**, 57 (1978)/ "Effects of Phase Change Energy Storage on the Performance of Air-based and Liquid-based Solar Heating Systems."

Shurcliff, W. A., *Solar Heated Buildings of North America*, Brick House Publishing Co., Harrisville, NH (1978).

Speyer, E., *Solar Energy*, **3**(4), 24 (1959). "Optimum Storage of Heat with a Solar House."

Szokolay, S. V., *World Solar Architecture*, Architectural Press, London (1980).

Telkes, M., *Proc Workshop on Solar Energy Storage Subsystems for the Heating and Cooling of Buildings*, Charlottesville, VA, p. 17, (1975). "Thermal Storage for Solar Heating and Cooling"

Telkes, M. and E. Raymond, *Heating and Ventilating*, 80 (1949). "Storing Solar Heat in Chemicals - A Report on the Dover House."

Terrell, R. E., *Solar Energy*, **23**, 451 (1979). "Performance of a Heat Pump Assisted Solar Heated Residence in Madison, Wisconsin."

Tybout, R. A. and G. O. G. Löf, *Natural Resources J.*, **10**, 268 (1970). "Solar Energy Heating."

Ward, D. S., G. O. G. Löf, C. C. Smith, and L. L. Shaw, *Solar Energy*, **19**, 79 (1977). "Design of a Solar Heating and Cooling System for CSU House II."

Yanagimachi, M., *Trans. of the Conference on Use of Solar Energy*, **3**, 32, University of Arizona Press (1958). "How to Combine: Solar Energy, Nocturnal Radiational Cooling, Radiant Panel System of Heating and Cooling, and Heat Pump to Make a Complete Year-Round Air Conditioning System."

Yanagimachi, M., *Proc. of the UN Conference on New Sources of Energy*, **5**, 233 (1964). "Report on Two and a Half Years' Experimental Living in Yanagimachi Solar House II."

Chapter 14

BUILDING HEATING: PASSIVE AND HYBRID METHODS

The active systems described in the previous chapter are based on collectors and storage systems that are not necessarily integrated into a building structure. Passive systems can be distinguished from active systems on either of two bases. The first distinction lies in the degree to which the functions of collection and storage are integrated into the structure of the building; windows and the rooms behind them can serve as collectors, with storage provided as sensible heat of the building structure and contents as they change temperature. Second, many passive systems require no mechanical energy for moving fluids for their operation; fluids and energy move by virtue of the temperature gradients established by adsorption of radiation (and hence the term *passive*). (Mechanical energy may be used to move insulation for loss control or to move fluids to distribute absorbed energy from one part of a building to another.)

By nature, passive heating is intimately concerned with architecture, as the building itself functions as collector and storage unit and as the enclosure in which people live, work, and are protected from an often harsh exterior environment. In this chapter we discuss in a largely qualitative way the factors that affect the thermal performance of a passive building (i.e., its ability to provide an acceptable level of human comfort). The engineering basis for thermal performance calculations is (as for active systems) the subject of Chapters 1 to 11. This chapter serves as an introduction to Chapter 22 on estimation of annual thermal performance of passive and hybrid systems.

The thrust of this book is the estimation of the long-term thermal performance of solar energy systems, to predict how much fuel will be needed to keep a building within a reasonably fixed span of interior temperatures. The esthetic features of a building that make it a pleasant place to live and work can be very much enhanced (or in some cases degraded) by admission of solar energy into the building spaces. Daylighting, the use of the visible part of the solar spectrum rather than artificial lighting, is closely coupled with passive heating. These matters are also of importance and relate to aspects of the overall process of building design that are treated in other books.

Some solar heating systems are combinations of active and passive systems. In this chapter and in Chapter 22 we discuss examples of hybrid systems, including systems with active collectors and passive storage, and combination of direct-gain passive and active systems. Other combinations are possible.

In recent years, there has developed a very substantial body of literature on passive solar processes. The *Passive Solar Journal* was published by the American Solar Energy Society. Since 1976 the society has run National Passive Solar Conferences which have resulted in conference proceedings. Numerous books on the subject have appeared, as have papers in other journals such as *Solar Energy* and architectural publications. Included in these publications have been works on quantitative aspects of passive heating processes, architecture, building (system) performance, daylighting, shading, heat transfer within buildings by natural and forced convection, cooling, energy storage, commercial aspects, and others.

14.1   CONCEPTS OF PASSIVE HEATING

Several concepts for passive and hybrid solar heating have developed that are sufficiently distinct as to provide a useful basis of discussion of the principles and functions of passive systems. These are direct gain, collector-storage wall, and sunspace. (Many other types and combinations of types have been noted, including convective loop, solar chimney, etc.) The common features are means of absorbing solar energy in the building, storing it in parts of the structure, and transferring it to the spaces to be heated.

Direct gain of energy through windows can meet part of building heat loads. The window acts as a collector and the building itself provides some storage. Overhangs, wingwalls, or other architectural devices are used to shade the windows during times when heating is not wanted. It is also necessary, in cold climates, to insulate the windows during periods of low solar radiation to prevent excessive losses. Direct gain can provide energy to the south side of a building; means may have to be provided to distribute energy to rooms not having south windows.

Collector-storage wall[1] combines the functions of collection and storage into a single unit that is part of a building structure. Part of a south wall may be single or double glazed; inside the glazing is a massive wall of masonry material or water tanks, finished black to absorb solar radiation. Heat is transferred from the storage wall to the room by radiation and convection from the room side of the wall and by forced or natural convection of room air through the space between glazing and wall. Room air may enter this space through openings (vents) in the bottom of the wall and return to the room through openings in the top. The storage unit could also be part of the roof and ceiling. Movable insulation may have to be provided in other than mild climates to control losses at times of low solar radiation.

Sunspace attachments to buildings have been used as solar collectors, with storage in walls, floors, or pebble beds. Forced air circulation to the rooms is an option to improve storage and utilization of absorbed energy. In cold climates, energy losses from sunspaces, which are greenhouse-like structures, can exceed the absorbed energy, and care must be used to assure that net gains accrue from such a

---

[1] Also referred to as a Trombe wall or Trombe-Michel wall.

system. The uses to which a sunspace is put, for human occupancy or for plants, will impose limits on the allowable temperature swings in the sunspace and affect the energy balances on it.

A survey of passive solar buildings has been prepared by AIA (1978). Many passive buildings are included in broader surveys of solar-heated buildings by Szokolay (1975) and by Shurcliff (1978). Many of the concepts now being studied and developed were set forth by Olgyay (1963). Anderson (1977) reviews passive heating concepts. Books on passive heating include those of Mazria (1979), Lebens (1980), and many others.

14.2 COMFORT CRITERIA AND HEATING LOADS

Active solar heating systems can be designed to provide the same levels of control of conditions in the heated (or cooled) spaces as conventional systems. With indoor temperatures essentially fixed at or a little above a minimum, load estimations can be done by conventional methods such as those outlined in Chapter 9. Passively heated buildings in many cases are not controlled within the same narrow temperature ranges; the thermal storage capacity of the building structure or contents is usually significant, and it may be necessary to use load calculation methods that take into account variable internal temperatures and capacitance of building components.

For residential buildings, there are limits in the variation of indoor conditions (temperature and humidity) considered comfortable by occupants. The limits are not well defined and are to a degree subjective. They depend on the activities of the individuals in the buildings and on their clothing. Air movement is also important, as are the temperatures of the interior surfaces with which an occupant exchanges radiation. Extensive discussions of comfort under various conditions of temperature, humidity, air movement, radiant exchange, activity levels, and clothing are provided by ASHRAE (1989) and Fanger (1972).

The concept of solar fraction, which is useful for systems having loads substantially independent of the size of the solar energy system, is not useful in passive heating. For systems in which energy is absorbed in part of the structure (i.e., in direct gain, collector-storage walls, and sunspace systems) and where the loss coefficient of the solar aperture is different than that of the insulated wall it replaces, the total loads (that are to be met by solar plus auxiliary) will be a function of the design of the solar energy system. The significant number resulting from performance calculations for these systems is the annual amount of auxiliary energy required.

14.3 MOVABLE INSULATION AND CONTROLS

Passive elements such as direct-gain and storage wall elements can lose energy at excessive rates if measures are not taken to control the losses. An example of this situation, a water heater equivalent of a storage wall using water tanks, was shown in Example 12.7.1 to have significant night losses. Hollingsworth (1947) and Dietz and Czapek (1950) noted the same problems based on measurements on an early MIT solar house. In passive heating, storage elements are also energy-absorbing elements and thus have large, relatively uninsulated areas, and steps must be taken to control losses in any but mild climates.

Movable insulation is the evident possibility. Movable insulation can take several forms. Drapes, shades, screens, and shutters provide nominal levels of insulation. More extensive insulation can be provided by movable foam plastic or glass wool panels or by such devices as "beadwall" panels described by Harrison (1975), in which lightweight plastic beads are pneumatically moved into or out of space between glazings. Many movable insulation systems useful in passive heating are now on the market.[2]

Insulation may be moved manually or automatically. If automatic, controls are needed to move it into or out of place, and detectors and control strategies must be arranged to maximize the net gains from the system while keeping indoor conditions within acceptable (and pleasant) limits. Controls must also be provided for mechanical devices that may be used for transferring heat from storage to rooms, such as fans used to circulate air around storage walls. Sebald et al. (1979), for example, have used simulation methods to assess the effects of air circulation control in collector-storage walls.

14.4   SHADING: OVERHANGS AND WINGWALLS

Overhangs and wingwalls are used with passive heating to reduce gains during times when heat is not wanted in the building. As these shading devices may partially shade absorbing surfaces during periods when collection is desired, it is necessary to estimate their effect on the absorbed radiation. In this section we show how this can be formulated for any point in time and also how monthly average effects of overhangs can be estimated. This discussion treats overhangs of finite length and is a summary of a detailed treatment by Utzinger (1979) and Utzinger and Klein (1979). For overhangs of infinite length (very long compared to the width of the receiver), the method of Jones (1980), discussed in Section 1.9, can be used.

A horizontal overhang shading device with its outer extremity parallel to the wall can be represented by one perpendicular to the wall as shown schematically in Figure 14.4.1. Its geometry is described by a set of dimensions: the projection P, the gap between the top of the receiver and the overhang G, and the left and right extensions E_L and E_R. The receiver (the window) height is H and width is W. Using the receiver height H as a characteristic dimension, the other dimensions can be expressed in dimensionless form as ratios to the height. Thus the relative width is

---

[2] Loss coefficients for windows with and without insulation are shown in Section 9.2.

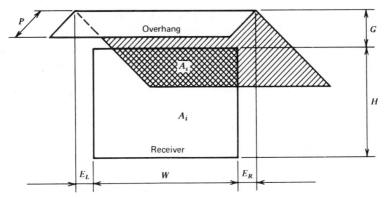

Figure 14.4.1 Diagrams of the shading of a vertical receiver by a horizontal overhang. From Utzinger (1979).

w, the relative projection is p, the relative gap is g, and the relative extension (with the left and right values the same) is e.

The ratio f_i of beam radiation received by the shaded receiver to that received by the unshaded receiver is the same as the fraction of the receiver area A_r which is exposed to direct beam radiation:

$$f_i = \frac{A_i}{A_r} \tag{14.4.1}$$

The value of f_i at any time depends on the dimensions of the overhang and receiver and on the angle of incidence of the beam radiation on the wall. An algorithm for computing this fraction has been developed by Sun (1975).

The area-average radiation on the partially shaded receiver at any time is the sum of beam, diffuse, and ground-reflected radiation. Assuming that diffuse and ground-reflected radiation are isotropic,[3]

$$I_r = I_b R_b f_i + I_d F_{r\text{-}s} + I \rho_g F_{r\text{-}g} \tag{14.4.2}$$

The three terms have the same general significance as those in Equation 2.15.1. The first term includes f_i to account for shading from beam radiation. The view factor of the receiver for radiation from the sky $F_{r\text{-}s}$ is reduced by the overhang from its value without an overhang, $(1 + \cos 90)/2$. Values of $F_{r\text{-}s}$ are shown in Table 14.4.1. The third term is the ground-reflected radiation. For vertical receivers, neglecting secondary reflections from the underside of the overhang, $F_{r\text{-}s}$ is $(1 - \cos 90)/2$, or 0.5.

---

[3] Anisotropic models can also be used with little modification in the analysis.

Table 14.4.1　Receiver Radiation View Factor of the Sky, $F_{r\text{-}s}$ [a]

e	g	w	0.10	0.20	0.30	0.40	0.50	0.75	1.00	1.50	2.00
			$F_{r\text{-}s}$ at $p =$								
0.00	0.00	1.0	0.46	0.42	0.40	0.37	0.35	0.32	0.30	0.28	0.27
		4.0	0.46	0.41	0.38	0.35	0.32	0.27	0.23	0.19	0.16
		25.0	0.45	0.41	0.37	0.34	0.31	0.25	0.21	0.15	0.12
	0.25	1.0	0.49	0.48	0.46	0.45	0.43	0.40	0.38	0.35	0.34
		4.0	0.49	0.48	0.45	0.43	0.40	0.35	0.31	0.26	0.23
		25.0	0.49	0.47	0.45	0.42	0.39	0.34	0.29	0.22	0.18
	0.50	1.0	0.50	0.49	0.49	0.48	0.47	0.44	0.42	0.40	0.38
		4.0	0.50	0.49	0.48	0.46	0.45	0.41	0.37	0.31	0.28
		25.0	0.50	0.49	0.47	0.46	0.44	0.39	0.35	0.27	0.23
	1.00	1.0	0.50	0.50	0.50	0.49	0.49	0.48	0.47	0.45	0.43
		4.0	0.50	0.50	0.49	0.49	0.48	0.46	0.43	0.39	0.35
		25.0	0.50	0.50	0.49	0.48	0.47	0.44	0.41	0.35	0.30
0.30	0.00	1.0	0.46	0.41	0.38	0.35	0.33	0.28	0.25	0.22	0.20
		4.0	0.46	0.41	0.37	0.34	0.31	0.26	0.22	0.17	0.15
		25.0	0.45	0.41	0.37	0.34	0.31	0.25	0.21	0.15	0.12
	0.25	1.0	0.49	0.48	0.46	0.43	0.41	0.37	0.34	0.30	0.28
		4.0	0.49	0.47	0.45	0.42	0.40	0.34	0.30	0.24	0.21
		25.0	0.49	0.47	0.45	0.42	0.39	0.33	0.29	0.22	0.18
	0.50	1.0	0.50	0.49	0.48	0.47	0.45	0.42	0.39	0.35	0.33
		4.0	0.50	0.49	0.47	0.46	0.44	0.39	0.34	0.27	0.26
		25.0	0.50	0.49	0.47	0.46	0.44	0.39	0.34	0.27	0.22
	1.00	1.0	0.50	0.50	0.49	0.49	0.48	0.47	0.45	0.42	0.40
		4.0	0.50	0.50	0.49	0.48	0.48	0.45	0.43	0.38	0.34
		25.0	0.50	0.50	0.49	0.48	0.47	0.44	0.41	0.35	0.30

[a] From Utzinger and Klein (1979).

Although in principle it is possible to calculate I_r at any time from Equation 14.4.2, the determination of f_i is tedious. For design purposes, we are not normally concerned with what happens at any particular time, but rather with monthly means. A monthly mean fraction of the beam radiation received by the shaded receiver relative to that on the receiver if there were no overhang, $\bar{f_i}$, can be calculated by integrating (or summing) beam radiation with and without shading over a month:

$$\bar{f_i} = \frac{\int G_b R_b f_i \, dt}{\int G_b R_b \, dt} \tag{14.4.3}$$

Then with $\bar{f_i}$, an equation analogous to Equation 14.4.2 can be written for the time and area average daily radiation on the shaded vertical receiving surface:

$$\overline{H}_r = \overline{H}\left[\left(1 - \frac{\overline{H}_d}{\overline{H}}\right)\overline{R}_b \bar{f_i} + \frac{\overline{H}_d}{\overline{H}} F_{r\text{-}s} + \frac{\rho g}{2}\right] \tag{14.4.4}$$

This is analogous to Equation 2.19.1 for monthly average radiation on a tilted, unshaded surface. The $\overline{H}_d/\overline{H}$ and $\overline{R}_b$ terms are found by the methods described in Chapter 2 for an isotropic sky. The Klein and Theilacker method can also be applied to this situation by multiplying the first term in Equation 2.20.5a (or the first two terms in Equation 2.20.4a) by $\bar{f}_i$, which results in

$$\overline{H}_r = \overline{H}\left[D\bar{f}_i + \frac{\overline{H}_d}{\overline{H}}F_{r-s} + \frac{\rho_g}{2}\right] \tag{14.4.5}$$

Utzinger and Klein present plots of f_i as a function of e, g, w, and p for various latitudes. One of these plots, for a latitude of $35°$, w of 4, and zero extension e, is shown in Figure 14.4.2. A set of the plots for latitudes of $35°$, $45°$, and $55°$ [Utzinger (1979)}is in Appendix I. At a latitude of $35°$, a projection of 0.3, a gap of 0.2, and no extensions, from Figure 14.4.2, $\bar{f}_i$ is 0.40 in August and 0.94 in October.

The nature of these curves dictates that a particular interpolation method be used. The general procedure to obtain an $\bar{f}_i$ consists of a set of steps to account for gap, width, and latitude; this procedure is modified in some cases to account for discontinuities in the relationships. The curves are shown as plots of $\bar{f}_i$ versus p for months. The first interpolation is a linear interpolation for gap g. The second step is another linear interpolation for extension e. Next, interpolation for relative width w is done by

$$\bar{f}_i = \bar{f}_{i2} + (\bar{f}_{i1} - \bar{f}_{i2})\left[\frac{\frac{1}{w} - \frac{1}{w_2}}{\frac{1}{w_1} - \frac{1}{w_2}}\right] \tag{14.4.6}$$

Then, linear interpolation is used to account for intermediate values of latitude.

This procedure is modified by two circumstances that are clear on examination of the plots. First, if the projection is large enough, $\bar{f}_i$ is independent of p for some months. The value of p at which $\bar{f}_i$ becomes independent of p is a linear function of the relative gap. Care must be used in interpolating for gap when on or near the horizontal portion of the curves. Second, during winter months when the receiver area is not shaded at all, $\bar{f}_i$ has a limiting value of 1. The intersection of $\bar{f}_i$ curves for a given month with the $\bar{f}_i = 1$ axis is also a linear function of the relative gap. Care must be used in interpolating for gap when near these intersections.

All of the computations of $\bar{f}_i$ described above are for surface azimuth angles of zero. Utzinger and Klein have shown that for surface azimuth angles within $\pm15°$, there is a negligible difference in the estimated radiation on shaded receivers from that on a south-facing surface. When the surface azimuth angles exceed $\pm30°$, the calculation for south-facing surfaces underpredicts summer radiation and overpredicts the winter radiation on the receiver by substantial amounts, and a more detailed computation is needed. If the surface azimuth angles exceed $\pm15°$, wingwalls may be more important in shading than overhangs.

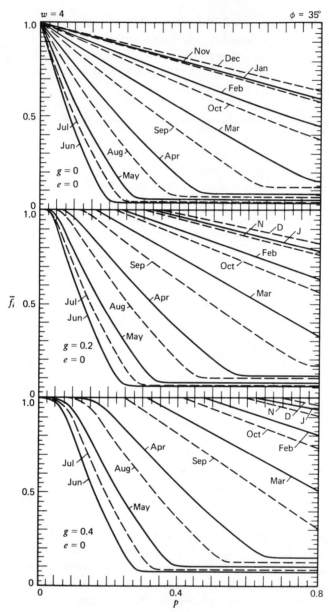

Figure 14.4.2 Monthly mean fraction of a vertical, south-facing receiver area receiving beam radiation as a function of relative overhang dimensions for latitude 35° and $w = 4$. For north-facing receivers in the southern hemisphere, interchange months as shown in Figure 1.8.2. From Utzinger (1979).

Example 14.4.1

Determine $\bar{f}_i$ for a window receiver with overhang with relative dimensions $w = 12$, $g = 0.4$, $p = 0.5$, and $e = 0$. The latitude is $40°$ and the month is March.

Solution

Interpolate w and ϕ in that order. The values of $\bar{f}_i$ from which the interpolations are done, using the appropriate plots in Appendix I, are as follows:

At $\phi = 35°$, $e = 0$ and $g = 0.4$: if $w = 4$, $\bar{f}_i = 0.80$, and if $w = 25$, $\bar{f}_i = 0.78$.

At $\phi = 45°$, $e = 0$ and $g = 0.40$: if $w = 4$, $\bar{f}_i = 0.96$, and if $w = 25$, $\bar{f}_i = 0.96$.

Equation 14.4.5 is used to interpolate for width. The inverse widths are $1/w = 0.0833$, $1/w_1 = 0.25$, and $1/w_2 = 0.04$. At a latitude of $35°$

$$\bar{f}_{i,35} = 0.80 + (0.78 - 0.80)\left[\frac{0.0833 - 0.25}{0.04 - 0.25}\right] = 0.79$$

At a latitude of $45°$, both $\bar{f}_{i1}$ and $\bar{f}_{i2}$ are 0.96, so $\bar{f}_{i,45} = 0.96$. We can now linearly interpolate between 0.79 and 0.96 for a latitude of $40°$:

$$\bar{f}_i = 0.79 + 0.5(0.96 - 0.79) = 0.87$$

Note that if the gap ratio had been 0.3 (for which curves are not shown), the four initial values would each have been the result of linear interpolations between $g = 0.2$ and $g = 0.4$. ■

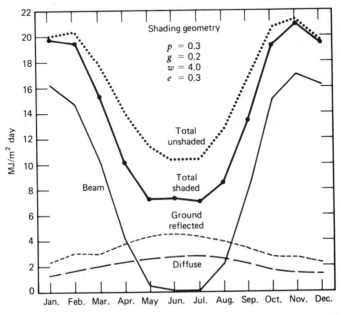

Figure 14.4.3 Monthly estimated radiation on vertical, south-facing receivers at Albuquerque, shaded and unshaded. From Utzinger and Klein (1979).

Figure 14.4.3 shows the results of month-by-month calculation of radiation on shaded and unshaded south-facing receivers at Albuquerque, NM ($\phi = 35°$), for a particular overhang geometry, calculated by Equation 14.4.5. The overhang is sufficiently large that beam radiation is essentially eliminated during the summer months, yet the total radiation on the shaded receiver in the summer is approximately two-thirds of that on the unshaded receiver. A similar computation for Minneapolis, MN ($\phi = 45°$) shows summer month radiation on the shaded receiver is more than half of that on the unshaded receiver.

14.5   DIRECT-GAIN SYSTEMS

A direct-gain passive system includes south-facing windows, very often with movable insulation and overhang shading to reduce incident summertime radiation, with the rooms behind the windows and their contents serving as cavity absorbers and having sufficient thermal capacitance to provide energy storage. The windows and room are, in effect, vertical, no-flow collectors with thermal capacitance. Loss coefficients are analogous to those of Section 6 but are generally based on room air temperature rather than on a plate temperature.[4] Movable insulation which is in place for part of a day means that the loss coefficients will have different values at different times of day.

The net rate of energy transfer across a vertical window (receiver) of area A_r at any time can be written as

$$Q_r = A_r \left[\alpha\, F_c \left(G_b R_b \tau_b f_i + G_d \tau_d F_{r\text{-}s} + \rho_g G \tau_g / 2 \right) - U_L F_I (T_r - T_a) \right]$$
(14.5.1)

The effective absorptance of the room, α, can be calculated with Equation 5.11.1. If the (vertical) window is unshaded by overhangs or wingwalls, f_i is unity and $F_{r\text{-}s}$ is 0.5. If there is shading by architectural features such as overhangs, the methods of Section 14.4 must be used to estimate f_i and $F_{r\text{-}s}$. The room temperature T_r will rise during periods when there is net gain through the windows and will be prevented from dropping below a set temperature by an auxiliary energy source.

The control function F_c is unity when there is no movable insulation covering the window and zero (or near zero) if movable insulation is in place. The three radiation terms in parentheses are the beam, diffuse, and ground-reflected radiation. The second term in the brackets represents the thermal losses. The loss coefficient U_L is for the uninsulated window, and $T_r - T_a$ is the temperature difference between room and ambient. Another control function F_I is unity when $F_c = 1$, and when insulation is in place it is the ratio of U_L with insulation to U_L without insulation.

[4] See Section 22.3.

The monthly average net gains (or losses) from a direct-gain system can be estimated by considering the solar gain into the rooms independently of the thermal losses through the window. The calculation of both solar gain and thermal losses must account for movable insulation. Thus the gain and loss terms of Equation 14.5.1 can be written in terms of hourly irradiation (I, I_b, and I_d) and summed for the month:

$$\Sigma\, Q_r = A_r\alpha\,\Sigma\, F_c\!\left(f_i\tau_b\, I_b\, R_b + \tau_d I_d F_{r\text{-}s} + \rho_g\tau_g\, I/2\right) - A_r\Sigma\, U_L F_I(T_r - T_a) \tag{14.5.2}$$

These calculations can be more conveniently done using monthly averages. The f_i for the individual hours can be replaced by a single value of $\bar f_i$ obtained from Figure 14.4.2 or Appendix I. Each of the solar gain terms includes the transmittance corresponding to the monthly mean angle of incidence determined by methods of Chapter 5. If movable insulation is used, $\bar F_c$ (a radiation-weighted value of the control function F_c) must be used. If the insulation is used only when no significant radiation is incident on the receiver, $\bar F_c$ will be unity. If this is not the case, information must be available on which to estimate $\bar F_c$. The average daily radiation absorbed in the room can then be written:

$$A_r\bar S = A_r\alpha\,\bar F_c\!\left(\bar f_i\bar\tau_b\bar H_b\bar R_b + \bar\tau_d F_{r\text{-}s}\bar H_d + \rho\bar\tau_g\bar H/2\right) \tag{14.5.3}$$

The average thermal losses for a day are

$$\bar Q_l = 2A_r\bar U_L\!\left(\bar T_r - \bar T_a\right) \tag{14.5.4}$$

where $\bar U_L$ is a monthly mean loss coefficient. If night insulation with thermal resistance R_N is used for a fraction f_N of the time,

$$\bar U_L = (1 - f_N)U_L + f_N U_L\!\left(\frac{U_L}{1 + R_N U_L}\right) \tag{14.5.5}$$

Example 14.5.1

A direct-gain system is to be used at a latitude of 40°. The south-facing, double-glazed window, with glass of $KL = 0.0370$ per sheet and without insulation, has an overall loss coefficient (inside air to outside air) of 3.2 W/m²C.

The window is 15.0 m wide and 1.25 m high. It is shaded by a rectangular overhang 0.5 m above the top edge of the window which projects out 0.625 m and is the same width as the window. The room surface area is 440 m² and it has a mean solar reflectance of 0.6.

For monthly average March conditions, the interior of the building is at an average temperature of 20 C, and the average ambient temperature is 3 C. The monthly average radiation $\bar H$ is 19.93 MJ/m², $\bar H_d$ is 6.26 MJ/m², $\bar H_b$ is 13.67 MJ/m², and ground reflectance is 0.2. What will be the net gain (or loss) from the window?

Solution

The average day of the month for March is the 16th. The average transmittance of the glazing can be estimated from Figure 5.3.1. The angle of incidence of the beam radiation $\bar{\theta}_b$ is 57°, and from Figure 5.3.1, $\tau_b = 0.71$. From Figure 5.4.1 the mean angle of incidence of both diffuse and ground-reflected radiation for a vertical surface is 59° so the same value of transmittance is used for the diffuse and ground-reflected components. (Note that Figure 5.4.1 does not give an exact equivalent angle for diffuse radiation for a shaded receiver, but uncertainties such as those due to assumption of isotropic diffuse radiation are probably greater than a correction for overhang.) Thus $\tau_d = \tau_g = 0.70$ and $\bar{R}_b$ from Figure 2.19.1(d) is 0.92.

For this shading overhang, $p = 0.625/1.25 = 0.5$, $g = 0.5/1.25 = 0.4$, and $e = 0$. The width w is $15/1.25 = 12$. These are the dimensions assumed in Example 14.4.1, and from that example $\bar{f}_i = 0.87$. From Table 14.4.1, $F_{r\text{-}s}$ is 0.42.

The room area is $15.0 \times 1.25 = 18.75$ m^2. The absorptance of the room-window combination is estimated by Equation 5.11.1.

$$\alpha = \frac{1 - 0.6}{(1 - 0.6) + 0.6 \times 0.70 \times 18.75/440} = 0.96$$

The estimated energy absorbed in the rooms, using Equation 14.5.3, is

$$A_r\bar{S} = 18.75 \times 0.96(0.87 \times 0.71 \times 13.67 \times 0.92 + 0.70 \times 6.26 \times 0.42$$

$$+ 0.2 \times 0.70 \times 19.93/2) = 198 \text{ MJ}$$

The thermal loss from the window over a 24 hour period (with $F_l = 1$) is

$$18.75 \times 24 \times 3.2 \times (20 - 3) \times 3600 = 88 \text{ MJ}$$

The average net gain is $198 - 88 = 110$ MJ/ day for the month.   ∎

Thus this window gives an estimated net gain of 110 MJ/day. By varying F_l, taken as unity in this example, the effects of control strategy on performance can be shown. A control method that excludes significant radiation will affect both $\bar{f}_i$ and $\bar{\tau}_b$, and hourly calculations would be needed. If very effective movable insulation were to reduce window losses to near zero whenever there would be a net loss from the wall, the direct-gain system would then produce the maximum possible net input to the rooms.

It is of interest to consider the energy balance on the building as a whole. There will be four major energy flow terms across the boundaries of direct-gain system, as shown schematically in Figure 14.5.1. The two input streams are the solar energy absorbed in the building and the auxiliary energy added L_A. The outputs are the excess energy that cannot be used or stored without driving the building to unacceptably high temperatures (the "dumped" energy Q_D) and the "load" (here defined as the skin losses and infiltration losses less internal energy generation).

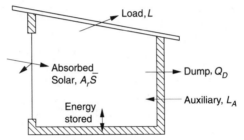

Figure 14.5.1 Monthly energy balance on a direct-gain building.

The auxiliary energy required for a month, neglecting differences in stored energy at the beginning and end of the month, will be

$$L_A = A_r \overline{S} - Q_D - L \tag{14.5.6}$$

The absorbed solar energy $\overline{S}$ is readily calculated. The load L, including the losses through the direct-gain window, can be estimated by standard building load calculation methods. The difficult term to estimate is Q_D, the dumped energy. Methods for estimating Q_D and L_A are presented in Chapter 22.

The loads on a direct-gain system are a function of the receiver area, as in general the loss coefficient U_L for a direct-gain window will be greater than that for an insulated wall. This is illustrated in Figure 14.5.2 for a building having a $(UA)_h$ of 200 W/C when the area of direct-gain receiver is zero. The term $(UA)_h$ is shown as a function of A_r for a double-glazed receiver with heavy drapes ($U_{L,r} = 1.3$ W/m²C) and for a receiver with 20 mm of tight-fitting foam insulation ($U_{L,r} = 0.95$ W/m²C).

Examples of the effects of several design parameters of direct-gain systems on auxiliary energy requirements are shown in Figure 14.5.3. The building without the

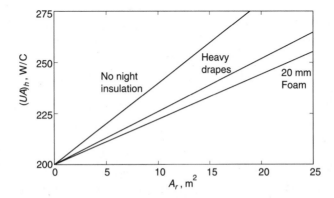

Figure 14.5.2 Dependence of $(UA)_h$ on the apertures of direct-gain receivers for $U_{wall} = 0.3$ W/m²C, $U_{window} = 4$ W/m²C, for heavy drapes with $U_N = 0.95$ W/m²C. Insulation is assumed to be in place an average of 12 hours a day.

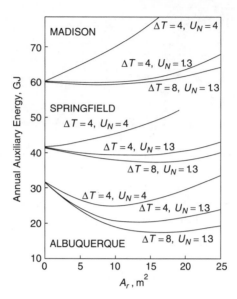

Figure 14.5.3 Examples of effects of receiver area, allowable temperature swing, and night insulation on the annual auxiliary energy requirements for direct-gain passive heating in Madison, WI, Springfield, IL, and Albuquerque, NM.

direct-gain window has a $(UA)_h$ of 200 W/C. The addition of direct-gain aperture, replacing corresponding areas of well-insulated wall, results in the increases in $(UA)_h$ shown in Figure 14.5.2 and affects the annual requirements for auxiliary energy in degrees that depend on system parameters and locations. The calculations were based on the following fixed parameters: minimum noon temperature = 17 C; $\beta = 90°$; $\gamma = 0°$; $(\tau\alpha)_n = 0.75$; and $C_b = 25$ MJ/C. The auxiliary energy required L_A is shown as a function of receiver area for three locations and for combinations of allowable temperature swing in the building and night insulation.

Note that there can be thermal optima in these systems. As the receiver area increases, solar gains increase but loads also increase. A point is reached at which either collector area or storage capacity begins to limit solar gains and incremental losses exceed incremental gains. The thermal optima in general will not coincide with economic optima.

The thermal advantage of direct-gain heating, based on these results, are most impressive in milder climates with high $\overline{K}_T$, with systems using night insulation, and with higher allowable temperature swings. Other parameter sets would give different Q_A requirements, but the trends would be the same.

The occupants of a building may use the system in different ways at different times of the year. For example, auxiliary heat may be turned off during parts of swing months, and higher than usual temperature swings tolerated. Operation of a system in this way can result in further reductions in auxiliary energy requirements.

Methods for estimating the annual performance (i.e., annual auxiliary energy requirements) of direct-gain systems are given in Sections 22.2 and 22.3.

14.6 COLLECTOR-STORAGE WALLS AND ROOFS

A collector-storage wall is essentially a high-capacitance solar collector coupled directly to the spaces to be heated. A diagram is shown in Figure 14.6.1. Solar radiation is absorbed on the outer surface of the wall. Energy is transferred from the room side of the wall to the spaces to be heated by convection and radiation. Energy can be transferred to the room by air circulating through the gap between the wall and glazing through openings at the top and bottom of the wall. Circulation can be by natural convection controlled by dampers on the vent openings (not shown) or by forced circulation by fans.

Radiation transmitted by the glazing and absorbed by the wall is calculated for any time period by the same methods as for the direct-gain component noted in Section 14.5, with the additional consideration that absorptance must be evaluated at the appropriate angles of incidence:

$$A_r S = A_r[f_i F_c\, I_b R_b(\tau\alpha)_b + I_d(\tau\alpha)_d F_{r\text{-}s} + I\rho_g(\tau\alpha)_g/2] \qquad (14.6.1)$$

and $$A_r \overline{S} = A_r \overline{F}_c[\overline{H}_b \overline{R}_b \overline{f}_i(\overline{\tau\alpha})_b + \overline{H}_d(\overline{\tau\alpha})_d F_{r\text{-}s} + \overline{H}\rho_g(\overline{\tau\alpha})_g/2] \qquad (14.6.2)$$

where $\overline{F}_C$ is an energy-weighted control function that will be less than unity if movable insulation is in place during times when significant radiation is available. An example of calculation of $\overline{S}$ for a system of this type (but without overhang) was shown in Section 5.10.

If a collector-storage wall uses water to provide thermal capacitance so that there are negligible temperature gradients in the wall and if there is no air circulation through the gap, then the situation is analogous to a flat-plate collector with high capacity. Equation 6.12.7 can then be used to predict the wall temperature as a function of time. Here U_L includes only front losses, as losses out through the glazing are indeed losses but losses out the back are gains to the space to be heated. Hourly calculations are essentially the same as Example 6.12.1, with S calculated from Equation 14.6.1, except that convection and radiation from the back must be

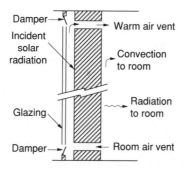

Figure 14.6.1 Section of a collector-storage wall. Dampers may be used to ventilate the gap in summer.

included. (Heat transfer to the rooms by these mechanisms is also needed for use in the energy balances on the rooms.)

A solid storage wall [e.g., a wall of concrete as used by Trombe et al. (1977)] can serve both structural and thermal purposes. A section of such a wall is shown in Figure 14.6.1. There will be temperature gradients through these walls, variable with time, which can be determined from energy balances as outlined in Section 8.6.

Storage walls may be either vented or unvented. Utzinger et al. (1980) have concluded that the annual auxiliary energy required by a building with a vented collector-storage wall is essentially the same as for one with an unvented wall, although the time distribution through the day of the delivered energy to the room will depend in part on venting. Balcomb et al. (1984) indicate that optimum venting can lead to performance improvements of several percent relative to unvented walls. A detailed discussion of flow in gaps is presented by Akbarzadeh et al. (1982).

A collector-storage wall supplying energy to a room and the auxiliary energy supply to the room may be controlled by several alternative strategies. If forced circulation through the gap is used, a thermostat must turn on the fan when the rooms need energy or call for auxiliary energy if the storage wall is unable to provide the needed energy. If natural circulation is used, it may be desirable to control addition of energy to the rooms by means of backdraft dampers on the vents. If summer venting to the outside is used, dampers will have to be positioned; this may be a single seasonal change which can be done manually. It may be necessary to move insulation into the air gap to control losses, and controls and mechanisms for accomplishing this must be provided. The equations for both the collector-storage wall and the rooms must reflect the modes of operation at any time. An example of the results of alternative control strategies is provided by Sebald et al. (1979).

The variations with time of day of energy flow into rooms from direct-gain windows and collector-storage walls are quite different. Figure 14.6.2 shows this contrast for south-facing receivers for a clear day. The energy absorbed in the rooms from the window occurs when the radiation enters the spaces. Energy is added to the rooms from the collector-storage wall by a combination of mechanisms: by heating of air flowing through the gap and by radiation and convection from the room side of the storage wall. The energy added by air flow lags that of the direct-gain system and is in turn lagged by the energy flow through the wall.

The effects of wall area, thickness, and night insulation on the performance of collector-storage wall systems in several locations is illustrated in Figures 14.6.3 and 14.6.4. In this example, the $(UA)_h$ of the building without a collector-storage wall is 200 W/C. As with direct-gain systems, the addition of receiver area increases $(UA)_h$ when losses out through the glazing of the collector storage wall are greater than that of the insulated wall it replaces. The $(UA)_h$ is shown as a function of A_r for collector-storage walls of various thicknesses, with and without night insulation. The walls are assumed to be concrete and the night insulation has $R = 1$ m^{2}C/W (corresponding to about 25 mm of foam insulation). The loss coefficient is a weak function of wall thickness for the double-glazed storage wall.

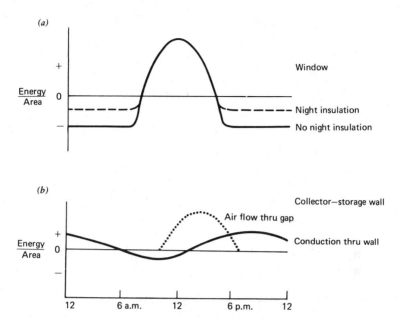

Figure 14.6.2 Energy flows through direct-gain and collector storage wall systems as a function of time of day. From Utzinger (1979).

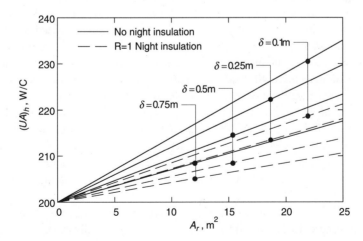

Figure 14.6.3 An example of the variation of a building UA for a building with a collector-storage wall, for no night insulation and for night insulation with $R = 1$ m^{2}C/W, for four thicknesses of concrete storage wall. U_L from wall surface to ambient is 2.0 W/m^{2}C.

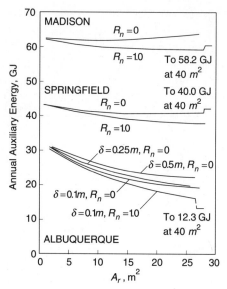

Figure 14.6.4   Examples of effects of location, collector-storage wall area, wall thickness and night insulation on auxiliary energy requirements. The wall is double glazed with $(\tau\alpha) = 0.705$ and is 0.25 m thick unless otherwise noted. $C = 12.5$ MJ/C, $\beta = 90°$, ,and $\gamma = 0°$.

Figure 14.6.4 shows examples[5] of effects of location, receiver area, wall thickness, and night insulation for collector-storage wall systems. In Madison, this system shows an optimum collector area (i.e., minimum Q_A) at about 10 m², but the effect of the wall on auxiliary energy requirements is essentially negligible under the assumptions made here. In Springfield, the uninsulated wall produces minor savings, with a minimum Q_A at an area less than 40 m². The insulated wall shows additional small savings.

In the climate of Albuquerque the impacts of the solar wall are more significant. In the range of areas up to 40 m² the auxiliary energy requirement diminishes with increasing A_r. The wall thickness of 0.1 m shows better performance than either of the thicker walls, and use of night insulation results in more improvement with the thin wall than it does with the thicker walls. The Albuquerque curves do not show a minimum out to 40 m² area; the loss coefficient of the collector-storage wall is higher than that of the insulated wall, but in this range of areas the increase in energy supplied to the building more than compensates for the increased losses from the building.

Again, other assumptions could lead to other results, and comparisons of the systems of Figure 14.6.4 should not be viewed as generalizations. Also, cost considerations have not been included.

---

[5] These are computed by the unutilizability method presented in Chapter 22.

This discussion has dealt with collector-storage walls. Similar concepts have been developed for flat roofs with storage capacity by Hay and Yellott (1970) and Hay (1973). The performance of a building using this method is described by Niles (1976). The principles are similar, but energy transport into the rooms is primarily by radiation from the storage ceiling. Collector-storage units are uncovered, and movable insulation is placed over them when collection is not occurring.

In closely related developments, transparent insulation of various types has been applied to the outside of opaque walls to reduce losses and at times produce net gains into the buildings. This is discussed in the proceedings of workshops on Transprent Insulation Technology (1988, 1989).

Methods for estimation of long-term performance of collector-storage walls are given in Sections 22.2 and 22.4.

14.7 SUNSPACES

An attached sunspace (also termed conservatory or greenhouse) is a glazed extension of the south side of a building, designed to provide a combination of energy gain and space which itself may be useful. Storage may be provided in the thermal mass of the floor and/or wall of the sunspace structure itself, or it may be provided separately, for example, in the form of a pebble bed using forced or natural circulation. Circulation of warm air from the sunspace to the building can be by natural convection or with the aid of fans and associated controls. Energy will also be transferred into the building through the walls separating the sunspace from the rooms. If a sunspace has a heavy wall between it and the building, it can be thought of as a collector-storage wall with an enlarged space between the glazing and the wall. If its glazing is primarily on the south side and storage is in the rooms to be heated (i.e., the sunspace itself), it can be considered as a direct-gain room.

The design and construction of sunspaces varies widely.[6] They can have open or closed ends, single or multiple slopes, and various arrangements of storage mass in the floor and wall. Representative configurations of sunspace are shown in Figure 14.7.1.

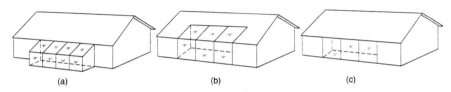

(a)   (b)   (c)

Figure 14.7.1 Common sunspace configurations. Type (a) can have opaque or glazed ends. Adapted from Balcomb et al. (1983).

[6] See Section 22.2.

Simulation models of sunspaces are related to those of collector-storage walls. A diagram of a thermal network for a sunspace with negligible end effects is shown in Figure 14.7.2. Schwedler (1981) studied networks of this kind for sunspaces with storage in the back walls and floor. He concluded that multiple nodes were needed in both vertical and horizontal directions in the floor, but multiple nodes were not needed in the horizontal direction in the storage wall. There are many possible variations of these networks to account for venting to the outside, air circulation of the building, end effects, use of water walls instead of masonry walls, use of pebble beds for storage, and sunspace configuration. The sunspace components in simulation programs such as TRNSYS (1990) are based on these networks.

Temperature excursions in sunspaces affect their utility and can be calculated by simulations. If minimum temperatures for protection of plants are maintained, through design or by use of auxiliary energy, sunspaces can serve as greenhouses. Figure 14.7.3, from Littler (1984), shows computed temperature-time history for a 2 month span for a sunspace in the winter climate of Kew, England. This sunspace is double glazed and has low thermal capacitance. In colder climates, temperatures will be more variable than those shown; sunspaces with more storage capacity will show less variation.

Sunspaces require the same thermal considerations as collector-storage walls or direct-gain systems, but the glazing area may be significantly larger than the receiver (collector) area. Losses occur continuously, although fans for heat transfer from sunspace to building would be shut off during nonsunny periods. Thus it may be necessary to use movable insulation to control losses, or the losses may exceed gains.

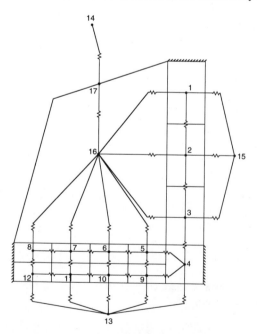

Figure 14.7.2 A thermal network for a two-dimensional sunspace. From Schwedler (1981).

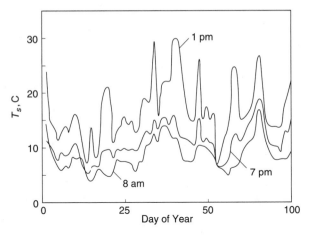

Figure 14.7.3 Sunspace air temperatures for January and February for Kew (London) England. Air is blown from the sunspace to the house when the sunspace temperature is above the room temperature. The three temperatures are at 8 AM, 1 PM, and 7 PM. From Littler (1984).

14.8 ACTIVE COLLECTION-PASSIVE STORAGE HYBRID SYSTEMS

Systems with active collection and passive storage (ACPS) can be of particular interest for closely spaced buildings where shading problems may rule out direct-gain or collector-storage wall systems. These hybrid ACPS systems can also be considered for retrofit, particularly to buildings of masonry construction. The basic concept is that active collectors are fed with fluid from the building and return the heated fluid to the building where it serves to provide heat needed at the time and also to raise the temperature of the structure. Storage is provided by the structure itself, as in a direct-gain system. These systems have the advantage that losses from the collector during nonoperating periods are eliminated by turning off the fluid flow to the collectors. The solar contribution to meeting the heating loads of the building, however, is usually limited by the storage capacity of the structure and thus by the allowable temperature swings of the interior spaces.

ACPS systems can be based on either air or liquid systems. The simplest use air collectors which heat room air which is recirculated to the building. This is shown schematically in Figure 14.8.1(a). It is also possible to use liquid heating collectors with a liquid-air heat exchanger as shown in Figure 14.8.1(b); the inlet temperature to the collector is higher than in the air systems, with the increase in temperature a function of the effectiveness of the heat exchanger. With the addition of another heat exchanger (air-water or liquid-water) downstream from the collector, either of these types can be arranged to provide service hot water; these heat exchangers are not shown on the diagrams. The following discussion is based on air systems, but with modification to allow for increased T_i, the same principles also apply to liquid systems.

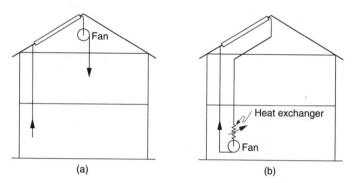

Figure 14.8.1   Active collect-passive store heating system schematics for systems based on (a) air and (b) liquids. The building structure provides storage.

For air systems, the inlet temperature of the collector is as low as it can be, i.e., room temperature. For given collector characteristics $[F_R(\tau\alpha)_n$ and $F_R U_L]$ the output of the collector will be very close to the maximum possible (differing only in that the room temperature will rise to some degree as energy is stored in the structure). Thus collector performance is comparable to that of active air systems during winter months and somewhat better during swing months.

Figure 14.8.2 shows an example of comparison of solar contributions to meeting heating loads as a function of collector area for ACPS systems for an allowable temperature swing of 5 C for residences of conventional and medium-weight construction in Springfield, IL. Also shown on the plot is the solar contribution of an active air heating system with pebble bed storage. In this example, at collector areas below solar fractions of about 0.4, the performance of the ACPS and that of the active systems are for practical purposes identical. Above solar

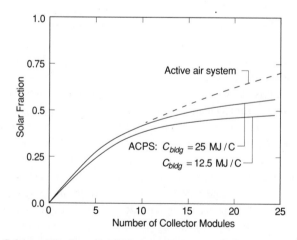

Figure 14.8.2   Solar contributions of ACPS systems of two building thermal capacitances and an active air system. $(UA)_h = 200$ W/C, based on weather data for Springfield, IL.

contributions (for this particular example) of about 0.4 for the lighter building and 0.5 for the heavier, the increase of solar contribution with increasing collector area levels off. (The temperature variations in the building with the full active systems would be about half that in the ACPS building with the lower temperature swing.)

ACPS systems should be designed with collector areas no larger than that corresponding to the knee in curves of $\mathcal{F}$ vs. A_c. At larger areas, the annual contributions are limited by storage capacity, and little is to be gained by increasing collector area. At collector areas below the knee, collector area will limit solar contribution in peak heating seasons, and storage capacity will limit solar contribution in spring and fall.

An economic analysis of an ACPS system shows life cycle solar savings vs. A_c curves of the same general shape as those for active systems, but with improved solar savings at low collector areas (due to lower system costs) and sharper drop-off of life cycle savings as collector area reaches the point at which storage is the limiting factor (i.e., the knee in the $\mathcal{F}$ vs. A_c curve).

14.9 OTHER HYBRID SYSTEMS

Many active solar heating systems are on buildings that have significant contributions to their heating loads from passive (usually direct-gain) solar heating. The nature of the active and passive systems dictates that the direct gains will be realized when they are available, and the active system will meet some part of the balance of the loads. Thus the time distribution of the loads being met by the active system will differ from those of systems in which no significant passive contribution is present, as the direct-gain system meets much of the daytime and evening load and the active system meets nighttime and early morning loads. The performance of these hybrid systems can be estimated by simulations. It is also possible to use design methods such as the f-chart method by applying corrections to the active systems performance; the auxiliary energy requirement resulting from this calculation is then the load that the passive system sees, and the passive performance is then estimated by one of the available methods. [See Evans and Klein (1984a).][7]

Another hybrid system that may be encountered is the combination of active collection-passive storage systems with direct gain [Evans and Klein (1984b)]. Energy collection occurs by two routes, the collectors and the direct-gain windows, but storage is provided only in the form of the thermal capacity of the structures. The inputs coincide in time, so analyzing the two systems independently and adding the solar gains leads to overprediction of the solar contribution and underprediction of auxiliary energy requirements. The energy balances on the building are similar to those of Figure 14.5.1, with the solar input being the sum of the inputs by the two mechanisms noted. If the building becomes overheated by the combination of inputs, venting of energy will become necessary.

---

[7] This will be shown in Section 22.6.

14.10   PASSIVE APPLICATIONS

A great variety of passively heated buildings have been built, operated, and enjoyed by their occupants. The performance of a few of these have been measured. Many innovative ideas and architectural approaches have evolved. In this section we present a brief picture of the operation of several buildings as illustrations of the range of possibilities.

The section diagrams in Figure 14.10.1 show several possible design concepts for passively heated buildings. Part (a) shows a direct-gain system, the Wallasey school in England. Part (b) is a section drawing of a house in Princeton, NJ, with a collector storage wall two stories high, and also includes a sunspace. Part (c) shows a building with clerestory windows, which provide both direct-gain heating and daylighting for rooms on the north side of the building. Part (d) is a section of a school in Rome, Italy, which includes sunspaces at three levels on the south side of the building [CEC (1988a)]. The sunspaces are separated from the main part of the building by both venetian blinds and sliding windows, so light and warm air can be admitted to or kept from the building independently [Funaro (1986)]. Daylighting is also provided for the corridors. Part (e) is a schematic section of a building with a central storage wall which is heated by air circulated from the top of the building by a small fan. Part (f) shows a house with air heating collectors below floor level with air flow through a pebble bed store and/or the building by natural circulation.

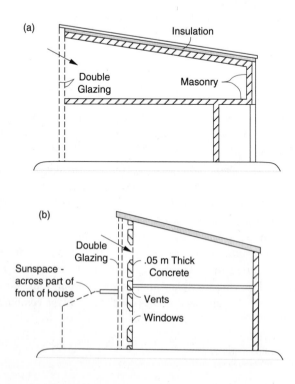

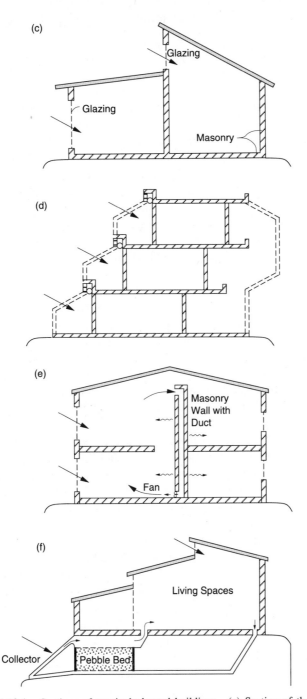

Figure 14.10.1 Sections of passively heated buildings. (a) Section of the Wallasey School direct-gain system. (b) Section of the Kelbaugh collector-storage wall system. (c) Building with clerestory windows to admit radiation to north rooms. (d) Part of a school building in Rome, showing sunspaces, classrooms, and corridors, designed by C. Greco. Adapted from CEC (1988). (e) A building with central storage core and air circulation. (f) A natural circulation system with separate air heating collectors and pebble bed storage.

One of the best known passive heating applications is the group of small single-family residential buildings at Odeillo, France. The collector-storage wall in the first of these buildings consists of double glazings, spaced 0.12 m from the concrete storage wall which is 0.60 m thick. The wall is painted black, and has vent openings 0.11 by 0.56 m at top and bottom which are 3.5 m apart. The collector occupies most of the south wall of the building. The auxiliary energy supply is electricity. The system was estimated by Trombe et al. (1977) to have supplied between 60 and 70% of the energy for an average house in the Odeillo climate. Based on extensive studies of this system, a building with three apartments was constructed that is better insulated, has smaller collector area per cubic meter of heated space (0.1 rather than 0.16 m^2/m^3), and utilizes storage walls of concrete 0.37 m thick. One of the original building is shown in Figure 14.10.2(a).

An example of a residence using a combined sunspace-direct-gain concept is the Balcomb house in Santa Fe, NM, shown in Figure 14.10.2(b). This two-story house has 0.35 m thick adobe walls that provide thermal capacitance and also a pebble bed storage unit to provide additional storage [see AIA (1978)]. This building operated with very little auxiliary energy input in the Santa Fe climate.

The Sargent house, built in the mid-1980s in Golden, CO, is shown in Figure 14.10.2(c). This direct-gain system provides about half of the total heating requirements of the building. Figure 14.10.2(d) shows a monastery building in New Mexico with clerestory windows admitting radiation to north rooms. All windows are double glazed, and solar energy provides most of the heating loads on the building [see AIA (1978)].

Many passive buildings have been built and operated in Europe. The Commission of the European Communities (CEC) publishes a series of case studies illustrating how passive principles can be applied to produce attractive and energy-efficient buildings. The waiting area of the Vielha Hospital in Catalonia, Spain, described in a CEC report (1988c), is shown in Figure 14.10.2(e). It is a direct-gain building with double-glazed windows covering the major part of the south facade. Shading control awnings are used to avoid summer overheating. The hospital waiting area is on the south side, is two floors high, and the first and second floors open off this space. The building is three floors high, is of compact shape, and the concrete and ceramics used in floors provide about 40 m^3 of direct storage. Passive solar gains contributed 33% of the total space heating load, 44% was supplied by auxiliary energy, and the balance was provided by internal energy generation. There are heating loads on this building throughout the year. All heating loads are met by solar in the summer months and an average of about a quarter of the loads are met by solar in the balance of the year. The monthly ambient temperatures range from about 8 to 18 C.

A development in Villefontaine, France, includes 42 single-family residences that include a combination of direct-gain and sunspace passive heating [CEC (1988b)]. Figure 14.10.2(f) shows three of these buildings. The sunspaces are on walls facing within 45° of south, have floor areas of 6.5 m, and are recessed into the building. A combination of fixed and movable shading devices control absorbed

Figure 14.10.2 (a) The Odeillo single family residence, utilizing collector-storage walls. (b) Residence in Santa Fe, NM. (c) Residence in Golden, CO. (d) Monastery in Pecos, NM. (e) Waiting room of the Vielha hospital, Catalonia, Spain. (f) Apartment buildings in Villefontaine, France.

(d)

(e)

(f)

radiation. Ventilation of the sunspaces is accomplished by windows and vents. Mechanical circulation of warm air from sunspaces to the rest of the building is used. All controls are operated manually. The space heating energy required by these buildings is 37% less than conventional designs in the same area. Occupant response to the homes has been positive, and turnover of occupants is low.

Warehouse heating, where temperature control may not be as critical as in buildings for human occupancy, is another application of interest. Thermal storage may be provided by the contents of a warehouse with direct-gain solar heating. Translucent plastic siding has been used on the south wall of such warehouses.

Direct-gain solar heating systems may have some problems associated with them. The contents of the room are, by definition, subjected to solar radiation, which can cause fabric degradation and fading of dyes. If floors are to be built with high capacitance (i.e., of masonry), there may be limits on the placement of rugs and furniture that would reduce the addition of heat to or removal from the storage. An experimental building at MIT is being used in a study of possible answers to these and other problems. Beam radiation incident on the windows is deflected upward by narrow-slat venetian blinds which have slats shaped to be concave upward and which have reflective materials on their upper surfaces. These blinds reflect the radiation upward to the ceiling rather than admitting it to the floor. The heat storage medium is a phase change material, $Na_2SO_4 \cdot 10H_2O$, in thin layers in ceiling panels. Heat transfer from the ceiling to the rooms is primarily by radiation [see Mahone (1978)].

14.11 HEAT DISTRIBUTION IN PASSIVE BUILDINGS

There is implicit in the discussion of energy storage in buildings, and the characterization of building conditions by a single temperature, that energy is distributed throughout the building by natural convection ("passively" in the complete sense of the word), or by forced convection. The basic passive heating systems discussed in this chapter all result (in their simplest forms) in solar energy addition to the south rooms. The transport of heat from the heated rooms to other rooms has been the subject of experimental and theoretical studies, including model studies using water tanks and full-scale experiments in laboratories. A review of this work has been presented by Anderson (1986).

Energy moves from one room to another by a combination of mechanisms. The dominant mechanism, if there are openings between them, is by convection. Thus convective flow through doorways is of importance in energy transport in passive buildings. Natural convection flows in buildings have been the subject of several studies. For example, Balcomb et al. (1984) report theoretical and experimental measurements of air flow in occupied buildings, and Kirkpatrick et al. (1986) have measured convective flow through doorways that connect a two-story sunspace and the rooms behind the sunspace. A temperature difference from one room to another means that there is a corresponding density difference, which leads to convective flow at the bottom of the doorway from the colder room to the warmer room and flow in the reverse direction at the top.

Any forced convection will be superimposed on the natural convection exchange between rooms and can dominate the natural convection processes. In a system such as that shown in Figure 14.10.1(e), air is moved from the warmer upper zones of a building to lower zones by air circulated downward through open cores in a storage wall. The forced convection requires relatively little mechanical energy and makes it possible to maintain comfortable temperature gradients throughout the building.

As pointed out by Anderson, an understanding of the natural convection processes occurring in a passive building can be essential in assuring that acceptable levels of comfort can be maintained.

14.12   COSTS AND ECONOMICS OF PASSIVE HEATING

The first costs, or investments, in passive heating of a building are those incremental costs associated with providing for solar heating compared to a building without solar heating. For direct-gain systems, two major costs must be considered. The first is for additional glazing area, any associated night insulation, and extra mass of the building to provide storage. Windows generally cost more than walls, and the total incremental cost of providing the windows, insulation, drapes, etc., must be considered. The replacement of a wood frame floor by a concrete and tile floor will entail an increment in cost. The second set of costs includes equipment costs of fans for heat distribution, controls for fans, and/or movable insulation. Some of these costs will be dependent on receiver area and others may be fixed. Equation 11.1.1 can usually be used to represent systems costs.

Collector-storage walls or roofs also involve increments in first cost compared to nonsolar buildings. The costs of the mass wall glazing, movable insulation, controls, vents, etc., should be included. If a mass wall occupies significant living space floor area or requires additional foundations, the costs of the space occupied and foundation should be included. The same considerations hold for sunspaces, where glazing and insulation expenses may be important.

For active collection-passive storage systems, the incremental costs will be the installed costs of the collector (less credits for wall or roof replaced), blower, pump, controls, ducting or piping, plus any costs of increments to the mass of the building designed to increase its storage capacity.

The economic gain from passive heating is in reduction of purchased (auxiliary) energy. Thus an economic analysis of passive heating must be based on an analysis of the thermal performance of a system to estimate the expected reduction in auxiliary energy needs. With the reduction in cost of purchased energy, the increase in cost of the structure, and the usual set of economic parameters such as interest, discount, and inflation rates, tax data, periods of mortgages, and economic analyses, the economic principles set forth in Chapter 11 can be applied to passive solar processes.

However, the loads on the building will usually increase significantly as receiver (window) area increases, in contrast to active systems where the loads are nearly independent of collector area. Thus the important results to be obtained from the thermal analysis are the annual amount of auxiliary energy needed and the reduction in annual energy purchased, and not the solar fraction. As noted in Section 10.10, a modified fractional solar contribution can be defined as

$$\mathcal{F}_c = L_S/L_0 \tag{14.12.1}$$

where L_S is the solar energy supplied, i.e., the reduction in purchased energy of the solar building compared to the nonsolar building, and L_0 is the load on the nonsolar building (the same building but with $A_r = 0$ and a standard wall replacing the receiver) that is to be met by purchased energy. With $\mathcal{F}_c$ so defined, the analysis of Chapter 11 needs no modification. (Note that the definitions of solar fraction used by many authors differ from this.) The application of the P_1, P_2 method leads to results that are similar to those for active systems.

For example, consider a residence to be constructed in Albuquerque, NM. The building has the thermal performance characteristics (i.e., the auxiliary energy requirements) of the building of Figures 14.5.2 and 14.5.3. The allowable temperature swing is 4 C, and night insulation is to be used on the direct-gain windows resulting in $U_L = 1.3$ W/m²C. The economic parameters are those of Example 11.8.2 (i.e., $C_{F1} = \$10/GJ$, $P_1 = 22.169$ and $P_2 = 1.035$). The fractional solar contributions as defined for this purpose are $\mathcal{F}_c = (L_{A,0} - L_A)/L_{A,0}$. The annual load on the building with $A_r = 0$ is 31 GJ, so $\mathcal{F}_c = (31 - L_A)/31$. Figure 14.12.1 shows life cycle savings vs. receiver area curves for four cost increments of

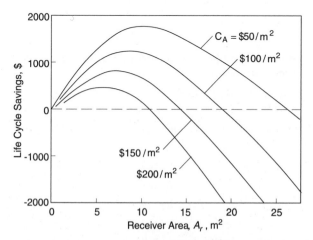

Figure 14.12.1 An example of the life cycle savings (LCS) of a direct-gain passive system in Albuquerque, for four levels of the incremental cost per unit area of direct-gain aperture over that of the insulated wall. Area-independent costs have been assumed negligible.

the direct-gain window and drapes over the insulated wall. The economic optimum area is shown to be less than the thermal optimum area, with the difference increasing as the incremental cost of the direct-gain aperture rises. These curves are generally the same shape as those encountered in active systems, and depending on location and system design, the curves may or may not show positive savings.

REFERENCES

AIA Research Corporation, Report to the Department of Housing and Urban Development HUD-PDR-287, Washington, DC (1978). "A Survey of Passive Solar Buildings."

Akbarzadeh, A., W. W. S. Charters, and D. A. Leslie, *Solar Energy*, **28**, 461 (1982). "Thermocirculation Characteristics of a Trombe Wall Passive Test Cell."

Anderson, B., *Solar Energy, Fundamentals in Building Design,* McGraw-Hill, New York (1977).

Anderson, R., *Passive Solar J.*, **3**, 33 (1986). "Natural Convection Research and Solar Building Applications."

ASHRAE *Fundamentals Handbook*, Am. Soc. Htg. Refrig. & Aircond. Engrs., Atlanta (1989).

Balcomb, J. D., R. W. Jones, R. D. McFarland, and W. O. Wray, *Passive Solar J.*, **1**, 67 (1983). "Expanding the SLR Method."

Balcomb, J. D., G. G. Jones, and K Yamaguchi, *Proc. Ninth National Passive Solar Conf.*, Columbus, OH. American Solar Energy Society (1984). "Natural Convection Air-Flow Measurements and Theory."

CEC (Commission of the European Communities), Research and Development Programmes on Solar Energy Applications to Buildings, Research Digest No. 1 (1988a).

CEC Project Monitor, Case Study 28 (1988b). "Les Tournesols, Villefontaine, France."

CEC Project Monitor, Case Study 34 (1988c). "Vielha Hospital, Catalonia, Spain."

Dietz, A. G. H. and E. L. Czapek, *Heating, Piping and Air Conditioning*, **118**, (March, 1950). "Solar Heating of Houses by Vertical South Wall Storage Panels."

Evans, B. L. and S. A. Klein, paper presented at Las Vegas ASME meeting (1984a). "A Design Method for Active Collection-Passive Storage Space Heating Systems."

Evans, B. L. and S. A. Klein, paper presented at Anaheim ASES meeting (1984b). "Combined Active Collection-Passive Storage and Direct Gain Hybrid Space Systems."

Fanger, P. O., *Thermal Comfort Analysis and Applications in Environmental Engineering,* McGraw-Hill, New York (1972).

Funaro, G., *Passive Solar J.*, **3**, 291 (1986). "Passive Solar Buildings in Italy."

Harrison, D., *Solar Energy*, **17**, 317 (1975). "Beadwalls."

Hay, H. R., and J. I. Yellott, *Mech. Engr.*, **92**, 19 (1973). "A Naturally Air-Conditioned Building."

Hay, H. R., *Mech. Engr.*, **95**, 18 (1973). "Energy, Technology, and Solarchitecture."

Hollingsworth, F. N., *Heating and Ventilating* **76** (May 1947). "Solar Heat Test Structure at MIT."

Jones, R. E., *Solar Energy*, **24**, 305 (1980). "Effects of Overhang Shading of Windows Having Arbitrary Azimuth."

Kirkpatrick, A., D. Hill, and K. Stokes, *Passive Solar J.*, **3**, 277 (1986). "Natural Convection in a Passive Solar Building."

Lebens, R. M., *Passive Solar Heating Design*, Applied Science Publishers, London (1980).

Littler, J., personal correspondence (1984)

Mahone, D., *Solar Age*, **3**(9), 20 (1978). "Three Solutions for Persistent Passive Problems."

Mazria, E., *The Passive Solar Energy Book*, Rondale Press, Emmaus, PA. (1979).

Niles, P. W. B., *Solar Energy*, **18**, 413 (1976). "Thermal Evaluation of a House Using a Movable Insulation Heating and Cooling System."

Olgyay, V., *Design with Climate*, Princeton University Press, Princeton NJ (1963).

Schwedler, M. C. A., M.S. Thesis, Mechanical Engineering, U. of Wisconsin-Madison (1981). "The Simulation of Sunspace Passive Solar Energy Systems."

Sebald, A. V., J. R. Clinton, and F. Langenbacher, *Solar Energy*, **23**, 479 (1979). "Performance Effects of Trombe Wall Control Strategies."

Shurcliff, W. A., *Solar Heated Buildings of North America*, Brick House Publishing Co., Harrisville, NH (1978).

Sun, Y. T., *ASHRAE Task Group on Energy Requirements*, 48 (1975). "SHADOW 1, Procedure for Determining Heating and Cooling Loads for Computerized Energy Calculations."

Szokolay, S. V., *Solar Energy and Building*, Architectural Press, London (1975).

Transparent Insulation Technology for Solar Energy Conversion, Proc. of the Third International Workshop (1989), and *The 2nd International Workshop on Transparent Insulation in Solar Energy Conversion for Buildings and other Applications* (1988), available from the The Franklin Co., 192 Franklin Road, Birmingham B30 2HE, UK.

TRNSYS Users Manual, Version 13.1, University of Wisconsin Solar Energy Laboratory (1990). [First public version was 7.0 (1976).]

Trombe, F., J. F. Robert, M. Cabanot, and B. Sesolis, *Solar Age*, **2**, 13 (1977). "Concrete Walls to Collect and Hold Heat."

Utzinger, D. M., M.S. Thesis, Mechanical Engineering, University of Wisconsin-Madison (1979). "Analysis of Building Components Related to Direct Solar Heating of Buildings."

Utzinger, D. M. and S. A. Klein, *Solar Energy*, **23**, 369 (1979). "A Method of Estimating Monthly Average Solar Radiation on Shaded Receivers."

Utzinger, M. D., S. A. Klein, and J. W. Mitchell, *Solar Energy*, **25**, 511 (1980). "The Effect of Air Flow Rate in Collector-Storage Walls."

Chapter 15

SOLAR COOLING

The use of solar energy to drive cooling cycles has been considered for two related purposes, to provide refrigeration for food preservation and to provide comfort cooling. In Section 15.1 we briefly review some of the literature relating to both of the applications, since there is a common underlying technology. From then on, we concentrate on problems relating to solar air conditioning. In particular, for application in temperate climates, we address questions of the use of flat-plate collectors for both winter heating and summer cooling.

Solar cooling of buildings is an attractive idea. Cooling is important in space conditioning of most buildings in warm climates and in large buildings in cooler climates. Cooling loads and availability of solar radiation are approximately in phase. The combination of solar cooling and heating should greatly improve use factors on collectors compared to heating alone. Solar air conditioning can be accomplished by three classes of systems: absorption cycles, desiccant cycles, and solar-mechanical processes. Within these classes there are many variation, using continuous or intermittent cycles, hot or cold side energy storage, various control strategies, various temperature ranges of operation, different collectors, etc. Each of these methods is reviewed in this chapter, with emphasis on absorption and desiccant cooling.

The future of many of the methods will depend on developments beyond the cooling process itself. Temperature constraints in the operation of collectors limit what can be expected of solar cooling processes. As collector operating temperatures are pushed upward, storage may then become a critical problem. The relationship of collector and storage characteristics to cooling performance will be evident in the following discussions.

Cooling is expensive, as is heating. Reduction in cooling loads through careful building design and insulation will certainly be warranted and, within limits, will be less expensive than providing additional cooling. Good building design and construction are needed to minimize loads on any air conditioning and heating system. Our concern is with cooling loads that cannot be avoided by building design.

15.1 SOLAR ABSORPTION COOLING

Two approaches have been taken to solar operation of absorption coolers. The first is to use continuous coolers, similar in construction and operation to conventional gas-

or steam-fired units, with energy supplied to the generator from the solar collector-storage-auxiliary system whenever conditions in the building dictate the need for cooling. The second is to use intermittent coolers similar in concept to that of commercially manufactured food coolers used many years ago in rural areas (the Crosley "Icyball") before electrification and mechanical refrigeration were widespread. Intermittent coolers have been considered for refrigeration, but most work in solar air conditioning has been based on continuous cycles.

Continuous absorption cycles can be adapted to operation from flat-plate collectors. The principles of these cooling cycles are described in ASHRAE (1985). A diagram of one possible arrangement is shown in Figure 15.1.1. The present temperature limitations of flat-plate collectors restrict consideration among commercial machines to lithium bromide-water systems. LiBr-H_2O machines require cooling water for cooling the absorber and condenser, and in most applications a cooling tower will be required. Operation with flat-plate collectors of ammonia-water coolers such as those now marketed for steam- or gas-fired operation is difficult because of the high generator temperatures required. Coolers based on other refrigerant-absorbent systems may be possible candidates for solar operation.

A commercial lithium bromide-water air conditioner, modified to allow supplying the generator with hot water rather than steam, was operated from a flat-plate water heater by Chung et al. (1963). An analytical study of solar operation of a LiBr-H_2O cooler and flat-plate collector combination by Duffie and Sheridan (1965) identified critical design parameters and assessed the effects of operating conditions on integrated solar operation. Under the assumptions made in their study, design of the sensible heat exchanger between absorber and generator, cooling water temperature, and generator design are important; the latter is more critical here than in fuel-fired coolers because of the coupled performance of the collector and cooler. An experimental program was also developed at the University of Queensland, Australia, in a specially designed laboratory house [Sheridan (1970)].

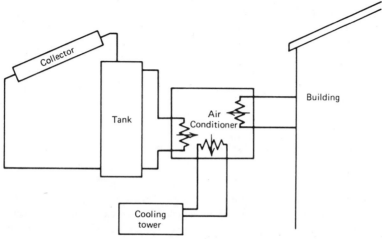

Figure 15.1.1 Simplified schematic of a solar absorption air conditioning system.

From these and other experiments, it was clear that LiBr-H$_2$O absorption air conditioners could be adapted for solar operation. Without modification, the commercial machines operated at reduced capacities, but they could be modified to operate at nominal capacities with energy supply to the generator by hot water. Part-load operation could be accomplished at little loss of the coefficient of performance (COP), with reduced dehumidification. Generator temperatures required for these air conditioners would be in a range suitable for flat-plate collectors (with the collectors operating at about the same temperature levels above ambient as those for winter heating operation).

If cooling requirements rather than heating loads fix collector size requirements, it is advantageous to use coolers with high COP. For example, double-effect evaporators can be used to decrease energy input requirements [Whitlow and Swearingen (1959) or Chinnappa (1973)]. The conditions and constraints of solar operation lead to cooler designs different than those for fuel operation.

Intermittent absorption cooling may be an alternative to continuous systems. Most work to date on these cycles has been directed at food preservation rather than comfort cooling. These cycles may be of interest in air conditioning because they offer potential solutions to the energy storage problem. In these cycles, distillation of refrigerant from the absorbent occurs during the regeneration stage of operation, and the refrigerant is condensed and stored. During the cooling portion of the cycles, the refrigerant is evaporated and reabsorbed. A schematic of the simplest of these processes is shown in Figure 15.1.2. "Storage" is in the form of separated refrigerant and absorbent. Modifications of this basically simple cycle may result in an essentially continuous cooling capacity and improved performance.

Refrigerant-absorbent systems used in intermittent cycles have been H$_2$SO$_4$-H$_2$O, NH$_3$-H$_2$O and NH$_3$-NaSCN. In the latter system, the absorbent is a solution of NaSCN in NH$_3$, with NH$_3$ the refrigerant. This system has been studied by Blytas and Daniels (1962) and by Sargent and Beckman (1968), and it appears to have good thermodynamic properties for cycles for ice manufacture. Williams et al. (1958) reported an experimental study of an intermittent NH$_3$-H$_2$O cooler using concentrating collector for regeneration.

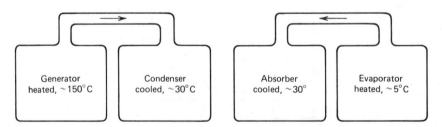

Figure 15.1.2 Schematic of an intermittent absorption cooling cycle. On the left is the regeneration cycle. On the right is the refrigeration cycle. The generator-absorber is a single vessel performing both functions, and the condenser-evaporator is also a single vessel performing both functions.

Chinnappa (1961, 1962) and Swartman and Swaminathan (1970) experimentally studied the operation of intermittent NH_3-H_2O machines in which flat-plate collectors provided the energy supply. The absorber and generator are separate vessels. The generator was an integral part of the collector, with refrigerant-absorbent solution in the tubes of the collector circulated by a combination of thermosyphon and a vapor lift (bubble) pump. Using approximately equal cycle times for the regeneration and refrigeration steps (5 to 6 hours each), overall COPs were found to be approximately 0.06 at generator temperatures rising from ambient to approximately 99 C during regeneration. Evaporator temperatures were below 0 C. With cooling water available at approximately 30 C, the effective cooling per unit area of collector surface per day for the experimental machine was in the range of 50 to 85 kJ/m^2 for clear days.

This is an incomplete history of developments of solar absorption cooling, but it provides an indication of the basis for interest in the combination of solar energy supply and absorption air conditioning. In the following section, we discuss some aspects of the theory of absorption cooling, particularly as it relates to solar operation, and in Sections 15.4 and 15.5 show data on solar cooling system performance.

15.2 THEORY OF ABSORPTION COOLING

Operation of absorption air conditioners with energy from flat-plate collector and storage systems is the most common approach to solar cooling. A schematic of a solar absorption cooling system is shown in Figure 15.2.1; this system (or variations using other methods of energy storage, auxiliary energy input, multiple-stage coolers, etc.) has been the basis of most of the experience to date with solar air conditioning.

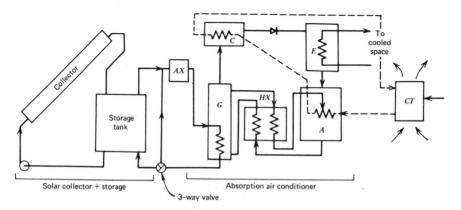

Figure 15.2.1   Schematic of a solar operated absorption air conditioner. The essential components of the cooler are: A, absorber; B, generator; C, condenser; E, evaporator; HX, heat exchanger to recover sensible heat; CT, cooling tower. AX is auxiliary energy source.

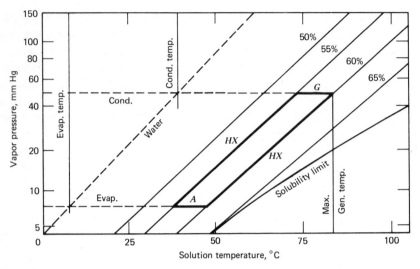

Figure 15.2.2 Pressure-temperature-concentration diagram for LiBr-H$_2$O, showing an idealized cooling cycle with letters on the segments corresponding to the process components of Figure 15.2.1.

The coolers used in most experiments are LiBr-H$_2$O machines with water-cooled absorbers and condensers. A pressure-temperature-concentration equilibrium diagram for LiBr and H$_2$O is shown in Figure 15.2.2. The idealized operation of a cycle is indicated on the diagram. The pressure in the condenser and generator is fixed by the condenser fluid coolant temperature. The pressure in the evaporator and absorber is fixed by the temperature of the cooling fluid to the absorber. The letters on the lines representing the cycle correspond to the processes occurring in the components indicated in Figure 15.2.1. The generation process is one of increasing the concentration from 55 to 60% while the equilibrium temperature of the solution rises from 72 to 82 C at the pressure of the condenser. In the absorber, the solution concentration drops from 60 to 55% as the solution temperature drops from 48 to 38 C, all at the evaporator pressure. In a real cycle, some sensible heat will have to be transferred in the generator and absorber (the amount dependent on the effectiveness of exchanger HX), there will be pressure changes through the generator due to hydrostatic head, and there will be temperature differences across all heat exchangers. Exact pressures, temperatures, and concentrations will vary with the machine and operation conditions; the numbers used here are for illustration of the nature of the process.

The maximum solution temperature in the generator is shown in Figure 15.2.2; the temperature of the heated fluid to the generator must be above the maximum generator temperature, which is determined by the condenser pressure and the concentration of the solution leaving the generator. The generator temperatures must be kept within the limits imposed by the characteristics of flat-plate collectors. The critical design factors and operational parameters include solution concentrations, effectiveness of the heat exchangers, and coolant temperature.

The pressure differences between the high- and low-pressure sides of LiBr-H_2O systems are small enough that these systems can use a vapor-lift pump and gravity return from absorber to generator as an alternative to mechanical pumping to move the solution from the low-pressure to the high-pressure side. Early absorption machines used the vapor-lift pump, but more recent designs use a mechanical pump because of improved performance.

An overall steady-state energy balance on the absorption cooler indicates that the energy supplied to the generator and to the evaporator must equal the energy removed from the machine via the coolant flowing through the absorber and condenser plus whatever net losses may occur to the surroundings

$$Q_G + Q_E = Q_A + Q_c + Q_{\text{Losses}} \qquad (15.2.1)$$

The thermal coefficient of performance COP is defined as the ratio of energy into the evaporator Q_E to the energy into the generator Q_G:

$$\text{COP} = \frac{Q_E}{Q_G} \qquad (15.2.2)$$

The coefficient of performance is a useful index of performance in solar cooling, where collector costs (and thus costs of Q_E) are important. Many LiBr-H_2O machines have nearly constant COP as the generator temperatures vary over the operating range, as long as the temperatures are above a minimum. The thermal COP is usually in the range of 0.6 to 0.8, and the major effect of variation in the solar energy temperature to the generator is to vary Q_E, the cooling rate.

Other types of COP can be defined [Mitchell (1986)]. A COP_e is the ratio of cooling to electrical energy used to provide air and liquid flows, operate controls, etc.

$$\text{COP}_e = \frac{Q_E}{\text{Electric input}} \qquad (15.2.3)$$

With water used as a coolant in the absorber and condenser, the generator temperatures are in the range 70 to 95 C. The temperature of the fluid supplied to the generator must be higher than this, which means that there is a small temperature range over which an unpressurized water storage tank can operate. Operation of most flat-plate collectors near 100 C is marginal. In addition, cooling towers are needed. These are three major problem areas in solar application of LiBr-H_2O coolers.

The schematic diagram of an ammonia-water cooler is similar to that of the cooler of Figure 15.2.1 except that a rectifying section must be added to the top of the generator to reduce the amount of water vapor going to the condenser. Pressure-temperature-concentration data for the ammonia-water system are shown in Figure 15.2.3. The basic solution processes are similar to those of the LiBr-H_2O system, but the pressures and pressure differences are much higher and mechanical pumps are needed to return solutions from the absorber to the generator. In many applications

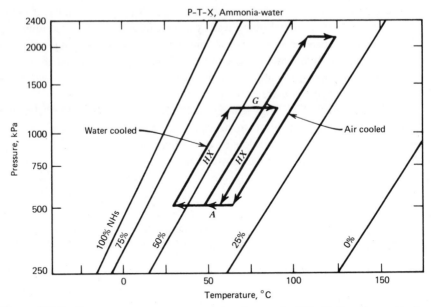

Figure 15.2.3 Pressure-temperature-concentration diagram for NH_3-H_2O, with two idealized cycles corresponding to water-cooled condenser and absorber, and air-cooled condenser and absorber.

the condenser and absorber are air cooled with generator temperatures in a range of 125 to 170 C. In applications where water cooling is used, generator temperatures may be in the range of 95 to 120 C. Both air-cooled and water-cooled cycles are shown in Figure 15.2.3. The condensing temperatures for the air-cooled condenser correspond to much higher generator temperatures than those for liquid-cooled systems.

There are two approaches to the calculation of performance of absorption coolers. It is possible to write for each component in the cooler a full set of energy balances, material balances, rate equations, and equilibrium relationships. These are solved simultaneously to determine the operating conditions and energy rates. This approach has been taken, for example, by Allen et al. (1973). The solution of the equations is expensive in computer time, and there is additional difficulty that real absorption coolers may include other components such as concentration adjusters (small containers that retain volumes of refrigerant or absorbent, depending on pressures in the machine, and so change the solution concentrations), which are difficult to include in the analysis.

The second alternative is to devise empirical models based on operating data from specific machines. These are much easier to use, particularly in simulations where repetitive computations are done. An early model of a LiBr-H_2O cooler was devised by Butz (1973) and used in an early simulation of absorption air conditioning [Butz et al. (1974)]. This model was modified by Ward and Löf (1975) and Oonk et al. (1975) to reflect data on later machines and by Blinn (1979), who has introduced a method of modeling transient operation of the cooler.

As a result of the on-off control strategy most commonly used in residential applications, chillers often do not operate at steady state. The room thermostat calls for cooling when the room temperature rises above a small control range and shuts off the chiller when room temperature drops below the range. If the cooling capacity is significantly greater than the building cooling load, the thermostat will cycle and the chiller will spend part of its operating time in a transient mode.

When a cooler is started up, the lithium bromide-water solution will begin circulating between the generator and absorber. No vapor will be evolved until the generator and all the solution held up in the generator have been heated to a temperature T_{min}. This temperature is the boiling point of a lithium bromide solution whose concentration corresponds to the chiller's initial charge at a pressure determined by the condenser temperature.

If the generator, sensible heat exchanger, and absorber are all modeled as constant effectiveness heat exchangers during start-up, if the generator is modeled as a single-node thermal capacitance, and if it is assumed that the absorber and sensible heat exchanger respond much more rapidly than the generator, then the generator temperature during start-up will vary exponentially:

$$T_G = T_{G,ss} + (T_{G,o} - T_{G,ss}) \exp(-t/t_H) \tag{15.2.4}$$

where T_G = the generator temperature
              $T_{G,ss}$ = the steady-state generator temperature
              $T_{G,o}$ = the initial generator temperature
              t_H = the generator time constant for start-up

Measurements of the time-temperature history of the generator of a 3-ton capacity chiller (Arkla model WF-36) located at Colorado State University (CSU) Solar House I show that during start-up the generator is described well by Equation 15.2.4, with a time constant t_H of about 8.0 minutes.

During cooldown, after the chiller has been shut off, solution will drain to the lowest point of the chiller (usually the absorber) and lose heat to the surrounding air by conduction and natural convection. In that case

$$T_S = T_a' + (T_{G,o} - T_a') \exp(-t/t_C) \tag{15.2.5}$$

where T_S = the solution temperature
              T_a' = the temperature of the surroundings
              t_C = the cooldown time constant

The cooldown time constant t_C will tend to be much larger than the start-up time constant, since it depends on free rather than forced convection. The cooldown time constant measured for the 3-ton chiller at CSU was about 63 minutes.

To complete the transient model, it is assumed that the instantaneous cooling delivered is a unique function of generator temperature, condensing water temperature (which fixes condenser and absorber temperatures), and evaporator temperature. The

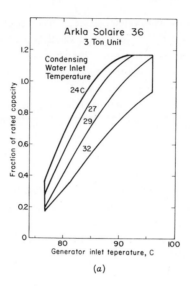

(a)

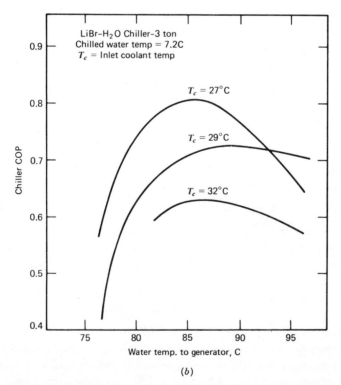

(b)

Figure 15.2.4 Manufacturer's data for 3 ton Arkla LiBr-H$_2$O water chiller, (a) showing capacity as a function of water inlet temperatures to generator and condenser at chilled water outlet temperature of 7.2 C, and (b) showing COP as a function of water inlet temperatures to generator and condenser. From Blinn (1979).

evaporator temperature is assumed constant, and manufacturer's performance data are used to determine cooling capacity and coefficient of performance as empirical functions of generator and condensing water temperatures.

When the generator temperature is above T_{min}, cooling and generator heat uptake are given by

$$Q_E = f_1(T_G, T_{CW}) \tag{15.2.6}$$

$$Q_G = f_2(T_G, T_{CW}) \tag{15.2.7}$$

where f_1 and f_2 are curve fits to manufacturer's data in terms of the generator temperature and the cooling water temperature. When the generator temperature is less than T_{min}, $Q_E = 0$ and $Q_G = (UA)_G(T_s - T_G)$.

Figure 15.2.4 shows the measured steady-state operation of a LiBr-H_2O chiller, indicating fraction of rated capacity and thermal COP as functions of inlet temperature of water to the generator and the coolant temperature.

The average COP over a long term can be evaluated as

$$\overline{COP} = \frac{\int Q_E \, dt}{\int Q_G \, dt} \tag{15.2.8}$$

Simulations using meteorological data for Charleston, SC, over a cooling season indicate that the transient operation of the chiller leads to approximately 8% lower COP than would be expected if transients were neglected. The months with the lowest cooling loads are the months with the highest differences, as those are the months when the cooler is oversized and cycles most frequently.

15.3   COMBINED SOLAR HEATING AND COOLING

Many applications of solar air conditioning will be done in conjunction with solar heating, with the same collector, storage, and auxiliary energy system serving both functions and supplying hot water. Figure 15.3.1 shows a combined heating and cooling system based on absorption cooling for air conditioning.

An important consideration in combined heating and cooling systems is the relative importance of the summer and winter loads. Either one may dictate the needed capacity of the collector and consequently its size and design. Climate is a major determining factor, and cooling requirements will dominate in climates like those of Phoenix and Miami. Commercial buildings are likely to have design fixed by cooling loads, even in cool climates. Also important are building design features that can affect relative energy requirements for the two loads. These include fenestration, shading by overhangs, wingwalls and foliage, and building orientation. Less obvious is the performance of the heating and cooling system; a poor

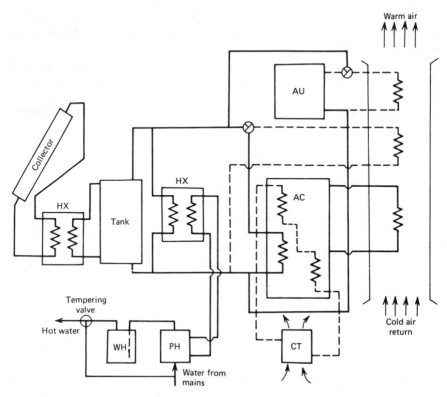

Figure 15.3.1 Schematic of a combined solar heating, air conditioning and hot water system using a LiBr-H$_2$O absorption cooler. Solid flow lines are for cooling, dashed lines are for heating, and dotted lines are air conditioner coolant flow. AU is auxiliary energy source, AC is air conditioner, CT is cooling tower, PH is preheat tank, and WH is water heater.

absorption cooler would require a larger collector area than one with a high COP and thus could shift the determination of collector needs from winter heating to summer cooling.

The location of storage, whether inside or outside the building, will have an effect on heating or cooling loads. If heat is to be stored and if the storage unit is inside the structure, heat losses from storage become uncontrolled gains during the heating season and additional loads during the cooling season. If collectors are part of the envelope of the building, back losses from the collector will also become uncontrolled gains during heating and additional loads during cooling.

Collector orientation may be affected by which load dominates; optimum orientation is approximately $\beta = \phi + 15°$ for winter use, $\beta = \phi - 15°$ for summer use, and $\beta = \phi$ for all-year use. Heating loads are likely to be higher in the morning, suggesting that the surface azimuth angle γ should be negative, while cooling loads peaking in the afternoon suggest that γ should be positive. Simulations can be used to answer these questions [see Oonk et al. (1975) for an example]; fortunately collector orientation is usually not critical.

As with solar heating alone, the major design problem is the determination of optimum collector area, with underdesign leading to excessive use of auxiliary energy and overdesign leading to low use factors on the capital-intensive solar energy system. Absorption air conditioners are more expensive than mechanical air conditioners. In climates where annual cooling loads are low, the use of absorption coolers will lead to higher cooling costs because of low use factors on the coolers.

15.4 A SIMULATION STUDY OF SOLAR AIR CONDITIONING

Simulations provide useful information on effects of design changes on the long-term performance of solar coolers. In this section we show the results of a simulation study of the system of Figure 15.3.1, a liquid system with a LiBr-H_2O chiller modeled as described in Section 15.2. The simulations were done using meteorological data for Albuquerque, NM.

The building simulated has a floor area of 150 m^2, is well insulated with $(UA)_h$ of 232 W/C, has infiltration of one-half air change per hour, and has reasonable levels of internal heat generation and building capacitance. Latent cooling loads are estimated as 0.3 of the sensible cooling loads [ASHRAE (1977)]. The desired room temperature range is 19 to 25 C. The hot water load is 300 kg/day heated from 11 to 55 C, and a tempering valve prevents delivery of water at temperatures above 55 C. Storage capacity of the preheat tank is 0.35 m^3.

The collectors have $F_R(\tau\alpha)_n = 0.72$ and $F_R U_L = 4.94$ W/m^{2}C, and are sloped toward the south at $\beta = 36°$ at the Albuquerque latitude of 35° N. The value of F'_R/F_R is 0.94. Simulation results are shown for five collector areas, from 5 to 50 m^2. An antifreeze solution with $C_p = 3900$ J/kgC is circulated at $\dot{m}/A_c = 0.0015$ kg/m^{2}s. The storage tank volume is 0.10 m^3/m^2, it has a loss coefficient of 10.5 W/m^{2}C and its temperature is limited to 100 C. The minimum tank temperature for solar heating is 30 C.

The chiller has a nominal capacity of 4.2 kW, a start-up time constant of 0.133 hours, and a cooldown time constant of 1.05 hours. The minimum useful source temperature is 77 C. Auxiliary energy in this set of simulations is provided in either of two ways. The primary method is by addition of heat to the generator of the chiller when needed. If the temperature in the building continues to be above the control temperature, parallel cooling by a separate (mechanical) cooler is computed to provide an approximation of energy required to meet any cooling loads not met by the absorption chiller.

Figure 15.4.1 shows the variation of the fraction of the annual energy needs (for heating, hot water, and cooling) met by this system as a function of collector

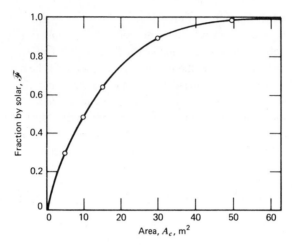

Figure 15.4.1   Annual fraction of energy supplied by solar energy for the Albuquerque simulations.

area. The curve has the same general shape as those for space heating. In the Albuquerque climate, the solar fraction reaches 0.98 at a collector area of 50 m² in the particular year used in these simulations. The monthly fractions are shown in Figure 15.4.2(a) where the inability of smaller collectors to meet large winter heating loads or summer cooling loads is evident. As with solar heating, the larger systems show poorer integrated efficiencies as they are oversized during spring and fall. The larger systems run at higher temperatures and have more thermal losses from the collectors. Figure 15.4.2(b) shows a plot of monthly efficiency (the ratio of the collected energy to the total radiation incident on the collector for the month) for the five collector areas. It is clear that high collector efficiencies are associated with low solar fractions in these systems with seasonally varying loads.

Figure 15.4.3 shows, for a collector area of 15 m², the cumulative heating, cooling, and hot water loads and the contributions by solar to meeting those loads. The year starts in September, between the heating and cooling seasons.

The annual results of these simulations are summarized in Table 15.4.1. The space heating and cooling loads increase slightly with collector area, as the control scheme maintains the building at higher mean temperatures in the winter and lower mean temperatures in the summer as the size of the solar energy system increases. (Different controls would have made some difference in these loads, but the general trends and conclusions would not change.) Annual efficiencies are shown; they decrease as collector area increases. Annual energy delivered per square meter of collector is shown; it is an approximate index of energy delivered per unit cost of the system (insofar as system cost is proportional to collector area).

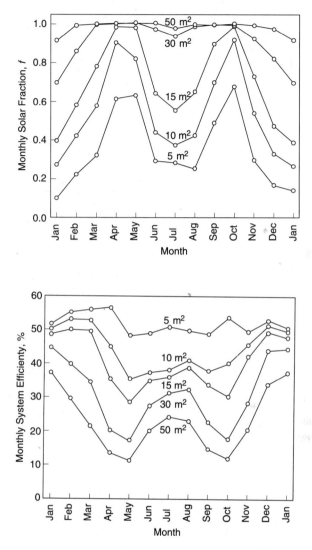

Figure 15.4.2 (a) Monthly fractions of loads met by solar energy for five collector areas for the Albuquerque simulation. (b) Monthly collector efficiencies for five collector areas for the Albuquerque simulations.

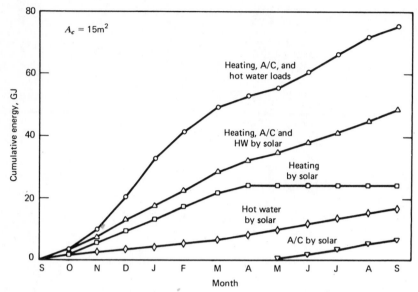

Figure 15.4.3 Cumulative loads and solar contributions for the 15 m^2 collector area for the Albuquerque simulations.

Table 15.4.1 Summary of Annual Performance for the Albuquerque Simulations of Solar Heating, Hot Water, and Cooling

Collector area, m^2	5	10	15	30	50
Space heating load, GJ	38.6	40.2	41.0	42.1	43.3
Air conditioning load, GJ	13.9	13.6	13.2	14.3	16.9
DHW load, GJ	20.2	20.2	20.2	20.2	20.2
Total load, GJ	72.8	74.0	74.5	76.6	80.4
Solar to space heating, GJ	8.7	17.9	24.5	36.1	42.5
Solar to air conditioning, GJ	0.0	5.0	6.7	13.8	16.8
Solar to DHW, GJ	12.9	15.2	16.6	18.6	19.6
Total solar to load, GJ	21.6	35.6	47.8	68.4	78.9
Fraction of load met by solar	0.30	0.28	0.64	0.89	0.98
Collector efficiency, %	52	44	39	30	22
Energy delivered, GJ/m^2	4.32	3.56	3.19	2.28	1.58

15.5 OPERATING EXPERIENCE WITH SOLAR COOLING

Detailed performance measurements of several solar cooling systems have been published. These systems are best described as experiments; the data provide an indication of the technical feasibility of solar absorption air conditioning.

The solar collector and heating system used on CSU House III in 1983 to 1984 [see Karaki et al. (1984)] was described in Section 13.3. The building is shown in Figure 13.2.5. The system was also equipped with an Arkla 3-ton LiBr-H$_2$O

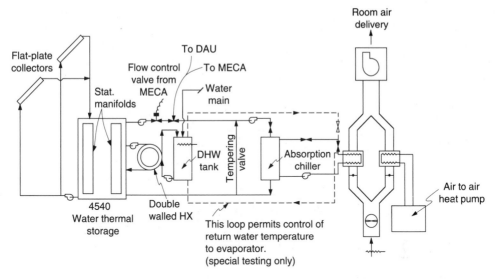

Figure 15.5.1 Schematic of the solar cooling system on CSU House III. From Karaki et al. (1984).

absorption chiller. A schematic of the complete system is shown in Figure 15.5.1. The 48.7 m^2 of collectors were double glazed with selective surfaces, and storage was in a water tank of 4530 kg capacity that was fitted with inlet and outlet manifolds designed to enhance stratification. Chilled water from the absorption machine was pumped through a heat exchanger in the air supply ducts for the building.

The chiller was a stand-alone unit located outside of the conditioned space. Heat rejection was accomplished by direct evaporative cooling by water trickling down over the outside walls of the condenser and absorber, so the functions of chiller and cooling tower were integrated into a single unit. The chiller had included in it a number of features designed to make it perform well under the transient operating conditions experienced in the solar operation and to avoid crystallization of LiBr in the generator during low-temperature operation. The effects of on-off cycling were minimized by controls that called for cooling to be continued during shutdown transients. [See Karaki et al. (1984) for details of the chiller, system, and controls.]

A diagram of the energy flows in the cooling operation is shown in Figure 15.5.2. A complete energy analysis of the system would involve determination of all of the flows; the experiments at CSU included measurement or calculation of all but a few noncritical flows.

This system carried about one-quarter of the summer cooling loads. The building houses offices, and internally generated cooling loads from office machines and computers started early in the day hours before the collectors turned on. Solar cooling performance by months is shown in Table 15.5.1. Table 15.5.2 provides information on 3 months of solar operation and shows the incident solar radiation, energy collected, energy supplied to the generator, energy removed at the evaporator,

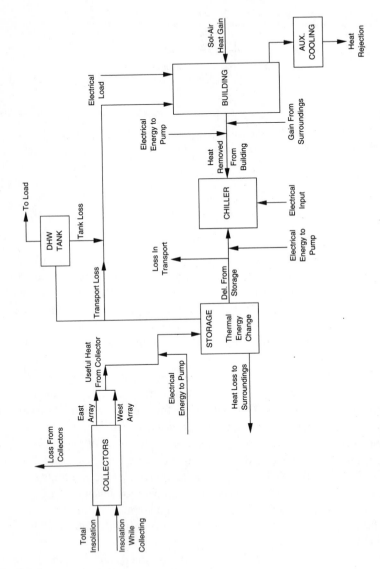

Figure 15.5.2 Energy flows in cooling operation at CSU House III. Adapted from Karaki et al. (1984).

Table 15.5.1   1983 Solar Cooling Performance of CSU House III.   Energy Quantities are in MJ/day (Average)[a]

Month	Total Cooling Load	Heat Pump Cooling	Solar Cooling	Percent by Solar
July	476	380	96	20
August	511	369	142	28
September	413	325	88	21

[a] Data are from Karaki et al. (1984).

Table 15.5.2   Incident Radiation and System Performance.   Energy Quantities are in MJ/day (Average)[a]

Month	Solar Radiation	Energy Collected	Energy to Generator	Q_E	Cooling	SCPF
July	904	272	183	126	96	0.11
August	964	300	232	159	141	0.15
September	1068	238	199	159	88	0.08

[a] Data from Karaki et al. (1984).

amount of cooling provided in the building, and a solar cooling performance factor (SCPF) (the ratio of heat removed at the cooling coil to the radiation incident on the collector).   The maximum SCPF would be the product of the daily collection efficiency and the chiller COP.   The SCPF measured at CSU were a half to three-quarters of the maximum.   Losses from storage (and other parts of the system) were substantial contributors to the difference;   the tank cooled off at night and considerable collection of energy was required to bring it back up to the threshold operating temperature.   Redesign of the system to reduce losses and changes in operating strategy could improve system performance.

15.6   APPLICATIONS OF SOLAR ABSORPTION AIR CONDITIONING

Under the auspices of various research, development, and demonstration programs, a number of solar-operated cooling systems have been installed and operated by companies and government agencies.   Many are 3-ton or 25-ton LiBr-H_2O units, and applications have been made to residential- and commercial-scale buildings.   Flat-plate collectors have been used on most of the systems;   several have been equipped with evacuated tube collectors that will operate at higher temperatures.   According to Mitchell (1986), the performance of these systems has ranged from very bad to very good.   Major problems included high parasitic power requirements for operation of fans, pumps, and cooling towers; inappropriate controls and control strategies; and use of oversized chillers that results in frequent cycling and performance degradation.

Single-effect LiBr-H$_2$O chillers operate with thermal COP that are limited to about 0.7, and actual operating coefficients may be very much less due to cycling and other problems. Double-effect chillers, in which two generators in series are used, can have COP values in the range 1.0 to 1.5; these improvements are obtained at the expense of considerable complication of the machines and will probably require the use of collectors operating at temperatures beyond the range of flat-plate collectors.

In contrast to solar heating, there are two major additional factors to be considered in evaluating solar air conditioning. First, there are substantial additional items of cost for the air conditioner and its associated piping, controls, auxiliary energy supply, etc. Second, in many climates there will be a substantial annual increment in the useful energy supplied from the collector, as it will not be oversized in the cooling season as it is for heating alone. The same methods of economic analysis that have been applied to other applications can be applied to cooling, and system sizing can be optimized. However, the factors that determine overall system performance (cycling, storage losses, etc., as noted in the CSU system development experiments) must be considered in the thermal performance calculations.

15.7   SOLAR DESICCANT COOLING

The second class of solar air conditioners is based on open-cycle dehumidification-humidification processes. These systems take in air from outside or from the building, dehumidify it with a solid or liquid desiccant, cool it by exchange of sensible heat, and then evaporatively cool it to the desired state. The desiccant is regenerated with solar energy. The components used include heat exchangers, heat and mass exchangers for dehumidification, and evaporative coolers. Many cycles have been studied; several of these are briefly noted in this section, and two are discussed in more detail in the following section.

Löf (1955) suggested solar operation of a system as shown in Figure 15.7.1. In this system, the drying agent is liquid triethylene glycol. The glycol is sprayed into an absorber where it picks up moisture from the building air. It is then pumped though a sensible heat exchanger to a stripping column where it is sprayed into a stream of solar-heated air. The high-temperature air removes water from the glycol, which then returns to the heat exchanger and absorber. Heat exchangers are provided to recover sensible heat, maximize the temperature in the stripper, and minimize the temperature in the absorber. Eliminators remove glycol spray from the air streams. This type of cycle, operated by steam, is marketed commercially and used in hospitals and other large installations. Solar operation has been studied by Lodwig et al.

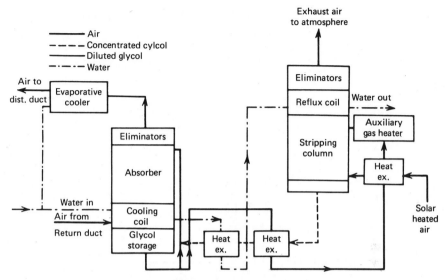

Figure 15.7.1 Schematic of a triethylene glycol open-cycle air conditioning system. From Löf (1955).

(1977), and a variation using the collector as the stripper is described by Collier (1979).

Dunkle (1965) showed a cycle designed for air conditioning in humid tropical or subtropical areas that is based on use of desiccant beds such as silica gel for drying air. The desiccants are regenerated by solar-heated air, and a pebble bed energy storage unit is included to allow operation during times of inadequate solar radiation. Rotary heat exchangers are used. Figure 15.7.2(a) shows a schematic of Dunkle's cycle, and Figure 15.7.2(b) shows the cycle on a psychrometric chart with the state numbers corresponding to conditions on the cycle schematic.

Baum et al. (1973) worked with a related system for solar cooling. Their cycle is shown schematically in Figure 15.7.3. The absorbing liquid is a solution of lithium chloride in water. Starting at the absorber (1), dilute LiCl solution is transferred by a pump (2) to a heat exchanger-distributor-header (3) and then to an open flat-plate collector (4) where water evaporates. Concentrated solution returns, via a heat exchanger (3) for recovery of sensible heat, to the absorber. Water is cooled in the evaporator (5), with water vapor from the evaporator going to the absorber. The chilled water from the evaporator is moved by a pump (6) to the air-to-water heat exchanger (7), which cools the building air (in this case, without direct contact with the absorbent solution). Means are provided to deaerate the solutions, recover sensible heat, and add makeup water. The absorber is cooled with a separate cooling coil.

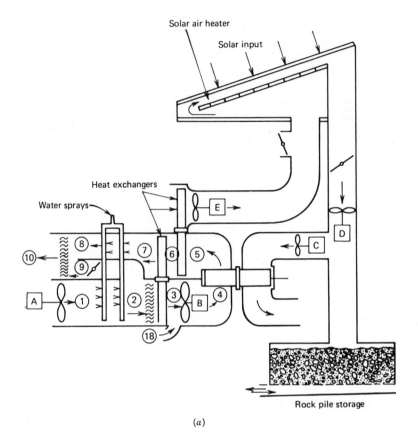

(a)

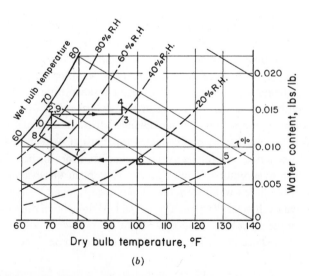

(b)

Figure 15.7.2 (a) Schematic of open-cycle solar air conditioning system using a rotary desiccant bed and heat exchangers. (b) The cycle on a psychrometric chart. From Dunkle (1965).

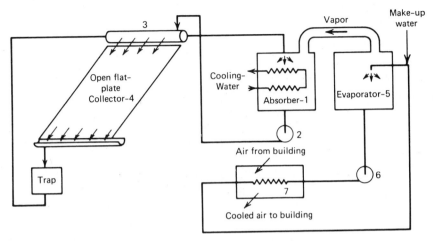

Figure 15.7.3 LiCl-H_2O open cycle cooling system. From Baum et al. (1973).

15.8 VENTILATION AND RECIRCULATION DESICCANT CYCLES

In the 1980s there has been considerable research on two related desiccant cycles that may have applications for operation with solar energy. The ventilation cycle is that used by Munters in the Munters Environmental Control (MEC) system; the other is the recirculation cycle.

The ventilation cycle (the MEC cycle) is shown in Figure 15.8.1(a), and is traced on the psychrometric chart of Figure 15.8.1(b) where the state points are numbered to correspond to the points on the process schematic. Ambient air is dried and heated by a dehumidifier from 1 to 2, regeneratively cooled by exhaust air from 2 to 3, evaporatively cooled from 3 to 4 (allowing control of both temperature and humidity), and introduced into the building. Exhaust air at state 5 moves in the countercurrent direction and is evaporatively cooled to 6, heated to 7 by the energy removed from the supply air in the regenerator, heated by solar or other source to 8, and then passed through the dehumidifier (desiccant) where it regenerates the desiccant.

The recirculation cycle is shown in Figure 15.8.2(a), and the corresponding psychrometric chart is in Figure 15.8.2(b). The same components are used, but building air is recirculated and ambient air is used only for regeneration. Both of the cycles use rotary dehumidifiers with solid desiccants and rotary heat exchangers.

The conditions shown in the figures are based on reasonable assumptions of the effectivenesses of the various components. In operation, the ambient conditions change over time and the response to changing conditions must be estimated. For solar operation, energy would be supplied (for example, by a water to air heat exchanger from water heated by flat-plate collectors) to heat the air from state 7 to state 8 in the ventilation cycle or state 3 to state 4 in the recirculation cycle. Nelson

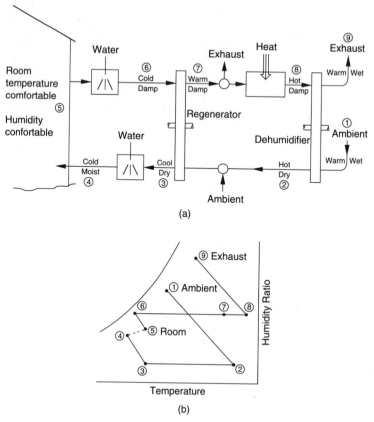

Figure 15.8.1 (a) Schematic of a solar-MEC system based on the ventilation cycle. (b) The solar-MEC cycle on a psychrometric chart. From Jurinak (1982).

(1976), Jurinak (1982), and Jurinak et al. (1984) have reported simulations of these systems. These studies have led to several generalizations about their solar operation.

Desiccant bed design is critical to solar operation if regeneration is to be accomplished at the temperatures of 60 to 80 C that can be produced by flat-plate collectors. High heat and mass transfer coefficients in the bed and low diffusion resistance in the desiccant particles are needed. The objective is to minimize the temperature difference driving forces needed to accomplish regeneration.

There are new degrees of freedom in the control of desiccant systems. Optimum meeting of variable latent and sensible cooling loads can be accomplished through control of air flow rates, dehumidifier wheel rotation rates, regeneration temperatures, and other process variables. The systems operate most efficiently at the highest supply temperature and humidity (states 4 of the ventilation cycle and 5 of the recirculation cycle) that allow the total load to be met.

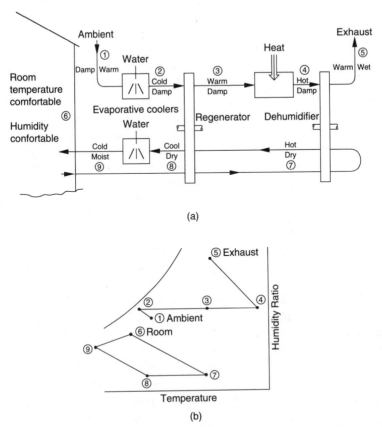

Figure 15.8.2   (a) Schematic of the recirculation cycle desiccant air conditioner.   (b) The recirculation cycle on a psychrometric chart. From Jurinak (1982).

These cycles, and others related to them, are potentially of interest for solar operation. The theory of their operation and how they can be integrated with solar energy supplies is the subject of research. Economic evaluations cannot yet be made.

15.9   SOLAR-MECHANICAL COOLING

The third type of solar cooling system that has received some attention couples a solar-powered Rankine-cycle engine with a more or less conventional air conditioning system. The design of conventional air conditioning systems is well established; the problems associated with solar operation are basically those associated with generating mechanical energy from solar energy and adaptation of air conditioning equipment for part-load operation.

Studies of Rankine-cycle solar air conditioning systems have concentrated on theoretical analysis [e.g., Teagan and Sargent (1973), Beekman (1975), and Olson (1977)]. An experimental system in a mobile laboratory [Barber (1974) and Prigmore and Barber (1974)] was in operation for several years, but no long-term operating data are available. The National Security and Resources Study Center at Los Alamos National Laboratory is partially cooled by a 77-ton solar vapor-compression system. Biancardi and Meader (1976), Eckhard and Bond (1976), and others describe development studies of engine-compressor systems.

The problems associated with solar Rankine air conditioning systems are substantial. Generation of mechanical power from solar radiation has not been shown to be economical on the scale of air conditioning operations, and it is difficult to envision a household-scale system that will convert solar to mechanical energy at less expense than conventional systems.

A simple Rankine-cycle cooling system is shown in Figure 15.9.1(a). Energy from the storage tank is transferred through a heat exchanger to a heat engine. The heat engine exchanges energy with the surroundings and produces work. As shown in Figure 15.9.1(b), the efficiency of the solar collector decreases as the operating temperature increases while the efficiency of the heat engine will increase as the operating temperature increases. Figure 15.9.1(c) shows overall system efficiency for converting solar energy to mechanical work and indicates an optimum operating temperature exists for steady-state operation.

A schematic diagram of a Rankine heat engine is shown in Figure 15.9.2(a). The corresponding temperature-entropy diagram is shown in Figure 15.9.2(b). This cycle is somewhat different from the conventional power plant cycle using water as a

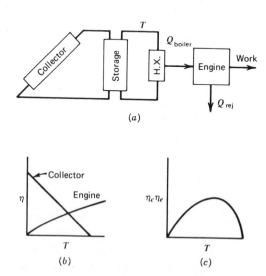

(a)

(b)   (c)

Figure 15.9.1 (a) Schematic of a solar-operated Rankine cycle cooler. (b) Collector and power cycle efficiencies as a function of operating temperature. (c) Overall system efficiency.

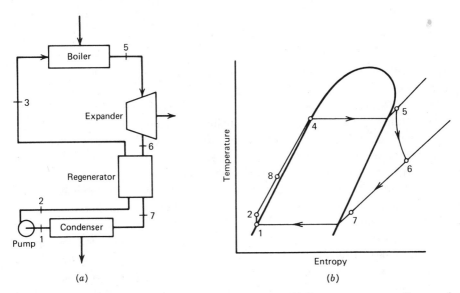

Figure 15.9.2 (a) Rankine power cycle with regeneration. (b) Temperature-entropy diagram of Rankine power cycle.

working fluid. In a conventional power plant cycle superheating and extraction are used to increase the cycle efficiency and to prevent moisture from eroding the turbine blades. In a solar energy system superheating is not desirable due to the increased temperature requirements of the collector, and extraction is not economical in small systems. To avoid moisture in the turbine, a fluid such as R 114 must be selected which has a positive slope to the saturated vapor line on a *T-S* diagram. With R 114, the outlet temperature of the turbine is significantly higher than the temperature of condensation, so a regenerator can be used to preheat the fluid leaving the condenser before it enters the boiler [see Olson (1977)].

The steady-state thermodynamic analysis of such a cycle is not difficult. However, the prediction of component performance at off-design conditions and the matching of components into a complete system so that the overall performance is optimized are not easy. For example, as the storage tank temperature changes through the day, the temperature to the boiler will change. This will result in variable energy being added to the working fluid. To ensure that fluid does not enter the turbine from the two-phase region, either auxiliary energy must be added at the boiler or the circulation rate of the fluid must be reduced. Both of these options result in design problems.

The expander that drives the air conditioner can be either a turbine (which will require a gearbox speed reduction unit) or a piston engine. It is not now possible to determine which system is best for solar operation. Some designs use a single working fluid for both the expander and the air conditioner compressor.

When a Rankine heat engine is coupled with a constant-speed air conditioner, the output of the engine will seldom match the required input to the air conditioner.

Consequently, a control system is needed to ensure matching of the engine and air conditioner. When the engine output is greater than needed, matching can be accomplished by throttling the energy supply to the engine, which means that the engine will work at an off-design condition and available energy will be wasted. Or excess energy from the engine can be used to produce electrical energy for other purposes. When the engine output is less than that required by the air conditioner, auxiliary energy must be supplied. This auxiliary can be supplied in the form of heat to the Rankine cycle or in the form of mechanical work to drive the air conditioner. The system can be designed to operate at variable speed. However, the operation of the air conditioner will be off-design with subsequent reduction of output.

As with all solar processes, steady-state conditions cannot be used to find optimum designs. It is necessary to evaluate design options based on estimates of integrated yearly performance. The effects of design variables are not intuitively obvious. The design and control of Rankine engine-vapor compression cooling has been studied by Olson (1977), who found that for a given location and collector size, there are optimum sizes of both the engine and the storage tank which maximize the annual solar contribution to meeting the cooling load.

Two other routes to solar-operated coolers are to be noted. There are now in existence solar power plants[1] that feed electrical energy into utility grids to help meet peak power demands; much of these peak demands are for air conditioning. In a more direct application, electrical energy from photovoltaic generators is used to operate mechanical refrigerators for storage of medicines and other valuable items.[2]

15.10   SOLAR-RELATED AIR CONDITIONING

Some components of systems installed for the purpose of heating a building can be used to cool the building, but without the direct use of solar energy. In this section we note three examples: (1) night cold storage system, (2) sky radiation systems, and (3) heat pump systems. These are referred to as "solar-related" methods for cooling, and the economics of the processes are interrelated with the economics of solar heating.

Night chilling of pebble beds to store "cold" for use the following day can provide some cooling capacity. In climates where night temperatures and humidities are low, pebble bed storage units of solar air heating systems can be cooled by passing outside air through an evaporative cooler, through the pebble bed, and to exhaust. This system was used on part of a laboratory building in Melbourne, Australia [see Close et al. (1968)]. During the air conditioning season the rock pile storage unit was cooled with evaporatively cooled air at night when, in that climate, the ambient wet-bulb temperatures were at most 15 to 20 C.

---

[1] See Sections 17.3 and 18.3.

[2] See Chapter 23.

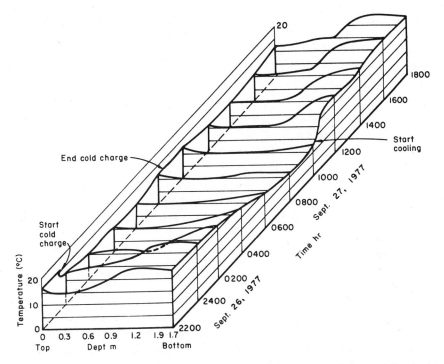

Figure 15.10.1 Temperature profiles in the pebble bed in CSU House II for a day late in the cooling season. From Karaki et al. (1977).

Essentially the same system was used on CSU House II [see Karaki et al. (1977)]. During the night, air flow was through the evaporative cooler, down through the pebble bed, and to exhaust. When cooling the building, flow was from the rooms, up through the pebble bed, and returned to the rooms. The pebble bed stratified, as it does in the heating mode, and sample profiles are shown in Figure 15.10.1. Karaki et al. found that in the climate of Fort Collins there was very little charging of the bed on the warmest summer nights, and the bed was underdesigned for cooling purposes during the height of the cooling season. Early and late in the cooling season, when nighttime temperatures are low, there was increased cooling capacity, with the bed chilled to a minimum temperature of 10 C by air entering at 7.5 C from the evaporative cooler.

Radiation to the night sky has been used to dissipate energy in several experimental systems. The nearly horizontal radiators are also used as collectors for solar heating. Bliss (1964), Yanagimachi (1958 and 1964), Hay (1973), and Hay and Yellott (1970) describe systems that use nocturnal radiation to chill water for subsequent cooling. Bliss found that on a monthly average the uncovered collectors on his laboratory at Tucson could dissipate at most approximately 4.1 MJ/m² per night. For the 93 m² collector-radiator on the Tucson laboratory, the monthly average cooling was equivalent to about 3.5 kW of continuous cooling.

Two problem areas are encountered in the design and operation of systems of this type. First, the characteristics that make a good collector are not those that make a good radiator. Neither covers nor selective surfaces can be used (unless movable covers are provided, as by Hay), as either one effectively limits nocturnal radiation to very low levels. Thus collectors are limited to the kinds that can be used at temperatures close to ambient, for example, collectors that could be used as sources for series heat pump systems. The second limitation is a climatic one. These systems could only be used where night sky temperatures are low, that is, where atmospheric moisture and dust content are low. Also, it is necessary that the nighttime wind speeds are low.

Heat pumps used as part of solar heating systems can also be used for cooling. If the cost of the heat pump can be justified by its use for air conditioning, the result will be to make the economics of the solar energy-heat pump systems described in Section 13.6 more favorable.

15.11   PASSIVE COOLING

Passive cooling depends on transfer of heat by natural means from a building to environmental sinks, including clear skies, the atmosphere, the ground, or water. Mechanisms of transport include radiation to clear skies, naturally occurring wind, air flow induced by temperature differences, conduction to ground, and conduction or convection to bodies of water. The options available are dependent on climate type, i.e., whether arid or humid. Many of these topics are discussed in the *Passive Cooling Handbook* (1980), in all of the recent preceedings of the passive solar conferences of the American Solar Energy Society, and elsewhere.

As with any cooling processes, the first approach is to minimize the cooling loads. In residential applications, major heat gains will often be solar gains through windows; these can be minimized through shading and by use of glass with high transmittance for the visible spectrum to provide daylighting and low transmittance in the infrared part of the solar spectrum. Shading can be provided by architectural features such as overhangs and wingwalls, or in many cases by proper placement of deciduous trees. Double roofs, with space between for air circulation, can keep the temperature of the ceilings of living spaces at reduced levels. Building temperatures may also be stabilized by provision of thermal capacitance if there are large diurnal ambient temperature variations.

Ventilation of building interiors is effective in cooling if ambient temperatures are moderate. Techniques for maximizing natural ventilation range from placement of windows and doors to take maximum advantage of prevailing winds to use of appropriate vents in attic and roof spaces to induce thermal circulation of air upward through a building. Gaps between wall and glazing in collector-storage walls can be vented to the outside during cooling seasons.

Earth tempering is a term applied to use of the ground as a heat sink. Ground temperatures below the surface normally do not vary much and are low enough that conduction to the ground can be a means of removal of energy from a building interior. Thermal coupling to ground can also provide some thermal capacitance.

Other methods of cooling and cooling load reduction can be considered, including hybrid methods that are in part operated with mechanical energy and in part by passive means. Control of humidity can be an important question for passively heated buildings. Dessicant coolers and systems with mechanical dehumidification can be considered hybrid systems and are discussed in this context in the *Passive Cooling Handbook*.

REFERENCES

Allen, R. W., R. H. Morse, et al., Reports on project NSF/RANN/SE/GI 39117, U. of Maryland, Mech. Engr. Dept. (1973-76). "Optimization Study of Solar Absorption Air Conditioning Systems."

ASHRAE *Handbook – Fundamentals*, American Society of Heating, Refrigerating, and Air Conditioning Engineers, Atlanta, GA (1985).

ASHRAE *Handbook and Product Directory/Equipment*, American Society of Heating, Refrigerating and Air Conditioning Engineers, New York (1975).

Barber, R. E., report from Barber-Nichols to Honeywell (1974). "Final Report on Rankine Cycle Powered Air Conditioning System."

Baum, V. A., A. Kakabaaef, A. Khandurdyev, O. Klychiaeva, and A. Rakhmanov, paper at International Solar Energy Congress, Paris (1973). "Utilization de L'energie Solair dans Les Conditions Particulieres des Regions a Climate Torrride et Aride Pour la Climatisation En Ete."

Beekman, D. M., M.S. Thesis, Mechanical Engineering, U. of Wisconsin-Madison (1975). "The Modeling of a Rankine Cycle Engine for Use in a Residential Solar Energy Cooling System."

Biancardi, R. and M. Meader, Paper in *Proc. of the Second Workshop on the Use of Solar Energy for the Cooling of Buildings*, 137 (1976). "Demonstration of a 3 ton Rankine Cycle Powered Air Conditioner."

Blinn, J. C., M.S. Thesis, Chemical Engineering., U. of Wisconsin-Madison (1979). "Simulation of Solar Absorption Air Conditioning"

Bliss, R. W., *Proc. of the UN Conf. on New Sources of Energy*, **5**, 148 (1964). "The Performance of an Experimental System Using Solar Energy for Heating, and Night Radiation for Cooling a Building."

Blytas, G. C. and F. Daniels, *J. Am. Chem. Soc.*, **84**, 1075 (1962). "Concentrated Solutions of NaSCN in Liquid Ammonia: Solubility, Density, Vapor Pressure, Viscosity, Thermal Conductance, Heat of Solution, and Heat Capacity."

Butz, L. W., M.S. Thesis, Mechanical Engineering, U. of Wisconsin-Madison (1973). "Use of Solar Energy for Residential Heating and Cooling."

Butz, L. W., W. A. Beckman, and J. A. Duffie, *Solar Energy*, **16**, 129 (1974). Simulation of a Solar Heating and Cooling System."

Chinnappa, J. C. V., *Solar Energy*, **5**, 1 (1961). "Experimental Study of the Intermittent Vapour Absorption Refrigeration Cycle Employing the Refrigerant-Absorbent System of Ammonia Water and Ammonia Lithium Nitrate."

Chinnappa, J. C. V., *Solar Energy*, **6**, 143, (1962). "Performance of an Intermittent Refrigerator Operated by a Flat-Plate Collector."

Chinnappa, J. C. V., Paper at International Solar Energy Society Congress, Paris (1973). "Computed Year-Round Performance of Solar-Operated Multi-State Vapour Absorption Air Conditioners at Georgetown, Guyana, and Colombo, Ceylon."

Chung, R., J. A. Duffie, and G. O. G. Löf, *Mech. Engr.*, **85**, 31 (1963). "A Study of a Solar Air Conditioner."

Close, D. J., R. V. Dunkle, and K. A. Robeson, *Mech. and Chem. Engr. Trans., Inst. Engs. Australia*, **MC4**, 45 (1968). "Design and Performance of a Thermal Storage Air Conditioning System."

Collier, R. K., *Solar Energy*, **23**, 357 (1979). "The Analysis and Simulation of an Open cycle Absorption Refrigeration System."

Duffie, J. A. and N. R. Sheridan, *Mech. and Chem. Engr. Trans., Inst. Engrs. Australia*, **MC1**, 79 (1965). "Lithium Bromide-Water Refrigerators for Solar-Operation."

Dunkle, R. V., *Mech. and Chem. Engr. Trans., Inst. Engrs. Australia*, **MC1**, 73 (1965). "A Method of Solar Air Conditioning."

Eckard, S. E. and J. A. Bond, paper in *Proc. of the Workshop on the Use of Solar Energy for the Cooling of Buildings*, 110 (1976). "Performance Characteristics of a 3-Ton Rankine Powered Vapor-Compression air Conditioner."

Hay, H. R. and J. I. Yellott, *Mech. Engr.*, **92**, (1), 19 (1970). "Naturally Air Conditioned Building."

Hay, H. R., *Mech. Engr.*, **95** (11), 18 (1973). "Energy Technology and Solararchitecture."

Jurinak, J. J., Ph.D. Thesis, Mechanical Engineering, U. of Wisconsin-Madison (1982). :Open Cycle Solid Desiccant Cooling - Component Models and System Simulations."

Jurinak, J. J., J. W. Mitchell, and W. A. Beckman, *Trans. ASME, J. Solar Energy Engrg.*, **106**, 252 (1984). "Open-Cycle Desiccant Air Conditioning as an Alternative to Vapor Compression Cooling in Residential Applications."

Karaki, S., P. R. Armstrong, and T. M. Bechtel, Colorado State University Report COO-2868-3 to U. S. Deptartment of Energy (1977). "Evaluation of a Residential Solar Air Heating and Nocturnal Cooling System."

Karaki, S, T. E. Brisbane, and G. O. G. Löf, Colorado State University Report SAN 11927-15 to U. S. Deptartment of Energy (1984). "Performance of the Solar Heating and Cooling System for CSU House III - Summer Season 1983 and Winter Season 1983-84."

Lodwig, E., D. A. Wilkie, and J. Bressman, Paper in *Proc. of the Am. Solar Energy Soc. Meeting*, Orlando, **1**, 7-1 (1977). "A Solar Powered Desiccant Air Conditioning System."

Löf, G. O. G., Paper in *Solar Energy Research*, U. of Wisconsin Press, Madison, p. 33 (1955). "House Heating and Cooling with Solar Energy."

Mitchell, J. W., in *Proc. 1986 Annual Mtg., Am. Solar Energy Soc.*, Boulder, CO., 63 (1986). "State of the Art of Active Solar Cooling."

Nelson, J. S., M. S. Thesis, Mechanical Engineering, U. of Wisconsin-Madison (1976). "An Investigation of Solar Powered Open-Cycle Air Conditioners."

Nelson, J. S., W. A. Beckman, J. W. Mitchell, and D. J. Close, *Solar Energy*, **21**, 273 (1978). "Simulations of the Performance of Open-Cycle Desiccant Systems Using Solar Energy."

Olson, T. J., M. S. Thesis, Mechanical Engineering, University of Wisconsin-Madison (1977). "Solar Source Rankine cycle Engine For Use in Residential Cooling."

Oonk, R. L., W. A. Beckman, and J. A. Duffie, *Solar Energy*, **17**, 21 (1975). "Modeling of the CSU Heating/Cooling System."

Passive Cooling Handbook, Report prepared for U.S. Department of Energy (1980).

Prigmore, D. R. and R. E. Barber, 9th IECEC Conference (1974). "A Prototype Solar Powered, Rankine Cycle System Providing Residential Air Conditioning and Electricity."

Sargent, S. L. and W. A. Beckman, *Solar Energy*, **12**, 137 (1968). "Theoretical Performance of an Ammonia-Sodium Thiocyanate Intermittent Absorption Refrigeration Cycle."

Sheridan, N. R., Paper at Int'l. Solar Energy Soc. Conference, Melbourne (1970). "Performance of the Brisbane Solar House."

Swartman, R. K. and C. Swaminathan, Paper at the Intl. Solar Energy Soc Conference, Melbourne (1970). "Further Studies on Solar-Powered Intermittent Absorption Refrigeration."

Teagan, W. P. and S. L. Sargent, Paper EH-94-1 at Intl. Solar Energy Soc Paris Conference, (1973). "A Solar Powered Combined Heating.Cooling System With the Air Conditioning Unit Driven by an Organic Rankine Cycle Engine."

Ward, D. S. and G. O. G. Löf, *Solar Energy*, **17**, 13 (1975). "Design and Construction of a Residential Solar Heating And Cooling System."

Whitlow, E. P. and J. S. Swearingen, Paper at Southern Texas Am. Inst. Chem. Engr. meeting (1959). "An Improved Absorption-Refrigeration Cycle."

Williams, D. A., R. Chung, G. O. G. Löf, D. A. Fester, and J. A. Duffie, *Refrigeration Engr.*, **66**, 33 (November 1958). "Cooling Systems Based on Solar Regeneration."

Yanagimachi, M., *Proc. of the UN Conference on New Sources of Energy*, **5**, 233 (1964). "Report on Two and a Half Years' Experimental Living in Yanagimachi Solar House II."

Yanagimachi, M., *Trans. of the Conference on Use of Solar Energy*, **3**, 32, U. of Arizona Press, Tucson (1958). "How to Combine: Solar Energy, Nocturnal Radiational Cooling, Radiant Panel System of Heating and Cooling, and Heat Pump to Make a Complete Year-Round Air Conditioning System."

Chapter 16

SOLAR INDUSTRIAL PROCESS HEAT

Very large amounts of energy are used for low-temperature process heat in industry, for such diverse applications as drying of lumber or food, cleaning in food processing, extraction operations in metallurgical or chemical processing, cooking, curing of masonry products, paint drying, and many others. Temperatures for these applications can range from near ambient to those corresponding to low-pressure steam, and energy can be provided from flat-plate collectors or concentrating collectors of low concentration ratios.

The principles of operation of components and systems outlined in earlier chapters apply directly to industrial process heat applications. The unique features of these applications lie in the scale on which they are used, system configurations and controls needed to meet industrial requirements, and the integration of the solar energy supply system with the auxiliary energy source and the industrial process. Process simulations similar to those outlined in Chapter 10 and chapters on heating and cooling are useful tools in the study of industrial applications. The design methods outlined in Chapters 20 (if energy is to be delivered at about 20 C) and 21 (for systems delivering energy at temperatures other than 20 C) are applicable to many solar industrial process heating operations. Chapter 17 treats solar applications for power generation and Chapter 18 is on solar ponds and evaporative processes; these are subjects closely related to industrial process applications.

Other than traditional crop drying and solar evaporation, which have been practiced over centuries, most solar applications for industrial process heat have been on a relatively small scale and are experimental in nature. A few large systems are in use. In this chapter we outline some general design and economic considerations and then briefly describe several examples of industrial applications to illustrate the potential utility of solar energy to industry and the kinds of special problems that industrial applications can present. Experiments of the 1970s are described in the *Proceedings of Solar Industrial Process Heat Symposium* (1977) and *Proceedings of Solar Industrial Process Heat Conference* (1978).

16.1 INTEGRATION WITH INDUSTRIAL PROCESSES

Two primary questions to be considered in a possible industrial process application concern the use to which the energy is to be put and the temperature at which it is to

620

be delivered. If a process requires hot air for direct drying, an air-heating system is probably the solar energy system best matched to the need. If steam is needed to operate an autoclave or indirect dryer, the solar energy system must be designed to produce steam and concentrating collectors will probably be required. If hot water is needed for cleaning in food processing, the solar energy system will be liquid heater. An important factor in determination of the best system for a particular use is the temperature of the fluid to the collector. The generalizations of building heating applications (e.g., that return air to collectors in air systems is usually at or near room temperatures) do not necessarily carry over to industrial processes, as the system configurations and energy uses may be quite different.

The energy may be needed at particular temperature or over a range of temperatures. If low-pressure steam is condensed in an indirect dryer, the condensate will probably be recirculated, and the solar process system will be called on to deliver essentially all of the energy at a constant temperature level. A once-through cleaning process may call for fresh water to be heated from supply temperature to some useful minimum level, so energy can be added to the water over a range of temperatures. A system in which a working fluid is recirculated back to the tank will probably operate over an intermediate range of temperatures.

The temperature at which energy is used in many industrial processes has not been limited by the characteristics of conventional energy supplies. The partial replacement of conventional sources by solar energy in retrofit applications is usually limited to operations in the same temperature ranges as that of the sources replaced. In new applications, since solar collectors operate more efficiently at lower temperatures, the industrial process itself should be examined to see if the temperature of energy delivery can be optimized.

Storage would usually be used in industrial processes, except where the maximum rate at which the solar energy system can deliver energy is not appreciably larger than the rate at which the process uses energy. In these cases the annual fraction of the energy needs delivered by solar energy will be small if the process operates in other than daylight hours. If storage is used, the energy balances and rate equations of Chapter 8 can be used to describe the storage subsystem.

The investments in industrial processes are generally large, and the transient and intermittent characteristics of solar energy supply are so unique that the study of options in solar industrial applications can be done by simulation methods at costs that are very small compared to the investments.

16.2 MECHANICAL DESIGN CONSIDERATIONS

Many industrial processes use large amounts of energy in small spaces. If solar is to be considered for these applications, the location of collectors can be a problem. It may be necessary to locate collector arrays on adjacent buildings or grounds, resulting in long runs of pipe or duct. Collector area may be limited by building roof area and orientation. Existing buildings are generally not designed or oriented to accommodate arrays of collectors, and in many cases structures to support collector arrays must be added to the existing structures. New buildings can be readily designed, often at little or no incremental cost, to allow for collector mounting and access.

Interfacing with conventional energy supplies must be done in a way that is compatible with the process. If air to dryers is to be preheated, it must be possible to get the solar-preheated air into the dryer air supply. In food processing, sanitation requirements of the plant must be met. It is not possible to generalize on these matters; the engineering of the solar energy process and the industrial process must be mutually compatible. In most of the examples in the following sections, solar heating was retrofitted, it supplied a relatively small part of the plant loads, and in varying degrees the range of problems noted here have been encountered.

16.3   ECONOMICS OF INDUSTRIAL PROCESS HEAT

Economic analyses of industrial processes are the same as outlined in Chapter 11, but in contrast to non-income-producing applications, deduction of fuel costs as a business expense, investment tax credits, and depreciation must be taken into account.[1] And, industry usually requires a higher return on investment (i.e., shorter payback time) than do individuals.

The investments in solar industrial process systems are as outlined in Section 11.1. For a retrofit system, mounting a collector array may be a major item of investment; in some experimental installations the cost of providing mountings exceeded the cost of the collectors. In new construction, provision for collector mounting may mean only nominal changes in the design of the structure. The operating costs include the same kinds of items as for other solar applications, such as parasitic power, maintenance, insurance, and real estate taxes.

The generalization made for solar heating that collector area can be considered as the primary design variable does not necessarily extend to industrial process applications. There is no general relationship between time dependence of loads and energy supply, and both collector area and storage capacity (and possibly others) may be important design parameters. In general, the design of a solar industrial process will have to be based on a study of the appropriate ranges of variables of all the important parameters, done in view of the characteristics of the energy-using process.

As high reliability is needed for industrial applications, solar will normally be combined with a conventional energy supply. The useful energy produced by the solar energy system serves to reduce the fuel consumption of the conventional (auxiliary) energy supply. Industry buys fuels in quantities that are large compared to homeowners, and the price of that fuel is generally less than is paid by the small

---

[1] This section is written with U.S. tax laws in mind; other treatments will be necessary where tax laws are substantially different from those of the United States.

buyer. In addition, funds spent for purchase of fuel are business expenses and therefore deductible from corporate income taxes. With a basic corporate income tax rate of 50%, the effective cost of energy would be halved. As with nonindustrial applications, it is difficult to build solar heating systems that can deliver energy at a sufficiently low cost to compete with inexpensive fuels. Differences in taxation of fuels, investment tax credits, and depreciation of equipment can make very substantial difference in the economic viability of these processes and are matters of legislation.

Miscellaneous expenses for parasitic power, insurance, and maintenance are also business expenses and therefore tax deductible. Investment tax credits effectively reduce first costs and thus encourage investments in equipment. Equipment costs can be depreciated, which has the effect of decreasing corporate taxes. Depreciation can be by straight line, declining balance, sum of digits, or other methods. Most industries will have established methods for depreciation and discount rates.

Seasonal industrial processes, which can only utilize the output of solar heating systems over part of the year, will show less favorable economics than will processes that operate all year. Food processing applications in temperate climates provide examples. This disadvantage can, in part, be minimized by collector orientation to maximize output over the period of use. It may also be possible to develop other off-season uses for energy, such as building heating. One of the examples in a following section shows such a combination of uses.

16.4 OPEN CIRCUIT AIR HEATING APPLICATIONS

Heated outside air is used in many industrial applications where recirculation of air is not practical because of contaminants. Examples are drying, paint spraying, and supplying fresh air to hospitals. Heating of ambient air is an ideal operation for a collector, as it operates very close to ambient temperature. In this case, Equation 6.7.6 becomes simply

$$Q_u = A_c F_R(\tau\alpha)S \tag{16.4.1}$$

where F_R may be essentially fixed if flow rate is fixed or it may vary if, for example, system controls are arranged to provide a fixed outlet temperature by adjusting the flow rate as inlet (ambient) temperature varies.

Equation 16.4.1 is evaluated for the hours of operation of the solar system. All of the methods of Chapters 2, 4, and 5 in determining S and Chapter 6 in calculating $F_R U_L$, $F_R(\tau\alpha)_n$ and incidence angle effects apply directly to these calculations. The critical radiation level of a collector heating ambient air is zero and utilizability is unity. For a full month's operation of such a system, the average daily useful gain can be approximated by

$$\overline{Q}_u = A_c F_R(\tau\alpha)_n \frac{\overline{(\tau\alpha)}}{(\tau\alpha)_n} \overline{H}_T \tag{16.4.2}$$

where $(\overline{\tau\alpha})/(\tau\alpha)_n$ can be estimated by the methods of Section 5.10. This calculation of monthly gains will work for systems that operate whenever the incident radiation is high enough to justify operation of the fan.

Two systems in which heating of ambient air is carried out are described here. The first is an experimental application to food drying, entailing several unique problems illustrative of special considerations that must be taken into account by a system designer. The second is a space heating system in which large quantities of outside air must be supplied to a bus maintenance facility in order to meet indoor air quality requirements.

An experimental application of air heating to drying of soybeans at the Gold Kist plant at Decatur, Alabama, was one of a series of demonstrations of solar industrial process heating sponsored in part by the U.S. Department of Energy. The system, described by Guinn (1978), consisted of 1200 m^2 (672 modules) of air heating collectors supplying warmed air to the dryer. The air was mixed with additional ambient air, heated by oil to the desired temperature, and used in the drying operation. Since the maximum output from the collectors was less than the energy needs of the dryer, all collected energy could be used (during the drying season) and no storage was provided. Figure 16.4.1 is a schematic of the process, and Figure 16.4.2 shows the collector array.

In this system, the filter and blower were on the upstream side of the collectors, to put the collectors under positive pressure and avoid ingestion of dusty air into the system where it might become an explosive mixture. The size of the collector array was essentially fixed by the dimensions of a parking lot over which the array was mounted. The collector slope was 15°, chosen as a compromise between performance and greater structural costs associated with larger slopes. The atmosphere in the plant area was dusty, and keeping the collectors reasonably clean was a significant operational problem.

Systems of this type, which are designed for small contributions by solar in relation to the total loads, can be operated without energy storage. No energy will be dumped as long as the maximum output of the collector is less than the energy needs of the application at the time the collector maximum occurs. It may be that the time of collector operation would be determined by the process itself (e.g., times when paint spraying is going on or when materials are in the dryer ready to dried), and under these circumstances storage may be needed.

The second once-through air system is on a bus maintenance facility in Denver, CO, in which diesel powered buses are stored, serviced, and maintained. Large quantities of fresh air are required in the facility to assure adequate air quality and remove exhaust fumes. The facility serves dual purposes, as a storage facility and as a maintenance base. The minimum allowable temperature in the storage area is 7 to 10 C, and the minimum in the maintenance area is 13 to 19 C. There is also an office area that is maintained at 21 C or above. The facility requires 12 to 15 air changes per hour, and the bulk of the heating load is heating the ventilation air. The building floor area is 32,500 m^2. A view of the roof of the building and the collector system is shown in Figure 16.4.3.

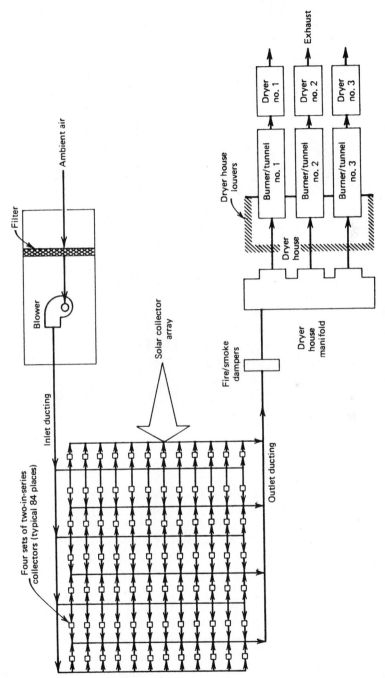

Figure 16.4.1 Layout of collector array and soybean drying installation. Adapted from Guinn (1978).

Figure 16.4.2 The 1200-m^2 air heating collector array at the Gold Kist soybean plant at Decatur, Alabama. Courtesy of Solaron Co.

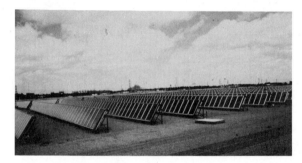

Figure 16.4.3 View of the roof mounted collectors on the Denver bus maintenance facility. Photo courtesy of Colorado State University.

There are five independent solar heating systems, each heating ambient air for supply to the building. One of them also supplies service hot water through an air-to-water heat exchanger in the duct from the collector to the building; this heated water is used for washing buses. Total collector area is 4300 m^2 of flat-plate air heaters of the type shown in Figure 6.14.1(b). They have one cover and a flat black absorbing surface and are sloped 55° to the south. The performance of two of these systems has been measured by Marlatt and Smith (1988). A schematic of the systems, including provision of heat with steam, is shown in Figure 16.4.4.

There are significant differences between these systems and the "standard" air systems noted in Chapter 13. They are once-through systems, and there is no air recirculation either from the building or from the storage units. Air to the stores is exhausted to the outside regardless of its temperature, and outside air is heated in the storage units when the controls call for this mode of heating.

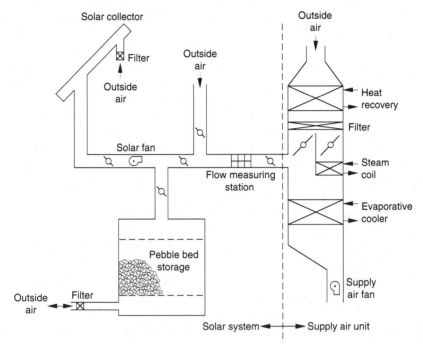

Figure 16.4.4 Schematic of the solar and auxiliary (steam) heating systems on the Denver RTD facility. From Wirsum (1988).

There are three basic modes of operation of these systems. The first is direct delivery of air from the collectors to the supply air units and thence to the building. Excess energy from the collectors is diverted to storage. In the second mode, outside air is drawn through the storage units to the supply air units and the building. In either of these two modes, auxiliary heat is supplied by steam coils in the supply air units. The third mode is the summer ventilation of the collector arrays. Wirsum (1988) has made a detailed simulation study of the operation of this system, its controls, modeling of the collectors, the utility of storage in this system in the Denver climate, and the factors that limit system performance.

This application represents an ideal one for air heaters. Performance of the collectors is as good as it can be expected to be. Solar contribution is limited primarily by the area of collector that can be readily accommodated on the roof of the building; the annual solar contribution is estimated to be 20% of the total heating load.

16.5 RECIRCULATING AIR SYSTEM APPLICATIONS

Indirect drying, with solar-heated air supplied to a drying chamber, has been applied to a variety of materials, including foods, crops, and lumber. In most such operations, adequate control of temperature and humidity can lead to improved product quality by controlling the rate of drying; this is usually accomplished by recycling to the collectors part of the air that has passed through the dryer.

Solar drying of lumber, to reduce its moisture content from that of the green wood from the trees to levels acceptable for use in building and manufacturing, has been the subject of several experiments [e.g., Read et al. (1974) and Duffie and Close (1978)]. Based on experiments at the U.S. Forest Products Laboratory [Tschernitz and Simpson (1979) and Simpson and Tschernitz (1985)], a production dryer has been constructed at a furniture manufacturing plant in Sri Lanka and is in routine use processing rubberwood for use in making furniture and laminated beams. The performance of the kiln and its energy supplies has been described by Simpson and Tschernitz (1989).

A schematic of the Sri Lanka dryer is shown in Figure 16.5.1. The major components of the system, designed for use at low latitudes, are a flat-plate collector built essentially horizontally on the ground, the kiln, an auxiliary waste wood burner, and controls, dampers, blowers, etc. The kiln is designed to have a capacity of 14 m^3 of 25 mm thick boards, with 19 mm sticks (spacers to allow air circulation). Four collectors are used; they have a total area of 132 m^3 and are sloped at about 0.5° to allow rain water runoff. A cross section of the collectors is shown in Figure 16.5.2. A single glass cover is used, with air circulating between the cover and a bed of charcoal that absorbs radiation and provides both extended heat transfer surface and insulation to minimize energy losses to the ground. The collectors operate at negative pressure so that leaks are into the system. They are fed with a mixture of recycled air from the dryer and ambient air.

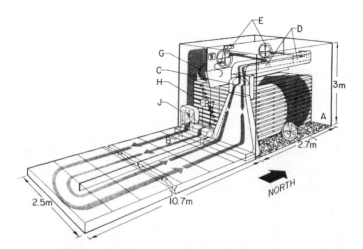

Figure 16.5.1 Schematic of a section of the Sri Lanka solar lumber dryer. (A) Drying chamber. (B) Collectors. (C) Blower to induce air flow through the collectors. (D) Hot air discharge to internal fans. (E) Internal fans. (G) Disk humidifier. (H) Damper motor for collector air flow. (J) Fresh air intake. Adapted from Simpson and Tschernitz (1985).

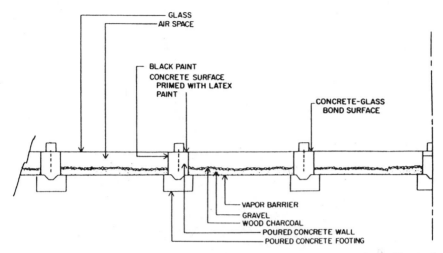

Figure 16.5.2 Cross section of part of the collectors of the Sri Lanka lumber dryer. The single glazing layer is sealed to the concrete frames with silicone sealant. Adapted from Simpson and Tschernitz (1985)

The kiln is of concrete block construction with inside dimensions 3.0 × 10.4 × 3.3 m. The roof is arranged to serve as a preheater for air entering the collectors. Automatic controls on the dryer are arranged to control the temperature and relative humidity of the drying air in acceptable ranges. Air flow through the collector is modulated on and off depending on whether the collector temperature is above a predetermined setpoint, which depends on the stage of dryness of the lumber. Four fans are available to move air through the dryer, with four used in early stages of drying and two in later stages when drying is slower and air requirements are less.

The dryer has been in use since 1984. Its performance has been measured over two periods, one during the monsoon (wet) season and one during a period of clearer skies. In a 2-week drying operation on a load of lumber in the wet season, the solar contribution to the total energy requirements of the process was 18%. During this period the solar radiation on the horizontal surface averaged 7.57 MJ/m², the dryer efficiency was 33%, and the moisture content of the wood was reduced from 73% to 18% (dry basis). During the 9 day drying operation on a load of lumber during better weather, the average daily radiation on the collectors averaged 14.7 MJ/m², the collectors supplied about 34% of the energy required, and the dryer efficiency was estimated to be 51%. The maximum estimated solar contributions in this dryer, in periods of minimum clouds, are estimated at 50%, and collector efficiencies are estimated at 35 to 40%. Air temperature rises through the collector are a maximum of about 30 C in the midday hours of clear days. Maximum temperatures in the dryer were about 35 C early in drying cycles to 50 C late in the cycles. Drying cycles are 7 to 14 days in length.

A solar kiln design for use at high latitudes and its operation in Ontario for drying jack pine studs is described by Yang (1980). Figure 16.5.3 shows a cross

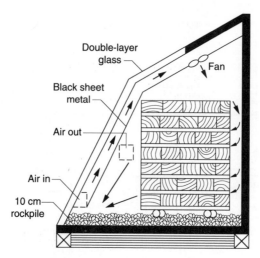

Figure 16.5.3 Cross section of an experimental solar lumber dryer at Thunder Bay, Ontario (latitude 48°N). From Yang (1980).

section of this dryer. Drying cycles in the summer are are about 12 days and in the winter about 100 days, compared to air drying cycles of about 240 days. The quality of the dried lumber was better than that for air drying.

A detailed review of indirect solar wood drying has been prepared by Steinmann (1989), who has made detailed measurements and simulations of drying lumber in an experimental kiln that is similar in general design to that of Figure 16.5.3.

16.6 ONCE-THROUGH INDUSTRIAL WATER HEATING

Large quantities of water are used for cleaning in the food processing industries, and recycling of used water is not practical because of contaminants picked up by the water in the cleaning processes. These processes involve heating of water from mains supply temperature to the desired temperature level for the process. As with other systems, if the solar collector output is always smaller than the loads on a process, the solar heating operation can be carried out without storage, but solar contributions relative to total loads will be small. Storage should usually be used with system schematics similar in concept to the domestic water system configurations shown in Figure 12.1.1. Systems are also designed in which supply water passes through the collector once and then to storage or the industrial process. If there is no storage or if the amount of water in storage is constant, flow through the collector is fixed by draw-off at the point of use. If variable-capacity storage is provided, flow though the collector can be controlled independently of the loads on the system.

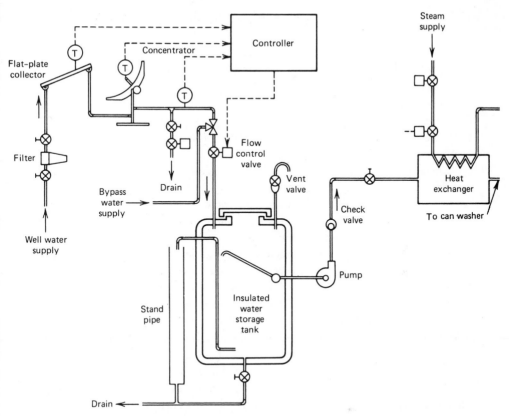

Figure 16.6.1 Schematic of the experimental water heating system in the Campbell Soup Co. plant, Sacramento, CA. Adapted from Vindum and Bonds (1978).

An example of this application is the experimental water heater that was constructed for the Campbell soup plant at Sacramento, CA, shown in Figure 16.6.1. Water supply from a well was pumped in series through a flat-plate collector and a concentrating collector and then to an insulated storage tank. The quantity of water in the tank was variable with time. The temperature of this water was boosted as needed by an auxiliary steam heater to bring it to 85 to 90 C as required in the can washer. This solar-heated water was mixed with heated water from the conventional system, used for can washing, treated, and discarded. The system, described by Vindum and Bonds (1978), was designed to provide 75% of the energy to one can line of 20 in the plant, so its output was small compared to the total energy requirement for hot water for the plant. This system, while it produced hot water in the same temperature range as that of flat-plate collectors used in absorption cooling, uses concentrating collectors for part of the heating process.

16.7   RECIRCULATING INDUSTRIAL WATER HEATING

An experimental solar heating system to preheat air and boiler feedwater in an onion drying plant in Gilroy, CA, [Graham et al. (1978)], was based on the use of evacuated tubular collectors delivering hot water at temperatures of 70 to 100 C. The gas-fired burner received preheated air from the solar system and delivered 93 C air to the dryer. The capacity of the collectors to deliver energy was small compared to the energy requirements of the plant, and no storage was used. A schematic of the solar heating process is shown in Figure 16.7.1.

In normal operation of the system, water was circulated through the collectors and the air-water preheat exchanger to heat drying air before it went to the gas burners. It was also possible to heat water in the condensate tank which is part of the plant boiler system, so during months when there was no drying to be done, the output from the collector could still be used.

The collector elements were evacuated collector tubes similar to the type shown in Figure 7.5.1(a) and were fitted with vee-shaped specular reflectors behind the tubes. The steady-state thermal characteristics of the collectors are shown in Figure 16.7.2, with the normal range of operating conditions indicated. Tubes were assembled into modules, with each 1.36 m^2 module including eight tubes and reflectors with appropriate manifold and mounting units. Four hundred and two modules were used for a total collecting area of 548 m^2. The collector array was mounted on the roof of a warehouse near the building housing the dryers.

A large area, industrial-scale water heating installation is in operation in Green Bay, WI, supplying hot water to a meat packing plant [Lane (1990)]. The system

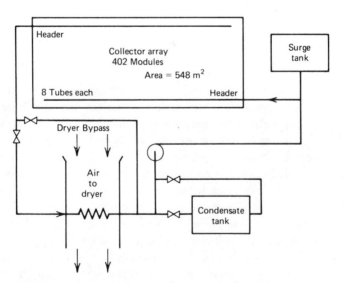

Figure 16.7.1   Schematic of the water heating system for the Gilroy Foods onion drying facility. Adapted from Graham et al. (1978).

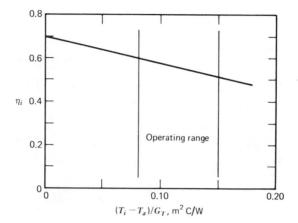

Figure 16.7.2 Characteristics of the evacuated tubular collectors used on the Gilroy Foods onion dryer. Adapted from Graham et al. (1978).

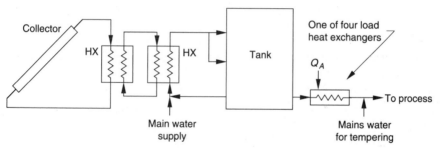

Figure 16.7.3 Schematic of the layout of the Green Bay water heating facility that supplies heated water to a meat packing plant.

uses recirculation of fluid in the collectors and partial recirculation of water in the storage tank. The basic system layout is shown in Figure 16.7.3; it includes arrays of flat-plate collectors heating propylene glycol and two heat exchangers with an intermediate water loop to transfer collected energy to water from mains or recirculated from storage. Water from the tank is pumped to the plant where it is heated in one of four load heat exchangers or mixed with mains water to obtain the desired temperatures. The collector arrays consist of 5256 modules, each with absorber area 2.77 m². The storage tank has a capacity of 1250 m³; it contains variable quantities of water, allowing the collector heat exchangers to be fed with mains water during periods when the tank is not full, thus partially decoupling the collector operation from the water demands of the plant. The requirements of the plant are for water heated to four temperatures (32, 35, 50, and 82 C); a maximum of about 1300 m³ of heated water is required each working day.

Figure 16.7.4   Collectors at the Green Bay water heating plant. Photo courtesy of R. Lane.

The solar heating system was built in 1984, suffered significant damage in its first winter of operation, and has since been redesigned and rebuilt. Figure 16.7.4 shows the collector arrays. Since reconstruction, the system has been in routine use. Its performance has been monitored; over the first 9 months of 1989, in 188 days of operation, it delivered a total of 21 TJ to the packing plant. During this period the collector panels were available 99% of the time. The average temperature increase of the water delivered was 33 C. The plant delivers hot water to the packing house under the terms of a 10-year contract.

16.8   SHALLOW POND WATER HEATERS

Large horizontal collectors consisting of plastic envelopes to contain water, rigid covers supported on concrete curbs, and underlying foam glass and sand insulation have been studied for possible application to provide large quantities of hot water for mineral leaching operation, for providing service hot water to a military base, and for heating water used for various purposes in a poultry processing plant [see Dickenson (1976) and Guinn and Hall (1977)]. A cross section of a collector module of this type is shown in Figure 16.8.1. Dimensions of single collector units are of the order of 3 or 4 m wide and 50 m long, with water depths of the order of 0.1 m. Details of construction of such a collector are provided by Casamajor and Parsons (1979).

The deeper the pond is, the less the temperature will rise and the higher the integrated efficiency. Pond depth is a function of the desired water temperature and cost of energy; it is also influenced by the practical considerations of preparation of large areas of very level ground for support of the collectors. Operation of the shallow pond collectors is usually in a batch mode. In a batch operation the collectors are filled in the morning, the temperature rises through the day as energy is absorbed, and when losses exceed absorbed energy, the heated water is drained to an insulated storage tank for subsequent use.

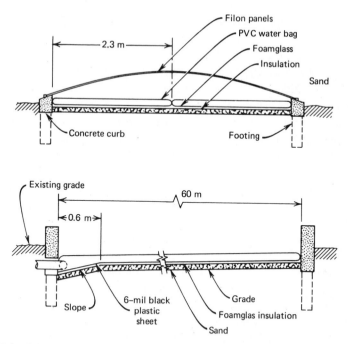

Figure 16.8.1 Schematic cross sections of a shallow pond water heater. Adapted from Guinn and Hall (1977).

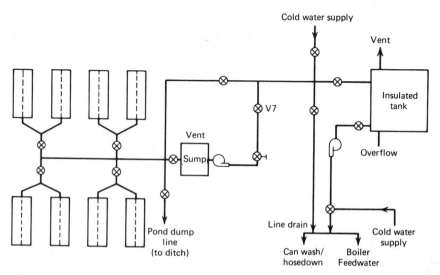

Figure 16.8.2 Schematic intermittent shallow pond solar water heating system for a poultry processing plant. Adapted from Guinn and Hall (1977).

A schematic for a system proposed for preheating water for a poultry processing plant is shown in Figure 16.8.2. This plant require a total of 150,000 liters of water per day, 40% of which is for boiler feedwater and 60% for can washing and plant cleanup operations. In this design, essentially all of the water from the insulated tank is used each day, and the collectors are fed with fresh water at supply temperature each morning.

Collectors of this type have been studied experimentally at Livermore, CA, and at Grants, NM, by Dickenson (1976) and Casamajor et al. (1977). Some practical problems have been encountered. The collectors must be very carefully sloped to drain, so as to diminish the amount of water left in them at night. Snow loads on the wide, flat spans of the covers have resulted in structural problems. The durability of the system is not yet established.

Collectors of this type are restricted to horizontal orientation, so there is a latitude and seasonal variation of the incident energy on them that is a function of latitude. The monthly average radiation data from Appendix G are directly applicable to these collectors. Hour-by-hour estimates of the performance of these collectors can be made by the methods of Section 6.12, as illustrated by Example 12.7.1. Long-term performance estimates can be made by the $\bar{\phi}$ method (see Chapter 21).

16.9   SUMMARY

Use of solar energy for industrial process heat represents a range of potential applications that could replace large quantities of fossil fuels. These applications face formidable competition. Industry purchases fuel in large quantities at relatively low cost. These costs are effectively further reduced by their deduction as an operating expense from corporate income with a resulting reduction of corporate income taxes. On the other hand, investment tax credits and depreciation of equipment tend to make the economics of solar application more favorable. There are not many industrial applications now being made (although the Luz power plants and the Israeli solar ponds described in the next chapters could be considered to be industrial processes).

REFERENCES

Casamajor, A. B. and R. E. Parsons, Lawrence Livermore Laboratory Report UCRL 52385 Rev 1 to U.S. Deptartment of Energy (1979). "Design Guide for Shallow Solar Ponds."

Casamajor A. B., T. A. Trautt, and J. M. Flowers, *Proc. of the Solar Industrial Process Heat Symposium* (1977), p. 141. "Shallow Solar Pond Design Improvements."

Dickenson, W. W., Report SAN/1038-76/1 to NATO Committee on Challenges of Modern Society for U.S. Energy Research and Development Administration (1976). "Performance of the Sohio Solar Water Heating System Using Large Area Plastic Collectors."

Duffie, N. A. and D. J. Close, *Solar Energy*, **20**, 405 (1978). "The Optimization of a Solar Timber Dryer Using an Absorbent Energy Store."

Graham, B. J., M. Sarlas, and P. D. Sirer, *Proc. of the Solar Industrial Process Heat Conference* (1978), p. 79. "Application of Solar Energy to the Dehydration of Onions."

Guinn, G. R., *Proc. of the Solar Industrial Process Heat Conference* (1978), p. 63. "Process Drying of Soybeans Using Heat from Solar Energy."

Guinn, G. R. and B. R. Hall, *Proc. of the Solar Industrial Process Heat Symposium* (1977), p. 161. "Solar Production of Industrial Process Hot Water Using Shallow Solar Ponds."

Lane, R., *Solar Today*, **4**, No. 5, 18 (1990). "Successful Large Scale Solar Hot Water."

Marlatt, W. P. and C. C. Smith, *IEA Status Seminar on Solar Air Systems*, March (1988). "Seasonal Performance Monitoring of a Large Solar Air-Heating System."

Proc. of the Solar Industrial Process Heat Symposium, College Park, MD, Report No. Conf. 770966. to U.S. Energy Research and Development Administration (1977), Govt. Printing Office Stock 061-000-00109-3.

Proc. of the Solar Industrial Process Heat Conference, Report No. Conf. 781015 to U. S. Department of Energy (1978), available from National Technical Information Service.

Read, W. R., A. Choda, and P. I. Cooper, *Solar Energy*, **15**, 309 (1974). "A Solar Timber Kiln."

Simpson, W. T., and J. L. Tschernitz, Forest Products Laboratory report FPL-44 (1985), U.S. Deptartment of Agriculture, Forest Service. "FPL Design for Lumber Dry Kiln Using Solar/Wood Energy in Tropical Latitudes."

Simpson, W. T. and J. L. Tschernitz, *Forest Products J.*, **39**, 23 (January, 1989). "Performance of a Solar/Wood Energy Kiln in Tropical Latitudes."

Steinmann, D. E., Ph.D. Thesis, Univ. of Stellenbosch, South Africa, (1989). "Solar Kiln Drying of Wood."

Tschernitz, J. L., and W. T. Simpson, *Solar Energy*, **22**, 563 (1979). "Solar Heated Forced Air Lumber Dryer."

Vindum, J. O. and L. P. Bonds, *Proc. of the Solar Industrial Process Heat Conference* (1978), p. 15. "Solar Energy for Industrial Process Hot Water."

Wirsum, M. C., M.S. Thesis, Mechanical Engineering, U. of Wisconsin-Madison (1988). "Simulation Study of a Large Solar Air Heating System."

Yang, K. C., *Forest Products J.*, **30**, 37 (March, 1980). "Solar Kiln Performance at a High Latitude, 48°N."

Chapter 17

SOLAR THERMAL POWER SYSTEMS

Conversion of solar to mechanical and electrical energy has been the objective of experiments for over a century. In 1872, Mouchot exhibited a steam-powered printing press at the Paris Exposition, in 1902 the Pasadena Ostrich Farm system was exhibited, and in 1913 a solar-operated irrigation plant started its brief period of operation at Meadi, Egypt. These and other developments utilized concentrating collectors to supply steam to heat engines. An interesting historical review of these experiments is provided by Jordan and Ibele (1956).

Much of the early attention to solar thermal-mechanical systems was for small-scale applications, with outputs ranging up to 100 kW, and most of them were designed for water pumping. In the past three decades, the possibilities of operating vapor compression air conditioners has been studied analytically and experimentally. Since 1975 there have been several large-scale experimental power systems constructed and operated, commercial plants of 30 and 80 MW electric (peak) generating capacity are operating, with additional 80 MW electric plants are under construction. Several of these will be noted in this chapter.

The processes for conversion of solar to mechanical and electrical energy by thermal means are fundamentally similar to other thermal processes, and the principles treated earlier on radiation, materials, collectors, storage, and systems form the basis of estimating the performance of solar thermal power systems. In addition, a new set of practical considerations are important, as these processes operate at higher temperatures than those treated in earlier chapters.

This chapter is concerned with generation of mechanical and electrical energy from solar energy by processes based on flat-plate or concentrating collectors and heat engines. The use of solar ponds for power production is noted in Chapter 18. There are two other kinds of solar power systems. The first includes photovoltaic processes for the direct conversion of solar to electrical energy by solid state devices; these will be discussed in Chapter 23. The second includes solar-biological processes that produce fuels for operation of conventional engines or power plants.

17.1 THERMAL CONVERSION SYSTEMS

The basic process for conversion of solar to mechanical energy is shown schematically in Figure 17.1.1. Energy is collected by either flat-plate or

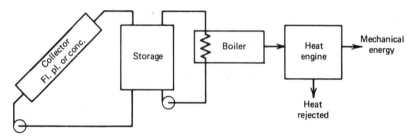

Figure 17.1.1 Schematic of a solar-thermal conversion system.

concentrating collectors, stored (if appropriate), and used to operate a heat engine. Problems of these systems lie in the fact that the efficiency of a collector diminishes as its operating temperature rises, while the efficiency of the engine rises as its operating temperature rises. The maximum operating temperatures for flat-plate collectors are low relative to desirable input temperatures for heat engines, and system efficiencies are low if flat-plate collectors are used.

Hottel (1955) pointed out that the availability of cover materials with high transmittance for solar radiation may make it possible to consider the use of flat-plate collectors to supply energy to heat engines. Recent developments in collector loss control by honeycombs, CPCs, concentrating collectors, or use of evacuated tubular collectors may open up additional systems for thermal conversion of solar to mechanical energy.

Methods of estimation of system performance are based on analysis of the components in the system. The analysis of the collectors has been covered in Chapters 6 and 7. If arrays are large, two factors must be taken into account. First, if collector modules are arranged in series, the methods of Section 7.6 can be used to calculate the effective $F_R U_L$ and $F_R(\tau\alpha)$. Second, large arrays inevitably mean long runs of piping, and the methods of Section 10.5 can be used or adapted to account for these losses. If pipe runs are very long, their heat capacity and that of the fluid in them can represent an appreciable factor in the energy balances during transient operation.

Models are also needed for engines. These can be derived from basic thermodynamic relationships or they can be graphical or analytical representations of operating data. Olson (1977) derives and uses one such model.

17.2   THE GILA BEND PUMPING SYSTEM

The Gila Bend solar irrigation pumping system described by Alexander et al. (1979) is shown schematically in Figure 17.2.1. It was based on an array of cylindrical parabolic collectors that was designed to produce hot water at 150 C (pressurized at about 7 atm). The hot water went to a heat exchanger which serves as a Freon boiler where its temperature dropped about 10 C in vaporizing the Refrigerant 113 working fluid. The hot water then went through the preheater where it preheated the fluid

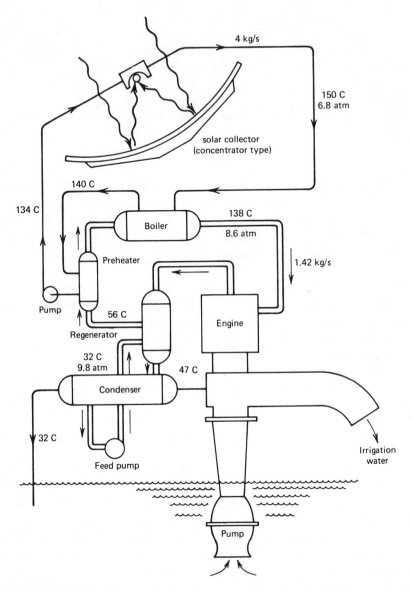

Figure 17.2.1   Schematic of the Gila Bend solar pumping system. Design flow and conditions are shown. Adapted from Alexander et al. (1979).

returning to the boiler. The water returned to the collector at about 134 C. There was no thermal energy storage in this system. Water was pumped into the irrigation canals whenever possible, and storage was in the form of water at the desired higher elevation.

The working fluid vapor went from the boiler through an entrainment separator and to a 50 hp turbine. The turbine drove a 630 l/s (10,000 gpm) pump through a gearbox. After leaving the turbine, the vapor went through a regenerator to partially preheat the boiler feed liquid and then to the condenser where it was condensed at approximately 32 C by part of the irrigation water. Condensate was returned by a feed pump to the boiler via the regenerator and preheater.

The collector array included nine rows of concentrating collectors, each 24.4 m long, with an aperture of 2.45 m and focal length of 0.91 m, for a total collecting area of 537 m^2. The rows were rotated about horizontal north-south axes. The receivers were copper tubes 41 mm in diameter, with nonselective black coatings, and insulated on the side away from the concentrator. First-year experiments were done with a glass half-cylinder cover, and in second-year experiments the cover was removed. Each of the nine rows of collectors was individually oriented to track the sun.

Alexander et al. (1979) describe a range of practical design and operating problems that were encountered in this experimental system. These include maintenance of specular reflectance of the reflector, tracking accuracy on various kinds of days, maintenance of mechanical and electronic components, integration of the solar energy system with an electric drive system to increase the use factor on the pump, and others.

This system, even though it underwent frequent and prolonged developmental changes, operated for 323 hours in the first year at an average capacity of 240 l/s and a maximum of 570 l/s. In its second year, it operated for 188 hours and delivered 1.24×10^5 m^3 of irrigation water.

17.3 THE LUZ SYSTEMS

Nine commercial Solar Electric Generating Systems (SEGS), designed, constructed and operated by Luz International, are operating in the Mojave desert of southern California. These plants are based on large parabolic trough concentrators providing steam to Rankine power plants. They generate peaking power which is sold to the Southern California Edison utility. The first of these plants is a 14 MW electric (MWe) plant, the next six are 30 MWe plants, and the two latest are 80 MWe. Additional 80 MWe plants were under contract in 1991. Basic data on seven operational plants are shown in Table 17.3.1. An aerial view of SEGS III, IV, and V is shown in Figure 17.3.1.

The design and economics of these plants are substantially influenced by U.S. federal law. The plants qualify under the Public Utilities Regulatory Policies Act (PURPA) as small power producers, and are allowed under this law to supplement solar output of the plant by fuels (natural gas) to the extent of 25% of the annual output. The plants can supply peaking power, using all solar energy, all natural gas, or a combination of the two, regardless of time or weather, within the constraint of the annual limit on gas use. The most critical time for power generation and delivery (and the time in which the selling price of the power per kwh is highest) is between noon and 6 PM in the months of June through September. Operating strategy is designed to maximize solar energy use and depends on gas to provide power during

Table 17.3.1   Characteristics of SEGS I to VII[a]

Plant	First Year	Turbine Size (MWe)	Solar Temp (C)	Field Area (m²)	Turbine Efficiency Solar	Gas	Annual Output (MWh)
I	1985	13.8	307	82,960	31.5[b]	–	30,100
II	1986	30	316	165,376	29.4	37.3	80,500
III	1987	30	349	230,300	30.6	37.4	92,780
IV	1987	30	349	230,300	30.6	37.4	92,780
V	1988	30	349	233,120	30.6	37.4	91,820
VI	1989	30	390	188,000	37.5	39.5	90,850
VII	1989	30	390	194,280	37.5	39.5	92,646

[a] From Jensen et al. (1989).
[b] Includes natural gas superheating.

Figure 17.3.1   Aerial view of SEGS III, IV, and V, three of the Luz plants located on the Mojave desert of southern California. Photo courtesy of Luz International, Inc.

cloudy periods early in the year and late in the summer when solar output drops off before the end of the peak power period. The turbine-generator efficiency is best at full load, and use of gas supplement to allow full-load operation maximizes plant output. The plants do not have energy storage facilities. A schematic of a typical plant is shown in Figure 17.3.2; the solar and natural gas loops are in parallel to allow operation with either or both of the energy sources. Data and experience with these plants have been reported by Jaffe et al. (1987), Kearney and Gilon (1988), Kearney et al. (1988), Jensen et al. (1989), and Harats and Kearney (1989).

   The major components in the systems are the collectors, the fluid transfer pumps, the power generation system, the natural gas auxiliary subsystem, and the controls. Most of the information included here is for SEGS VI and VII.

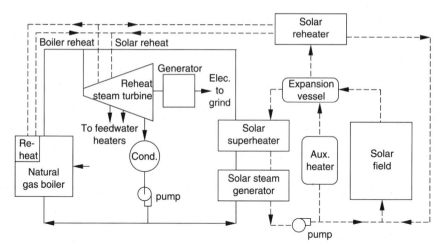

Figure 17.3.2 Schematic process diagram for SEGS systems. Adapted from Kearney and Gilon (1988).

Three collector designs have been used in these plants, LS-1 in SEGS I, LS-2 in II to VII, and LS-3 in part of VII and in subsequent plants. Data for LS-2 and LS-3 are shown in Table 17.3.2. The reflectors are made up of back-silvered, low-iron float-glass panels which are shaped over parabolic forms. Metallic and lacquer protective coatings are applied to the back of the silvered surface, and no measurable degradation of the reflective surface has been observed. The glass is mounted on truss structures, with the position of large arrays of modules adjusted by hydraulic drive motors. The receivers are 70 mm diameter steel tubes with cermet selective surfaces surrounded by a vacuum glass jacket. The surfaces have an absorptance of 0.96 and an emittance of 0.19 at 350 C.

The reflectance of the mirrors is 0.94 when clean. Maintenance of high reflectance is critical to plant operation. With a total of $1.35 \times 10^6 \, m^2$ of mirror area in the first seven plants, mechanized equipment has been developed for cleaning the reflectors, which is done regularly at intervals of about 2 weeks.

Table 17.3.2   Characteristics of the LS-2 and LS-3 Collector Modules[a]

	LS-2	LS-3
Area, m^2	235	545
Mirror segments	120	224
Aperture, m	5.0	5.76
Length, m	47.1	95.2
Focal length, m	1.84	2.12
Concentration ratio	71	82
Distance between rows, m	12.5-15	17.3
Optical efficiency	0.74-0.76	0.80

[a] Adapted from Haratz and Kearney (1989).

Tracking of the collectors is controlled by sun sensors that utilize an optical system to focus radiation on two light-sensitive diodes, with imbalance causing corrections in the positioning of the collectors. There is a sensor and controller on each collector assembly; the resolution of the sensor is about 0.5°. The collectors rotate about horizontal north-south axes, an arrangement which results in slightly less energy incident on them over a year but which favors summertime operation when peak power is needed and its sale brings the greatest revenue.

A synthetic heat transfer fluid is heated in the collectors and is piped to the solar steam generator and superheater where it generates the steam to be supplied to the turbine. Reliable high-temperature circulating pumps are critical to the success of the plants, and significant engineering effort has gone into assuring that pumps will stand the high fluid temperatures and temperature cycling. The normal temperature of the fluid returned to the collector field is 304 C and that leaving the field is 390 C. Recent experience indicates that availability of the collector fields is about 99%.

The power generation system consists of a conventional Rankine cycle reheat steam turbine with feedwater heaters, deaerators, etc. The condenser cooling water is cooled in forced draft cooling towers.

The Luz plants are designed with the help of a simulation program that uses hourly radiation data. Plant performance guarantees based on that program are provided to the investors in the projects. SEGS III to VII show performance slightly exceeding the warranted performance. Peak hour capacity factors, the ratio of available output to rated (warranted) output over all of the peak hours of the year, exceeded 100% by a small margin. Annual overall capacity factors have been about 30%. The projected cost of delivered peak hour power from SEGS VIII and IX is $0.07 to $0.08 per kwh.

17.4   CENTRAL RECEIVER SYSTEMS

In Chapter 8, some of the optical properties of large "power tower" or central receiver concentrators were noted. The potential applications of these collectors is in large-scale conversion processes with outputs in the megawatt range. Large-scale systems must collect energy from large areas. In these systems beam radiation from a large array of (relatively) small heliostats is focused on a central receiver, thus collecting by optical rather than thermal means. A photograph of an experimental system is shown in Figure 7.15.1.

An early paper on a central receiver concept by Baum (1956) outlined a design in which heliostats were to be mounted on semicircular railway tracks, with the assemblies of mirrors moved through the day to keep beam radiation focused on the central receiver. More recent design concepts call for heliostats on mounts at fixed locations and movable about two axes of rotation to accomplish concentration. Major research and development efforts in the United States are aimed at solving the range of optical, thermal, and mechanical problems associated with the development of electric power generation systems based on these concepts. These early efforts were reported in the proceedings of the Department of Energy workshop on central solar thermal electric systems (1978). A review of general design considerations is provided by Vant-Hull in a paper in the workshop proceedings.

The major components in the system are the heliostat field, the heliostat controls, the receiver, the storage system, and the heat engine which drives the generator. Heliostat design concepts were briefly outlined in Section 7.9. The objective of heliostat design is to deliver radiation to the receiver at the desired flux density at minimum cost. For an extensive discussion of optical problems, see the proceedings of the ERDA workshop on optical analysis (1977).

A range of receiver shapes has been considered, including cavity receivers and cylindrical receivers. The optimum shape is a function of intercepted and absorbed radiation, thermal losses, receiver cost, and design of the heliostat field. Vant-Hull suggests that for a large heliostat field a cylindrical receiver has advantages when used with Rankine cycle engines, particularly for radiation from heliostats at the far edges of the field. If higher temperatures are required for operation of Brayton cycle turbines, it may be necessary to use cavity receivers with larger tower height to heliostat field area ratios.

It has been observed in many solar power studies that the solar collector represents the largest cost in the system. Under these circumstances, an efficient engine is justified to obtain maximum useful conversion of the collected energy. Several possible thermodynamic cycles have been considered. First, Brayton or Stirling gas cycle engines operated at inlet temperatures of 800 to 1000 C provide high engine efficiencies but are limited by low gas heat transfer coefficients, by the need for recouperators, and by the practical constraints on collector design (i.e., the need for cavity receivers) imposed by the requirements of 1000 C temperatures. Second, turbines driven from steam generated in the receiver would operate at 500 to 550 C and have several advantages over the Brayton cycle. Heat transfer coefficients in the steam generator are high, allowing the use of high energy densities and smaller receivers with energy absorption on the outer surface. Cavity receivers are not needed, and cylindrical receivers permit larger heliostat fields to be used. Use of reheat cycles would improve steam turbine performance but entails mechanical design problems. It is also possible to use steam turbines with steam generated from an intermediate heat transfer fluid circulated through collector and boiler. The fluids could be molten salts or liquid metals, and cylindrical receivers could be operated at around 600 C with such systems. These indirect systems are the only ones that readily lend themselves to the use of storage. Low-temperature steam or organic fluid turbines may also be used, with collector temperatures around 300 C in receivers with lower solar radiation flux densities.

In the next section, a very large central receiver system, Solar One, will be described. Other experimental plants have been built and operated, including a 5 MWe plant in the Crimea, USSR and 1 MWe plants in Italy, Spain, and France.

17.5   SOLAR ONE POWER PLANT

The construction of Solar One, a large central receiver plant with a peak electric generating capacity of over 10 MWe, was completed in 1982. Located at Barstow, California, the plant underwent 2 years of operation in an experimental test and evaluation phase and 3 years in a power production phase. Power generated by the plant was supplied to the grid of Southern California Edison. Photographs of the collector, the receiver, and heliostats are shown in Figure 17.5.1. A summary of experience with the plant during the 3 years of power production operation is provided by Radosevich and Skinrood (1989).

The heliostat system consists of 1818 individually oriented reflectors, each consisting of 12 concave panels with a total area of 39.13 m^2, for a total array area of 71,100 m^2. The reflecting material is back-silvered glass.

The receiver is a single pass superheat boiler, generally cylindrical in shape, 13.7 m high, 7 m in diameter, with the top 90 m above ground. It is an assembly of 24 panels, each 0.9 m wide and 13.7 m long. Six of the panels on the south side (which receives the least radiation) are used as feedwater preheaters and the balance are used as boilers. Each panel is made up of parallel 0.069 m diameter alloy tubes welded together along their total length. The panels are coated with a nonselective flat black paint which was heat cured in place with solar radiation. Average absorptance for solar radiation, after a second coating was applied, was about 0.96. The receiver was designed to produce 50,900 kg/hr of steam at 516 C with the absorbing surface operating at a maximum temperature of 620 C.

(a)

Figure 17.5.1 Solar One, at Barstow, CA. (a) Overall view of the heliostat field and receiver tower, when system was not operating. (b) A heliostat. (c) The receiver, photographed from above, when in operation. Photographs courtesy of Sandia National Laboratories.

The thermal storage system is in a fluid loop that is separated from the steam cycle loop by heat exchangers. Steam generated in excess of that required for power generation can be diverted to the charging heat exchanger, in which oil is heated for transfer to a bed of oil, sand, and gravel. Since the plant is basically a peaking plant, most steam is used in the turbine as it is generated.

The power generation system includes the turbine generator and its auxiliaries and the cooling system for heat rejection.

The control systems include the computers that monitor and control the plant and the beam characterization system that corrects the alignment of the heliostats and evaluates heliostat performance.

The plant support system includes all of the auxiliaries necessary for the operation of the plant, such as water conditioners, compressed gas supplies, liquid waste disposal, and electrical systems.

During its 3 years of power production operation, measurements were made on the performance of the various components in the systems. Radosevich and Skinrood (1989) have summarized the main conclusions of the operation; highlights from their summary serve to illustrate the potential and the problems of operation of large central receiver systems.

The heliostat field uses mirrors with an average reflectance of 90.3%. Dirt on the mirrors reduced this reflectance, with the reduction a function of frequency of washing. Based on experience with Solar One, it appears that the reflectance of the mirrors can be maintained at about 93% of its nominal value if heloistats are washed effectively at biweekly intervals. Some corrosion of the silver backing was experienced; this can be held to a few percent per year by proper design of the modules to minimize the contact of the silvering with moisture. A third consideration with the large heliostat field is reliability. During the third year of the power production operation, an average of 98.8% of the 1818 heliostats were operational.

The receiver was a source of more problems. The panels experienced warping, presumably due to temperature gradients caused by uneven distribution of incident radiation. In addition, leaks developed in tubes and repairs were required. Warpage did not itself affect operation of the receiver other than to expose support structure to radiation entering through openings that developed between the panels, but it probably contributed to leaks and would reduce the useful life of the receiver. The black paint on the receiver had an initial high absorptance of 0.96 to 0.97; this diminished at the rate of about 0.02 per year over a 3-year span.

During the power production operation, the plant availability averaged 82% of the time in which beam radiation was available to operate it. Thus the turbogenerator was connected to the grid or the store was being charged 0.82 of the time for which beam radiation was available. (The balance of the time, the down-time, was attributable to scheduled or unscheduled maintenance, stowing of the heliostats because of excessive winds, etc.) In the Barstow climate, the annual average availability, based on total daylight hours, averaged 55%. The plant more than met its design goal of 10 MWe; its maximum power output was a net (after auxiliary energy for plant operation was deducted) of 11.7 MWe.

The annual system efficiencies, the ratio of net electrical energy produced (based on 24 hour plant energy requirements) to annual incident beam radiation, were 4.1, 5.8, and 5.7% during the 3 years of power production operation. The maximum monthly efficiency was 8.7%. (Maximum instantaneous efficiencies would be higher, probably close to the design maximum of 15%.) Capacity factors, the ratio of the net electrical output to the rated net output over 24 hours, averaged 8, 12, and 11% over the 3 years of power production operation, with a maximum monthly factor of 24% in a summer month.

Solar One was a large-scale experiment that provided operating data and experience on a range of component and system problems while at the same time providing peaking power to a utility. In future plants, receiver design, storage subsystem, and use of molten salt or liquid sodium in the receiver would be considered.

REFERENCES

Alexander, G., D. F. Busch, R. D. Fischer, and W. A. Smith, Battelle Columbus Laboratories report SAND 79-7009 (1979). "Final Report on the Modification and 1978 Operation of the Gila Bend Solar-Powered Irrigation Pumping System."

Baum, V. A., in *Proc. of the World Symposium on Applied Solar Energy*, Stanford Research Institute, Menlo Park, CA., p. 289 (1956). "Prospects for the Application of Solar Energy, and Some Research Results in the USSR."

Harats, Y. and D. Kearney, paper presented at ASME meeting, San Diego (1989). "Advances in Parabolic Trough Technology in the SEGS plants."

Hottel, H. C., paper in *Solar Energy Research*, University of Wisconsin Press, Madison, p. 85 (1955). "Power Generation with Solar Energy."

Jaffe, D., S. Friedlander, and D. W. Kearney, in *Proc. ISES World Congress*, Hamburg, FRG (1987). "The Luz Solar Electric Generating Systems in California."

Jensen, C., H. Price, and D. Kearney, paper presented at ASME meeting, San Diego (1989). "The SEGS Power Plants: 1988 Performance."

Jordan, R. C. and W. C. Ibele, in *Proc. of the World Symposium on Applied Solar Energy*, Stanford Research Institute, Menlo Park, CA, p. 81 (1956). "Mechanical Energy from Solar Energy."

Kearney, D. W., H. Price, Y. Gilon, and D. Jaffe, paper at ASME meeting, Denver (1988). "Performance of the SEGS Plants."

Kearney, D. W., and Y. Gilon, paper presented at VDI Solarthermische Kraftwerke, Koln, Germany meeting (1988). "Design and Operation of the Luz Parabolic Trough Solar Electric Plants."

Olson, T. J. M.S. Thesis, Mechanical Engineering, University of Wisconsin-Madison (1977). "Solar Source Rankine Cycle Engines for use in Residential Cooling."

Proc. of the ERDA Solar Workshop on Methods for Optical Analyses of Central Receiver Systems, University of Houston - Sandia Laboratories (1977). Available from the U.S. Deptartment of Commerce.

Proc. of the 1978 DOE Workshop on Systems Studies for Central Solar Thermal Electric, University of Houston - Sandia Laboratories (1978). Available from the U.S. Deptartment of Commerce.

Radosevich, L. G., and A. C. Skinrood, Trans. *ASME, J. Solar Energy Engrg.,* **111**, 144 (1989). "The Power Production Operation of Solar One, the 10 MWe Solar Thermal Central Receiver Pilot Plant."

Central Receiver Technology Status and Assessment, report SERI/SP-220-3314 (1989), Solar Energy Research Institute, Golden, CO.

Chapter 18

SOLAR PONDS; EVAPORATIVE PROCESSES

In this chapter we start with a discussion of solar ponds. These are large, shallow bodies of salt water that are arranged so that the temperature gradients are reversed from the normal, i.e., so the hottest layers are at the bottom of the ponds. This allows their use for collection and storage of solar energy, which may under ideal conditions be delivered at temperatures of 40 to 50 C above ambient. The first section is a description of ponds and their operation. The next is an introduction to analysis of the thermal performance of ponds (and thus might have been placed in Part I), and in the third section three applications are discussed.

At least two other methods have been noted for maintaining inverted temperature gradients by suppressing convection in ponds through the use of horizontal, vertical, or honeycomb membranes [see Hull (1980)] and polymer gel layers [see Wilkins and Lee (1987)]. These are not treated here.

We then note three evaporative processes: distillation of salt water to produce fresh water, evaporation of salt brines to produce salt, and drying to remove moisture from solids. Evaporation for salt production and drying of crops are many centuries old and are the basis of commercial and agricultural enterprises in many parts of the world. A substantial body of know-how and design information has been developed on evaporation and drying, and in the discussion here only the principles involved are noted. Distillation developments have occurred over the past century and are treated in more detail. In all cases our concern is with direct distillation, evaporation, and drying in which solar energy is absorbed in the apparatus where evaporation occurs rather than in a separate collector; indirect processes, in which solar energy is delivered from collectors to evaporation, drying, or distillation processes, are covered, by implication, in the chapter on industrial processes.

18.1   SALT-GRADIENT SOLAR PONDS

Temperature inversions have been observed in natural lakes, with higher temperatures in the bottom layers than near the surface. These lakes have high concentrations of dissolved salts in the bottom layers and much more dilute solutions at the surface. This phenomenon suggested the possibility of constructing and using ponds as large-scale horizontal solar collectors. Tabor (1964), Weinberger (1964), and Tabor and Matz (1965) reported a series of theoretical and experimental studies of these salt-gradient ponds.[1] Salt-gradient ponds have been proposed for power production, salt production, and providing thermal energy for buildings. Tabor and Weinberger (1981) review theory and operation of ponds. See Nielsen (1988) for a recent and useful review of the theory and applications. For a comprehensive treatment of all aspects of the topic, see Hull et al. (1989).

Typical gradients of salinity and temperature in an operating solar pond are shown in Figure 18.1.1. Three layers are present. At the top is a homogeneous layer of solution of low concentration, an upper convecting zone (UCZ). The next layer is a thick nonconvecting or gradient zone (GZ), which provides insulation as heat is transferred up out of the heated layer below it only by conduction. The bottom layer is another homogeneous layer, the lower convecting zone (LCZ), with a high concentration of salt. The basic criterion for solar pond operation is that the density of the concentrated solution in the bottom zone must be higher at its maximum temperature than the density of the more dilute layers above it. Much of the solar radiation is transmitted through the upper layers, and the bottom layers become heated.

Several phenomena occurring in the operation of ponds are briefly described below. These include absorption of radiation, diffusion processes that tend to drive ponds toward homogeneity, stability, pond construction, salt and water requirements, and heat extraction.

The reflectance of a smooth water surface for solar radiation is of the order of 5% at near normal incidence. As radiation is transmitted through water, the longer wavelengths are absorbed near the surface. Mesasurements by Nielson (1988) indicate that little radiation at $\lambda > 0.7$ μm is transmitted through a meter of clean water, but over 95% of the radiation in the wavelength range 0.3 to 0.6 μm is transmitted through this much water. At the time of these measurements, the zenith angle of the sun was approximately 12°, so radiation was nearly normal to the water surface. In this pond there was significant radiation transmitted to the bottom layers.

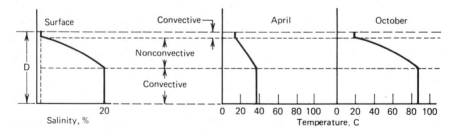

Figure 18.1.1 Salinity and temperature profiles for a salt-gradient pond. From Nielsen (1978), with permission of the American Section of the International Solar Energy Society.

[1] These are often called simply "solar ponds." As there are other types of ponds called by the same name, the term salt-gradient pond avoids confusion.

The concentration gradients that exist in ponds lead to diffusion of salt from the concentrated layer at the bottom to the dilute layer at the top. In order for pond stability to be maintained, salt must be added to the lower layer and removed from the upper layer. As this must be done by injecting and removing solutions, there is an additional upward flow due to water addition and removal. Nielsen notes that the rates of salt transport into the upper zone are of the order of 3 kg/m^2 per month. Thus the quantities of salt which must be added and removed are significant. The corresponding quantities of dilute solution must be discarded and concentrated solution obtained, or the dilute solution processed to allow recycling. Water evaporation from the open surfaces of ponds must also be replaced. [See Batty et al. (1987).]

Solar ponds are of the order of 1 to 3 m deep. They are constructed on level ground by a combination of excavation and embankments, and membrane liners are used to make the basins leak proof. Layers of clay may be used over the liners to protect them and improve their durability. Leaks from ponds can prevent their functioning as useful energy collection systems. Filling ponds must be done in such a way that the necessary salinity gradients are established by addition of fresh water or brines at appropriate levels.

Surface waves caused by winds tend to cause mixing and increases in thickness of the upper convective layer, ultimately leading to degradation of pond performance. Floating nets or arrays of floating pipes are used as wave barriers. The larger the ponds, the more critical is the use of the barriers.

There are several advantages to large ponds. The larger the pond, the smaller will be the ratio of perimeter to area. As edge losses can be significant, performance of larger ponds is better than that of smaller ponds. Also, wall effects such as development of convection in the gradient layer due to temperature gradients and conduction in the wall are minimized. Costs per unit area of ponds will decrease as pond size increases. (Ponds that have been built have ranged in size from about 300 m^2 experiments to a 210,000 m^2 production pond.)

Removal of heat from the bottom layer of a pond can be accomplished either by means of a large area-heat exchanger submerged in the pond or by simultaneous slow removal of hot brine from the lower level at one part of the pond and return of (cooled) brine at the lower level at another location. Hull et al. (1986) describe experiments on a 1000 m^2 pond in which both methods were studied; properly done, either one can be an effective method for extracting energy without destabilizing the pond.

Cleanliness of the ponds may be a problem if they are not large in extent as contaminants can reduce transmittance. Some "dirt" will float at the surface, some will sink to the bottom, and some will float at intermediate levels where its density matches that of the solution, and removal of these contaminants may have to be done by filtration or skimming. The effects of dirt may be of diminishing importance as pond size increases, as the effects will be more pronounced at edges. In very large ponds, significant effects of dirt may be caused only by major events such as dust storms.

18.2   POND THEORY

Numerous studies of various aspects of the theory of solar ponds have been published; Nielsen (1988) and Hull et al. (1986, 1989) provide useful introductions to these publications. In this section some aspects of the basic theory of ponds are outlined.

Equations for the steady-state operation of ponds can be written that are the same in form as the Hottel-Whillier equation (Equation 6.7.6) [Kooi (1979)], where heat removal from the pond is by brine extraction and injection at the LCZ. The factor F' is unity. The temperature gradient in the flow (horizontal) direction from injection point to extraction point is accounted for by an F_R, which has the same functional form as that for a flat-plate collector; its value for practical ponds is close to unity. The loss coefficient U_L includes contributions for upward losses through the nonconvective (gradient) zone and for ground and edge losses. Hull et al. (1986) have used the Kooi analysis and work on ground heat losses to derive an equation for a steady-state pond with F_R near unity as

$$Q_u = A\left[I\tau\alpha - \left(\frac{k}{\Delta x} + g\right)(T_{LCZ} - T_{UCZ})\right] \tag{18.2.1}$$

where k is the thermal conductivity of the solution in the gradient zone (nearly that of water), Δx is the thickness of the gradient zone, and g is a ground loss coefficient. The coefficient g is estimated by an equation of the form

$$g = (a/x_g + bP/A)/k_g \tag{18.2.2}$$

where P is the perimeter of a pond of area A, k_g is the effective thermal conductivity of the ground under the pond, and x_g is the distance from the bottom of the pond down to the water table. The constants a and b are for a particular pond. This assumes that both upward and ground losses are to a single temperature T_{UCZ}, which will be close to mean ambient temperature. See Hull et al. (1984) for a detailed discussion of ground losses.

Radiation is absorbed in a pond at various depths and wavelengths, as the optical extinction coefficient is a strong function of wavelength in the solar energy spectrum. Transmission in ponds has been represented in two ways, as the sum of four exponentials representing four parts of the spectrum by Rabl and Nielsen (1975) or as a logarithmic function by Bryant and Colbeck (1977), both being based on early experiments by oceanographers. Afeef and Mullett (1989) review these experiments and describe new experiments indicating that the transmittance of unfiltered salt solutions was substantially less than that of water alone, but that filtering or prolonged settling (such as would occur in solar ponds) increases transmittances of 20% solutions of sodium and magnesium chlorides to close to that of water.

Kooi has calculated η/F_R vs. $\Delta T/I$ for ponds of varying depths and compared them to flat-plate collectors, as shown in Figure 18.2.1. Curves B and C are for

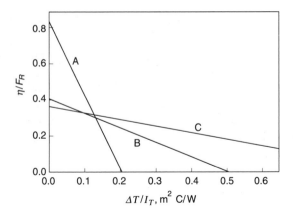

Figure 18.2.1 Calculated η/F_R vs $\Delta T/I$ curves for: (A) a one cover selective flat-plate collector, (B) a solar pond with the top of the lower convective zone one meter from the surface, and (C) a pond with the top of the LCZ two meters below the surface. Adapted from Kooi (1979).

ponds with the top of the lower nonconvecting zone 1 and 2 m below the surface. The difference in intercepts represents additional radiation absorption by the additional meter of solutions in the UCZ and GZ. The lower slope of C reflects the additional meter of insulation (i.e., solution) over the LCZ. Curve A is for a one-cover, selective flat-plate collector. At high operating points ($\Delta T/I$), the performance of ponds appears to be better than the flat-plate collector. The utility of this comparison is limited, however.   It does not account for the increases in incident radiation on the flat-plate collector gained by orienting it at more favorable angles nor the fact that ponds lose heat continuously and collectors lose heat only when they are operating. Costs and applicability to specific energy supply problems must also be considered.

18.3   APPLICATIONS OF PONDS

Several dozen ponds have been constructed and operated. Most of them are designed for experimental purposes, but as of 1989 several are production ponds delivering energy routinely to meet energy loads. The ponds briefly described in the following paragraphs are each designed for a different type of application.

At El Paso, TX, an existing 3350 m² water storage pond near a food canning plant was converted to a solar pond for providing process heat to the plant and for experiments on power generation and desalination. The operation started in 1986 with production of heat for the plant and experimental operation of a Rankine cycle turbine using an organic working fluid. The long-term intention is to use the pond for process water heating only.

A 2000 m² pond was constructed at Miamisburg, OH, in 1978 for supplying heat to a municipal swimming pool. Difficulties were initially encountered with leaks in the liner and corrosion of a submerged copper pipe heat extraction exchanger. These were overcome, and the pond is now operated with an external heat exchanger. It supplies all of the heat needed for the pool during the swimming season and also provides heat to a bath house during other parts of the year. The water supply used to fill the pool in the spring is at 11 C; after the pool is filled, the pond provides the energy to heat it to the desired 27 C operating temperature and maintain that temperature until the pool is closed in the fall [Wittenberg and Etter (1982)].

The largest solar ponds have been built in Israel. After years of research and experience with experimental ponds (by Tabor and his colleagues) a series of large ponds have been constructed in the Dead Sea area which have been designed for power production. The first of these was a 6250 m^2 pond at En Boqek that operated a 150 kW power plant. This has been superseded by two much larger ponds, one of 40,000 m^2 and the other of 210,000 m^2, which together supply energy for a 5 MW Rankine cycle peaking plant. These ponds operate at substantially higher temperatures than does the Miamisburg pond; bottom layer temperatures of 70 to 80 C are used. Heat is extracted by withdrawing concentrated brine from the lower convective layer, pumping it through the heat exchanger (boiler) where it supplies heat to the

Figure 18.3.1(a) The 3350 m^2 El Paso pond. Photo courtesy of University of Texas at El Paso. (b) The 210,000 m^2 and 40,000 m^2 ponds at the Dead Sea, Israel. Photo courtesy of Ormat.

organic working fluid, and returning it to the lower layer through low-velocity injectors.

Photographs of the pond at El Paso and the Israeli Dead Sea installation are shown in Figure 18.3.1.

18.4 SOLAR DISTILLATION

The first known application of solar distillation was in 1872 when a still at Las Salinas on the northern deserts of Chile started its three decades of operation to provide drinking water for animals used in nitrate mining. The still utilized a shallow black basin to hold the salt water and absorb solar radiation; water vaporized from the brine, condensed on the underside of a sloped transparent cover, ran into troughs, and was collected in tanks at the end of the still. Most stills built and studied since then have been based on the same concepts, though many variations in geometry, materials, methods of construction, and operation have been employed. This section is concerned with these basin-type stills. Other configurations have been used based on evaporation from wicks or from brines cascading over weirs to allow other than horizontal energy-absorbing surfaces. A very comprehensive summary of literature on all aspects of solar distillation is provided by Talbert et al. (1970).

The basin-type still is shown in section in Figure 18.4.1. A still may have many bays side by side, each of the type shown. The covers are usually glass; they may also be air-supported plastic films. The basin may be on the order of 10 to 20 mm deep (referred to as shallow basins) or they may be 100 mm or more deep (referred to as deep basins). The widths are of the order of 1 to 2 m, with lengths widely variable up to 50 to 100 m.

In operation, solar radiation is transmitted through the cover and absorbed by the salt water and the basin. The solution is heated, water evaporates, and vapor rises to the cover by convection where it is condensed on the underside of the cover. Condensate flows by gravity into the collection troughs at the lower edges of the cover; the covers must be at sufficient slope that surface tension of the water will cause it to flow into the troughs without dropping back into the basin. The trough is constructed with enough pitch along its length so that condensate will flow to the

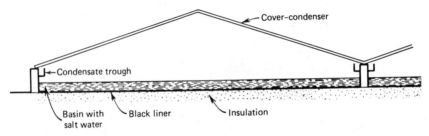

Figure 18.4.1 Schematic cross section of a bay of a basin-type solar still.

lower end of the still, where it drains into a product water collection system. Operation of a still may be continuous or batch. If seawater (approximately 3.5% salt) is used as feed, the concentration is usually allowed to double before the brine is removed, so about half of the water in the feed is distilled off.

Figure 18.4.2 shows the major energy flows in a still while it is operating. The objective of still design is to maximize Q_{evap}, the transport of absorbed solar radiation to the cover-condenser by water vapor, as this is directly proportional to still productivity. All other energy transfer from basin to surroundings should be suppressed as far as is possible. Most energy flows can be evaluated from basic principles, but terms such as leakage and edge losses are difficult to quantify and may be lumped together in a catch-all term determined experimentally for a particular still.

The basic concepts of solar still operation have been set forth by Dunkle (1961) and others. The thermal network for a basin-type still is similar in principle to Figure 6.4.1 for a flat-plate collector, but with three differences. Energy transfer from basin to cover occurs by evaporation-condensation in addition to convection and radiation. The losses from the back of the still are to the ground. The depth of the water in the still is usually such that its capacitance must be taken into account. There is no useful energy gain in the sense of that of a flat-plate collector; still output is measured by the evaporation-condensation transfer from basin to cover. A thermal network is shown in Figure 18.4.3, where the resistances correspond to the energy flows in Figure 18.4.2. Terms for leakage, edge losses, entering feedwater, and leaving brine or product are not shown.

An energy balance on the water in the basin (and the basin itself), per unit area of basin, can be written as

$$G\tau_c\alpha = q_e + q_{r,b\text{-}g} + q_{c,b\text{-}g} + q_k + (mC_p)_b\frac{dT_b}{dt} \qquad (18.4.1)$$

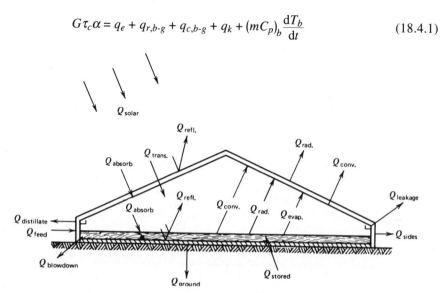

Figure 18.4.2 The major energy transport mechanisms in a basin-type still. From Talbert et al. (1970).

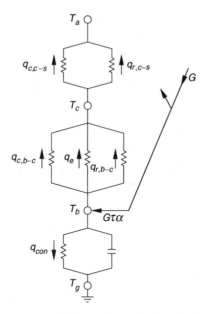

Figure 18.4.3 Basic thermal network for a basin-type still.

where the subscripts, e, r, c, and k represent evaporation-condensation, radiation, convection, and conduction, respectively. The subscripts b and g refer to basin and glazing (cover), and τ_c is the transmittance of the cover and the water film or droplets on its underside.

The capacitance of the glazing will normally be small compared to that of the water and basin. In most modern still designs the pitch of the covers is small and the cover area is approximately the same as the basin area. An energy balance on the cover, neglecting its capacitance and solar energy absorbed by it, can be written as

$$q_e + q_{r,b\text{-}g} + q_{c,b\text{-}g} = q_{c,g\text{-}a} + q_{r,g\text{-}a} \tag{18.4.2}$$

These two equations are analogous to Equations 6.12.1 and 6.12.2 for flat-plate collectors. Here q_e is not a linear function of the temperature difference between the plate and the glazing, and the two equations need to be solved simultaneously to find T_b, T_c, and q_e as functions of time.

Dunkle (1961) provides convenient ways of estimating the terms for internal heat transfer in the still for use in these equations. The radiation exchange between basin and cover $q_{r,b\text{-}g}$ is calculated by Equation 3.8.3. The cover is usually glass, and during operation a thin layer of condensate forms on most of its undersurface. Dunkle recommends that the term be written in the form

$$q_{r,b\text{-}g} = 0.9\sigma\left(T_b^4 - T_g^4\right) \tag{18.4.3}$$

For estimating the convection energy transfer from basin to cover, $q_{c,b\text{-}g}$, he suggests that the normal Rayleigh number must be modified to account for buoyancy effects due to the fact that heat and mass transfer occur simultaneously. The buoyancy term in the Grashof number is modified by the density gradient caused by the composition gradient (in addition to the temperature gradient). In a horizontal enclosed air gap, a relationship between Nusselt and Rayleigh numbers is

$$Nu = 0.075(Ra)^{1/3} \tag{18.4.4}$$

where the temperature difference in the Rayleigh number, $\Delta T'$, is an equivalent temperature difference accounting for density differences due to water vapor concentration differences. For air and water,

$$\Delta T' = (T_b - T_g) + \left(\frac{p_{wb} - p_{wg}}{2016 - p_{wb}}\right) T_b \tag{18.4.5}$$

where p_{wb} and p_{wg} are the vapor pressures of water in mm Hg of the solution in the basin at T_b and of water at the cover temperature T_g. Temperatures are in degrees Kelvin.

From Equations 18.4.4 and 18.4.5 the convection coefficient in a still is

$$h_c' = 0.884\left[(T_b - T_g) + \left(\frac{p_{wb} - p_{wg}}{2016 - p_{wb}}\right) T_b\right]^{1/3} \tag{18.4.6}$$

and the heat transfer between the basin and cover is

$$q_{c,b\text{-}g} = h_c'(T_b - T_g) \tag{18.4.7}$$

By analogy between heat and mass transfer, the mass transfer rate can be written as

$$\dot{m}_D = 9.15 \times 10^{-7} h_c'(p_{wb} - p_{wg}) \tag{18.4.8}$$

and the heat transfer by evaporation-condensation is

$$q_e = 9.15 \times 10^{-7} h_c'(p_{wb} - p_{wg}) h_{fg} \tag{18.4.9}$$

where $\dot{m}_D$ is the mass transfer rate in kg/m²s, and h_{fg} is the latent heat of water in J/kg.

The heat transfer terms from cover to ambient are formulated in the same way as for flat-plate collectors. If the still has insulation under the basin, heat loss to the ground can be written as

$$q_k = U_G(T_b - T_a) \tag{18.4.10}$$

where U_G is an overall loss coefficient to ground assuming the ground to be at a temperature equal to ambient. This term should be small in a large, well-designed still.

If the basin is very shallow and well insulated, the heat capacity term in Equation 18.4.1 can be neglected and steady-state solutions found. However, for practical reasons, most stills will have sufficient depth that the capacitance of the basin should be considered. If the still is not well insulated, an effective ground capacitance will also have to be considered by writing another differential equation relating energy stored in a layer of ground to heat flows into and out of that layer.

This set of equations, with data on radiation, temperature, and wind speed and with design parameters of the still, can be solved for T_b as a function of time, and the productivity is then calculated from Equation 18.4.8. This analysis does not include capacitance rates of feedwater or leaving brine or product nor edge effects and leakage, which are difficult to formulate; these are often lumped together in a term determined experimentally as that required to make energy balances close. There is evidence for the existence of temperature gradients in still basins that make the surface temperature different from the bulk temperature of the salt water.

The instantaneous efficiency of a still at any time is defined as the ratio of the heat transfer in the still by evaporation-condensation to the radiation on the still:

$$\eta_i = \frac{q_e}{A\,G} \qquad (18.4.11)$$

This is usually integrated over some extended period (e.g., day or month) to indicate long-term performance. If there is any loss of product water back into the still (by dripping from the cover or by evaporation or leakage from collecting troughs), less product would be available than is indicated by this equation. Efficiency from experimental measurements is

$$\eta_i = \frac{\dot{m}_p h_{fg}}{A\,G} \qquad (18.4.12)$$

where $\dot{m}_D$ is the rate at which distillate is produced from the still (which may be less than $\dot{m}_D$) and h_{fg} is the latent heat of vaporization.

The objective of still design is to maximize q_e (i.e., $\dot{m}_D$), which is proportional to the vapor pressure difference between basin and cover. Thus it is desirable to have the basin temperature as high as possible, which will increase the ratio of heat transfer by evaporation-condensation to that by convection and radiation. This has led to designs with shallow basins with small heat capacity which heat up more rapidly than deep basins and operate at higher mean temperatures.

Many practical considerations govern still design and operation. Shallow basins require precise leveling of large areas, which is costly. Crystals of salt build up on dry spots in basins, leading to reduced overall absorptance and reduced effective basin area. Leakage can cause problems in three ways: distillate can leak back into the basin; salt water can leak out of the basin; and humid air from inside the still can leak out through openings in the cover. Occasional flushing of still basins has been found necessary to remove accumulations of salt and organisms such as algae which grow in the brines. Algal growths can be controlled by additions of algicides.

A wide variety of experiment basin-type stills have been built and studied. Two design trends have evolved. Large-area deep basin stills as shown in Figure 18.4.4 can be built by standard construction techniques, are durable, and are relatively inexpensive. Part of the distillate production from these deep-layer stills is obtained at night, when the basin remains warm and the cover temperature goes down as the ambient temperature drops.

Modular shallow basin stills, as shown in Figure 18.4.5, have lower thermal capacitance, produce somewhat more water, and yield most of their product water in the daytime. They may be more expensive to construct than deep-layer stills.

Details of these and many other designs are included in Talbert et al. (1970). A less extensive report on the potential of distillation applications in developing economies was published by the United Nations (1970). Cooper, in a series of papers (e.g., 1973), has made a detailed simulation study of solar still performance. Proctor (1973) has experimentally investigated the possibility of augmenting solar still output with waste heat.

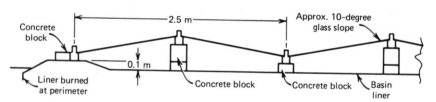

Figure 18.4.4   Schematic section of an experimental deep layer still at Daytona Beach, FL. Brine depth in still is 0.1 to 0.15 m. From Talbert et al. (1970).

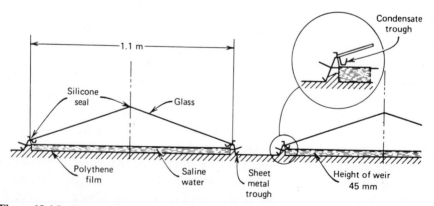

Figure 18.4.5   Schematic section of an experimental shallow basin still at Muresk, W. A. Brine depth is 0.01 to 0.03 m. Adapted from Morse and Read (1968).

18.5   EVAPORATION

Solar evaporation to produce salts from sea water or other brines is a large-scale industrial operation, and on a worldwide basis approximately one-third of the annual salt production is from solar processes. In the United States, solar salt production facilities are concentrated in the Great Salt Lake and San Francisco Bay areas. This is a well-established technology.

A description of the history and operation of salt manufacture by solar evaporation of sea water is provided by See (1960). The process is a fractional crystallization. Shallow pans or ponds are filled with the brine to be evaporated and exposed to radiation. As long as evaporation exceeds rainfall, concentration will increase with time, and crystallization will occur when saturation is reached. Pond operation depends on the source of feedwater. Sea water contains approximately 3.5% salts, of which approximately two-thirds is NaCl (the usual desired product). Sea water evaporation can be considered to consist of three steps. The first is brine concentration to bring the solution to saturation (approximately 23% salts) and is the largest in area. This solution is transferred to second-stage pans, where the first salts to crystallize are calcium, magnesium, and iron carbonates and sulfates. Then the solution goes to the crystallizing ponds where NaCl crystallizes out. More water is allowed to evaporate until the solution nears the saturation point for the magnesium and potassium chlorides and bromides and other salts (the "bittern" salts), at which point the solution is removed from the pond. The salt precipitated in the crystallizer pond is NaCl in a reasonably pure form and is harvested mechanically.

Solar evaporation is used with feeds other than sea water. Sodium sulfate solutions are concentrated in 440,000 m^2 ponds on the Atacama desert of Chile for the manufacture of anhydrous sodium sulfate [Suhr (1970)]. The Caracol, a solar evaporating process near Mexico, D. F., concentrates brines from the Lake Texcoco area to feed a fuel-fired evaporator plant producing sodium carbonate decahydrate.

Solar evaporation has been of substantial interest for salt production and also for estimation of evaporative losses from water storage reservoirs. A number of studies of the process have been published, including reports of measurements from salt pans [e.g., Bloch et al. (1951)] and lakes and reservoirs [e.g., Geological Survey (1954)]. (Salt solutions in evaporators will have lower vapor pressures than lake water; otherwise the processes are similar.)

An energy balance over a fully mixed pond is very much like Equation 18.4.1 for the basin of a solar still, but heat is transferred to the atmosphere by evaporation, convection, and radiation rather than to a cover:

$$G\alpha = q_e + q_{r,p\text{-}a} + q_{c,p\text{-}a} + q_k + (mC_p)\frac{dT_b}{dt} \qquad (18.5.1)$$

where q_e, q_k, and capacitance terms have the same significance as for a still, and the subscript p refers to the pond. (For deep ponds that are thermally stratified, the heat capacity term requires more detailed treatment.)

For evaporation from open water surfaces, the ratio of energy transport from the water by convection to that by evaporation has been shown by Bowen (1926) to be nearly constant regardless of wind speed. The ratio is given by

$$\frac{q_{c,p\text{-}a}}{q_e} = \frac{0.46(T_p - T_a)}{p_{wp} - p_a}\frac{p}{760} \tag{18.5.2}$$

where T_p and T_a are the temperatures of the brine surface and ambient, p_{wp} and p_a are the vapor pressure of the surface solution and the ambient vapor pressure of water, and p is the barometric pressure (with all pressures in mm Hg).

These equations have been applied to evaporation from lakes by Cummings (1946) and others and to evaporation from salt pans by Bloch et al. (1951) [see also Spiegler (1955)]. Bloch et al., in an experimental study of evaporation from brines, measured diurnal temperature variations of solution, temperature gradients in solution during evaporation and at night, and effects of dyes which increase absorptance of the brines for solar radiation.

Net annual evaporation from salt pans in the San Francisco area is usually in the range of 0.9 to 1.1 m with an annual rainfall in the range of 0.25 to 0.5 m, and in dry years the net annual evaporation was 1.2 m [See (1960)]. Thus in this area 0.5 to 1 m³ of water is evaporated per year per square meter of pond.

18.6   DIRECT SOLAR DRYING

Direct crop drying is another well-established use of solar energy, in which the standard practice is to spread the materials to be dried in thin layers to expose them to radiation and wind. (Indirect drying, by supply of solar-heated air to more or less conventional dryers, was noted in Chapter 16.) The practices of solar drying are based on long experience. Questions regarding further applications concern possible improvements in the process or changes that might be made to improve product quality or dry additional kinds of materials.

The equations describing a radiant drying process can be written in forms similar to those for evaporation from pans. However, several additional phenomena may have to be considered. If the material to be dried is opaque to solar radiation, drying occurs at the surface and moisture is transported to the surface by capillary action or diffusion. As with other drying processes, there may be constant-rate periods when energy balances determine rates of drying (when the surface of the material is always saturated with moisture) and falling-rate periods when moisture transport through the drying material to its surface controls the rate of evaporation. The determination of the vapor pressure of water at the surface of the drying material may be difficult.

In addition to field crops, drying of grapes in tiers of trays has been reported by Wilson (1965), and an analysis of this type of dryer is presented by Selcuk et al. (1974). Experiments on direct drying of an oil shale to reduce its moisture content prior to retorting have been described by Talwalkar et al. (1965). Taylor and Weir (1985) describe an analytical and experimental study of a timber-drying kiln with forced air circulation in which some of the radiation is absorbed in the drying lumber and some in the adjoining structure.

18.7   SUMMARY

In this chapter we have briefly described some of the major considerations regarding solar ponds, distillation, evaporation, and direct drying. Much of the information needed in estimating the performance of these processes can be found by methods outlined in earlier chapters. Most are horizontal collectors, some of which have integral storage, and evaporation of water is either an objective or a problem in each of the processes.

REFERENCES

Afeef, M. and L. B. Mullett, *Solar & Wind Technology*, 6, 1 (1989). "Solar Transmission in Salt Solutions with Reference to Solar Ponds."

Batty, J. C., J. P. Riley, and Z. Panahi, *Solar Energy*, 39, 483 (1987). "A Water Requirement Model for Salt-Gradient Solar Ponds."

Bloch, M. R., L. Farkas, and K. S. Spiegler, *Ind. and Engr. Chemistry*, 43, 1544 (1951). "Solar Evaporation of Salt Brines."

Bowen, I. S., *Physical Review*, 27B, 779 (1926). "The Ratio of Heat Loss by Conduction and by Evaporation from Any Water Surface."

Bryant, H. C., and I. Colbeck, *Solar Energy*, 19, 321 (1977). "A Solar Pond for London?"

Cooper, P. I., *Solar Energy*, 14, 451 (1973). "Digital Simulation of Experimental Solar Still Data."

Cummings, N. W., *Trans. of the American Geophysical Union*, 27, 81 (1946). "The Reliability and Usefulness of the Energy Equations for Evaporation."

Dunkle, R. V., *International Developments in Heat Transfer*, Conference at Denver Part 5, 895 (1961). "Solar Water Distillation: The Roof Type Still and a Multiple Effect Diffusion Still."

Geological Survey Professional Paper 269, U. S. Government Printing Office, Washington, DC (1954). "Water Loss Investigations: Lake Hefner Studies, Technical Report."

Hull, J. R., *Solar Energy*, 25, 317 (1980). "Membrane Stratified Solar Ponds."

Hull, J. R., K. V. Liu, W. T. Sha, J. Kamal, and C. E. Nielsen, *Solar Energy*, 33, 25 (1984). "Dependence of Ground Heat Loss upon Solar Pond Size and Perimeter Insulation: Calculated and Experimental Results."

Hull, J. R., A. B. Scranton, J. M. Mehta, S. H. Cho, and K. E. Kasza, report ANL 86-17, Argonne National Laboratory, Argonne, IL. (1986). "Heat Extraction from the ANL Research Salt-Gradient Solar Pond."

Hull, J. R., C. E. Nielsen, and P. Golding, *Salinity Gradient Solar Ponds*, CRC Press, Boca Raton, FL (1989).

Kooi, C. F., *Solar Energy*, **23**, 37 (1979). "The Steady State Salt-Gradient Solar Pond."

Morse, R. N. and W. R. Read, *Solar Energy*, **12**, 5 (1968). "A Rational Basis for the Engineering Development of a Solar Still."

Nielsen, C. E., paper in *Proc. of the 1978 Annual Meeting of the American Section of the ISES*, **2**(1), 932 (1978). "Equilibrium Thickness of the Stable Gradient Zone in Solar Ponds."

Nielsen, C. E., paper in *Advances in Solar Energy*, (K. Boer, ed.) **4**, 445 (1988), American Solar Energy Society and Plenum Press. "Salinity-Gradient Solar Ponds"

Proctor, D., *Solar Energy*, **14**, 433 (1973). "The Use of Waste Heat in a Solar Still."

Rabl, A., and C. E. Nielsen, *Solar Energy*, **17**, 1 (1975). "Solar Ponds for Space Heating."

See, D. S., in *Sodium Chloride*: The Production and Properties of Salt and Brine, D. W. Kaufman (ed.), American Chemistry Society, Washington, DC, Monograph #145, Chapter 6 (1960). "Solar Salt."

Selcuk, M. K., O. Ersay, and M. Akyurt, *Solar Energy*, **16**, 81 (1974). "Development, Theoretical Analysis and Performance Evaluation of Shelf Type Solar Dryers."

Spiegler, K. S., in *Solar Energy Research*, University of Wisconsin Press, Madison, WI 119(1955). "Solar Evaporation of Salt Brines in Open Pans."

Suhr, H. B., Paper No. 4/27 presented at the International Solar Energy Society Conference, Melbourne (1970). "Energy-Balance Calculations on the Production of Anhydrous Sodium Sulfate with Solar Energy and Waste Heat."

Tabor, H., *Proc. of the UN Conference on New Sources of Energy*, **4**, 59 (1964). "Large Area Solar Collectors (Solar Ponds) for Power Production."

Tabor, H. and R. Matz, *Solar Energy*, **9**, 177 (1965). "Solar Pond Project."

Tabor, H., and Z. Weinberger, in *Solar Energy Handbook* (J. Kreider and F Kreith, eds), McGraw-Hill. New York (1981) "Nonconvecting Solar Ponds."

Talbert, S. B., J. A. Eibling, and G. O. G. Löf, *Manual on Solar Distillation of Saline Water*, Office of Saline Water, U. S. Dept. of Interior, Research and Development Progress Report No. 546 (1970).

Talwalkar, A. T., J. A. Duffie, and G. O. G. Löf, *Proc. of the UN Conference on New Sources of Energy*, **5**, 284 (1965). "Solar Drying of Oil Shale."

Taylor, K. J., and A. D. Weir, *Solar Energy*, **34**, 249 (1985). "Simulation of a Solar Timber Dryer."

United Nations Department of Ecomonic and Social Affairs, *Solar Distillation*, Publication Sales No. E. 70. II. B. 1 (1970).

Weinberger, H., *Solar Energy*, **8**, 45 (1964). "The Physics of the Solar Pond."

Wilkins, E. S., and T. K. Lee, *Solar Energy*, **39**, 33 (1987). "Development of the Gel Pond Technology."

Wilson, B. W., *Proc. of the UN Conference on New Sources of Energy*, **5**, 296 (1965). "The Role of Solar Energy in the Drying of Vine Fruit."

Wittenberg, L. J., and D. E. Etter, paper presented at ASME meeting, Phoenix (1982). "Heat Extraction from a Large Solar Pond."

Part III

DESIGN METHODS

In Part I we treated fundamentals of solar processes and in Part II we described applications of these fundamental ideas to several practical processes. In Part III, we consider methods of estimation of long-term thermal performance of solar energy processes, an essential step in their design.

Chapter 19 is on simulations. These are detailed transient thermal performance calculations. Parallels are drawn between these numerical experiments and the physical experiments that are necessary in order to have confidence in the calculations. A brief description of a transient process simulation program TRNSYS is provided and examples of its use are included. Simulations are shown to be clearly warranted in the design of large, one of a kind, or experimental systems.

Chapters 20 to 22 are discussions of methods for substituting monthly calculations for the detailed calculations of simulations. These "design methods" are based on correlations of the results of many numerical experiments (simulations) or on utilizability concepts. A range of active and passive system types are treated in these chapters.

Chapter 23 is unique in that it treats a process that is not thermal in nature, i.e., photovoltaic processes, but it builds on the same background information and systems approach that we use for thermal processes. A brief, qualitative description of solar cells is provided, followed by discussion of generator and system models, and then a design procedure that is applicable for determining the long-term output of many PV systems.

Chapter 19

SIMULATIONS IN SOLAR PROCESS DESIGN

In Chapter 10 the basic concepts of simulation were developed, i.e., the simultaneous solutions of the sets of equations (models) that describe the performance of solar processes. Example 10.9.1 is a simulation, albeit a simple one amenable to hand calculation. In Chapters 12 through 18, the parametric studies of applications that were presented were developed using simulations. This chapter carries the use of simulations a step further and treats the application of simulations to the design of specific processes.

Simulations, like any other calculations, are only as good as the models that are the basis of the programs and the skill with which they are used. There has been over two decades of experience with simulations, and considerable effort has been placed on validations. There have been detailed intercomparisons of simulation programs that were developed independently by authors in various countries. There have been experimental studies of component performance that are (with theory) the basis of confidence in component models. There have been detailed comparisons of simulations with measurements on the operation of systems. The conclusion to be drawn from these validations is that if simulations are properly used, they can provide a wealth of information about solar processes and thermal design.

This chapter includes a brief description of TRNSYS, a transient process simulation program developed for study of solar processes, and its applications. Examples are shown of the kinds of information that simulations and experiments can provide and illustrate the use of a simulation program to determine dynamic system behavior and integrated performance. Comparisons of physical measurements and simulations are discussed, and problems and limitations of simulations are noted.

19.1 SIMULATION PROGRAMS

The use of simulation methods in the study of solar processes is a relatively recent development. Sheridan et al. (1967) used an analog computer in simulation studies of operation of solar water heaters. Gupta and Garg (1968) developed a model for thermal performance of a natural circulation solar water heater with no load, represented solar radiation and ambient temperature by Fourier series, and were able to predict a day's performance in a manner that agreed substantially with experiments. Close (1967) used numerical modeling and a factorial design method to determine which water heater system design factors are most important. Gupta (1971) used a response factor method that is amendable to hand calculation for short-term process operation. Buchberg and Roulet (1968) developed a thermal model of a house heating system, simulated its operation with a year's hourly meteorological data, and applied a pattern search optimization procedure in finding optimum designs. Other process simulations have been done by Löf and Tybout (1973), Butz et al. (1974) and Oonk et al. (1975). Since these publications, solar process simulation programs have come into widespread use.

Some of the programs that have been applied to solar processes have been written specifically for study of solar energy systems. Others were intended for nonsolar applications but have had models of solar components added to them to make them useful for solar problems. The common thread in them is the ability to solve the combinations of algebraic and differential equations that represent the physical behavior of the equipment. Simulation programs fall into two general categories. The first includes those that are "special purpose" programs that represent the performance of specific types of systems. In these programs, the equations for the components are combined algebraically to simplify the computations; they are generally easy to use but are not flexible. Programs in the second category, the general purpose programs, are more flexible and can be applied to a wide range of systems, but are more difficult to use. In these programs, the equations representing components (collectors, storage units, pumps, etc.) are kept separate and are solved simultaneously rather than being combined algebraically.

WATSUN, a program developed at the University of Waterloo in Canada (1989), models a large number of active systems and includes stratified and nonstratified tanks, an economics program, and many other features. MINSUN, developed under International Energy Agency auspices [Chant (1985)], is specifically designed for simulation of systems with seasonal storage where daily temperature changes are small; one-day time steps are used. G^3, from Switzerland [Guisan et al. (1988)], is a simplified program that is based on daily input-output diagrams; it is designed for fast and simple evaluation of active systems. ISFH, from West Germany [Schreitmüeller (1988)], is a versatile program based on combining detailed component subroutines; it includes a means of generating a "synthetic climate" from monthly average weather data.

EMPG2 (European Modelling Group Program 2) is a modular simulation program that has the capability of handling a wide variety of transient thermal systems, system control strategies, and load types. It is extensively documented [Dutre (1985)].

Simulation programs have been written for design of passive systems. Examples are PASOLE and SERIRES. A review of use of computers in the design of passive systems [Klein (1983)] outlines many of these efforts to that date.

These programs evolve with time, and their exact nature can only be recorded as of a particular time. TRNSYS, the program to be described below, is a public domain program that in 1990 was available in version 13.1; it has undergone eight major revisions and many more minor changes and is a far more powerful program than it was when it first became available in 1977.

19.2   THE UTILITY OF SIMULATIONS

Simulations are numerical experiments and can give the same kinds of thermal performance information as can physical experiments. They are, however, relatively quick and inexpensive and can produce information on effect of design variable changes on system performance by series of experiments all using exactly the same loads and weather. These design variables could include selectivity of the absorbing surface, number of covers on the collector , collector area, and so on. With cost data and appropriate economic analysis, simulation results can be used to find least cost systems.

Simulations are uniquely suited to parametric studies and thus offer to the process designer the capability to explore the effects of design variables on long-term system performance. They offer the opportunity to evaluate effects of system configuration and alternative system concepts. They have the advantage that the weather used to drive them is reproducible, allowing parametric and configuration studies to be made without uncertainties of variable weather. By the same token, a system can be "operated" by simulations in a wide range of climates to determine the effects of weather on design.

Simulations are complementary to physical experiments. Component-scale experiments are necessary to understand component behavior and lend confidence to the corresponding mathematical models. System-scale experiments are necessary to bring to light the many practical problems inherent in any complicated system that simulations cannot fully model. Careful comparisons of experiments and simulations lead to improved understanding of each. Once simulations have been verified with experiments, new systems can be designed with confidence using simulation methods.

In principle, all of the physical parameters of collectors, storage, and other components are the variables that need to be taken into account in the design of solar processes. The number of parameters that must be considered may be quite small, as there is a backlog of experience which indicates that the sensitivity of long-term performance or process economics to many parameters is small. For application to heating of a building, for example, the primary system design variable is collector area, with storage capacity and other design variables being of secondary importance provided they are within reasonable bounds of good design practice (as noted in Chapter 13). Simulations can provide information on effects of collector area (or other variables) which is essentially impossible to get by other means.

19.3   INFORMATION FROM SIMULATIONS

The parallels between numerical experiments (simulations) and physical experiments are strong. In principle, it is possible to compute what it is possible to measure. In practice, it may be easier to compute than to measure some variables (for example, temperatures in parts of a system that are inaccessible for placement of temperature sensors). Simulations can be arranged to subject systems to extremes of weather, loads, or other external forces. They also can be used to interject faults in a system (e.g., a failure in a circulating pump) to see what the effects of the fault would be.

There are two basic kinds of data that can be obtained from simulations. First, integrated performance over extended periods can be determined. This is normally wanted for a year that represents the long-term average climate in which a proposed process would operate. A year is the time base of most economic studies, but information may be needed for other periods from days to spans of many years. These data are readily obtained by integrating the desired quantities (collector output, tank losses, pump or blower on time, heat exchange across an exchanger, amount of auxiliary energy required, etc.) over the appropriate time period.

Second, information on process dynamics is available. The selection of materials of construction may be affected by the temperature excursions through which the component goes in either normal or abnormal operation. Pressure on liquid systems is determined by temperatures in the various parts of the liquid loops. For example, temperatures in the tops of storage tanks in systems with collector heat exchangers may be moderate, but temperatures in the collector outlets may be well above the boiling point, and pressure relief valves may be needed on collector outlets as well as on tanks.

A year's operation of a solar heating system for a residence in the Madison WI climate has been simulated. The building has 150 m² floor area, is well insulated, and is to have a solar heating and hot water system to supply part of the heating load. Collector area is the primary design variable. Values of other design parameters were selected to represent good design practice, including collector characteristics and ratio of storage capacity to collector area. Integration of delivered solar energy and auxiliary energy required at several collector areas give information on the solar fraction $\mathcal{F}$, the fraction of annual heating and hot water loads carried by solar energy, that is, on the amount of energy usefully delivered to the building from the solar energy system. This is shown in Figure 19.3.1(a). A similar curve for water heating only is shown in Figure 19.3.1(b); it is more nearly linear over a wide range of collector areas (primarily because the loads and solar radiation are more evenly matched throughout the whole year in this application).

An example of temperature variation with time is shown in Figure 19.3.2 for a selected 2-week period for two collector areas for a space and water heating system. These data, if examined for the whole year (or for those periods in the year when temperatures are likely to go to extremes), will provide indication of the magnitude of safety problems which must be addressed by providing means to "dump" energy.

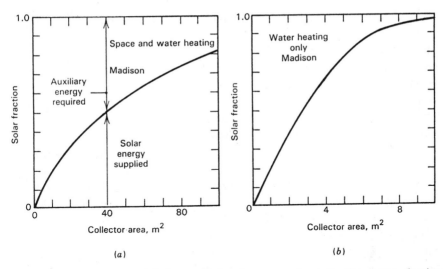

Figure 19.3.1 (a) The thermal performance of a solar space and water heating system, showing the fraction of annual heating and hot water loads carried by solar energy as a function of collector area. (b) The thermal performance of a solar water heating system, showing the fraction of the annual hot water loads carried by solar energy as a function of collector area.

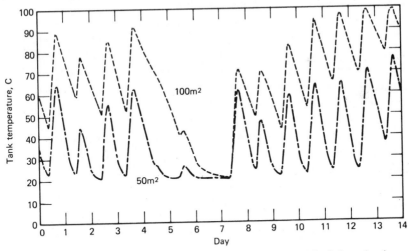

Figure 19.3.2 Tank temperature vs time for a 2-week period in March for a heating and hot water system for Madison, WI, assuming a fully mixed tank, for two collector areas and fixed storage-collector area ratio.

Other kinds of information can also be obtained. It is possible to estimate the times at which auxiliary energy is needed and the relationship of those times to meteorological conditions. It is possible to estimate storage unit losses. In general, any variable that appears in the set of system equations can be investigated.

19.4 TRNSYS, A THERMAL PROCESS SIMULATION PROGRAM

TRNSYS [Klein et al. (1975, 1976, 1990)] is a widely used, modular thermal process simulation program. Originally developed for solar energy applications, it now is used for simulation of a wider variety of thermal processes. Subroutines are available that represent the components in typical solar energy systems. A list of the components and combinations of components in the TRNSYS library (as of 1990) is shown in Table 19.4.1. Users can readily write their own component subroutines if they are not satisfied with those provided. By a simple language, the components are "connected" together in a manner analogous to piping, ducting, and wiring in a physical system. The programmer also supplies values for all of the parameters describing the components to be used. The program does the necessary simultaneous solutions of the algebraic and differential equations which represent the components and organizes the input and output. Varying levels of complexity can be used in the calculation. For example, a flat-plate collector can be represented by constant values of $F_R U_L$ and $F_R(\tau\alpha)$, or it can be represented by values of U_L and $(\tau\alpha)$ which are calculated each time step for the conditions that change through time as the simulation proceeds. The user of the program must determine how detailed he wishes his simulation to be; the more detail, the higher the cost in terms of programming effort and computer time. Most of the simulation studies reported in preceding chapters on specific solar application were done with TRNSYS.

Current versions of TRNSYS have, in the executive program, convergence promoters and other means of speeding computations. There are three integration algorithms in TRNSYS; the user can choose the one best suited to the problem at hand. The one that is extensively used is the Modified-Euler method. It is essentially a first-order predictor-corrector algorithm using Euler's method for the predicting step and the trapezoid rule for the correcting step. The advantage of a predictor-corrector integration algorithm for solving simultaneous algebraic and differential equations is that the iterative calculations occurring during a single time step are performed at a constant value of time. (This is not the case for the Runge-Kutta algorithms.) As a result, the solutions to the algebraic equations of the system converge, by successive substitution, as the iteration required to solve the differential equation progresses. The calculation scheme can be described in the following manner.

At time t, the values of the dependent variables T^p are predicted using their values and the values of their derivatives, $(dT/dt)_0$, from the previous time step:

$$T^p = T_0 + \Delta t \left(\frac{dT}{dt}\right)_0 \tag{19.4.1}$$

where T^p is the predicted value of all of the dependent variables at time t (note that this prediction step is exactly the method used to integrate Example 19.3.1); T_0 is the value of the dependent variables at time $t - \Delta t$; Δt is the time step interval at which solutions to the equations of the system model will be obtained; and $(dT/dt)_0$ is the value of the derivative of the dependent variables at time $t - \Delta t$.

Table 19.4.1 Components in the Standard Library of TRNSYS 13

Utility Components	Building Loads and Structures
Data reader	Energy/(degree-hour) house
Time-dependent forcing function	Detailed zone (transfer function)
Algebraic operator	Roof and attic
Radiation processor	Overhang and wingwall shading
Quantity integrator	Window
Psychrometrics	Thermal storage wall
Load profile sequencer	Attached sunspace
Collector array shading	Multizone building
Convergence promoter	
Weather data generator	**Hydronics**
	Pump/fan
Solar Collectors	Flow diverter/mixing valve/tee piece
	Pressure relief valve
Linear thermal efficiency data	Pipe
Detailed performance map	
Single or biaxial incidence angle modifier	**Controllers**
Theoretical flat-plate	
Theoretical CPC	Differential controller with hysteresis
	Three-stage room thermostat
Thermal Storage	Microprocessor controller
Stratified liquid storage (finite difference)	
Algebraic tank (plug flow)	**Output**
Rockbed	
	Printer
Equipment	Plotter
	Histogram plotter
On/off auxiliary heater	Simulation summarizer
Absorption air conditioner	Economics
Dual-source heat pump	
Conditioning equipment	**User-Contributed Components**
Cooling coil	
Cooling tower	PV/thermal collector
Chiller	Storage battery
	Regulator/inverter
Heat Exchangers	Electrical subsystem
Heat exchanger	
Waste heat recovery	**Combined Subsystems**
Utility Subroutines	Liquid collector-storage system
	Air collector-storage system
Data interpolation	Domestic hot water system
First order differential equations	Thermosyphon solar water heater
View factors	
Matrix inversion	
Least squares curve fitting	
Psychrometric calculations	

The predicted values of the dependent variables, T^p, are then used to determine corrected values, T^c, by evaluating their derivatives, (dT/dt), as a function t, T^p, and the solutions to the algebraic equations of the model:

$$\frac{dT}{dt} = f(t, T^p, \text{algebraic solutions}) \tag{19.4.2}$$

The corrected values of the dependent variables, T^c, are obtained by applying the trapezoidal rule:

$$T^c = T_0 + \frac{\Delta t}{2}\left[\left(\frac{dT}{dt}\right)_0 + \frac{dT}{dt}\right] \tag{19.4.3}$$

If $\dfrac{2(T^c - T^p)}{T^c + T^p} > \varepsilon$ $\tag{19.4.4}$

where ε is an error tolerance, T^p is set equal to T^c, and application of Equations 19.4.2 and 19.4.3 is repeated. When the error tolerance is satisfied, the solution for that time step is complete and the whole process is repeated for the next time step.

As an illustration of the use of the general program TRNSYS for simulating a solar energy system and the nature of the results that can be obtained by simulations, consider the following example.

Example 19.4.1

A hot water load of 3000 kg water/day at a minimum temperature of 60 C is evenly distributed between the hours of 0700 and 2100. This load is to be met in substantial part by a solar collector assembly of total effective area 65.0 m². The two-cover collector has the following characteristics:

> Tilt $= \beta = 40°$ to south
> $U_L = 4.0$ W/m²C
> $(\tau\alpha) = 0.77$
> $F' = 0.95$
> Flow rate through collector $= \dot{m}_c = 0.903$ kg/s

The tank has the following characteristics:

> V = 3.9 m³
> Height-diameter ratio = 3
> Loss coefficient $U_L = 0.40$ W/m²C
> Ambient temperature at tank = 21 C
> The supply water to the tank = 15 C

The auxiliary heater is controlled such that if the temperature of the water from the tank is less than 60 C, it will heat the water from the storage tank temperature to 60 C. If T_s exceeds 60 C, the water delivered to meet the load will be at a temperature above 60 C.

The system shown in the diagram that follows is to be operated at Boulder, CO, latitude 40°N, for a week in January. Hourly values of solar radiation and ambient temperature are as shown in the plots and are the same data as given in Table 2.5.2. Assuming that the initial tank temperature is 60 C, compute the percentage of the load that is carried by solar energy.

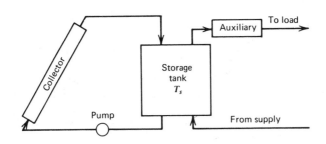

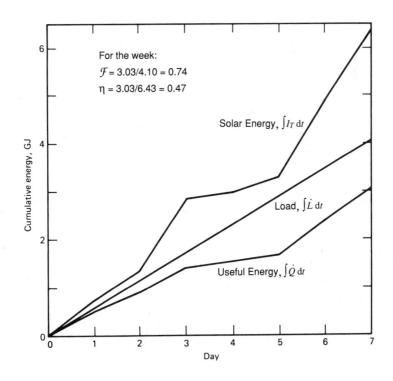

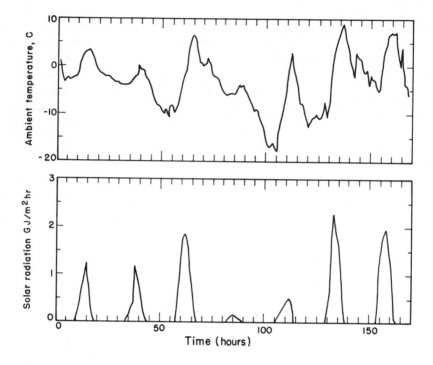

Solution

Solution to this problem were obtained using TRNSYS. A summary of the results is given in the table below. Two solutions are shown, one for an unstratified storage tank and one for a three-node approximation to a stratified storage tank. The load requirement is a fixed amount of water at a temperature of at least 60 C; the two energy deliveries are slightly different, as the stratified tank delivers water at slightly higher temperature. The minimum total load is 3.96 GJ, and both systems slightly exceed this since the delivery temperature sometimes exceeds 60 C. The percentages of the total energy supplied to the water by solar energy for the two cases are $(2.74/4.04)100 = 68\%$ and $(3.03/4.10)100 = 74\%$. The amount of auxiliary energy that is required in the two cases is 1.30 GJ for the unstratified tank and 1.07 GJ for the three-node tank. If a conventional water heater were designed to deliver exactly the minimum energy required (3.96 GJ), the savings in purchased energy would have been 2.66 and 2.89 GJ, respectively.

Plots of cumulative energy quantities versus time for the week are shown for the case of the fully mixed tank. The points for the totals at the end of each day are connected by straight lines; smoother plots could have been obtained by using smaller time intervals. The tables are summaries of integrated energy quantities. All energy quantities are in GJ.

Case 1   Unstratified Storage Tank

End of Day	Cumulative Energy			Change of Energy of Tank	Cumulative Energy		
	Incident Solar	Useful Gain	Tank Loss		Supplied from Tank	Supplied from Auxiliary	Total Load
1	0.71	0.21	0.02	−0.27	0.46	0.10	0.56
2	1.34	0.43	0.03	−0.38	0.78	0.35	1.13
3	2.84	1.17	0.04	−0.11	1.24	0.47	1.71
4	2.95	1.17	0.05	−0.45	1.57	0.71	2.28
5	3.34	1.28	0.05	−0.53	1.76	1.00	2.76
6	4.87	2.09	0.07	−0.13	2.15	1.26	3.41
7	6.43	2.81	0.08	−0.01	2.74	1.30	4.04

Case 2   Three-section Storage Tank

End of Day	Cumulative Energy			Change of Energy of Tank	Cumulative Energy		
	Incident Solar	Useful Gain	Tank Loss		Supplied from Tank	Supplied from Auxiliary	Total Load
1	0.71	0.24	0.02	−0.30	0.52	0.05	0.57
2	1.34	0.50	0.03	−0.42	0.89	0.25	1.14
3	2.84	1.28	0.04	−0.15	1.39	0.34	1.73
4	2.95	1.28	0.05	−0.58	1.81	0.50	2.31
5	3.34	1.42	0.05	−0.60	1.97	0.90	2.87
6	4.87	2.27	0.06	−0.18	2.39	1.06	3.45
7	6.43	3.05	0.08	−0.06	3.03	1.07	4.10

19.5   SIMULATIONS AND EXPERIMENTS

The extent to which simulations represent the operation of real physical systems depends on the level of detail included in the numerical experiment. Component models can vary in complexity, as can systems. In principle, simulations can be as detailed as the user wishes. In practice, there may be factors in system operation which are difficult to simulate, such as leaks in air systems, and operation of real systems may be less ideal than the simulation indicates.

Many component experiments have been carried out, for example, the kind noted in Chapter 6 on flat-plate collectors. Some comparisons of simulations and experiments of building heating have been made of both short-term process dynamics and long-term integrated performance. For example, Mitchell et al. (1980) have used tank temperatures and integrated energy quantities as indicators of process operation and compared measured and computed tank temperatures and energy quantities over several 10-day periods from CSU House I. An example of a tank temperature comparison is shown in Figure 19.5.1.

Comparisons of the simulated and measured useful energy gain from the collector and solar auxiliary contribution are shown in Table 19.5.1. In this study, the simulation was very carefully done to closely represent the physical situation; by doing so, it was possible to obtain agreements that are comparable to the accuracy of the measurements on the physical system.

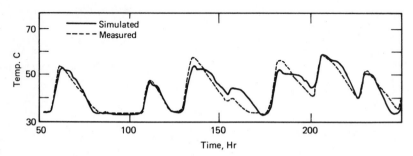

Figure 19.5.1 Comparisons of measured and predicted storage tank temperatures for Colorado State University House I. Adapted from Pawelski (1976).

Table 19.5.1 **Comparison of Simulated and Measured Energy Quantities from CSU House I from Three Different Time Periods[a]**

	Period	Measured	Simulated	Difference (%)
Collected solar	I	2388	2577	7.0
energy (MJ)	II	2419	2292	5.2
	III	2086	2012	3.5
Air heater heat	I	2076	2041	1.4
flow (MJ)	II	3243	3238	0.0
	III	1810	1736	3.6
Delivered solar	I			
energy (MJ)	II	1952	2025	3.1
	III	1517	1573	2.7
Auxiliary (MJ)	I	0	0	0.0
	II	1291	1213	3.3
	III	238	162	3.7
Preheater heat	I	398	303	3.8
flow (MJ)	II	132	116	0.7
	III	204	132	3.5

[a] From Mitchell et al. (1980).

19.6   METEOROLOGICAL DATA

Weather conditions and loads can be considered as forcing functions operating on the sets of equations that describe system performance. As noted in Chapter 9, loads may be functions of weather for building heating and cooling or they may be determined by factors not related to weather. Meteorological data, including solar radiation, ambient temperature, and wind speed, influence collector performance, and information such as that in Table 2.5.2 is needed to calculate system performance over time.

Most simulations are done with past meteorological data, and it is necessary to select a data set to use in simulations. For studies of process dynamics, data for a few days or weeks may be adequate if they represent the range of conditions of interest. For design purposes it is best to use a full year's data or a full season's data if the process is a seasonal one. Data are available for many years for some stations, and it is necessary to select a satisfactory set. Klein (1976) developed the concept of a **design year**. He used heating season data for 9 months for each of 8 years. For each of the months (e.g., the Januarys) the month was selected that had the radiation closest to the 8-year average. Monthly average temperatures were used as a secondary criterion where needed. The set of months so selected constituted the design year. Discontinuities between months normally cause no difficulty. The selection of the design year was evaluated by simulations. Table 19.6.1 shows the results of this procedure and indicates the annual fraction of heating loads carried by solar energy for two collector areas for a particular building, for the design year and for the full 8 years. The design year provides a good representation of the 8 year period, at least for purposes of simulating solar heating systems.

Table 19.6.1   The Design (Heating Season) Year for Madison, Based on 8 Years of Data (1948-1949 to 1955-1956)[a]

Month	Year	Monthly Solar Fractions Area 1	Area 2
Sept.	51-52	0.97	1.00
Oct.	55-56	0.79	0.98
Nov.	49-50	0.35	0.63
Dec.	49-50	0.26	0.49
Jan.	53-54	0.23	0.43
Feb.	54-55	0.36	0.66
Mar.	53-54	0.53	0.85
Apr.	55-56	0.72	0.96
May	52-53	0.77	0.98
Annual Contribution		0.47	0.69
Annual Contribution for 8 Years		0.47	0.67

[a] From Klein (1976).

A related but more detailed study by Hall et al. (1978) of 23 years of data for each of 26 stations in the U.S. solar radiation network has led to the generation of **typical meteorological year** (TMY) tapes for these locations. The TMY data for Madison have been used in heating system simulations and the results compared to those based on the full 23 years of data; in one comparison the TMY data indicated a solar contribution of 0.60, and the value for the full 23 year simulation was 0.58.

Caution must be exercised in the use of typical years if the process to be designed is to provide a high fraction of the loads by solar energy. Years that are far from the average years tend to include unusual sequences or extremes of weather, and the results of simulations based on the typical year data may be significantly different from the long-term average for systems with high $\mathcal{F}$. Also, year-to-year variation in weather will lead to significant year-to-year variation in a solar process output, and a performance prediction based on a typical year will not apply for a specific year unless that year happens to be nearly the same as the typical year.

Another approach to the use of existing meteorological data is to synthesize from it data for shorter periods which, when used in simulations, will provide information on longer periods of operation. For example, Anand et al. (1978) have worked out a two-step procedure for condensing data. First, they statistically arrange the data (insolation and temperature) to place it in bins (ranges of data pairs). For these data pairs, they curve fit expressions for diurnal variation of insolation and temperature to get a small number of synthesized days which are representative of a large number of actual days.

Another approach to development of meteorological data is to synthesize sequences of hours and days, starting from the means, with the synthesized sequences having the same means, the same distributions, the same autocorrelations, and (ultimately) the same cross correlations between one meteorological variable and another as the real data. This procedure has the advantage that it would allow generation of data for locations for which only monthly means are available. Based on the original work of Degelman (1976), Knight et al. (1991) used a combination of autoregression models for solar radiation and ambient temperature to generate data that have statistics comparing favorably with TMY data.

Information on time dependence of loads must also be available if simulations are to be done. The general approach to determination of loads is the same as other parts of a system, i.e., develop the set of equations or numerical data that relate energy rates and temperatures to time. These load characteristics are then solved as part of the overall set of equations that describe the solar process.

The usual situation is to have the loads on a nonsolar system the same as those on a solar energy system designed for the same task. However, there are many systems where the total amount of energy to be supplied is influenced by the solar energy system. For example, solar water heating systems with tanks larger than conventional tanks will have larger losses and consequently increased loads. Passively heated buildings will have loads that depend on the size of the solar aperture, as the heat losses per unit area through aperture will in general be more than through well-insulated walls.

19.7 LIMITATIONS OF SIMULATIONS

Simulations are powerful tools for process design, for study of new processes, and for understanding how existing systems function and might be improved. However, there are limits to what can be done with them.

First, there is implicit in this discussion of simulations the assumption that they are properly done. It is easy to make program errors, assume erroneous constants, neglect factors which may be important, and err in a variety of other ways. Like other engineering calculations, a high level of skill and judgment is required in order to produce useful results.

As noted above, it is possible, in principle, to model a system to whatever degree is required to extract the desired information. In practice, it may be difficult to represent in detail some of the phenomena occurring in a system. Physical world realities include leaks, plugged or restricted pipes, scale on heat exchangers, failure of controllers, poor installation of equipment, etc. The simulations discussed here are of the thermal processes; mechanical and other considerations can affect the thermal performance of systems.

There is no substitute for carefully conceived and carefully executed experiments. Such experiments will reveal whether or not the theory is adequate and where difficulties lie in design and operation of the systems. At its best, a combination of numerical experiment (simulation) and physical experiment will lead to better systems, better understanding of how processes work, better knowledge of what difficulties can be expected and what can be done about them, and what next logical steps should be taken in the evolution of new systems.

Simulations and development laboratory experiments are complementary. Comparisons of the results of measurements in the field of performance of purchased and installed systems with simulations has in some instances shown greater differences than those with experiments. The reasons are two. First, field measurements are often very difficult to make, and differences may be ascribed to poor measurements. Second, commercially installed systems are not always built and operated with the same care and knowledge as laboratory systems, and they may not work as well as laboratory systems.

In brief, simulations are powerful tools for research and development, for understanding how systems function, and for design. They must, however, be done with care and skill.

REFERENCES

Anand, D. K., I. N. Dief, and R. W. Allen, Paper 78-WA/Sol-16 , ASME, San Francisco (1978). "Stochastic Predictions of Solar Cooling System Performance."

Buchberg, H. and J. R. Roulet, *Solar Energy*, **12**, 31 (1968). "Simulation and Optimization of Solar Collection and Storage for House Heating."

Butz, L.W., W. A. Beckman, and J. A. Duffie, *Solar Energy*, **16**, 129 (1974). "Simulation of a Solar Heating and Cooling System.

Chant, V. *Central Solar Heating Plants with Seasonal Storage - the MINSUN Simulation and Optimization Program Application and User's Guide*, International Energy Agency (IEA) Solar Heating and Cooling Task VII, National Research Council of Canada, Ottawa, Document No. CENSOL2, (1985).

Close, D. J., *Solar Energy*, **11**, 112 (1967). "A Design Approach for Solar Processes."

Degelman, L. O., *ASHRAE Trans., Symposium on Weather Data*, Seattle, WA, p. 435 (1976). "A Weather Simulation Model for Building Energy Analysis."

Dutre, W. L., *A European Transient Simulation Model for Thermal Solar Systems*, D. Reidel Publishing, Dordrecht (1985) (for the Commission of the European Communities).

Guisan, O., B. Lachal, A. Mermouid, and O. Rudaz, in *Advances in Solar Energy Technology, the Proc. of the 1987 SES Congress*, Hamburg, FRG, **1**, 848 (1988). "G-3 Model, A Simple Model for the Evaluation of the Thermal Output of a Solar Collection System on a Daily Basis."

Gupta, C. L., *Solar Energy*, **13**, 301 (1971). "On Generalizing the Dynamic Performance of Solar Energy Systems."

Gupta, C. L. and H. P. Garg, *Solar Energy*, **12**, 163 (1968). "System Design in Water Heaters with Natural Circulation."

Hall, I. J., R. R. Prairie, H. E. Anderson, and E. C. Boes, *Proc. of 1978 Annual Meeting, American Section of ISES*, Denver, **2**(2), 669 (1978). "Generation of a Typical Meteorological Year.

Klein, S. A., Ph.D. Thesis, University of Wisconsin-Madison (1976). "A Design Procedure for Solar Heating Systems."

Klein, S. A., *Passive Solar J.*, **2**, 57 (1983). "Computers in the Design of Solar Energy Systems"

Klein, S. A. and W. A. Beckman, *ASHRAE Trans.*, **82**, 623 (1976). "TRNSYS – A Trasnsient Simulation Program."

Klein, S. A., P.I. Cooper, T. L. Freeman, D. M. Beekman, W. A. Beckman, and J. A. Duffie, *Solar Energy*, **17**, 29 (1975). "A Method of Simulation of Solar Processes and Its Application."

Klein, S. A. et al., *TRNSYS Users Manual*, Version 13, EES Report 38, U. of Wisconsin Engineering Experiment Station (1990). (The first public version was 7 (1976).)

Knight, K. M., S. A. Klein, and J. A. Duffie, *Solar Energy*, **46**, 109 (1991). "A Methodology for the Synthesis of Hourly Weather Data."

Löf, G. O. G. and R. A. Tybout, *Solar Energy*, **14**, 253 (1973). "Cost of House Heating with Solar Energy."

Mitchell, J. W., W. A. Beckman, and M. J. Pawelski, *Trans. ASME, J. Solar Energy Engrg.*, **102**, 192 (1980). "Comparisons of Measured and Simulated Performance for CSU Solar House I."

Oonk, R. L., W. A. Beckman, and J. A. Duffie, *Solar Energy*, **17**, 21 (1975). "Modeling of the CSU Heating/Cooling System."

Pawelski, M. J., M.S. Thesis, Mechanical Engineering, U. of Wisconsin-Madison (1976). "Development of Transfer Function Load Models and Their Use in Modeling the CSU House I."

Schreitmüeller, K. R., in *Advances in Solar Energy Technology, the proc. of the 1987 SES Congress*, Hamburg, FRG, **1**, 828 (1988). "Modeling of Active Solar Energy Systems with Holistic Characteristics."

Sheridan, N. R., K. J. Bullock, and J. A. Duffie, *Solar Energy*, **11**, 69 (1967). "Study of Solar Processes by Analog Computer."

WATSUN Users Manual and Program Documentation, Version 11.1, WATSUN Simulation Laboratory, University of Waterloo, Canada (1989).

Chapter 20

DESIGN OF ACTIVE SYSTEMS: f-CHART

The liquid and air system configurations described in Section 13.2 are common configurations, and there is considerable information and experience on which to base designs. For residential-scale applications, where the cost of the project does not warrant the expense of a simulation, performance predictions can be done with "short-cut" methods. Design procedures are available for many of these systems that are easy to use and provide adequate estimates of long-term thermal performance. In this chapter we briefly note some of these methods. The f-chart method, applicable to heating of buildings where the minimum temperature for energy delivery is approximately 20 C, is outlined in detail. Methods for designing systems delivering energy at other minimum temperatures, as are encountered solar absorption air conditioning or industrial process heat applications, are presented in Chapter 21.

20.1 REVIEW OF DESIGN METHODS

Design methods for solar thermal processes can be put in three general categories, according to the assumptions on which they are based and the ways in which the calculations are done. They produce estimates of annual useful outputs of solar processes, but they do not provide information on process dynamics.

The first category applies to systems in which the collector operating temperature is known or can be estimated and for which critical radiation levels can be established. The first of these, the utilizability methods, are based on analysis of hourly weather data to obtain the fraction of the total month's radiation that is above a critical level.[1] Another example in this category is the heat table method of Morse as described by Proctor (1975). This is a straightforward tabulation of integrated collector performance as a function of collector characteristics, location, and orientation, assuming fixed fluid inlet temperatures.

The second category of design methods includes those that are correlations of the results of a large number of detailed simulations. The f-chart method of Klein et al. (1976, 1977) and Beckman et al. (1977) is an example. The results of many numerical experiments (simulations) are correlated in terms of easily calculated

---

[1] This method and a further development are outlined in Chapter 21.

dimensionless variables. The results of the f-chart method have served as the basis for further correlations, e.g., by Ward (1976), who has used only January results to characterize a year's system operation; by Barley and Winn (1978), who used a two-point curve fit to obtain location dependent annual results; and by Lameiro and Bendt (l978), who also obtained location-dependent annual results with three-point curve fits. The SEU (Solar Energy Unit of University College Cardiff) methods of Kenna (1984a,b) are correlation methods which are applicable to designing open-loop and closed-loop heating systems. Another example in the second category is the method of Los Alamos Scientific Laboratory [Balcomb and Hedstrom (1976)], which is a correlation of the outputs of simulations for specific systems and two collector types.

The third category of design methods is based on short-cut simulations. In these methods, simulations are done using representative days of meteorological data and the results are related to longer term performance. The SOLCOST method [Connelly et al. (1976)] simulates a clear day and a cloudy day and then weights the results according to average cloudiness to obtain a monthly estimate of system performance.

20.2 THE f-CHART METHOD

This and the following sections outline the f-chart method for estimating the annual thermal performance of active heating systems for buildings (using either liquid or air as the working fluid) where the minimum temperature of energy delivery is near 20 C. The system configurations that can be evaluated by the f-chart methods are expected to be common in residential applications.

The f-chart method provides a means for estimating the fraction of a total heating load that will be supplied by solar energy for a given solar heating system. The primary design variable is collector area; secondary variables are collector type, storage capacity, fluid flow rates, and load and collector heat exchanger sizes. The method is a correlation of the results of many hundreds of thermal performance simulations of solar heating systems. The conditions of the simulations were varied over appropriate ranges of parameters of practical system designs. The resulting correlations give f, the fraction of the monthly heating load (for space heating and hot water) supplied by solar energy as a function of two dimensionless parameters. One is related to the ratio of collector losses to heating loads, and the other is related to the ratio of absorbed solar radiation to heating loads.

The f-charts have been developed for three standard system configurations, liquid and air systems for space (and hot water) heating and systems for service hot water only. A schematic diagram of the standard heating system using liquid heat transfer fluids is shown in Figure 20.2.1. This system normally uses an antifreeze solution in the collector loop and water as the storage medium. Collectors may be drained when energy is not being collected, in which case water is used directly in the collectors and a collector heat exchanger is not needed. A water-to-air load heat exchanger is used to transfer heat from the storage tank to a domestic hot water sub-

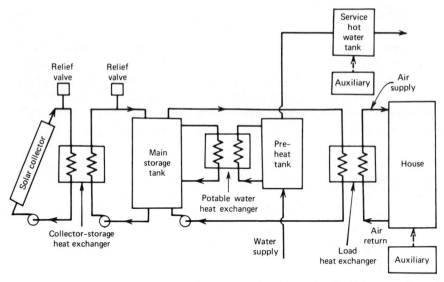

Figure 20.2.1   Schematic of the standard system configuration using liquid heat transfer and storage media.

Table 20.2.1   Ranges of Design Parameters Used in Developing the *f*-Charts for Liquid Systems[a]

0.6	<	$(\tau\alpha)_n$	<	0.9
5	<	$F'_R A_c$	<	120 m^2
2.1	<	U_L	<	8.3 W/m^{2}C
30	<	β	<	90°
83	<	$(UA)_h$	<	667 W/C

[a] From Klein et al. (1976).

system. Although Figure 20.2.1 shows a two-tank domestic hot water system, a one-tank system could be used as described in Section 12.4. An auxiliary heater is provided to supply energy for the space heating load when it cannot be met from the tank. The ranges for major design variables used in developing the correlations are given in Table 20.2.1.

The standard configuration of a solar air heating system with a pebble bed storage unit is shown in Figure 20.2.2. Other equivalent arrangements of fans and dampers can be used to provide the same modes of operation. Energy required for domestic hot water is provided by an air-to-water heat exchanger in the hot air duct leaving the collector. During summer operation, it is best not to store solar energy in the pebble bed, and a manually operated storage bypass is usually provided in this system to allow summer water heating. The ranges of design parameters used in developing the correlations for this system are shown in Table 20.2.2.

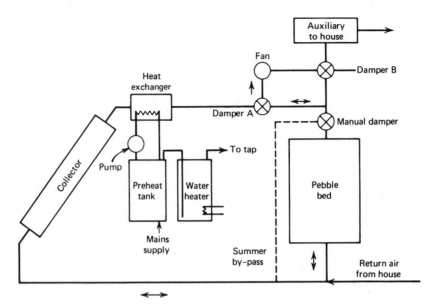

Figure 20.2.2 The standard air system configuration.

**Table 20.2.2   Ranges of Design
Parameters Used in Developing the
f-Chart for Air Systems[a]**

0.6	<	$(\tau\alpha)_n$	<	0.9
5	<	$F_R'A_c$	<	120 m^2
2.1	<	U_L	<	8.3 W/m^{2}C
30	<	β	<	90^0
83	<	$(UA)_h$	<	667 W/C

[a] From Klein et al. (1976).

The standard configuration for a solar domestic water heating system is shown in Figure 20.2.3. The collector may heat either air or liquid. The solar energy is transferred via a heat exchanger to a domestic hot water (DHW) preheat tank, which supplies solar-heated water to a conventional water heater or an in-line low-capacitance "zip" heater where the water is further heated to the desired temperature if necessary. A tempering valve may be provided to maintain the tap water below a maximum temperature. These changes in the system configuration do not have major effects on the performance of the system (see Section 12.4).

Detailed simulations of these systems have been used to develop correlations between dimensionless variables and f, the monthly fraction of loads carried by

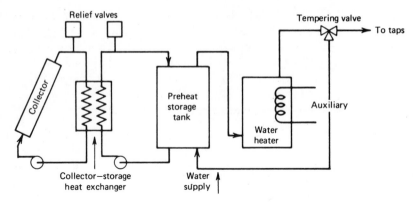

Figure 20.2.3   The standard system configuration for water heating only. The collector may heat air or water.

solar energy. The two dimensionless groups are

$$X = \frac{A_c F'_R U_L (T_{ref} - \overline{T}_a) \Delta t}{L} \tag{20.2.1}$$

$$Y = \frac{A_c F'_R (\overline{\tau\alpha}) \overline{H}_T N}{L} \tag{20.2.2}$$

where A_c =   collector area (m²)
F'_R =   collector-heat exchanger efficiency factor[2]
U_L =   collector overall loss coefficient (W/m²C)
Δt =   total number of seconds in the month
$\overline{T}_a$ =   monthly average ambient temperature (C)
T_{ref} =   an empirically derived reference temperature (100 C)
L =   monthly total heating load for space heating and hot water (J)
$\overline{H}_T$ =   monthly average daily radiation incident on the collector surface per unit area (J/m²)
N =   days in month
$(\overline{\tau\alpha})$ =   monthly average transmittance-absorptance product

Equations 20.2.1 and 20.2.2 can be rewritten as

$$X = F_R U_L \times \frac{F'_R}{F_R} \times (T_{ref} - \overline{T}_a) \times \Delta t \times \frac{A_c}{L} \tag{20.2.3}$$

$$Y = F_R (\tau\alpha)_n \times \frac{F'_R}{F_R} \times \frac{(\overline{\tau\alpha})}{(\tau\alpha)_n} \times \overline{H}_T N \times \frac{A_c}{L} \tag{20.2.4}$$

[2] Although we indicate only a modification to F_R to account for the collector-storage heat exchanger, both $F_R(\tau\alpha)$ and $F_R U_L$ can be modified to account for the collector heat exchanger, duct losses, and flow arrangements. (See Chapter 10.)

$F_R U_L$ and $F_R(\tau\alpha)_n$ are obtained from collector test results by the methods noted in Chapter 6. The ratio F'_R/F_R corrects for various temperature drops between the collector and the storage tank and is calculated by methods summarized in Chapter 10. The ratio $(\overline{\tau\alpha})/(\tau\alpha)_n$ is estimated by the methods noted in Section 5.10. The average air temperature $\overline{T}_a$ is obtained from meteorological records for the month and location desired, and $\overline{H}_T$ is found from the monthly average daily radiation on the surface of the collector as outlined in Chapter 2. The calculation of monthly loads L is discussed in Chapter 9. (There is no requirement in *f*-Chart that any particular method be used to estimate the loads.) The collector area is A_c. Thus all of the terms in these two equations are readily determined from available information.

Example 20.2.1

A solar heating system is to be designed for Madison, WI (latitude 43°N), using one-cover collectors with $F_R(\tau\alpha)_n = 0.74$ and $F_R U_L = 4.00$ W/m²C as determined from standard collector tests. The flow rate in use will be the same as that in the tests. The collector is to face south with a slope of 60° from the horizontal. The average daily radiation on a 60° surface for January in Madison is 11.9 MJ/m² (from Example 2.19.1) and the average ambient temperature is –7 C (from Appendix G). The heating load is 36.0 GJ for space and hot water. The collector-heat exchanger correction factor F'_R/F_R is 0.97. For all months the ratio of the monthly average to normal incidence transmittance-absorptance product $(\overline{\tau\alpha})/(\tau\alpha)_n$ is taken to be 0.96 for one-cover collectors, as noted in Section 5.10. (This ratio can be calculated month by month, if desired.) Calculate X and Y for these conditions for collector areas of 25 and 50 m².

Solution

From Equations 20.2.3 and 20.2.4 with $A_c = 25$ m²,

$$X = (4.0)0.97[100 - (-7)](31 \times 86{,}400)(25)/(36.0 \times 10^9) = 0.77$$

$$Y = 0.74 \times 0.97 \times 0.96 \left(12.9 \times 10^6\right)31(25)/(36.0 \times 10^9) = 0.18$$

For 50 m², the values of X and Y are proportionally higher:

$$X = 0.77 \times 50/25 = 1.54$$

$$Y = 0.18 \times 50/25 = 0.35 \qquad ■$$

As will be shown in later sections, the variables X and Y are used to determine f_i, the monthly fraction of the load supplied by solar energy. The energy contribution for the month is the product of f_i and the monthly heating and hot water load L_i. The fraction of the annual heating load supplied by solar energy $\mathscr{F}$ is the sum of the monthly solar energy contributions divided by the annual load:

$$\mathcal{F} = \frac{\sum f_i L_i}{\sum L_i} \qquad (20.2.5)$$

20.3   THE *f*-CHART FOR LIQUID SYSTEMS

For systems of the configuration shown in Figure 20.2.1, the fraction f of the monthly total load supplied by the solar space and water heating system is given as a function of X and Y in Figure 20.3.1.   The relationship of X, Y, and f in equation form is

$$f = 1.029Y - 0.065X - 0.245Y^2 + 0.0018X^2 + 0.0215Y^3 \qquad (20.3.1)$$

Because of the nature of the curve fit of Equation 20.3.1, *it should not be used outside of the range shown by the curves of Figure 20.3.1.* If a calculated point falls outside of this range, the graph can be used for extrapolation with satisfactory results.

Example 20.3.1

The solar heating system described in Example 20.2.1 is to be a liquid system. What fraction of the annual heating load will be supplied by the solar energy for a collector

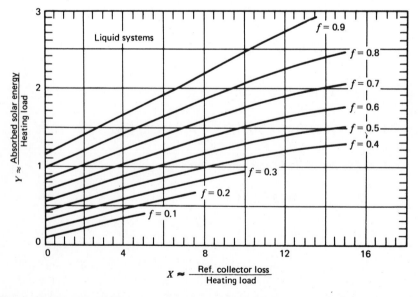

Figure 20.3.1   The *f*-chart for systems using liquid heat transfer and storage media. From Beckman et al. (1977).

area of 50 m²? The monthly combined loads on the system are indicated in the table below. (See Example 2.19.1 for $\overline{H}_T$ values.)

Solution

From Example 20.2.1, the values of X and Y for 50 m² are 1.54 and 0.35, respectively, in January. From Figure 20.3.1 (or Equation 20.3.1), $f = 0.24$. The total heating load for January is 36.0 GJ. Thus, the energy delivery from the solar heating system in January is

$$fL = 0.24 \times 36.0 = 8.6 \text{ GJ}$$

The fraction of the annual heating load supplied by solar energy is determined by repeating the calculation of X, Y, and f for each month and summing the results as indicated by Equation 20.2.5. The table shows the results of these calculations. From Equation 20.2.5 the annual fraction of the load supplied by solar energy is

$$\mathcal{F} = 85.9/203.2 = 0.42$$

Monthly and Annual Performance of a Liquid Heating System in Madison

Month	$\overline{H}_T$, MJ/m²	$\overline{T}_a$	L, GJ	X	Y	f	fL, GJ
Jan.	11.9	−7	36.0	1.54	0.35	0.24	8.6
Feb.	15.5	−6	30.4	1.64	0.49	0.35	10.5
Mar.	15.8	0	26.7	1.95	0.63	0.44	11.7
Apr.	14.5	7	15.7	2.98	0.96	0.60	9.4
May	15.4	13	9.2	4.91	1.73	0.88	8.1
June	15.9	19	4.1	9.93	4.01	[a]1.00	4.1
July	16.3	21	2.9	14.15	6.01	[a]1.00	2.9
Aug.	16.6	20	3.4	12.23	5.22	[a]1.00	3.4
Sept.	15.8	15	6.3	6.78	2.59	1.00	6.3
Oct.	14.9	10	13.2	3.54	1.21	0.71	9.4
Nov.	9.6	1	22.8	2.18	0.44	0.27	6.2
Dec.	8.5	−5	32.5	1.68	0.28	0.16	5.3
Total			203.2				85.9

[a] These points have coordinates outside of the range of the f-chart correlation.

■

To determine the economic optimum collector area, the annual load fraction corresponding to several different collector areas must be determined. The annual load fraction is then plotted as a function of collector area, as shown in Figure 20.3.2. The information in this figure can then be used for economic calculations as shown in Chapter 11.

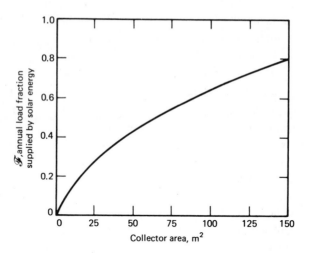

Figure 20.3.2   Annual load fraction vs. collector area.

For liquid systems, f-chart calculations can be modified to estimate changes in long-term performance due to changes in storage tank capacity and load heat exchanger characteristics. This is done by modifying the values of X or Y as described below.

Storage Capacity

Annual system performance is relatively insensitive to storage capacity as long as capacity is more than approximately 50 liters of water per square meter of collector. When the costs of storage are considered, there are broad optima in the range of 50 to 200 liters of water per square meter of collector.

The f-chart was developed for a storage capacity of 75 liters of stored water per square meter of collector area. The performance of systems with storage capacities in the range of 37.5 to 300 l/m² can be determined by multiplying the dimensionless group X by a storage size correction factor X_c/X from Figure 20.3.3 or Equation 20.3.2:

$$\frac{X_c}{X} = \left(\frac{\text{Actual storage capacity}}{\text{Standard storage capacity}}\right)^{-0.25} \qquad (20.3.2)$$

for

$$0.5 \leq \left(\frac{\text{Actual storage capacity}}{\text{Standard storage capacity}}\right) \leq 4.0$$

where the standard storage capacity is 75 liters of water per square meter of collector area.

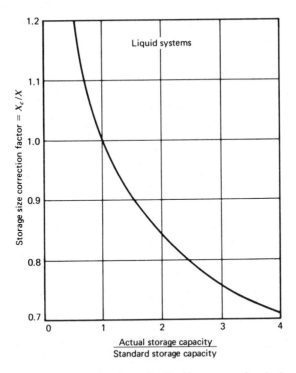

Figure 20.3.3 Storage size correction factor for liquid systems. Standard storage size is 75 $1/m^2$.

Example 20.3.2

For the conditions of Example 20.3.1, what would be the annual solar contribution if the storage capacity of the tank were doubled, to 150 $1/m^2$?

Solution

To account for changes in storage capacity, the value of X calculated in the previous examples must be modified using Figure 20.3.3 or Equation 20.3.2. The ratio of actual storage size to standard storage size is 2.0, so

$$\frac{X_c}{X} = (2.0)^{-0.25} = 0.84$$

For January the corrected value of X is then

$$X_c = 0.84 \times 1.54 = 1.29$$

The value of Y remains 0.35. From the f-chart, $f = 0.25$. The solar contribution for January is

$$fL = 0.25 \times 36.0 \text{ GJ} = 9.1 \text{ GJ}$$

Repeating these calculations for the remaining 11 months gives an annual solar load fraction of 0.44 (versus 0.42 for the standard storage size).   ∎

Load Heat Exchanger Size

As the heat exchanger used to heat the building air is reduced in size, the storage tank temperature must increase to supply the same amount of heat, resulting in higher collector temperatures and reduced collector output. A measure of the size of the heat exchanger needed for a specific building is provided by the dimensionless parameter $\varepsilon_L C_{min}/(UA)_h$ where ε_L is the effectiveness of the water-air load heat exchanger, C_{min} is the minimum fluid capacitance rate (mass flow rate times the specific heat of the fluid) in the load heat exchanger and is generally that of the air, and $(UA)_h$ is the overall energy loss coefficient-area product for the building used in the degree-day space heating load model.

From thermal considerations, the optimum value of $\varepsilon_L C_{min}/(UA)_h$ is infinity. However, system performance is asymptotically dependent upon the value of this parameter, and for values of $\varepsilon_L C_{min}/(UA)_h$ greater than 10, performance will be essentially the same as that for the infinitely large value. The reduction in performance due to an undersized load heat exchanger will be significant for values of $\varepsilon_L C_{min}/(UA)_h$ less than about 1. Practical values of $\varepsilon_L C_{min}/(UA)_h$ are generally between 1 and 3 when the cost of the heat exchanger is considered. See Beckman et al. (1977) for further discussion.

The f-chart for liquid systems was developed with $\varepsilon_L C_{min}/(UA)_h = 2$. The performance of systems having other values of $\varepsilon_L C_{min}/(UA)_h$ can be estimated from the f-chart by modifying Y by a load heat exchanger correction factor Y_c/Y, as indicated in Equation 20.3.3 or Figure 20.3.4:

$$\frac{Y_c}{Y} = 0.39 + 0.65 \exp[-0.139(UA)_h/\varepsilon_L C_{min}] \tag{20.3.3}$$

for $0.5 \le \left(\frac{\varepsilon_L C_{min}}{(UA)_h}\right) \le 50$

Example 20.3.3

For the conditions of Example 20.3.1, what will be the solar contribution if the load heat exchanger is used under the following circumstances: the air flow rate is 520 l/s, the water flow rate is 0.694 l/s, and the heat exchanger effectiveness at these flow rates is 0.69. The building overall energy loss coefficient-area product $(UA)_h$ is 463 W/C.

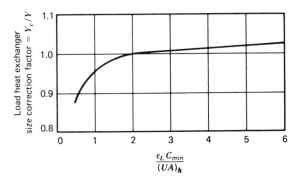

Figure 20.3.4   Load heat exchanger size correction factor.

Solution

First, the value of C_{min} is determined. This usually is the capacitance rate of the air, which in this example is

$$C_{min} = 520 \times 1.20 \times 1010 \times 1000 = 630 \text{ W/C}$$

The capacitance rate of the water is 2910 W/C so that of the air is lower. Then

$$\frac{\varepsilon_L C_{min}}{(UA)_h} = 0.69 \times \frac{630}{463} = 0.94$$

This heat exchanger is smaller than the standard value of 2 used in developing Figure 20.3.1. The correction factor from Figure 20.3.4 or Equation 20.3.3 is

$$\frac{Y_c}{Y} = 0.95 \quad \text{and} \quad Y_c = 0.95 \times 0.35 = 0.34$$

From Figure 20.3.1, $f = 0.22$ for January, and the solar energy contribution for the month is

$$fL = 0.22 \times 36.0 \text{ GJ} = 8.0 \text{ GJ}$$

The annual solar load fraction is 0.40. ∎

  If both the storage and load heat exchanger sizes differ from the standards used to develop the f-chart, the correction factors discussed in Examples 20.3.2 and 20.3.3 would both be applied to find the appropriate values of X_c and Y_c for determination of f. Thus if the storage correction of Example 20.3.2 and the load heat exchanger correction of Example 20.3.3 are both needed, f would be determined at $X_c = 1.29$ and $Y_c = 0.34$ where $f = 0.24$.

20.4   THE f-CHART FOR AIR SYSTEMS

The monthly fraction of total heating load supplied by the solar air heating system shown in Figure 20.2.2 has been correlated with the same dimensionless parameters X and Y as were defined in Equations 20.2.1 and 20.2.2. The correlation is given in Figure 20.4.1 and Equation 20.4.1. It is used in the same manner as the f-chart for liquid-based systems:

$$f = 1.040Y - 0.065X - 0.159Y^2 + 0.00187X^2 - 0.0095Y^3 \qquad (20.4.1)$$

This equation is not to be used outside of the range of values of X and Y shown in Figure 20.4.1.

Example 20.4.1

A solar heating system is to be designed for a building in Madison, WI with two-cover collectors facing south at a slope of 60°. The air heating collectors have the following characteristics: $F_R U_L = 2.84$ W/m²C, and $F_R(\tau\alpha)_n = 0.49$. For this application of the two cover collector $(\overline{\tau\alpha})/(\tau\alpha)_n = 0.93$. The total space and water heating load for January is 36.0 GJ (as in Examples 20.3.1 to 20.3.3). What fraction of the load would be supplied by solar energy with a system having a collector area of 50 m²?

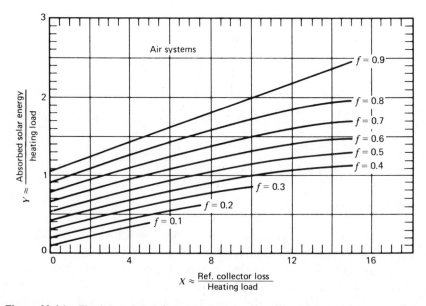

Figure 20.4.1   The f-chart for air systems of the configuration shown in Figure 20.2.2. From Beckman et al. (1977).

Solution

For air systems there will be no correction factor for the collector heat exchanger, and duct losses are assumed to be small, so $F_R'/F_R = 1$. From Equations 20.2.3 and 20.2.4,

$$X = 2.84[100 - (-7)](31 \times 86,400)\frac{50}{36.0 \times 10^9} = 1.13$$

$$Y = 0.49 \times 1 \times 0.93(12.9 \times 10^6)(31)\frac{50}{36.0 \times 10^9} = 0.23$$

Then f for January, from Figure 20.4.1 or Equation 20.4.1, is 0.16. The solar energy supplied by this system in January is

$$fL = 0.16 \times 36.0 \text{ GJ} = 5.8 \text{ GJ}$$

As with the liquid systems, the annual system performance is obtained by summing the energy quantities for all months. The result of the calculation is that 34% of the annual load is supplied by solar energy. ∎

Air systems require two correction factors, one to account for effects of storage size if it is other than 0.25 m³/m², and the other to account for air flow rate that affects stratification in the pebble bed. In addition, care must be exercised to be sure that the values of $F_R(\tau\alpha)_n$ and $F_R U_L$ from collector tests are obtained for the same air flow rates as will be used in an installation. The corrections shown in Chapter 10 can be used to account for duct losses, air flow rate, etc. The correction factors for storage capacity and air flow rate are outlined below. There is no load heat exchanger in air systems.

Air Flow Rate

An increase in air flow rate tends to improve system performance by increasing F_R and tends to decrease performance by reducing the thermal stratification in the pebble bed. The f-chart for air systems is based on a standard collector air flow rate of 10 l/s of air per square meter of collector area. The performance of systems having other collector air flow rates can be estimated by using the appropriate values of F_R and Y and then modifying the value of X by a collector air flow rate correction factor X_c/X (as indicated in Figure 20.4.2 or Equation 20.4.2) to account for the degree of stratification in the pebble bed:

$$\frac{X_c}{X} = \left(\frac{\text{Actual air flow rate}}{\text{Standard air flow rate}}\right)^{0.28} \tag{20.4.2}$$

for

$$0.5 \le \left(\frac{\text{Actual air flow rate}}{\text{Standard air flow rate}}\right) \le 2.0$$

where the standard air flow rate is 10 l/s per square meter of collector area.

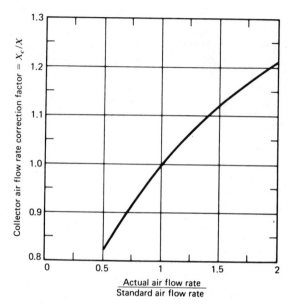

Collector air flow rate correction factor = X_c/X

Actual air flow rate
Standard air flow rate

Figure 20.4.2 Correction factor for air flow rate to account for stratification in the pebble bed. The standard flow rate is 10 l/m²s.

Example 20.4.2

The system of Example 20.4.1 is to be designed using a collector air flow rate of 15 l/s per square meter of collector. Estimate the change in annual performance of the system resulting from the increased air flow.

Solution

Increasing the air flow rate affects F_R and stratification in the pebble bed. The effects of air flow rate on F_R and thus on $F_R U_L$ and $F_R(\tau\alpha)_n$ must either be determined either by collector tests at the correct air flow rate or be estimated by the methods of Section 6.20. In this case, $F_R(\tau\alpha)_n = 0.52$ and $F_R U_L = 3.01$ W/m²C at 15 l/m²s. The corrected X to account for pebble bed stratification is found from Equation 20.4.2 or Figure 20.4.2:

$$\frac{X_c}{X} = \left(\frac{15}{10}\right)^{0.28} = 1.12$$

Thus the X to be used is the value from Example 20.4.1 corrected for the increased air flow rate both through the collector and through the pebble bed:

$$X_c = 1.13 \times \frac{3.01}{2.84} \times 1.12 = 1.34$$

Correcting Y for the new value of F_R [i.e., $F_R(\tau\alpha)_n$]

$$Y_c = 0.23 \times \frac{0.52}{0.49} = 0.24$$

From the air *f*-chart, $f = 0.16$ and $fL = 5.9$ GJ for January. The calculation for the year indicates that 34% of the annual load is supplied by solar energy. (This is essentially the same as at the standard air flow rate; there will be increased fan power required at the higher air flow rate.) ∎

Pebble Bed Storage Capacity

The performance of air systems is less sensitive to storage capacity than that of liquid systems. Air systems can operate in the collector-load mode, in which the storage unit is bypassed. Also, pebble beds are highly stratified, and additional capacity is effectively added to the cold end of the bed, which is seldom heated and cooled to the same extent as the hot end.

The *f*-chart for air systems is for a storage capacity of 0.25 cubic meters of pebbles per square meter of collector area, which corresponds to 350 kJ/m^{2}C for typical void fractions and rock properties. The performance of systems with other storage capacities can be determined by modifying X by a storage size correction factor X_c/X, as indicated in Figure 20.4.3 or Equation 20.4.3:

$$\frac{X_c}{X} = \left(\frac{\text{Actual storage capacity}}{\text{Standard storage capacity}} \right)^{-0.30} \tag{20.4.3}$$

for

$$0.5 \le \left(\frac{\text{Actual storage capacity}}{\text{Standard storage capacity}} \right) \le 4.0$$

where the standard storage capacity is 0.25 m^3/m^2.

Example 20.4.3

If the system of Example 20.4.1 has storage capacity which is 60% of the standard capacity, what fraction of the annual heating load would the system be expected to supply?

Solution

The storage size correction factor, from Figure 20.4.3 (or Equation 20.4.3), is 1.17. Then for January

$$\frac{X_c}{X} = 1.17 \quad \text{and} \quad X_c = 1.17 \times 1.13 = 1.32$$

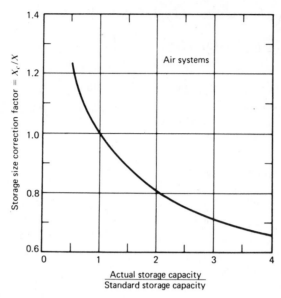

Figure 20.4.3 Storage size correction factor for air systems. The standard storage capacity is $0.25\ \text{m}^3/\text{m}^2$.

Here Y remains 0.23. From Figure 20.4.1 or Equation 20.4.1, $f=0.15$ and $fL=0.15 \times 36.0$ GJ $= 5.5$ GJ. The fraction of the annual load supplied by solar energy $\mathcal{F}$ is 0.32 (compared to 0.34 for the standard storage size).   ■

   If both air flow rate and storage size are not standard, there will be two corrections on X to be made (in addition to any corrections due to changes in F_R) and the final X will be the product of the uncorrected value and the two correction factors.

   If a phase change energy storage unit is used in place of the rock bed, Equation 8.8.9, an empirical equation for the equivalent rock bed storage capacity, can be used to predict system performance. The properties and mass of the phase change material are used to estimate the size of an equivalent rock bed, which is then used in the air f-chart correlations. Some phase change material properties are given in Table 8.8.1.

20.5   SERVICE WATER HEATING SYSTEMS

Figure 20.3.1, the f-chart for liquid heating systems, can be used to estimate the performance of solar water heating systems having the configuration shown in Figure 20.2.3 by defining an additional correction factor on X. The mains water temperature T_m and the minimum acceptable hot water temperature T_w both affect the performance of solar water heating systems. Both T_m and T_w affect the average system operating temperature level and thus also the collector energy losses. The dimensionless group X, which is related to collector energy losses, can be corrected to include these effects. If monthly values of X are multiplied by a water heating correction factor X_c/X in Equation 20.5.1, the f-chart for liquid-based solar space and water heating systems (Equation 20.3.1 or Figure 20.3.1) can be used to estimate monthly values of f for water heating systems. All temperatures are in Celsius.

$$\frac{X_c}{X} = \frac{11.6 + 1.18T_w + 3.86T_m - 2.32\overline{T}_a}{100 - \overline{T}_a} \qquad (20.5.1)$$

This method of estimating water heater performance is based on storage capacity of 75 l/m^2 and on the typical day's distribution of hot water use occurring each day as shown in Figure 9.1.2. Other distributions of use in a day have small effect on system performance; however, day-to-day variations in loads can have a substantial effect on performance. [See Buckles and Klein (1980).]

The water heating correction factor is based on the assumption of a well-insulated solar preheat tank, and losses from an auxiliary tank were not included in the f-chart correlations. For systems supplying hot water only, loads on the system should also include losses from the auxiliary tank. (These are normally included in the energy supplied to a conventional water heater.) Tank losses can be estimated from the insulation and tank area, but this frequently leads to their underestimation as losses through connections, mounting brackets, etc., can be significant. It is recommended that tank loss calculations be based on the assumption that the entire tank is at the water set temperature T_w.

The use of a tempering valve on the supply line to mix cold supply water with solar-heated water above the water set temperature has little effect on the overall output of the solar system, as noted in Section 12.4, and the method indicated here can be used for systems either with or without the tempering valve.

Example 20.5.1

A solar water heating system is to be designed for a residence in Madison, WI (latitude 43°N). The collectors considered for this purpose have two covers with $F_R'(\tau\alpha)_n = 0.64$ and $F_R'U_L = 3.64$ W/m^{2}C. The collectors are to face south at a slope of 45°. The water heating load is 400 liters per day heated from 11 to 60 C. The storage capacity of the preheat tank is to be 75 liters of water per square meter of collector area. The auxiliary tank has a capacity of 225 liters and a loss coefficient of 0.62 W/m^{2}C. The tank is a cylinder 0.50 m diameter and 1.16 m high. Estimate the fraction of the January heating load supplied by solar energy for this system with a collector area of 10 m^2. The radiation on the collector $\overline{H}_T$ is 11.1 MJ/m^2 and $(\overline{\tau\alpha})/(\tau\alpha)_n$ is 0.94.

Solution

The monthly load is the energy required to heat the water from T_m to T_w plus the auxiliary tank losses. For January, the energy to heat the water is

$$400 \times 1 \times 4190(60 - 11)\, 31 = 2.55 \text{ GJ}$$

The loss rate from the auxiliary tank is $UA(T_w - T_a')$. The tank area is 2.21 m², so the loss rate for T_a' of 20 C is

$$2.21 \times 0.62(60 - 20) = 55 \text{ W}$$

The energy required to supply this loss for the month is

$$55 \times 31 \times 24 \times 3600 = 0.15 \text{ GJ}$$

The total load to be used in calculation of X and Y is then

$$2.55 + 0.15 = 2.70 \text{ GJ}$$

We now calculate X_c and Y:

$$X_c = X\frac{X_c}{X} = \frac{10 \times 3.64 \times [100 - (-7)] \times 31 \times 24 \times 3600}{2.70 \times 10^9}$$

$$\times \frac{11.6 + 1.18(60) + 3.86(11) - 2.32(-7)}{100 - (-7)} = 5.10$$

$$Y = 0.64 \times 0.94 \times 11.1 \times 10^6 \times 31 \frac{10}{2.70 \times 10^9} = 0.76$$

From Figure 20.3.1 or Equation 20.3.1, $f = 0.37$. When this process is repeated for all 12 months, the annual solar contribution is estimated to be 59%. ∎

20.6 *f*-CHART RESULTS

In the original development of *f*-charts [Klein (1976)], it was necessary to make a number of assumptions about systems and their performance. Several of these are worth noting as they are useful in interpreting results obtained from this method.

First, all liquid storage tanks were assumed to be fully mixed, both for main storage tanks for liquid systems and preheat tanks for all water heating. This assumption, as shown in Chapter 10, tends to lead to conservative estimates of long-term performance by overestimating collector inlet temperature. Second, for reasons of economy in simulations, all days were considered symmetrical about solar noon. This also leads to conservative estimates of system outputs. For water heating only, it has been noted that energy in water above the set temperature is not considered useful. Thus, the computations tend to be conservative in their predictions. On the other hand, very well insulated storage tanks were assumed for liquid systems, and it was assumed that there are no leaks in systems; most air systems leak to some extent, which will tend to degrade performance below predicted levels.

There are implicit assumptions in this method. Systems are well built, flow distribution to collectors is uniform, flow rates are as assumed, system configurations are close to those for which the correlations were developed, and control strategies used are nearly those assumed in the *f*-chart development. If systems do not meet these conditions, they cannot be expected to perform as estimated by the *f*-chart method.

Three steps have been used to check the results of *f*-chart. The first is comparisons with detailed simulations in many locations (simulations themselves having been compared with measurements). Second, laboratory measurements on experimental systems have been compared to *f*-chart results. Third, measurements made on operating systems in the field have also been used for comparisons. The comparisons must be based on use of measured meteorological data in the *f*-chart calculations and on measured loads; variability in weather is large enough that the weather experienced by a system in any one year may be substantially different from the average data ordinarily used in the calculations.

The results obtained with *f*-chart have been compared to results of detailed simulations for a variety of locations. Agreement is generally to within 3% for most U.S. locations and within 11% for Seattle, the worst case. Agreement of monthly solar fractions is not nearly as good as annual fractions, and the *f*-chart method should be used to estimate annual performance only.

Fanney (1979) did a year-long experimental study of solar domestic hot water systems at the U.S. National Bureau of Standards. The annual fraction of the total load (water draw-off plus auxiliary tank losses) supplied by solar from measurements and from *f*-chart predictions is given in Table 20.6.1. Although there are some

Table 20.6.1   Measured and Predicted Values of Annual Solar Fraction[a]

Number of Tanks	Type	Heat Exchanger	Measured	Predicted
1	Liquid	External	0.36	0.37
2	Liquid	External	0.37	0.40
1	Liquid	Internal	0.45	0.43
2	Liquid	Internal	0.33	0.30
1	Air	External	0.20	0.21

[a] From Fanney (1979).

differences between measured and predicted performance, the results agree reasonably well.

Two experimental building heating systems supply data for comparisons. MIT House IV (Section 13.5) was 52% heated by solar energy over 2 years, while *f*-chart estimates (based on measured meteorological conditions) indicate a solar fraction of 57%. (The system configuration is close to, but not the same as, the *f*-chart system.) CSU House II, an air system (Section 13.4), was supplied with 72% of its heat by solar energy; *f*-chart predicts 76% for the period.

The U.S. National Solar Data Network (NSDN) program and other related programs provide additional data on operation of systems in routine use on buildings. The quality of the data and the operation of the systems does not match that of laboratory experiments, but the circumstances under which the operations are carried out are "real world." Duffie and Mitchell (1983) have summarized some of these data for systems that are reasonably close in design to the standard configurations. The results were shown in terms of comparisons of measured solar fraction $\mathcal{F}_M$ with predicted solar fraction $\mathcal{F}_P$. A set of these comparisons is shown in Figure 20.6.1. The agreement is generally within ±15%. Three air systems show better performance than predicted; this may reflect the difficulty of making good measurements on air systems.

Data for some systems, when plotted on the coordinates of Figure 20.6.1, lie in the upper left portion of the plot. This is an indication that there is a problem in the design or construction of the systems and that corrective action should be taken to improve performance.

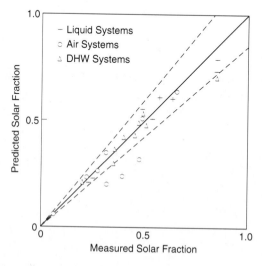

Figure 20.6.1 Comparison of predicted and measured annual or seasonal performance of solar heating systems. From Duffie and Mitchell (1983).

20.7 PARALLEL SOLAR ENERGY-HEAT PUMP SYSTEMS

For the parallel solar energy-heat pump system shown in Figure 20.7.1, Anderson (1979) and Anderson et al. (1980) have developed a design method based on a combination of the "bin" method and the f-chart method. In the parallel mode of operation the solar system is the primary energy source, and its operation is unaffected by the presence of a heat pump, that is, the heat hump system acts as the solar system auxiliary energy source. Consequently, the f-chart method can be used to determine the solar contribution to the heating load. The remaining portion of the load is met by a combination of the energy delivered by the heat pump and auxiliary energy. Although the performance of the heat pump is affected by the presence of the solar system, the Anderson et al. study observed that this interaction is small and can be neglected. This means that the only effect of the solar system on the heat pump is a reduction of the load that the heat pump will meet. The results of bin method calculations for a heat-pump-only system can then be modified to predict heat pump performance in the presence of a solar system.

A typical set of heat pump and load characteristics is shown in Figure 20.7.2 as a function of ambient temperature. When the ambient temperature is above the balance point, the heat pump can supply more energy Q_{del} than is needed by the load L. When the ambient temperature is below the balance point, auxiliary energy must be used in addition to the heat pump.

The bin method for estimating the monthly energy usage of a stand-alone heat pump system is described in ASHRAE (1976). The method uses long-term weather data to determine the number of hours in which the ambient temperatures were within 2.8 C (5 F) temperature ranges called bins. The number of hours in each temperature bin for a particular month can be used to estimate the monthly purchased energy. For example, suppose a month has 15 hours in a temperature bin centered around 10 C. To meet the load during this 15 hours, the system needs to run only 15 hours times

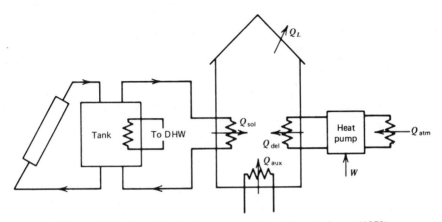

Figure 20.7.1 Parallel solar heat pump system. From Anderson (1979).

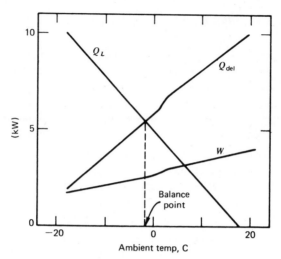

Figure 20.7.2   Typical heat pump and load characteristics as a function of ambient temperature. From Anderson (1979).

the ratio of L to Q_{del} or 15(2.2/8.1) = 4.1 hours, and in this 4.1 hours the energy required by the heat pump is 4.1 × 3.4 = 13.9 kWh (50 MJ). This calculation must be repeated for each temperature bin above the balance point.

At temperatures below the balance point the heat pump alone cannot meet the load and auxiliary energy must be used to make up the deficit. If 12 hours are in the bin centered around −10 C, the heat pump will run continuously for the 12 hours at a rate of 2.1 kW for a total electrical requirement of 2.1 × 12 = 25.2 kWh (91 MJ). In addition, auxiliary energy must make up the difference between the load of 7.7 kW times 12 hours and the delivered energy of 4.6 kW times 12 hours, or 37.2 kWh (134 MJ). By repeated application of these calculations, the monthly purchased energy can be estimated from which annual values can be calculated.

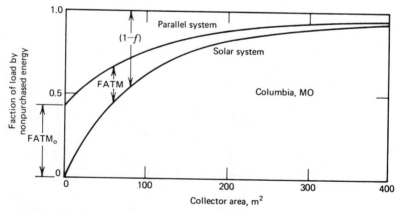

Figure 20.7.3   Parallel solar heat pump system performance for January in Columbia, MO. From Anderson et al. (1980).

For a house with a parallel solar-heat pump system in Columbia, MO the combined system performance is shown in Figure 20.7.3 as a function of collector area. For zero collector area the system is a stand-alone heat pump and the fraction of the load supplied by nonpurchased energy from the air is $FATM_0$. At any finite collector area, some energy is supplied by solar and some is from the ambient air. The design procedure assumes that on a monthly basis

$$FATM = FATM_0 (1 - f) \qquad (20.7.1)$$

where f is the monthly fraction by solar. This equation is a result of assuming that the only effect of the solar system on the heat pump performance is a reduction in the load.

With monthly results from f-chart, a bin method calculation, and Equation 20.7.1, the monthly fraction of the load supplied by nonpurchased energy can be estimated. The remainder of the load is supplied by a combination of compressor work supplied to the heat pump and auxiliary energy. If the auxiliary energy is electricity, there is no need to separate the purchased energy into fractions by compressor work and by auxiliary. However, if the auxiliary is not electricity, it is

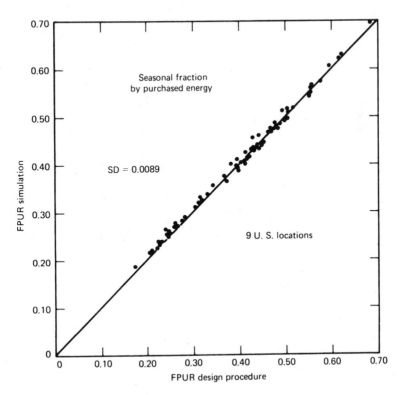

Figure 20.7.4 Comparison of purchased energy fractions calculated by design procedure and by detailed simulations. From Anderson et al. (1980).

necessary to know each of the two fractions to do an economic assessment. Equation 20.7.2 is recommended by Anderson et al. to find the work fraction,

$$FW = FW_0 (1 - f) \tag{20.7.2}$$

where FW_0 is the work fraction for the stand alone heat pump. The auxiliary fraction is then

$$FAUX = (1 - FATM_0 - FW_0)(1 - f) \tag{20.7.3}$$

The results of using the design procedure given by Equations 20.7.1 through 20.7.3 have been compared to detailed computer simulations, and typical results are shown in Figure 20.7.4. In this figure the fraction of the total load supplied by purchased energy calculated by the two methods compare very well. A similar conclusion can be made concerning the work fraction, the fraction from the atmosphere, and the fraction by solar.

20.8   SUMMARY

The f-chart method provides a means of quickly estimating the long-term performance of solar heating systems of standard configurations. The data needed are monthly average radiation and temperature, the collector parameters available from standard collector tests, and estimates of loads.

It should be recognized that there are uncertainties in the estimates obtained from the f-chart procedure. The major uncertainties arise from several sources. First, the meteorological data can be in error by as much as 5 to 10%, particularly when the horizontal data are converted to radiation on the plane of the collector. Second, average data are used in the calculations, and any particular year may vary widely from that average. Third, it is extremely difficult to predict what building heating loads will be as they are dependent on the habits of the occupants. Fourth, systems must be carefully engineered and constructed, with minimal heat losses, leakage, and other mechanical and thermal problems. Finally (and probably least important), there are some differences between the f-chart correlation and individual data points.

It is difficult to quantitatively assess the impacts of these uncertainties on the results obtained from the method. However, two generalizations can be made. First, the relative effects of design changes can be established. For example, the effects on annual performance of a change in plate absorptance and emittance can be shown. The second decimal place is significant in this context. Second, the method will predict the performance of a given system, but because of the uncertainties, only the first decimal should be considered as significant.

The calculations of the f-chart method can easily be done by hand, but they can be tedious. The method has been programmed in combination with life cycle economic analysis. [See, for example, FCHART (1985).]

REFERENCES

Anderson, J. V., M. S. Thesis, Mechanical Engineering, U. of Wisconsin-Madison (1979). "Procedures for Predicting the Performance of Air-to-Air Heat Pumps in Stand-Alone and Parallel Solar-Heat Pump Systems."

Anderson, J. V., J. W. Mitchell, and W. A. Beckman, *Solar Energy*, **25**, 155 (1980). "A Design Method for Parallel Solar-Heat Pump Systems."

ASHRAE *Systems Handbook*, American Society of Heating, Refrigerating, and Air-Conditioning Engineers, New York (1976).

Balcomb, J. D. and J. C. Hedstrom, *Proc. of the International Solar Energy Society Conference*, Winnipeg, **4**, 281 (1976). "A Simplified Method for Sizing a Solar Collector Array for Space Heating."

Barley, C. D. and C. B. Winn, *Solar Energy*, **21**, 279 (1978). "Optimal Sizing of Solar Collectors by the Method of Relative Areas."

Beckman, W. A., S. A. Klein, and J. A. Duffie, *Solar Heating Design by the f-Chart Method*, Wiley-Interscience, New York (1977).

Buckles, W. E., and S. A. Klein, *Solar Energy*, **25**, 417 (1980). "Analysis of Solar Domestic Hot Water Heaters."

Connelly, M., R. Giellis, G. Jenson, and R. McMordie, *Proc. of the International Solar Energy Society Conference*, Winnipeg, **10** 220 (1976). "Solar Heating and Cooling Computer Analysis--A Simplified Sizing Method for Non-Thermal Specialists."

Duffie, J. A., and J. W. Mitchell, *Trans. ASME, J Solar Energy Engrg.*, **105**, 3 (1983). "*f*-Chart: Predictions and Measurements."

Fanney, A. H., personal communication on the National Bureau of Standards Solar Domestic Hot Water Test Facility (1979).

FCHART Users Manual, f-Chart Software, Middleton, WI (1985).

Kenna, J. *Solar Energy*, **32**, 687 (1984a). "A Parametric Study of Open Loop Solar Heating Systems - I," and Solar Energy **32**, 707 (1984b). "A Parametric Study of Open Loop Solar Heating Systems - II."

Klein, S. A., Ph.D. Thesis, Chemical Engineering, U. of Wisconsin-Madison (1976). "A Design Procedure for Solar Heating Systems."

Klein, S. A., W. A. Beckman, and J. A. Duffie, *Solar Energy*, **18**, 113 (1976). "A Design Procedure for Solar Heating Systems."

Klein, S. A., W. A. Beckman, and J. A. Duffie, *Solar Energy*, **19**, 509 (1977). "A Design Procedure for Solar Air Heating Systems."

Lameiro, G. and P. Bendt, paper in *Proc.of the 1978 Meeting of the American Section of the International Solar Energy Society*, Denver, **2** (1), 113 (1978). "The GFL Method for Designing Solar Energy Space Heating and Domestic Hot Water Systems."

Proctor, D., paper presented at International Solar Energy Society Meeting, Los Angeles (1975). "Methods of Predicting the Heat Production of Solar Collectors for System Design."

Ward, J. C., *Proc. of the International Solar Energy Society Conference*, Winnipeg, **4**, 336 (1976). "Minimum-Cost Sizing of Solar Heating Systems."

Chapter 21

DESIGN OF ACTIVE SYSTEMS BY UTILIZABILITY METHODS

In Chapter 20 the f-chart method was presented as a design tool for solar systems that deliver energy to a load at minimum temperatures near 20 C. There are, however, many applications that use energy at temperatures higher or lower than 20 C. A warehouse heating system may be required to keep the building above freezing so that all energy delivered above 0 C is useful. A solar-operated absorption cycle air conditioning system may be able to use all energy above 75 C. Industrial process heat temperature requirements can be at almost any level. This chapter describes design methods for active systems for which the f-chart method does not apply.

The design of active solar systems can be done with detailed computer simulations, as noted in Chapter 19. The expense of simulations can be significant, but the procedure should be considered in the final analysis for all large systems. Preliminary designs and designs for small systems require inexpensive methods for predicting long-term performance, and the methods presented here (and the f-chart method) are in this category.

The first of the utilizability methods is monthly average hourly utilizability for flat-plate collectors (the ϕ method). In Chapter 2, the concepts of utilizability were introduced, and methods were shown for calculating hourly and daily utilizability without reference to how critical radiation levels are determined. In this chapter, we briefly review the initial development of Whillier (1953) and Hottel and Whillier (1958), the generalized ϕ method of Liu and Jordan (1963), and the $\overline{\phi}$ method of Klein (1978) and Collares-Pereira and Rabl (1979a,b). We then show how Klein and Beckman (1979) combined daily utilizability with the f-chart concept to account for finite storage capacity (the $\overline{\phi}, f$-chart method). In each case we show how critical radiation levels are established and illustrate the use of the methods.

Design methods are available for many solar-thermal systems, but not all. The utilizability methods all require knowledge of the collector inlet fluid temperature, which often is not known. The $\overline{\phi}, f$-chart method allows collector inlet temperature to vary with the storage tank temperature but requires that the load be a closed loop with the fluid returned to the tank at or above a minimum temperature. Also, any device between the solar system and the load must have a conversion efficiency which is independent of the temperature level at which energy is delivered (as long as it is above the minimum temperature). This rules out solar-to-mechanical systems.

21.1   HOURLY UTILIZABILITY

Utilizability can be defined as the fraction of the incident solar radiation that can be converted to useful heat. It is the fraction utilized by a collector having $F_R(\tau\alpha) = 1$ and operating at a fixed inlet to ambient temperature difference. Although this collector has no optical losses and has a heat removal factor of unity, the utilizability is always less than 1 since the collector does have thermal losses.[1]

As shown in Section 6.8, an analytical expression for utilizability in terms of the hourly radiation incident on the plane of the collector can be derived from Equation 6.7.6:

$$Q_u = A_c F_R[I_T(\tau\alpha) - U_L(T_i - T_a)]^+ \qquad (21.1.1)$$

The radiation level must exceed a critical value before useful output is produced. This critical level is found by setting Q_u in Equation 21.1.1 equal to zero:

$$I_{Tc} = \frac{F_R U_L(T_i - T_a)}{F_R(\tau\alpha)} \qquad (21.1.2)$$

The useful output of the collector can be expressed in terms of the critical radiation level as

$$Q_u = A_c F_R(\tau\alpha)(I_T - I_{Tc})^+ \qquad (21.1.3)$$

If the critical radiation level is constant for a particular hour (say 10 to 11) for a month (N days), then the monthly average hourly collector output for this hour is given by

$$\overline{Q}_u = \frac{A_c F_R(\tau\alpha)}{N} \sum_{}^{N} (I_T - I_{Tc})^+ \qquad (21.1.4)$$

The monthly average radiation in this particular hour is $\bar{I}_T$, so the average useful output can be expressed as

$$\overline{Q}_u = A_c F_R(\tau\alpha)\bar{I}_T\phi \qquad (21.1.5)$$

where the utilizability ϕ is defined as

$$\phi = \frac{1}{N} \sum_{}^{N} \frac{(I_T - I_{Tc})^+}{\bar{I}_T} \qquad (21.1.6)$$

The procedure for calculating ϕ was shown in Example 2.22.1 for a vertical surface at Blue Hill, MA. The result of the procedure, when the critical radiation

---

[1] If a collector heat exchanger is present, F_R' can be used in place of F_R.

level was varied, was the utilizability curve of Figure 2.22.3. Whillier (1953) and later Liu and Jordan (1963) have shown that in a particular location for a 1-month period, ϕ is essentially the same for all hours. Thus, although the curve of Figure 2.22.3 was derived for the hour-pair 11 to 12 and 12 to 1, it is valid for all hour-pairs in Blue Hill.

The ϕ curve of Figure 2.22.3 is specific to a particular location and orientation, and in order that the method be more generally useful, Liu and Jordan developed the generalized ϕ curves presented in Section 2.23. These are independent of location and orientation. Figures 2.23.1 or Equations 2.23.5 are used to determine ϕ at the dimensionless critical radiation level X_c, which was defined as $I_{Tc}/\bar{I}_T$. We can now write X_c in terms of collector parameters:

$$X_c = \frac{I_{Tc}}{\bar{I}_T} = \frac{F_R U_L (T_i - T_a)}{F_R (\tau\alpha)_n \frac{(\tau\alpha)}{(\tau\alpha)_n} \bar{I}_T} \tag{21.1.7}$$

where $(\tau\alpha)/(\tau\alpha)_n$ is determined for the mean day of the month and the appropriate hour angle. This ratio can be estimated from b_o. With ϕ determined, the utilizable energy $\phi\bar{I}_T$ follows.

The main utility of hourly utilizability is in estimating the output of processes which have a critical radiation level that changes significantly through the day. This change may be due to a regular and pronounced diurnal temperature variation or to shifts in collector inlet temperature caused by characteristics of the load. The following example is an industrial process in which the water returned to the collector varies in a regular manner through the day, allowing critical radiation levels (and thus useful energy) to be calculated for each hour.

Example 21.1.1

An industrial process is supplied water from an array of collectors. The temperature of the return water from the process to the collectors varies from hour to hour, but for a given hour is nearly constant through the month. The system is located in Albuquerque NM, at latitude 35°, and the collectors are sloped 35° to the south. For the month of March, the temperatures at which water returns to the collectors are shown in the second column of the table. The average ambient temperatures for the hours are shown in the next column. Data for $\bar{I}_T$ are shown in the fourth column; they are calculated from $\bar{H}$ from Appendix G by Equation 2.23.3. The angles of incidence of beam radiation on the collector (which are nearly the same as the hour angles for the south-facing collectors with $\beta = \phi$ and δ close to zero) are shown in the fifth column.

The collector characteristics are $F_R(\tau\alpha)_n = 0.726$; $F_R U_L = 3.89$ W/m²C; and $b_o = -0.11$. The ground reflectance can be taken as 0.2.

Estimate the March output of the collector operating under these circumstances.

Solution

The data in columns 6 through 9 in the table show the results of the hour-by-hour calculations for March. The calculations for the hour 8 to 9 are as follows.

At $\theta_b = 52.5°$, from Equation 6.17.10,

$$K_{\tau\alpha} = (\tau\alpha)/(\tau\alpha)_n = 1 - 0.11\left(\frac{1}{\cos 52.5} - 1\right) = 0.929$$

The dimensionless critical radiation level X_c is calculated from Equation 21.1.7:

$$X_c = \frac{3.89(23 - 3)3600 \times 10^{-6}}{0.726 \times 0.929 \times 1.74} = 0.239$$

For the month of March, $\overline{K}_T$ is 0.68. It is thus necessary to interpolate between $\overline{K}_T$ of 0.6 and 0.7 in Figures 2.23.1(d). For March with $\beta = \phi$, $\overline{R}_b = 1.28$ from Table D-7. From Figure 2.23.1(d), ϕ is very close to 0.77 at both values of $\overline{K}_T$. Alternatively, Equations 2.23.5 can be used, which lead to ϕ of 0.769. The useful gain for the collector for this hour for the 31 days of the month is thus

$$F_R(\tau\alpha)_n \frac{(\tau\alpha)}{(\tau\alpha)_n} N\overline{I}_T\phi = 0.726 \times 0.929 \times 31 \times 1.74 \times 0.769$$

$$= 27.9 \text{ MJ/m}^2$$

This process is repeated for each hour of the day. The conditions of operation for the hours 11 to 12 and 12 to 1 happen to be the same, so the calculations are done once for that hour-pair. The sum of the outputs for all hours is the expected gain from the collector for the month. Thus the month's useful gain is 378 MJ per square meter of collector.

Hour	T_r C	T_a C	$\overline{I}_T$ MJ/m^2	θ_b deg	$K_{\tau\alpha}$	X_c	ϕ	NQ_u/A MJ/m^2
8-9	23	3	1.74	52.5	0.929	0.239	0.769	27.9
9-10	23	3	2.49	37.6	0.971	0.160	0.843	45.9
10-11	23	3	3.08	22.6	0.991	0.126	0.875	60.1
11-12	28	4	3.41	7.9	1.000	0.136	0.865	66.3
12-1	28	4	3.41	7.9	1.000	0.136	0.865	66.3
1-2	47	9	3.08	22.6	0.991	0.240	0.764	52.5
2-3	47	9	2.49	37.6	0.971	0.303	0.706	38.4
3-4	47	9	1.74	52.5	0.929	0.454	0.575	20.9
								Total = 378

■

The ϕ curves are very powerful design tools, but they can be misused. They cannot be directly applied to many liquid-based building heating systems since the critical level varies considerably during the month due to the finite storage capacity. Two limiting cases of solar heating fit into the ϕ curve restrictions: air heating systems in midwinter and systems with seasonal storage. For the case of air systems, the inlet air temperature to the collector during much of the winter will be the return air from the house. This is because little excess energy is available for storage and what is stored is effectively stratified by the rock bed. Only in the spring and fall will the collector inlet temperature rise much above the room temperature. In the case of annual (long-term) storage, the storage tank temperature varies slowly during a month so that a monthly average tank temperature and the critical level can be found by trial and error. This is illustrated in the next section. As shown in Example 21.1.1, some industrial process heat applications can also be analyzed with ϕ curves if a critical level for each hour for the month can be determined.

The line on the ϕ charts labeled "limiting curve of identical days" is the ϕ curve that would be obtained if all days in a month were identical. Only if K_T is high or if the critical level is very low do all ϕ curves approach this limit, and only then can average days be used to predict long-term performance.

21.2   DAILY UTILIZABILITY

As noted in Section 2.24, the use of ϕ curves involves calculations for each hour or hour-pair. This means that for the most common problems, four to eight calculations will be needed per month. This situation led Klein (1978), and also Collares-Pereira and Rabl (1979a,b), to simplify the calculations for those systems for which a single critical radiation level can be defined for all hours for a month.

Daily utilizability was defined in Section 2.24 as the sum for a month, over all hours and all days, of the radiation on a tilted surface that is above a critical level divided by the monthly radiation. In equation form

$$\bar{\phi} = \sum_{days} \sum_{hours} \frac{(I_T - I_{Tc})^+}{\bar{H}_T N} \tag{21.2.1}$$

The critical level I_{Tc} is similar to that defined by Equation 21.1.2 except that monthly average transmittance-absorptance must be used in place of $(\tau\alpha)$, and T_i and $\bar{T}_a$ are representative inlet and daytime temperatures for the month:

$$I_{Tc} = \frac{F_R U_L (T_i - \bar{T}_a)}{F_R (\overline{\tau\alpha})} = \frac{F_R U_L (T_i - \bar{T}_a)}{F_R (\tau\alpha)_n \dfrac{(\overline{\tau\alpha})}{(\tau\alpha)_n}} \tag{21.2.2}$$

The value of $(\overline{\tau\alpha})/(\tau\alpha)_n$ can be calculated with Equation 5.10.5. The monthly average daily useful energy gain is then given by

$$\sum \bar{Q}_u = A_c F_R (\overline{\tau\alpha}) \bar{H}_T \bar{\phi} \tag{21.2.3}$$

It will be recalled from Section 2.24 that Klein developed a method for calculating $\bar{\phi}$ as a function of $\bar{K}_T$, the geometric factor $\bar{R}/R_n$, and the dimensionless critical radiation level $\bar{X}_c$; $\bar{R}$ is calculated by the methods of Sections 2.19 and 2.20, and R_n by Equation 2.24.2.

The monthly average critical radiation ratio $\bar{X}_c$ is the ratio of the critical radiation I_{Tc} to the noon radiation level for a day of the month in which the total radiation for the day is the same as the monthly average. Writing this ratio for the noon hour, we have

$$\bar{X}_c = \frac{I_{Tc}}{r_{t,n} R_n \bar{H}} = \frac{F_R U_L (T_i - \bar{T}_a)/F_R(\overline{\tau\alpha})}{r_{t,n} R_n \bar{K}_T \bar{H}_o} \tag{21.2.4}$$

Equations[2] 2.24.4 or Figure 2.24.2 will then give $\bar{\phi}$. The procedure, except for the calculation of $\bar{X}_c$, was illustrated in Example 2.24.1.

In the next example a problem involving a large capacity seasonal storage unit will be used to illustrate an application of the $\bar{\phi}$ method.

Example 21.2.1

Estimate the annual performance of a liquid solar heating system in Madison having a very large storage volume (seasonal storage). The collector area is 50 m², and its slope is 60° to the south. The storage tank contains 250,000 liters of water. The losses from the tank are to the ground at an average temperature of 7 C. The storage tank area-loss coefficient product $(UA)_s$ is 30 W/C. The building $(UA)_h$ is 200 W/C and the characteristics of the collector-heat exchanger-piping combination are $F_R'(\tau\alpha)_n = 0.78$ and $F_R' U_L = 4.55$ W/m²C. The monthly average weather data are given in the table that follows.

Solution

Since the storage tank is very large, its temperature will not change significantly through each month so an average collector inlet temperature can be assumed, and the $\bar{\phi}$ method can be used. An energy balance on the tank for a 1-month period is

$$(mC_p \Delta T)_s = \text{Collector output} - \text{Tank losses} - \text{Energy to load}$$

---

[2] For convenience, Equations 2.24.4 are repeated here:

$$\bar{\phi} = \exp\left\{ \left[a + b(R_n/\bar{R}) \right] \left[\bar{X}_c + c\bar{X}_c^2 \right] \right\}$$

$$a = 2.943 - 9.271\bar{K}_T + 4.031\bar{K}_T^2$$

$$b = -4.345 + 8.853\bar{K}_T - 3.602\bar{K}_T^2$$

$$c = -0.170 - 0.306\bar{K}_T + 2.936\bar{K}_T^2$$

In this solution, it will be assumed that all of the energy necessary to meet the heating needs of the building for all months can be supplied across a load heat exchanger, i.e., that all of the loads are met by solar energy.

The initial step is to guess a tank temperature at the beginning of March. March is conveniently used as the beginning of the year since it usually has the lowest tank temperature and is therefore easiest to estimate. Once the tank temperature is known at the beginning of the month, the tank temperature at the end of the month is guessed. The average tank temperature for the month is used to calculate the collector output and tank losses.[3] The tank energy balance equation is used to determine the tank temperature at the end of the month. This calculated temperature is compared with the assumed value, and if they agree, the calculation continues with the next month. If they disagree, another monthly average temperature is estimated and a new final tank temperature is calculated. The procedure is repeated for all 12 months, and the final tank temperature is compared to the initial guess made for March. If they agree, the calculations are complete; if they disagree, the calculations must be repeated. The details of the March calculations follow, and the results for all 12 months are given in the table. The month's heating load is found with Equation 9.2.6

$$(UA)_h(DD) = 200 \times 3600 \times 24 \times 599 = 10.4 \text{ GJ}$$

The tank losses depend on the monthly average tank temperature $\overline{T}_s$, which is unknown. An initial tank temperature of 24.6 and an average temperature of 24.9 C will be assumed (and later checked):

$$(UA)_s(\overline{T}_s - T_{\text{ground}}) = 30 \times 3600 \times 24 \times 31 \, (24.9 - 7) = 1.4 \text{ GJ}$$

The monthly average daily useful energy gain will be estimated first, using the assumed 24.9 C tank temperature. For the month $\overline{H}$ is 12.89 MJ/m². From Example 2.19.1 the monthly average radiation on the collector is 15.8 MJ/m². The monthly average ratio $(\overline{\tau\alpha})/(\tau\alpha)_n$ will be assumed to be a constant at 0.96. The critical radiation level I_{Tc} is

$$I_{Tc} = \frac{4.55(24.9 - 1)}{0.78 \times 0.96} = 145 \text{ W/m}^2$$

Here the daytime average temperature was estimated to be 2 C higher than the 24-hour average.

Details of the radiation and utilizability calculations are given in Example 2.24.1. The dimensionless critical radiation level is calculated by Equation 21.2.4:

$$\overline{X}_c = \frac{145 \times 3600}{0.146 \times 1.12 \times 12.89 \times 10^6} = 0.25$$

Using Equations 2.24.4, with $\overline{K}_T = 0.50$, $\overline{\phi}$ is 0.68. The monthly energy gain is then calculated by multiplying Equation 21.2.3 by 31, the number of days in the

---

[3] An Euler type integration could also have been used.

month.

$$N \sum \bar{Q}_u = 31 \times 50 \times 0.78 \times 0.96 \times 15.8 \times 10^6 \times 0.68 = 12.4 \text{ GJ}$$

The change in tank temperature during the month of March is then

$$\Delta T = (12.4 - 1.4 - 10.4) \frac{10^9}{4190 \times 0.25 \times 10^6} = 0.6 \text{ C}$$

Thus the tank temperature at the end of the month is 0.6 C higher than at the beginning. The average temperature is equal to the assumed 24.9 C, and the final temperature is 25.2 C. This final March temperature is taken as that for the beginning of April and is the basis for estimation of the April average temperature. This process is repeated for all months. At the end of February (i.e., after a year's calculation) the tank temperature is back to 24.6 C, which is the same as the assumed starting temperature. The results of the year's calculations are shown in the table. If there

Month	ρ_g	$\bar{T}_a + 2$ C	DD C-day	$\bar{H}$ MJ/m²	$\bar{K}_T$	$\bar{R}$	R_n
Mar.	0.4	1	602	12.89	0.500	1.23	1.12
Apr.	0.2	9	340	15.88	0.475	0.91	0.93
May	0.2	15	194	19.79	0.507	0.78	0.86
June	0.2	21	79	22.11	0.535	0.72	0.83
July	0.2	23	50	21.96	0.547	0.74	0.85
Aug.	0.2	22	60	19.39	0.545	0.86	0.92
Sept.	0.2	17	140	14.75	0.518	1.07	1.03
Oct.	0.2	12	282	10.34	0.500	1.44	1.21
Nov.	0.2	4	508	5.72	0.395	1.69	1.18
Dec.	0.4	−4	743	4.42	0.374	1.92	1.28
Jan.	0.7	−6	828	5.85	0.442	2.03	1.46
Feb.	0.7	−5	695	9.13	0.490	1.69	1.38

Month	$\bar{X}_c$	$\bar{\phi}$	$N\sum \bar{Q}_u$ GJ	Q_{loss} GJ	Q_{load} GJ	T C
Mar.	0.25	0.68	12.4	1.4	10.4	24.6
Apr.	0.21	0.72	11.7	1.6	5.9	25.2
May	0.19	0.72	12.9	2.1	3.4	29.2
June	0.20	0.71	12.6	2.6	1.4	36.4
July	0.25	0.64	12.1	3.3	0.8	44.6
Aug.	0.32	0.56	10.8	3.9	1.0	52.1
Sept.	0.43	0.47	8.3	4.0	2.4	57.8
Oct.	0.53	0.42	7.3	4.2	4.9	59.6
Nov.	0.92	0.32	3.5	3.6	8.8	57.9
Dec.	0.99	0.34	3.3	2.9	12.8	49.4
Jan.	0.55	0.48	6.6	2.1	14.3	37.5
Feb.	0.34	0.60	9.8	1.4	12.0	28.2

had been differences in the initial and final temperatures, the calculations for the year would be repeated until they agreed.

Nothing has been specified in this problem about the characteristics of the heat exchanger between the tank and the building; it is assumed that it is large enough so that the needed energy can be transferred to the building even late in the winter when the tank temperature drops to levels that are not much higher than room temperature. There is also an implicit assumption that the collector area and storage capacity are large enough so that the tank is always hot enough to meet the load. A more complete analysis would include the characteristics of the load heat exchanger, and energy delivered to the load would be calculated as the heat transferred rather than the heating loads on the building.

∎

There are many other applications of $\bar{\phi}$. These include processes such as swimming pool heating, passive and hybrid heating systems, and photovoltaic systems. Some of these will be discussed in the following chapters. In the next section a combination of $\bar{\phi}$ with a correlation method is described and applied to additional problems.

21.3   THE $\bar{\phi}, f$-CHART METHOD

The utilizability design concept is useful whenever the collector operates at a known critical radiation level throughout a month. In a more typical situation the collector is connected to a tank so that the monthly sequence of weather and the load time distribution result in a fluctuating storage tank temperature and consequently a variable critical radiation level. The f-chart method was developed to overcome the restriction of constant critical level, but it is limited to systems which deliver energy to a load near 20 C. The method presented here is not restricted to loads at 20 C.

In this section, the $\bar{\phi}$ concept is combined with the f-chart idea to produce a design method for closed-loop solar systems shown in Figure 21.3.1. In these systems energy supplied to the load must be above a specified minimum useful temperature, and it must be used at a constant coefficient of performance or thermal

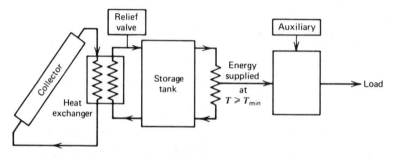

Figure 21.3.1   Schematic of a closed-loop solar energy system.

efficiency so that the load on the solar system can be calculated. (If the load is, for example, a heat engine that performs better as the temperature increases, then the thermal load on the solar system cannot be calculated from knowledge of the heat engine output.)

The storage tank is assumed to be pressurized so that energy dumping does not occur. The return temperature from the load is always at or above T_{min}. A separate auxiliary system is in parallel with the solar system and makes up any energy deficiency of the solar system.

The maximum monthly average daily energy that can be delivered by such a system is given by

$$\sum \bar{Q}_{u,max} = A_c F_R (\overline{\tau\alpha}) \bar{H}_T \bar{\phi}_{max} \tag{21.3.1}$$

which is similar to Equation 21.2.3 except that $\bar{\phi}$ is replaced by $\bar{\phi}_{max}$. The maximum daily utilizability is calculated from the minimum monthly average critical radiation ratio:

$$\bar{X}_{c,min} = \frac{F_R U_L (T_{min} - \bar{T}_a)/F_R(\overline{\tau\alpha})}{r_{t,n} R_n \bar{H}} \tag{21.3.2}$$

The method of calculating $\bar{\phi}_{max}$ is exactly as illustrated in Example 21.2.1.

For a particular storage size-collector area ratio, Klein and Beckman (1979) correlated the results of many detailed simulations of the system of Figure 21.3.1 with two dimensionless variables. These variables are similar to the f-chart variables but are not exactly the same. The f-chart ordinate Y is replaced by $\bar{\phi}_{max}Y$.

$$\bar{\phi}_{max}Y = \bar{\phi}_{max}\frac{A_c F_R(\overline{\tau\alpha})\bar{H}_T N}{L} \tag{21.3.3}$$

and the f-chart abscissa X is replaced by a modified variable X'.

$$X' = \frac{A_c F_R U_L (100)\Delta t}{L} \tag{21.3.4}$$

The change in X is that $(100 - \bar{T}_a)$ has been replaced by an empirical constant 100 (or 180 if English units are used).

Figures 21.3.2(a-d) are $\bar{\phi}$, f-charts for four different storage volume-collector area ratios. The information in these figures can be represented analytically by

$$f = \bar{\phi}_{max}Y - 0.015(e^{3.85f} - 1)(1 - e^{-0.15X'})(R_s)^{0.76} \tag{21.3.5}$$

where R_s is the ratio of the standard storage heat capacity per unit of collector area of 350 kJ/m^2 C to the actual storage capacity. Although f is given implicitly by Equation 21.3.5, it is easy to solve for f by Newton's method or by trial and error.

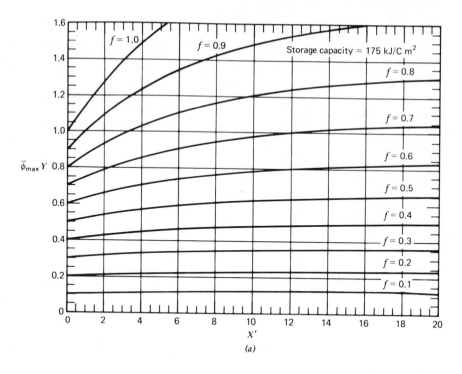

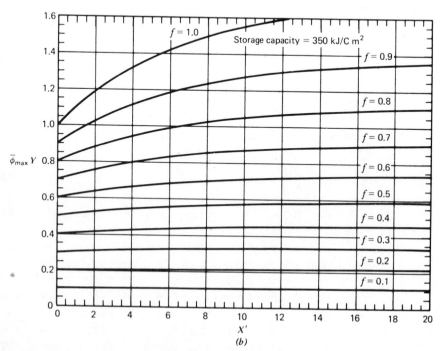

Figure 21.3.2 $\overline{\phi}, f$-charts for various storage capacity-collector area ratios. From Klein and Beckman (1979)

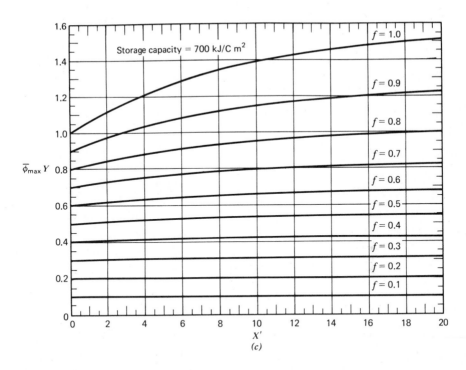

(c)

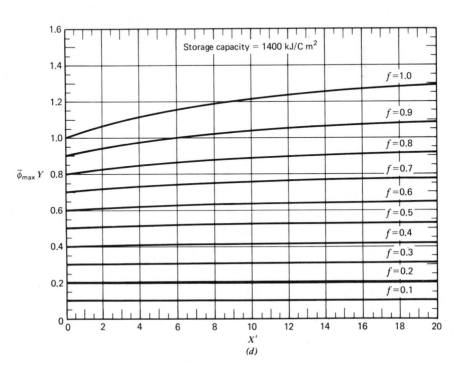

(d)

The $\bar{\phi}, f$-charts are used in the same manner as the f-charts. Values of $\bar{\phi}_{max} Y$ and X' are calculated from long-term radiation data and load patterns for the location in question. The value of f is then determined from the figures or from Equation 21.3.5. The product fL is the average monthly contribution of solar energy to meeting the load. The calculations are repeated for each month, from which the annual fraction $\mathcal{F}$ can be determined.

Example 21.3.1

An industrial solar energy system in Omaha, NE (latitude 41°), requires energy above 60 C at a rate of 12 kW for a 12-hour period each day. The average ambient temperature and the monthly average daily radiation on a horizontal surface are given in the table. The collector heat exchanger characteristics are $F'_R(\tau\alpha)_n = 0.72$; $F'_R U_L = 2.63$ W/m^{2}C; $F_R = 0.8$; $(\overline{\tau\alpha})/(\tau\alpha)_n = 0.94$; tilt $= 40°$; $\gamma = 0$; and $A_c = 50$ m^2. The storage tank holds 4180 liters of water.

Solution

The detailed calculations for January will be illustrated. Results of intermediate calculations for all months are given in the table. The radiation incident on the tilted surface is calculated by the methods of Section 2.19. The ground reflectance will be assumed to be 0.2. $\overline{H}_o$ is 14.6 MJ/m^2 for January, $\overline{K}_T$ is 0.59, and $\overline{H}_d/\overline{H}$ is 0.31. From Equation 2.19.2, $\overline{R}$ is 1.90, and $r_{t,n}$ and $r_{d,n}$ are 0.176 and 0.164.

The beam radiation conversion factor at noon, $R_{b,n}$, is 1.97 from Equation 1.8.3a. For an average day with $K_T = 0.59$, the ratio of diffuse to total radiation H_d/H is 0.38 from Figure 2.11.2. From Equation 2.24.2, the noon ratio of radiation on a tilted surface to that on a horizontal surface for the average day of the month is

$$R_n = \left(1 - \frac{0.164 \times 0.38}{0.176}\right)1.97 + \frac{0.164 \times 0.38}{0.176}\left(\frac{1 + \cos 40}{2}\right) + 0.2\left(\frac{1 - \cos 40}{2}\right) = 1.61$$

The ratio $R_n/\overline{R}$ is then 0.85.

The monthly average transmittance-absorptance product divided by the normal incidence value is 0.94 from the method given in Section 5.10. The critical level at the minimum useful temperature is found from Equation 21.3.2:

$$\overline{X}_c = \frac{2.63[60 - (-5)]/(0.72 \times 0.94)}{0.176 \times 1.61 \times 8.6 \times 10^6/3600} = 0.37$$

The value of $\bar{\phi}_{max}$ is found from Figure 2.24.2 or Equation 2.24.4, with $a = -1.120$, $b = -0.379$, and $c = 0.669$:

$$\bar{\phi}_{max} = \exp\left[(-1.17 - 0.33 \times 0.85)(0.37 + 0.704 \times 0.37^2)\right] = 0.51$$

The value of Y is found from Equation 21.3.3:

$$Y = \frac{50 \times 0.72 \times 0.94 \times 1.90 \times 8.6 \times 10^6 \times 31}{12,000 \times 12 \times 3600 \times 31} = 1.07$$

and $\bar{\phi}_{max}Y$ is then $0.51 \times 1.07 = 0.54$.
    From Equation 21.3.4

$$X' = \frac{50 \times 2.63 \times 100 \times 31 \times 24}{12,000 \times 12 \times 31} = 2.19$$

and it remains constant for the whole year. The storage capacity per unit of collector

1	2	3	4	5	6	7	8	9	10	11	12
Average Day	$\bar{H}_o$	$\bar{T}_a$	$\bar{H}$	$\bar{R}_b$	$\bar{K}_T$	$\dfrac{\bar{H}_d}{\bar{H}}$	$\bar{R}$	$r_{t,n}$	$r_{d,n}$	$R_{b,n}$	R_n
	MJ/m^2	C	MJ/m^2								
Jan. 17	14.6	-5	8.6	2.32	0.59	0.31	1.90	0.176	0.164	1.97	1.61
Feb. 16	20.0	-3	11.6	1.83	0.58	0.32	1.55	0.160	0.148	1.65	1.39
Mar. 16	27.0	3	14.9	1.40	0.55	0.38	1.23	0.145	0.134	1.37	1.16
Apr. 15	34.4	10	19.4	1.08	0.56	0.37	1.03	0.131	0.121	1.16	1.06
May 15	39.6	17	21.5	0.89	0.54	0.39	0.91	0.122	0.112	1.03	0.98
June 11	41.8	22	23.6	0.81	0.56	0.37	0.86	0.118	0.109	0.97	0.96
July 17	40.6	25	23.8	0.84	0.59	0.35	0.88	0.120	0.110	1.00	0.97
Aug. 16	36.3	23	21.8	0.99	0.60	0.34	0.98	0.127	0.117	1.10	1.04
Sept. 15	29.6	19	16.6	1.26	0.56	0.37	1.14	0.139	0.128	1.28	1.12
Oct. 15	22.0	12	12.3	1.67	0.56	0.37	1.40	0.154	0.143	1.55	1.26
Nov. 14	15.8	4	8.3	2.17	0.52	0.37	1.72	0.171	0.159	1.88	1.42
Dec. 10	13.1	-2	7.0	2.50	0.53	0.36	1.94	0.181	0.169	2.09	1.55

	13	14	15	16	17	18	19	20	21	22
Average Day	$\dfrac{R_n}{\bar{R}}$	$\dfrac{(\overline{\tau\alpha})}{(\tau\alpha)_n}$	$\bar{X}_c$	$\bar{\phi}$	Y	$\bar{\phi}Y$	X'	f	L	fL
			(min)	(max)		(max)			GJ	GJ
Jan. 17	0.85	0.94	0.37	0.51	1.07	0.54	2.19	0.52	16.1	8.3
Feb. 16	0.90	0.94	0.34	0.55	1.17	0.64	2.19	0.60	14.5	8.7
Mar. 16	0.95	0.93	0.32	0.57	1.18	0.68	2.19	0.63	16.1	10.2
Apr. 15	1.03	0.93	0.26	0.63	1.29	0.81	2.19	0.74	15.6	11.6
May 15	1.08	0.91	0.24	0.66	1.23	0.81	2.19	0.74	16.1	11.9
June 11	1.11	0.90	0.21	0.70	1.27	0.88	2.19	0.80	15.6	12.4
July 17	1.10	0.91	0.18	0.73	1.33	0.97	2.19	0.86	16.1	13.8
Aug. 16	1.06	0.92	0.18	0.73	1.36	0.99	2.19	0.88	16.1	14.1
Sept. 15	0.98	0.93	0.22	0.69	1.22	0.84	2.19	0.76	15.6	11.9
Oct. 15	0.90	0.94	0.28	0.62	1.12	0.70	2.19	0.65	16.1	10.5
Nov. 14	0.82	0.93	0.39	0.53	0.92	0.49	2.19	0.46	15.6	7.2
Dec. 10	0.80	0.94	0.44	0.48	0.89	0.42	2.19	0.41	16.1	6.6
								Totals	189.6	129.2

area is

$$\frac{mC_p}{A_c} = \frac{4180 \times 4190}{50} = 350 \text{ kJ/m}^2\text{C}$$

so Figure 21.3.2(b) can be used to find f. Alternatively, Equation 21.3.5 can be used but a trial-and-error solution is necessary. With X' and $\bar{\phi}_{max} Y$ equal to 2.19 and 0.54 respectively, f is 0.52. The load for the month of January supplied by solar is then $0.52 \times 16.1 = 8.3$ GJ. The other months are shown in the table. The annual fraction by solar is then the sum of column 22, the energy from solar, divided by the sum of column 21, the yearly load:

$$\mathcal{F} = 127.1/189.2 = 0.67 \qquad\qquad\blacksquare$$

The $\bar{\phi}, f$-chart calculations overestimate f due to the assumptions that there are no losses from the tank and that the load heat exchanger is infinite in size. Corrections can be applied to eliminate both of these assumptions.

The rate at which energy is lost from the storage tank to the surroundings at T_a' is given by

$$\dot{Q}_{st} = (UA)_s(T_s - T_a') \qquad\qquad (21.3.6)$$

If T_a' and $(UA)_s$ are both constant for a month, then integration of Equation 21.3.6 over a month yields the month's tank losses:

$$Q_{st} = (UA)_s(\bar{T}_s - T_a')\Delta t \qquad\qquad (21.3.7)$$

where $\bar{T}_s$ is the monthly average tank temperature.

The total load on the solar system is the useful load plus the energy required to meet the tank losses. If the tank losses are modest so that the tank seldom drops below the minimum temperature, the solar system cannot tell the difference between the energy withdrawn to supply the load and the tank losses. In equation form, the fraction of the total load supplied by solar (including the tank losses) is

$$f_{TL} = \frac{L_S + Q_{st}}{L_0 + Q_{st}} \qquad\qquad (21.3.8)$$

where L_S is the solar energy supplied to the load. Once Q_{st} is known, f_{TL} can be calculated from the $\bar{\phi}, f$-charts in exactly the same manner as illustrated in Example 21.3.1. The usual interpretation for the fraction of the load supplied by solar energy is the ratio L_S/L_0, the solar energy supplied to the load divided by the useful load. If we use the symbol f for this fraction, Equation 21.3.8 can be written as

$$f = f_{TL}\left(1 + \frac{Q_{st}}{L_0}\right) - \frac{Q_{st}}{L_0} \qquad\qquad (21.3.9)$$

The tank losses cannot be calculated exactly but two limiting values can be determined which should bracket the actual losses. A low estimate for tank losses is to assume the tank remains at T_{min} all month. An upper bound for tank losses is to assume the average tank temperature is the same as the monthly average collector inlet temperature $\overline{T}_i$. The actual average tank temperature will be lower than this value since the collector does not operate 24 hours a day. An estimate for $\overline{T}_i$ can be found using the $\overline{\phi}$ charts. The average daily utilizability is

$$\overline{\phi} = \frac{f_{TL}}{Y} \tag{21.3.10}$$

With this value of $\overline{\phi}$, the monthly average operating critical level can be found from the $\overline{\phi}$ charts and $\overline{T}_i$ can then be found. Klein and Beckman (1979) recommend that the arithmetic average of T_{min} and $\overline{T}_i$ be used to evaluate tank losses from Equation 21.3.7. A more conservative estimate is to use $\overline{T}_i$ for estimating tank losses.

The process is iterative. An estimate is first made of the monthly average tank temperature from which Q_{st} is found from Equation 21.3.7. This estimate of tank losses is used as part of the total load and the $\overline{\phi}, f$-chart method is used to estimate f_{TL}. Then $\overline{\phi}$ is calculated from Equation 21.3.10 and $\overline{X}_c$ is found from the $\overline{\phi}$ charts. This dimensionless critical level is used to find I_{Tc} and $\overline{T}_i$ from Equations 21.3.5 and 21.3.2. The tank temperature is compared with the initial guess and the process is repeated if necessary. With the final value of Q_{st}, Equation 21.3.9 is used to find f, the fraction of the load supplied by solar.

Example 21.3.2

Consider the solar system of Example 21.3.1 but include the effect of tank losses. The tank has a $(UA)_s$ of 5.9 W/C and tank losses are to surroundings at 20 C. Do the calculations for January.

Solution

Only the month of January will be considered in this example. For January the average tank temperature will be assumed to be 62 C so that the tank losses, from Equation 21.3.7, are

$$Q_{st} = 5.9(62 - 20) \times 3600 \times 24 \times 31 = 0.7 \text{ GJ}$$

The total load is then $16.1 + 0.7 = 16.8$ GJ. The values of $\overline{\phi}_{max}Y$ and X' are then 16.1/16.8 times the values from Example 21.3.1. Thus

$$\overline{\phi}_{max}Y = 0.54 \times 16.1/16.8 = 0.52$$

$$X' = 2.19 \times 16.1/16.8 = 2.10$$

From the $\bar{\phi}, f$-charts (or Equation 21.3.5) we obtain

$$f_{TL} = 0.50$$

From Equation 21.3.10,

$$\bar{\phi} = 0.50/1.02 = 0.49$$

and X_c is 0.39 from Figure 2.24.2 or Equation 2.24.4 (by trial and error). Since the original value of X_c was 0.37, the temperature difference of 65 C must be increased by the ratio 0.39/0.37. The estimate of $\bar{T}_i$ is then

$$\bar{T}_i = 65\frac{0.39}{0.37} - 5 = 63.5 \text{ C}$$

and the average tank temperature is estimated to be $(63.5 + 60)/2 = 61.7$ C, which is close to the initial guess so no iterations are necessary. The fraction by solar is then found from Equation 21.3.9:

$$f = 0.50\left(1 + \frac{0.7}{16.1}\right) - \frac{0.7}{16.1} = 0.48$$

The tank losses have reduced the fraction of the load by solar in January from 52 to 48% and result in $0.48 \times 16.1 = 7.7$ GJ being supplied by solar rather than 8.3 GJ. On an annual basis the solar fraction is reduced from 67 to 64%. ■

The load heat exchanger adds thermal resistance between the storage tank and the load. This resistance elevates the storage tank temperature, which results in reduced useful energy collection and increased tank losses. In the development of the $\bar{\phi}, f$-charts, the load heat exchanger was assumed to be infinite in size, and consequently the value of f will be optimistic and a correction is necessary.

The average rate of solar energy supplied to the load is found by dividing L_S by the number of seconds during the month in which the load was required, Δt_L. The average increase in tank temperature necessary to supply the required energy rate is

$$\Delta T = \frac{Q_{st}/\Delta t_L}{\varepsilon_L C_{min}} = \frac{fL/\Delta t_L}{\varepsilon_L C_{min}} \tag{21.3.11}$$

where ε_L is the load heat exchanger effectiveness and C_{min} is the smaller of the two fluid capacity rates in the heat exchanger. This temperature difference is added to T_{min} to find the monthly average critical radiation ratio from Equation 21.4.2. Since f is unknown at the beginning, it is necessary to first estimate ΔT and then follow the procedure illustrated in Example 21.3.2 to find f. This value of f is used in Equation 21.3.11 to check the estimate of ΔT. The calculations are illustrated in the following example.

Example 21.3.3

Include the effect of a load heat exchanger on the performance of the system described in Examples 21.3.1 and 21.3.2. The heat exchanger effectiveness is 0.45 and the minimum capacitance rate is 3000 W/C.

Solution

From Example 21.3.1 for January: $\bar{R} = 1.90$, $R_n = 1.61$, and $R_n/\bar{R} = 0.85$; $r_{t,n} = 0.176$, $\bar{K}_T = 0.59$, $\bar{H} = 8.6$ MJ/m², $\bar{T}_a = -5$ C, and $(\overline{\tau\alpha})/(\tau\alpha)_n = 0.94$. As a first estimate of ΔT use 4 C, so from Equation 21.3.2 the minimum critical radiation level is

$$\bar{X}_{c,\min} = \frac{\dfrac{2.63[60 + 4 - (-5)]}{0.72 \times 0.94}}{0.176 \times 1.61 \times 8.6 \times 10^6/3600} = 0.40$$

From Figure 2.24.2 or Equations 2.24.4, $\bar{\phi}_{\max} = 0.49$.

Since we wish to consider tank losses, a guess of the tank temperature is necessary to determine the total load. With a tank temperature of 66 C, $Q_{st} = 0.7$ MJ. The total load is $16.1 + 0.7 = 16.8$ MJ. Then

$$Y = \frac{50 \times 0.72 \times 0.94 \times 1.90 \times 8.6 \times 10^6 \times 31}{16.8 \times 10^9} = 1.02$$

Thus $\bar{\phi}_{\max}Y = 0.49 \times 1.02 = 0.50$. The value of X' is

$$X' = \frac{50 \times 2.63 \times 100 \times 3600 \times 24 \times 31}{16.8 \times 10^9} = 2.11$$

From Equation 21.3.5 or Figure 21.3.2(b), $f_{TL} = 0.48$.

We must now check the tank loss approximation by evaluating $\bar{X}_c$ at $\bar{\phi} = 0.48/1.02 = 0.47$. From Figure 2.24.2 or Equation 2.24.4, $\bar{X}_c$ is 0.41. From Equation 21.3.2

$$\bar{T}_i - \bar{T}_a = \frac{0.41 \times 0.176 \times 1.61 \times 8.6 \times 10^6 \times 0.72 \times 0.94}{2.63 \times 3600} = 73$$

so that $\bar{T}_i = 73 - 5 = 68$. The average temperature for tank losses is then

$$\bar{T} = (64 + 68)/2 = 66$$

Since this is the same as the guess, no iteration is necessary for tank losses. From

Equation 21.3.9, the fraction by solar is

$$f = 0.48\left(1 + \frac{0.7}{16.1}\right) - \frac{0.7}{16.1} = 0.46$$

This value of f is used in Equation 21.3.11 to find ΔT:

$$\Delta T = \frac{0.46 \times 16.1 \times 10^9/(12 \times 3600 \times 31)}{0.45 \times 3000} = 4 \text{ C}$$

Since this is the same as the initial guess, the calculations for January are complete. The load met by solar is then $0.46 \times 16.1 = 7.3$ GJ. ∎

It is interesting to compare the January results of the last three examples. With no tank losses or load heat exchanger, the contribution by solar is 8.3 GJ. With tank losses considered, this was reduced to 7.7 GJ, and with the addition of the load heat exchanger, the energy supplied by solar is 7.3 GJ. These are not insignificant reductions.

Comparisons have been made of $\bar{\phi}, f$-chart predictions with detailed performance predictions from the TRNSYS simulation. In Table 21.3.1, the results for space heating systems which deliver energy to the load with a minimum temperature of 20 C are compared in six climates. Also, the f-chart results are presented. The estimates from all three methods are in good agreement.

In Table 21.3.2, the monthly results of an industrial process heating example are compared for two climates. The process used energy above a minimum temperature of 60 C at a constant rate between 6 AM and 6 PM, 7 days a week. The agreement between TRNSYS and $\bar{\phi}, f$-chart results is excellent in both locations. Differences between the TRNSYS results and $\bar{\phi}, f$-chart results on a monthly basis are affected by energy carryover from month to month, and the radiation data used in the TRNSYS simulations may be too small a sample to adequately represent the long-term statistical distribution.

Table 21.3.1   Comparison of TRNSYS, f-chart, and $\bar{\phi}, f$-chart Results for T_{min} = 20 C[a]

	Space Heating Annual Solar Load Fractions		
	TRNSYS	f-chart	$\bar{\phi}, f$-chart
Albuquerque, NM (1959)	0.79	0.78	0.81
Blue Hill, MA (1958)	0.49	0.50	0.52
Boulder, CO (1956)	0.67	0.68	0.72
Madison, WI (1948)	0.45	0.47	0.47
Medford, OR (1969)	0.55	0.53	0.56
Seattle, WA (1960)	0.57	0.56	0.59

[a] From Klein and Beckman (1979).

Table 21.3.2   Solar Load Fractions for a Process Heating
Application   (T_{min} = 60 C).

	Albuquerque, NM		New York, NY	
	TRNSYS	$\bar{\phi},f$-chart	TRNSYS	$\bar{\phi},f$-chart
Jan.	0.89	0.86	0.77	0.76
Feb.	0.96	1.00	0.65	0.67
Mar.	0.80	0.82	0.58	0.63
Apr.	0.90	1.00	0.49	0.52
May	0.75	0.78	0.30	0.32
June	0.66	0.64	0.20	0.16
July	0.69	0.66	0.26	0.21
Aug.	0.74	0.71	0.39	0.37
Sept.	0.95	0.93	0.73	0.68
Oct.	0.92	0.89	0.58	0.56
Nov.	0.91	0.90	0.45	0.40
Dec.	0.94	0.93	0.66	0.66
Year	0.84	0.84	0.50	0.49

[a] From Klein and Beckman (1979).

The $\bar{\phi},f$-chart method has potential for misuse, and as a result its limitations need to be emphasized. The method is intended for applications in which the load can be characterized by a single temperature, T_{min}. The load must be relatively uniform on a day-to-day basis as it will product inaccurate results for processes in which the load distribution is highly irregular. The method is not applicable for systems in which energy supplied to the load is used at an efficiency or COP that depends upon temperature, such as the solar-Rankine engine.

21.4   SUMMARY

In this chapter three design methods have been presented: ϕ, $\bar{\phi}$, and $\bar{\phi},f$-chart. The ϕ method can usually be replaced by the $\bar{\phi}$ method with approximately a fourfold reduction in calculation. The hourly and daily utilizability methods require that the collector critical radiation level remain constant for a month, which means that the difference between the collector inlet temperature and ambient temperature is nearly fixed. In practice this applies best to systems with very large storage tanks or to systems with no storage in which the collector inlet fluid is from a constant temperature source. Neither of these two situations is very common. The $\bar{\phi},f$-charts were developed for closed-loop systems with finite storage where the load is characterized by a single minimum useful temperature. This is a common system, but it does not cover all practical applications. Systems that are not covered by these methods must be designed using detailed simulations.

REFERENCES

Collares-Pereira M., and A. Rabl, *Solar Energy*, **23**, 223 (1979a). "Derivation of Method for Predicting Long Term Average Energy Delivery of Solar Collectors."

Collares-Pereira, M. and A. Rabl, *Solar Energy*, **23**, 235 (1979b). "Simple Procedure for Predicting Long Term Average Performance of Nonconcentrating and of Concentrating Collectors."

Hottel, H. C. and A. Whillier, *Trans. of the Conference on the Use of Solar Energy*, Vol. 2, part I, U. of Arizona Press, 74, (1958). "Evaluation of Flat-Plate Collector Performance."

Klein, S. A., *Solar Energy*, **21**, 393 (1978). "Calculation of Flat-Plate Collector Utilizability."

Klein, S. A. and W. A. Beckman, *Solar Energy*, **22**, 269 (1979). "A Generalized Design Method for Closed-Loop Solar Energy Systems."

Liu, B. Y. H. and R. C. Jordan, *Solar Energy*, **7**, 53 (1963). "A Rational Procedure for Predicting the Long-Term Average Performance of Flat-Plate Solar-Energy Collectors."

Whillier, A., Sc.D. Thesis, Mechanical Engineering, Massachusetts Institute of Technology (1953). "Solar Energy Collection and Its Utilization for House Heating."

Chapter 22

DESIGN OF PASSIVE AND HYBRID HEATING SYSTEMS

The principles underlying passive (and active) solar processes are outlined in Chapters 1 to 11, and passive heating and cooling processes and phenomena associated with them are described in Chapters 14 and 15. In this chapter we deal with questions of estimating the annual performance of several types of passive building-heating systems. The solar-load ratio correlation method developed at Los Alamos Scientific Laboratory is first introduced. The application of utilizability methods to direct-gain and collector-storage wall systems is shown. Then design methods for two hybrid systems are outlined, for active-collection, passive-storage systems and for systems having significant fractions of annual loads carried by both active and passive processes. This combination of methods will allow the annual performance of a wide variety of passive and hybrid systems to be estimated.

22.1 APPROACHES TO PASSIVE DESIGN

Thermal design of passive buildings is closely interrelated to architectural design, as the collection of solar energy and its storage are accomplished in elements of the structure itself. There is a spectrum of methods for estimating long-term thermal performance of these buildings that ranges from practical experience ("... do it this way in this climate and it will work ...") to use of charts and tables that are based on combinations of experience and calculations to correlation and utilizability methods that are the counterparts of the methods for active systems.

Estimation of solar energy absorbed in a building is not difficult and can be done by the methods outlined in Sections 5.10 and 5.11. Load calculations are more uncertain but can be done by the methods outlined in Chapter 9 or other suitable means that the designer may wish to use. An important problem in estimating passive system performance is the estimation of how much of the absorbed energy cannot be used because it is available at a time when it exceeds the loads and the capacity of the building to store it (i.e., how much energy is "dumped"). The thermal problem that is addressed in this chapter is coupled with the aesthetic problem, the challenge of making the building attractive and functional.

Mazria (1979) presents specific recommendations on sizing of solar apertures and providing storage capacity. For example, for a direct-gain system in cold climates, the ratio of area of south-facing glass to floor area should be 0.19 to 0.38; if masonry storage is to be used, the interior walls and floors should be a minimum of 0.1 m (4 in.) thick; movable insulation should provide a tight cover for the glazing. A design based on these specific recommendations can then be "fine tuned" by calculation of heat losses, solar gains, average indoor temperature, indoor temperature fluctuations, auxiliary energy requirements, and economics.

A "Method 5000" developed in France and described by Lebens and Myer (1984) is a three-part method: first, calculation of heat loss from the building; second, calculation of the total solar energy absorbed by the passive solar components; third, calculation of the utilized solar and internal gains and thus of the auxiliary energy requirements. The critical step is the third. Utilization factors are obtained from correlations as a function of absorbed solar and internal gains, $\overline{T}_a$, a heat loss coefficient of the building, and the thermal mass of the building.

These design methods, and the solar-load ratio and unutilizability methods to be described in more detail in the following sections, have a common basic feature, i.e., they all use correlations of results of simulations to determine long-term performance. The correlating variables, the definitions of terms, the ways in which the correlations are used, and the generality of the correlations are, however, very different from one method to another. Care must be exercised in the use of these methods to see that the terminology and correlations are understood and properly used.

22.2   THE SOLAR-LOAD RATIO METHOD

The solar-load ratio (SLR) method is widely used for designing direct-gain, collector-storage wall, and sunspace systems; it was developed by Balcomb et al. (1980, 1983a,b). It is a method of calculating annual requirements for auxiliary energy, based on extensive simulation studies of performance of many passive heating systems done with the simulation program PASOLE [McFarland (1978)]. The simulations were in turn compared to experiments on test cells. A total of 94 standard system configurations were simulated and correlations were developed for their performance. In addition, sensitivity curves were developed to assess the impact on performance of a number of design and climatic variables. Background information on this method is in volume 1 of the *Passive Design Handbook*, and details of the method and sensitivity analyses are in volumes 2 and 3 [Balcomb et al. (1980, 1983a,b)]. The method is presented in *Passive Solar Heating Analysis* [Balcomb et al. (1984)] and summarized in Balcomb et al. (1983b). In this section we show the essentials of the SLR method for selected systems of each of the three types and examples of the sensitivity curves. The significant end result of SLR calculations is the annual requirement for auxiliary energy; individual months may be substantially in error, but annual results of the method are generally within ±3% of results of detailed simulations.

The SLR method is based on a set of defined terms that must be properly used in order for the correlations to work. All energy terms and calculations are on a monthly basis. Some of the terms in the SLR method have dimensions, and care must be exercised to see that appropriate units are used. In particular, in SI units, energy quantities in this method are in watt-hours (Wh) rather than in joules. In the discussion to follow, loss coefficient-area products are in W/C, and loads and auxiliary energy quantities are in Wh. Important definitions and symbols follow.

The **solar wall** is the glazed building wall that provides the solar gains that are to be estimated. The **solar aperture** is that portion of the wall that is glazed to admit solar radiation. The **net glazing area** A_r is the area of the solar aperture after mullions, framing, etc., have been subtracted. The **projected area** A_{rp} is the projection of the net glazing area on a vertical plane normal to the azimuth of the glazing. (Note that for direct-gain and collector storage walls, A_r and A_{rp} are the same; for sunspaces, A_r will be substantially larger than A_{rp}.)

The SLR correlations are based on loads that do not include losses[1] through the solar aperture. The **net reference load** L_{ns} is the monthly heat loss from the nonsolar portions of the building, i.e., from all of the building envelope except the solar aperture. The **net load coefficient** is the net reference load per degree of temperature difference between indoors and outdoors, i.e., a UA of the nonsolar parts of the building.[2] The net reference load in Wh for the month, if $(UA)_{ns}$ is in W/C and DD is C-days for the month, is

$$L_{ns} = 24(UA)_{ns} (DD) \tag{22.2.1}$$

where DD is the number of degree-days in the month, using the appropriate base temperature. (See Chapter 9 for a method for calculating degree-days to any desired base temperature.)

The **gross reference load** L is the heat loss from the building, including both solar and nonsolar portions of the building. The **total load coefficient** is the gross reference load per degree of difference between indoor and outdoor temperature, i.e., a UA for the whole building. In a form analogous to Equation 22.2.1, the load in Wh is

$$L = 24(UA)(DD) \tag{22.2.2}$$

The building load coefficients $(UA)_{ns}$ and UA can be calculated by standard methods [e.g., ASHRAE (1989)].

The **solar savings fraction** f_{ns} is the fraction of the net reference load that is met by solar energy.[3] Thus

$$f_{ns} = 1 - L_A/L_{ns} \tag{22.2.3}$$

---

[1] Losses through the solar aperture are, however, included in the calculation of auxiliary energy required, as will be shown.

[2] It can be thought of as the UA of the building if the solar wall were adiabatic.

[3] The fraction f_{ns} is not comparable to solar fraction as defined and used in active systems, nor to that in the unutilizability passive design methods described in Sections 22.3 and 22.4. Since f_{ns} is the fraction of a non-existant load, it should not be viewed as having physical significance.

The **load collector ratio** LCR is the ratio of the net load coefficient to the projected area of the solar aperture.

$$LCR = 24(UA)_{ns}/A_{rp} \qquad (22.2.4)$$

The **load collector ratio for the solar aperture** LCR_s is the ratio of the loss coefficient of the solar aperture (based on the projected area) to the projected area. The 94 standard systems are so defined that LCR_s is determined for each of them, and appropriate values are tabulated.

The monthly average daily solar radiation absorbed in the building per unit of projected area $\overline{S}'$ is calculated by the methods of Section 5.10, and Section 14.4 if there are shading overhangs. In equation form,

$$\overline{S}' = \overline{S}A_r/A_{rp} \qquad (22.2.5)$$

The general correlation for solar savings fraction that is applicable to all three types of systems (direct-gain, collector-storage wall, and sunspace) is

$$f_{ns} = 1 - (1 - F)K \qquad (22.2.6a)$$

$$K = 1 + G/LCR \qquad (22.2.6b)$$

$$F = \begin{cases} B - Ce^{-DX} & \text{when } X > R \\ AX & \text{when } X < R \end{cases} \qquad (22.2.6c)$$

In either case, $F \le 1$. Here X is a generalized solar-load ratio given by

$$X = \frac{\dfrac{N\overline{S}'}{DD} - LCR_s \times H}{LCR \times K} \qquad (22.2.6d)$$

In these equations A, B, C, D, G, H, LCR_s, and R are constants that have been determined for each of the 94 passive systems types and N is the number of days in the month.

The auxiliary energy required for each month is obtained by rearranging Equation 22.2.3:

$$L_A = L_{ns}(1 - f_{ns}) \qquad (22.2.7)$$

The calculation is done for each month of the year or heating season to provide an estimate of the annual auxiliary energy required.

Original development of the SLR method was done in English units, and some of the defined quantities are not dimensionless. In Balcomb et al. (1983b), constants and examples are given in both English and SI units. A summary of units that should be used in calculations using either set of units is in Table 22.2.1. The table also shows the symbols used by Balcomb et al. and those used in this book.

Tables 22.2.2 and 22.2.3 indicate design parameters for a set of 22 passive systems designs selected from the 94 designs for which correlations have been

Table 22.2.1   Units Used in the SLR Design Method[a]

Quantity	Symbols		Units	
	This book	Balcomb	SI	English
Net glazing area	A_r	-	m^2	ft^2
Projected area	A_{rp}	A_p	m^2	ft^2
Net reference load	-	-	Wh	Btu
Net load coefficient	$24(UA)_{ns}$	NLC	Wh/C-day	Btu/F-day
Gross reference load	L	-	Wh	Btu
Total load coefficient	$24(UA)$	TLC	Wh/C-day	Btu/F-day
Load collector ratio	LCR	LCR	Wh/m^2C-day	Btu/ft^2F-day
Solar savings fraction	f_{ns}	SSF	-	-
Degree days	DD	DD	C-day	F-day
Absorbed solar radiation	$\overline{S}'$	S'	Wh/m^2	Btu/ft^2
Dimensional correlation constant	G	G	$[Wh/m^2C$-day$]^{-1}$	$[Btu/ft^2F$-day$]^{-1}$

[a] Adapted from Balcomb et al. (1983a).

Table 22.2.2   Design Parameters for Direct-Gain and Collector-Storage Wall Passive Systems[a]

Type	Thermal Storage Capacity KJ/m^2C	Storage[b] Mass Thickness m	Number of Glazings	Storage Mass-Glazing Area Ratio	Night[c] Insulation
Direct-Gain Systems					
A1	613	0.051	2	6	No
A2	613	0.051	3	6	No
A3	613	0.051	2	6	Yes
C1	1227	0.102	2	6	No
C2	1227	0.102	3	6	No
C3	1227	0.102	2	6	Yes
Vented Collector-Storage Wall Systems					
A1	306	0.0152	2	1	No
A2	459	0.0229	2	1	No
A3	613	0.0305	2	1	No
A4	919	0.0457	2	1	No
D2	613	0.0305	3	1	No
D3	613	0.0305	1	1	Yes
E2[d]	613	0.0305	3	1	No
E3[d]	613	0.0305	1	1	Yes

[a] Adapted from Balcomb et al. (1983b).
[b] Mass (or wall) thickness assuming a volumetric heat capacity of 620 kJ/m^3C.
[c] Night insulation, when used is "R9" ($U_N = 0.63$ W/m^2C).
[d] The E systems have a selective surface with $\alpha = 0.95$, $\varepsilon = 0.10$ on the glazing side of the wall. For nonselective surfaces, $\alpha = 0.95$ and $\varepsilon = 0.90$.

Table 22.2.3   Design Parameters for Sunspace Passive Systems[a]

Type	Geometry[b]	Glazing[c] Slope	Common[d] Wall	End Walls	Night Insulation
A1	Attached	50	Masonry	Opaque	No
A2	Attached	50	Masonry	Opaque	Yes
B3	Attached	90/30	Masonry	Glazed	No
B4	Attached	90/30	Masonry	Glazed	Yes
B5	Attached	90/30	Insulated	Opaque	No
B6	Attached	90/30	Insulated	Opaque	Yes
E3	Semi-enclosed	90/30	Insulated	Common	No
E4	Semi-enclosed	90/30	Insulated	Common	Yes

[a] Adapted from Balcomb et al. (1983b).

[b] See Figure 15.7.1.

[c] Double-glazing, south-facing, slopes as indicated and shown on Figure 15.7.1.

[d] Common wall separates sunspace from balance of building. If masonry, 0.30 m thick, $k = 1.73$ W/mC. If insulated, conductance is 0.28 W/m^{2}C.

Table 22.2.4   Solar Load Ratio Correlation Parameters[a]

Type	A	B	C	D	R	G	LCR_s	H
Direct-Gain Systems								
A1	0.5650	1.0090	1.0440	0.7175	0.3931	53.1	0	0
A2	0.5906	1.0060	1.0650	0.8099	0.4681	29.9	0	0
A3	0.5442	0.9715	1.1300	0.9273	0.7086	15.0	0	0
C1	0.6344	0.9887	1.5270	1.4380	0.8632	54.5	0	0
C2	0.6763	0.9994	1.400	1.3940	0.7604	29.9	0	0
C3	0.6182	0.9859	1.5660	1.4370	0.8990	13.6	0	0
Vented Collector-Storage Wall Systems								
A1	0	1	0.9194	0.4601	−9	0	73.7	1.11
A2	0	1	0.9680	0.6318	−9	0	73.7	0.92
A3	0	1	0.9964	0.7123	−9	0	73.7	0.85
A4	0	1	1.0190	0.7332	−9	0	73.7	0.79
D2	0	1	1.0150	0.8994	−9	0	52.2	0.80
D3	0	1	1.0346	0.7810	−9	0	50.5	1.08
E2	0	1	1.0476	1.0050	−9	0	49.3	0.66
E3	0	1	1.0919	1.0739	−9	0	31.2	0.61
Sunspace Systems								
A1	0	1	0.9587	0.4770	−9	0	105.5	0.83
A2	0	1	0.9982	0.6614	−9	0	59.0	0.77
B3	0	1	0.9689	0.4685	−9	0	109.5	0.82
B4	0	1	1.0029	0.6641	−9	0	55.0	0.76
B5	0	1	0.9408	0.3866	−9	0	92.5	0.97
B6	0	1	1.0068	0.6778	−9	0	48.2	0.84
E3	0	1	0.9565	0.4827	−9	0	111.2	0.81
E4	0	1	1.0214	0.7694	−9	0	61.3	0.79

[a] From Balcomb et al. (1983b).

developed. This limited set covers a variety of combinations of direct-gain, collector-storage wall, and sunspace systems; various glazings; storage characteristics; and systems with and without night insulation. Table 22.2.4 gives the necessary constants for Equations 22.2.4 for this set. Note that all of the entries in this table are dimensionless except LCR_s which has units of Wh/m²C-day, and G, which has units of (Wh/m²C-day)⁻¹.

In the development of the correlations a number of assumptions were made in addition to those indicated in the tables. The room temperatures were assumed to be in the range of 18.3 to 23.9 C and sunspace temperatures in the range of 7.2 to 35.0 C. The collector-storage wall top and bottom vents were each assumed to have areas of 3% of the projected area, with the vertical distance between vents 2.4 m. Night insulation, when used, was assumed to be in place from 5:30 PM to 7:30 AM throughout the year. Shading was taken as negligible, and ground reflectance was taken as 0.3. Sketches of the three types of sunspaces included in the table are shown in Figure 22.2.1. Glazing is assumed to be glass 3.2 mm thick with 12.7 mm spacing between panes.

The SLR correlations (in addition to those shown here) included unvented collector-storage walls, water walls, and designs of sunspaces similar to those shown but with different design parameters.

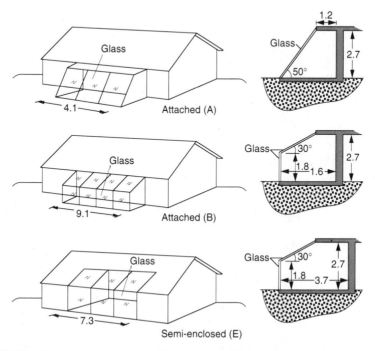

Figure 22.2.1 The geometry of sunspaces included in Tables 22.2.3 and 22.2.4. Adapted from Balcomb et al. (1983a).

Example 22.2.1

A residential building in Tulsa (latitude 36°) has a south-facing vented concrete collector-storage wall of 15 m² area. The wall is 0.305 m thick, and the thermal storage capacity is 610 kJ/m²C. Triple glazing is used, a flat black paint is applied to the outer surface of the masonry, and no night insulation is provided. The building exclusive of the solar wall has a UA of 200 W/C, including effects of infiltration and internal generation. The average January daily absorbed solar radiation and degree-days to the appropriate base are 5.67 MJ/m² per day and 491 C-days; these are calculated from the data in Appendix G using the methods of Chapters 2 and 5. Here $(\tau\alpha)_n$ was taken as 0.70.

Estimate the January requirement for auxiliary energy to heat this building.

Solution

This system is essentially the same as the vented collector-storage wall system D2 in Table 22.2.4. From the table the constants for this system are $A = 0$, $B = 1$, $C = 1.015$, $D = 0.8994$, $R = -9$, $G = 0$, $LCR_s = 52.2$ Wh/m²C-day, and $H = 0.80$.

$$\text{Net load coefficient} = 200 \times 24 = 4800 \text{ Wh/C-day}$$

$$LCR = 24(UA)_{ns}/A_{rp} = 4800/15 = 320 \text{ Wh/m}^2\text{C-day}$$

$$K = 1 + G/LCR = 1$$

From Equation 22.2.6d,

$$X = \left(\frac{31 \times 5.67 \times 10^6}{3600 \times 491} - 52.2 \times 0.8\right)(320)^{-1} = 0.180$$

From Equation 22.2.4c,

$$F = 1 - 1.015e^{-0.8994 \times 0.180} = 0.137$$

From Equation 22.2.4a,

$$f_{ns} = 1 - (1 - F) = 0.137$$

From Equation 22.2.7 the auxiliary energy is then

$$L_A = 4800 \times 3600 \times 491(1 - 0.137) = 7.3 \text{ GJ} \qquad \blacksquare$$

Figure 22.2.2 is an example of the inter-relationships of the LCR, $N\overline{S}'/DD$, and f_{ns}. It is for direct-gain system A1; the shapes of other curves are generally similar. As the load-collector ratio diminishes at a fixed ratio $N\overline{S}'/DD$, the solar

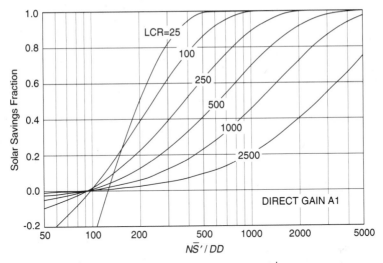

Figure 22.2.2 A sample plot of f_{ns} as a function of LCR and S'/DD, Adapted from Balcomb et al. (1980).

savings fraction rises and L_A (the ultimate criterion of a system's worth) will decrease. At a fixed LCR, as the ratio $N\bar{S}'/DD$ increases, the solar savings fraction will increase but the increase is sharper at low values of LCR than at high.

The SLR method includes correlations for 94 designs. However, systems other than those of the 94 designs will need to be evaluated. Thus the sensitivity of the solar savings fraction to several design parameters has been computed using simulations. Two types of sensitivities are shown. The first is a set of general relationships of sensitivity of all systems in all climates to design parameters; this is illustrated here by the response of f_{ns} to the amount of night insulation used. The second type includes sensitivities of specific types of systems in specific locations to design parameter changes; many of these have been determined, and a single illustration is included here.

Night insulation can have very substantial effects on passive heating performance. The reference insulation conductance used in development of the SLR correlations for systems with night insulation is 0.63 W/m²C ($R9$), a value that may be difficult to achieve, so it is necessary to assess the effects of lesser levels. Figure 22.2.3 shows a correction factor to be applied to the difference in f_{ns} for the reference insulation ($R9$) and no insulation for direct-gain and collector-storage wall systems. The factor is given by the equation

$$Y = \frac{1 + R_0/1.59}{1 + R_0/R} \tag{22.2.8}$$

where R_o is 0.32 m²C/W for direct-gain systems, 0.56 m²C/W for water wall, and

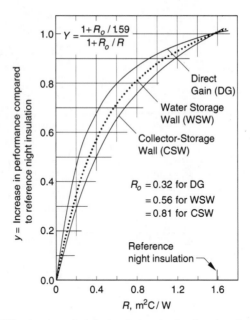

Figure 22.2.3   Effectiveness of night insulation for levels other than the reference levels. Adapted from Balcomb et al. (1980).

0.81 m²C/W for masonry collector-storage wall systems. The f_{ns} for a system with an intermediate level of night insulation is

$$f_{ns} = f_{ns,\text{no ins}} + Y(f_{ns,\text{ref ins}} - f_{ns,\text{no ins}})$$

(22.2.9)

Example 22.2.2

Calculation of the solar savings fraction of a direct-gain system in a particular location and month shows $f_{ns} = 0.46$ with no insulation and $f_{ns} = 0.78$ with the reference night insulation. What would be the f_{ns} if the system were to be insulated at night with $R3$ insulation having a resistance R of 0.53 m²C/W?

Solution

From Figure 22.2.3 or Equation 22.2.8,

$$Y = (1 + 0.32/1.59)/(1 + 0.32/0.53) = 0.75$$

Then for the insulation with $R = 0.53$ m²C/W,

$$f_{ns} = 0.46 + 0.75(0.78 - 0.46) = 0.70 \qquad \blacksquare$$

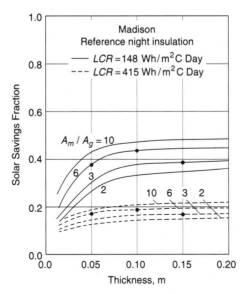

Figure 22.2.4 An example of the SLR method sensitivity curve: the effects of thickness of storage mass and ratio of mass area to glazing area for double-glazed, direct-gain systems with the reference night insulation at two values of *LCR*, in Madison. Adapted from Balcomb et al. (1983a).

An example of a specific sensitivity curve, the dependence on storage mass, thickness, and ratio of mass area to glazing area of direct-gain systems with particular values of *LCR* in Madison, WI, is shown in Figure 22.2.4. The reference system performance is shown by heavy dots. The solid curves are for systems with load collector ratio of 148 Wh/m^{2}C-day, and the dashed curves are for *LCR* of 415 Wh/m^{2}C-day. In both cases night insulation with an *R* value of 1.59 m^{2}C/W is used.

The sensitivity of f_{ns} to storage mass and absorptance, number of glazings, and temperature limits has been computed for direct-gain systems. Effects of storage mass and properties, glazing properties, wind speed, wall to room heat transfer coefficient, and several other parameters has been worked out for collector-storage wall systems. See the *Passive Solar Handbook* (volumes 2 and 3) or *Passive Solar Heating Analyses* for details.

22.3 THE UNUTILIZABILITY DESIGN METHOD: DIRECT-GAIN

The unutilizability method (the "double-U method") for design of direct-gain and collector-storage wall systems, developed by Monsen et al. (1981, 1982), is based on the concept that a passively heated building can be viewed as a collector with finite heat capacity. Analyses are developed for estimating the auxiliary energy requirements of two limiting cases. The first is for an infinite capacitance structure that can store all energy in excess of loads. The second is for a zero capacitance structure than can store no energy. Correlations are used to determine where between these limits the auxiliary energy requirements of a real structure must lie.

As is the case with f-chart and SLR design methods, the UU method calculations are done on a monthly basis. The significant end result of the calculations is the annual auxiliary energy required. The method requires calculation of loads; these calculations can be done by any method the designer wishes. The simplest load calculation method is the degree-day method; allowance can be made for variable base temperature T_b, night thermostat setback, night insulation on the direct-gain windows, and internal generation (other than solar gains through the direct-gain receiver). In contrast to the solar-load ratio method, loads as defined for this method are for the whole building including the direct-gain aperture.

Figure 22.3.1 shows the monthly energy streams entering and leaving a direct-gain structure. Solar energy absorbed in the room is $\overline{H}_T N(\overline{\tau\alpha})A_r = N\overline{SA}_r$. The energy lost through the building shell by conduction, by infiltration, etc., is shown as the load. At times there will be insufficient solar energy to meet loads and auxiliary energy L_A must be supplied. There will also be times when there is excess solar energy absorbed that is not used to meet losses and cannot be stored, and this Q_D must be vented, or "dumped." The diagram does not show energy stored; at any time during a month sensible heat may be stored in or removed from the structure if it has thermal capacitance.

Consider first a hypothetical building with **infinite storage capacity**. During a month, all absorbed energy in excess of the load is stored in the structure. The temperature of the conditioned space remains constant, as infinite capacitance for storage implies zero temperature change. This stored energy can be used at any time to offset auxiliary energy needs. However, month-to-month carryover is not considered.

A monthly energy balance gives the auxiliary energy requirement for the infinite capacity building, $L_{A,i}$:

$$L_{A,i} = (L - N\overline{SA}_r)^+ \qquad\qquad (22.3.1)$$

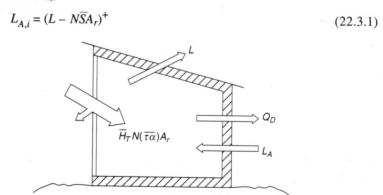

Figure 22.3.1 Monthly energy streams for a direct-gain building. Energy storage is not indicated.

where L is the load for the entire building, i.e., the auxiliary energy that would be required if the transmittance of the glazing were zero. The superscript plus sign indicates that only positive values of the difference of monthly energy quantities in the bracket are considered.

Next consider a hypothetical **zero storage capacity** building. Since there is no storage capacity, energy deficits must be made up as they occur by use of auxiliary energy and excess solar energy is dumped as it is absorbed. The temperature in the building is again fixed, but by addition or removal of energy rather than storage. An instantaneous energy balance on the structure gives the rate at which energy must be removed from the structure, $\dot{Q}_D$:

$$\dot{Q}_D = [I_T(\tau\alpha)A_r - (UA)_h(T_b - T_a)]^+ \tag{22.3.2}$$

As with collectors in active systems, a critical radiation level can be defined as that level at which the gains just offset the losses:

$$I_{Tc} = (UA)_h (T_b - T_a)/(\tau\alpha) A_r \tag{22.3.3}$$

In this case absorbed radiation above the critical level must be dumped and is "unutilizable." The dumped energy for the month Q_D can then be written as

$$Q_D = A_r(\overline{\tau\alpha}) \int_{mo} (I_T - I_{Tc})^+ \, dt \tag{22.3.4}$$

Figure 22.3.2 shows operation over a sequence of days of a zero-capacitance system. It is assumed that over a month I_{Tc} can be considered to be constant at its monthly average value given by Equation 22.3.5:

$$I_{Tc} = (UA)_h(T_b - \overline{T_a})/(\overline{\tau\alpha})A_r \tag{22.3.5}$$

Energy above I_{Tc} must be dumped, and energy below I_{Tc} is useful. Thus the shaded area represents dumped energy, the white areas under the curves represent useful solar energy and the cross-hatched areas represent auxiliary energy for the zero-capacitance building.

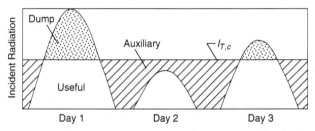

Figure 22.3.2 Dumped, useful, and auxiliary energy for a zero-capacitance direct-gain system. From Monsen et al. (1981).

Equation 22.3.4 can be expressed in terms of the monthly average utilizability $\bar{\phi}$, as defined in Equation 2.24.1.

$$\bar{\phi} = \frac{1}{\bar{H}_T N} \sum_{\text{days}} \sum_{\text{hours}} \left(I_T - I_{Tc} \right)^+ \tag{22.3.6}$$

So Q_D is given by

$$Q_D = N \, \bar{S} A_r \bar{\phi} \tag{22.3.7}$$

The amount of auxiliary energy required by the zero-capacitance building can now be determined from a monthly energy balance. It is the load plus dumped energy less the absorbed solar energy.

$$L_{A,z} = L - (1 - \bar{\phi}) N \bar{S} A_r \tag{22.3.8}$$

Equations 22.3.1 and 22.3.8 give bounds on the amount of auxiliary energy required. A correlation based on simulations indicates where within these limits the auxiliary energy for a real system will lie. The correlations have been developed in terms of the fraction of the load supplied by solar energy, which is defined in the same way as for active systems (i.e., as $f = 1 - L_A/L$). A solar-load ratio is defined as

$$X = \frac{\bar{H}_T N (\overline{\tau \alpha}) A_r}{L} = \frac{N \bar{S} A_r}{L} \tag{22.3.9}$$

For the infinite-capacitance system, from Equation 22.3.1, X is the solar fraction f_i:

$$f_i = (1 - L_{A,i}/L) = X \tag{22.3.10}$$

For the zero-capacitance case

$$f_z = (1 - \bar{\phi}) X \tag{22.3.11}$$

In the simulation study of systems with finite capacitance, the building interior was assumed to consist of a single zone at uniform temperature with an effective thermal capacitance (mass times heat capacity) C. The temperature was allowed to float between minimum and maximum thermostat set temperatures. Auxiliary energy was provided to keep the temperature at the minimum setpoint temperature, and ventilation was used to dump energy and keep the temperature below the specified maximum.

The monthly solar fraction f was found to correlate with the solar-load ratio X and a storage-dump ratio:

$$Y = \frac{C_b \Delta T_b}{\bar{H}_T (\overline{\tau \alpha}) A_r \bar{\phi}} = \frac{C_b \Delta T_b}{\bar{S} A_r \bar{\phi}} = \frac{N C_b \Delta T_b}{Q_D} \tag{22.3.12}$$

where C_b is the effective thermal capacitance, ΔT_b is the difference in upper and lower temperatures (i.e., the range over which building temperature is allowed to float), and the units are such that Y is dimensionless. Values of C_b are available from Table 9.5.1. (C_b is approximately 120 kJ/m²C based on floor area for typical residential frame construction.) The parameter Y is the monthly ratio of the maximum storage capacity of the building to the solar energy that would be dumped if the building has zero thermal capacitance. The zero- and infinite-capacitance buildings thus have values of Y of zero and infinity, respectively.

A correlation for a monthly solar fraction f, in terms of X, Y, and $\bar{\phi}$, is

$$f = \min\{PX + (1 - P)(3.082 - 3.142\bar{\phi})[(1 - \exp(-0.329X)], 1\}$$
(22.3.13a)

where $\qquad P = [1 - \exp(-0.294Y)]^{0.652}$ (22.3.13b)

The auxiliary energy required is then calculated from f by

$$L_A = L(1 - f)$$
(22.3.14)

Note that since the load varies with aperture, values of f calculated this way for different solar apertures are not comparable with one another. It is appropriate to compare auxiliary energy requirements L_A for different apertures.

An example of an f vs. X plot for various values of Y for $\bar{\phi} = 0.6$ is shown in Figure 22.3.3.

The annual auxiliary energy requirements predicted with Equations 22.3.12 and 22.3.13 have been compared to results of detailed simulations for a wide variety of system parameters and climates. The standard deviation of the differences of annual solar fractions was 3%. (The standard deviation of monthly differences was 6%.)

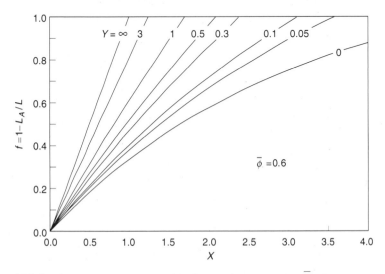

Figure 22.3.3 $(1 - L_A/L)$ vs. X and Y for direct-gain system, for $\bar{\phi}$ of 0.6. (Note: use Equations 22.3.13 for other values of $\bar{\phi}$). From Monsen et al. (1981).

Example 22.3.1

A well-insulated residence in Springfield, IL, is to be heated by direct gain. There are four rooms having direct-gain south-facing windows; the characteristics of each of the energy absorption structures are as outlined in Example 5.10.1. The area of each of the direct-gain windows is 4.5m² and the absorptance of the cavities is 0.87. In the table that follows $\bar{S}$ and $(\overline{\tau\alpha})$ are reproduced from Example 5.10.1.

The effective heat capacity of the building (which is of moderately heavy construction) is estimated to be 41.4 MJ/C. The allowable temperature swing is 6 C and the low setpoint temperature is 18.3 C. The UA of the building exclusive of the direct-gain windows is 138 W/C, and the U of the windows (which do not have night insulation) is 4.17 W/m²C.

Estimate the annual auxiliary energy requirements for heating this building.

Solution

The calculations for January are outlined in detail. The results for the year are summarized in the table. The degree-day data are from Appendix G. (The low set-point is 18.3 C, the same as the base temperature for the degree-day data in the table.)

First calculate the load. The UA of the entire building, including the four direct-gain windows of area $4 \times 4.5 = 18$ m², is

$$(UA)_h = 138 + 18 \times 4.17 = 213 \text{ W/C}$$

The January load is then

$$L = 213 \times 657 \times 24 \times 3600 = 12.09 \text{ GJ}$$

Second, calculate the solar-load ratio X from Equation 22.3.9:

$$X = 7.60 \times 10^6 \times 31 \times 18/(12.09 \times 10^9) = 0.351$$

Third, calculate I_{Tc} with Equation 22.3.5; $\bar{T}_a = -3$ C, so

$$I_{Tc} = 213(18.3 - (-3))/(0.70 \times 18) = 360 \text{ W/m}^2$$

Fourth, $\bar{\phi}$ is to be calculated. The procedure is essentially the same as that of Example 2.24.1. The quantities needed and their values for the mean day of January are as follows:

$$\omega_s = 71.3°, \quad r_{t,n} = 0.174, \quad r_{d,n} = 0.162$$

$$H_d/H = 0.70 \text{ at } K_T = 0.436, \quad R_{b,n} = \frac{\cos|\phi - \delta - \beta|}{\cos|\phi - \delta|} = 1.80$$

$$R_n = 1.108, \quad \bar{R} = \bar{H}_T/\bar{H} = 10.84/6.63 = 1.63$$

$$\bar{R}/R_n = 1.63/1.108 = 1.47$$

We can now calculate $\overline{X}_c$

$$\overline{X}_c = \frac{I_{Tc}}{r_{t,n}R_n\overline{H}} = \frac{360 \times 3600}{0.174 \times 1.108 \times 6.63 \times 10^6} = 1.014$$

With these, get $\overline{\phi}$ from Equation 2.24.4, with $\overline{K}_T = 0.436$, $a = -0.332$, $b = -1.171$, and $c = 0.254$.

$$\overline{\phi} = \exp\{(-0.332 - 1.171 \times 0.678)[1.014 + 0.254(1.014)^2]\} = 0.239$$

Fifth, Y can now be calculated from its defining equation:

$$Y = \frac{41.4 \times 10^6 \times 6.0}{7.60 \times 10^6 \times 18 \times 0.239} = 7.60$$

Sixth, a solar fraction for January is calculated from Equation 22.3.13:

$$P = [1 - \exp(-0.294 \times 7.60)]^{0.652} = 0.929$$

$$f = 0.929 \times 0.351 + (1 - 0.929)(3.082 - 3.142 \times 0.239)$$

$$[1 - \exp(-0.329 \times 0.351)] = 0.344$$

The auxiliary energy is then calculated with Equation 22.3.14.

$$L_A = 12.09 \times 10^9(1 - 0.344) = 8.0 \text{ GJ}$$

Finally, the significant result of the calculations is the sum of the monthly auxiliary energy requirements. From the table, the annual auxiliary required is 29 GJ.

Month	$\overline{S}$ MJ/m^2	$(\overline{\tau\alpha})$	L GJ	X_c	I_{Tc} W/m^2	$\overline{\phi}$	Y	f	L_A GJ
Jan.	7.60	0.70	12.09	1.01	360	0.24	7.60	0.34	7.93
Feb.	8.47	0.67	9.90	0.79	340	0.28	5.93	0.42	5.79
Mar.	7.32	0.63	8.15	0.64	269	0.34	5.62	0.48	4.27
Apr.	6.36	0.58	3.90	0.30	129	0.59	3.66	0.74	1.00
May	5.84	0.55	1.56	0.07	28	0.90	2.64	1.00	0.00
June	5.63	0.54	0.33	0.00	0	1.00	2.45	1.00	0.00
July	5.66	0.53	0.18	0.00	0	1.00	2.44	1.00	0.00
Aug.	6.42	0.55	0.26	0.00	0	1.00	2.15	1.00	0.00
Sep.	7.98	0.60	0.87	0.00	0	1.00	1.73	1.00	0.00
Oct.	9.35	0.66	3.07	0.15	77	0.80	1.86	1.00	0.00
Nov.	8.44	0.70	7.12	0.51	208	0.48	3.39	0.55	3.20
Dec.	6.61	0.70	10.89	1.08	327	0.27	7.81	0.33	7.28
Totals			58.3						29.5

22.4 THE UNUTILIZABILITY DESIGN METHOD: COLLECTOR-STORAGE WALLS

The unutilizability concept developed for direct-gain systems has been applied, with modifications, to collector-storage wall systems [Monsen et al. (1982)]. The presence of the wall introduces significant differences in the manner in which the energy balances on the building are written, in the definition of the critical radiation level and solar-load ratio, and in the form of the correlations for solar contribution. As with the direct-gain system, the UU method establishes the limiting cases of zero- and infinite-capacitance buildings; where a real building lies between the limits is determined by correlations based on simulations. The calculations are again done monthly, with the significant final result being the annual amount of auxiliary energy needed for the passively heated structure.

The monthly energy flows in a collector-storage wall building are shown schematically in Figure 22.4.1. The heating loads are shown in two parts. Load L_{ad} is that which would be experienced if the collector-storage wall were replaced by an adiabatic wall. It is the same as the net reference load in the SLR method, and can be calculated by conventional means such as the degree-day method (Equation 22.4.1a). The load L_w is the monthly energy loss from the building through the collector-storage wall that would be experienced if the transmittance of the glazing for solar radiation were zero, and can be estimated by Equation 22.4.1b:

$$L_{ad} = (UA)_{ad}(DD) \tag{22.4.1a}$$

$$L_w = U_w\, A_r(DD) \tag{22.4.1b}$$

The wall conductance is

$$U_w = \cfrac{1}{\dfrac{1}{\overline{U_L}} + \dfrac{1}{U_i} + \dfrac{\delta}{k}} \tag{22.4.2}$$

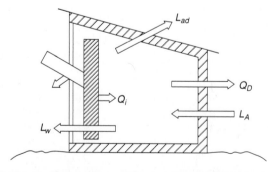

Figure 22.4.1 Monthly energy flows for a collector-storage wall building. Storage of energy in the wall or structure is not indicated. From Monsen et al. (1982).

where U_i is the inside heat transfer coefficient between the inner wall surface and the air in the room. ASHRAE (1989) recommends a value of 8.3 W/m²C for U_i. The wall thickness is δ and its thermal conductivity is k; $\overline{U}_L$ is the average heat loss coefficient from the outer wall surface through the glazing to ambient. $\overline{U}_L$ is conceptually the same as loss coefficients from plates of flat-plate collectors, and has typical values for one to three glazings of 3.7, 2.5, and 1.9 W/m²C, respectively. If night insulation is used, $\overline{U}_L$ can be estimated as a time-average of day and night conductances:

$$\overline{U}_L = (1 - i)U_L + i\left(\frac{U_L}{1 + R_N U_L}\right) \tag{22.4.3}$$

where U_L is the loss coefficient without night insulation, R_N is the thermal resistance of the night insulation, and i is the fraction of the 24-hour day in which the insulation is in place.

The energy dump stream, Q_D in Figure 22.4.1, can be considered as the sum of two energy flows. The loads L_{ad} and L_w are calculated assuming the indoor temperature to be at the low thermostat set temperature. In general, the room temperature can be expected to rise above this temperature, and the actual losses from the building will be greater than $L_{ad} + L_w$; this increase represents one contribution to Q_D. The other contribution is energy which must be removed from the building to keep the indoor temperature from exceeding the high thermostat set temperature.

Let Q_n be the net monthly heat transfer through the collector-storage wall into the heated space. It is at any time a complex function of wall parameters and meteorological conditions. The wall can be modeled as a distributed network of resistors, capacitors, and current sources, as shown in Figure 8.6.2. However, by assuming (a) that there are no vents in the wall to allow circulation between the room and the gap between the wall and the glazing[4] and (b) that the thermal storage capacity of a wall (the maximum change in internal energy of a wall) over a month is negligible compared to the total energy flow through the wall over the month,[5] it can be shown that the temperature profile through the wall is linear. Thus, for monthly average computations, the network of Figure 8.6.2 is replaced by the simple resistance network shown in Figure 22.4.2. An estimation of Q_n for a month can be based on this network.

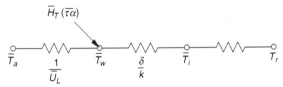

Figure 22.4.2 Monthly average resistance network for collector-storage walls. From Monsen et al. (1982).

---

[4] Utzinger et al. (1980) show that there is little difference in the annual performance of vented and unvented collector-storage walls.
[5] Typically, the storage capacity of a collector-storage wall is less than the heating loads on the building for a single day.

A monthly average daily energy balance on the outer (energy-absorbing) surface of the wall shows how the absorbed energy is distributed to losses to ambient through the glazing and by conduction through the wall with subsequent transfer into the heated space. If T_a and $\bar{U}_L$ are assumed constant at their monthly average values,

$$\bar{S} = U_k\left(\bar{T}_w - T_r\right)\Delta t + \bar{U}_L\left(\bar{T}_w - \bar{T}_a\right)\Delta t \qquad (22.4.4)$$

where U_k is the conductance from the outer wall surface to the room given by

$$U_k = \frac{U_i\,k}{k + U_i\,\delta} \qquad (22.4.5)$$

Here $\bar{T}_w$ is the monthly average outer wall temperature and T_r is the room temperature which is assumed to be at the low thermostat set temperature; additional losses through the wall occurring as a result of the average room temperature being above T_r will be accounted for in Q_D. The number of time units in a day is Δt.

The average outer wall temperature can be found from Equation 22.4.4:

$$\bar{T}_w = \frac{\bar{S} + \left(U_k T_r + \bar{U}_L \bar{T}_a\right)\Delta t}{\left(U_k + \bar{U}_L\right)\Delta t} \qquad (22.4.6)$$

This can be used to calculate the net monthly heat transfer to the building:

$$Q_n = U_k A_r\left(\bar{T}_w - T_r\right)\Delta t N \qquad (22.4.7)$$

We turn now to the question of establishing the limits of performance of buildings with collector-storage walls. For a hypothetical building with infinite thermal capacitance, all of the net gain Q_n can be used. Allowing no month-to-month carryover of stored energy, a monthly energy balance on this infinite-capacitance building is

$$L_{A,i} = \left(L_{ad} - Q_i\right)^+ \qquad (22.4.8)$$

A hypothetical building having zero thermal capacitance in the collector-storage wall and in the rest of the structure will have the maximum auxiliary energy requirements. A collector-storage wall with zero mass can be considered to be a radiation shield which alters the amplitude but not the time of the solar gains to the building. A monthly energy balance on this zero-capacitance building gives

$$L_{A,z} = \left(L_{ad} - Q_n + Q_D\right)^+ \qquad (22.4.9)$$

where Q_D, the energy dumped, is the month's energy entering the building through the collector-storage wall that does not contribute to reduction of the auxiliary energy requirement. Here Q_D can be determined by integrating $\dot{Q}_D$, the rate at which excess energy must be removed to prevent the indoor temperature from rising above the low thermostat set temperature. At any time, $\dot{Q}_D$ is the difference between the rate of heat transfer through the collector-storage wall into the building and the rate of energy loss from the rest of the building:

$$\dot{Q}_D = [U_k A_r (T_w - T_r) - (UA)_{ad}(T_b - T_a)]^+ \tag{22.4.10}$$

An energy balance on the absorbing surface of the hypothetical zero thermal capacity collector-storage wall at any time gives

$$I_T(\tau\alpha)A_r = U_L A_r(T_w - T_a) + U_k A_r(T_w - T_r) \tag{22.4.11}$$

This can be solved for T_w:

$$T_w = \frac{I_T(\tau\alpha) + U_L T_a + U_k T_r}{U_L + U_k} \tag{22.4.12}$$

which when substituted into Equation 22.4.10 yields

$$\dot{Q}_D = \left[U_k A_r \left(\frac{I_T(\tau\alpha) - U_L(T_r - T_a)}{U_L + U_k} \right) - (UA)_{ad}(T_b - T_a) \right]^+ \tag{22.4.13}$$

Assuming $(\tau\alpha)$ and T_a to be constant at their monthly mean values of $\overline{(\tau\alpha)}$ and $\overline{T}_a$, Equation 22.4.13 can be integrated over the month to give Q_D for this zero-capacitance building.

$$Q_D = \frac{U_k A_r \overline{(\tau\alpha)}}{U_L + U_k} \sum (I_T - I_{Tc})^+ \tag{22.4.14}$$

where the critical radiation level I_{Tc} which makes $\dot{Q}_D$ zero is given by

$$I_{Tc} = \frac{1}{\overline{(\tau\alpha)}A_r} \left[(UA)_{ad}\left(\frac{U_L}{U_k} + 1\right)\frac{T_b - \overline{T}_a}{T_r - \overline{T}_a} + U_L A_r \right](T_r - \overline{T}_a) \tag{22.4.15}$$

The summation in Equation 22.4.14 is the same as that in the definition of the daily utilizability $\overline{\phi}$, Equation 2.24.1. Thus

$$Q_D = \frac{U_k A_r \overline{S}N\overline{\phi}}{U_L + U_k} \tag{22.4.16}$$

Equations 22.4.8 and 22.4.9 provide estimates of the limits of the performance of collector-storage wall systems. The solar fractions corresponding to the limits are defined as

$$f_i = 1 - \frac{L_{A,i}}{L_{ad} + L_w} = \frac{L_w + Q_i}{L_{ad} + L_w} \qquad (22.4.17)$$

and $\qquad f_z = 1 - \frac{L_{A,z}}{L_{ad} + L_w} = f_i - \frac{U_k}{U_L + U_k} \overline{\phi X} \qquad (22.4.18)$

where the solar-load ratio X is defined as

$$X = \frac{\overline{S} N A_r}{L_{ad} + L_w} \qquad (22.4.19)$$

    To establish where between these limits a real system will operate, correlation methods are used. A parameter is needed that describes the storage capacity of the system; this will include storage capacity of the building, S_b, and that of the wall, S_w. Building thermal storage capacity for a month (i.e., with N cycles) is

$$S_b = C_b(\Delta T_b)N \qquad (22.4.20)$$

where C_b is the effective building storage capacitance and ΔT_b is the allowable temperature swing, i.e., the difference between the high and low thermostat set-points. Values of C_b can be estimated from information in Table 9.5.1. The storage capacity of the wall for the month (i.e., with N cycles) is

$$S_w = C_p \delta A_r \rho \left(\Delta T_w \right) N \qquad (22.4.21)$$

where C_p, ρ, δ, and A_r are the heat capacity, density, thickness, and area of the wall, and ΔT_w is one-half of the difference of the monthly average temperatures of the outside and inside surfaces of the wall (the monthly average gradient through the wall being linear). The heat transfer through the wall into the heated space Q_i can be expressed in terms of ΔT_w as

$$Q_i = \frac{2kA_r}{\delta} \left(\Delta T_w \right) \Delta t N \qquad (22.4.22)$$

Thus

$$S_w = \frac{\rho C_p \delta^2}{2k \, \Delta t} Q_i \qquad (22.4.23)$$

    A correlation for f [here defined as $1 - L_A/(L_{ad}+L_w)$] has been developed from simulations. f is a function of f_i and a dimensionless storage-dump ratio defined as the ratio of a weighted storage capacity of the building and wall to the energy that would be dumped by a building having zero capacitance.

$$Y = \frac{S_b + 0.047 S_w}{Q_D} \tag{22.4.24}$$

The correlation for f is shown in Figure 22.4.3. The equation for solar fraction is[6]

$$f = \min\{Pf_i + 0.88(1 - P)[1 - \exp(-1.26f_i)], 1\} \tag{22.4.25a}$$

$$P = \left[1 - \exp(-0.144Y)\right]^{0.53} \tag{22.4.25b}$$

(Monsen et al. did not use the zero capacitance limit in the correlation. However, as shown in Example 22.4.1, Equations 22.4.25 and 22.4.18 predict approximately the solar fraction at zero thermal storage.)

To estimate the annual performance of a collector-storage wall system, a set of steps for each month are to be followed. First, the month's absorbed radiation is calculated by the methods of Section 5.10. Second, the loads L_{ad} and L_w are estimated, taking into account internal generation if necessary. Third, the month's heat gain across the collector-storage wall Q_i is estimated using Equations 22.4.5 to 22.4.7. Fourth, the energy dump that would occur in a zero-capacitance system Q_D is calculated from Equation 22.4.16. This requires that daily utilizability be known; it

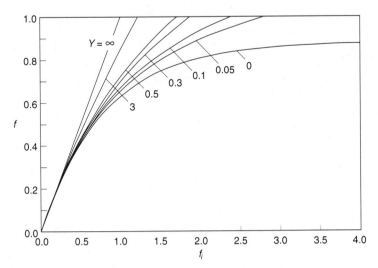

Figure 22.4.3 Correlation for f as a function of f_i and the storage–dump ratio Y. From Monsen et al. (1982).

[6] In Equation 22.4.18 f_i can exceed unity and should not be limited to the range 0 to 1 when used in Equation 22.4.25a.

is based on the critical radiation level given by Equation 22.4.15. Fifth, a series of additional parameters are needed to calculate f from Equation 22.4.25. These are f_i (from Equation 22.4.17), S_b and S_w (the building and wall effective thermal capacitances from Equations 22.4.20 and 22.4.23), and Y (the storage-dump ratio, Equation 22.4.24). The last step is to calculate the auxiliary energy required for the month:

$$Q_A = (L_{ad} + L_w)(1 - f) \qquad\qquad (22.4.26)$$

The monthly auxiliary energy requirements are then summed to get the annual auxiliary energy needs.

Example 22.4.1

A collector-storage wall system is to be designed for Springfield, IL. The daily average energy absorbed by this wall in MJ/m^2, $(\overline{\tau\alpha})$, the ambient temperature, and the number of degree days to base 18.3 C, are given in the table for Example 22.3.1. The south-facing vertical collector-storage wall is double glazed. Night insulation is not to be used.

The building has the following characteristics. The minimum allowable inside temperature T_r is 18.3 C; the allowable temperature swing ΔT is 6 C; and internal energy generation is small. The $(UA)_h$ is 138 W/C and the effective heat capacity of the building C_b is 23.5 MJ/C. The overall heat transfer coefficient from the storage wall to the interior is 8.3 W/m^{2}C.

The wall has the following characteristics. It is masonry with a density of 2105 kg/m^3, and its heat capacity is 0.95 kJ/kgC. The wall is 0.46 m thick and has an area of 18 m^2 (the same as the solar aperture). The loss coefficient for the double-glazed wall to the outside is estimated by the methods of Section 6.4 to be 2.50 W/m^{2}C. Thus: $\delta_w = 0.46$ m, $\rho_w = 2105$ kg/m^3, $C_{p,w} = 0.95$ kJ/kgC, $A_r = 18$ m^2, $U_L = 2.50$ W/m^{2}C, $U_i = 8.3$ W/m^{2}C, and $k_w = 1.73$ W/mC.

What is the estimated need for auxiliary energy for this building for the year?

Solution

The solution for January is shown in detail. The results of the calculations for all months are tabulated. The steps are as outlined above.

Step 1: The values of $\overline{S}$ and $(\overline{\tau\alpha})$ are reproduced from Examples 5.10.1 and 22.3.1.

Step 2: Calculate the loads L_{ad} and L_w with Equation 22.4.1 and 22.4.2, using degree-day data from Appendix G:

$$L_{ad} = 138 \times 657 \times 24 \times 3600 = 7.83 \text{ GJ}$$

$$U_w = \left(\frac{1}{2.50} + \frac{1}{8.3} + \frac{0.46}{1.73}\right)^{-1} = 1.27 \text{ W/m}^2\text{C}$$

$$L_w = 1.27 \times 18 \times 657 \times 24 \times 3600 = 1.30 \text{ GJ}$$

Step 3: Estimate Q_n from Equations 22.4.5 to 22.4.7. The conductance from the outer wall to the room is

$$U_k = \frac{8.3 \times 1.73}{1.73 + 8.3 \times 0.46} = 2.59 \text{ W/m}^2\text{C}$$

The average wall temperature is

$$\bar{T}_w = \frac{7.60 \times 10^6 + \left[2.59 \times 18.3 + 2.5(-3)\right]86,400}{(2.59 + 2.5)86,400} = 25.1 \text{ C}$$

Then the net heat transferred into the rooms through the storage wall is

$$Q_n = 2.59 \times 18(25.1 - 18.3)86,400 \times 31 = 0.849 \text{ GJ}$$

Step 4: Estimate Q_D that would occur if the capacitance of the system were zero using Equation 22.4.16. This requires the calculation of I_{Tc} and $\bar{\phi}$.

$$I_{Tc} = \frac{1}{0.70 \times 18}\left[138\left(\frac{2.5}{2.59} + 1\right)\frac{18.3 - (-3)}{18.3 - (-3)} + 2.5 \times 18\right][18.3 - (-3)] = 535 \text{ W/m}^2$$

(Note that T_r and T_b are the same in this example. As shown in Chapter 9, where internal generation $\dot{g}$ is significant T_b will be lower than T_r.)

Next calculate $\bar{\phi}$. The procedure is like that outlined in Example 22.3.1. The following intermediate results are obtained: $\omega_s = 71.3°$, $r_{t,n} = 0.174$, $r_{d,n} = 0.162$, H_d/H (at $K_T = 0.436$) = 0.70, $R_{b,n} = 1.80$, $R_n = 1.108$, $\bar{R} = 1.63$, $\bar{R}/R_n = 1.47$.

$$\bar{X}_c = \frac{535 \times 3600}{0.174 \times 1.108 \times 6.63 \times 10^6} = 1.50$$

Now calculate $\bar{\phi}$ from Equation 2.24.4. As in Example 22.3.1, $a = -0.332$, $b = -1.171$, $c = 0.254$:

$$\bar{\phi} = \exp\left\{[-0.332 - 1.171(0.678)][1.50 + 0.254\,(1.50)^2]\right\} = 0.097$$

We can now calculate Q_D from Equation 22.4.16:

$$Q_D = \frac{2.59 \times 18 \times 7.60 \times 10^6 \times 31 \times 0.097}{2.5 + 2.59} = 0.208 \text{ GJ}$$

Step 5: Calculate f_i, S_b, S_w, and Y from Equations 22.4.17 to 22.4.23:

$$f_i = \frac{1.30 + 0.849}{7.83 + 1.30} = 0.235$$

$$S_b = 23.5 \times 10^6 \times 6 \times 31 = 4.37 \text{ GJ}$$

$$S_w = \frac{2105 \times (0.46)^2 \times 950 \times 0.849 \times 10^9}{2 \times 1.73 \times 86,400} = 1.20 \text{ GJ}$$

$$Y = \frac{(4.37 + 0.047 \times 1.20) 10^9}{0.208 \times 10^9} = 21.3$$

(Note: f_z is not used in the calculation of L_A, but it is interesting to compare it with f_i. From Equation 22.4.18 and 22.4.19,

$$f_z = 0.235 - \left[\left(\frac{2.59}{2.50 + 2.59}\right) 0.097\right]\frac{7.60 \times 10^6 \times 31 \times 18}{(7.83 + 1.30) 10^9} = 0.212$$

The zero capacitance limit can also be calculated from Equation 22.4.25. With $Y = 0$, P is also zero and

$$f_z = 0.88\left(1 - e^{-1.26 \times 0.235}\right) = 0.266$$

Thus both methods give nearly the same results.

In this example for this month the reduction in solar fraction on going from infinite to zero capacitance is about 10%.)

Step 6: Calculate the solar fraction f and L_A from Equations 22.4.25 and 22.4.26.

$$P = \left[1 - \exp(-0.144 \times 21.3)\right]^{0.53} = 0.975$$

$$f = 0.971 \times 0.235 + 0.88(1 - 0.975)\left[1 - \exp(-1.26 \times 0.235)\right] = 0.234$$

$$L_A = (7.83 + 1.30)10^9 \times (1 - 0.234) = 6.99 \text{ GJ}$$

The process is repeated for the other months of the heating season. The annual requirement for auxiliary energy for this building is 27 GJ. The results are shown in the table.

Mo.	$\bar{S}$ MJ/m^2	$(\overline{\tau\alpha})$	L_w+L_{ab} GJ	Q_n GJ	$I_{T,c}$ W/m^2	$\bar{\phi}$	Q_D GJ	f_i	f	L_A GJ
Jan.	7.60	0.70	9.13	0.85	535	0.097	0.21	0.24	0.23	7.0
Feb.	8.47	0.67	7.48	1.10	504	0.120	0.26	0.29	0.29	5.3
Mar.	7.32	0.63	6.16	1.20	400	0.172	0.36	0.34	0.33	4.1
Apr.	6.36	0.58	2.95	1.37	192	0.442	0.77	0.61	0.57	1.3
May	5.84	0.55	1.18	1.58	41	0.849	1.41	1.48	1.00	0
June	5.63	0.54	0.25	1.83	0	1.000	1.55	7.44	1.00	0
July	5.63	0.52	0.14	2.02	0	1.000	1.61	14.66	1.00	0
Aug.	6.42	0.55	0.20	2.17	0	1.000	1.82	11.31	1.00	0
Sep.	7.98	0.61	0.65	2.29	0	1.000	2.19	3.65	1.00	0
Oct.	9.35	0.66	2.32	2.39	115	0.703	1.87	1.17	0.94	0.2
Nov.	8.44	0.70	5.38	1.59	309	0.315	0.73	0.44	0.43	3.1
Dec.	6.61	0.70	8.23	0.69	485	0.120	0.23	0.23	0.23	6.4
Total										27.4

22.5 HYBRID SYSTEMS: ACTIVE COLLECTION WITH PASSIVE STORAGE

Active air or liquid collectors may be used in heating of buildings in which storage is provided in the structure of the building. This hybrid system has the advantages of control of thermal losses from the receiver (the collector) by shutdown when the radiation is below the critical level and elimination of the cost and complexity of a separate storage system. It has the disadvantages of the significant temperature swings needed in the building to provide storage and, for given building structure and allowable temperature swing, upper limits on the amount of solar energy that can be delivered to the building. Active-collection, passive-storage systems may be applied as retrofits to buildings that cannot accommodate large active systems, particularly in areas of high building density where shading limits the utility of direct-gain or collector-storage wall passive systems but roof areas are available for collection.

The systems are relatively simple. Air heating collectors or liquid heating collectors with liquid-air heat exchangers may be used to supply energy that is distributed by air flow to the heated spaces. Controls are required to turn on collector fluid flow when collection is possible and to turn off the collector or otherwise dump energy when the building interior temperature reaches its maximum allowable value.

A method for estimation of annual performance of these systems based on utilizability concepts has been devised by Evans and Klein (1984a). It is similar to that of Monsen et al. (1981) for direct-gain passive systems (Section 22.3), with estimation of active collector performance by $\bar{\phi}$ methods. Two critical radiation levels are defined, one for the collector and one for the building. Limits of performance are again established by consideration of infinite- and zero-capacitance buildings, with the performance of real buildings determined by correlations based on simulations.

As shown in Section 21.2, the output of an active collector for a month can be expressed as

$$\Sigma Q_u = A_c \bar{H}_T F_R (\overline{\tau\alpha}) N \bar{\phi}_c = A_c F_R \bar{S} N \bar{\phi}_c \qquad (22.5.1)$$

where the monthly average utilizability associated with collecting energy is designated as $\bar{\phi}_c$. The heat removal factor in Equation 22.5.1 and subsequent equations should include the modifications discussed in Chapter 10 (i.e., F_R' should be used for F_R). The critical radiation level to be used in determinating $\bar{\phi}_c$ is

$$I_{Tc,c} = \frac{F_R U_L (\bar{T}_i - \bar{T}_a)}{F_R (\overline{\tau\alpha})} \qquad (22.5.2)$$

where all symbols are the same as for Equation 21.2.2 and $\bar{T}_i$ is the monthly average inlet temperature, in this case the building temperature during collection. For the limiting cases of zero and infinite building capacitance, T_i will be constant. For real cases, T_i will vary above the minimum building temperature, but this variation normally does not significantly affect ΣQ_u.

For a building with infinite storage capacity the monthly energy balance is

$$L_{A,i} = (L - \Sigma Q_u)^+ \qquad (22.5.3)$$

For a building with zero capacitance, energy will have to be dumped if at any time the solar gain exceeds the load. The analysis is exactly parallel to that in Section 22.3 for direct-gain systems. The intensity of radiation incident on the collector that is just enough to meet the building heating load without dumping is called the dumping critical radiation level, $I_{Tc,d}$. On a monthly average basis it is given by

$$I_{Tc,d} = \frac{(UA)_h (\bar{T}_b - \bar{T}_a) + A_c F_R U_L (\bar{T}_i - \bar{T}_a)}{A_c F_R (\overline{\tau\alpha})} \qquad (22.5.4)$$

where $(UA)_h$ is again the overall loss coefficient-area product of the building, $\bar{T}_b$ is the average building base temperature, and $\bar{T}_i$ is the average building interior temperature.

Thus for the zero-capacitance building, on a monthly average basis, a radiation level of $I_{Tc,c}$ is necessary to permit collection. A radiation level of $I_{Tc,d}$ is necessary for the collector to just meet the building loads without dumping. This is shown in Figure 22.5.1 for a sequence of 3 days. Energy above $I_{Tc,d}$ must be dumped and

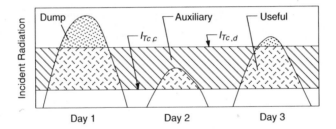

Figure 22.5.1 A sequence of days showing the collecting critical radiation level $I_{Tc,c}$ and the dumping critical radiation level $I_{Tc,d}$ for the zero-capacitance house.

can be estimated as

$$Q_D = A_c F_R \overline{S} N \overline{\phi}_d \tag{22.5.5}$$

where $\overline{\phi}_d$ is the monthly average utilizability (more accurately, unutilizability) based on $I_{Tc,d}$. The energy supplied by the collector that is useful in meeting the heating load of this zero-capacitance building is then the difference between the total energy collected and the energy dumped:

$$Q_{u,b} = \Sigma Q_u - Q_D = A_c F_R \overline{S} N \left(\overline{\phi}_c - \overline{\phi}_d \right) \tag{22.5.6}$$

The auxiliary energy required for the month for this zero-capacitance building is then

$$L_{A,z} = \left(L - \Sigma Q_u + Q_D \right)^+ \tag{22.5.7}$$

Equations 22.5.3 and 22.5.7 establish the limits of auxiliary energy requirements. The auxiliary energy requirements for a building with finite capacitance can be estimated by correlations of f, the fraction of the month's energy supplied from solar energy, with a solar-load ratio X and a storage-dump ratio Y, where X and Y must be properly defined for the active-collection, passive-storage system. The solar-load ratio is

$$X = A_c F_R \overline{S} N / L \tag{22.5.8}$$

The storage-dump ratio is

$$Y = \frac{C_b \Delta T_b}{A_c F_R \overline{S} \, \overline{\phi}_d} = \frac{N C_b \Delta T_b}{Q_D} \tag{22.5.9}$$

The building capacitance is the product of the passive energy storage capacity C_b and the allowable building temperature swing ΔT_b, as in Equation 22.3.12.

The correlation for monthly solar fraction is in terms of $\bar{\phi}_c$, the monthly collecting utilizability, and $\bar{\phi}_d$, the monthly dumping utilizability:[7]

$$f = PX\bar{\phi}_c + (1 - P)(3.082 - 3.142\bar{\phi}_u)[1 - \exp(-0.329X)] \qquad (22.5.10a)$$

$$P = [1 - \exp(-0.294Y)]^{0.652} \qquad (22.5.10b)$$

$$\bar{\phi}_u = 1 - \bar{\phi}_c + \bar{\phi}_d \qquad (22.5.10c)$$

The quantity $\bar{\phi}_u$ is the zero capacitance system unutilizability resulting from energy loss from collectors $(1 - \bar{\phi}_c)$ and energy loss by dumping $\bar{\phi}_d$.

For the month, the auxiliary energy requirement is then

$$L_A = L(1 - f) \qquad (22.5.11)$$

As with other design methods, this method is applied each month and the auxiliary energy requirements summed to obtain annual energy needs, the significant result. Comparisons of annual energy requirements obtained by this method with those obtained from detailed simulations show good agreement for a wide range of system design parameters and locations.

Systems with active storage (evaluated with the f-chart method) can be compared with the same systems except for passive storage (evaluated with the utilizability techniques of this section). Figure 22.5.2 shows an example of solar fraction as a function of collector area for each type. For the systems with active storage, the storage capacity increases in proportion to collector area. For the system with passive storage, the capacity is fixed by the design of the building. For this example, the solar contribution is the same for each system up to an area of about 25

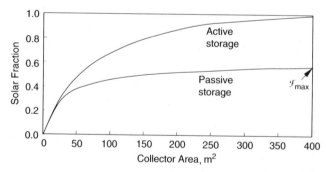

Figure 22.5.2 Annual solar fraction as a function of collector area for systems in Madison, WI. Adapted from Evans and Klein (1984a).

---

[7] Equation 22.5.10 reduces to Equation 22.3.13 for direct-gain systems. For direct-gain systems, $I_{Tc,c}$ is zero and $\bar{\phi}_c$ is unity. Then $\bar{\phi}_u$ is the same as $\bar{\phi}_d$, F_R is unity, and X and Y will be the same as in Equations 22.3.9 and 22.3.12.

m² and an annual solar fraction of 0.3. Evans and Klein show that the collector area at which the differences in performance became significant can be characterized by a ratio of active storage capacity to passive storage capacity. If it is assumed that the active storage capacity is $A_c(mC_p)_s \times 75$, i.e., that the store can change temperature over a 75 C range, and if the passive capacity is the numerator in Equation 22.5.9, i.e., $C_b \Delta T_b$, then the ratio of the two storage capacities is

$$\frac{Q_{s,a}}{Q_{s,p}} = \frac{75 A_c (mC_p)_s}{C_b \Delta T_b} \qquad (22.5.12)$$

If this ratio is less than 2.5, the active storage annual performance advantage will be less than 5%. If the ratio is greater than 2.5, the advantage may or may not exceed 5%, and analyses of both types of system should be made to evaluate relative performance.

Passive storage implies an upper limit to storage capacity and thus to solar contributions to annual loads. Addition of collector area beyond about 100 m² to the passive storage system of Figure 22.5.2 adds little to the solar contribution. A maximum annual solar contribution $\mathcal{F}_{max}$ is shown; it can be estimated from the passive storage capacity per unit floor area Q_{sp} and the yearly heating load per unit floor area L_a:

$$\mathcal{F}_{max} = 0.655 \, (Q_{sp}/L_a)^{0.36} \qquad (22.5.13)$$

The resulting estimated maximum annual solar fraction is within ±6% for a 90% confidence interval compared to the design method. Figure 22.5.3 shows $\mathcal{F}_{max}$ as a function of the annual load per square meter of floor area, for light, medium, and heavy construction as defined in Table 9.5.1, for an allowable temperature swing of 7 C. This figure is based on a wide range of heating loads for uninsulated to heavily insulated buildings. It serves to quickly estimate the best that can be done with large collector areas, i.e., the limits imposed on annual performance by storage considerations alone.

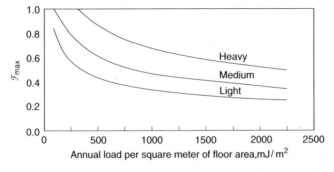

Figure 22.5.3 $\mathcal{F}_{max}$ as a function of annual load per square meter of floor area and building construction for an allowable temperature swing of 7 C. From Evans and Klein (1984a).

Example 22.5.1

A building in Springfield, IL, is to be heated by an active-collection, passive-storage system. The collector is to be located on the roof facing south at a slope of 55°. The incident radiation and $(\overline{\tau\alpha})$ (calculated by the methods of Chapters 2 and 5 using a ground reflectance of 0.3) are shown in the table.

The collector is an air heater with $F_R(\tau\alpha)_n = 0.62$ and $F_R U_L = 3.10$ W/m²C. Absorptance at normal incidence is 0.93 and the collector has one cover with $KL = 0.0125$, resulting in $(\tau\alpha)_n = 0.846$. Air goes from the building interior into the collector and the heated air is returned directly via the warm air duct system to the building. Collector area is 20 m².

By the notation of Table 9.5.1 the structure is of medium construction and is estimated to have a thermal capacitance of 22.9 MJ/C. It has a floor area of 150 m², the overall loss coefficient-area product $(UA)_h = 175$ W/C, and the loads are adequately represented by use of degree-days to base 18.3 C when the interior temperature is 21 C. The allowable temperature range in the building is 21 to 26 C.

Solution

The calculations are done by months in the sequence of the equation numbers. First the load is calculated. Then the critical radiation levels $I_{Tc,c}$ and $I_{Tc,d}$ are determined from Equations 22.5.2 and 22.5.4, and the utilizabilities ϕ_c and ϕ_d are calculated by the methods of Section 2.24. The solar-load ratio X and the storage-dump ratio Y are then determined by Equations 22.5.8 and 22.5.9. Then f is calculated by Equation 22.5.10 and L_A by Equation 22.5.11. This is illustrated in detail for January, and the results for the year are summarized in the table.

For January,

$$F_R(\overline{\tau\alpha}) = 0.62 \times 0.812/0.846 = 0.595$$

$$L = 175 \times 657 \times 86400 = 9.93 \text{ GJ}$$

Month	$\overline{S}$ MJ/m²	$(\overline{\tau\alpha})$	L GJ	$I_{Tc,c}$ W/m²	$I_{Tc,d}$ W/m²	$\overline{\phi}_c$	$\overline{\phi}_{cb}$	f	L_A GJ
Jan.	9.50	0.81	9.93	125	438	0.68	0.20	0.27	7.2
Feb.	11.65	0.80	8.13	116	404	0.73	0.28	0.36	5.2
Mar.	11.74	0.78	6.70	92	311	0.78	0.39	0.46	3.6
Apr.	12.38	0.76	3.21	50	149	0.88	0.66	0.75	0.8
May	12.97	0.75	1.29	23	43	0.95	0.90	1.00	0.0
June	13.46	0.75	0.27	0	0	1.00	1.00	1.00	0.0
July	13.66	0.75	0.15	0	0	1.00	1.00	1.00	0.0
Aug.	13.85	0.76	0.21	0	0	1.00	1.00	1.00	0.0
Sep.	14.16	0.79	0.71	5	0	0.99	1.00	1.00	0.0
Oct.	13.59	0.80	2.53	37	101	0.92	0.78	0.90	0.2
Nov.	11.11	0.84	5.89	75	250	0.81	0.36	0.47	3.1
Dec.	8.16	0.81	8.95	115	399	0.68	0.22	0.26	6.6
Total									26.7

From Equation 22.5.2, the collecting critical radiation level is

$$I_{Tc,c} = 3.10 \, [21 - (-3)]/0.595 = 125 \text{ W/m}^2$$

From Equation 22.5.4, the dumping critical radiation level is

$$I_{Tc,d} = \frac{175[18.3 - (-3)] + 20 \times 3.10[21 - (-3)]}{20 \times 0.595} = 438 \text{ W/m}^2$$

For this month and location, from the methods of Chapter 2, $n = 17$, $\delta = -20.9°$, $\omega_s = 71.3°$, $\overline{H}_o = 15.21$ MJ/m^2, $r_{t,n} = 0.174$, $r_{d,n} = 0.162$, $\overline{K}_T = 0.436$ and H_d/H (at $K_T = 0.436$) is 0.70. Then $R_{b,n} = 2.05$, $R_n = 1.30$, $\overline{R} = 1.764$, $R_n/\overline{R} = 0.734$, $a = -0.332$, $b = -1.171$, and $c = 0.254$. To calculate $\overline{\phi}_c$ we first calculate $\overline{X}_{c,c}$ from Equation 2.24.3:

$$\overline{X}_{c,c} = \frac{125 \times 3600}{0.174 \times 1.30 \times 6.63 \times 10^6} = 0.300$$

and then, from Equation 2.24.4, $\overline{\phi}_c = 0.680$.

To calculate $\overline{\phi}_d$, we need $\overline{X}_{c,d}$:

$$\overline{X}_{c,d} = \frac{438 \times 3600}{0.174 \times 1.30 \times 6.63 \times 10^6} = 1.051$$

and $\overline{\phi}_d = 0.204$.

The absorbed radiation per unit area of collector $\overline{S}$ is 9.50 MJ. Next calculate the solar-load ratio X by Equation 22.5.8:

$$X = 20 \times 0.733 \times 9.50 \times 10^6 \times 31/(9.93 \times 10^9) = 0.435$$

Calculate Y, the storage-dump ratio, by Equation 22.5.9:

$$Y = 22.9 \times 10^6 \times 5/(20 \times 0.733 \times 9.50 \times 10^6 \times 0.204) = 4.030$$

We can now use Equations 22.5.10 to get P, $\overline{\phi}_u$ and f:

$$P = \left[1 - \exp(-0.294 \times 4.030)\right]^{0.652} = 0.788$$

$$\overline{\phi}_u = 1 - 0.680 + 0.204 = 0.524$$

$$f = 0.788 \times 0.435 \times 0.680 + (1 - 0.788)(3.082 - 3.142 \times 0.524)$$

$$\times \left[1 - \exp(-0.329 \times 0.435)\right] = 0.273$$

Then $\quad L_A = 9.93 \, (1 - 0.273) = 7.22$ GJ

A check of the solar fraction f for May and October shows 100% solar, and no auxiliary will be required for those and the intervening months. The annual auxiliary energy required is the sum of the monthly values and is estimated to be 27 GJ.

∎

22.6 OTHER HYBRID SYSTEMS

Buildings may have both passive and full active heating systems. The nature of these systems is such that the passive system will meet whatever part of the loads it can, and the active system, plus auxiliary, will meet the balance of the loads. The passive part of this hybrid system will function in essentially the same way as it would without the active part, and its performance can be estimated by the methods of Sections 22.2 to 22.4. The loads that the active system plus the auxiliary source must meet are as a first approximation the L_A for the passive system.

Corrections to this load on the active plus auxiliary system can be made for two effects. First, the time distribution of the loads on the active system is different. Passive gains tend to be in the daytime and evening, so the loads seen by the active system tend to be in the night and morning, resulting in increased storage temperature and diminished performance. Second, the strategy of control of these hybrid systems will affect mean building temperature and thus loads. The set temperature for operation of the active solar system will be higher than that for the auxiliary to assure that solar is used before auxiliary is supplied. Thus the higher the solar fraction, the higher the mean building temperature and the greater the loads (Chapter 13), and thus the actual load on the active system will be higher than that estimated as L_A by the passive method.

Evans et al. (1985) have used simulations to evaluate the extent of these two effects on the performance of the active part of this kind of hybrid system. The results of temperature elevation induced by controls are shown in terms of an increase in active system load relative to a zero-active-collector-area load; it is a function of the active and passive contributions. The effects of time distribution of loads are shown in terms of overprediction of active system performance by the f-chart method and are a function of active and passive contributions and of active system storage capacity. Neither of these effects is large; typically they are each a few percent.

Evans and Klein (1984b) have also used simulation methods to analyze the combination of active-collection, passive-storage systems and direct-gain passive systems, both of which use only the capacitance of the building as storage. Both of these systems deliver energy to the building, and thus to storage, at the same time. Consequently, energy dump from the two systems also occurs at the same time, and the flow to the collector must be shut off and the direct-gain aperture blocked by curtains or shades. It is not practical to calculate the contributions of the two systems independently, as the performance of each is degraded from what it would be if the other were not there. The alternative is to assume that the active and passive collectors deliver energy to the building in essentially the same way and can be considered as a single system. Under this assumption, the unutilizability methods can be applied to this hybrid if the critical radiation levels for dumping energy from the building are properly defined.

REFERENCES

ASHRAE *Handbook - Fundamentals*. American Society of Heating, Refrigerating, and Air-Conditioning Engineers, Atlanta, GA (1989).

Balcomb, J. D., D. Barley, R. D. McFarland, J. Perry, W. D. Wray and S. Noll, *Passive Solar Design Handbook*, Vol. 2, U.S. Department of Energy, Washington, DC, Report No. DOE/CS-0127/2 (1980).

Balcomb, J. D., R. W. Jones, C. E. Kosiewicz, G. S. Lazarus, R. D. McFarland and W. D. Wray, *Passive Solar Design Handbook*, Vol. 3, American Solar Energy Society, Boulder, CO (1983a).

Balcomb, J. D., R. W. Jones, R. D. McFarland and W. O. Wray, *Passive Solar J.*, **1**, 67 (1983b). "Expanding the SLR Method."

Balcomb, J. D., R. W. Jones, R. D. McFarland, W. O. Wray, D. Barley, C. Kosiewicz, G. Lazarus, J. Perry and G. Schoenau, *Passive Solar Heating Analysis - A Design Manual*, ASHRAE, Atlanta, GA (1984).

Evans, B. L. and S. A. Klein, paper presented at Las Vegas ASME meeting (1984a). "A Design Method of Active Collection-Passive Storage Space Heating Systems."

Evans, B. L. and S. A. Klein, paper presented at Anaheim American Solar Energy Society meeting (1984b). "Combined Active Collection-Passive Storage and Direct-Gain Hybrid Space Systems."

Evans, B. L., S. A. Klein and J. A. Duffie, *Solar Energy*, **35**, 189 (1985). "A Design Method for Active-Passive Hybrid Space Heating Systems."

Lebens, R. M. and A. Myer, paper presented at United Kingdom/International Solar Energy Society Workshop on Microcomputers in Building Energy Design, Birmingham, UK (1984). "An Outline of the Methode 5000."

McFarland, R. D., Los Alamos National Laboratory Report LA-7433-MS (1978). PASOLE: A General Simulation Program for Passive Solar Energy."

Mazria, E., *The Passive Solar Energy Book*, Rodale Press, Emmaus, PA (1979).

Monsen, W. A., S. A. Klein and W. A. Beckman, *Solar Energy*, **27**, 143 (1981). "Prediction of Direct Gain Solar Heating System Performance."

Monsen, W. A., S. A. Klein and W. A. Beckman, *Solar Energy*, **29**, 421 (1982). "The Unutilizability Design Method for Collector-Storage Walls."

Total Environmental Action, Inc., *Passive Solar Design Handbook*, Vol. 1, U.S. Department of Energy, Washington, DC, Report No. DOE/CS-0127/1 (1980)

Utzinger, D. M., S. A. Klein and J. W. Mitchell, *Solar Energy*, **25**, 511 (1980). "The Effect of Air Flow Rate in Collector-Storage Walls."

Chapter 23

DESIGN OF PHOTOVOLTAIC SYSTEMS

This chapter is an anomaly in that the processes it treats are not primarily thermal in nature. However, the equations that are encountered in design of many photovoltaic (PV) systems are very similar to those describing passive heating processes, and design methods based on utilizability concepts have been developed that are analogous to those in Chapter 22 for passive heating. Radiation calculations developed for thermal processes are directly applicable to PV converters. This chapter includes a brief description of PV converters (solar cells), models that describe their electrical and thermal characteristics, a short treatment of models of batteries and other components in the systems, and notes on applications. These serve as introductions to the design method and to comments on simulation of PV systems.

Photovoltaic converters are semiconductor devices that convert part of the incident solar radiation directly into electrical energy. A brief history of photovoltaics is presented by Wolf (1981). An early public demonstration of solar cells occurred in 1955 in Americus, GA, where a small panel of experimental silicon cells provided energy to charge a battery and power telephone equipment. Since then, tremendous progress has been made in the development of cell technology and applications. Early handmade cells were 5% efficient, were 1 or 2 cm^2 in area, and had outputs of a few milliwatts. In 1990, laboratory cells have been reported with efficiencies of over 30%, cells are being manufactured with areas of many square meters, and the cells produced in the world in the year had an aggregate peak generating capacity of about 50 MW. Solar cells are being used in a range of applications, from powering watches and calculators to charging batteries for boats and communications systems to medium-scale systems for power generation. Photovoltaics have been the means of supplying power to most of the satellites that have been launched since the beginnings of space programs.

There is a very substantial body of literature on photovoltaics. Maycock and Stirewalt (1981) present a layman's treatment of the subject; examples of technical books are Green (1982) on semiconductors and cells, Rauschenbach (1980) on cell array design, and Kenna and Gillett (1985) on a particular application (water pumping). In addition, there are proceedings of many conferences and meetings (such as the Photovoltaic Specialists Conferences that are held at approximately two-year intervals), reviews that have appeared in *Advances in Solar Energy* and elsewhere, and papers in scientific journals.

23.1   PHOTOVOLTAIC CONVERTERS

Incident solar radiation can be considered as discrete "energy units" called photons. From Sections 3.1 and 3.2, the product of the frequency v and wavelength λ is the speed of light:

$$C = \lambda v \qquad (23.1.1)$$

The energy of a photon is a function of the frequency of the radiation (and thus also of the wavelength) and is given in terms of Planck's constant h by

$$E = hv \qquad (23.1.2)$$

Thus the most energetic photons are those of high frequency and short wavelength.

The most common photovoltaic cells are made of single-crystal silicon. An atom of silicon in the crystal lattice absorbs a photon of the incident solar radiation, and if the energy of the photon is high enough, an electron from the outer shell of the atom is freed. This process thus results in the formation of a hole-electron pair, a hole where there is a lack of an electron, and an electron out in the crystal structure. These normally disappear spontaneously as electrons recombine with holes. The recombination process can be reduced by building into the cells a potential barrier, a thin layer or junction across which a static charge exists. This barrier is created by doping the silicon on one side of the barrier with very small amounts (of the order of one part in 10^6) of boron to form p-silicon, which has a deficiency of electrons in its outer shell, and that on the other side with phosphorus to form n-silicon, which has an excess of electrons in its outer shell. The barrier inhibits the free migration of electrons, leading to a buildup of electrons in the n-silicon layer and a deficiency of electrons in the p-silicon. If these layers are connected by an external circuit, electrons (i.e., a current) will flow through that circuit. Thus free electrons created by absorption of photons are in excess in the n-silicon and flow through the external circuit to the p-silicon. Electrical contacts are made by metal bases on the bottom of the cell and by metal grids or meshes on the top layer (which must be largely uncovered to allow penetration of photons). A schematic section of a cell of this type and a schematic of a cell in a circuit are shown in Figure 23.1.1.

There are many variations on cell material, design, and methods of manufacture. Amorphous or polycrystalline silicon (Si), cadmium sulfide (CdS), gallium arsenide (GaAs), and other semiconductors are used for cells.

The output of cells is limited by several factors. There is a minimum energy level (and thus a maximum wavelength) of photons that can cause the creation of a hole-electron pair. For silicon, the maximum wavelength is 1.15 μm. Radiation at higher wavelengths does not produce hole-electron pairs but heats the cell. Each photon causes the creation of a single hole-electron pair, and the energy of photons in excess of that required to create hole-electron pairs is also converted to heat. From these considerations alone, the maximum theoretical efficiency of silicon cells is 23%.

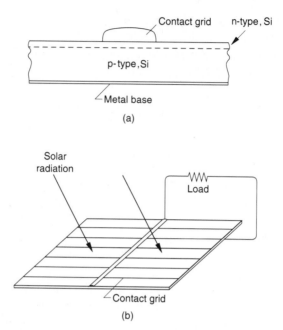

Figure 23.1.1   (a) Section of a silicon solar cell.   (b) Schematic of a cell, showing top contacts.

In addition, there are reflection losses at the top surface of the cells (which can be reduced by antireflective coatings). Part of the top layer of a cell must be covered by the contact grid, which reduces active cell area. There are also electrical effects such as sheet resistance (resistance to the flow of electrons across the top layer to the grid) that limit cell output.

Solar cell technology is evolving rapidly, in developing new and more efficient cells and in reducing costs of manufacture. Modules are available with many cells connected in series and parallel to provide convenient currents and voltages. Cell modules can be purchased on the market in 1990 that are over 12% efficient and have design lifetimes of over 10 years. Experimental cells with multiple junctions (i.e., two or more layers of materials with varying spectral response) have been constructed that have efficiencies of more than 30%.

23.2   PV GENERATOR CHARACTERISTICS AND MODELS

An assessment of the operation of solar cells and the design of power systems based on solar cells must be based on the electrical characteristics, i.e., the voltage-current relationships of the cells under various levels of radiation and at various cell temperatures. Many cell models have been developed, ranging from simple idealized models to detailed models that reflect the details of the physical processes occurring in the cells. For system design purposes, the model must provide the means to calculate current, voltage, and power relationships of cell arrays over the range of operating conditions to be encountered, and the data on cell or module characteristics needed to use the model must be available from the manufacturer. For this purpose detailed models do not appear to be needed, and simple models work well. Rauschenbach (1980), Townsend (1989), and Eckstein (1990) reviewed several models and their utility for system design purposes.

Figure 23.2.1 is an equivalent circuit that can be used either for an individual cell or for a module consisting of several cells or for an array consisting of several modules. This circuit requires that five parameters be known: the light current I_L, the diode reverse saturation current I_o, the series resistance R_s, the shunt resistance R_{sh}, and a curve fitting parameter a. At a fixed temperature and solar radiation, the current-voltage characteristic of this model is given by

$$I = I_L - I_D - I_{sh} = I_L - I_o\{\exp\left[(V + IR_s)/a\right] - 1\} - \frac{V + IR_s}{R_{sh}} \qquad (23.2.1)$$

The power is given by

$$P = IV \qquad (23.2.2)$$

All of the five parameters may be functions of cell temperature; this is discussed further in the next section.

Current-voltage (I-V) characteristics of a PV module are shown in Figure 23.2.2. The current axis (where $V = 0$) is the short circuit current I_{sc}, and the intersection with the voltage axis (where $I = 0$) is the open circuit voltage V_{oc}. For this module the current is nearly constant up to about 15 V, at which point the diode current becomes increasingly important. At open circuit conditions at about 20 V, all of the generated light current is passing through the diode and shunt resistance. For comparison, a single 1 cm^2 silicon cell at a solar radiation level of 1000 W/m^2 has an open circuit voltage of about 0.6 V and a short circuit current of about 20 to 30 mA.

The power as a function of voltage is also shown in Figure 23.2.2. The maximum power that can be obtained corresponds to the rectangle of maximum area under the I-V curve. At the maximum power point the power is P_{mp}, the current is I_{mp}, and the voltage is V_{mp}. Ideally, cells would always operate at the maximum

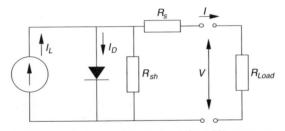

Figure 23.2.1   The equivalent circuit for a PV generator.

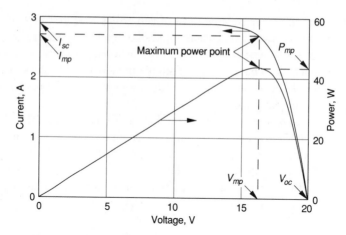

Figure 23.2.2 Typical *I-V* and *P-V* curves for a PV module.

power point, but practically cells operate at a point on the *I-V* curve that matches the *I-V* characteristic of the load. This will be discussed further in what follows.

Current-voltage curves are shown in Figure 23.2.3 for a module operating at a fixed temperature and at several radiation levels. The locus of maximum power points is shown. For this module, the short circuit current increases in proportion to the solar radiation while the open circuit voltage increases logarithmically. As long as the curved portion of the *I-V* characteristic does not intersect the current axis (where $V = 0$), the short circuit current is nearly proportional to the incident radiation.[1]

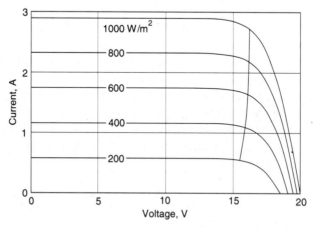

Figure 23.2.3 *I-V* curves for a PV generator at several radiation levels. The locus of maximum power points is shown.

---

[1] Thus if the incident radiation is assumed to have a fixed spectral distribution, the short circuit current can be used as a measure of incident radiation. It is also possible to impose a bias voltage across the cell, in effect move the voltage axis to the left, and extend the range of radiation fluxes that can be measured. [See Schöffer et al. (1964).]

Cells are usually mounted in modules, and multiple modules are used in arrays. Individual modules may have cells connected in series and parallel combinations to obtain the desired voltage. Arrays of modules may likewise be arranged in series and parallel. For identical modules or cells connected in series, the voltages are additive, and when connected in parallel, the currents are additive. If the cells or modules are not identical, a more complicated analysis is required. Figure 23.2.4 shows *I-V* characteristics of one, two, and four identical modules connected in various ways.

All of the *I-V* curves of this section were generated using Equation 23.2.1. The five parameters in the model were obtained indirectly using measurements of the current and voltage characteristics of a module at reference conditions. Traditionally, measurements of PV electrical characteristics are made at a reference incident radiation of 1000 W/m^2 and an ambient temperature of 25 C. Measurements of current and voltage at these or other known reference conditions are often available at open circuit conditions, short circuit conditions, and maximum power conditions. It is not possible to determine all five model parameters if measurements are made at only these three current-voltage pairs. However, as pointed out by Rauschenbach (1980), the shunt resistance of most modern cells is very large and the last term of Equation 23.2.1 often can be neglected. If, at low voltages, the *I-V* curve exhibits a pronounced negative slope instead of the nearly horizontal line as shown in Figure 23.2.2, then it is necessary to consider the shunt resistance. The remainder of this chapter is based on the assumption that shunt resistance is infinite.

With the shunt resistance taken as infinity, four model parameters remain with only three measured points on the *I-V* curve available to determine the four parameters. The problem of obtaining an additional independent equation will be discussed later in this section.

At short circuit conditions, the diode current is very small and the light current is equal to the short circuit current:

$$I_L = I_{sc} \qquad\qquad (23.2.3)$$

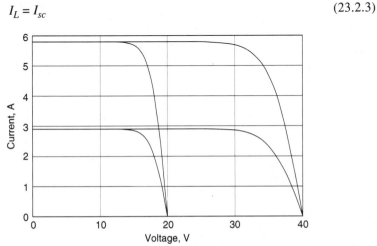

Figure 23.2.4 *I-V* curves for PV generators connected in various series and parallel arrangements.

At open circuit conditions the current is zero and the 1 in Equation 23.2.1 is small compared to the exponential term so that

$$I_o = I_L \exp(-V_{oc}/a) \qquad (23.2.4)$$

The measured I-V pair at maximum power conditions can be substituted into Equation 23.2.1 along with I_o from Equation 23.2.4 and I_L from Equation 23.2.3. Again neglecting the 1, the result is:[2]

$$R_s = \frac{a\ln\left(1 - \frac{I_{mp}}{I_L}\right) - V_{mp} + V_{oc}}{I_{mp}} \qquad (23.2.5)$$

Equation 23.2.5 not only relates R_s to a but also limits their maximum values. If both R_s and a are to be positive, as would be expected from the model, then the maximum value of R_s occurs when a approaches zero and the maximum value of a occurs when R_s approaches zero. In order to determine values for these two parameters another independent condition is needed. Manufacturers traditionally do not supply an additional measurement of an I-V pair. An alternative is to use the condition at the maximum power point where the derivative of power with respect to voltage is zero. In either case the resulting equations sometimes produce disturbing results like negative values for the series resistance. The problem lies with the simplified circuit of Figure 23.2.1; it cannot reproduce all of the complicated phenomena that occur when photons interact with a semiconductor. In spite of these problems the model is adequate for systems design and can even be extended to approximate the observed temperature dependence. The fourth equation comes from temperature considerations.

Figure 23.2.5 shows the effect of temperature on a module I-V characteristic; at a fixed radiation level, increasing temperature leads to decreased open circuit voltage and slightly increased short circuit current. In order for the model to reproduce these effects, it is necessary to know how the model parameters I_o, I_L, and a vary with temperature. The series resistance R_s is assumed independent of temperature in this model. Townsend (1989) showed that the following equations are good approximations for many PV modules:

$$\frac{a}{a_{ref}} = \frac{T_c}{T_{c,ref}} \qquad (23.2.6)$$

$$I_L = \frac{G_T}{G_{T,ref}}[I_{L,ref} + \mu_{I,sc}(T_c - T_{c,ref})] \qquad (23.2.7)$$

$$\frac{I_o}{I_{o,ref}} = \left(\frac{T_c}{T_{c,ref}}\right)^3 \exp\left[\frac{\varepsilon N_s}{a_{ref}}\left(1 - \frac{T_{c,ref}}{T_c}\right)\right] \qquad (23.2.8)$$

---

[2] It is not necessary to neglect the 1 in deriving expressions for I_o and R_s (i.e. Equations 23.2.4 and 23.2.5) but the algebra is considerably simplified and the errors introduced are very small.

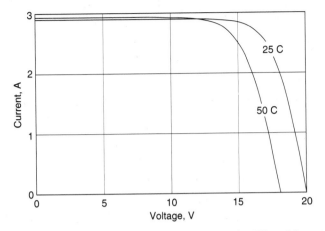

Figure 23.2.5 *I-V* characteristics and maximum power points of a PV module at temperatures of 25 and 50 C.

where ε is the material bandgap energy, 1.12 eV for silicon and 1.35 eV for gallium arsenide, and N_s is the number of cells in series in a module times the number of modules in series. All of the quantities with the subscript *ref* are from measurements at reference conditions. The temperature coefficient of the short circuit current $\mu_{I,sc}$ is obtained from measurements at the reference irradiation from

$$\mu_{I,sc} = \frac{\mathrm{d}I_{sc}}{\mathrm{d}T} \cong \frac{I_{sc}(T_2) - I_{sc}(T_1)}{T_2 - T_1} \tag{23.2.9}$$

where T_2 and T_1 are two temperatures centered around the reference temperature. Similarly, the temperature coefficient of the open circuit voltage can be obtained from measurements:

$$\mu_{V,oc} = \frac{\mathrm{d}V_{oc}}{\mathrm{d}T} \cong \frac{V_{oc}(T_2) - V_{oc}(T_1)}{T_2 - T_1} \tag{23.2.10}$$

It is now possible to obtain the fourth model equation. If the temperature coefficients of the short circuit current and open circuit voltage are both known, then Townsend (1989) showed that an additional independent equation can be found by equating the experimental value of $\mu_{V,oc}$ with the value determined from the analytical expression for the derivative, $\mathrm{d}V_{oc}/\mathrm{d}T$. Differentiating V_{oc} in Equation 23.2.4 with respect to T and using Equations 23.2.6, 23.2.8 and the definition of $\mu_{I,sc}$, it can be shown that

$$a_{ref} = \frac{\mu_{V,oc}T_{c,ref} - V_{oc,ref} + \varepsilon N_s}{\dfrac{\mu_{I,sc}T_{c,ref}}{I_{L,ref}} - 3} \tag{23.2.11}$$

If the value of a_{ref} determined from Equation 23.2.11 is greater than zero but less than the maximum value obtained from Equation 23.2 5 (i.e., with R_s equal to zero), then the I-V cell/module/array model is complete. With measurements at the reference condition of V_{oc}, I_{sc}, I_{mp}, V_{mp}, $\mu_{I,sc}$, and $\mu_{V,oc}$, Equations 23.2.3 through 23.2.5 and 23.2.11 are used to determine I_L, I_o, R_s, and a at the reference condition. Equations 23.2.6 through 23.2.8 are then used to determine I_o, I_L, and a at any cell temperature. The process is as follows: find $I_{L,ref}$ from Equation 23.2.3, find a_{ref} from Equation 23.2.11, find $I_{o,ref}$ from Equation 23.2.4, find $R_{s,ref}$ from Equation 23.2.5, and then find a, I_L, and I_o at the cell temperature using Equations 23.2.6 through 23.2.8. The current-voltage-power characteristics are then found from Equations 23.2.1 and 23.2.2 with R_{sh} set to infinity. The next section shows how to determine the cell temperature.

If the value of a_{ref} obtained from Equation 23.2.11 is negative or greater than the maximum value obtained from Equation 23.2.5, then the choice is to either use these values (which are often adequate for predicting observed cell behaviour) or to modify the model presented here. One possibility is to change the meaning of N_s in Equation 23.2.11 and let it represent another model parameter. Of course, another condition is then necessary.

Example 23.2.1

A PV module with 36 cells in series and an area of 0.427 m² has the following measured characteristics at reference conditions: $I_{sc} = 2.9$ amperes, $V_{oc} = 20$ volts, $I_{mp} = 2.67$ amperes, and $V_{mp} = 16.5$ volts. The temperature coefficient of the short circuit current $\mu_{I,sc}$ and the open circuit voltage $\mu_{V,oc}$ are 1.325×10^{-3} A/K and -0.0775 V/K, respectively. Determine the values of I_o, I_L, R_s, and a at a cell temperature of 67.2 C and the maximum power and efficiency predicted by the model at the reference conditions.

Solution

From Equation 23.2.3,

$$I_{L,ref} = I_{sc} = 2.9 \text{ A}$$

From Equation 23.2.11

$$a_{ref} = \frac{-0.0775 \times 298 - 20 + 1.12 \times 36}{\dfrac{0.001325 \times 298}{2.9} - 3} = 0.969$$

From Equation 23.2.4

$$I_{o,ref} = 2.9 e^{-20/0.969} = 3.15 \times 10^{-9} \text{ A}$$

From Equation 23.2.5

$$R_{s,ref} = \frac{0.969\ln\left(1 - \frac{2.67}{2.90}\right) - 16.5 + 20}{2.67} = 0.391\ \Omega$$

Equations 23.2.6 through 23.2.8 are used at the cell temperature of 67.2 C to determine I_L, A, and I_o. The series resistance in this model is assumed to be independent of temperature so $R_{s,ref}$ is equal to R_s. •

$$I_L = \left(\frac{1000}{1000}\right)[2.90 + 0.001325(340.2 - 298)] = 2.96\ A$$

$$a = \frac{0.969 \times 340.2}{298} = 1.106$$

$$I_o = 3.15 \times 10^{-9}\left(\frac{340.2}{298}\right)^3 \exp\left[\frac{1.12 \times 36}{0.969}\left(1 - \frac{298}{340.2}\right)\right] = 8.16 \times 10^{-7}\ A$$

It is possible to differentiate the equation for power, Equation 23.2.2, with respect to V and set the result equal to zero. However, the result is a transcendental equation and thus must be solved numerically. It is just as easy to numerically solve for the maximum power by varying V until the maximum power is found. The result of such a numerical search is $I_{mp} = 2.73$ A, $V_{mp} = 16.2$ V, and $P_{mp} = 44.2$ W. These compare favorably with the measured maximum power conditions used to find the model constants. The cell efficiency is $44.2/(0.427 \times 1000) = 0.104$. ∎

The temperature dependence of the maximum power point efficiency of a module is an important parameter in estimating system performance as presented in Section 23.4. The maximum power point efficiency of a module is given by

$$\eta_{mp} = \frac{I_{mp}V_{mp}}{A_c G_T} \tag{23.2.12}$$

The temperature dependence of this efficiency can be expressed in terms of a maximum power point efficiency temperature coefficient $\mu_{P,mp}$ as[3]

$$\eta_{mp} = \eta_{mp,ref} + \mu_{P,mp}(T_c - T_{ref}) \tag{23.2.13}$$

A value of $\mu_{P,mp}$ can be obtained by solving the cell model for the maximum power efficiency over a range of temperatures. Figure 23.2.6 shows the results of these calculations for the module used in the examples of this section. The maximum power point efficiency is seen to be a linear function of temperature and, for this and many modules, it is also independent of solar radiation.

---

[3] $\mu_{P,mp}$ is generally a negative number.

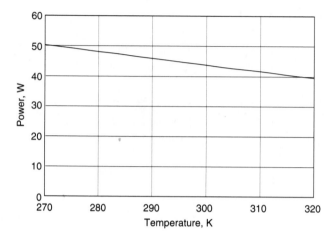

Figure 23.2.6 Maximum power point efficiency as a function of module temperature.

Fortunately, it is not necessary to repeat the calculations that generated Figure 23.2.6 to estimate $\mu_{P,mp}$. From Equation 23.2.12,

$$\mu_{P,mp} = \frac{d\eta_{mp}}{dT} = \left(I_{mp}\frac{dV_{mp}}{dT} + V_{mp}\frac{dI_{mp}}{dT}\right)\frac{1}{A_cG_T} \tag{23.2.14}$$

For many modules, $\mu_{I,sc}$ is small which indirectly results in a value of dI_{mp}/dT of approximately zero and dV_{mp}/dT to be approximately equal to dV_{oc}/dT. With these approximations, the temperature coefficient of maximum power efficiency is approximated by

$$\mu_{P,mp} = \frac{I_{mp}}{A_cG_T}\frac{dV_{oc}}{dT} = \eta_{mp,ref}\frac{\mu_{Voc}}{V_{mp}} \tag{23.2.15}$$

Example 23.2.2

For the module and reference conditions of Example 23.2.1, determine $\mu_{P,mp}$ from Equation 23.2.15 and compare it with the results obtained from Figure 23.2.6. What is the module maximum power point efficiency at 340.2 K?

Solution

At the reference conditions the maximum power voltage is 16.2 V (Example 23.2.1). Thus, from Equation 23.2.15,

$$\mu_{P,mp} = -0.0775 \times 0.104/16.2 = -0.00050 \ (1/K)$$

From Figure 23.2.6,

$$\mu_{P,mp} = \frac{\eta(330) - \eta(270)}{330 - 270} = \frac{0.087 - 0.118}{330 - 270} = -0.00052 \ (1/K)$$

The maximum power point efficiency at 340.2 K is found from Equation 23.2.13 using the efficiency of 0.104 from Example 23.2.1:

$$\eta_{P,mp} = 0.104 - 0.00050(340.2 - 298)) = 0.083$$

When the complete model is used to find the maximum power point efficiency at these conditions, the result is 0.082. ∎

23.3 CELL TEMPERATURE

The temperature of operation of a PV module is determined by an energy balance. The solar energy that is absorbed by a module is converted partly into thermal energy and partly into electrical energy which is removed from the cell through the external circuit. The thermal energy must be dissipated by a combination of heat transfer mechanisms; the upward losses occur by the same mechanisms as losses from the covers of flat-plate collectors, as detailed in Section 6.4. Back losses will usually be more important than in thermal collectors, since heat transfer from the module should be maximized so the cells will operate at the lowest possible temperature.

In some array designs, cells are operated at elevated radiation levels by use of linear or circular concentrators. Cell temperature control may be enhanced by water cooling; in these cases, the loss coefficient must be modified to account for the additional mechanism for heat transfer, and the sink temperature to which this heat transfer occurs may not be the ambient temperature.

Cell arrays have been designed and built to produce combinations of electrical and thermal energy. The University of Delaware Solar One, a residential-scale experimental building, was equipped with PV cells that were air cooled; the cooling air was used for space heating in the building. A system based on a similar concept is installed on a house in Providence, RI [Loferski et al. (1988)]. For best operation of the cells, they should be at the minimum possible temperature; this limits the possible applications of combined systems to situations where thermal energy is needed at low temperatures.

An energy balance on a unit area of module which is cooled by losses to the surroundings can be written as[4]

$$\tau\alpha G_T = \eta_c G_T + U_L(T_c - T_a) \tag{23.3.1}$$

---

[4] These energy balances can also be written for an hour in terms of I_T.

where τ is the transmittance of any cover that may be over the cells, α is the fraction of the radiation incident on the surface of the cells that is absorbed, and η_c is the efficiency of the module in converting incident radiation into electrical energy. This efficiency will vary from zero to the maximum module efficiency depending on how close to the maximum power point the module is operating. The loss coefficient U_L will include losses by convection and radiation from top and bottom and by conduction through any mounting framework that may be present, all to the ambient temperature T_a.

The nominal operating cell temperature (NOCT) is defined as the cell or module temperature that is reached when the cells are mounted in their normal way at a solar radiation level of 800 W/m², a wind speed of 1 m/s, an ambient temperature of 20 C, and no load operation (i.e., with $\eta_c = 0$). Measurements of the cell temperature, ambient temperature, and solar radiation can be used in Equation 23.3.1 to determine the ratio $\tau\alpha/U_L$:

$$\tau\alpha/U_L = (T_{c,NOCT} - T_a)/G_{T,NOCT} \tag{23.3.2}$$

The temperature at any other condition, assuming $\tau\alpha/U_L$ to be constant, is then found from

$$T_c = T_a + (G_T\tau\alpha/U_L)(1 - \eta_c/\tau\alpha) \tag{23.3.3}$$

The $\tau\alpha$ in the last term of Equation 23.3.3 is not generally known, but an estimate of 0.9 can be used without serious error since the term $\eta_c/\tau\alpha$ is small compared to unity.

Example 23.3.1

For the module of Examples 23.2.1 and 23.2.2, determine the module temperature in a desert location where the ambient temperature is 40 C and the incident solar radiation is 920 W/m². The module is operating at 12.91 V and has an area of 0.427 m². At NOCT conditions, the module temperature is 46 C.

Solution

This is a trial and error solution. First guess a module efficiency and find the cell temperature from Equation 23.3.3. With this cell temperature and with values of $I_{o,ref}$, L_L, $L_{L,ref}$, and a_{ref} from Example 23.2.1, use Equations 23.2.6 through 23.2.8 to find values for I_o, I_L, and a at this temperature. At the given voltage, calculate the current from Equation 23.2.1 and the power from 23.2.2. Evaluate the module efficiency and compare with the initial guess. Since the module efficiency is not a strong function of temperature, the process will converge rapidly. Only the final iteration will be shown.

From NOCT conditions, Equation 23.3.2 gives $\tau\alpha/U_L = (46 - 20)/800 = 0.0325$. With a trial module efficiency of 0.0818, Equation 23.3.3 gives

$$T_c = 313 + 920 \times 0.0325 \, (1 - 0.0818/0.9) = 340.2 \text{ K}$$

This is the same temperature that was used in Example 23.2.1 so that $a = 1.106$ and $I_o = 8.16 \times 10^{-7}$. The value of I_L must be evaluated with the incident radiation of 920 W/m².

$$I_L = \frac{920}{1000}[2.9 + 0.001325(340.2 - 298)] = 2.719 \text{ A}$$

With these values and a voltage of 12.91, the current from Equation 23.2.1 is given by

$$I = 2.719 - 8.16 \times 10^{-7}\left[\exp\left(\frac{12.91 + 0.391I}{1.106} - 1\right)\right]$$

which is transcendental and must be solved numerically. The solution is $I = 2.49$ A and the power is $P = 2.49 \times 12.91 = 32.1$ W. The efficiency is $32.1/(920 \times 0.427) = 0.082$, which is what was assumed at the beginning. This particular current and voltage is also the maximum power point for this module under these operating conditions. The efficiency at the maximum power point was estimated in Example 23.2.2 to be 0.083. ∎

23.4 LOAD CHARACTERISTICS AND DIRECT-COUPLED SYSTEMS

Photovoltaic cells will operate at the voltage and current at which its characteristics match that of the load to which it is connected. Examples of load characteristics are a resistive load, a battery, a series motor, and the power grid. The examples in this section are of direct-connected systems. In the following section, note is made of power conditioning equipment (maximum power point trackers) that can be used where there is serious mismatch between generator and load characteristics.

Figure 23.4.1 shows the characteristics of resistive loads and I-V curves for a solar cell module of 45 W nominal capacity. The I-V curves are typical of those for three times of day for modules with fixed orientation. The resistive load curves shown each intersect the module I-V curves at the maximum power points. At any time, the combination of the module and the load will operate at the intersection of the characteristic curves of the two components. The optimum load resistance, i.e., the slope of the resistive load curve, will vary through a day. A fixed resistive load is thus not an optimum load for a PV generator. Khouzam et al. (1991) studied the problem of determining the optimum fixed resistance for clear days.

A typical storage battery charging characteristic is shown in Figure 23.4.2(a), together with PV generator characteristics and a load. The internal resistance of a

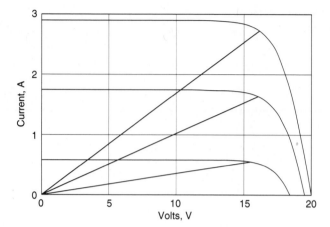

Figure 23.4.1   PV generator characteristics at three incident radiation levels and resistive load lines corresponding to the maximum power points for each radiation level.

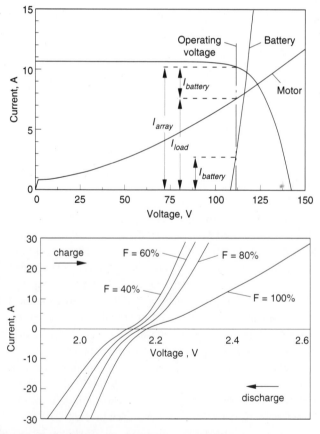

Figure 23.4.2   (a) Characteristics of a lead-acid storage battery and a PV generator array.   (b) Voltage vs. state of charge curve for a typical 12 V battery.

battery depends on the rate of charging, and the battery characteristic thus has a finite slope. Also, the voltage required to charge a battery (and the voltage produced by a battery on discharge) is a function of the state of charge, as shown in Figure 23.4.2(b); as the state of charge nears 100%, voltage rises rapidly. In effect, the load characteristic in Figure 23.4.2(a) moves to the right as charging proceeds to completion. The battery charging characteristic can more nearly match the locus of maximum power points than that of a resistive load. A typical discharge characteristic of a storage battery is also shown.

Photovoltaic generators may be used to drive machines such as electric pumps, refrigerators, and other devices. Direct current motors are of several types, each having different characteristics. The speed at which motors run is a function of voltage and torque, which in turn depend on the characteristics of the load on the motor. Thus an analysis of the operation of PV generators supplying energy to motors must be a simultaneous solution of the characteristics of all of the components. Townsend (1989) and Eckstein (1990) have compared series, shunt, and permanent magnet DC motors, including as loads ventilating fans, centrifugal pumps, and positive displacement pumps. Figure 23.4.3 shows the characteristic of two different DC motors (connected to a water pumping system) and PV generator characteristics at three radiation levels. The system with the separately exited motor begins delivering water at a lower incident radiation than the system with the series motor. Both systems operate far from the maximum power point at both low and high radiation levels. Thus this load is not well matched to PV cell characteristics, direct connection is not an efficient arrangement, and power conditioning equipment[5] would probably be justified.

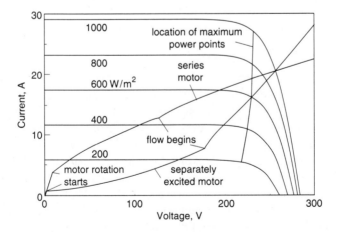

Figure 23.4.3 Characteristics of an array of cell modules at five radiation levels coupled to a water pumping system using two different DC motors. Adapted from Townsend (1989).

[5] See Section 23.5.

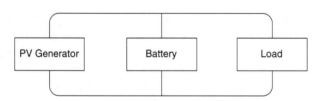

Figure 23.4.4   PV array with battery and load in parallel.

Electrical loads that are combinations of components may be used. A battery may be wired so as to charge when there is current from the PV generator that exceeds load current and discharge when the generator does not produce enough. A common circuit to accomplish this is shown in Figure 23.4.4. In such a parallel circuit the voltage across the battery and the load is the same. The load current is the sum of the generated current and the battery current; the battery current is considered positive if it is discharging and negative if it is charging.

The effective use of photovoltaic generators in many applications will depend on system arrangements more complex than those of the direct-coupled arrangements inferred here. Devices for controlling voltage (with some sacrifice of power) can be used for tracking maximum power points. Some of these control devices and strategies are noted in the following section.

23.5   CONTROLS;   MAXIMUM POWER POINT TRACKERS

Solar cells are expensive, and in most installations controls are used to maximize the output of arrays and protect electrical components from damage. Kirpich et al. (1980) provide a general discussion of some of these problems. Two examples of control mechanisms are described here to illustrate the kind of considerations that may be important.

Systems that consist of PV generators, storage batteries, and loads need controls to protect the battery from overcharge or deep discharge. Overcharge will damage the storage batteries used in these systems, and high-voltage cut-off or power shunting devices are used to interrupt the current to batteries after full charge has been achieved. The voltage across a battery is a function of the state of charge of the battery. Excessive discharge will also damage the batteries, and low-voltage cut-off devices are used to detect low voltages and disconnect the batteries. This results in an interruption of power to the loads and necessitates the use of either very large batteries or an additional source of electric energy if very high degrees of reliability are required. Stand-alone systems requiring very high degrees of reliability have been studied by Klein and Beckman (1987) and Gordon (1987).

Cell output can be maximized by operating near or at the maximum power point. It is clear from the descriptions of load characteristics in the previous section that except in special cases this cannot be achieved without decoupling of the cells and the load (and batteries). Maximum power trackers are devices that keep the impedance of the circuit of the cells at levels corresponding to best operation, and also convert the resulting power from the PV array so that its voltage is that required by the load. There are some power losses associated with the tracking process; efficiencies greater than 90% are possible.

If AC power is needed, DC/AC inverters will be required. This may be the case if AC machinery is to be operated or if the PV system is to be tied into a utility grid. Inverters are available that will generate frequencies to match utility frequencies. Grid-connected systems may have batteries and be controlled so as to maximize the contribution of the PV system to meeting the load and thus minimize the energy purchased. Or, they may "float" on the grid without batteries, with the system adding to or drawing from the grid depending on the loads and available radiation. For practical purposes, this type of system functions as one with infinite storage capacity if the capacity of the grid is large compared to the capacity of the PV system.

23.6 APPLICATIONS

There are many thousands of installed PV systems. A handbook on stand-alone PV systems produced by Sandia (1988) notes that applications as of 1987 included 3100 refrigerators and freezers; 17,500 remote monitoring and data communications stations; 1100 communications systems; 17,000 systems for residences, boats, recreational vehicles, etc.; and 21,000 water pumping installations. In addition there are innumerable very small scale applications for calculators, watches, toys, etc. In this section, brief descriptions are provided of four applications, to illustrate the range of energy problems that can be approached with PV generators.

A 1320 W generator provides energy to drive an AC submersible pump to provide a community water supply at Bendals, Antigua (West Indies). The cell array consists of twenty-eight 12 V modules, with four parallel subarrays each with seven modules in series. The array is mounted at a fixed slope of 17° on a tracking mount. Batteries are not used, and pumped water is stored. A 1500 W three-phase inverter is used. The multistage pump is driven by a 105 V induction motor, with the pump-motor assembly in a 15 cm borehole. The head on the pump is 50 m, and an average of 19,000 liters of water per day is needed to supply the community with water.

In the high, dry environment of central Oregon, a VHF radio repeater station is powered by a 350 W system which includes battery storage. The array is ground mounted at a slope of 59°, and consists of eleven 12 V modules in three sub-arrays, two of 180 W and one of 135 W nominal capacity. Each subarray has its own controller to avoid overcharge or excessive discharge of the batteries. The batteries used have a capacity of 1800 Ah, and the loads on the system are 58 Ah per day. Thus the system has a nominal storage capacity of energy to meet the loads for 30 days. Restrictions on charging and deep discharging put the actual capacity at 14 days; a very high degree of reliability is needed in this application.

A 50 kWp (peak) system provides electrical energy to a dairy farm on Fota Island, Cork, Ireland [McCarthy and Wrixon (1986)]. The loads on the system vary seasonally, from peak loads of 115 kWh per day in summer to minimum loads of 5 kWh per day in winter. In the Irish climate, the generator output exceeds 200 kWh per day in summer but provides only 30 kWh per day in winter. The array includes 2775 modules each of nominal 19 W capacity, mounted facing south at a 45° slope. The maximum recorded output of the array was 47 kW. The maximum recorded module temperature was 50 C when the ambient temperature was 22 C. Storage is provided in lead-acid batteries having aggregate capacity of 600 Ah. Inverters are used in the supply to the farm, and a large line-commutated inverter conditions excess electric energy that is supplied to the utility grid. The system was installed in 1983; it operates automatically with a weekly visual check and a general maintenance check each 3 months.

At Gardner, MA (near Boston), thirty 2.2 kW generators were installed in 1985 and 1986 on neighboring residences in an experimental study of the impact of large, distributed PV systems on utility grid loading, design, and protection [Gulachenski et al. (1988)]. Each of the thirty systems included a 2 kW inverter, and the outputs of the inverters were directly connected to the existing distribution panels in the houses. The systems were all in a small area and represent a possible future "saturation" of PV generators in a utility service area. Problems of concern include: the effects of slow transients caused by normal variation of PV generator output; effects of fast transients due to short circuits, lightning, reduced utility voltage, switching surges, etc.; and the performance of the inverters and their interactions with household appliances. In aggregate, the thirty systems have a 50.4 kW capacity at a solar irradiance of 1000 W/m^2. The output of the systems peaks at times of utility peak loads during normal working hours in the summer. The output exceeds the loads on the residences during the peak hours, and residential systems such as these can feed power to the grid at peak hours and reduce the utility's peak generating capacity needs.

23.7   DESIGN PROCEDURES

There are many parallels between the design of thermal systems and the design of PV systems. The equipment is sufficiently expensive that investment costs dominate life cycle costs, and overdesign of the systems is to be avoided. Thus there is need for means to predict the annual useful electrical output from PV systems. There is a substantial body of experience now available on which to base the sizing of these systems. The methods available for predicting long-term output range from detailed simulations to design methods based on utilizability concepts to hand calculation methods shown in the Sandia handbook (1988). Eckstein (1990) has developed PV component models for the simulation program TRNSYS (Chapter 19). In this section, a design method developed by Siegel et al. (1981) and Evans (1981) and extended by Clark et al. (1984) is presented. (It is similar in concept to the unutilizability design method of Monsen et al. for direct-gain passive heating systems presented in Chapter 22.) It can be applied to systems meeting loads with any monthly average hourly load profile but is limited to systems using maximum power point control.

The method includes two major steps. First, systems without storage are considered, and then the effects of finite (or infinite) storage are taken into account. The simplest system to analyze is one in which the loads always exceed the generating capacity. If at times the capacity of the generator is larger than the load, there is an additional problem of determining how much energy must be dumped, stored, or sold to a utility. If a system includes storage batteries of finite capacity, it is also necessary to estimate the amount of excess energy that will be produced at times when the batteries are fully charged. Evans et al. (1980a,b) have addressed all of these problems in preparing graphs for predicting the fraction of loads carried by solar energy for a wide range of daily load patterns. The method presented here is an analytical method that provides the same information.

For the ith hour of the day, the power output of a PV array with maximum power point tracking is

$$P_i = A_c G_{T,i} \eta_{mp} \eta_e \tag{23.7.1}$$

where A_c is the array area, $G_{T,i}$ is the incident solar radiation on the array, η_{mp} is the maximum power point efficiency of the array as found from Equation 23.2.13, and η_e is the efficiency of any power conditioning equipment. Substituting the cell temperature from Equation 23.3.3 into 23.2.13 yields

$$\eta_{mp} = \eta_{mp,ref} \left[1 + \frac{\mu_{mp}}{\eta_{mp,ref}} (T_a - T_{ref}) + \frac{\mu_{mp} G_T}{\eta_{mp,ref}} \frac{\tau\alpha}{U_L} \left(1 - \frac{\eta_{mp}}{(\tau\alpha)} \right) \right] \tag{23.7.2}$$

The term $\eta_{mp}/\tau\alpha$ is always small compared to unity and thus can be approximated by $\eta_{mp,ref}$ without introducing significant error. The power is then

$$P_i = A_c G_{T,i} \eta_{mp,ref} \eta_e \left[1 + \frac{\mu_{mp}}{\eta_{mp,ref}} (T_a - T_{ref}) + \frac{\mu_{mp} G_T}{\eta_{mp,ref}} \frac{\tau\alpha}{U_L} (1 - \eta_{mp,ref}) \right] \tag{23.7.3}$$

The monthly average array electrical energy output E_i for the hour i is found by integrating Equation 23.7.3 over the hourly period i and then summing over the month (e.g. from 10:00 to 11:00 for the month of June). The monthly average array efficiency $\overline{\eta}_i$ for the hour i is the ratio $\overline{E}_i/A_c\overline{I}_T$. Carrying out the summation yields[6]

$$\overline{\eta}_i = \eta_{mp,ref} \eta_e \left[1 + \frac{\mu_{mp}}{\eta_{mp,ref}} (\overline{T}_{a,i} - T_{ref}) + \frac{\mu_{mp} \overline{I}_T}{\eta_{mp,ref}} \frac{\tau\alpha}{U_L} (1 - \eta_{mp,ref}) Z_i \right] \tag{23.7.4}$$

where

$$Z_i = \frac{1}{N \overline{I}_T^2} \sum_{n=1}^{N} I_{T,n}^2 \tag{23.7.5}$$

---

[6] The instantenous power output of the cell and the monthly average hourly ambient temperature are assumed to remain constant for the hour.

and is a function of geometry and $\bar{k}_T$. Although Z_i can be found using the long-term distribution of hourly radiation, the process is time consuming. An empirical fit for Z_i is

$$Z_i = \left(\frac{\bar{I}_o}{\bar{I}_T}\right)^2 (a_1\, b_1 + a_2\, b_2 + a_3\, b_3) \tag{23.7.6a}$$

where

$$a_1 = R_b^2 + \rho\left(1 - \cos \beta\right) R_b + \rho^2 \left(1 - \cos \beta\right)^2/4 \tag{23.7.6b}$$

$$a_2 = R_b\left(1 + \cos \beta - 2\, R_b\right) + \rho\left(1 + \cos \beta - 2R_b\right)\left(1 - \cos \beta\right)/2 \tag{23.7.6c}$$

$$a_3 = \left[\left(1 - \cos \beta\right)/2 - R_b\right]^2 \tag{23.7.6d}$$

$$b_1 = -\,0.1551 + 0.9226\bar{k}_T \tag{23.7.6e}$$

$$b_2 = 0.1456 + 0.0544\, \ln\bar{k}_T \tag{23.7.6f}$$

$$b_3 = \bar{k}_T\left(0.2769 - 0.3184\bar{k}_T\right) \tag{23.7.6g}$$

A dimensionless critical radiation is defined as the ratio of a critical radiation level to the monthly average radiation where the critical level is that at which the output of the array is equal to the monthly average hourly load for that hour, $\bar{L}_i/A_c\bar{\eta}_i$

$$X_c = \frac{I_{Tc,i}}{\bar{I}_{T,i}} = \frac{\bar{L}_i}{A_c\,\bar{\eta}_i\,\bar{I}_{T,i}} \tag{23.7.7}$$

The fraction of the incident radiation that exceeds this critical radiation level is ϕ_i and can be estimated by the Liu and Jordan generalized utilizability functions, or (more easily) by the Clark et al. utilizability functions (Equations 2.23.5 to 2.23.8).

The monthly average hourly generation that exceeds the load can be written as

$$\bar{E}_{ex,i} = \bar{E}_i\,\phi_i \tag{23.7.8}$$

and the energy to the load is

$$\bar{E}_{L,i} = \bar{E}_i\left(1 - \phi_i\right) \tag{23.7.9}$$

Monthly average daily excess energy and energy to the load are obtained by summing over all hours of the day:

$$\bar{E}_L = \sum_{i=1}^{24} \bar{E}_{L,i} \tag{23.7.10}$$

and

$$\bar{E}_{ex} = \sum_{i=1}^{24} \bar{E}_{ex,i} \tag{23.7.11}$$

Thus the monthly average fraction of the load carried by the PV system with no storage is

$$f_o = \frac{\overline{E}_L}{\sum_{i=1}^{24} \overline{L}_i} = \frac{\overline{E}_L}{\overline{L}} \qquad (23.7.12)$$

Clark and Evans have both studied the effects of load variation on a time scale smaller than an hour and found that the variation causes little difficulty and that use of an $I_{T,c}$ that is constant for an hour is a satisfactory procedure.

From Equation 23.7.11 $\overline{E}_{ex}$ is the monthly average daily amount of energy that cannot be utilized by the load when it is generated and so must be stored, dumped, or fed to a utility grid. If storage is provided, the fraction of the load met by solar will increase by an amount $\Delta f_s = f - f_o$, where f is the solar fraction with storage and f_o is that with no storage. From physical considerations it is possible to put bounds on Δf_s. To do so, it is convenient to define a term that is the value of Δf_s that would be found if all of the excess energy of a system without storage could be stored. This dumped energy without storage is

$$d_o = \eta_b \overline{E}_{ex}/\overline{L} \qquad (23.7.13)$$

where η_b is the the the battery storage efficiency.

If d_o is much smaller than $B_c/\overline{L}$, the ratio of battery capacity to the average daily load, the battery will never become completely charged and no energy will be dumped. Thus at the lower limit, as d_o approaches 0, Δf_s approaches 0.

If d_o is very large, i.e., if the excess energy that is available for storage is large compared to the average load, Δf_s approaches its upper limit. This limit is $1 - f_o$, since the fraction of the load supplied by solar energy cannot exceed unity. In addition to this limit, there is another upper limit on Δf_s. If d_o is large, all loads in daytime will be met by the array and the battery will only be discharged at night. In this situation Δf_s may be limited by the ratio of the battery capacity to the load, $B_c/\overline{L}$. Thus the upper limit on Δf_s can be expressed as

$$\Delta f_{max} = \min[(1 - f_o), B_c/\overline{L}] \qquad (23.7.14)$$

Clark et al. (1984) developed an empirical equation for Δf_s which satisfies these constraints for both high and low values of d_o.

$$\Delta f_s = \frac{1}{2P}\left\{d_o + \Delta f_{max} - \left[(d_o + \Delta f_{max})^2 - 4Pd_o \Delta f_{max}\right]^{1/2}\right\} \qquad (23.7.15)$$

where $\qquad P = 1.315 - 0.1059\dfrac{f_o\overline{L}}{B_c} - \dfrac{0.1847}{\overline{K}_T} \qquad (23.7.16)$

The correlation was developed by computing values of Δf_s for a range of diurnal load types, three climates, and battery capacities from 0 to $2\overline{L}$. The simulation program TRNSYS, including PV generator, regulator inverter, and battery models based on those of Evans et al. (1978), was used.

Clark et al. compared the results of performance calculations using the procedures outlined above with the TRNSYS simulations and found that the standard deviation of the annual differences was 2.4%. (The standard deviation of 672 monthly load fractions was 3.9%.)

Example 23.7.1

Determine the monthly electrical output and solar fraction for a PV power system array that uses four modules having the following characteristics: NOCT temperature = 46 C, reference temperature = 25 C, maximum power efficiency = 10.4%, single module area = 0.427 m^2, temperature coefficient of maximum power point efficiency = –0.00050, efficiency of maximum power point electronics = 0.9, array slope = 35°, array azimuth = 0°, usable battery capacity = 1200 Wh, battery efficiency = 0.8. The system is located at a latitude of 40°, the month is March, the monthly average daily horizontal radiation is 13.2 MJ/m^2, the ground reflectance is 0.2, and the load is constant at 100 W between 9 AM and 3 PM and zero at other times. The 12 monthly average hourly temperatures, beginning at 6 AM, are 4, 5, 5, 6, 8, 10, 11, 12, 12, 11, 10, and 9 C.

Solution

The details for only the hour 10 to 11 will be shown. Intermediate values for the 12 hours of interest are shown in the table. It is first necessary to calculate solar radiation incident on the array for the hour. For day 75, the average day in March, δ = –2.4°, H_o = 27.4 MJ, and ω_s = 88°. Thus, $\overline{K}_T$ = 13.2/27.4 = 0.481. From Equations 2.13.2 and 2.13.4 for the hour 10 to 11, ω = –22.5, r_t = 0.129 and r_d = 0.123. From Equation 1.8.2, R_b = 1.347, and from Equation 2.12.1, $\overline{H}_d/\overline{H}$ = 0.448, so that from Equation 2.23.4,

$$\overline{I}_T = 13.2 \times 10^6[(0.129 - 0.448 \times 0.123)1.347 +$$

$$0.448 \times 0.123 \times (1 + \cos 35)/2 + 0.2 \times 0.129\,(1 - \cos 35)/2]$$

$$= 2.00 \times 10^6 \text{ J}$$

The extraterrestrial radiation for the hour, $\overline{I}_o$, from Equation 1.10.4 is 3.37 MJ. From Equation 23.7.6, Z_i = 1.514. From Equation 23.3.2 at NOCT conditions, $U_L/\tau\alpha = 800/(49 - 20) = 30.8$ W/m^{2}K. The module efficiency for the hour can now be calculated from Equation 23.7.4:

$$\bar{\eta}_i = 0.104 \times 0.9 \left[1 - \frac{0.0005}{0.104} (8 - 28) \right.$$

$$\left. - \frac{0.0005 \times 2 \times 10^6}{0.104 \times 30.7 \times 3600} (1 - 0.104)\, 1.514 \right] = 0.090$$

The monthly average hourly electrical output is then $\bar{E}_i = 0.090 \times 4 \times 0.427 \times 2 \cdot 10^6/3600 = 86$ W.

From 2.23.8, $R_h = \bar{I}_T / (\bar{H}_T\, r_T) = 2.00/(13.2 \times 0.129) = 1.17$. The dimensionless critical radiation is the ratio of the critical radiation level from Equation 23.7.7 to the incident radiation:

$$X_{c,i} = \frac{L_i}{\bar{I}_T A_c \eta_i} = \frac{100 \times 3600}{2 \times 10^6 \times 4 \times 0.427 \times 0.090} = 1.166$$

From Equation 2.23.5, the utilizability ϕ is 0.159 and from Equation 23.7.8 the hourly excess is $\bar{E}_{ex,i} = 86 \times 0.159 = 14$ W. The monthly energy to the load is $\bar{E}_i - \bar{E}_{ex,i} = 72$ W.

The table shows the results for all of the hours of interest. The monthly average electric generation is 664 W and the excess is 242 W. With no storage, useful energy is available only during the middle 6 hours of the days. The total energy to the load for the month is 422 Wh,[7] so the monthly average fraction of the load carried by the PV system without batteries f_o is $422/600 = 0.703$.

Hour	T_a C	$\bar{I}_T$ MJ	Z_i	$\bar{\eta}_i$	$\bar{E}_i$ Wh	$\bar{L}_i$ Wh	$X_{c,i}$	ϕ	$\bar{E}_{ex,i}$ Wh	$\bar{E}_{L,i}$ Wh
6-7	4	0.15	2.204	0.102	7	0	0.000	1.000	7	0
7-8	5	0.60	1.890	0.099	28	0	0.000	1.000	28	0
8-9	5	1.11	1.715	0.096	50	0	0.000	1.000	50	0
9-10	6	1.61	1.592	0.093	71	100	1.412	0.087	6	65
10-11	8	2.00	1.514	0.090	86	100	1.166	0.159	14	72
11-12	10	2.22	1.475	0.088	93	100	1.072	0.193	18	75
12-1	11	2.22	1.475	0.088	93	100	1.077	0.191	18	75
1-2	12	2.00	1.514	0.088	84	100	1.189	0.149	13	71
2-3	12	1.61	1.592	0.090	69	100	1.454	0.075	5	64
3-4	11	1.11	1.715	0.093	49	0	0.000	1.000	49	0
4-5	10	0.60	1.890	0.096	27	0	0.000	1.000	27	0
5-6	9	0.15	2.204	0.100	7	0	0.000	1.000	7	0
Totals					664	600			242	422

[7] These calculations are shown in terms of rates, in watts. Since these rates are taken as constant over hours, the energy quantities in watt-hours are numerically the same as the rates.

To find the contribution by the battery system, Δf_s must be determined. From Equation 23.7.13, $d_o = 0.8 \times 242/600 = 0.323$, and from Equation 23.7.14, Δf_{max} = min[1 – 0.703), 0.323] = 0.297. From Equation 23.7.16, $P = 1.315 - 0.1059 \times$ $0.703 \times 600/1200 - 0.1847/0.481 = 0.894$, and from Equation 23.7.15, $\Delta f_s =$ 0.232. The monthly fraction of the load supplied by the PV system is then 0.703 + 0.232 = 0.935, and the monthly average daily electrical output is $0.935 \times 600 = 561$ Wh.   ∎

A study by Menicucci (1986) compared predictions of long-term PV generator output using the Clark et al. method (in the form of the program PV-FCHART) with measurements on operating systems and found very good agreement. Differences were between 1 and 4% on an annual basis when the radiation measured on the plane of the array was used in the performance calculation.

23.8   HIGH-FLUX PV GENERATORS

Cell array costs are for most systems an important part of the total cost of the system. The contribution of array costs to the total cost ranged from 12 to 60% in the 15 examples cited in the Sandia Handbook (1988), with the fraction tending to be larger in larger systems. The predominance of cell costs has led to consideration of operation of cells at high solar radiation flux levels on the basis that concentrators cost less than cells. A consideration in the design of high-flux systems is the maintenance of uniform radiation intensity on all cells in an array. The current in a series array will be limited by the current in the cell subjected to the lowest flux; this can degrade performance if one or more cells are underilluminated.

An early high-flux generator was built and tested by Beckman and Schöffer (1966); it utilized water cooling from extended heat transfer area on the backs of cells that operated at about 200 times ordinary solar flux. The output of the cells was about 100 times larger than that of the same cells in nonconcentrated radiation. Uniform radiation on the array was achieved by using an assembly of flat mirrors, each reflecting a uniform image onto the focal plane larger than the PV array. The cells were p-n silicon cells with 20 gridlines/cm; these were optimized for the 150 to 350 kW/m^2 radiation fluxes used in the experiments.

One and two axis tracking concentrators are used with PV generators. Refracting and reflecting systems have been applied. Fresnel refractors have been used with concentration ratios (ratio of aperture area to cell area) of 40 to 400, linear reflecting concentrators with ratios of 25 to 40, and CPC-type optics at ratios of about 9. Cells for use with concentrators have closely spaced front contacts to minimize series resistance in the cells, and as they operate at higher fluxes and thus higher temperatures, the materials in the cells and coatings must be able to withstand higher temperatures than cells in normal operation. Cell cooling can be passive or active; passive cooling involves natural convection over extended heat transfer areas on the backs of the cells, and active cooling depends on forced circulation of coolant. A system with active cooling must be fitted with safety devices to defocus the concentrator in case of coolant flow failure, as permanent cell damage will result from severe overheating.

23.9 SUMMARY

This discussion of photovoltaic cells and their applications is an extremely brief treatment of a subject about which many books have been written and one in which new technology appears at a rapid rate. Applications are increasing almost daily. Thus this chapter is only an introduction to a very large and dynamic topic. However, the same considerations of incident radiation on fixed and tracking surfaces apply here as in strictly thermal processes. There is an added consideration that the spectral response of PV cells is such that spectral distribution of solar radiation is more significant than in thermal devices. The same energy balance concepts that apply to collectors apply to PV modules, with the modification that some energy is removed as electrical energy. And the application of utilizability concepts to the estimation of monthly performance of PV generators is parallel to that for passive thermal processes. Thus the earlier chapters provide the groundwork for understanding both thermal and PV processes.

REFERENCES

Beckman, W. A. and P. Schöffer, *Proc. 20th Annual Power Sources Conf.*, 190 (1966). "An Experimental High Flux Solar Power System."

Clark, D. R., S. A. Klein, and W. A. Beckman, *Solar Energy*, **33**, 551 (1984). "A Method for Estimating the Performance of Photovoltaic Systems."

Eckstein, J. H., M.S. Thesis, Mechanical Engineering, U. of Wisconsin-Madison (1990). "Detailed Modelling of Photovoltaic Systems Components"

Evans, D. L., *Solar Energy*, **27**, 555 (1981). "Simplified Method for Predicting Photovoltaic Array Output."

Evans, D. L., W. A. Facinelli, and R. T. Otterbein, Sandia National Laboratories, Report No. SAND 78-7031 (1978). "Combined Photovoltaic/Thermal System Studies."

Evans, D. L., W. A. Facinelli, and L. P. Koehler, Sandia National Laboratories, Report No. SAND 80-7013 (1980a). "Simulation and Simplified Design Studies of Photovoltaic Systems."

Evans, D. L., W. A. Facinelli, and L. P. Koehler, Sandia National Laboratories, Report No. SAND 80-7185 (1980b). "Simplified Design Guide for Estimating Photovoltaic Flat Array and System Performance."

Gordon, J., *Solar Cells*, **20**, 295 (1987). "Optimal Sizing of Stand-Alone Photovoltaic Power Systems."

Green, M. A., *Solar Cells*, Prentice-Hall, Englewood Cliffs, NJ (1982).

Gulachenski, E. M., et al., *Proc. 1988 Annual Meeting,* American Solar Energy Society, p.63 (1988). "The Gardner, MA, 21st Century Residential Photovoltaic Project."

Kenna, J. P. and W. B. Gillett, *Solar Water Pumping - A Handbook*, I. T. Publications, London (1985).

Khouzam, K., L. Khouzam, and P. Groumpos, *Solar Energy*, **46**, 101 (1991) "Optimum Matching of Ohmic Loads to the Photovoltaic Array."

Kirpich, A., G. O'Brien, and N. Shepard, in *Solar Energy Technology Handbook*, (W. C. Dickinson and P. N. Cheremisnoff, eds.), Marcel Dekker, New York **B**, 313 (1980). "Electric Power Generation: Photovoltaics."

Klein, S. A. and W. A. Beckman, *Solar Energy*, **39**, 499 (1987). "Loss-of-Load Probabilities for Stand-Alone Photovoltaic Systems."

Loferski, J. J., J. M. Ahmad, and A. Pandey, *Proc. 1988 Annual Meeting, Am. Solar Energy Soc.*, p.427 (1988). "Performance of Photovoltaic Cells Incorporated into Unique Hybrid Photovoltaic/Thermal Panels of a 2.8 kW Residential Solar Energy Conversion System."

McCarthy, S., and G. T. Wrixon, in *Energy for Rural and Island Communities, IV*, (J. Twidell, I. Hounam, and C. Lewis, eds.), Pergamon Press, Oxford, p 9 (1986). "The Fotavoltaic Project - a 50 kWp Photovoltaic Array on Fota Island."

Maycock, P. D., and E. N. Stirewalt, *Photovoltaics*, Brick House Publishing, Andover, MA (1981).

Menicucci, D. F., *Solar Cells*, **18**, 383 (1986). "Photovoltaic Array Performance Simulation Models."

Rauschenbach, H. S., *Solar Cell Array Design Handbook: the Principles and Technology of Photovoltaic Energy Conversion*, Van Nostrand Reinhold, New York (1980).

Sandia National Laboratories, Report No. SAND 87-7023 (1988). *Stand-Alone Photovoltaic Systems - A Handbook of Recommended Design Practices*. Available from National Technical Information Service, U.S. Department of Commerce.

Schöffer, P., P. Kuhn, and C. M. Sapsford, *Proc. UN Conf. on New Sources of Energy*, United Nations, New York, **4**, 444 (1964). "Instrumentation for Solar Radiation Measurements."

Siegel, M. D., S. A. Klein, and W. A. Beckman, *Solar Energy*, **26**, 413 (1981). "A Simplified Method for Estimating the Monthly-Average Performance of Photovoltaic Systems."

Townsend, T. U., M.S. Thesis, Mechanical Engineering, U. of Wisconsin-Madison (1989). "A Method for Estimating the Long-Term Performance of Direct-Coupled Photovoltaic Systems."

Wolf, M., in *Solar Energy Handbook*, (J. F. Kreider and F. Kreith, eds.), McGraw-Hill, New York (1981). "Photovoltaic Solar Energy Conversion Systems."

Appendix A

PROBLEMS

The problems below are arranged by chapters. Most of them have quantitative answers. A few of them are descriptive. They are selected to be representative of interesting and significant problem types; many of them can be rearranged or broadened in scope to provide a wider range of problem solving experience. Time is assumed to be solar time unless otherwise stated.

Chapter 1

1.1   From the diameter and effective surface temperature of the sun, estimate the rate at which it emits energy. What fraction of this emitted energy is intercepted by Earth? Estimate the solar constant, given the mean Earth-sun distance.

1.2   What would be the "solar constant" for Venus? Mean Venus-sun distance is 0.72 times the mean Earth-sun distance. Assume the sun to be a blackbody emitter at 5777 K.

1.3   What fraction of the extraterrestrial radiation is at wavelengths below 0.5 μm? 2 μm? What fraction is included in the wavelength range 0.5 to 2.0 μm?

1.4   Divide the extraterrestrial solar spectrum into 10 equal increments of energy. Specify a characteristic wavelength for each increment.

1.5   Calculate the angle of incidence of the beam solar radiation at 1400 (2 PM) solar time on February 10 at latitude $43.3°$ on surfaces of the following orientations:
  a   Horizontal
  b Sloped to south at $60°$
  c Slope of $60°$, facing $40°$ west of south
  d Vertical, facing south
  e Vertical, facing west

1.6   **a** When it is 2 PM Mountain Standard Time (MST) on Feb. 3 in North Platte, NE ($L = 101°$W), what is the solar time?
  b When it is 2 PM MST in Boise, ID ($L = 116°$W), on Feb. 3, what is the solar time?
  c What EDT corresponds to solar noon on July 31 for Portland, ME ($L = 70.5°$W)?
  d What CDT corresponds to 10:00 AM on July 31 for Iron Mountain, MI ($L = 90°$W)?

1.7   Determine the sunset hour angle and day length for Madison and for Miami for the following dates: **a** January 1, **b** March 22, **c** July 1. **d** the mean day of February.

1.8   A concentrating collector is located at $\phi = 27°$ and is rotated about a single axis so as to always minimize the angle of incidence of beam radiation on it. On April 5, at solar times of 9 AM and noon, calculate the angle of incidence:
  a If the axis is horizontal and east-west
  b If the axis is horizontal and north-south
  c If the axis is parallel to the earth's axis

1.9 Estimate R_b for a collector at Madison ($\phi = 43^\circ$):
    a Sloped 60° from horizontal, with $\gamma = 0^\circ$, at 2:30 on March 5
    b Sloped 45°, with surface azimuth angle of 15°, at 10:30 solar time on March 5

1.10 For Minneapolis ($\phi = 45^\circ$) on Feb. 8:
    a What is H_o?
    b For the month, what is $\overline{H}_o$?
    c What is I_o for the hour ending at 11:00 AM on the 8th?

1.11 A window on a building in Kansas City faces 15° west of south. An "ell" of the building is 10 ft away from the east edge of the window, and projects out 20.2 ft from the plane of the window. If the vertical dimension of the edge of the ell is large compared to that of the window, plot the times of day at which the window will start to be shaded by the ell as a function of the time of year.

1.12 An overhang over a south-facing window has dimensions as follows (see Figure 14.4.1):
$G = 0.25$ m, $H = 1.75$ m, $W = 3.25$ m, $P = 0.75$ m, $E_L = E_R = 0.50$ m.
    a For a location at $\phi = 35^\circ$ for a point at the middle of the window, plot a shading diagram.
    b Will this point receive beam radiation at 1 PM solar time on February 16? At 3 PM on July 17? At 5 PM on August 16?

Chapter 2

2.1   **a** For the clear day illustrated in Figure 2.5.1, determine the hourly and daily solar radiation on a horizontal surface.
    b For the cloudy day illustrated in Figure 2.5.1, determine the hourly and daily solar radiation on a horizontal surface.

2.2 Do Problem 1.4 for a typical terrestrial distribution.

2.3 Estimate the monthly average radiation on a horizontal surface in January and June in Madison, starting with the average hours of sunshine data from Table 2.7.1. Compare the result with data from Appendix G.

2.4   **a** Solar radiation on a horizontal surface integrated over the hour 11 to 12 AM on Jan. 9 at Boulder, CO ($\phi = 40^\circ$) is 402 kJ/m². What is the clearness index k_T for that hour? Estimate the fraction of the hour's radiation that is diffuse.
    b Solar radiation on a horizontal surface integrated over the day of Jan. 9 at Boulder ($\phi = 40^\circ$ is 4.48 MJ/m². What is the clearness index K_T for that day? What is the estimated fraction of the day's energy which is diffuse?

2.5 The average daily solar radiation on a horizontal surface in Lander, WY ($\phi = 42.8^\circ$) is 18.9 MJ/m² in March. How much of this is beam and how much is diffuse?

2.6 For Madison, 43°N, on January 20, the total radiation on a horizontal surface is 8.0 MJ/m².
    a Estimate the total horizontal radiation for 10 to 11.
    b Estimate the beam and diffuse radiation for 10 to 11.
  For a surface sloped 60° to the south:
    c What is R_b for that hour?
    d If all radiation is treated as beam, what is total radiation on the tilted surface I_T for that hour?
    e If the diffuse and ground-reflected radiation together are considered to be independent of orientation, what is I_T for that hour?
    f If the diffuse radiation is uniform over sky and ground reflectance is considered, what is I_T for that hour? Let ρ_g be 0.2 (bare ground) and 0.7 (snow).

g If the Hay and Davies model is assumed, with circumsolar diffuse added to the beam radiation, what is I_T for that hour? Let ρ_g be 0.2 (bare ground) and 0.7 (snow).

h If the HDKR model is assumed, accounting for circumsolar diffuse and horizon brightening, what is I_T for that hour? Let ρ_g be 0.2 (bare ground) and 0.7 (snow).

i If the Perez model is assumed, what is I_T for that hour? Let ρ_g be 0.2 (bare ground) and 0.7 (snow).

2.7 The day's radiation on a horizontal surface in Madison ($\phi = 43^\circ$) on Dec. 22 is 8.80 MJ/m^2. There is a fresh snow cover. For the hour 11 to 12, estimate the diffuse radiation, the ground-reflected radiation, and the total radiation on a south-facing vertical surface.

2.8 Estimate the average daily radiation in March on a surface sloped 35° to the south for Albuquerque:

a Using the isotropic model

b Using the Klein-Theilacker method

2.9 For Albuquerque, plot the monthly average daily radiation as a function of month for collectors with the following orientations: Assume ground reflectance is 0.2 for all months.

a Horizontal surface

South-facing:

b Collector tilt of 20°

c Collector tilt equal to latitude (35°)

d Collector tilt of 50°

e Vertical collector

2.10 A window faces south at a location with latitude 36°. Estimate the January average radiation on the window if $\overline{K}_T$ for the month is 0.47.

2.11 For Madison in October, for a south-facing surface sloped at 58°, estimate $\overline{R}$ and $\overline{H}_T$. What assumptions have you made?

2.12 Estimate the monthly average daily radiation incident on a south-facing vertical surface in July in Madison. Take the ground reflectance as 0.2.

a Using Equation 2.19.2

b Using Equation 2.20.4

2.13 Estimate the standard clear sky (i.e., 23 km visibility) beam and diffuse radiation on a horizontal surface for Dec. 23 for Minneapolis ($\phi = 46.5^\circ$, elevation 432 m) for each of the hours from sunrise to noon.

2.14 Estimate the hourly beam radiation on the aperture of a collector which rotates continuously about a horizontal north-south axis so as to track the sun. The data base is measurements of normal incidence radiation measured with a pyrheliometer and integrated over the hour, and is indicated in the table. The latitude is 38° and the date is January 7.

Hour	8-9	9-10	10-11	11-12	12-1	1-2	2-3	3-4
I_{bn}	0.35	0.70	2.66	3.05	3.30	3.19	1.80	1.42

2.15 For Minneapolis ($\phi = 45^\circ$) in February:

a The radiation integrated over the hour from 10:00 to 11:00 on Feb. 8 is 1.57 MJ/m^2. What is k_T? What are I_b and I_d for the hour?

b For Feb. 8, the day's integrated radiation is 10.80 MJ/m^2. What are K_T, H_b, and H_d?

c For the month of February, the average daily radiation is 8.67 MJ/m^2. What are $\overline{K}_T$, $\overline{H}_b$, and $\overline{H}_d$?

2.16 On Feb. 8 in Minneapolis, the total radiation for the hour 10:00 to 11:00 is 1.57 MJ/m^2. Estimate the beam, diffuse, and ground-reflected radiation and the total radiation on a south-facing surface having a slope of 60°:
    a   By the Liu and Jordan model
    b   By the HDKR model
    c   By the Perez model

2.17 For the circumstances of Problem 2.16a, estimate the beam radiation on a surface that is rotated continuously about a north-south axis so as to minimize the angle of incidence of the beam radiation.

2.18 For Madison, in February, assume $\rho_g = 0.6$. With $\gamma = 0°$, for each of two cases, $\beta = 58°$ and $\beta = 90°$, estimate H_T and its three parts.

2.19 A building in Springfield, IL, has a south-facing vertical surface that is designed to absorb solar radiation. Calculate the monthly radiation on the surface and the three components of the radiation
    a   Using the Liu and Jordan method
    b   Using the Klein and Theilacker method
Use the following values for ground reflectance for the 12 months, starting with January: 0.7, 0.7, 0.4, 0.2, 0.2, 0.2, 0.2, 0.2, 0.2, 0.2, 0.4, 0.7.

2.20 A window 2.1 m high and 4.2 m long. A horizontal diffuse reflector is placed in front of the window. It is hinged at the bottom edge of the window so that when raised to a vertical position against the window it just covers the window. During daylight hours, the reflector is lowered to its horizontal position. Radiation data for an hour are: I is 1.61 MJ/m^2, I_d is 0.33 MJ/m^2, I_b is 1.28 MJ/m^2, and R_b is 2.31.
    a   If the (diffuse) reflectance of the reflector is 0.85, what will be the total radiation per unit area on the window?
    b   If the diffuse reflectance drops to 0.7, what will be the total radiation per unit area on the window?

2.21 A cylindrical concentrator like that shown in Figure 7.9.1 is installed at latitude 40°. The date is Feb. 20. Radiation for two hour-long periods is shown in the table.

Hour	I_{bn}, MJ/m^2
11-12	2.31
3-4	1.67

    a   If the collector is rotated about a horizontal east-west axis to minimize the angle of incidence of beam radiation on the aperture, what is the estimated beam radiation on the aperture for each hour?
    b   If the collector is rotated on a horizontal north-south axis, what is the estimated beam radiation on the aperture for each hour?

2.22 A south-facing window is on a building at Blue Hill, MA. For the month of January, the critical radiation level is 0.72 MJ/m^2 (based on an indoor temperature of 20 C and an outdoor temperature of –1 C). The average January solar radiation on a vertical surface at Blue Hill is 1.52, 1.15 and 0.68 MJ/m^2 for the hour pairs centered at 0.5, 1.5 and 2.5 hours from solar noon. Calculate the utilizable energy. Use the generalized utilizability method.

2.23 A collector with $\beta = 60°$ and $\gamma = 0°$ is located at Minneapolis. The critical radiation level is 536 W/m^2. For the month of February, for the hour-pair 10 to 11 and 1 to 2, what is the utilizability? What is the utilizable energy?
    If the loss coefficient of the collector is reduced to 60% of the stated value (i.e., the critical radiation level is reduced to 322 W/m^2), what are the utilizability and the utilizable energy? (See Appendix G for data. Assume ρ_g is 0.60.)

2.24 A collector is used in a swimming pool heating system at latitude 34°S and is sloped 20° to the north. For the month of December the average daily radiation on a horizontal surface is 28.7 MJ/m^2. The critical radiation level for the collector for this month in this application is 124 W/m^2. Calculate the daily utilizability and the month's utilizable energy.

2.25 Using the shading plane concept, calculate the beam radiation incident on a window 2 m high and 3 m long, located 0.5 m below an infinitely long horizontal overhang with a projection of 0.7 m. The latitude is 38°, the window faces south, and the hour is 1 to 2 PM. The normal beam radiation is 800 W/m^2. Do the calculation for **a** Jan. 17, **b** April 15.

Chapter 3

3.1 Verify the values of the blackbody spectral emissive power shown in Figure 3.4.1 for: **a** $T = 1000$ K and $\lambda = 10$ μm; **b** $T = 400$ K and $\lambda = 5$ μm; **c** $T = 6000$ K and $\lambda = 1$ μm.

3.2 What is the percentage of the blackbody radiation from a source at 300 K in the wavelength region from 8 to 14 μm? (This is the so-called "window" in the earth's atmosphere.)

3.3 Write a computer subroutine to calculate the fraction of the energy from a blackbody source at T in the wavelength interval a to b.

3.4 Calculate the energy transfer per unit area by radiation between two large parallel flat plates. The temperature and emittance of one plate are 500 K and 0.45 and for the other plate are 300 K and 0.2. What is the radiation heat transfer coefficient?

3.5 Calculate an overall heat loss coefficient for a plate at 50 C, facing up, when exposed to an ambient temperature of 10 C. The plate emittance is 0.88, the dew point is 3 C, and the wind heat transfer coefficient is 25 W/m^{2}C.

3.6 Consider two large flat-plates spaced l mm apart. One plate is at 100 C and the other is at 50 C. Determine the convective heat transfer between the plates for the following conditions:
a Horizontal, heat flow up, $l = 20$ mm
b Horizontal, heat flow up, $l = 50$ mm
c Inclined at 45°, heat flow up, $l = 20$ mm

3.7 Compute the equilibrium temperature of a thin, polished copper plate 1 m by 1 m by 1 mm, under the following conditions:
a In earth orbit, with solar radiation normal to a side of the plate. Neglect the influence of the earth. See Table 4.5.1.
b Just above the earth's surface, with solar radiation normal to the plate and the sun directly overhead. See Table 4.5.1.
Assume the following: the sky is clear and transmits 0.80 of the solar radiation; the equivalent blackbody sky temperature is 10 C less than the ambient temperature; the ambient air temperature is 25 C; the wind heat transfer coefficient is 23 W/m^{2}C; and the earth's surface is effectively a blackbody at 15 C.

3.8 Consider two thin circular disks, thermally isolated from each other, and suspended horizontally, side by side in the same plane, inside a glass sphere on low conductance mounts. The sphere is filled with an inert gas, such as dry nitrogen, to prevent deterioration of the surfaces. The dimensions of the disks are identical. One disk is painted with black paint ($\alpha_b = 0.95$, $\varepsilon_b = 0.95$) and the other with white paint ($\alpha_w = 0.35$, $\varepsilon_w = 0.95$). The glass has a transmittance for solar radiation τ_g of 0.90 and an emittance for long-wave radiation of 0.88. The convection coefficient h between each of the disks and the glass cover is 16 W/m^{2}C. (Note that the disks have two sides and that the edges can be neglected.) When exposed to an unknown solar radiation on a horizontal surface G, the temperature of the white disks T_w is 5 C and temperature of the black disc T_b is 15 C. T_a is 0 C.

    a   Write the energy balances for the black and white disk assuming the glass cover is at a uniform temperature T_c.

    b   Derive an expression for the combined convection and radiation heat transfer coefficient.

    c   Using the result of **b**, derive an expression giving the incident solar radiation as a function of the difference in temperature between the black and white disks.

    d   What is the incident solar radiation G for the conditions stated?

3.9   Determine the convection heat transfer between two large flat-plates (covers in a collector) separated by a distance of 20 mm and inclined at an angle of 60°. The temperature of the lower plate is 110 C and that of the upper plate is at 60 C.

3.10   What is the convective heat transfer for the conditions of Problem 3.9 when a slat-type honeycomb is inserted in the space between the plates? The slat spacing is 10 mm.

3.11   Determine the heat transfer coefficient for air flowing by forced convection in a 1 m wide by 2 m long by 15 mm deep channel. The flow rate is 0.012 kg/s. What is the heat transfer coefficient if the plate spacing is halved? What is the heat transfer coefficient if the mass flow rate is doubled? $T_{air} = 25$ C.

3.12   Estimate the pressure drop in a pebble bed with 3×4 m flow area and with 2 m flow length. The flow rate through the bed is 1.1 kg/s. The pebbles are 0.02 m diameter river washed gravel with a void fraction of 0.45. Use an average air temperature of 40 C.

3.13   Estimate the rock diameter required for a pressure drop of 55 Pa for the conditions of Problem 3.12. Assume the pebble void fraction remains at 0.45 for all pebble sizes.

3.14   A flat-plate collector with one cover is inclined at a slope of 50° from the horizontal. The plate temperature is 100 C and its emittance is 0.10. The cover temperature is 31.5 C, and the (glass) cover has an emittance of 0.88.

    a   Calculate the radiative transfer from plate to cover.

    b   Calculate a radiation heat transfer coefficient from plate to cover.

    c   Calculate the convective transfer from plate to cover.

    d   Given an effective surroundings temperature of 10 C, estimate the net radiative exchange between the cover and the surroundings.

3.15   The ambient temperature around the collector in Problem 3.14 is 10 C, and the wind speed is 5.0 m/s. Estimate the wind coefficient h_w

    a   If the collector is free standing and has dimensions 2.5×10 m.

    b   If the collector is flush mounted on a building with a volume of 564 m^3.

3.16   A counterflow heat exchanger has water entering one side at a rate of 3.75 kg/s and a temperature of 49 C. Glycol enters the other side at a variable flow rate and a temperature of 65 C. The C_p of the glycol is 3780 J/kgC. The UA of the exchanger is 2.10×10^5 W/C. What are the outlet temperatures of the two streams?

Chapter 4

4.1   Consider a surface that has been prepared for use in outer space and has the following spectral characteristics:

$$\rho_\lambda = 0.10 \quad \text{for} \quad 0 < \lambda < \lambda_c \; \mu m$$
$$\rho_\lambda = 0.90 \quad \text{for} \quad \lambda_c < \lambda < \infty \; \mu m$$

For λ_c of 1, 2, and 3 μm, calculate the equilibrium temperature of the plate. Assume the sun can be approximated by a blackbody at 6000 K and that the solar flux on the plate is 1367 W/m². Also assume that the back side of the plate is perfectly insulated.

4.2   A selective surface for solar collector absorber plates has the characteristics

$$\alpha_\lambda = \varepsilon_\lambda = 0.95 \quad \text{for} \quad 0 < \lambda < 1.8 \; \mu m$$
$$\alpha_\lambda = \varepsilon_\lambda = 0.05 \quad \text{for} \quad 1.8 < \lambda < \infty \; \mu m$$

Assume the sun to be a blackbody emitter at 5777 K. Calculate the absorptance of the surface. If the surface is at 150 C, calculate its emittance.

4.3   For the black chrome surface (after the humidity test) of Figure 4.6.3:
- **a**   Determine the absorptance for extraterrestrial solar radiation for the surface.
- **b**   What is the absorptance of the surface of the surface for the terrestrial solar radiation distribution of Table 2.6.1?
- **c**   What is the emittance of the surface at a temperature of 350 C?

4.4   Estimate the emittance of a surface having characteristics of curve C of Figure 4.6.2, if its temperature is: **a** 175 C and **b** 30 C.

4.5   A surface is designed for use in space. It has wavelength-dependent properties as shown in the table. The surface (on a spacecraft) is at 350 K. Calculate the ratio α/ε.

Wavelength Range, micron	α_λ	ε_λ
0-0.6	0.80	0.20
0.6-2.6	0.25	0.75
2.6-100	0.10	0.90

4.6   The absorptance of the surface of Problem 4.4b can be assumed to be that for radiation at normal incidence. Assuming that Figure 4.11.1 adequately represents the characteristics of the surface, what is α at 30°? 45°? 60°?

Chapter 5

5.1   Calculate the reflectance of one glass surface for angles of incidence of **a** 10°; **b** 30°; **c** 50°; and **d** 70°. The index of refraction is 1.526.

5.2   Calculate the transmission of three nonabsorbing glass covers at angles of 10° and 70° and compare your results to Figure 5.1.3.

5.3   For glass with $K = 20 \; m^{-1}$ and 2.0 mm thick, calculate the transmission of two covers: **a** at normal incidence and **b** at 50°.

5.4   Calculate the $(\tau\alpha)$ product for a two-glass-cover collector ($KL = 0.0370$ per plate) with a flat black collector plate surface for radiation incident on the collector at an angle of **a** 25° and **b** 60°. $\varepsilon_p = 0.96$.

5.5   Estimate the transmittance of a single glass cover with $KL = 0.0370$ for diffuse radiation from the sky and for radiation reflected from the ground. The slope of the cover is **a** 45° and **b** 90°.

5.6   What is the transmittance for solar radiation of a collector cover with index of refraction 1.60 at an angle of incidence of 58°? The cover is 2 mm thick and the extinction coefficient is 10 m^{-1}. If the refractive index is 1.40, what will be the transmittance?

5.7   A glass to be used for a cover on a solar collector has $K = 25$ m^{-1} and is 2.5 mm thick. Solar radiation is incident on the glass at an angle of incidence of 55°.
   a   What is the reflectance of a single surface?
   b   What is the transmittance of a single cover?
   c   What is the transmittance of two covers?
   d   For **b**, compare the results obtained with the exact method, the approximate method, and Figure 5.3.1.

5.8   Estimate the radiation absorbed by a collector under the following conditions: $I = 3.0$ MJ/m^2, $I_{bT} = 4.1$ MJ/m^2, $I_d = 0.4$ MJ/m^2, $\theta_{bT} = 25°$, $\beta = 45°$; τ is given in Figure 5.3.1 for $KL = 0.037$ and one cover; $\alpha = 0.93$ and is independent of angle; and ground reflectance is 0.2.

5.9   A collector has a single cover of low-iron glass. It is at a latitude of 48° and a slope of 63°. For a particular hour, the angle of incidence of beam radiation on the collector (taken as that at the half-hour point) is 36.20° and the zenith angle is 71.5°. I is 1.03 MJ/m^2 and I_o is 1.58 MJ/m^2. Ground reflectance is 0.4.
   The cover is glass with $KL = 0.013$. The absorptance of the plate for total radiation at normal incidence is 0.955, and it has an angular dependence as shown in Fig. 4.11.1.
   a   What is $(\tau\alpha)_b$?   c   What is $(\tau\alpha)_g$?
   b   What is $(\tau\alpha)_d$?   d   Calculate S for the hour.

5.10   A vertical, south-facing collector in Madison has an angular dependence of absorptance as shown in Figure 4.11.1 and $\alpha_n = 0.93$. The collector has one cover of low-iron glass with $KL = 0.013$ per sheet. For the month of January average ground reflectance is estimated to be 0.60. Using the isotropic sky model
   a   What are the three components of absorbed radiation?
   b   What is the total absorbed radiation?
   c   What is the average transmittance-absorptance product?

5.11   For the "Tedlar" cover of Figure 5.6.2, estimate the transmittance for blackbody radiation from a source at **a** 45 C and **b** 145 C.

5.12   A building in Madison has as part of the south surfaces a collector-storage wall with double glazing ($KL = 0.0125$ per sheet) and a black absorbing surface behind it that has $\alpha_n = 0.92$ and an angular dependence as shown in Figure 4.11.1. Calculate the absorbed beam, diffuse, ground-reflected, and total radiation for the month of January. Assume that the ground reflectance averages 0.7 for the month. State any assumptions you make.

Chapter 6

6.1   Compare the value of U_L calculated with Equation 6.4.7 to the graphs of Figure 6.4.4 at
   a $h_w = 5$ W/m^{2}C, $\varepsilon_p = 0.95$, $T_p = 60$ C, $T_a = 10$ C, for 2 covers
   b $h_w = 20$ W/m^{2}C, $\varepsilon_p = 0.1$, $T_p = 100$ C, $T_a = 40$ C, for 1 cover

6.2   Calculate the overall loss coefficient for a flat-plate solar collector located in Madison and sloped toward the equator with a slope equal to the latitude. Assume a single glass cover 25 mm above the absorber plate, a wind speed of 6.5 m/s, an absorber long-wave emittance of 0.11, and 70 mm of rock wool back insulation having a conductivity of 0.034 W/mC. The mean plate temperature is 100 C, and the ambient temperature is 25 C. Neglect edge effects and absorption of solar radiation by the glass. The collectors are mounted flush on the surface of a house having a volume of 300 m^3.

6.3 **a** Verify the convective heat flows shown in Figure 6.4.3(a), using the convection equation of Sections 3.11.

  b What will be the cover temperature, total loss rate, and individual convection and radiation transfers if a set of convection suppression slats with aspect ratio 0.3 is used between plate and cover?

6.4 A flat-plate solar collector has 2 glass covers, a black absorber with $\varepsilon_p = 0.95$, mean plate temperature of 110 C at an ambient temperature of 10 C, and a wind loss coefficient of 10 W/m^{2}C. Estimate its top loss coefficient. If the back of the collector is insulated with 50 mm of mineral wool insulation of k of 0.035 W/mC, what is its overall loss coefficient? (Neglect edge effects.) The slope is 45°.

6.5 Consider a flat-plate collector with a fin and tube type absorber plate. Assume $U_L = 8.0$ W/m^{2}C, the plate is 0.5 mm thick, the tube center-to-center distance is 100 mm, and the heat transfer coefficient inside the 20 mm diameter tubes is 300 W/m^{2}C. Assume bond conductance is high. Calculate the collector efficiency factor F' for **a** copper fins, **b** aluminum fins, and **c** steel fins.

6.6 A flat-plate water heating collector absorber plate is copper 1.00 mm thick. Tubes 10 mm in diameter are spaced 160 mm apart. The collector overall loss coefficient is 3.0 W/m^{2}C and the inside heat transfer coefficient is 300 W/m^{2}C. The solder bond between the plate and tubes is 5 mm wide and averages 2 mm thick. The solder has a conductivity of 20 W/mC. What is F' for the collector?

6.7 Derive the F' expression for the air heater shown in Figure 6.14.1(b). Assume U_b is negligible.

6.8 Estimate the useful output of a solar collector when $F_R U_L = 6.3$ W/m^{2}C, $F_R(\tau\alpha) = 0.83$, $T_i = 56$ C, $T_a = 14$ C, and $I_T = 3.4$ MJ/m^2.

6.9 Calculate F', F'', and F_R for an air heater of the type shown in Figure 6.14.1(b) having characteristics as listed. Evaluate all properties at 70 C.

$A_c = 2$ m^2	$T_1 = 100$ C
$U_L = 5.0$ W/m^{2}C	$T_2 = 40$ C
$\varepsilon_1 = \varepsilon_2 = 0.95$	Air flow rate = 0.028 kg/s
Plate spacing = 0.02 m	Length of flow channel = 2 m
Cross-sectional area of flow channel = 0.02 m^2	

6.10 The solar energy absorbed by a solar collector S and the ambient temperatures are given in the table below. The collector has $U_L = 5.2$ W/m^{2}C and $F_R = 0.92$. Determine the useful output of the collector for the day in question if the inlet temperature is constant at 35 C.

Hour	S, MJ/m^2	T_a, C	Hour	S, MJ/m^2	T_a, C
7-8	0.01	-3	12-1	3.42	9
8-9	0.40	0	1-2	3.21	11
9-10	1.90	4	2-3	1.54	5
10-11	2.85	5	3-4	1.07	1
11-12	3.02	7	4-5	0.52	-4

6.11 A "rule of thumb" for solar air heaters is that the air flow rate (at 20 C) should be 10 liters per second per square meter of collector area. For an air heater with this flow rate for which $F' = 0.75$ and $U_L = 6.0$ W/m^{2}C, calculate F_R.

6.12 A solar water heater, when operating, has $F_R U_L = 5.5$ W/m^{2}C. In an experiment, the pump is turned off, and the plate temperature is measured to be 118 C. The pump is then started, with the inlet water at 30 C. What is the useful gain from the collector if the solar radiation does not change?

6.13 Determine the mean plate and fluid temperatures for a water heating collector operating under the following conditions: $U_L = 4.0$ W/m^{2}C; $F' = 0.90$; $\dot{m}/A_c = 0.015$ kg/m^{2}s; tube diameter = 10 mm; inside heat transfer coefficient = 300 W/m^{2}C; inlet temperature = 55 C; ambient temperature = 15 C; incident radiation = 1000 W/m^2; $(\tau\alpha)_e = 0.85$; dust and shading coefficient = 1.5%.

6.14 What is the critical radiation level for the circumstances of Problem 6.13?

6.15 If the circulating pump in Problem 6.13 were to fail, what would you expect the plate temperature to be? What do you expect would happen to U_L under these circumstances, and how would it affect an estimate of the plate temperature?

6.16 Calculate $(\tau\alpha)_e$ for diffuse radiation incident on a single-cover selective-surface collector. The absorptance of the plate at normal incidence is 0.9, and the ratio of α/α_n for the surface is as given in Figure 4.11.1. The plate emittance is 0.10. Do the calculation for **a** a KL product of 0.0524 and **b** KL product of 0.0125.

6.17 For November 2 in Madison the total solar radiation on horizontal surface at 11:30 AM is 8.0 MJ/m^2 and the air temperature is 5 C. Estimate the steady-state efficiency and exit temperature of an air heater of the type of Figure 6.14.1(b) with parameters as follows:
Absorber plate $\alpha = 0.95$, $\varepsilon = 0.2$
Single glass cover, $\varepsilon_c = 0.88$, $KL = 0.0125$
Wind speed = 5 m/s, building volume = 400 m^3
Air mass flow rate = 0.016 kg/m^{2}s
Air entering temperature = 38 C
Plate-to-cover spacing = 20 mm
Air passage depth = 10 mm
Collector width = 1.3 m
Collector length = 3 m
Polyurethane foam back insulation thickness = 60 mm
Collector tilt = 53° (south-facing)

6.18 Estimate the energy collection and efficiency of a module of a two-cover water heating flat-plate collector operating at a latitude of 30° in a space heating process. The date is March 11 and the hour is 1500 to 1600.
Assume: $T_s = T_a$; $(\tau\alpha)_e = 1.02\tau\alpha$; all incident solar is beam radiation. Ignore dust and shading and ignore edge losses.
The design parameters of the module are as follows:
0.75 m wide, 5 m long
Two glass covers, $KL = 0.0125$ per cover
Metal plate 1.0 mm thick; $k = 50$ W/mC
Tubes are 10 mm diameter, on 100 mm centers
$\alpha = 0.95$, independent of incident angle
Back insulation is 80 mm thick; $k = 0.048$ W/mC
Water flow rate is 0.03 kg/s (per module)
Heat transfer coefficient inside riser is 300 W/m^{2}C

Slope from horizontal = 50°
Cover spacing = 25 mm
Plate emittance $\varepsilon_p = 0.95$

For this hour, the operating conditions are as follows:
Solar radiation for the hour on a horizontal surface = 1.40 MJ/m^2
Ambient temperature = 2 C
Wind heat transfer coefficient = 10 W/m^{2}C
Temperature of water entering collector = 35 C

6.19 Estimate the hour-by-hour useful gain and day's efficiency for a flat-plate solar collector located in Boulder, CO, and tilted toward the equator with a slope equal to the latitude. Use the hourly radiation data of Jan. 10 from Table 2.5.1. The collector is the same as that in Problem 6.18 and the inlet water temperature is 25 C.

6.20 In an industrial application for ventilating a paint-spraying operation, ambient air is heated with flat-plate solar air heating collectors. The test results of the air heaters show that $F_R U_L = 4.0$ W/m^{2}C and $F_R(\tau\alpha) = 0.61$. The ambient temperature is 12 C and the solar radiation incident on the collector is 665 W/m^2. Calculate the rate of useful energy collection and the collector efficiency.

6.21 Based on Equation 6.7.6, would it be better to have an absorber surface in a one cover collector with: **a** $\alpha = 0.95$ and $\varepsilon = 0.20$, or **b** $\alpha = 0.90$ and $\varepsilon = 0.10$? What is the basis of the choice?

6.22 A flat-plate water heating collector has the following characteristics:
Plate: Copper, 0.026 cm thick, 15 mm diameter tubes on 80 mm centers,
$\quad \alpha_n = 0.93$, $\varepsilon = 0.11$, $\dot{m}/A_c = 0.0080$ kg/m^{2}s
Covers: 1 glass, KL for the cover $= 0.0125$, spaced 25 mm from the plate
Insulation on back and edge: $U_{be} = 0.82$ W/m^{2}C
Collector orientation: 60° slope to south
Conditions of operation: $T_a = 10$ C, wind heat transfer coefficient $= 20$ W/m^{2}C,
$\quad T_i = 35$ C
$\quad h_{fi} = 300$ W/m^{2}C
For an hour in which $I_T = 2.60$ MJ/m^2 and the radiation is nearly normal to the plane of the collector, estimate Q_u/A.

6.23 Hour-long collector test gives the results shown in the table. For this collector, what are $F_R(\tau\alpha)_n$ and $F_R U_L$? The collector area is 3.0 m^2.

Q_u, MJ	I_T, MJ/m^2	T_{ci}, C	T_{co}, C
6.80	3.10	16.0	12.0
1.49	3.22	86.1	12.0

6.24 A collector has $F_R(\tau\alpha)_n = 0.73$ and $F_R U_L = 6.08$ W/m^{2}C. When $T_i = T_a$, what is the instantaneous efficiency? If the instantaneous efficiency is 0, the ambient temperature is 30 C, and the inlet temperature is 150 C, what is the radiation on the collector?

6.25 The experimental data in the table are from a standard NBS/ASHRAE test on an air heating collector. The module is 7.83×4.00 ft, for a gross area of 31.33 ft^2. The test flow rate is 562.5 lb/hr. What are $F_R(\tau\alpha)_n$ and $F_R U_L$ (in SI units)?

Run	I_T, Btu/ft^{2}hr	T_a, F	$T_{f,i}$, F	$T_{f,o}$, F	ΔT_f	η	$\Delta T_f/I_T$
1	303.6	77.39	182.51	197.32	14.81	21.0	0.346
2	308.8	78.78	185.21	199.59	14.38	20.1	0.345
3	317.9	78.74	142.26	167.99	25.74	34.9	0.200
4	317.7	79.30	142.25	166.52	24.27	32.9	0.198
5	285.4	57.73	63.43	99.05	35.62	53.8	0.020
6	312.3	58.69	64.21	103.28	39.07	53.9	0.018
7	317.1	63.02	90.12	123.85	33.74	45.8	0.085
8	306.0	63.59	89.57	121.48	31.91	44.9	0.085

6.26 An indoor experimental evaluation of a solar water heating collector produced the data in the table below. The collector has a selective coating with α of 0.87 to 0.92 and ε of 0.10 to 0.20. The single glass cover has $\tau = 0.92$ at normal incidence. The collector dimensions are 0.914×2.133 m overall for a gross area of 1.95 m^2. Glass (aperture) area is 1.76 m^2

and the effective absorber area is 1.72 m^2. For all of the data in the table, the wind speed was 3.4 m/s. The average water flow rate is 0.0359 kg/s. For the conditions of these tests, what are $F_R(\tau\alpha)_n$ and $F_R U_L$ base on gross collector area?

T_a, C	23.9	24.4	25.0	26.7	26.7	24.4	25.6	26.1
T_i, C	23.9	24.7	46.1	46.7	52.6	53.0	78.5	78.9
T_o, C	30.7	32.8	51.4	53.3	57.8	58.9	80.9	82.4
G_T, W/m^2	789	947	789	947	789	947	789	947
$\dot{m}$, kg/s	0.0363	0.0385	0.0357	0.0362	0.0357	0.0364	0.0348	0.0359

6.27 The water heating collector of Problem 6.26 is to be operated at reduced flow rates. Estimate new values of $F_R(\tau\alpha)_n$ and $F_R U_L$ based on the gross collector area for flow rates of **a** 0.020 kg/s and **b** 0.010 kg/s.

6.28 The incidence angle modifier coefficient b_o is normally found experimentally. However, it is possible to estimate it; this can be useful for accounting for incidence angle effects in simulations. For a one-cover collector with a cover with KL of 0.0125 and $\alpha_n = 0.93$, estimate b_o based on a calculation of $(\tau\alpha)$ at normal incidence and at $\theta = 60°$.

6.29 A collector with $F' = 0.88$, $U_L = 4.0$ W/m^{2}C, and $(\tau\alpha)_{av} = 0.75$ is operated as a water heater with a flow rate per unit area of 0.012 kg/m^{2}s. Water enters at 17 C, ambient temperature is 8 C, and I_T for the hour is 2.26 MJ/m^2. The collector area is 11.2 m^2. Calculate F_R, Q_u, and η for the hour.

6.30 A liquid heating collector has test data from standard collector tests indicating $F_R U_L = 4.32$ W/m^{2}C and $F_R(\tau\alpha)_n = 0.81$ when operated at a flow rate of 0.012 kg/m^{2}s. The heat capacity of the liquid is 3200 J/kgC. If the flow rate is cut in half, what would be the appropriate values of the two parameters to use in a process design?

6.31 Calculations of the performance of a flat-plate collector with one cover and a selective surface show that $(\tau\alpha)_b$ at 4:30 PM is 0.71. At 4:30, the angle of incidence of beam radiation is 57°. At normal incidence, the absorptance of the plate is 0.93 and the transmittance of the cover is 0.91. Estimate b_o for this collector. What assumptions do you have to make?

6.32 Collector tests produce the following results: $F_R(\tau\alpha)_n = 0.673$, $F_R U_L = 3.07$ W/m^{2}C, and $b_o = -0.17$. This collector operates for a day under the conditions listed in the table. (I_T is in kJ/m^2 and temperatures are in C.) Complete the table.

Hour	I_T	T_a	T_i	θ_b	S	Q_u/A_c	η
8-9	314	-8.3	31.0	58			
9-10	724	-1.7	29.6	45			
10-11	1809	1.7.	27.1	34			
11-12	2299	3.3	30.5	26			
12-1	1926	5.6	35.7	26			
1-2	420	7.2	34.3	34			

From 2 PM on, radiation is negligible. What is η for the day?

Chapter 7

7.1 A concentrating collector is to have a tubular receiver with a plug in it so that the liquid being heated flows through an annulus. The inside and outside diameters of the steel outer tube are 0.054 and 0.059 m respectively, and the outer diameter of the plug is 0.045 m. The assembly is surrounded by a concentric glass tube cover. A collector module is 0.30 m in aperture width and 3.10 m long with the receiver length equal to the reflector length. The water flow rate through the tube is 0.0168 kg/s. If U_L is 7.5 W/m^{2}C (based on the absorbing surface area) and the average fluid temperature is 100 C, what are F' and F_R for this collector?

7.2 In Example 7.3.1 and 7.3.2 the receiver surfaces has an emittance of 0.91. If a selective surface with the same absorptance but with $\varepsilon = 0.18$ were to be used, what would be the useful gain from the collector?

7.3 A CPC collector array has an acceptance half-angle θ_c of 15°. It is oriented along as east-west axis in a fixed position so that the slope of the array is 60°. The application is for heating at a location with latitude 43°. For the hour 10 to 11 on Jan. 26, the radiation on a horizontal plane is 0.36 MJ/m^2 diffuse and 1.20 MJ/m^2 beam. The CPC is not truncated. It is covered by a single glass cover with $KL = 0.0370$.

    a Estimate the absorbed radiation if the specular reflectance of the concentrator is 0.85. At normal incidence, α is 0.95 and its angular dependence is as shown in Figure 4.11.1.

    b Estimate the output of the collector per unit of aperture area for this hour if U_L is 9.0 W/m^{2}C of absorbing area, F_R is 0.92, the inlet fluid temperature is 55 C, and T_a is –5 C.

7.4 A CPC collector array at a location with $\phi = 45^\circ$ is sloped at an angle of 65° to the south. The acceptance half-angle of the CPC elements is 9°, and they are truncated to a height-aperture ratio of 2.5. The array is covered with a single glass cover with $KL = 0.0125$. The reflectance of the CPC is 0.85. Assume that an average absorptance of the receiver can be taken as 0.88 independent of the angle of incidence. For the hour 10 to 11 the incident beam normal radiation as measured with a pyrheliometer is 1.13 MJ/m^2 and the diffuse on a horizontal surface I_d is 0.37 MJ/m^2.

    a On Jan. 2, what would be the absorbed radiation?

    b On Mar. 15, what would be the absorbed radiation?

7.5 A linear parabolic concentrator of width 1.80 m, length 10 m, and focal length 0.92 m is fitted with a flat receiver located in the focal plane. It is arranged to be continuously adjusted about a horizontal north-south axis. The latitude is 32° and the date is March 16. The image in the focal plane, when the beam radiation is normal to aperture, can be approximated as shown on the figure. The receiver is centered on the centerline of the parabola and is 0.022 m wide and 10.5 m long. The loss coefficient will be very near that of the Suntec collector (Figure 7.13.2). For this receiver $\tau\alpha$ is 0.78 and ρ of the concentrator is 0.87.

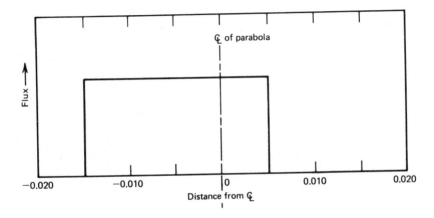

    a   Estimate the absorbed energy for the hours 9 to 10 and 11 to 12 for this date, if the I_{bn} for the hours are 2.06 and 3.14 MJ/m², respectively.

    b   If the collector is used to boil water at 120 C, what will be the energy collected during these hours? The ambient temperatures are 2 and 9 C, respectively.

7.6   The mean sun-Venus distance is 67×10^6 miles. On Venus, what would be the maximum concentration ratio for **a** a linear concentrator and **b** a circular concentrator.

7.7   The figures below show variation of incidence angle modifier, $K_{\tau\alpha}$ with θ_T (the incidence angle in the direction transverse to the tubes) and an efficiency curve measured at near normal incidence for a CPC collector array.

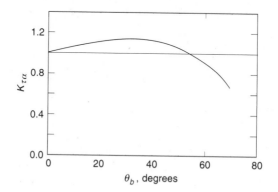

A collector of this type is located at $\phi = 38^\circ$, with $\beta = 40^\circ$ and $\gamma = 0^\circ$. The tubes are oriented in the north-south (up and down) direction. The data in the table are for March 21. Complete the table, assuming that the conditions of operation are close to those under which the test were made.

Time	T_i	T_a	G_T, W/m²	$K_{\tau\alpha}$	Q_u, W/m²
noon	75 C	20 C	925		
2 PM.	75	22	760		
4 PM.	75	22	385		

7.8   A linear parabolic concentrator of length of 10 m has an aperture of 1.80 m and a focal length of 1.26 m. It is mounted on a horizontal east-west axis with continuous adjustment to minimize the angle of incidence on the aperture. It is at a latitude of 35°. The date is March 21. The receiver is flat and is located at the focal plane. Assume the sun to be the non-uniform solar disk.

    a   Determine the size of the receiver to intercept 0.95 of the image at 2:45 PM solar time: (1) for a perfect reflector and (2) for a reflector with a standard deviation of angular error $\sigma = 0.15^\circ$.

    b   The loss coefficient of the receiver is estimated to be 14 W/m²C based on absorber area. $\tau\alpha$ is estimated to be 0.78 and ρ of the concentrator is 0.87. Estimate Q_u for this collector at the time indicated in **a** if $\sigma = 0.15^\circ$; G_{bn} is 2.14 MJ/m², $T_i = 230$ C, and $T_a = 7$ C.

7.9   It has been suggested that a solar-thermal power plant should be based on a concentrator similar to the type shown in Figures 7.15.1 and 17.5.2. Sketch such an arrangement, showing schematically the following dimensions: the distance L from a flat reflecting element to a spherical receiver of diameter D; and the width of the flat reflector W. If reflector size is limited by wind loading to $W = 10$ m and if (in a large system) $L_{max} = 1$ km, what (in terms of D) must be the pointing accuracy of the reflector?

Chapter 8

8.1 Rework Example 3.1, but with a tank containing 500 kg of water and with $(UA)_s$ of 5.56 W/C. Comment on your results.

8.2 Develop a set of equations analogous to those in Section 8.4 for a two-node (partially stratified) tank in which water to meet a load is replaced by water at a constant temperature from the mains.

8.3 Estimate the pressure drop for Example 8.5.1. Does the pressure drop satisfy the minimum recommended pressure drop of Table 13.2.1?

8.4 Estimate the equivalent thermal capacity of a phase change energy storage unit made with paraffin wax (see Table 13.7.1 for properties). The total mass of PCES material is 1300 kg and the temperature change is 40 to 55 C.

Chapter 9

9.1 A water heating system serving a commercial building is set up with a recirculation loop that assures quick availability of hot water at 38 C at all parts of the system. The estimated UA of this loop is 38.1 W/C. The UA of the hot water tank is 11.0 W/C. Cold water enters the heater from the mains at 12 C. Hot water use in the building averages 610 l/day monday through Friday, and 210 l/day on weekends.
   a How much energy is required per week to heat the water (without considering losses)?
   b Estimate the total weekly energy to be supplied to this hot water system if recirculation is continuous.
   c If the recirculation system is shut off from 6 PM to 8 AM each night, if the total mass of pipe and water is equivalent to the water in 122 m of 19 mm diameter tubing, and if there is no hot water use during the hours when the recirculation pump is off, estimate the total weekly energy requirements of the system.

9.2 A building has $(UA)_h = 335$ W/C and is to be maintained at 19 C. The ambient temperature is –3 C. Internal generation in the building is equivalent to 1.5 kW. What is the balance temperature? How much energy must be supplied to the building from fuel or solar heating to maintain the building at 19 C, assuming steady conditions?

9.3 The building of Problem 9.2 is located in Madison, WI. Estimate the annual energy that must be delivered to the building to keep it at 19 C under the following assumptions.
   a The $(UA)_h$ is 335W/C.
   b The building has additional insulation to reduce $(UA)_h$ to 224 W/C.
   c The building has additional insulation to reduce $(UA)_h$ to 112 W/C.

Chapter 10

10.1 A water heating system has a fully mixed tank of capacity 400 liters. The tank is located in a building having a temperature of 19 C. The collector area is 5.0 m^2. The tank area-loss coefficient product $(UA)_s$ is 0.81 W/C, $F_R(\tau\alpha)$ is 0.76, and $F_R U_L$ is 4.80 W/m^{2}C. In a particular hour when the ambient temperature is 12 C, the radiation on the plane of the collector is 18.0 MJ. The load on the system is 5.20 MJ. If the tank temperature is 43 C at the beginning of the hour, estimate the temperature at the end of the hour.

10.2 A fully mixed storage tank containing 1500 kg of water has a loss coefficient-area product of 11.1 W/C. The tank start a particular 24 hour period at a temperature of 45 C. Q_u is added from a collector, and L_S is removed to a load. The loads L_S are indicated in the table below. The tank is located inside a building where the temperature is 20 C.
   Collector data are: $A_c = 30$ m^2; $F_R(\tau\alpha) = 0.78$; $F_R U_L = 7.62$ W/m^{2}C.

For this purpose, assume that $(\tau\alpha)$ is independent of angle of incidence. The incident radiation and outdoor ambient temperature are also given in the table. Times shown are the end of the hour.

Time	1 AM	2	3	4	5	6	7	8	9	10	11	12
Load, MJ	12	12	11	11	13	14	18	21	20	20	18	16
I_T, MJ/m^2									1.09	1.75	2.69	3.78
T_a, C	−4	−4	−5	−3	−1	0	0	1	0	2	4	10

Time	1 PM	2	3	4	5	6	7	8	9	10	11	12
Load, MJ	14	14	13	18	22	24	18	20	15	11	10	9
I_T, MJ/m^2	3.87	3.41	2.77	1.82	1.53							
T_a, C	10	8	8	6	4	4	3	5	6	6	7	6

a   Estimate the useful gain from the collector for the day.
b   Check the 24 hour energy balance on the tank.

10.3   Derive Equation 10.2.4.

10.4   Calculate the value of the collector heat exchanger correction factor F_R'/F_R for the following conditions.

     Heat exchanger effectiveness = 0.60
     Collector fluid flow rate = 0.70 kg/s
     Collector fluid heat capacity = 3350 J/kgC
     Tank heat exchanger flow rate = 0.70 kg/s
     $F_R U_L = 3.75$ W/m^{2}C
     Collector area = 50 m^2

10.5   A 5.2 m^2 water heating collector is connected by 20 m of 19 mm diameter piping on each of the inlet and outlet sides to the storage tank. The pipe is insulated with 18 mm of foam rubber insulation with conductivity 0.050 W/mC. The heat transfer coefficient on the outside of the insulated pipe is 25 W/m^{2}C. F_R is 0.92 and $(\dot{m}C_p)_c$ is 295 W/C. The collector has $F_R(ta) = 0.72$ and $F_R U_L = 4.65$ W/m^{2}C. If the insulated piping is exposed to ambient temperature, what will be the equation for useful output of the collector?

10.6   An industrial process air heating system is shown in the sketch. The air being heated by the combination of solar plus auxiliary enters the heating section at 35 C, is partially heated by solar energy from the storage tank via the load heat exchanger, and then is further heated by the auxiliary exchanger to 55 C. The collector has an area of 80 m^2. $F_R'(\tau\alpha) = 0.78$, and can be considered as constant for this problem. $F_R' U_L = 4.45$ W/m^{2}C. The fully mixed tank has a capacity of 5000 kg. It is well insulated, and $(UA)_s$ is 140 kJ/hrC. The tank is outdoors. The load heat exchanger has an effectiveness near unity (i.e., it is very large) so the return temperature to the tank is always the same as the inlet air temperature. The air

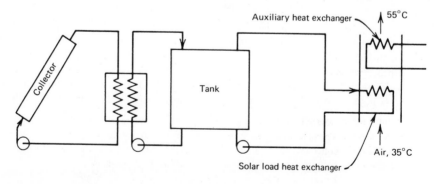

to be heated flows at constant rate for 24 hours a day at a rate of 0.296 kg/s and has a heat capacity of 1013 J/kgC. The flow rate of the water through the heat exchanger is 0.056 kg/s.

The radiation on the collector and ambient temperature data for a 12-hour period are shown in the table. The initial tank temperature is 47 C. For this period, how much energy is supplied to the load from the tank? How much auxiliary energy is needed?

Time	T_a, C	I_T, MJ/m^2	Time	T_a, C	I_T, MJ/m^2
6-7	7	0	12-1	15	3.87
7-8	8	0	1-2	14	3.41
8-9	9	1.09	2-3	12	2.77
9-10	9	1.75	3-4	12	1.82
10-11	9	2.69	4-5	12	1.53
11-12	12	3.78	5-6	11	0

10.7 The system of Problem 10.6 has a load heat exchanger effectiveness of 0.58. Calculate the solar energy supplied and the auxiliary energy required for the same conditions of operation.

10.8 A liquid heating system with a one-cover selective collector with emittance of 0.2 is set up to preheat water from the mains and supply it at T_s to a boiler, as shown in the diagram.

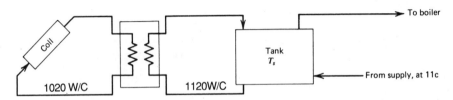

The characteristics of the system are as follows:
$F_R(\tau\alpha)_n = 0.73$
$F_R U_L = 5.10$ W/m^{2}C
Heat exchanger effectiveness = 0.60
Capacitance rate on collector side = 1020 W/C
Capacitance rate on tank side = 1120 W/C
Collector area = 20 m^2
Tank capacity = 1000 liters
Tank UA = 6.0 W/C
Supply water temperature = 11 C
Initial tank temperature = 35 C

The meteorological data and load flow rate for a four-hour period are shown in the table.

Hour	I_T, MJ/m^2	T_a, C	Load Flow, kg
10-11	0.09	14	150
11-12	1.75	17	150
12-1	3.45	18	0
1-2	2.75	20	150

Make the following assumptions:
The controller turns the pump on whenever energy can be collected.
The $(\tau\alpha)$ for the hours in question are the same as $(\tau\alpha)_n$.
All quantities in the table are constant at their averages over each hour.
The tank is fully mixed.
The tank is in a room at a constant temperature of 20 C.

Calculate

a The tank temperature at 2 PM.

b The integrated energy balance (over the 4 hours) on the tank.

10.9 For Problem 10.8 any one of the following might occur (each independently of the others). For each, indicate how the 2 PM T_s would be different from that of Problem 10.8, and explain why.

a The effectiveness of the collector heat exchanger increases to 0.90.

b The absorber surface deteriorates resulting in an increase in emittance to 0.5.

c The pump runs at half speed.

d The supply water temperature increases to 14 C.

10.10 The manufacturer of a solar collector panel supplies the technical data shown in the figures. You are designing a solar domestic hot water heating system that uses six of these panels. The panels are connected in three parallel circuits each having two panels in series. Assume the same flow through each parallel circuit. A 50-50 mixture of propylene glycol and water is circulated through the panels.

A heat exchanger with an effectiveness of 0.45 separates the collector loop from the tank loop. The heat exchanger is located near the storage tank. (See Figure 20.2.3.) The 12 mm diameter copper piping connecting the collectors to the tank and exposed to T_a is

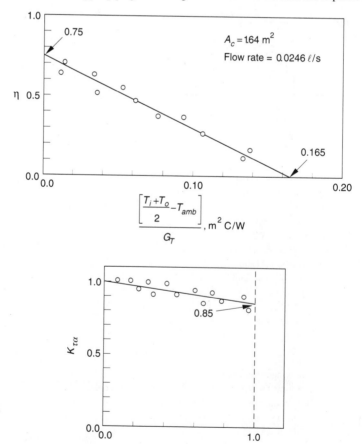

12 m long on both the inlet and outlet of the collector. The heat transfer coefficient on the outside of the pipe insulation is very large. The pipe insulation has an outside diameter of 30 mm and has a thermal conductivity of 0.043 W/mC. The collector loop pump delivers 0.040 l/s and the tank loop pump delivers 0.050 l/s.

Calculate the numerical values of the constants a and b, and an equation for the function $f(\theta)$ in the following equation which gives the useful gain of the collector system in megajoules for an hour period when I_T is in MJ/m^2 for an hour:

$$Q_u = a\left[I_T f(\theta) - b\left(T_{\text{tank}} - T_a\right)\right]^+$$

10.11 For the north row of the collector array of Example 1.9.3, estimate the output of the collector row for an hour in which the beam radiation on the plane of the (unshaded) collectors is 2.95 MJ/m^2 and the diffuse radiation on this plane is 0.88 MJ/m^2. (Note: Assumptions must be made about the angular distribution of the diffuse and the reflected radiation on the collector row. One approximation would be to assume that the combination of the diffuse from the sky and reflected radiation from the surroundings are in sum isotropic. Other assumptions are possible.)

For this collector, $F_R(\tau\alpha)_n = 0.83$, $F_R U_L = 4.07$ W/m^{2}C, and $b_o = 0.14$. For the time in question, $T_i = 40$ C and $T_a = 14$ C.

10.12 An experimental solar heating system is constructed with collectors facing south and arranged in two banks, each of area 50 m^2, the first of which is at a slope of 20^0 and the second at a slope of 70^0. For the collectors $F_R(\tau\alpha)_n$ is 0.83 and $F_R U_L$ is 4.07 W/m^{2}C. Flow is in series from the first to the second bank. At a latitude of 40^0 at 3:30 PM on March 16 the collector inlet temperature is 40 C, the ambient temperature is 14 C, and G is 580 W/m^2. What will be the output of the 100 m^2 array?

Chapter 11

11.1 Calculate the present worth of a cost that is expected to be $5700 in 10 years if the market discount rate is **a** 8%/year and **b** 12%/year.

11.2 What is the annual loan payment on a loan of $20,000 which is borrowed at 8.5% for 15 years if all payments are uniform?

11.3 What is the present worth factor for a series of payments over 18 years if the inflation rate is 7.8% per year and the market discount rate is 9.5% per year?

11.4 Calculate the present worth of a 10 period series of costs the first of which is $1000 payable at the end of the first period. The costs inflate at 5%/period. The discount rate d is **a** 3%/period, and **b** 7%/period.

11.5 A collector is to be installed which has plastic glazing. It is expected that the glazing will have to be replaced every 3 years. The cost of its replacement now is $15/m^2; this is expected to inflate at 7%/year. What is the present worth of the cost of maintaining the glazing on a 75 m^2 collector array over a period of 20 years, if the market discount rate is 10%.

11.6 A solar energy system is to be paid for with a loan in the amount of $2700. The interest rate on the loan is 10.5%/year, and the period is 8 years. The market discount rate is 9.5%/year.
a   Calculate the annual payment to the lender.
b   Calculate the monthly payment for the same loan (assuming that $i = 10.5/12$ per month, compounded monthly).
c   What is the present worth of the interest payments on the loan of part **a**?

11.7 **a** What is the present worth of a cost that is expected to be $4500 seven years hence, if the discount rate is 8% per year?

b Fuel for an operation now costs $700. If the inflation rate is 5% per year, what will it cost at the end of 7 years?

c The cost of fuel for an operation will be billed at the end of a year and will be $700. If the inflation rate is 5% per year, what will be its cost at the end of 7 years?

d A cost for maintenance of solar equipment will be $175 for the first year, payable at the end of the year. If the cost is expected to inflate at 5% per year and the discount rate is 8% per year, what is the present worth of the maintenance cost in the 10th year?

e What is the present worth of the *series* of maintenance costs of **d** for the 10 years?

11.8 In Examples 11.6.3 and 11.6.4, a fixed effective income tax rate of 45% was assumed. What would be the effect on the answer to theses examples if the income tax rate was 45% during the first 10 years, and 55% in the last 10 years? (Note: This could be done quantitatively. Indicate from qualitative considerations what change you expect.)

11.9 The first cost of a solar heating system (after tax credits) is $6000. Of this, area-dependent costs are $96/m^2, and the area-independent cost is $120. It is paid for by a down payment of $1000 and the balance by a 5 year, 10% per year mortgage. The load on the system is 124 GJ/yr, of which 64% is met by solar energy. C_{F1} is $12/GJ and i_F is 7% per year. The discount rate is 9% per year. Assume that maintenance and cost of parasitic power are negligible and there are no property taxes on the solar heating equipment. The owner's effective tax bracket is 0.42. Assume that all payments are made at the end of the year in which they are incurred. Prepare a 10 year table of annual costs (as in Example 11.6.3).

a. If resale value is 0 and N_e is 10 years, what is the *LCS*?

b. If resale value is 0 and N_e is five years, what is the *LCS*?

c. If resale value is $2000 and N_e is 5 years, what is the *LCS*?

11.10 Redo Problem 11.9 by the P_1, P_2 method.

11.11 If in Problem 11.9/11.10 the fuel inflation rate is uncertain to ±3% (i.e., $i_F = 7 \pm 3\%$), what is the uncertainty in the *LCS*? If the calculation of the annual solar fraction is uncertain to ±2% (i.e., $\mathcal{F} = 64 \pm 5\%$), what is the uncertainty in the *LCS*? If both of these uncertainties exist, what is the uncertainty in the *LCS*?

11.12 For the conditions of Problem 11.9/11.10:

a What is payback time B of Section 11.7 without discounting fuel costs?

b What is payback time B is fuel costs are discounted?

11.13 For the system of Problem 11.9/11.10, thermal performance calculations produce the $\mathcal{F}$ vs A_c curve shown below. Estimate the optimum collector area and the life cycle savings at the optimum area.

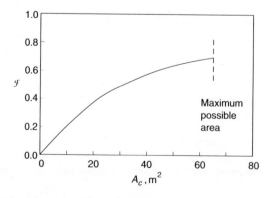

11.14 A solar process is to deliver 70% of an annual energy requirement of 300 GJ. The solar process equipment cost (installed) is $18,000. The equipment is to be paid for by a 20% down payment and 80% loan at 9% per year over 10 years. The market discount rate is 10%/yr. The resale value is expected to be small. This is not an income producing facility. Other costs are as follows: Insurance, maintenance and operating power, $250 in first year, expected to inflate at 6% per year. The owner's federal income tax rate is 40% and the installation is in a state where there is no state income tax. Auxiliary energy supplied is 90 GJ/yr, its initial cost is $8.80/GJ, and it is expected to rise at 10% per year. What are the life cycle savings over a 15 year period?

11.15 A life cycle cost economic analysis is to be made for a solar process. The first costs is $11,000, and it supplies 110 GJ/yr of solar energy to the process (which is not an income producer). The equipment is durable, and it is expected that it can be resold at the end of the period of the analysis for 80% of its installed first cost. 20% of the first cost is paid as cash, and the balance is financed by a loan at 9% interest over 10 years. Insurance, maintenance, parasitic power, and real estate taxes are negligible.

Real estate taxes are 2% of investment, expected to inflate at 6% per year. The effective income tax bracket is 40%. The discount rate is 15% per year.

Fuel cost in the first year (for auxiliary or for fuel-only operation) is $10/GJ and is expected to inflate at 14% per year.

a   What are the life cycle savings of this system over 15 years, compared to a fuel-only system?

b   What are the life cycle savings of this system over 20 years, compared to a fuel-only system? (Note: the change is in the period of the analysis, not the period of the mortgage.)

c   What is the return on investment? (This is a long calculation.)

11.16 A state's legislation provides tax credits for solar energy systems. The administrative rules state that the life cycle fuel savings must exceed the total present worth of the costs of the system within a 25-year period. A solar water heater installed on a house cost a total of $1550; it is expected to meet 60% of an annual load of 19 GJ. The back-up energy source is electricity at 0.045 $/kwh average. The rules specify that an inflation rate of 12% per year shall be used in the calculations. Will this water heater meet the state's criterion?

Chapter 12

12.1 What would be the effect on the amount of solar energy delivered by a collector and tank if the same water heating load were concentrated between the hours of 1700 and 1800, rather than spread out uniformly through the 24 hours of the day?

12.2 Consider the water heater and storage tank of Example 10.9.1. Discuss qualitatively what will happen to the system performance if the following operation or design changes are made. Consider each independently. (Note: Whenever possible, use equations to justify qualitative conclusions.)

a   The collector cover glass is removed (e.g., by breakage).

b   The area of the collector is doubled.

c   The same total load is applied for each day but it is all required between 6 PM and midnight.

d   The storage tank design is changed so that water in the tank is thermally stratified rather than mixed.

Chapter 14

14.1 For a vertical collector-storage wall located in Albuquerque and having dimensions and overhang as described in Problem 2.25, estimate the monthly average absorbed radiation for the month of April. The wall is double glazed with glass of $KL = 0.0370$ per sheet. The α_n of the wall is 0.89.

14.2 A direct-gain system has the same geometry as that used to calculate Figure 14.4.3, but is located in Minneapolis (latitude 44.9°). The ground reflectance is 0.6 in December and January, 0.4 in November and February, and 0.2 for other months.

    a For the month of March, estimate the beam, the diffuse, the ground-reflected, and the total radiation on the shaded receiver, and also the total radiation if there were no shading.

    b Repeat for each month of the year and prepare a plot like Figure 14.4.3.

14.3 Redo Example 14.5.1 with these changes: The location is Madison, WI, the gap is 0.25 m, and the month is February $(\overline{H} = 9.12 \text{ MJ/m}^2, T_a = -6 \text{ C})$. What is the net gain (or loss) from the window?

14.4 What would be the effect of adding night insulation to the collector-storage wall of Problem 14.3? The resistance of the insulation is 1.6 $\text{m}^2\text{C/W}$ and it is to be placed over the windows from 6 PM to 6 AM every day during the heating season.

Chapter 18

18.1 A basin-type solar still is operating under conditions such that the water temperature is 60 C and the ambient temperature is 24 C. The glass temperature is 57 C. Estimate the heat fluxes from basin to cover by convection, by evaporation-condensation, and by radiation.

Chapter 19

19.1 Determine the collector area to supply 75% of the hot water requirements of a residence of a family of four in Boulder, CO, based on the meteorological date for the week of January 8 through 14 as given in Table 2.5.2. Figure 12.1.1(b) is the system schematic. Each person requires 45 kg of water per day at a temperature of 60 C or above. Assume that the load is uniformly distributed over the day from 7 AM to 9 PM. Whenever the storage temperature is less than 60 C, the auxiliary energy source supplies sufficient energy to heat the water coming from the storage tank to 60 C. Use the following collector and storage parameters in your analysis:

Storage mass to collector area	60 kg/m^2
Collector tilt	40° to the south
Collector loss coefficient,	4.0 W/m^{2}C
Cover transmittance-absorptance product $(\tau\alpha)_n$	0.77
Collector efficiency factor F'	0.95
Flow rate through the collector	50 kg/m^{2}hr
Temperature of supply water to bottom of tank	15 C
Loss coefficient of tank	1.05 W/m^{2}C
Height to diameter ratio of cylindrical tank	3
Ambient temperature at tank	21 C

Notes: This problem is best done by use of a simulation program such as TRNSYS. Or, a new program can be written. For a time period as short as a week, the initial tank temperature can affect system performance. You may wish to find the steady periodic solution, in which the final tank temperature is the same as the initial tank temperature.

19.2   Do problem 19.1 with a system like that of Figure 12.1.1(d), where the effectiveness of the collector heat exchanger is **a** 0.78 and **b** 0.35. Assume that the capacitance rates on the two sides of the heat exchanger are the same.

Note: At the end of Appendix A there is an extensive simulation problem typical of those used by the authors as term projects.

Chapter 20

20.1   A well-insulated residence located at a latitude of $43°$ has an area-loss coefficient product of 145 W/C. It has on it a collector of $A_c = 22$ m^2 facing south at a slope of $70°$. The collector-heat exchanger parameters from the standard collector tests are $F_R'(\tau\alpha)_n = 0.79$ and $F_R'U_L = 3.80$ W/m^2C. The collector has one cover. The storage tank capacity is 1650 liters. For the load heat exchanger $\varepsilon_L C_{min}/(UA)_h$ is 2.00. $(\tau\alpha)/(\tau\alpha)_n$ is 0.96. Radiation on the collector $\overline{H}_T$, calculated by the methods of Chapter 2, is 10.7 MJ/m^2. Appendix G provides information on average temperature and degree days. The water heating load for the building is 2.06 GJ per month. For the month of January:
   a   What are the total loads?
   b   What are X and Y?
   c   What is the fraction carried by solar?
   d   If storage volume is halved, what is f?
   e   If $\varepsilon_L C_{min}/(UA)_h$ is reduced to 0.5 (in addition to halving the storage volume), what is f?

20.2   A space heating system in Madison, WI, uses liquid heating collectors. The house and system parameters and weather for a particular month are as follows:

Days in month	30
Radiation on the plane of the collector, $\overline{H}_T$	15 MJ/m^2
Average ambient temperature	–2 C
Degree-days	553 C-days
House area-loss coefficient product, $(UA)_h$	350 W/C
$F_R'(\tau\alpha)_n$	0.78
$F_R'U_L$	7.0 W/m^2 C
$(\tau\alpha)/(\tau\alpha)_n$	0.95
Storage volume per unit collector area	180 l/m^2
Load heat exchanger	Standard size

   a   For a collector area of 100 m^2, calculate the corrected value of X for use in f-chart.
   b   For a collector area of 100 m^2, calculate the corrected value of Y.
   c   For this collector area, what is f?
   d   If the collector area is 50 m^2, what is f?

20.3   A residential building in Madison, WI is to be fitted with a solar heating system of the standard configuration for air systems shown in Figure 20.2.2. The UA of the building is 400 W/C. The storage capacity is the standard storage capacity in the f-chart correlation. Air flow rate is also the standard value. The collector faces south with a slope of $50°$. The collector to be used has the test results shown on the plot; these were measured at the same flow rate as is to be used in the application. For January, water heating loads are expected to be 2.15 GJ, and space heating loads correspond to 830 degree-days. Average ambient temperature is –7.0 C. $\overline{H}_T$ on the collector is 10.5 MJ/m^2 per day. Assume $(\tau\alpha)/(\tau\alpha)_n$ is 0.96.
   a   What collector area is required to meet 0.5 of the January loads?
   b   What is the annual solar fraction for the collector area of part **a**?

20.4 An air system in Madison, WI has $A_c = 40$ m^2, $F_R(\tau\alpha)_n = 0.76$, $F_R U_L = 4.85$ W/m^{2}C, 2 covers, $\beta = 50°$, $\gamma = 30°$, air flow rate $= 12$ l/m^{2}s, storage capacity $= 0.15$ m^3/m^2 of rock of bulk (apparent) density $= 1700$ kg/m^2. The building has $(UA)_h = 300$ W/C. Calculate $\mathcal{F}$ for this system.

20.5 A residential building in Madison has an area-loss coefficient product of 300 W/C. A liquid solar heating system is to be used on this building, with its collector mounted at $60°$ slope and facing $30°$ west of south. A constant monthly hot water load of 1.9 GJ is expected. The collector area is to be 50 m^2, and the collector-heat exchanger parameters are $F_R'(\tau\alpha)_n = 0.72$ and $F_R' U_L = 3.5$ W/m^{2}C. The collector has 2 covers. The storage capacity is to be 5000 liters. The load heat exchanger is to be sized so $\varepsilon_L C_{min}/(UA)_h = 2.0$. Estimate the annual solar fraction.

20.6 For the heat pump characteristics shown in Figure 20.7.2 and the bin data for the month of January given in the table below, determine the monthly work and auxiliary energy requirements for a house with UA of 300 W/C. (The bins cover 2 C and are identified by their central temperatures.)

Bin	-26	-24	-22	-20	-18	-16	-14	-12	-10
Hours	5.5	4	11.5	12	33.5	34	40	44	43

Bin	-8	-6	-4	-2	0	2	4	6	8
Hours	77	76	92	110	103.5	46.5	8.5	3	0

20.7 If a solar system is added in parallel with the system of Problem 20.6, and supplies 30% of the heating load in January, what are the month's heat pump work and auxiliary energy requirements?

Chapter 21

21.1 For the conditions of Problem 20.4 (but with $\gamma = 0$), use the generalized ϕ-charts to estimate the fraction by solar for the month of January. Assume that the collector receives air at a temperature of 20 C for the whole month. The following information may be useful in making the calculations.

For January: the average day is 17; $\delta = -20.9°$; $\omega_s = 69.12°$; $R_b = 2.67$; $K_T = 0.44$; $\bar{H}_o = 13.23$ MJ/m^2; $\bar{H} = 5.85$ MJ/m^2; $T_a = -7$ C; $T_{day} = -3$ C; $H_d/H = 0.39$; $\rho_g = 0.2$.

Hour from Noon	r_t	r_d	θ_z	θ	$(\tau\alpha)/(\tau\alpha)_n$
0	0.179	0.167	63.9	13.9	0.96
0.5	0.176	0.165	64.3	15.7	0.96
1.5	0.152	0.147	67.2	25.8	0.95
2.5	0.108	0.113	72.6	38.8	0.94
3.5	0.055	0.065	80.1	52.6	0.92

21.2 Repeat Problem 21.1 using daily utilizability. Assume $\overline{(\tau\alpha)}/(\tau\alpha)_n$ is 0.94.

21.3 In Example 14.5.1, the direct-gain window was left uncovered during the night, which resulted in a significant heat loss and a reduced net energy gain. Estimate the average net gain for the month of March that could be obtained with a perfect control system that would cover the window with a perfect insulator ($U = 0$) whenever the rate of energy gain is negative.

21.4   Using the generalized ϕ method, calculate the total energy collection for January in a location at 40°N latitude under the following conditions:

$\overline{K_T} = 0.5$

$F_R U_L = 4.2$ W/m²C

$F_R(\tau\alpha) = 0.75$ (constant)

Collector tilt = 55° (facing due south)

$\overline{H_o} = 15.06$ MJ/m²

$\overline{T_a} = 0$ C

$T_i = 50$ C

21.5   Repeat Problem 21.4 using the $\overline{\phi}$ method.

21.6   For Miami, estimate the contribution in July by solar for an absorption air conditioning system that has a COP of 0.7 when the temperature to the generator is greater than 70 C. The solar system characteristics are as given in Example 21.3.1. The collector tilt is 25° to the south. The collector area is 25 m². The cooling load in July is 5.1 GJ. Neglect storage losses.

21.7   For Problem 21.6, the tank has an area-loss coefficient product of 7 W/C. Estimate the reduction in performance due to tank losses if **a** the tank is outside the building and **b** the tank is inside the building.

Chapter 22

22.1   A building at Chattanooga, TN (latitude 35°), has 16.1 m² of south-facing windows. The area-loss coefficient product for the entire building is 200 W/C. The $(\tau\alpha)$ of the window-room combination is 0.79. The loss coefficient of the window is 3.45 W/m²C. The room thermostat is set at 18.3 C.

For the month of February at this location, $\overline{H}$ is 9.23 MJ/m² and $\overline{K_T}$ is 0.41. For the mean day in February, the day length is 10.8 hours and the zenith angle of the sun at noon is 48°. The average ground reflectance is 0.40, $\overline{H_T}$ is 10.61 MJ/m², and $\overline{T_a}$ is 6 C.

a   If the building has negligible thermal capacitance, what fraction of the February solar radiation on the window is useful in heating the building?

b   The building has an effective thermal capacitance of 20.2×10^6 J/C. Estimate how much auxiliary energy would be required if (1) the allowable temperature swing is 5 C and (2) the allowable temperature swing is 10 C.

22.2   A building in Springfield, IL, is to be heated in part by an ACPS system. The collector is to be located on the roof, facing south, at a slope of 55°. The incident radiation $\overline{H_T}$, the absorbed solar radiation $\overline{S}$, and $(\overline{\tau\alpha})$ for the month of January are estimated to be 11.78 MJ/m², 8.60 MJ/m², and 0.73, respectively.

The collector is an air heater with $F_R(\tau\alpha)_n = 0.62$ and $F_R U_L = 3.10$ W/m²C. $(\tau\alpha)_n$ is 0.846 and $A_c = 20.0$ m². Air goes from the building interior through the collector and back to the interior.

The building is of medium weight construction (Table 9.5.1), and is estimated to have a thermal capacitance of 22.9×10^6 J/C. The floor area is 150 m², and the loads are adequately represented by use of degree-day loads with a base temperature of 18.3 C when the interior is at 21 C. The allowable temperature range in the building is 21 to 26 C.

Estimate the auxiliary energy required to heat this building in January.

Chapter 23

23.1 A PV module has the following characteristics: $I_L = 13.6$ A, $I_o = 0.008$ A, $a = 23.6$V, $R_s = 0.9$ ohm, $R_{sh} = \infty$ ohm. If this module is connected to a resistive load of 20 ohms, what is the power?

23.2 A manufacturer gives the following "name plate" data for a silicon module at 1000 W/m^2 and 25 C: $V_{oc} = 186.6$ V, $I_{sc} = 13.6$ A, $I_{mp} = 11.26$ A, $P_{mp}=1480$ W, $N_s = 300$, $\mu_{I,sc} = 0.02$ A/K, and $\mu_{V,oc} = -0.716$ V/K. Estimate the maximum power at an incident solar radiation of 800 W/m^2 and a cell temperature of 30 C.

23.3 Eight photovoltaic modules are available, each with the I-V characteristics of Example 23.2.1 (which is the same as the 25 C curve of Figure 23.2.5). At a radiation level of 1000 W/m^2 and a cell temperature of 25 C, what is the best series/parallel connection plan to maximize power delivered to a resistive load of **a** 6 ohms, **b** 7 ohms, **c** 12 ohms?

23.4 A PV array has the following characteristics: $\alpha = 0.92$, $\tau = 0.94$, and $U_L = 25$ W/m^{2}C. **a** If the incident radiation is 1000W/m^2 and the ambient temperature is 35 C, estimate the cell temperature. **b** What is T_{NOCT} (the nominal operating cell temperature)? What important assumptions are made in your solutions?

23.5 The following properties describe a solar cell module at reference conditions (i.e., solar radiation of 1000 W/m^2 and cell temperature of 20 C): $V_{oc} = 15$ V, $I_{sc} = 1.2$ A, $I_{mp} = 1.0$ A, $V_{mp} = 12.5$ V, $\mu_{I,sc} = 0.0005$ A/K, $\mu_{V,oc} = -0.05$ V/K, and $T_{NOCT} = 46$ C. The module operates at a solar radiation level of 750 W/m^2 and a cell temperature of 15 C. What will be the power delivered to a resistive load of 13.48 ohms? (This resistive load was chosen to correspond to the maximum power point.)

23.6 At an incident solar radiation of 1000 W/m^2 a PV solar system with an area of 0.1 m^2 has a maximum power output of 12.2 W at a cell temperature of 18.6 C and 10.0 W and 59.2 C. The system NOCT temperature is 40 C. The efficiency of the maximum power point tracking electronics is 0.92. Estimate the maximum power produced at a radiation level of 800 W/m^2 and an ambient temperature of 30 C.

23.7 Write a computer program to reproduce the results of Example 23.7.1.

Semester Project

The following is an example of a system simulation problem given as a semester project. This requires the integration and application to systems of many of the ideas used in short problems in this appendix. The project is open-ended in the sense that the student must decide what variables will be treated and how the results will be presented. Appropriate meteorological data must be available. There are many variations on problems of this type; this one includes some complications that can be omitted if a shorter project is wanted.

INDUSTRIAL AIR HEATER SIMULATION

The Application

For an industrial operation, a continuous supply of heated air is required at a rate of 1.40 kg/s, at a temperature of at least 65 C. The air supply is at a constant temperature of 22 C. The location is Albuquerque, NM. A liquid solar heating system has been proposed to supply part of the needed energy. The major system components are: collector, collector heat exchanger, storage tank, load heat exchanger, and auxiliary heat exchanger. They are arranged as shown in the diagram.

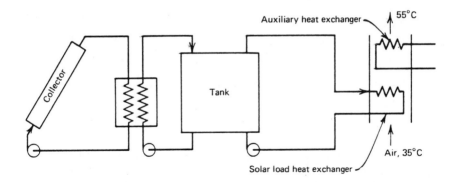

You are assigned the task of evaluating this system and reporting to your management on the technical and economic feasibility of the system. To do this, you are to write a simulation program that will allow you to do year-long simulations of the system, and allow you to explore the effects of several critical variables.. The results of the thermal analysis will then be used as the basis for an economic analysis.

In the proposed process the air enters the heating section at 22 C, is heated first by solar energy, and is then heated in the auxiliary heat exchanger. It is to leave the heater section at a temperature no lower than 65 C. The auxiliary heat exchanger has the capacity to provide all of the needed energy if necessary. (Thus its cost is common to all systems and does not enter into economic comparisons of systems.)

Freeze protection is to be provided by use of an antifreeze solution in the collector loop. You should consider what, if anything, should be done about boiling.

The Equipment

Some of the component design parameters can be fixed, based on past experience and on equipment that is available. For purposes of these simulations, assume the following:

Collector: $F_R(\tau\alpha)_n = 0.81$ and $F_R U_L = 4.45$ W/m^{2}C at $\dot{m}/A_c = 0.015$ kg/m^{2}s
$b_o = -0.12$
$\gamma = 0$ and $\beta = $ latitude
A_c is variable

Collector heat exchanger: The UA of the exchanger and thus also its effectiveness ε_c can be specified at any level.

Storage: A wide range of storage tank sizes is available. Based on past experience it is recommended that the ratio of storage tank capacity to collector area be kept at 100 l/m^2. Assume that the tanks are cylindrical with a ratio of height to diameter of 1. Assume a loss coefficient for the tanks of 5.5 W/m^{2}C.

Load heat exchanger: Cross flow water-to-air heat exchangers are available in a wide range of sizes. It is recommended that the flow rate on the water side be fixed at 0.30 kg/s. The heat capacity of air under the conditions of this operation can be taken as 1013 J/kgC. Use the effectiveness of this exchanger ε_L to describe its operation.

Other Components: The other components, such as piping, ductwork, pumps, controls, etc., are standard. Any assumptions about them should be clearly stated.

The Thermal Analysis

Some thought about this problem will lead you to the following: There are three design variables that are or may be important in determining thermal performance and thus economic feasibility. They are the collector area A_C, the collector heat exchanger effectiveness ε_c, and the load heat exchanger effectiveness ε_L. You should explore the effects of these design variables on:

L_A, the auxiliary energy required

L_S, the solar energy delivered across the load heat exchanger

Q_L, the energy losses from the storage tank.

It is also of interest to explore process dynamics, i.e., the short-term variations of energy flows and temperatures. It is suggested that for a time period of your choice (a period which would be important in process design) you consider what happens with time to the following variables:

T, the tank temperature

T_{As}, the temperature of the air leaving the load heat exchanger

T_{Ao}, the temperature of the air leaving the auxiliary heat exchanger

L_S, the rate of solar energy delivery to the load

Q_u, the rate of solar energy delivery to the tank

I_T, the radiation incident on the collector

L_A, the rate of delivery of auxiliary energy to the load.

You will have to make assumptions in order to write the simulation program. In particular, you will have to decide how you wish to model the storage tank. Considering it to be a fully-mixed tank leads to the simplest program, and has the advantage of producing conservative results.

You will need some guidance in selecting collector area ranges to explore. This can be obtained by use of the design methods described in Chapters 20 and 21. (These methods provide a first estimate only; they are not a substitute for the simulations.) As a first suggestion, you should cover a range of at least 100 to 400 m² (but this may not be an adequate range). Also note that you can calculate the total minimum amount of energy required to heat the air from 22 to 65 C, you can estimate roughly what the annual efficiency of the collector might be, and thus you can get a rough first estimate of the solar contribution of a system.

The Economic Analysis

Based on the thermal analysis, an economic analysis is to be prepared. This does not need to be done in detail, but should be sufficient to indicate to your management what the economic feasibility of this project might be. In order to do this, assume the following economic parameters:

Cost of collectors = $ 250/m²

Cost of storage = $ 0.50/liter

Fixed costs = $ 20,000

Cost now of delivering auxiliary energy = $ 10/GJ

Cost of heat exchangers = $ 5 per unit of W/C

Your Procedure

Prepare a simulation program in the language of your choice. You will need two meteorological data files, one a one-day file of hourly data for debugging your program and a second a TMY (Typical Meteorological Year) file of hourly data which you will use for the simulations.

Your Report

An executive summary should be included in your report. It should be one or two pages in length and should contain the essential facts of what you did and what conclusions you reached. It is intended for managers who are not familiar with the details of solar processes but who have strong engineering backgrounds

The technical report itself should follow standard technical report format. It should include such items as: the basis of the calculations, assumptions made, tables or charts of data, analysis of your results, notes of what might be done about practical problems such as boiling and freezing, and whatever else may be appropriate. It should be the kind of report you would prepare for your supervisor who has some familiarity with the technology.

Appendix B

NOMENCLATURE

Table B-1

This Table of Nomenclature is a partial listing of symbols. Those that are used infrequently or in limited parts of the book and are defined locally do not appear on this list. In some cases section references are provided to the section where a symbol is defined or where there might be confusion about its significance. A special table of radiation nomenclature is provided (Table B-2).

A	area, auxiliary, altitude
A_a	aperture area
A_c	collector area
A_i	anisotropy index (2.16)
A_r	receiver area
a	photovoltaic model curve fitting parameter (23.2)
a,b	coefficients in empirical relationships
b_o	incidence angle modifier coefficient (6.17)
C	cost, concentration ratio (7.2), capacitance rate (3.17), building thermal capacitance (9.5)
C_A	cost per unit of collector area (11.1)
C_E	cost of equipment (11.1)
C_F	cost of energy from fuel (11.1)
C_p	specific heat
C_1, C_2	Planck's first and second radiation constants
C_b	bond conductance
D	diameter (defined locally), down payment fraction
d	dust factor, market discount rate
E	energy, equation of time
e	emissive power, base of natural logarithm
F	fin efficiency factor (6.5), control function (defined locally)
F'	collector efficiency factor (6.5)
F''	collector flow factor (6.7)
F_R	collector heat removal factor (6.7)
F'_R	collector heat exchanger factor (10.2)
$F_{i\text{-}j}$	diffuse energy leaving surface i that is incident on surface j without reflection/diffuse energy leaving surface i (view factor)
f	fraction, modulating factor (2.16), monthly solar fraction (10.10)
f_i	fraction shaded (14.4)
$\mathcal{F}$	annual fraction by solar energy (10.10)
g	gravitational constant
G_{sc}	solar constant, W/m^2
G	irradiance (see Table B-2)
H	daily irradiation (see Table B-2)
h	heat transfer coefficient, Planck's constant

h_{fi}	heat transfer coefficient inside tube or duct
h_r	radiation heat transfer coefficient (3.10)
h_w	wind heat transfer coefficient (3.15)
I	hourly irradiation (see Table B-2), current (23.2)
I_{Tc}	hourly critical radiation on tilted surface
i	inflation rate
K	extinction coefficient
$K_{\tau\alpha}$	incidence angle modifier (6.17)
k	thermal conductivity, Boltzmann's constant
k_T	hourly clearness index (2.10)
$\bar{k}_T$	monthly average hourly clearness index
K_T	daily clearness index (2.11)
$\bar{K}_T$	monthly average daily clearness index (2.12)
l	length, thickness
L	longitude, length, distance, loss, load
$\dot{L}$	load rate
L_A	load met by auxiliary energy
L_0	load with zero area solar energy system
L_S	load met by solar energy
m	mass, air mass, mean, mortgage interest rate
$\dot{m}$	flow rate
n	day of year (1.6), index of refraction, hours of bright sunshine in a day
N	number of covers, day length, term of mortgage or economic analysis
N_p	payback time
P	power
p	pressure, vapor pressure
P_1	ratio of life cycle fuel savings to first year fuel energy cost (11.8)
P_2	ratio of owning cost to initial cost (11.8)
PWF	present worth factor (11.8)
q	energy per unit time per unit length or area
Q	energy, energy per unit time
R	ratio of total radiation on a tilted plane to that on the plane of measurement (usually horizontal) (2.15), heat transfer resistance
$\bar{R}$	monthly average R (2.19)
R_b	ratio of beam radiation on a tilted plane to that on the plane of measurement (usually horizontal) (1.8)
$\bar{R}_b$	monthly average R_b (2.19)
R_d	ratio of diffuse radiation on a tilted plane to that on the plane of measurement (usually horizontal)
R_n	noon radiation ratio (2.24)
ROI	return on investment (11.7)
r	radius
r_d	ratio of diffuse radiation in an hour to diffuse in a day
r_r	mirror radius (7.9)
r_t	ratio of total radiation in an hour to total in a day
S	absorbed solar radiation per unit area
$\bar{S}$	monthly average absorbed solar radiation per unit area (5.10)
s	shade factor
T	temperature
T_{dp}	dew point temperature

t	time
U	overall heat transfer coefficient
U_L	collector overall heat loss coefficient
v	specific volume
V	volume, velocity, voltage (23.2)
W	distance between tubes
X	dimensionless collector loss ratio (20.2)
X'	modified dimensionless loss ratio (21.4)
X_c	dimensionless critical radiation level (2.22)
$\overline{X}_c$	monthly average critical radiation ratio (2.23)
Y	dimensionless absorbed energy ratio (20.2, 21.3)

Greek

α	absorptance, thermal diffusivity
α_s	solar altitude angle
α_o	object altitude angle
α_p	profile angle (1.9)
β	slope
γ	surface azimuth angle, bond thickness, intercept factors
γ_s	solar azimuth angle (8.5)
γ_o	object azimuth angle (1.9)
δ	declination, thickness (defined locally), dispersion
δ_{ij}	delta-function: $\delta_{ij} = 1$ when $i = j$ and $\delta_{ij} = 0$ when $i \neq j$
ε	emittance, effectiveness (3.17)
η	efficiency (defined locally)
θ	angle (defined locally), angle between surface normal and incident radiation
θ_c	acceptance half-angle of CPC
θ_z	zenith angle
λ	wavelength
λ_c	cutoff wavelength of selective surface
μ	absolute viscosity, cosine of polar angle
ν	kinematic viscosity, frequency
ρ	reflectance, density
σ	Stefan-Boltzmann constant, standard deviation
τ	transmittance
ϕ	latitude, angle (defined locally), utilizability (2.22)
ϕ_r	rim angle (7.9)
ψ	angle (defined locally), ground coverage of collector array
ω	hour angle, solid angle
ω_s	sunset (or sunrise) hour angle

Subscripts

A	auxiliary
a	air, ambient, absorbed, aperture, annual
b	blackbody, beam back, bond, bed
c	collector, critical, cover, corrected, cold
d	diffuse, day
e	effective, equivalent, edge
f	fin, fluid, fuel

g	glass, ground, glazing
h	hot
i	incident, inlet
l	loss
m	moving, mean
n	normal, noon
o	overall, out, extraterrestrial
p	plate
r	radiation, reflected, receiver
s	storage, sunset, specular, scattered
T	tilted
t	top
u	useful
w	wind
z	zenith
λ	wavelength

Table B-2   RADIATION NOMENCLATURE (SEE SECTION 1.5)

G	irradiance (W/m^2)
H	irradiation for a day (J/m^2)
I	irradiation for an hour (J/m^2); can also be considered as an hourly average rate

Subscripts

b	beam
d	diffuse
n	normal
T	on tilted plane
o	extraterrestrial

The absence of subscript b or d means total radiation. The absence of subscript n or T means radiation on a horizontal surface. A "bar" above H or I means monthly average.

Examples

G_o	extraterrestrial irradiance on horizontal plane
I_d	an hour's diffuse radiation on a horizontal plane
$\bar{H}_b$	a monthly average daily beam radiation on a horizontal surface
$\bar{H}_T$	monthly average daily radiation (beam plus diffuse) on a tilted plane

INTERNATIONAL SYSTEM OF UNITS

BASIC UNITS

meter, m	length
kilogram, kg	mass
second, s	time
kelvin, K	temperature

DERIVED UNITS

All other units are derived from basic and supplementary units. Some derived units have special names.

The following decimal prefixes are recommended for use with SI units:

tera	T	10^{12}	giga	G	10^9
mega	M	10^6	kilo	k	10^3
milli	m	10^{-3}	micro	μ	10^{-6}
nano	n	10^{-9}	pico	p	10^{-12}
femto	f	10^{-15}	atto	a	10^{-18}

The use of the following prefixes should be limited:

hecto h 10^2	deca da 10
deci d 10^{-1}	centi c 10^{-2}

SOME CONVERSIONS OF UNITS

Exact conversion factors are indicated by an asterisk(*)

Length: m, m/s

1 ft = 0.304 8*m
1 in = 25.4* mm
1 mile = 1.609 km
1 ft/min = 0.005 08* m/s
1 mile/hr = 0.477 0 m/s
1 km/hr = 0.277 78 m/s

Area: m^2

1 ft^2 = 0.092 903 04* m^2
1 in^2 = 0.000 645 16* m^2
1 $mile^2$ = 2.590 km^2

Volume: m^3, m^3/kg, m^3/s

(Note: 1 liter = 10^{-3} m^3)
1 ft^3 = 28.32 liters
1 U.K. gal = 4.546 liters
1 U.S. gal = 3.785 liters
1 ft^3/lb = 0.0062 43 m^3/kg
1 cfm = 0.471 9 liters/s
1 U.K. gpm = 0.075 77 liters/s
1 U.S. gpm = 0.063 01 liters/s
1 cfm/ft^2 = 5.080 liters/m^2s

Mass: kg, kg/m^3, kg/s, kg/m^2s

1 lb = 0.453 592 37* kg
1 oz = 28.35 g
1 lb/ft^3 = 16.02 kg/m^3
1 g/cm^3 = 10^3 kg/m^3
1 lb/h = 0.000 125 6 kg/s
1 lb/ft^2h = 0.001 356 kg/m^2s

Force: newton N = kg m/s^2, N/m

pascal Pa $\equiv$ N/m^2
1 lbf = 4.448 N
1 lbf/ft = 14.59 N/m
1 dyne/cm = 1 (mN)/m
1 mm H_2O = 9.806 65* Pa
1 bar = 10^5 Pa
1 psi = 6.894 kPa
1 in H_2O = 249.1 Pa
1 mm Hg = 133.3 Pa
1 in Hg = 3.378 kPa
1 atm = 101.325* kPa

**Energy: joule J $\equiv$ Nm = Ws,
J/kg, J/kgC**

1 kWh = 3.6* MJ
1 Btu = 1.055 kJ
1 therm = 105.5 MJ
1 kcal = 4.186 8* kJ
1 Btu/lb = 2.326* kJ/kg
1 Btu/lbF = 4.186 8* kJ/kgC
1 Btu/ft^2 = 0.01136 MJ/m^2
1 cal/cm^2 = 1 langley = 0.04187 MJ/m^2

**Power: watt $\equiv$ J/s = N m/s, W/m^2,
W/m^{2}C, W/mC**

1 Btu/h = 0.293 1 W
1 kcal/h = 1.163* W
1 hp = 0.745 7 kW
1 ton refr. = 3.517 kW
1 W/ft^2 = 10.76 W/m^2
1 Btu/h ft^{2}F = 5.678 W/m^{2}C
1 Btu/h ftF = 1.731 W/mC
1 Btu/h ft^2 (F/in) = 0.1443 W/mC
1 Btu/ft^{2}h = 3.155 W/m^2

Viscosity: Pa s = N s/m^2 = kg/m s

1 cP (centipoise) = 10^{-3} Pa s
1 lbf h/ft^2 = 0.172 4 MPa s

Appendix D

MONTHLY $\overline{R}_b$ AS FUNCTION OF ϕ AND $(\phi - \beta)$

These tables are calculated by Equation 2.19.3 and are the basis of Figure 2.19.1. In the northern hemisphere use months across the top. In the southern hemisphere use months across the bottom. The table may be used with little error for azimuth angles of $0 \pm 15^\circ$.

Table D-1 $\phi - \beta = 30^\circ$

ϕ	Jan	Feb	Mar	Apr	May	Jun	Jul	Aug	Sep	Oct	Nov	Dec
0	0.59	0.69	0.83	1.00	1.13	1.20	1.17	1.05	0.90	0.74	0.61	0.56
5	0.62	0.72	0.84	0.98	1.09	1.14	1.12	1.03	0.90	0.76	0.65	0.59
10	0.67	0.75	0.86	0.97	1.06	1.10	1.08	1.01	0.90	0.79	0.69	0.64
15	0.72	0.79	0.88	0.97	1.03	1.06	1.05	0.99	0.91	0.82	0.74	0.70
20	0.79	0.85	0.91	0.97	1.02	1.04	1.03	0.99	0.93	0.87	0.80	0.77
25	0.88	0.91	0.95	0.98	1.01	1.02	1.01	0.99	0.96	0.93	0.89	0.87
30	1.00	1.00	1.00	1.00	1.00	1.00	1.00	1.00	1.00	1.00	1.00	1.00
35	1.16	1.11	1.07	1.03	1.00	0.99	0.99	1.02	1.05	1.10	1.15	1.18
40	1.39	1.26	1.15	1.06	1.01	0.98	0.99	1.04	1.11	1.22	1.35	1.43
45	1.72	1.46	1.26	1.11	1.02	0.99	1.00	1.07	1.19	1.39	1.64	1.81
50	2.24	1.75	1.41	1.17	1.04	0.99	1.01	1.11	1.30	1.62	2.08	2.44
55	3.16	2.19	1.60	1.25	1.06	1.00	1.03	1.16	1.44	1.95	2.83	3.63
60	5.17	2.91	1.88	1.34	1.09	1.00	1.04	1.22	1.62	2.47	4.30	6.64
ϕ	Jul	Aug	Sep	Oct	Nov	Dec	Jan	Feb	Mar	Apr	May	Jun

Table D-2 $\phi - \beta = 25^\circ$

ϕ	Jan	Feb	Mar	Apr	May	Jun	Jul	Aug	Sep	Oct	Nov	Dec
0	0.67	0.76	0.88	1.02	1.13	1.19	1.16	1.07	0.93	0.80	0.69	0.64
5	0.71	0.79	0.89	1.00	1.09	1.13	1.11	1.04	0.93	0.82	0.73	0.68
10	0.76	0.82	0.90	0.99	1.06	1.09	1.07	1.02	0.94	0.85	0.77	0.74
15	0.82	0.87	0.93	0.98	1.03	1.05	1.04	1.00	0.95	0.89	0.83	0.80
20	0.90	0.93	0.96	0.99	1.01	1.02	1.02	1.00	0.97	0.94	0.91	0.89
25	1.00	1.00	1.00	1.00	1.00	1.00	1.00	1.00	1.00	1.00	1.00	1.00
30	1.14	1.09	1.05	1.02	0.99	0.98	0.99	1.01	1.04	1.08	1.12	1.15
35	1.32	1.22	1.12	1.05	1.00	0.97	0.98	1.02	1.09	1.18	1.29	1.35
40	1.57	1.38	1.21	1.08	1.00	0.97	0.98	1.05	1.16	1.32	1.51	1.63
45	1.93	1.60	1.33	1.13	1.02	0.97	0.99	1.08	1.24	1.50	1.83	2.06
50	2.51	1.91	1.48	1.19	1.03	0.97	1.00	1.12	1.35	1.75	2.32	2.76
55	3.53	2.38	1.69	1.27	1.06	0.98	1.01	1.17	1.49	2.11	3.14	4.09
60	5.76	3.16	1.98	1.37	1.09	0.99	1.03	1.23	1.68	2.66	4.76	7.44
ϕ	Jul	Aug	Sep	Oct	Nov	Dec	Jan	Feb	Mar	Apr	May	Jun

Table D-3 $\phi - \beta = 20^\circ$

ϕ	Jan	Feb	Mar	Apr	May	Jun	Jul	Aug	Sep	Oct	Nov	Dec
0	0.74	0.82	0.92	1.03	1.12	1.17	1.15	1.07	0.96	0.85	0.76	0.72
5	0.79	0.85	0.93	1.01	1.08	1.11	1.10	1.04	0.96	0.87	0.80	0.77
10	0.84	0.89	0.94	1.00	1.04	1.07	1.06	1.02	0.97	0.91	0.85	0.83
15	0.91	0.94	0.97	1.00	1.02	1.03	1.02	1.01	0.98	0.95	0.92	0.90
20	1.00	1.00	1.00	1.00	1.00	1.00	1.00	1.00	1.00	1.00	1.00	1.00
25	1.11	1.08	1.04	1.01	0.99	0.98	0.98	1.00	1.03	1.07	1.10	1.12
30	1.26	1.18	1.10	1.03	0.98	0.96	0.97	1.01	1.07	1.15	1.24	1.29
35	1.46	1.31	1.17	1.06	0.98	0.95	0.97	1.02	1.12	1.26	1.42	1.51
40	1.73	1.49	1.27	1.10	0.99	0.95	0.97	1.05	1.19	1.41	1.66	1.82
45	2.13	1.72	1.39	1.14	1.00	0.95	0.97	1.08	1.28	1.60	2.01	2.29
50	2.76	2.06	1.55	1.21	1.02	0.95	0.98	1.12	1.39	1.86	2.54	3.06
55	3.88	2.56	1.76	1.28	1.05	0.96	1.00	1.17	1.54	2.24	3.43	4.52
60	6.30	3.39	2.06	1.38	1.07	0.96	1.01	1.23	1.74	2.83	5.18	8.19
ϕ	Jul	Aug	Sep	Oct	Nov	Dec	Jan	Feb	Mar	Apr	May	Jun

Table D-4 $\phi - \beta = 15^\circ$

ϕ	Jan	Feb	Mar	Apr	May	Jun	Jul	Aug	Sep	Oct	Nov	Dec
0	0.82	0.87	0.95	1.03	1.10	1.14	1.12	1.06	0.98	0.90	0.83	0.80
5	0.86	0.91	0.96	1.01	1.06	1.08	1.07	1.03	0.98	0.92	0.87	0.85
10	0.92	0.95	0.97	1.00	1.03	1.04	1.03	1.01	0.99	0.96	0.93	0.92
15	1.00	1.00	1.00	1.00	1.00	1.00	1.00	1.00	1.00	1.00	1.00	1.00
20	1.10	1.07	1.03	1.00	0.98	0.97	0.98	0.99	1.02	1.06	1.09	1.11
25	1.22	1.15	1.08	1.01	0.97	0.95	0.96	0.99	1.05	1.13	1.20	1.24
30	1.38	1.26	1.14	1.03	0.96	0.93	0.95	1.00	1.09	1.22	1.34	1.42
35	1.59	1.40	1.21	1.06	0.96	0.92	0.94	1.02	1.15	1.33	1.54	1.66
40	1.89	1.58	1.31	1.10	0.97	0.92	0.94	1.04	1.22	1.48	1.80	2.00
45	2.32	1.83	1.44	1.15	0.98	0.92	0.95	1.07	1.31	1.68	2.17	2.50
50	3.00	2.18	1.60	1.21	1.00	0.92	0.96	1.11	1.42	1.96	2.74	3.34
55	4.20	2.72	1.82	1.29	1.03	0.93	0.97	1.16	1.57	2.36	3.69	4.91
60	6.80	3.60	2.13	1.39	1.05	0.94	0.99	1.23	1.77	2.98	5.57	8.87
ϕ	Jul	Aug	Sep	Oct	Nov	Dec	Jan	Feb	Mar	Apr	May	Jun

Table D-5 $\phi - \beta = 10^\circ$

ϕ	Jan	Feb	Mar	Apr	May	Jun	Jul	Aug	Sep	Oct	Nov	Dec
0	0.88	0.92	0.97	1.03	1.08	1.10	1.09	1.05	1.00	0.94	0.89	0.87
5	0.93	0.96	0.98	1.01	1.03	1.05	1.04	1.02	0.99	0.97	0.94	0.93
10	1.00	1.00	1.00	1.00	1.00	1.00	1.00	1.00	1.00	1.00	1.00	1.00
15	1.08	1.06	1.03	1.00	0.97	0.96	0.97	0.99	1.01	1.05	1.07	1.09
20	1.18	1.13	1.06	1.00	0.96	0.94	0.94	0.98	1.04	1.10	1.17	1.20
25	1.31	1.21	1.11	1.01	0.94	0.91	0.93	0.98	1.07	1.18	1.29	1.35
30	1.48	1.33	1.17	1.03	0.94	0.90	0.92	0.99	1.11	1.27	1.44	1.54
35	1.71	1.47	1.25	1.06	0.94	0.89	0.91	1.00	1.17	1.39	1.64	1.79
40	2.03	1.66	1.34	1.10	0.95	0.89	0.91	1.03	1.24	1.55	1.92	2.15
45	2.48	1.92	1.47	1.14	0.96	0.89	0.92	1.06	1.33	1.76	2.31	2.70
50	3.21	2.30	1.64	1.21	0.98	0.89	0.93	1.10	1.44	2.04	2.92	3.59
55	4.48	2.85	1.87	1.28	1.00	0.90	0.94	1.15	1.59	2.46	3.93	5.27
60	7.24	3.77	2.19	1.38	1.03	0.90	0.95	1.21	1.80	3.10	5.91	9.49
ϕ	Jul	Aug	Sep	Oct	Nov	Dec	Jan	Feb	Mar	Apr	May	Jun

Table D-6 $\phi - \beta = 5^{\circ}$

ϕ	Jan	Feb	Mar	Apr	May	Jun	Jul	Aug	Sep	Oct	Nov	Dec
0	0.94	0.96	0.99	1.02	1.04	1.05	1.05	1.03	1.00	0.97	0.95	0.94
5	1.00	1.00	1.00	1.00	1.00	1.00	1.00	1.00	1.00	1.00	1.00	1.00
10	1.07	1.05	1.02	0.99	0.97	0.96	0.96	0.98	1.01	1.04	1.06	1.08
15	1.16	1.10	1.04	0.99	0.94	0.92	0.93	0.97	1.02	1.08	1.14	1.17
20	1.26	1.18	1.08	0.99	0.92	0.89	0.91	0.96	1.04	1.14	1.24	1.29
25	1.40	1.27	1.13	1.00	0.91	0.87	0.89	0.96	1.07	1.22	1.36	1.44
30	1.58	1.38	1.19	1.02	0.91	0.86	0.88	0.97	1.12	1.32	1.52	1.64
35	1.82	1.53	1.27	1.05	0.91	0.85	0.88	0.98	1.17	1.44	1.74	1.91
40	2.15	1.73	1.37	1.08	0.91	0.85	0.88	1.01	1.24	1.60	2.03	2.30
45	2.63	2.00	1.50	1.13	0.93	0.85	0.88	1.04	1.33	1.82	2.44	2.87
50	3.39	2.39	1.67	1.19	0.94	0.85	0.89	1.08	1.45	2.11	3.07	3.81
55	4.73	2.97	1.90	1.27	0.97	0.86	0.90	1.12	1.60	2.54	4.13	5.59
60	7.63	3.92	2.23	1.37	0.99	0.86	0.92	1.18	1.81	3.20	6.20	10.03
ϕ	Jul	Aug	Sep	Oct	Nov	Dec	Jan	Feb	Mar	Apr	May	Jun

Table D-7 $\phi - \beta = 0^{\circ}$

ϕ	Jan	Feb	Mar	Apr	May	Jun	Jul	Aug	Sep	Oct	Nov	Dec
0	1.00	1.00	1.00	1.00	1.00	1.00	1.00	1.00	1.00	1.00	1.00	1.00
5	1.06	1.04	1.01	0.98	0.96	0.95	0.95	0.97	1.00	1.03	1.05	1.06
10	1.13	1.08	1.03	0.97	0.92	0.90	0.91	0.95	1.00	1.06	1.12	1.14
15	1.22	1.14	1.05	0.97	0.90	0.87	0.88	0.94	1.02	1.11	1.20	1.24
20	1.33	1.22	1.09	0.97	0.88	0.84	0.86	0.93	1.04	1.17	1.30	1.37
25	1.48	1.31	1.14	0.98	0.86	0.81	0.83	0.93	1.07	1.25	1.43	1.53
30	1.66	1.43	1.20	1.00	0.85	0.79	0.82	0.93	1.12	1.35	1.60	1.74
35	1.91	1.59	1.28	1.02	0.85	0.77	0.80	0.94	1.17	1.48	1.82	2.02
40	2.26	1.79	1.38	1.05	0.84	0.75	0.79	0.96	1.24	1.64	2.12	2.42
45	2.76	2.07	1.51	1.09	0.83	0.73	0.77	0.98	1.33	1.86	2.55	3.02
50	3.55	2.46	1.69	1.15	0.83	0.69	0.75	1.00	1.45	2.17	3.21	4.00
55	4.94	3.06	1.92	1.21	0.81	0.64	0.72	1.03	1.60	2.60	4.30	5.86
60	7.96	4.04	2.25	1.28	0.77	0.55	0.65	1.05	1.80	3.28	6.45	10.50
ϕ	Jul	Aug	Sep	Oct	Nov	Dec	Jan	Feb	Mar	Apr	May	Jun

Table D-8 $\phi - \beta = -5^{\circ}$

ϕ	Jan	Feb	Mar	Apr	May	Jun	Jul	Aug	Sep	Oct	Nov	Dec
0	1.05	1.03	1.00	0.97	0.95	0.94	0.94	0.96	0.99	1.02	1.04	1.05
5	1.11	1.06	1.01	0.96	0.91	0.89	0.90	0.94	0.99	1.05	1.10	1.12
10	1.18	1.11	1.03	0.94	0.88	0.85	0.86	0.92	1.00	1.08	1.16	1.20
15	1.28	1.17	1.06	0.94	0.86	0.82	0.84	0.90	1.01	1.13	1.25	1.31
20	1.39	1.25	1.09	0.94	0.84	0.80	0.82	0.90	1.03	1.20	1.35	1.44
25	1.54	1.34	1.14	0.96	0.83	0.78	0.80	0.90	1.06	1.27	1.49	1.60
30	1.73	1.47	1.20	0.97	0.83	0.76	0.79	0.91	1.11	1.37	1.66	1.82
35	1.99	1.63	1.28	1.00	0.83	0.76	0.79	0.92	1.16	1.50	1.89	2.11
40	2.35	1.83	1.38	1.03	0.83	0.75	0.79	0.94	1.23	1.67	2.20	2.53
45	2.87	2.12	1.51	1.08	0.84	0.75	0.79	0.97	1.32	1.89	2.64	3.15
50	3.68	2.52	1.69	1.14	0.86	0.76	0.80	1.01	1.44	2.20	3.31	4.17
55	5.12	3.13	1.92	1.21	0.88	0.76	0.81	1.05	1.59	2.65	4.44	6.09
60	8.23	4.12	2.25	1.30	0.90	0.77	0.82	1.11	1.79	3.33	6.65	10.89
ϕ	Jul	Aug	Sep	Oct	Nov	Dec	Jan	Feb	Mar	Apr	May	Jun

Table D-9 $\phi - \beta = -10^o$

ϕ	Jan	Feb	Mar	Apr	May	Jun	Jul	Aug	Sep	Oct	Nov	Dec
0	1.09	1.05	1.00	0.94	0.89	0.87	0.88	0.92	0.97	1.03	1.08	1.10
5	1.15	1.08	1.01	0.92	0.86	0.83	0.84	0.89	0.97	1.06	1.13	1.17
10	1.23	1.13	1.02	0.91	0.83	0.79	0.81	0.88	0.98	1.10	1.20	1.25
15	1.32	1.19	1.05	0.91	0.81	0.76	0.78	0.86	0.99	1.15	1.29	1.36
20	1.44	1.27	1.09	0.91	0.79	0.74	0.76	0.86	1.01	1.21	1.39	1.49
25	1.59	1.37	1.13	0.92	0.78	0.72	0.75	0.86	1.05	1.29	1.53	1.66
30	1.79	1.49	1.20	0.94	0.78	0.71	0.74	0.87	1.09	1.39	1.71	1.89
35	2.05	1.65	1.27	0.97	0.78	0.70	0.74	0.88	1.14	1.52	1.94	2.19
40	2.42	1.86	1.38	1.00	0.78	0.70	0.74	0.90	1.21	1.69	2.26	2.62
45	2.95	2.15	1.51	1.04	0.79	0.70	0.74	0.93	1.30	1.91	2.71	3.26
50	3.79	2.56	1.68	1.10	0.81	0.70	0.75	0.96	1.41	2.22	3.40	4.30
55	5.26	3.17	1.91	1.17	0.83	0.71	0.76	1.01	1.56	2.67	4.55	6.27
60	8.43	4.18	2.24	1.26	0.85	0.71	0.77	1.06	1.76	3.36	6.80	11.19
ϕ	Jul	Aug	Sep	Oct	Nov	Dec	Jan	Feb	Mar	Apr	May	Jun

Table D-10 $\phi - \beta = -15^o$

ϕ	Jan	Feb	Mar	Apr	May	Jun	Jul	Aug	Sep	Oct	Nov	Dec
0	1.12	1.06	0.98	0.90	0.83	0.80	0.81	0.87	0.95	1.03	1.11	1.14
5	1.18	1.10	0.99	0.88	0.80	0.76	0.77	0.85	0.95	1.06	1.16	1.21
10	1.26	1.14	1.01	0.87	0.77	0.72	0.74	0.83	0.95	1.10	1.23	1.29
15	1.36	1.21	1.04	0.87	0.75	0.70	0.72	0.82	0.97	1.15	1.32	1.40
20	1.48	1.28	1.07	0.87	0.74	0.68	0.70	0.81	0.99	1.21	1.42	1.54
25	1.63	1.38	1.12	0.88	0.73	0.66	0.69	0.81	1.02	1.29	1.56	1.71
30	1.83	1.51	1.18	0.90	0.72	0.65	0.68	0.82	1.06	1.39	1.74	1.94
35	2.10	1.67	1.26	0.92	0.72	0.64	0.68	0.83	1.11	1.52	1.98	2.25
40	2.47	1.88	1.36	0.96	0.73	0.64	0.68	0.85	1.18	1.69	2.30	2.68
45	3.01	2.17	1.49	1.00	0.74	0.64	0.68	0.88	1.27	1.91	2.76	3.34
50	3.86	2.58	1.66	1.05	0.75	0.64	0.69	0.91	1.38	2.22	3.45	4.40
55	5.36	3.19	1.89	1.12	0.77	0.65	0.70	0.95	1.52	2.67	4.62	6.40
60	8.58	4.20	2.21	1.21	0.79	0.65	0.71	1.00	1.72	3.36	6.90	11.41
ϕ	Jul	Aug	Sep	Oct	Nov	Dec	Jan	Feb	Mar	Apr	May	Jun

Table D-11 $\phi - \beta = -20^o$

ϕ	Jan	Feb	Mar	Apr	May	Jun	Jul	Aug	Sep	Oct	Nov	Dec
0	1.15	1.06	0.96	0.85	0.76	0.72	0.74	0.81	0.92	1.03	1.12	1.17
5	1.21	1.10	0.97	0.84	0.73	0.68	0.71	0.79	0.92	1.06	1.18	1.24
10	1.28	1.15	0.99	0.83	0.71	0.65	0.68	0.78	0.92	1.10	1.25	1.33
15	1.38	1.21	1.01	0.82	0.69	0.63	0.66	0.76	0.94	1.14	1.34	1.43
20	1.50	1.29	1.05	0.83	0.68	0.61	0.64	0.76	0.96	1.21	1.44	1.57
25	1.66	1.38	1.10	0.84	0.67	0.60	0.63	0.76	0.99	1.28	1.58	1.75
30	1.86	1.51	1.15	0.85	0.66	0.59	0.62	0.77	1.03	1.38	1.76	1.98
35	2.13	1.67	1.23	0.88	0.66	0.58	0.62	0.78	1.08	1.51	2.00	2.29
40	2.51	1.88	1.33	0.91	0.67	0.58	0.62	0.80	1.14	1.68	2.32	2.73
45	3.05	2.17	1.45	0.95	0.68	0.58	0.62	0.82	1.22	1.90	2.79	3.39
50	3.91	2.58	1.62	1.00	0.69	0.58	0.63	0.85	1.33	2.21	3.49	4.47
55	5.41	3.19	1.85	1.06	0.71	0.59	0.64	0.89	1.47	2.65	4.65	6.49
60	8.65	4.20	2.16	1.14	0.73	0.59	0.65	0.94	1.66	3.34	6.94	11.55
ϕ	Jul	Aug	Sep	Oct	Nov	Dec	Jan	Feb	Mar	Apr	May	Jun

Table D-12 $\phi - \beta = -25^{\circ}$

ϕ	Jan	Feb	Mar	Apr	May	Jun	Jul	Aug	Sep	Oct	Nov	Dec
0	1.16	1.06	0.93	0.80	0.69	0.64	0.66	0.75	0.88	1.02	1.13	1.19
5	1.22	1.10	0.94	0.78	0.66	0.61	0.63	0.73	0.88	1.05	1.19	1.26
10	1.30	1.14	0.96	0.78	0.64	0.58	0.61	0.72	0.88	1.08	1.26	1.35
15	1.40	1.20	0.98	0.77	0.62	0.56	0.59	0.71	0.90	1.13	1.34	1.45
20	1.52	1.28	1.02	0.78	0.61	0.54	0.57	0.70	0.92	1.19	1.45	1.59
25	1.67	1.38	1.06	0.78	0.60	0.53	0.56	0.70	0.94	1.27	1.59	1.77
30	1.88	1.50	1.12	0.80	0.60	0.52	0.56	0.71	0.98	1.37	1.77	2.00
35	2.15	1.66	1.19	0.82	0.60	0.52	0.55	0.72	1.03	1.49	2.01	2.31
40	2.52	1.87	1.29	0.85	0.61	0.51	0.55	0.74	1.09	1.66	2.33	2.76
45	3.07	2.15	1.41	0.89	0.61	0.51	0.56	0.76	1.17	1.88	2.79	3.42
50	3.93	2.56	1.57	0.93	0.63	0.52	0.56	0.79	1.28	2.18	3.49	4.50
55	5.43	3.16	1.79	0.99	0.64	0.52	0.57	0.82	1.41	2.61	4.66	6.53
60	8.67	4.16	2.10	1.07	0.66	0.52	0.58	0.87	1.59	3.29	6.93	11.59
ϕ	Jul	Aug	Sep	Oct	Nov	Dec	Jan	Feb	Mar	Apr	May	Jun

Table D-13 $\phi - \beta = -30^{\circ}$

ϕ	Jan	Feb	Mar	Apr	May	Jun	Jul	Aug	Sep	Oct	Nov	Dec
0	1.17	1.05	0.90	0.74	0.62	0.56	0.58	0.69	0.84	1.00	1.14	1.20
5	1.23	1.08	0.91	0.73	0.59	0.53	0.56	0.67	0.83	1.03	1.19	1.27
10	1.30	1.13	0.92	0.72	0.57	0.51	0.53	0.65	0.84	1.06	1.26	1.36
15	1.40	1.19	0.95	0.72	0.56	0.49	0.52	0.64	0.85	1.11	1.34	1.46
20	1.52	1.26	0.98	0.72	0.54	0.47	0.50	0.64	0.87	1.17	1.45	1.60
25	1.68	1.36	1.02	0.73	0.54	0.46	0.50	0.64	0.90	1.24	1.59	1.78
30	1.88	1.48	1.08	0.74	0.54	0.45	0.49	0.65	0.93	1.34	1.77	2.01
35	2.15	1.63	1.15	0.76	0.54	0.45	0.49	0.66	0.98	1.46	2.00	2.32
40	2.52	1.84	1.24	0.79	0.54	0.45	0.49	0.67	1.04	1.62	2.32	2.76
45	3.06	2.12	1.36	0.82	0.55	0.45	0.49	0.69	1.11	1.84	2.78	3.42
50	3.91	2.52	1.51	0.87	0.56	0.45	0.49	0.72	1.21	2.13	3.47	4.50
55	5.40	3.11	1.72	0.92	0.57	0.45	0.50	0.75	1.34	2.56	4.62	6.51
60	8.61	4.08	2.02	0.99	0.58	0.46	0.51	0.79	1.51	3.21	6.87	11.55
ϕ	Jul	Aug	Sep	Oct	Nov	Dec	Jan	Feb	Mar	Apr	May	Jun

Table D-14 Vertical Surface

ϕ	Jan	Feb	Mar	Apr	May	Jun	Jul	Aug	Sep	Oct	Nov	Dec
0	0.60	0.36	0.07	0.00	0.00	0.00	0.00	0.00	0.00	0.27	0.54	0.67
5	0.71	0.46	0.15	0.00	0.00	0.00	0.00	0.00	0.04	0.36	0.65	0.79
10	0.84	0.56	0.25	0.00	0.00	0.00	0.00	0.00	0.12	0.46	0.76	0.92
15	0.98	0.68	0.34	0.06	0.00	0.00	0.00	0.01	0.21	0.57	0.90	1.07
20	1.15	0.81	0.44	0.13	0.00	0.00	0.00	0.06	0.30	0.68	1.06	1.25
25	1.35	0.96	0.55	0.21	0.05	0.01	0.02	0.13	0.40	0.82	1.24	1.47
30	1.59	1.13	0.67	0.30	0.11	0.05	0.08	0.21	0.50	0.97	1.46	1.74
35	1.90	1.34	0.80	0.39	0.18	0.11	0.14	0.29	0.61	1.15	1.74	2.10
40	2.32	1.59	0.96	0.48	0.25	0.17	0.21	0.37	0.74	1.36	2.11	2.58
45	2.90	1.93	1.14	0.59	0.33	0.24	0.28	0.47	0.89	1.63	2.61	3.27
50	3.80	2.38	1.36	0.71	0.41	0.31	0.35	0.56	1.05	1.99	3.35	4.39
55	5.34	3.03	1.64	0.84	0.50	0.38	0.43	0.67	1.26	2.48	4.55	6.45
60	8.61	4.08	2.02	0.99	0.58	0.46	0.51	0.79	1.51	3.21	6.87	11.55
ϕ	Jul	Aug	Sep	Oct	Nov	Dec	Jan	Feb	Mar	Apr	May	Jun

Appendix E

PROPERTIES OF MATERIALS

Density, kg/m^3

Copper	8795
Steel	7850
Aluminum	2675
Glass, standard	2515
Concrete, typical	2400
Ice, −1 C	918
Gypsum plaster, dry, 23 C	881
Oak, 14% wet	770
Pine, 15% wet	570
Pine fiberboard, 24 C	256
Cork board, dry, 18 C	144
Silica aerogel, experimental	105
Ebonite, expanded, 10 C	64
Mineral wool, batts, −2 C	32
Polyurethane foam, rigid	24
Polystyrene, expanded, 10 C	16
Air, 20 C, p_o	1.204

Thermal Conductivity, W/mC

Copper	385
Aluminum	211
Steel	47.6
Ice, −1 C	2.26
Concrete, typical	1.73
Glass, standard	1.05
Water, 20 C	0.596
Gypsum plaster, dry, 23 C	0.170
Oak, 14% wet	0.160
Pine, 15% wet	0.138
Pine fiberboard, 24 C	0.051 9
Cork board, dry	0.041 8
Mineral wool, batts, −2 C	0.034 6
Polystyrene, expanded, 10 C	0.034 6
Ebonite, expanded, 10 C	0.030 3
Air, 20 C, p_o	0.026
Polyurethane foam, rigid	0.024 5
Silica aerogel, experimental	0.004

Specific Heat, kJ/kgC

Water, 20 C, p_o	4.19
Ice, −21 C to −1 C	2.10
Steam (C_p), 100 C, p_o	1.95
Air (C_p), 100 C, p_o	1.012
Limestone	0.91
Concrete, 18 C	0.84
Glass	0.82
Stone, typical	0.8
Steel	0.50

Heat of Vaporization, kJ/kg

Water, 20 C	2454
Water, 100 C	2257
R12, 0 C, sat.	151.5
R22, 0 C, sat.	205.4
R11, 0 C, sat.	188.9
R500, 0 C, sat.	183.0
R717, 0 C, sat.	1263.3

Momentum Diffusivity = Kinematic Viscosity, m^2/s

Air, 20 C, p_o	14.95×10^{-6}
Water, 20 C, p_o	1.01×10^{-6}

Heat (Thermal) Diffusivity, m^2/s

Air, 20 C, p_o	21.2×10^{-6}
Water, 20 C, p_o	0.142×10^{-6}

Mass Diffusivity, m^2/s

Water vapor in air, 20 C, p_o	26.1×10^{-6}

Surface Tension, N/m

Water/air, 20 C, p_o	0.072 8

Miscellaneous Information

NTP (Normal temperature T_o, pressure p_o):
  $T_o = 273.15$ K $= 0$ C
  $p_o = 101.325$ kPa $= 1$ atm
Standard Gravity g_o:
  $g_o = 9.806\,65$ m/s^2
Velocity of sound in air:
  344 m/s at p_o, 20 C, 50% R.H·

Gas constants:
  $R_u = 8314.4$ J/kmol (universal)
  $R_a = 287.045$ J/kgK (air)
  $R_v = 461.52$ J/kgK (water vapor)
Stefan-Boltzmann constant:
  $\sigma = 5.67 \times 10^{-8}$ W/m^{2}K^4

Properties of Saturated Water

T, C	ρ, kg/m^3	C_p, J.kgK	k, W/mK	μ, Pa s	α, m^2/s	Pr
0	1002	4218	0.552	17.9×10^{-4}	1.31×10^{-7}	13.06
20	1001	4182	0.597	10.1×10^{-4}	1.43×10^{-7}	7.02
40	995	4178	0.628	6.55×10^{-4}	1.51×10^{-7}	4.34
60	985	4184	0.651	4.71×10^{-4}	1.55×10^{-7}	3.02
80	974	4196	0.668	3.55×10^{-4}	1.64×10^{-7}	2.22
100	960	4216	0.680	2.82×10^{-4}	1.68×10^{-7}	1.74
120	945	4250	0.685	2.33×10^{-4}	1.71×10^{-7}	1.45
140	928	4283	0.684	1.99×10^{-4}	1.72×10^{-7}	1.24
160	910	4342	0.680	1.73×10^{-4}	1.73×10^{-7}	1.10
180	889	4417	0.675	1.54×10^{-4}	1.72×10^{-7}	1.00
200	867	4505	0.665	1.39×10^{-4}	1.71×10^{-7}	0.94
220	842	4610	0.652	1.26×10^{-4}	1.68×10^{-7}	0.89
240	816	4756	0.635	1.17×10^{-4}	1.64×10^{-7}	0.88
260	786	4949	0.611	1.08×10^{-4}	1.58×10^{-7}	0.87
280	753	5208	0.580	1.02×10^{-4}	1.48×10^{-7}	0.91
300	714	5728	0.540	0.96×10^{-4}	1.32×10^{-7}	1.02

Properties of Air at One Atmosphere

T, C	ρ, kg/m^3	C_p, J.kgK	k, W/mK	μ, Pa s	α, m^2/s	Pr
0	1.292	1006	0.0242	1.72×10^{-5}	1.86×10^{-5}	0.72
20	1.204	1006	0.0257	1.81×10^{-5}	2.12×10^{-5}	0.71
40	1.127	1007	0.0272	1.90×10^{-5}	2.40×10^{-5}	0.70
60	1.059	1008	0.0287	1.99×10^{-5}	2.69×10^{-5}	0.70
80	0.999	1010	0.0302	2.09×10^{-5}	3.00×10^{-5}	0.70
100	0.946	1012	0.0318	2.18×10^{-5}	3.32×10^{-5}	0.69
120	0.898	1014	0.0333	2.27×10^{-5}	3.66×10^{-5}	0.69
140	0.854	1016	0.0345	2.34×10^{-5}	3.98×10^{-5}	0.69
160	0.815	1019	0.0359	2.42×10^{-5}	4.32×10^{-5}	0.69
180	0.779	1022	0.0372	2.50×10^{-5}	4.67×10^{-5}	0.69
200	0.746	1025	0.0386	2.57×10^{-5}	5.05×10^{-5}	0.68
220	0.715	1028	0.0399	2.64×10^{-5}	5.43×10^{-5}	0.68
240	0.688	1032	0.0412	2.72×10^{-5}	5.80×10^{-5}	0.68
260	0.662	1036	0.0425	2.79×10^{-5}	6.20×10^{-5}	0.68
280	0.638	1040	0.0437	2.86×10^{-5}	6.59×10^{-5}	0.68
300	0.616	1045	0.0450	2.93×10^{-5}	6.99×10^{-5}	0.68

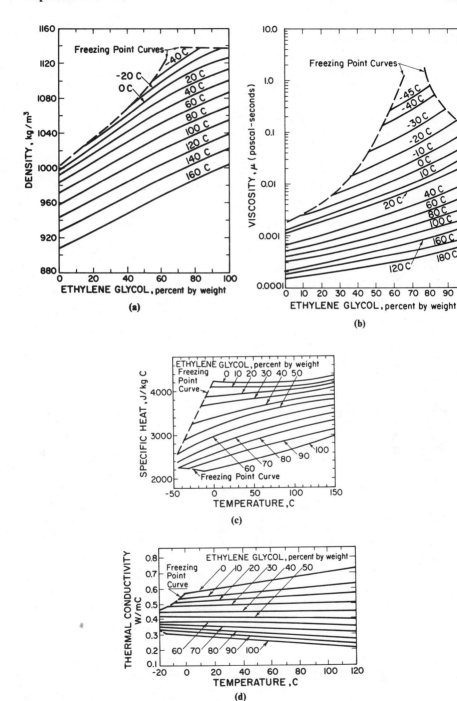

Figure E-1 (a-f) Properties of aqueous ethylene glycol solutions. Adapted with permission from information supplied by Union Carbide. The manufacturer recommends only the use of inhibited ethylene glycol ("UCAR Thermofluid 17").

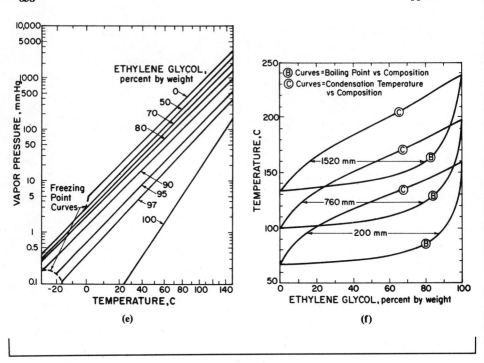

(e)   (f)

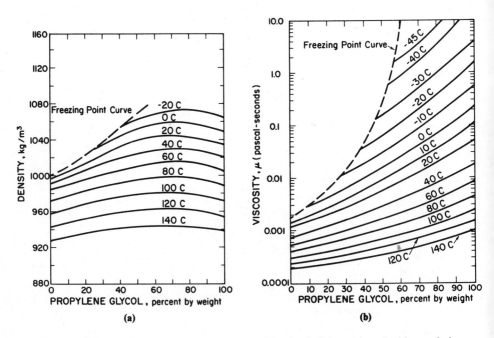

(a)   (b)

Figure E-2 (a-f) Properties of aqueous propylene glycol solutions. Adapted with permission from information supplied by Union Carbide. The manufacturer recommends only the use of inhibited propylene glycol ("UCAR Foodfreeze 35").

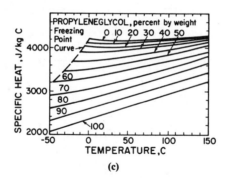

(c)

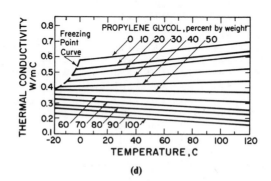

(d)

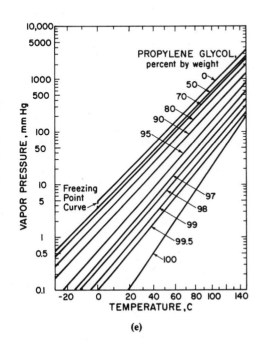

(e)

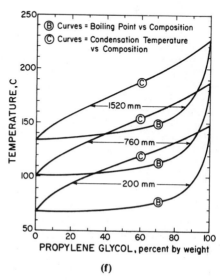

(f)

PRESENT WORTH FACTORS (See Section 11.5)

N = 5

d	0	1	2	3	4	5	*i* 6	7	8	9	10	11	12
0	5.000	5.101	5.204	5.309	5.416	5.526	5.637	5.751	5.867	5.985	6.105	6.228	6.353
1	4.853	4.950	5.049	5.150	5.253	5.358	5.466	5.575	5.686	5.799	5.915	6.033	6.153
2	4.713	4.807	4.902	4.999	5.098	5.199	5.302	5.407	5.514	5.623	5.734	5.847	5.962
3	4.580	4.669	4.761	4.854	4.950	5.047	5.146	5.246	5.349	5.454	5.561	5.669	5.780
4	4.452	4.538	4.626	4.716	4.808	4.901	4.996	5.093	5.192	5.293	5.395	5.500	5.606
5	4.329	4.413	4.497	4.584	4.672	4.762	4.853	4.947	5.042	5.139	5.238	5.338	5.441
6	4.212	4.292	4.374	4.457	4.542	4.629	4.717	4.807	4.898	4.992	5.087	5.183	5.282
7	4.100	4.177	4.256	4.336	4.418	4.501	4.586	4.673	4.761	4.851	4.942	5.036	5.131
8	3.993	4.067	4.143	4.220	4.299	4.379	4.461	4.545	4.630	4.716	4.804	4.894	4.986
9	3.890	3.961	4.035	4.109	4.185	4.263	4.342	4.422	4.504	4.587	4.672	4.759	4.847
10	3.791	3.860	3.931	4.003	4.076	4.151	4.227	4.304	4.383	4.464	4.545	4.629	4.714
11	3.696	3.763	3.831	3.900	3.971	4.043	4.117	4.191	4.268	4.345	4.424	4.505	4.586
12	3.605	3.669	3.735	3.802	3.870	3.940	4.011	4.083	4.157	4.231	4.308	4.385	4.464
13	3.517	3.580	3.643	3.708	3.774	3.841	3.909	3.979	4.050	4.122	4.196	4.271	4.347
14	3.433	3.493	3.555	3.617	3.681	3.746	3.812	3.879	3.948	4.018	4.089	4.161	4.235
15	3.352	3.410	3.470	3.530	3.592	3.655	3.719	3.784	3.850	3.917	3.986	4.056	4.127
16	3.274	3.331	3.388	3.447	3.506	3.567	3.629	3.691	3.755	3.821	3.887	3.954	4.023
17	3.199	3.254	3.309	3.366	3.424	3.482	3.542	3.603	3.665	3.728	3.792	3.857	3.924
18	3.127	3.180	3.234	3.288	3.344	3.401	3.459	3.518	3.577	3.638	3.700	3.764	3.828
19	3.058	3.109	3.161	3.214	3.268	3.323	3.379	3.436	3.493	3.552	3.612	3.673	3.736
20	2.991	3.040	3.091	3.142	3.194	3.247	3.301	3.357	3.413	3.470	3.528	3.587	3.647

N = 10

d	0	1	2	3	4	5	*i* 6	7	8	9	10	11	12
0	10.000	10.462	10.950	11.464	12.006	12.578	13.181	13.816	14.487	15.193	15.937	16.722	17.549
1	9.471	9.901	10.354	10.831	11.335	11.865	12.425	13.014	13.635	14.289	14.979	15.705	16.470
2	8.983	9.383	9.804	10.248	10.716	11.209	11.728	12.275	12.851	13.458	14.097	14.770	15.479
3	8.530	8.903	9.295	9.709	10.144	10.603	11.085	11.594	12.129	12.692	13.286	13.910	14.567
4	8.111	8.459	8.825	9.210	9.615	10.042	10.492	10.965	11.462	11.986	12.537	13.117	13.727
5	7.722	8.046	8.388	8.748	9.126	9.524	9.943	10.383	10.846	11.334	11.847	12.386	12.953
6	7.360	7.664	7.983	8.319	8.672	9.043	9.434	9.845	10.277	10.731	11.208	11.710	12.238
7	7.024	7.308	7.607	7.921	8.251	8.598	8.962	9.346	9.749	10.172	10.618	11.085	11.577
8	6.710	6.976	7.256	7.550	7.859	8.184	8.525	8.883	9.259	9.655	10.070	10.507	10.965
9	6.418	6.667	6.930	7.205	7.495	7.798	8.118	8.453	8.805	9.174	9.562	9.970	10.398
10	6.145	6.379	6.625	6.884	7.155	7.440	7.739	8.053	8.382	8.728	9.091	9.472	9.872
11	5.889	6.110	6.341	6.584	6.838	7.105	7.386	7.680	7.989	8.313	8.652	9.009	9.383
12	5.650	5.858	6.075	6.303	6.543	6.793	7.057	7.333	7.622	7.926	8.244	8.578	8.929
13	5.426	5.622	5.826	6.041	6.266	6.502	6.749	7.008	7.280	7.565	7.864	8.177	8.505
14	5.216	5.400	5.593	5.795	6.007	6.229	6.462	6.705	6.961	7.228	7.509	7.803	8.111
15	5.019	5.193	5.374	5.565	5.765	5.974	6.193	6.422	6.662	6.914	7.177	7.453	7.743
16	4.833	4.997	5.169	5.349	5.537	5.734	5.940	6.156	6.383	6.619	6.867	7.127	7.399
17	4.659	4.814	4.976	5.146	5.323	5.509	5.704	5.908	6.121	6.344	6.577	6.822	7.077
18	4.494	4.641	4.794	4.955	5.123	5.298	5.482	5.674	5.875	6.085	6.305	6.536	6.776
19	4.339	4.478	4.623	4.775	4.934	5.100	5.273	5.455	5.644	5.843	6.050	6.267	6.494
20	4.192	4.324	4.462	4.606	4.756	4.913	5.077	5.248	5.428	5.615	5.811	6.016	6.230

N = 15

d	0	1	2	3	4	5	6	7	8	9	10	11	12
							i						
0	15.000	16.097	17.293	18.599	20.024	21.579	23.276	25.129	27.152	29.361	31.772	34.405	37.280
1	13.865	14.851	15.926	17.098	18.375	19.767	21.285	22.942	24.748	26.718	28.867	31.212	33.770
2	12.849	13.738	14.706	15.759	16.906	18.156	19.517	21.000	22.616	24.377	26.297	28.389	30.669
3	11.938	12.741	13.614	14.563	15.596	16.719	17.942	19.273	20.722	22.300	24.017	25.888	27.925
4	11.118	11.845	12.634	13.492	14.423	15.436	16.536	17.733	19.035	20.451	21.991	23.667	25.491
5	10.380	11.039	11.754	12.530	13.372	14.286	15.279	16.357	17.529	18.802	20.187	21.691	23.327
6	9.712	10.311	10.960	11.664	12.426	13.254	14.151	15.125	16.182	17.329	18.575	19.929	21.399
7	9.108	9.654	10.244	10.883	11.575	12.325	13.138	14.019	14.974	16.010	17.134	18.354	19.677
8	8.559	9.057	9.595	10.177	10.807	11.488	12.225	13.024	13.889	14.826	15.842	16.943	18.137
9	8.061	8.516	9.007	9.538	10.111	10.731	11.402	12.127	12.912	13.761	14.681	15.678	16.757
10	7.606	8.023	8.473	8.958	9.481	10.046	10.657	11.317	12.030	12.802	13.636	14.539	15.516
11	7.191	7.574	7.986	8.430	8.908	9.425	9.982	10.584	11.233	11.935	12.694	13.514	14.400
12	6.811	7.163	7.541	7.949	8.387	8.860	9.369	9.919	10.511	11.151	11.842	12.587	13.393
13	6.462	6.786	7.135	7.509	7.912	8.345	8.812	9.314	9.856	10.440	11.070	11.749	12.483
14	6.142	6.441	6.762	7.107	7.477	7.875	8.303	8.764	9.260	9.794	10.370	10.990	11.659
15	5.847	6.124	6.420	6.738	7.079	7.445	7.839	8.262	8.717	9.206	9.733	10.300	10.911
16	5.575	5.831	6.105	6.399	6.714	7.051	7.413	7.803	8.220	8.670	9.153	9.672	10.231
17	5.324	5.561	5.815	6.087	6.378	6.689	7.024	7.382	7.767	8.180	8.623	9.100	9.612
18	5.092	5.312	5.547	5.799	6.069	6.357	6.665	6.996	7.351	7.731	8.139	8.577	9.048
19	4.876	5.081	5.300	5.533	5.783	6.050	6.336	6.641	6.969	7.320	7.696	8.099	8.532
20	4.675	4.867	5.070	5.288	5.519	5.767	6.032	6.315	6.618	6.942	7.289	7.661	8.059

N = 20

d	0	1	2	3	4	5	6	7	8	9	10	11	12
							i						
0	20.000	22.019	24.297	26.870	29.778	33.066	36.786	40.995	45.762	51.160	57.275	64.203	72.052
1	18.046	19.802	21.780	24.009	26.524	29.362	32.568	36.190	40.284	44.913	50.150	56.074	62.778
2	16.351	17.885	19.608	21.546	23.728	26.186	28.958	32.084	35.612	39.594	44.093	49.174	54.917
3	14.877	16.221	17.727	19.417	21.317	23.453	25.857	28.564	31.613	35.050	38.926	43.299	48.232
4	13.590	14.771	16.092	17.571	19.231	21.093	23.185	25.536	28.180	31.156	34.506	38.279	42.531
5	12.462	13.503	14.665	15.965	17.419	19.048	20.874	22.922	25.222	27.806	30.710	33.977	37.651
6	11.470	12.391	13.417	14.562	15.840	17.269	18.868	20.659	22.665	24.916	27.442	30.277	33.463
7	10.594	11.411	12.320	13.332	14.459	15.717	17.122	18.692	20.448	22.414	24.617	27.086	29.856
8	9.818	10.546	11.353	12.250	13.247	14.358	15.596	16.977	18.519	20.242	22.169	24.325	26.740
9	9.129	9.779	10.498	11.296	12.181	13.164	14.258	15.476	16.834	18.349	20.039	21.928	24.040
10	8.514	9.096	9.739	10.450	11.238	12.112	13.082	14.160	15.359	16.694	18.182	19.841	21.693
11	7.963	8.487	9.063	9.700	10.403	11.182	12.044	13.001	14.063	15.243	16.556	18.018	19.647
12	7.469	7.941	8.460	9.031	9.661	10.356	11.125	11.977	12.920	13.967	15.129	16.421	17.857
13	7.025	7.451	7.919	8.433	8.998	9.622	10.310	11.070	11.910	12.841	13.872	15.017	16.287
14	6.623	7.009	7.432	7.896	8.406	8.966	9.583	10.263	11.014	11.844	12.762	13.779	14.906
15	6.259	6.610	6.994	7.414	7.874	8.379	8.934	9.545	10.217	10.959	11.779	12.685	13.687
16	5.929	6.249	6.597	6.978	7.395	7.851	8.352	8.902	9.506	10.172	10.905	11.714	12.608
17	5.628	5.920	6.238	6.585	6.963	7.376	7.829	8.325	8.870	9.468	10.126	10.851	11.650
18	5.353	5.620	5.911	6.227	6.572	6.947	7.358	7.807	8.298	8.838	9.430	10.081	10.798
19	5.101	5.347	5.613	5.902	6.216	6.558	6.932	7.339	7.784	8.272	8.806	9.392	10.036
20	4.870	5.096	5.340	5.605	5.893	6.205	6.545	6.916	7.320	7.762	8.245	8.774	9.355

N = 25

d	0	1	2	3	4	5	6	7	8	9	10	11	12
							i						
0	25.000	28.243	32.030	36.459	41.646	47.727	54.865	63.249	73.106	84.701	98.347	114.413	133.334
1	22.023	24.752	27.929	31.633	35.958	41.014	46.933	53.869	62.003	71.550	82.762	95.935	111.419
2	19.523	21.832	24.510	27.622	31.245	35.470	40.401	46.164	52.906	60.800	70.051	80.897	93.621
3	17.413	19.375	21.644	24.272	27.322	30.867	34.994	39.804	45.417	51.974	59.639	68.606	79.104
4	15.622	17.298	19.229	21.459	24.038	27.028	30.498	34.531	39.224	44.693	51.071	58.516	67.213
5	14.094	15.532	17.184	19.085	21.277	23.810	26.740	30.137	34.079	38.660	43.990	50.197	57.431
6	12.783	14.024	15.444	17.072	18.943	21.098	23.585	26.458	29.784	33.639	38.112	43.308	49.350
7	11.654	12.729	13.954	15.356	16.961	18.803	20.923	23.364	26.183	29.440	33.210	37.578	42.645
8	10.675	11.611	12.674	13.885	15.269	16.851	18.666	20.750	23.148	25.912	29.103	32.791	37.058
9	9.823	10.641	11.568	12.620	13.817	15.182	16.743	18.530	20.580	22.936	25.648	28.774	32.382
10	9.077	9.796	10.607	11.525	12.566	13.749	15.097	16.636	18.396	20.412	22.727	25.388	28.452
11	8.422	9.056	9.769	10.574	11.482	12.512	13.682	15.012	16.530	18.264	20.248	22.523	25.134
12	7.843	8.405	9.035	9.743	10.540	11.440	12.459	13.615	14.929	16.425	18.133	20.086	22.321
13	7.330	7.830	8.388	9.014	9.716	10.506	11.398	12.406	13.548	14.846	16.322	18.005	19.926
14	6.873	7.320	7.817	8.372	8.993	9.689	10.473	11.356	12.353	13.483	14.764	16.220	17.878
15	6.464	6.865	7.309	7.803	8.355	8.971	9.662	10.439	11.314	12.301	13.417	14.683	16.119
16	6.097	6.457	6.856	7.298	7.790	8.338	8.950	9.636	10.406	11.272	12.249	13.353	14.602
17	5.766	6.092	6.451	6.848	7.288	7.776	8.321	8.929	9.609	10.372	11.230	12.197	13.288
18	5.467	5.762	6.086	6.444	6.839	7.277	7.763	8.304	8.907	9.582	10.339	11.189	12.146
19	5.195	5.463	5.758	6.081	6.437	6.830	7.266	7.749	8.286	8.886	9.556	10.306	11.148
20	4.948	5.192	5.460	5.753	6.075	6.430	6.822	7.255	7.735	8.269	8.864	9.529	10.272

N = 30

d	0	1	2	3	4	5	6	7	8	9	10	11	12
							i						
0	30.000	34.785	40.568	47.575	56.085	66.439	79.058	94.461	113.283	136.308	164.494	199.021	241.333
1	25.808	29.703	34.389	40.042	46.878	55.164	65.225	77.462	92.367	110.545	132.735	159.843	192.981
2	22.396	25.589	29.412	34.002	39.529	46.201	54.270	64.050	75.922	90.353	107.916	129.313	155.400
3	19.600	22.235	25.374	29.126	33.624	39.029	45.541	53.404	62.914	74.435	88.413	105.392	126.034
4	17.292	19.481	22.076	25.163	28.846	33.254	38.541	44.900	52.563	61.813	73.000	86.545	102.965
5	15.372	17.203	19.363	21.919	24.955	28.571	32.891	38.065	44.276	51.746	60.748	71.613	84.744
6	13.765	15.307	17.116	19.246	21.765	24.751	28.302	32.537	37.601	43.668	50.953	59.716	70.272
7	12.409	13.716	15.241	17.028	19.131	21.612	24.549	28.037	32.190	37.147	43.076	50.182	58.715
8	11.258	12.372	13.667	15.176	16.942	19.017	21.461	24.351	27.778	31.851	36.704	42.499	49.433
9	10.274	11.230	12.335	13.618	15.111	16.856	18.904	21.313	24.157	27.523	31.518	36.271	41.937
10	9.427	10.253	11.202	12.299	13.569	15.046	16.771	18.792	21.166	23.965	27.273	31.192	35.848
11	8.694	9.411	10.232	11.175	12.262	13.520	14.982	16.687	18.681	21.022	23.776	27.027	30.873
12	8.055	8.682	9.395	10.211	11.147	12.225	13.472	14.918	16.603	18.572	20.879	23.590	26.786
13	7.496	8.046	8.670	9.379	10.190	11.119	12.188	13.423	14.855	16.520	18.464	20.738	23.407
14	7.003	7.489	8.037	8.658	9.363	10.169	11.091	12.151	13.375	14.792	16.438	18.356	20.599
15	6.566	6.997	7.482	8.028	8.646	9.347	10.147	11.063	12.115	13.327	14.729	16.356	18.250
16	6.177	6.562	6.992	7.475	8.019	8.633	9.331	10.126	11.035	12.078	13.279	14.667	16.275
17	5.829	6.174	6.558	6.987	7.468	8.009	8.621	9.315	10.104	11.007	12.041	13.231	14.605
18	5.517	5.827	6.171	6.554	6.981	7.460	7.999	8.608	9.298	10.083	10.979	12.005	13.184
19	5.235	5.515	5.825	6.168	6.550	6.976	7.453	7.990	8.596	9.282	10.061	10.951	11.968
20	4.979	5.233	5.513	5.822	6.165	6.545	6.970	7.446	7.980	8.583	9.265	10.040	10.922

Appendix G

METEOROLOGICAL DATA

The meteorological data in this appendix are useful in designing systems by the methods presented in Part III. The data include: $\overline{H}$, the monthly average daily radiation on a horizontal surface, in MJ/m^2; $\overline{K}_T$, the monthly average clearness index; $\overline{T}$, the 24-hour monthly average ambient temperature, in C; and DD, the average number of degree days in the month to the base temperature 18.3 C.

The data for the United States stations are from the SOLMET program [see Cinquenami et al. (1978)]; the Canadian data are from the *Canadian Climate Normals* (1982); the European radiation data are from the CEC's *European Solar Radiation Atlas* (1982); the remaining world-wide radiation data are from the *Solar Radiation and Radiation Balance Data (The World Network)* (1970, 1976, 1982); and the temperature data are from *World Weather Records* (1965). All of the degree-days are calculated by the method presented in Section 9.3.

The table is arranged alphabetically by region: Africa, Asia, Europe, North America, Pacific and South America. Within each region stations are grouped by country (and state or province for the U.S. and Canada.)

These data are subject to change as additional measurements become available, as further refinements are made in the processing of existing data, and as the inevitable errors in compilations of this type are uncovered and corrected.

REFERENCES

Cinquemani, V., J. Owenby, and R. G. Baldwin, report prepared for U.S. Department of Energy by the National Oceanic and Atmospheric Administration (1978). "Input Data for Solar Systems."

Canadian Climate Normals, 1951-1980, Vol. 1, *Solar Radiation;* Vol. 2, *Temperatures,* Ministry of Supply and Services Canada, Ottawa (1982).

Commision of the European Communities (CEC), *European Solar Radiation Atlas*, Vol. 1, *Global Radiation on Horizontal Sufaces*, Vol. 2, *Inclined Surfaces* (W. Palz, ed.), Verlag TÜV Rheinland, Koln (1984).

World Weather Records, 1951-1960, Vol. 1-6, U.S. Government Printing Office (1965).

Solar Radiation and Radiation Balance Data (The World Network), 1964-1968 (1970), *1969-1973* (1976), *1974-1979* (1982) Hydrometeorlogical Publishing House, Leningrad.

AFRICA
-Angola-
  Luanda
-Egypt-
  Cairo
-Ethiopia-
  Addis Ababa
-Israel-
  Bet Dagan
-Kenya-
  Nairobi
-Morocco-
  Casablanca
-Nigeria-
  Benin City
-Senegal-
  Dakar
-South Africa-
  Pretoria
-Sudan-
  El Fasher
-Tunisia-
  Sidi-Bou-Said
-Uganda-
  Entebbe

ASIA
-India-
  Madras
  New Delhi
-Japan-
  Akita
  Kagoshima
  Shimizu
-Malaysia-
  Kuala Lumpur
-Pakistan-
  Karachi
  Lahore
-Singapore-
  Singapore
-Sri Lanka-
  Colombo
-Thailand-
  Bangkok

EUROPE
-Belgium-
  Brussels

-Denmark-
  Copenhagen
-Germany-
  Stuttgart
-Great Britian-
  London
-Greece-
  Athens
-Italy-
  Rome
-Netherlands-
  Vlissigen
-Portugal-
  Lisbon
-Poland-
  Warsaw
-Romania-
  Cluj
-Spain-
  Almeria
-Switzerland-
  Zurich
-USSR-
  Kiev
  Leningrad
  Moscow
  Odessa

NORTH AMERICA
Canada
-Alberta-
  Beaverlodge
  Edmonton
  Suffield
-British Columbia-
  Cape St. James
  Fort Nelson
  Mt. Kobau Observ.
  Nanaimo Depart. Bay
  Port Hardy
  Prince George
  Summerland
  Vancouver
-Manitoba-
  Churchill
  The Pas
  Winnipeg
-New Brunswick-
  Fredericton

-Newfoundland-
  St. John's West
-Northwest Territories-
  Alert
  Baker Lake
  Cambridge Bay
  Coral Harbour
  Eureka
  Frobisher Bay
  Hall Beach
  Inuvik
  Isachen
  Mould Bay
  Norman Wells
  Resolute
  Sachs Harbour
-Nova Scotia-
  Halifax Citadel
  Kentville Sable Island
-Ontario-
  Big Trout Lake
  Elora Research Station
  Guelph
  Kapuskasing
  Moosonee
  Ottawa
  Toronto
-Prince Edward Island-
  Charlottetown
-Quebec-
  Fort Chimo
  Inoucdjouac
  Montreal
  Nitchequon
  Normandin
  Sept-Iles
-Saskatchewan-
  Bad Lake
  Swift Current
-Yukon-
  Whitehorse

Caribbean
  San Juan, PR

Mexico
  Ciudad Univ.

United States
-Alabama-
  Birmingham

Mobile
Montgomery
-Alaska-
Annette
Barrow
Bethel
Bettles
Big Delta
Fairbanks
Gulkana
Homer
Juneau
King Salmon
Kodiak
Kotzebue
Mcgrath
Nome
Summit
Yakutat
-Arizona-
Phoenix
Prescott
Tucson
Winslow
Yuma
-Arkansas-
Fort Smith
Little Rock
-California-
Bakersfield
Daggett
Fresno
Long Beach
Los Angeles
Mount Shasta
Needles
Oakland
Red Bluff
Sacramento
San Diego
San Francisco
Santa Maria
-Colorado-
Colorado Springs
Denver
Eagle
Grand Junction
Pueblo
-Connecticut-
Hartford
-Delaware-
Wilmington

-District of Columbia-
Washington
-Florida-
Apalachicola
Daytona Beach
Jacksonville
Miami
Orlando
Tallahassee
Tampa
West Palm Beach
-Georgia-
Atlanta
Augusta
Macon
Savannah
-Hawaii-
Hilo
Honolulu
Lihue
-Idaho-
Boise
Lewiston
Pocatello
-Illinois-
Chicago
Moline
Springfield
-Indiana-
Evansville
Fort Wayne
Indianapolis
South Bend
-Iowa-
Burlington
Des Moines
Mason City
Sioux City
-Kansas-
Dodge City
Goodland
TopekaS
Wichita
-Kentucky-
Lexington
Louisville
-Louisiana-
Baton Rouge
Lake Charles
New Orleans
Shreveport

-Maine-
Caribou
Portland
-Maryland-
Baltimore
-Massachusetts-
Boston
-Michigan-
Alpena
Detroit
Flint
Grand Rapids
Sault Ste. Marie
Traverse City
-Minnesota-
Duluth
Internat'l Falls
Minneapolis
Rochester
-Mississippi-
Jackson
Meridian
-Missouri-
Columbia
Kansas City
St. Louis
Springfield
-Montana-
Billings
Cut Bank
Dillon
Glasgow
Great Falls
Helena
Lewistown
Miles City
Missoula
-Nebraska-
Grand Island
North Omaha
North Platte
Scottsbluff
-Nevada-
Elko
Ely
Las Vegas
Lovelock
Reno
Tonopah
Winnemucca
-New Hampshire-
Concord

-New Jersey-
  Newark
-New Mexico-
  Albuquerque
  Clayton
  Farmington
  Roswell
  Truth Or Consequences
  Tucumcari
  Zuni
-New York-
  Albany
  Binghamton
  Buffalo
  Massena
  New York (Lag)
  New York(Cen Pk)
  Rochester
  Syracuse
-North Carolina-
  Asheville
  Cape Hatteras
  Charlotte
  Greensboro
  Raleigh-Durham
-North Dakota-
  Bismarck
  Fargo
  Minot
-Ohio-
  Akron-Canton
  Cincinnati
  Cleveland
  Columbus
  Dayton
  Toledo
  Youngstown
-Oklahoma-
  Oklahoma City
  Tulsa
-Oregon-
  Astoria
  Burns
  Medford
  North Bend
  Pendleton
  Portland
  Redmond
  Salem
-Pennsylvania-
  Allentown

Erie
Harrisburg
Philadelphia
Pittsburgh
Wilkes-Barre
-Rhode Island-
  Providence
-South Carolina-
  Charleston
  Columbia
  Greenville
-South Dakota-
  Huron
  Pierre
  Rapid City
  Sioux Falls
-Tennessee-
  Chattanooga
  Knoxville
  Memphis
  Nashville
-Texas-
  Abilene
  Amarillo
  Austin
  Brownsville
  Corpus Christi
  Dallas
  Del Rio
  El Paso
  Fort Worth
  Houston
  Laredo
  Lubbock
  Lufkin
  Midland-Odessa
  Port Arthur
  San Angelo
  San Antonio
  Sherman
  Waco
  Wichita Falls
-Utah-
  Bryce Canyon
  Cedar City
  Salt Lake City
-Vermont-
  Burlington
-Virginia-
  Norfolk
  Richmond
  Roanoke

-Washington-
  Olympia
  Seattle-Tacoma
  Spokane
  Yakima
-West Virginia-
  Charleston
  Huntington
-Wisconsin-
  Eau Claire
  Green Bay
  La Crosse
  Madison
  Milwaukee
-Wyoming-
  Casper
  Cheyenne
  Rock Springs
  Sheridan

PACIFIC
-Australia-
  Aspendale
  Darwin
  Perth
-Fiji-
  Nandi
-New Zealand-
  Wellington
-Phillipines-
  Quezon City
-U.S. Pacific Territory-
  Koror Island
  Kwajalein Island
  Wake Island

SOUTH AMERICA
-Argentina-
  Buenos Aires
-Chile-
  Valparaiso
-Ecuador-
  Izobamba
-Peru-
  Huancayo
-Venezuela-
  Caracas
  Maracibo

	Jan	Feb	Mar	Apr	May	Jun	Jul	Aug	Sep	Oct	Nov	Dec

Africa

Angola

Luanda — Latitude −8.8

	Jan	Feb	Mar	Apr	May	Jun	Jul	Aug	Sep	Oct	Nov	Dec
$\overline{H}$	20.21	20.93	19.54	18.57	17.32	14.68	12.66	12.60	15.77	18.05	20.16	19.46
K_T	.52	.53	.52	.53	.55	.49	.42	.38	.43	.47	.52	.50
$\overline{T}$	26	27	27	27	25	22	20	21	22	25	26	26
DD	0	0	0	0	0	3	13	12	3	0	0	0

Egypt

Cairo — Latitude 30.0

	Jan	Feb	Mar	Apr	May	Jun	Jul	Aug	Sep	Oct	Nov	Dec
$\overline{H}$	11.84	15.60	19.32	23.15	26.32	27.95	27.10	25.23	21.97	17.86	13.24	11.12
K_T	.56	.60	.61	.63	.66	.68	.67	.66	.66	.65	.59	.56
$\overline{T}$	14	15	17	21	25	27	28	28	26	24	19	15
DD	143	104	69	12	1	0	0	0	0	2	32	116

Ethiopia

Addis Ababa — Latitude 9.0

	Jan	Feb	Mar	Apr	May	Jun	Jul	Aug	Sep	Oct	Nov	Dec
$\overline{H}$	19.14	20.96	21.27	20.38	19.28	16.56	13.64	14.05	17.05	21.74	22.27	20.12
K_T	.59	.60	.57	.54	.52	.45	.37	.38	.46	.61	.68	.64
$\overline{T}$	17	18	19	19	19	17	15	15	17	16	17	17
DD	56	33	25	29	21	50	102	113	69	79	62	60

Israel

Bet Dagan — Latitude 32.0

	Jan	Feb	Mar	Apr	May	Jun	Jul	Aug	Sep	Oct	Nov	Dec
$\overline{H}$	10.65	14.32	18.02	22.47	26.39	28.60	27.93	25.88	21.82	16.88	12.32	9.56
K_T	.53	.57	.58	.62	.66	.69	.69	.69	.67	.63	.58	.51
$\overline{T}$	12	13	14	17	21	24	25	26	25	22	18	14
DD	191	161	134	58	12	2	1	0	1	7	47	134

Kenya

Nairobi — Latitude −1.3

	Jan	Feb	Mar	Apr	May	Jun	Jul	Aug	Sep	Oct	Nov	Dec
$\overline{H}$	23.21	23.59	22.27	18.89	16.71	15.09	12.94	14.16	19.00	20.19	19.10	22.05
K_T	.63	.62	.59	.52	.49	.46	.39	.40	.51	.54	.52	.61
$\overline{T}$	18	18	19	19	18	16	15	16	17	19	18	18
DD	40	34	27	26	44	73	110	89	62	30	36	44

Morocco

Casablanca — Latitude 33.6

	Jan	Feb	Mar	Apr	May	Jun	Jul	Aug	Sep	Oct	Nov	Dec
$\overline{H}$	9.68	12.89	17.03	20.97	23.03	24.31	24.82	23.10	19.62	14.56	10.87	8.47
K_T	.51	.54	.56	.58	.58	.59	.61	.61	.61	.56	.54	.48
$\overline{T}$	12	13	15	16	18	20	22	23	22	19	16	13
DD	190	152	122	90	41	17	6	3	5	30	85	169

Nigeria

Benin City — Latitude 6.1

	Jan	Feb	Mar	Apr	May	Jun	Jul	Aug	Sep	Oct	Nov	Dec
$\overline{H}$	15.42	16.77	17.30	17.13	17.15	15.44	12.84	12.68	13.91	15.85	17.25	15.95
K_T	.46	.47	.46	.46	.47	.43	.36	.34	.37	.44	.51	.48
$\overline{T}$	27	28	28	28	28	26	25	25	26	27	28	27
DD	0	0	0	0	0	0	0	0	0	0	0	0

Senegal

Dakar — Latitude 14.7

	Jan	Feb	Mar	Apr	May	Jun	Jul	Aug	Sep	Oct	Nov	Dec
$\overline{H}$	18.43	22.02	24.34	25.42	24.93	23.01	20.25	19.06	19.18	20.19	18.36	17.02
K_T	.62	.67	.68	.67	.65	.60	.53	.50	.52	.60	.60	.59
$\overline{T}$	21	20	21	22	23	26	27	27	28	27	26	23
DD	8	11	11	6	2	0	0	0	0	0	0	2

	Jan	Feb	Mar	Apr	May	Jun	Jul	Aug	Sep	Oct	Nov	Dec
South Africa												
Pretoria				Latitude			−25.8					
$\overline{H}$	23.39	22.40	20.24	16.68	15.29	14.17	15.28	17.79	20.41	21.88	23.60	24.53
K_T	.55	.56	.57	.57	.65	.67	.69	.67	.63	.57	.57	.57
$\overline{T}$	21	21	20	17	13	10	10	13	16	20	20	21
DD	11	9	20	64	170	250	259	170	86	20	18	11
Sudan												
El Fasher				Latitude			13.6					
$\overline{H}$	20.47	23.27	25.35	26.06	25.66	24.46	23.04	23.12	23.52	22.81	21.73	20.49
K_T	.68	.70	.70	.68	.67	.64	.61	.61	.64	.67	.70	.70
$\overline{T}$	18	20	25	27	30	30	27	25	26	26	22	19
DD	46	16	0	0	0	0	0	0	0	0	5	31
Tunisia												
Sidi-Bou-Said				Latitude			36.9					
$\overline{H}$	8.81	11.25	15.94	19.59	23.99	26.18	27.03	23.79	19.07	14.28	10.59	8.29
K_T	.51	.51	.55	.55	.60	.63	.66	.64	.61	.59	.58	.53
$\overline{T}$	11	12	13	16	19	23	26	27	25	20	16	12
DD	230	188	160	96	33	3	0	0	1	19	91	189
Uganda												
Entebbe				Latitude			0.1					
$\overline{H}$	18.02	18.13	18.17	17.43	16.35	16.01	15.46	16.45	18.05	17.78	17.55	17.38
K_T	.50	.48	.48	.47	.47	.48	.46	.46	.48	.48	.48	.49
$\overline{T}$	22	22	22	22	22	21	21	21	21	22	22	22
DD	3	2	2	3	3	5	9	7	4	3	3	3

Asia

	Jan	Feb	Mar	Apr	May	Jun	Jul	Aug	Sep	Oct	Nov	Dec
India												
Madras				Latitude			13.0					
$\overline{H}$	18.20	22.11	24.18	24.57	22.69	20.21	19.57	19.96	19.73	16.63	15.53	14.31
K_T	.60	.66	.66	.65	.59	.53	.52	.53	.54	.49	.50	.48
$\overline{T}$	25	26	28	31	32	32	30	30	30	28	26	25
DD	0	0	0	0	0	0	0	0	0	0	0	0
New Delhi				Latitude			28.6					
$\overline{H}$	11.34	13.81	16.15	19.88	21.10	18.69	18.03	16.88	18.04	15.60	12.39	10.52
K_T	.51	.52	.50	.54	.53	.46	.45	.44	.53	.55	.54	.51
$\overline{T}$	15	18	23	29	33	35	31	30	30	26	20	16
DD	132	52	5	0	0	0	0	0	0	0	20	87
Japan												
Akita				Latitude			39.7					
$\overline{H}$	4.58	7.22	11.39	14.74	17.19	17.07	17.44	16.22	13.07	9.92	5.42	3.84
K_T	.30	.35	.41	.42	.43	.41	.43	.44	.43	.44	.33	.27
$\overline{T}$	−1	−0	3	9	14	18	22	24	19	13	7	2
DD	589	518	478	290	156	52	10	5	35	169	340	493
Kagoshima				Latitude			31.6					
$\overline{H}$	8.48	10.01	12.82	14.19	15.17	14.66	17.08	17.84	13.98	12.41	9.33	8.24
K_T	.42	.40	.41	.39	.38	.36	.42	.47	.42	.46	.44	.43
$\overline{T}$	7	9	12	16	19	23	27	28	25	19	14	9
DD	348	273	216	98	33	6	0	0	2	37	133	282
Shimizu				Latitude			32.7					
$\overline{H}$	9.50	11.07	14.11	14.93	16.22	15.53	18.90	18.36	14.40	12.06	9.92	9.02
K_T	.48	.45	.46	.41	.41	.38	.47	.49	.44	.46	.48	.49
$\overline{T}$	9	10	12	17	20	23	26	28	25	20	16	11
DD	305	243	198	79	27	6	0	0	1	21	83	219

	Jan	Feb	Mar	Apr	May	Jun	Jul	Aug	Sep	Oct	Nov	Dec

Malaysia

Kuala Lumpur — Latitude 3.1

	Jan	Feb	Mar	Apr	May	Jun	Jul	Aug	Sep	Oct	Nov	Dec
$\overline{H}$	17.68	19.08	19.44	18.77	17.70	16.77	17.13	17.43	17.15	18.33	15.50	16.74
K_T	.51	.52	.52	.50	.50	.48	.49	.48	.46	.50	.44	.49
$\overline{T}$	26	27	27	27	27	27	27	27	26	26	26	26
DD	0	0	0	0	0	0	0	0	0	0	0	0

Pakistan

Karachi — Latitude 24.8

	Jan	Feb	Mar	Apr	May	Jun	Jul	Aug	Sep	Oct	Nov	Dec
$\overline{H}$	16.36	18.74	21.10	22.75	23.89	23.50	20.19	18.59	20.80	19.77	17.27	15.37
K_T	.67	.66	.63	.61	.60	.58	.50	.49	.60	.66	.68	.67
$\overline{T}$	19	22	25	27	30	31	30	29	28	27	25	21
DD	25	6	0	0	0	0	0	0	0	0	0	8

Lahore — Latitude 31.5

	Jan	Feb	Mar	Apr	May	Jun	Jul	Aug	Sep	Oct	Nov	Dec
$\overline{H}$	10.06	13.71	17.90	20.44	22.67	22.20	20.03	18.96	18.36	15.60	12.19	10.01
K_T	.49	.55	.58	.56	.57	.54	.49	.50	.56	.58	.57	.53
$\overline{T}$	13	16	21	27	32	34	32	31	30	26	19	14
DD	188	87	16	0	0	0	0	0	0	1	38	141

Singapore

Singapore — Latitude 1.0

	Jan	Feb	Mar	Apr	May	Jun	Jul	Aug	Sep	Oct	Nov	Dec
$\overline{H}$	16.71	17.64	17.88	16.69	15.40	15.11	15.59	15.78	16.24	15.63	13.90	14.37
K_T	.47	.47	.47	.45	.44	.45	.45	.44	.44	.42	.39	.41
$\overline{T}$	26	26	27	27	27	28	27	27	27	27	26	26
DD	0	0	0	0	0	0	0	0	0	0	0	0

Sri Lanka

Colombo — Latitude 6.9

	Jan	Feb	Mar	Apr	May	Jun	Jul	Aug	Sep	Oct	Nov	Dec
$\overline{H}$	16.59	17.39	19.64	19.57	18.10	16.93	16.03	16.01	15.50	16.25	17.07	16.84
K_T	.50	.49	.53	.52	.49	.47	.44	.43	.42	.45	.50	.52
$\overline{T}$	26	28	29	30	30	29	28	28	28	28	27	25
DD	0	0	0	0	0	0	0	0	0	0	0	0

Thailand

Bangkok — Latitude 13.7

	Jan	Feb	Mar	Apr	May	Jun	Jul	Aug	Sep	Oct	Nov	Dec
$\overline{H}$	16.59	17.39	19.64	19.57	18.10	16.99	16.09	16.03	15.51	16.31	17.22	16.44
K_T	.55	.52	.54	.51	.47	.45	.42	.42	.42	.48	.56	.56
$\overline{T}$	26	28	29	30	30	29	28	28	28	28	27	25
DD	0	0	0	0	0	0	0	0	0	0	0	0

Europe

Germany

Stuttgart — Latitude 48.8

	Jan	Feb	Mar	Apr	May	Jun	Jul	Aug	Sep	Oct	Nov	Dec
$\overline{H}$	3.47	5.98	9.60	14.59	17.87	19.43	19.55	16.12	13.07	7.99	3.94	2.72
K_T	.35	.39	.42	.46	.46	.47	.49	.47	.50	.45	.35	.32
$\overline{T}$	–0	3	8	9	13	17	18	19	16	11	5	3
DD	576	425	324	284	161	66	54	35	95	239	402	471

Greece

Athens — Latitude 38.0

	Jan	Feb	Mar	Apr	May	Jun	Jul	Aug	Sep	Oct	Nov	Dec
$\overline{H}$	6.57	9.40	13.59	18.08	22.56	24.72	24.86	22.28	17.59	12.22	8.18	6.09
K_T	.40	.43	.48	.51	.57	.59	.61	.60	.57	.52	.46	.40
$\overline{T}$	10	10	11	15	20	25	28	28	23	19	14	11
DD	260	234	231	114	25	1	0	0	5	37	140	231

	Jan	Feb	Mar	Apr	May	Jun	Jul	Aug	Sep	Oct	Nov	Dec
Poland												
Warsaw					Latitude		52.3					
$\underline{H}$	1.90	3.48	8.32	11.94	16.42	19.30	18.18	16.47	10.41	5.09	2.12	1.32
K_T	.25	.26	.39	.39	.43	.47	.46	.49	.42	.33	.24	.21
$\underline{T}$	−2	−3	1	7	13	17	19	18	13	8	3	0
DD	641	604	549	331	177	66	44	59	165	315	459	555
Romania												
Cluj					Latitude		46.8					
$\underline{H}$	4.74	7.37	12.07	15.76	20.16	22.18	21.97	18.95	14.16	9.06	4.69	3.22
K_T	.43	.45	.50	.48	.52	.53	.55	.54	.52	.48	.38	.33
$\underline{T}$	−3	−3	3	9	14	18	20	19	15	9	4	−1
DD	660	596	475	282	148	54	29	41	117	291	429	598
Switzerland												
Zurich					Latitude		47.5					
$\underline{H}$	3.00	5.81	9.81	14.16	18.09	19.64	20.87	16.41	12.87	7.23	3.60	2.37
K_T	.28	.36	.41	.44	.47	.47	.52	.47	.48	.39	.30	.26
$\underline{T}$	−1	0	5	9	13	16	18	17	14	9	4	0
DD	601	504	428	293	181	93	60	72	140	303	438	564
USSR												
Kiev					Latitude		50.4					
$\underline{H}$	3.39	5.87	9.37	14.08	19.00	21.97	19.94	17.12	12.66	6.93	2.96	2.08
K_T	.38	.41	.42	.45	.49	.53	.50	.50	.50	.42	.29	.28
$\underline{T}$	−6	−5	−1	8	15	19	20	19	14	8	1	−3
DD	756	658	583	323	132	48	27	40	139	337	507	657
Leningrad					Latitude		60.0					
$\underline{H}$	1.16	3.37	7.48	12.11	18.71	21.06	19.00	14.37	8.43	3.60	1.21	0.61
K_T	.34	.40	.44	.44	.51	.51	.49	.46	.41	.33	.26	.26
$\underline{T}$	−8	−8	−4	3	10	15	18	17	11	5	−0	−4
DD	803	733	700	450	265	109	53	82	219	410	555	703
Moscow					Latitude		55.8					
$\underline{H}$	2.23	4.85	8.83	12.87	17.90	20.65	18.21	14.97	9.40	4.35	1.79	1.17
K_T	.39	.44	.46	.44	.48	.50	.46	.46	.41	.32	.26	.26
$\underline{T}$	−10	−10	−4	5	12	17	19	17	11	5	−2	−7
DD	874	778	697	409	208	81	46	79	220	429	606	778
Odessa					Latitude		46.5					
$\underline{H}$	3.44	5.41	9.23	14.42	19.34	22.30	21.58	18.88	14.27	8.60	3.79	2.61
K_T	.31	.32	.38	.44	.50	.53	.53	.54	.52	.45	.30	.27
$\underline{T}$	−2	−2	2	8	15	20	22	22	17	11	5	0
DD	635	565	515	299	125	31	11	16	74	230	388	555

North America

Canada

Alberta												
Beaverlodge					Latitude		55.2					
$\underline{H}$	2.90	6.30	12.13	17.27	20.07	22.12	21.17	17.18	11.26	6.68	3.44	2.08
K_T	.48	.55	.62	.59	.53	.54	.54	.53	.49	.48	.47	.44
$\underline{T}$	−16	−10	−6	3	9	13	15	14	10	4	−5	−12
DD	1060	798	756	471	281	169	121	146	269	432	702	926

	Jan	Feb	Mar	Apr	May	Jun	Jul	Aug	Sep	Oct	Nov	Dec
	Edmonton				Latitude		53.6					
H	3.74	7.06	12.61	17.49	20.94	22.54	23.20	18.08	13.07	8.06	4.20	2.78
K_T	.54	.57	.61	.58	.55	.54	.59	.55	.55	.54	.51	.49
$\overline{T}$	−15	−10	−5	4	11	15	17	16	11	6	−4	−10
DD	1032	781	722	424	226	120	75	99	226	389	660	889
	Suffield				Latitude		50.3					
H	4.89	8.50	13.92	17.74	21.66	23.73	24.60	20.44	14.47	9.32	5.14	3.73
K_T	.55	.59	.62	.57	.56	.57	.62	.60	.57	.56	.50	.49
$\overline{T}$	−14	−9	−4	5	12	16	19	18	13	7	−2	−9
DD	992	758	691	394	215	100	43	57	178	353	618	840
British Columbia												
	Cape St. James				Latitude		51.9					
H	2.70	5.17	9.34	14.35	19.43	20.47	18.97	16.68	12.10	6.59	3.31	2.05
K_T	.34	.39	.44	.47	.51	.49	.48	.50	.49	.42	.36	.31
$\overline{T}$	4	5	5	7	9	11	13	14	13	10	7	5
DD	446	378	415	354	298	233	178	146	166	262	342	412
	Fort Nelson				Latitude		58.8					
H	1.72	4.61	9.84	16.21	19.07	20.50	19.68	15.67	10.09	5.38	2.05	1.10
K_T	.43	.50	.56	.58	.52	.50	.50	.50	.48	.46	.39	.38
$\overline{T}$	−24	−17	−10	2	10	14	17	15	9	1	−12	−21
DD	1305	985	871	502	278	142	98	138	294	534	909	1218
	Mt. Kobau Observ.			Latitude		49.1					
H	4.77	7.97	12.25	17.49	21.40	23.18	24.42	21.15	15.33	9.28	4.73	3.66
K_T	.50	.53	.53	.55	.55	.56	.61	.61	.59	.53	.43	.44
$\overline{T}$	−9	−6	−5	−1	4	8	12	12	8	2	−5	−8
DD	830	680	734	588	441	305	196	199	316	499	687	802
	Nanaimo Depart. Bay			Latitude		49.2					
H	2.93	5.85	10.47	15.90	20.96	22.07	23.66	19.50	13.61	7.83	3.74	2.27
K_T	.31	.39	.46	.50	.54	.53	.59	.57	.52	.45	.34	.28
$\overline{T}$	3	5	6	9	12	15	18	18	15	11	6	4
DD	468	372	379	283	188	105	50	51	109	244	357	431
	Port Hardy				Latitude		50.7					
H	2.83	5.39	8.87	13.28	17.83	18.91	19.13	15.40	10.88	6.20	3.07	2.04
K_T	.33	.38	.40	.43	.46	.46	.48	.45	.43	.38	.31	.28
$\overline{T}$	2	4	4	7	9	12	14	14	12	9	5	4
DD	493	403	431	351	280	198	153	147	198	298	390	459
	Prince George				Latitude		53.9					
H	2.73	5.30	10.09	15.79	18.15	21.42	20.81	17.04	11.12	6.14	3.07	1.78
K_T	.40	.43	.50	.53	.48	.52	.53	.52	.47	.42	.38	.32
$\overline{T}$	−12	−6	−2	4	9	13	15	14	10	5	−3	−8
DD	942	683	623	421	283	172	120	146	262	419	636	812
	Sandspit				Latitude		53.3					
H	2.40	4.96	9.26	13.81	18.29	18.26	17.20	15.26	10.64	5.81	2.97	1.73
K_T	.33	.39	.45	.46	.48	.44	.43	.46	.44	.39	.35	.30
$\overline{T}$	2	4	4	6	9	12	14	15	13	9	6	3
DD	505	414	446	369	299	204	142	123	167	289	384	462
	Summerland				Latitude		49.6					
H	3.44	6.48	11.54	16.66	20.82	22.64	23.68	19.55	14.47	8.45	3.82	2.50
K_T	.37	.44	.51	.53	.54	.54	.59	.57	.56	.49	.36	.31
$\overline{T}$	−3	0	4	9	14	17	21	20	15	9	3	−1
DD	672	509	453	290	161	65	20	28	112	291	474	601

	Jan	Feb	Mar	Apr	May	Jun	Jul	Aug	Sep	Oct	Nov	Dec
	Vancouver				Latitude		49.3					
$\overline{H}$	2.94	5.53	10.03	15.09	20.15	21.78	22.95	18.62	13.22	7.38	3.59	2.28
K_T	.31	.37	.44	.48	.52	.52	.57	.54	.51	.43	.33	.28
T	3	5	6	9	12	15	17	17	14	10	6	4
DD	477	370	382	294	206	117	70	72	130	250	363	440

Manitoba

	Jan	Feb	Mar	Apr	May	Jun	Jul	Aug	Sep	Oct	Nov	Dec
	Churchill				Latitude		58.8					
$\overline{H}$	2.27	5.80	12.27	18.89	20.04	21.92	20.36	15.28	8.80	4.22	2.37	1.48
K_T	.56	.63	.70	.67	.54	.53	.52	.49	.41	.36	.45	.51
T	−28	−26	−20	−10	−2	6	12	11	5	−2	−12	−22
DD	1419	1237	1199	852	614	366	215	229	390	614	912	1255
	The Pas				Latitude		54.0					
$\overline{H}$	3.44	7.08	12.60	18.36	20.68	21.38	20.76	16.52	11.02	6.15	3.27	2.43
K_T	.51	.58	.62	.61	.55	.52	.52	.50	.46	.42	.41	.45
T	−23	−18	−11	0	8	14	18	16	10	4	−8	−18
DD	1271	1016	914	550	313	142	76	108	263	458	774	1112
	Winnipeg				Latitude		49.9					
$\overline{H}$	5.25	9.05	14.06	17.74	20.90	22.74	22.99	19.00	13.32	8.15	4.64	3.82
K_T	.57	.62	.62	.56	.54	.55	.57	.55	.52	.48	.44	.49
T	−19	−16	−8	3	11	17	20	18	12	6	−5	−14
DD	1165	949	821	448	229	88	45	65	191	381	684	1001

New Brunswick

	Jan	Feb	Mar	Apr	May	Jun	Jul	Aug	Sep	Oct	Nov	Dec
	Fredericton				Latitude		45.9					
$\overline{H}$	5.48	8.92	12.35	15.32	17.94	19.91	19.61	17.32	13.20	8.50	5.05	4.12
K_T	.47	.52	.50	.47	.46	.48	.49	.49	.48	.44	.39	.41
T	−9	−8	−3	4	11	16	19	18	13	8	2	−6
DD	849	744	645	430	245	97	44	59	160	331	498	765

Newfoundland

	Jan	Feb	Mar	Apr	May	Jun	Jul	Aug	Sep	Oct	Nov	Dec
	St. John's West				Latitude		47.5					
$\overline{H}$	4.14	7.08	10.42	13.60	16.54	19.60	19.94	15.72	12.02	6.82	4.11	3.03
K_T	.39	.44	.44	.42	.43	.47	.50	.45	.45	.37	.35	.33
T	−4	−4	−2	2	6	11	16	16	12	7	4	−1
DD	685	632	629	501	388	220	105	107	206	348	444	610

Northwest Territories

	Jan	Feb	Mar	Apr	May	Jun	Jul	Aug	Sep	Oct	Nov	Dec
	Alert				Latitude		82.5					
H	0.00	0.03	2.01	11.78	22.63	24.57	18.78	10.66	3.61	0.35	0.00	0.00
Ktbar	.00	.00	.60	.61	.61	.55	.46	.41	.48	–	.00	.00
T	−32	−34	−33	−25	−12	−1	4	1	−10	−20	−27	−30
DD	1562	1453	1596	1296	930	579	457	540	855	1178	1347	1497
	Baker Lake				Latitude		64.3					
H	0.78	3.33	10.28	17.75	22.33	21.69	19.47	13.83	7.66	3.55	1.24	0.32
Ktbar	.53	.58	.72	.69	.62	.53	.51	.47	.42	.43	.53	.56
T	−33	−33	−28	−17	−6	4	11	10	2	−8	−20	−28
DD	1590	1425	1432	1068	766	429	240	277	482	806	1158	1441
	Cambridge Bay				Latitude		69.1					
H	0.19	1.82	7.57	16.03	21.76	23.16	18.54	11.97	6.20	2.85	0.39	0.00
Ktbar	–	.58	.66	.68	.62	.55	.48	.43	.40	.53	.73	.00
T	−34	−34	−31	−22	−9	2	8	7	−1	−12	−24	−30
DD	1608	1475	1537	1206	859	505	329	371	571	930	1263	1497

	Jan	Feb	Mar	Apr	May	Jun	Jul	Aug	Sep	Oct	Nov	Dec
Coral Harbour					Latitude		64.2					
$\overline{H}$	0.78	3.55	9.84	18.04	23.43	22.88	17.97	13.76	7.98	3.95	1.24	0.36
Ktbar	.51	.61	.68	.70	.65	.56	.47	.47	.44	.48	.52	.60
$\overline{T}$	−30	−29	−25	−16	−6	2	9	7	1	−8	−18	−26
DD	1488	1335	1348	1038	762	487	304	343	523	809	1074	1357
Eureka					Latitude		80.0					
$\overline{H}$	0.00	0.15	2.81	11.94	23.15	24.79	18.82	10.52	4.10	0.58	3.74	2.27
Ktbar	.00	–	.59	.61	.63	.56	.46	.40	.46	–	.00	.00
$\overline{T}$	−36	−38	−37	−28	−11	2	5	3	−8	−22	−32	−35
DD	1695	1576	1726	1377	899	497	404	468	798	1252	1494	1646
Frobisher Bay					Latitude		63.8					
$\overline{H}$	0.84	3.52	9.19	17.61	21.20	19.74	16.42	12.81	7.42	3.31	1.14	0.39
Ktbar	.50	.58	.63	.68	.59	.48	.43	.43	.40	.39	.44	.53
$\overline{T}$	−26	−26	−23	−14	−3	3	8	7	2	−5	−13	−22
DD	1360	1237	1271	978	667	448	336	357	478	722	939	1243
Hall Beach					Latitude		68.8					
$\overline{H}$	0.19	1.99	7.64	16.37	21.97	23.87	18.95	13.11	6.14	2.64	0.42	0.01
Ktbar	–	.60	.66	.69	.62	.57	.49	.47	.39	.48	.69	.00
$\overline{T}$	−31	−32	−30	−21	−9	0	5	5	−1	−11	−22	−27
DD	1528	1411	1481	1176	849	550	403	427	567	893	1194	1416
Inuvik					Latitude		68.3					
$\overline{H}$	0.18	1.95	7.46	15.56	20.35	21.96	18.76	12.18	6.36	2.39	0.42	0.01
Ktbar	.70	.55	.63	.65	.58	.53	.49	.43	.40	.41	.56	.00
$\overline{T}$	−30	−29	−25	−14	−1	10	14	11	3	−8	−21	−27
DD	1484	1321	1342	978	593	256	171	249	458	818	1170	1410
Isachen					Latitude		78.8					
$\overline{H}$	0.00	0.09	3.19	12.08	22.11	23.31	16.95	9.19	4.11	0.52	0.00	0.00
Ktbar	.00	–	.59	.61	.61	.53	.42	.35	.43	.59	.00	.00
$\overline{T}$	−35	−37	−35	−26	−12	−1	3	1	−9	−19	−28	−33
DD	1643	1537	1643	1320	939	576	470	540	813	1168	1401	1584
Mould Bay					Latitude		76.2					
$\overline{H}$	0.00	0.45	4.11	13.53	22.56	22.81	17.37	9.64	4.81	0.90	0.03	0.00
Ktbar	.00	–	.58	.66	.63	.52	.43	.37	.43	.52	.00	.00
$\overline{T}$	−34	−35	−33	−24	−11	−0	4	1	−7	−18	−27	−31
DD	1605	1498	1584	1272	914	559	449	525	744	1112	1347	1534
Norman Wells					Latitude		65.3					
$\overline{H}$	0.47	2.73	8.41	15.48	20.14	23.09	20.26	14.41	8.11	2.81	0.83	0.17
Ktbar	.43	.52	.61	.61	.56	.56	.53	.50	.46	.37	.44	.60
$\overline{T}$	−29	−26	−20	−7	5	14	16	13	6	−5	−18	−27
DD	1463	1246	1181	765	404	156	110	177	370	710	1095	1388
Resolute					Latitude		74.7					
$\overline{H}$	0.00	0.63	5.24	14.72	23.05	24.91	18.69	11.03	5.22	1.28	0.07	0.00
Ktbar	.00	.79	.66	.70	.64	.58	.47	.42	.43	.54	.00	.00
$\overline{T}$	−32	−33	−31	−23	−11	−1	4	2	−5	−15	−25	−29
DD	1562	1442	1540	1242	905	567	442	494	702	1035	1284	1475
Sachs Harbour					Latitude		72.0					
$\overline{H}$	0.06	1.12	6.54	14.99	21.78	22.52	20.13	11.59	5.49	1.70	0.18	0.00
Ktbar	–	.64	.68	.67	.61	.53	.51	.43	.40	.45	–	.00
$\overline{T}$	−30	−31	−28	−20	−8	2	6	4	−2	−12	−22	−27
DD	1509	1377	1447	1149	818	493	388	451	609	927	1209	1416

	Jan	Feb	Mar	Apr	May	Jun	Jul	Aug	Sep	Oct	Nov	Dec
Nova Scotia												
Halifax Citadel				Latitude			44.7					
$\overline{H}$	5.11	8.14	12.09	14.54	17.38	20.00	19.11	17.94	13.98	8.96	5.32	3.85
K_T	.41	.46	.48	.44	.44	.48	.47	.50	.50	.45	.39	.35
T	-4	-4	-1	4	10	15	18	19	15	10	5	-1
DD	688	630	583	417	276	124	55	48	117	267	408	610
Kentville				Latitude			45.1					
$\overline{H}$	4.95	8.48	12.45	15.28	18.90	21.18	20.48	18.32	14.11	9.04	5.14	3.71
K_T	.41	.48	.50	.46	.48	.51	.51	.52	.51	.46	.38	.35
T	-5	-5	-1	4	10	16	19	18	14	9	4	-2
DD	722	658	598	418	250	97	39	51	135	289	429	641
Sable Island				Latitude			43.9					
$\overline{H}$	4.21	6.94	11.59	15.55	19.24	20.64	20.52	17.72	14.19	8.67	4.88	3.52
K_T	.33	.38	.45	.46	.49	.49	.51	.50	.50	.43	.35	.31
T	0	-1	1	3	7	11	16	18	16	12	7	3
DD	564	540	545	450	360	223	106	60	97	216	331	487
Ontario												
Big Trout Lake				Latitude			53.8					
$\overline{H}$	3.86	7.68	13.11	19.02	20.16	19.26	20.06	15.38	9.86	5.64	3.33	2.74
K_T	.57	.63	.64	.63	.53	.47	.51	.46	.41	.39	.41	.50
T	-25	-21	-15	-4	5	12	16	14	8	2	-9	-20
DD	1326	1111	1016	663	430	203	110	150	311	513	819	1184
Elora Research Station				Latitude			43.7					
$\overline{H}$	5.97	9.80	12.91	17.25	18.74	21.72	21.72	18.54	13.84	8.57	4.73	4.23
K_T	.46	.53	.50	.51	.48	.52	.54	.52	.49	.42	.33	.37
T	-8	-7	-3	5	11	17	19	18	14	9	2	-5
DD	821	716	651	397	222	74	43	59	139	307	492	728
Guelph				Latitude			43.5					
$\overline{H}$	6.00	10.21	13.32	16.32	20.26	22.72	22.08	19.02	13.98	9.32	4.88	4.49
K_T	.46	.55	.52	.49	.51	.54	.54	.53	.49	.45	.34	.39
T	-7	-7	-2	6	12	17	20	19	15	9	3	-4
DD	790	694	626	376	213	68	35	47	120	286	474	694
Kapuskasing				Latitude			49.4					
$\overline{H}$	4.56	8.56	13.87	17.15	19.52	23.29	20.87	16.29	12.03	6.97	4.03	3.59
K_T	.48	.57	.61	.54	.51	.56	.52	.47	.46	.40	.38	.45
T	-19	-16	-9	1	8	14	17	15	10	4	-4	-15
DD	1143	966	858	534	318	146	90	123	256	433	681	1023
Moosonee				Latitude			51.3					
$\overline{H}$	4.20	8.11	12.59	16.68	17.97	20.12	18.57	15.10	10.58	5.99	3.40	3.09
K_T	.50	.59	.58	.54	.47	.48	.47	.45	.42	.37	.35	.44
T	-20	-19	-12	-2	6	12	15	14	10	4	-5	-16
DD	1200	1030	948	618	393	204	123	147	270	442	684	1063
Ottawa				Latitude			45.5					
$\overline{H}$	5.74	9.44	13.61	16.75	19.88	21.37	21.28	18.11	13.36	8.58	4.72	4.33
K_T	.48	.54	.55	.51	.51	.51	.53	.51	.48	.44	.36	.41
T	-11	-9	-3	6	13	18	21	19	15	9	2	-7
DD	896	770	651	377	181	58	27	42	132	308	498	793
Toronto				Latitude			43.7					
$\overline{H}$	5.19	8.20	12.05	16.07	19.83	21.97	21.94	18.68	14.04	9.24	4.82	3.90
K_T	.40	.45	.47	.48	.50	.53	.54	.52	.49	.45	.34	.34
T	-5	-4	1	8	14	19	22	21	17	11	5	-2
DD	710	621	546	323	160	40	14	20	73	233	403	617

	Jan	Feb	Mar	Apr	May	Jun	Jul	Aug	Sep	Oct	Nov	Dec

Prince Edward Island
Charlottetown   Latitude   46.3

	Jan	Feb	Mar	Apr	May	Jun	Jul	Aug	Sep	Oct	Nov	Dec
$\bar{H}$	5.32	8.97	12.61	15.88	18.57	20.92	20.10	17.71	12.93	7.90	4.94	3.79
K_T	.47	.53	.52	.48	.48	.50	.50	.50	.47	.41	.39	.38
$\bar{T}$	−7	−7	−3	3	9	15	19	18	14	9	3	−4
DD	772	708	651	465	292	121	45	52	141	301	447	676

Quebec
Fort Chimo   Latitude   58.1

	Jan	Feb	Mar	Apr	May	Jun	Jul	Aug	Sep	Oct	Nov	Dec
$\bar{H}$	2.29	5.60	12.03	18.40	17.62	18.58	16.59	13.70	9.01	4.51	2.30	1.43
K_T	.52	.58	.67	.65	.48	.45	.42	.43	.42	.37	.41	.44
$\bar{T}$	−23	−22	−18	−9	0	7	11	10	5	−1	−8	−18
DD	1289	1139	1116	825	562	345	225	253	389	595	798	1137

Inoucdjouac   Latitude   58.5

	Jan	Feb	Mar	Apr	May	Jun	Jul	Aug	Sep	Oct	Nov	Dec
$\bar{H}$	2.43	6.08	12.88	19.55	20.48	20.98	17.88	13.99	9.75	4.30	1.80	1.38
K_T	.57	.64	.73	.70	.55	.51	.46	.44	.45	.36	.33	.45
$\bar{T}$	−25	−25	−21	−11	−2	4	9	9	5	−0	−7	−18
DD	1326	1212	1205	876	617	418	285	297	401	580	765	1122

Montreal   Latitude   45.5

	Jan	Feb	Mar	Apr	May	Jun	Jul	Aug	Sep	Oct	Nov	Dec
$\bar{H}$	5.30	8.80	12.51	15.87	19.07	20.25	20.96	17.23	13.45	8.04	4.61	3.92
K_T	.45	.51	.50	.48	.49	.49	.52	.49	.49	.41	.35	.38
$\bar{T}$	−10	−9	−2	6	13	18	21	20	15	9	2	−7
DD	871	761	635	383	176	54	24	37	117	296	483	772

Nitchequon   Latitude   53.2

	Jan	Feb	Mar	Apr	May	Jun	Jul	Aug	Sep	Oct	Nov	Dec
$\bar{H}$	3.71	7.83	12.98	18.80	19.63	19.01	16.98	14.46	9.30	5.21	3.28	2.72
K_T	.52	.62	.63	.62	.52	.46	.43	.43	.39	.35	.39	.46
$\bar{T}$	−23	−21	−15	−6	2	10	14	12	6	−0	−8	−19
DD	1280	1110	1019	723	506	262	165	214	363	574	798	1162

Normandin   Latitude   48.8

	Jan	Feb	Mar	Apr	May	Jun	Jul	Aug	Sep	Oct	Nov	Dec
$\bar{H}$	5.13	9.10	14.43	17.92	18.94	20.51	19.41	16.91	11.66	6.65	4.34	3.92
K_T	.52	.60	.62	.56	.49	.49	.48	.49	.44	.38	.39	.47
$\bar{T}$	−18	−16	−9	1	9	15	17	15	10	5	−3	−14
DD	1125	963	837	534	303	134	87	120	244	426	636	1004

Sept-Iles   Latitude   50.2

	Jan	Feb	Mar	Apr	May	Jun	Jul	Aug	Sep	Oct	Nov	Dec
$\bar{H}$	4.32	8.04	11.78	14.86	18.36	20.60	18.47	17.05	11.69	7.14	4.13	3.16
K_T	.48	.56	.53	.47	.48	.50	.46	.50	.46	.43	.40	.42
$\bar{T}$	−14	−13	−7	0	6	12	15	14	9	4	−3	−11
DD	1001	862	772	549	386	206	120	148	274	456	624	908

Saskatchewan
Bad Lake   Latitude   51.3

	Jan	Feb	Mar	Apr	May	Jun	Jul	Aug	Sep	Oct	Nov	Dec
$\bar{H}$	4.97	8.59	13.97	17.99	21.72	23.54	23.60	19.32	13.95	8.99	4.99	3.55
K_T	.60	.62	.64	.58	.57	.57	.59	.57	.56	.56	.52	.51
$\bar{T}$	−18	−14	−8	4	11	16	18	17	11	5	−5	−13
DD	1119	893	812	445	233	111	66	80	220	417	699	964

Swift Current   Latitude   50.3

	Jan	Feb	Mar	Apr	May	Jun	Jul	Aug	Sep	Oct	Nov	Dec
$\bar{H}$	5.05	8.72	14.05	17.86	21.58	23.14	24.35	20.13	14.36	9.45	5.27	3.83
K_T	.57	.60	.63	.57	.56	.56	.61	.59	.56	.56	.52	.51
$\bar{T}$	−15	−10	−5	4	11	16	19	18	12	6	−4	−10
DD	1026	803	713	430	246	111	56	70	205	387	666	886

	Jan	Feb	Mar	Apr	May	Jun	Jul	Aug	Sep	Oct	Nov	Dec

Yukon

Whitehorse   Latitude   60.7

	Jan	Feb	Mar	Apr	May	Jun	Jul	Aug	Sep	Oct	Nov	Dec
$\overline{H}$	1.30	4.01	9.21	15.74	19.59	21.04	18.75	14.98	9.02	4.36	1.62	0.74
$\overline{K}_T$	.42	.50	.56	.58	.54	.51	.48	.49	.45	.42	.39	.37
$\overline{T}$	−21	−13	−8	0	7	12	14	13	8	0	−9	−17
DD	1209	882	821	540	362	200	151	193	327	558	813	1081

Caribbean

Puerto Rico

San Juan   Latitude   18.4

	Jan	Feb	Mar	Apr	May	Jun	Jul	Aug	Sep	Oct	Nov	Dec
$\overline{H}$	15.05	17.44	20.30	21.47	20.58	20.63	21.27	20.87	19.00	17.20	15.53	14.03
$\overline{K}_T$	.54	.56	.58	.57	.53	.53	.55	.55	.53	.53	.54	.53
$\overline{T}$	24	24	25	25	26	27	27	27	27	27	26	25
DD	18	14	15	11	9	6	6	6	6	7	9	14

Mexico

Ciudad Univ.   Latitude   19.4

	Jan	Feb	Mar	Apr	May	Jun	Jul	Aug	Sep	Oct	Nov	Dec
$\overline{H}$	16.48	18.28	21.36	21.84	20.05	19.62	18.48	17.70	15.88	16.48	15.69	15.59
$\overline{K}_T$	.60	.59	.61	.58	.51	.50	.47	.46	.44	.51	.56	.60
$\overline{T}$	12	14	16	17	17	17	16	16	16	15	13	12
DD	194	130	82	57	53	59	87	87	91	119	154	191

United States

Alabama

Birmingham   Latitude   33.6

	Jan	Feb	Mar	Apr	May	Jun	Jul	Aug	Sep	Oct	Nov	Dec
$\overline{H}$	8.02	10.97	14.71	19.00	21.08	21.77	20.55	19.57	16.52	13.74	9.73	7.51
$\overline{K}_T$	.42	.46	.49	.53	.53	.53	.51	.52	.51	.53	.48	.42
$\overline{T}$	7	8	12	17	21	25	27	26	23	17	11	7
DD	359	283	211	71	19	2	1	1	7	72	219	344

Mobile   Latitude   30.7

	Jan	Feb	Mar	Apr	May	Jun	Jul	Aug	Sep	Oct	Nov	Dec
$\overline{H}$	9.40	12.48	15.98	19.55	21.25	21.21	19.47	18.63	16.45	14.74	10.84	8.62
$\overline{K}_T$	.45	.49	.51	.53	.53	.51	.48	.49	.50	.54	.49	.44
$\overline{T}$	11	12	15	20	24	27	28	28	25	21	15	12
DD	242	179	121	33	7	1	1	1	3	28	128	217

Montgomery   Latitude   32.3

	Jan	Feb	Mar	Apr	May	Jun	Jul	Aug	Sep	Oct	Nov	Dec
$\overline{H}$	8.53	11.49	15.22	19.63	21.54	22.39	20.90	19.82	16.66	14.32	10.39	8.17
$\overline{K}_T$	.43	.46	.50	.54	.54	.54	.51	.53	.51	.54	.49	.44
$\overline{T}$	9	10	14	18	22	26	27	27	24	19	13	9
DD	308	234	170	66	22	5	3	4	10	63	185	291

Alaska

Annette   Latitude   55.0

	Jan	Feb	Mar	Apr	May	Jun	Jul	Aug	Sep	Oct	Nov	Dec
$\overline{H}$	2.02	4.26	8.14	13.05	16.72	16.64	16.33	13.20	9.22	4.79	2.48	1.39
$\overline{K}_T$	.33	.37	.41	.44	.44	.40	.41	.40	.40	.34	.34	.28
$\overline{T}$	1	3	3	6	10	13	14	15	12	8	4	2
DD	544	440	461	372	275	187	152	143	198	317	420	500

Barrow   Latitude   71.3

	Jan	Feb	Mar	Apr	May	Jun	Jul	Aug	Sep	Oct	Nov	Dec
$\overline{H}$	0.00	0.84	5.57	11.93	12.94	17.34	16.56	9.71	4.70	1.43	0.05	0.00
$\overline{K}_T$	.00	.41	.56	.53	.37	.41	.42	.36	.33	.35	.00	.00
$\overline{T}$	−26	−28	−26	−18	−7	1	4	3	−1	−9	−18	−25
DD	1370	1299	1380	1097	792	534	458	477	578	855	1092	1330

	Jan	Feb	Mar	Apr	May	Jun	Jul	Aug	Sep	Oct	Nov	Dec
Bethel				Latitude			60.8					
$\overline{H}$	1.10	3.59	8.38	13.63	16.49	17.23	14.64	10.44	7.95	4.20	1.53	0.56
K_T	.36	.45	.51	.50	.45	.42	.38	.34	.39	.40	.37	.28
$\overline{T}$	−15	−13	−11	−4	4	11	13	11	7	−1	−8	−15
DD	1029	882	920	676	431	236	198	234	338	599	796	1041
Bettles				Latitude			66.9					
$\overline{H}$	0.11	1.95	6.99	13.95	19.29	21.08	17.74	12.20	7.63	2.86	0.45	0.00
K_T	.25	.44	.55	.57	.54	.51	.46	.43	.46	.43	.38	.00
$\overline{T}$	−25	−22	−17	−6	5	13	14	11	4	−7	−19	−25
DD	1345	1131	1090	741	404	170	152	237	420	776	1107	1330
Big Delta				Latitude			64					
$\overline{H}$	0.52	2.81	8.08	14.12	18.96	20.24	18.32	13.95	8.70	3.70	1.06	0.10
K_T	.34	.46	.56	.55	.53	.49	.48	.47	.47	.44	.43	.15
$\overline{T}$	−21	−16	−11	−1	8	14	15	13	6	−4	−14	−20
DD	1202	957	905	591	328	152	126	189	360	684	965	1190
Fairbanks				Latitude			64.8					
$\overline{H}$	0.34	2.51	7.66	13.55	18.21	19.89	17.51	12.69	8.05	3.32	0.84	0.02
K_T	.28	.45	.55	.54	.51	.48	.46	.43	.45	.42	.40	.05
$\overline{T}$	−24	−19	−13	−2	9	15	16	13	7	−4	−16	−24
DD	1323	1050	955	602	316	140	126	194	350	685	1035	1299
Gulkana				Latitude			62.2					
$\overline{H}$	0.83	3.25	8.60	14.81	18.32	19.96	18.30	14.21	9.02	4.43	1.32	0.32
K_T	.35	.45	.55	.56	.51	.49	.47	.47	.46	.46	.39	.23
$\overline{T}$	−22	−16	−10	−1	7	12	14	12	6	−3	−14	−21
DD	1242	949	869	581	371	205	176	226	365	658	981	1206
Homer				Latitude			59.6					
$\overline{H}$	1.39	3.79	8.62	14.18	17.97	19.87	18.14	13.49	8.98	4.96	1.99	0.73
K_T	.38	.43	.50	.51	.49	.48	.47	.43	.43	.45	.42	.29
$\overline{T}$	−6	−4	−2	2	6	9	11	11	8	3	−2	−6
DD	750	622	643	500	396	283	237	237	310	478	613	750
Juneau				Latitude			58.4					
$\overline{H}$	1.32	3.20	6.93	11.87	14.66	16.06	14.52	11.18	7.25	3.63	1.69	0.70
K_T	.31	.34	.39	.42	.40	.39	.37	.35	.34	.31	.31	.23
$\overline{T}$	−5	−2	−0	4	8	12	13	12	10	5	0	−3
DD	715	576	573	440	325	218	189	210	275	407	543	650
King Salmon				Latitude			58.7					
$\overline{H}$	1.66	4.28	9.08	13.69	16.82	17.49	15.71	11.87	8.83	5.38	2.32	1.03
K_T	.40	.46	.52	.49	.46	.43	.40	.38	.41	.46	.44	.35
$\overline{T}$	−10	−9	−6	−0	6	10	13	12	9	1	−6	−11
DD	886	753	765	558	390	249	201	211	302	541	714	917
Kodiak				Latitude			57.8					
$\overline{H}$	1.69	4.04	8.87	13.71	15.62	17.37	15.98	13.21	9.02	5.55	2.34	1.10
K_T	.37	.41	.49	.48	.42	.42	.41	.42	.41	.45	.41	.32
$\overline{T}$	−1	−0	0	3	6	10	12	13	10	5	2	−1
DD	595	520	565	469	378	264	203	191	257	422	502	606
Kotzebue				Latitude			66.9					
$\overline{H}$	0.09	1.86	6.75	13.40	18.65	20.85	17.34	11.86	7.36	2.91	0.37	0.00
K_T	.20	.42	.53	.55	.53	.50	.45	.41	.44	.43	.31	.00
$\overline{T}$	−20	−20	−18	−11	−1	6	12	10	5	−5	−14	−20
DD	1180	1078	1129	868	593	368	236	268	404	715	954	1184

	Jan	Feb	Mar	Apr	May	Jun	Jul	Aug	Sep	Oct	Nov	Dec
McGrath					Latitude		63					
$\overline{H}$	0.66	2.93	7.86	13.48	16.89	18.01	15.66	11.57	7.89	3.60	1.14	0.22
$\overline{K_T}$	.33	.44	.52	.52	.47	.44	.41	.39	.42	.40	.38	.21
T	−23	−18	−13	−3	7	13	15	12	7	−4	−15	−23
DD	1271	1013	963	643	365	176	149	218	356	683	999	1276
Nome					Latitude		64.5					
$\overline{H}$	0.34	2.54	7.17	13.46	17.86	19.90	16.05	11.27	7.64	3.48	0.74	0.03
$\overline{K_T}$	.26	.44	.51	.53	.50	.49	.42	.38	.42	.43	.33	.06
T	−14	−15	−14	−7	2	8	10	10	6	−2	−9	−15
DD	1013	929	993	768	520	333	270	284	387	627	822	1041
Summit					Latitude		63.3					
$\overline{H}$	0.64	2.85	7.93	14.07	18.54	18.54	16.02	11.85	7.99	3.91	1.22	0.18
$\overline{K_T}$	.35	.44	.53	.54	.52	.45	.42	.40	.43	.44	.43	.19
T	−17	−14	−12	−5	3	9	11	9	4	−4	−12	−16
DD	1091	907	927	691	478	279	241	294	421	704	920	1070
Yakutat					Latitude		59.5					
$\overline{H}$	1.14	3.01	7.08	11.93	14.39	15.27	13.71	10.70	7.20	3.91	1.53	0.58
$\overline{K_T}$	.31	.34	.41	.43	.39	.37	.35	.34	.34	.35	.32	.23
T	−4	−2	−1	2	6	10	12	12	9	5	0	−3
DD	700	575	596	482	377	264	214	222	283	422	546	657
Arizona												
Phoenix					Latitude		33.4					
$\overline{H}$	11.60	15.60	20.59	26.73	30.38	31.10	28.23	26.02	22.88	17.90	13.06	10.57
$\overline{K_T}$	.60	.65	.68	.74	.76	.75	.69	.69	.71	.69	.64	.59
T	11	13	15	20	25	29	33	32	29	22	15	11
DD	251	175	134	52	14	2	0	1	2	28	128	233
Prescott					Latitude		34.7					
$\overline{H}$	11.54	15.15	20.17	25.83	29.84	31.35	26.22	23.74	22.18	17.51	12.94	10.52
$\overline{K_T}$	.63	.65	.68	.72	.75	.75	.64	.63	.70	.69	.66	.62
T	3	5	7	11	16	21	24	23	20	14	8	4
DD	483	383	363	229	120	37	12	20	43	161	324	455
Tucson					Latitude		32.1					
$\overline{H}$	12.47	16.25	21.16	26.83	30.33	30.99	26.58	24.78	22.47	18.18	13.72	11.30
$\overline{K_T}$	.62	.65	.69	.74	.76	.75	.65	.66	.69	.68	.65	.61
T	11	12	14	19	23	28	30	29	27	21	15	11
DD	256	194	158	65	20	3	1	2	5	35	140	239
Winslow					Latitude		35					
$\overline{H}$	11.18	15.06	20.21	25.92	29.45	30.78	26.64	24.30	21.89	17.17	12.71	10.15
$\overline{K_T}$	.61	.65	.68	.72	.74	.74	.65	.65	.69	.68	.65	.60
T	0	4	7	12	17	22	26	25	21	14	6	1
DD	560	406	353	203	94	24	7	12	36	160	367	538
Yuma					Latitude		32.7					
$\overline{H}$	12.44	16.38	21.79	27.39	30.97	31.94	27.85	26.44	23.28	18.42	13.79	11.35
$\overline{K_T}$	.63	.67	.71	.75	.77	.77	.69	.70	.72	.70	.66	.62
T	13	15	18	22	26	30	34	34	31	24	18	14
DD	182	111	74	20	4	0	0	0	0	8	74	168
Arkansas												
Fort Smith					Latitude		35.3					
$\overline{H}$	8.44	11.34	14.89	18.34	21.71	23.72	23.44	21.31	17.05	13.63	9.67	7.74
$\overline{K_T}$	.47	.49	.50	.51	.54	.57	.58	.57	.54	.55	.50	.46
T	4	6	10	17	21	26	28	27	23	17	10	5
DD	448	339	259	87	26	4	1	2	11	82	250	406

	Jan	Feb	Mar	Apr	May	Jun	Jul	Aug	Sep	Oct	Nov	Dec
	Little Rock				Latitude		34.7					
$\overline{H}$	8.30	11.38	14.90	18.29	21.90	23.91	23.07	21.12	17.23	13.95	9.62	7.64
$\overline{K_T}$	.45	.49	.50	.51	.55	.58	.57	.56	.54	.55	.49	.45
$\overline{T}$	4	6	10	17	21	26	27	27	23	17	10	5
DD	444	349	272	116	51	14	9	10	30	113	262	412

California

	Jan	Feb	Mar	Apr	May	Jun	Jul	Aug	Sep	Oct	Nov	Dec
	Bakersfield				Latitude		35.4					
$\overline{H}$	8.70	12.51	18.11	23.77	28.49	31.21	30.46	27.48	22.61	16.55	10.70	7.69
$\overline{K_T}$	.48	.54	.61	.66	.71	.75	.75	.74	.71	.66	.56	.46
$\overline{T}$	9	11	14	17	21	25	29	28	25	19	13	9
DD	316	217	182	104	50	17	6	8	18	71	185	311
	Daggett				Latitude		34.9					
$\overline{H}$	10.88	14.54	20.12	25.82	29.42	31.41	29.55	27.04	22.80	17.21	12.31	9.94
$\overline{K_T}$	.59	.62	.68	.72	.74	.76	.73	.72	.72	.68	.63	.59
$\overline{T}$	9	11	14	18	22	27	31	30	26	20	13	9
DD	311	212	167	73	22	4	0	1	5	44	176	300
	Fresno				Latitude		36.8					
$\overline{H}$	7.45	11.49	17.78	23.75	28.19	31.02	30.47	27.51	22.54	16.22	10.09	6.52
$\overline{K_T}$	.43	.52	.62	.67	.71	.74	.75	.74	.72	.67	.55	.41
$\overline{T}$	7	10	12	16	20	23	27	26	23	18	12	8
DD	347	247	212	122	57	20	6	10	21	85	213	337
	Long Beach				Latitude		33.8					
$\overline{H}$	10.53	13.79	18.28	22.00	23.43	24.30	26.10	23.83	19.31	15.06	11.39	9.61
$\overline{K_T}$	.55	.58	.61	.61	.59	.59	.64	.63	.60	.59	.57	.55
$\overline{T}$	12	13	14	16	18	20	22	23	22	19	16	13
DD	214	171	171	124	93	60	34	29	33	67	124	193
	Los Angeles				Latitude		33.9					
$\overline{H}$	10.52	13.78	18.38	22.15	23.38	24.06	26.19	23.60	19.08	14.95	11.39	9.63
$\overline{K_T}$	.56	.58	.61	.61	.58	.58	.64	.63	.59	.58	.57	.55
$\overline{T}$	13	13	14	15	17	18	20	21	20	18	16	14
DD	209	172	181	145	115	83	56	49	51	83	126	177
	Mount Shasta				Latitude		41.3					
$\overline{H}$	6.36	9.73	14.19	19.93	24.82	27.66	29.26	25.12	19.70	13.11	7.49	5.74
$\overline{K_T}$	.44	.49	.53	.58	.63	.66	.72	.69	.67	.60	.48	.44
$\overline{T}$	1	3	5	8	12	16	20	19	16	11	5	2
DD	542	425	426	321	224	125	55	69	112	251	393	512
	Needles				Latitude		34.8					
$\overline{H}$	11.18	15.36	20.72	26.31	30.10	31.68	28.85	25.85	22.86	17.46	12.76	10.37
$\overline{K_T}$	.61	.66	.70	.73	.75	.76	.71	.69	.72	.69	.65	.61
$\overline{T}$	11	14	16	21	26	31	35	34	31	24	16	12
DD	248	157	114	36	8	1	0	0	1	21	119	231
	Oakland				Latitude		37.7					
$\overline{H}$	8.03	11.55	16.53	21.82	25.10	26.68	26.36	23.30	19.31	13.76	9.34	7.35
$\overline{K_T}$	.48	.53	.58	.62	.63	.64	.65	.63	.63	.58	.52	.48
$\overline{T}$	9	11	12	13	15	17	17	18	18	16	13	10
DD	288	209	205	164	133	90	81	77	63	102	177	268
	Red Bluff				Latitude		40.2					
$\overline{H}$	6.47	10.13	15.37	21.68	26.95	29.51	30.33	26.23	20.95	13.94	8.02	5.79
$\overline{K_T}$	.43	.50	.56	.63	.68	.71	.75	.72	.70	.62	.49	.42
$\overline{T}$	7	10	12	15	20	24	28	27	24	18	12	8
DD	347	240	218	124	48	10	2	4	11	70	201	326

	Jan	Feb	Mar	Apr	May	Jun	Jul	Aug	Sep	Oct	Nov	Dec
	Sacramento				Latitude		38.5					
$\overline{H}$	6.78	10.67	16.55	22.74	27.64	30.46	30.51	26.89	21.65	14.93	8.87	6.11
K_T	.42	.50	.59	.65	.69	.73	.75	.73	.71	.64	.51	.42
T	7	10	12	15	18	21	24	23	22	17	12	8
DD	347	243	219	139	75	27	12	15	23	85	211	334
	San Diego				Latitude		32.7					
$\overline{H}$	11.07	14.38	18.53	21.99	22.74	23.41	24.82	23.35	19.49	15.60	12.06	10.26
K_T	.56	.59	.61	.61	.57	.57	.61	.62	.60	.59	.58	.56
T	13	14	14	16	17	19	21	22	21	19	16	14
DD	200	159	164	125	101	76	49	39	44	76	123	179
	San Francisco				Latitude		37.6					
$\overline{H}$	8.03	11.46	16.52	21.80	25.26	26.98	27.15	24.02	19.78	13.91	9.33	7.29
K_T	.48	.52	.58	.62	.63	.65	.67	.65	.64	.58	.52	.48
T	9	11	12	13	15	16	17	17	18	16	13	10
DD	300	229	229	192	158	115	110	104	90	126	192	282
	Santa Maria				Latitude		34.9					
$\overline{H}$	9.69	12.95	17.96	21.81	24.30	26.66	26.58	23.90	19.64	15.37	11.05	9.12
K_T	.53	.55	.61	.61	.61	.64	.65	.64	.62	.61	.57	.54
T	10	11	12	13	14	15	17	17	17	16	13	11
DD	274	224	239	203	183	145	122	121	112	141	187	254

Colorado

	Jan	Feb	Mar	Apr	May	Jun	Jul	Aug	Sep	Oct	Nov	Dec
	Colorado Springs				Latitude		38.8					
$\overline{H}$	10.11	13.37	17.59	21.92	24.16	26.89	25.10	22.99	19.97	15.42	10.72	8.87
K_T	.63	.63	.63	.63	.61	.64	.62	.63	.66	.67	.62	.61
T	−2	−0	2	8	13	18	21	21	16	10	3	−1
DD	627	524	514	319	183	70	30	39	107	261	458	587
	Denver				Latitude		39.8					
$\overline{H}$	9.54	12.79	17.37	21.33	24.24	26.68	25.80	23.21	19.60	14.77	10.03	8.30
K_T	.62	.62	.63	.62	.61	.64	.63	.63	.65	.65	.61	.60
T	−1	0	3	9	14	19	23	22	17	11	4	0
DD	609	505	488	308	182	80	37	45	109	251	433	563
	Eagle				Latitude		39.7					
$\overline{H}$	8.56	12.23	17.05	21.93	25.60	28.48	27.07	23.66	20.06	14.83	9.86	7.84
K_T	.56	.59	.62	.63	.64	.68	.67	.65	.67	.65	.59	.56
T	−8	−5	−0	6	11	15	19	18	13	7	−1	−7
DD	808	646	583	386	245	129	56	76	172	351	568	769
	Grand Junction				Latitude		39.1					
$\overline{H}$	8.98	12.70	17.63	22.55	27.01	29.50	27.99	24.77	20.82	15.27	10.42	8.30
K_T	.57	.60	.63	.65	.68	.71	.69	.67	.69	.66	.61	.58
T	−3	1	5	11	17	22	26	24	20	13	4	−1
DD	661	488	413	235	99	26	7	13	49	194	424	611
	Pueblo				Latitude		38.3					
$\overline{H}$	10.15	13.30	17.75	22.21	24.55	27.64	26.24	23.86	20.20	15.45	10.82	8.88
K_T	.62	.62	.63	.63	.62	.66	.64	.65	.66	.66	.62	.60
T	−1	2	4	11	16	21	25	24	19	13	5	1
DD	603	472	435	236	111	29	11	16	58	201	406	550

Connecticut

	Jan	Feb	Mar	Apr	May	Jun	Jul	Aug	Sep	Oct	Nov	Dec
	Hartford				Latitude		41.9					
$\overline{H}$	5.42	8.11	11.11	14.93	17.81	19.14	18.72	16.14	13.11	9.68	5.65	4.37
K_T	.39	.42	.42	.44	.45	.46	.46	.45	.45	.45	.37	.35
T	−4	−3	2	9	15	20	23	21	17	11	5	−2
DD	691	593	507	294	143	41	18	29	84	227	395	633

	Jan	Feb	Mar	Apr	May	Jun	Jul	Aug	Sep	Oct	Nov	Dec
Delaware												
Wilmington					Latitude		39.7					
$\overline{H}$	6.49	9.39	13.04	16.80	19.41	21.37	20.70	18.33	14.96	11.17	7.31	5.54
K_T	.42	.45	.47	.48	.49	.51	.51	.50	.50	.49	.44	.40
$\overline{T}$	0	1	5	11	17	22	24	23	20	14	8	2
DD	569	489	410	229	105	32	17	22	53	168	329	524
District of Columbia												
Washington					Latitude		39					
$\overline{H}$	6.50	9.26	12.77	16.56	19.50	21.58	20.63	18.36	15.21	11.39	7.38	5.46
K_T	.41	.44	.46	.47	.49	.52	.51	.50	.50	.49	.43	.38
$\overline{T}$	0	1	5	12	17	22	24	23	19	13	7	1
DD	570	489	414	231	123	51	32	40	79	202	352	540
Florida												
Apalachicola					Latitude		29.7					
$\overline{H}$	9.68	12.78	16.73	21.33	23.73	22.68	20.58	19.15	17.42	15.57	11.80	9.28
K_T	.45	.49	.53	.58	.59	.55	.51	.50	.52	.56	.52	.46
$\overline{T}$	12	13	16	20	24	27	27	28	26	22	16	13
DD	206	157	111	35	10	2	2	2	4	24	99	185
Daytona Beach					Latitude		29.2					
$\overline{H}$	10.88	13.77	17.57	21.39	22.34	20.73	20.25	19.09	16.78	14.21	11.76	9.88
K_T	.50	.52	.55	.58	.56	.51	.50	.50	.50	.51	.52	.48
$\overline{T}$	15	15	18	21	24	26	27	27	26	23	18	15
DD	148	119	86	39	17	7	5	5	6	23	71	136
Jacksonville					Latitude		30.5					
$\overline{H}$	10.21	13.22	17.28	21.07	22.21	21.40	20.46	19.23	16.37	13.88	11.30	9.28
K_T	.49	.51	.55	.57	.55	.52	.51	.51	.49	.51	.51	.47
$\overline{T}$	13	14	16	20	24	26	27	27	26	21	16	13
DD	197	155	111	43	16	5	4	4	7	32	106	187
Miami					Latitude		25.8					
$\overline{H}$	12.01	14.91	18.20	21.10	20.92	19.39	20.01	18.50	16.53	14.79	12.70	11.57
K_T	.51	.53	.55	.56	.53	.48	.50	.48	.48	.50	.51	.52
$\overline{T}$	20	20	22	24	26	27	28	28	28	25	22	20
DD	56	44	32	15	9	4	4	3	4	10	26	48
Orlando					Latitude		28.6					
$\overline{H}$	11.35	14.12	17.96	21.55	22.58	20.79	20.45	18.99	16.99	14.81	12.44	10.52
K_T	.51	.53	.56	.58	.56	.51	.51	.50	.50	.53	.54	.51
$\overline{T}$	16	16	19	22	25	27	27	28	27	24	19	16
DD	120	92	62	25	10	4	3	3	4	15	53	106
Tallahassee					Latitude		30.4					
$\overline{H}$	9.95	12.91	16.80	20.70	21.98	21.38	19.84	19.01	16.95	14.96	11.45	9.22
K_T	.47	.50	.53	.56	.55	.52	.49	.50	.51	.55	.52	.47
$\overline{T}$	11	13	16	20	24	27	27	27	26	21	15	12
DD	223	166	109	33	7	1	1	1	2	26	124	212
Tampa					Latitude		28					
$\overline{H}$	11.47	14.30	18.09	21.66	22.68	20.97	19.90	18.76	16.94	15.29	12.57	10.62
K_T	.51	.53	.56	.58	.57	.51	.49	.49	.50	.54	.54	.50
$\overline{T}$	16	17	19	22	25	27	28	28	27	24	19	16
DD	121	90	63	23	9	3	3	3	4	15	54	108
West Palm Beach					Latitude		26.7					
$\overline{H}$	11.35	13.99	17.66	20.59	20.93	19.37	20.20	18.88	16.11	13.89	12.03	10.88
K_T	.49	.51	.54	.55	.52	.48	.50	.49	.47	.48	.50	.50
$\overline{T}$	19	19	21	23	25	27	28	28	28	25	22	19
DD	80	65	48	25	15	8	7	7	7	16	38	70

	Jan	Feb	Mar	Apr	May	Jun	Jul	Aug	Sep	Oct	Nov	Dec

Georgia

Atlanta — Latitude 33.7

	Jan	Feb	Mar	Apr	May	Jun	Jul	Aug	Sep	Oct	Nov	Dec
$\overline{H}$	8.14	11.00	14.80	19.14	21.05	21.73	20.57	19.39	16.14	13.62	10.02	7.66
K_T	.43	.46	.49	.53	.53	.52	.51	.52	.50	.53	.50	.43
T	6	7	11	16	21	24	26	25	22	17	11	6
DD	393	315	252	108	41	12	8	9	23	99	239	375

Augusta — Latitude 33.4

	Jan	Feb	Mar	Apr	May	Jun	Jul	Aug	Sep	Oct	Nov	Dec
$\overline{H}$	8.53	11.53	15.19	19.62	21.17	21.62	20.47	18.92	16.00	13.85	10.40	8.19
K_T	.44	.48	.50	.54	.53	.52	.50	.50	.49	.53	.51	.46
T	8	9	13	18	22	26	27	26	23	18	12	8
DD	337	266	201	84	30	9	6	8	19	87	207	329

Macon — Latitude 32.7

	Jan	Feb	Mar	Apr	May	Jun	Jul	Aug	Sep	Oct	Nov	Dec
$\overline{H}$	8.72	11.57	15.47	19.71	21.40	21.79	20.26	19.50	16.33	14.15	10.67	8.28
K_T	.44	.47	.51	.54	.53	.53	.50	.52	.50	.54	.51	.45
T	9	10	14	19	23	26	27	27	24	19	13	9
DD	310	241	181	73	28	10	8	8	19	78	192	300

Savannah — Latitude 32.1

	Jan	Feb	Mar	Apr	May	Jun	Jul	Aug	Sep	Oct	Nov	Dec
$\overline{H}$	9.02	11.85	15.88	19.99	21.02	20.93	20.24	18.40	15.48	13.81	10.69	8.55
K_T	.45	.48	.52	.55	.53	.51	.50	.49	.47	.52	.51	.46
T	10	11	14	19	23	26	27	27	25	20	14	10
DD	272	210	151	58	19	5	3	4	10	53	158	262

Hawaii

Hilo — Latitude 19.7

	Jan	Feb	Mar	Apr	May	Jun	Jul	Aug	Sep	Oct	Nov	Dec
$\overline{H}$	12.71	14.14	15.31	16.29	17.63	18.82	18.43	18.07	17.56	15.57	12.54	11.57
K_T	.47	.46	.44	.43	.45	.48	.47	.47	.49	.49	.45	.45
T	22	22	22	22	23	24	24	24	24	24	23	22
DD	16	13	16	12	9	6	5	5	5	6	8	14

Honolulu — Latitude 21.3

	Jan	Feb	Mar	Apr	May	Jun	Jul	Aug	Sep	Oct	Nov	Dec
$\overline{H}$	13.39	15.85	18.41	20.39	22.13	22.75	22.73	22.32	20.55	17.48	14.37	12.86
K_T	.51	.53	.54	.54	.56	.57	.58	.58	.58	.56	.53	.51
T	22	22	23	24	25	26	27	27	27	26	25	23
DD	15	12	13	8	5	2	2	1	1	3	5	11

Lihue — Latitude 22.0

	Jan	Feb	Mar	Apr	May	Jun	Jul	Aug	Sep	Oct	Nov	Dec
$\overline{H}$	12.52	14.75	16.75	18.63	20.71	21.21	21.14	20.64	19.78	16.45	13.10	11.95
K_T	.48	.50	.49	.49	.52	.53	.53	.54	.56	.53	.49	.48
T	22	22	22	23	24	25	26	26	26	25	24	23
DD	25	21	23	16	11	6	6	5	5	7	11	20

Idaho

Boise — Latitude 43.6

	Jan	Feb	Mar	Apr	May	Jun	Jul	Aug	Sep	Oct	Nov	Dec
$\overline{H}$	5.51	9.53	14.80	20.74	25.84	27.96	29.66	24.93	19.72	12.91	7.13	4.96
K_T	.42	.52	.57	.62	.66	.67	.73	.70	.69	.63	.50	.43
T	−2	2	5	9	14	18	24	22	17	11	4	0
DD	621	460	411	274	153	62	11	18	78	231	422	564

Lewiston — Latitude 46.4

	Jan	Feb	Mar	Apr	May	Jun	Jul	Aug	Sep	Oct	Nov	Dec
$\overline{H}$	3.85	6.92	11.57	16.29	20.91	22.88	26.51	21.92	16.29	9.76	4.69	3.25
K_T	.34	.41	.48	.50	.54	.55	.66	.62	.60	.51	.37	.33
T	−0	3	6	10	15	18	23	22	17	11	5	2
DD	583	423	390	264	168	89	35	45	104	254	416	523

Pocatello — Latitude 42.9

	Jan	Feb	Mar	Apr	May	Jun	Jul	Aug	Sep	Oct	Nov	Dec
$\overline{H}$	6.12	10.01	15.57	20.66	25.89	28.15	29.51	25.42	20.08	13.65	7.81	5.42
K_T	.46	.53	.60	.61	.66	.67	.73	.71	.70	.65	.53	.45
T	−5	−1	2	7	12	17	22	21	15	9	2	−3
DD	719	552	511	334	204	100	29	40	129	295	488	654

	Jan	Feb	Mar	Apr	May	Jun	Jul	Aug	Sep	Oct	Nov	Dec
Illinois												
Chicago				Latitude			41.8					
$\overline{H}$	5.76	8.62	12.56	16.56	20.31	22.79	22.07	19.51	15.37	11.00	6.42	4.55
$\overline{K}_T$	.41	.44	.47	.49	.51	.55	.54	.54	.53	.51	.42	.36
$\overline{T}$	−4	−3	3	10	16	21	24	23	19	13	5	−2
DD	700	586	486	264	127	32	17	20	64	189	411	627
Moline				Latitude			41.5					
$\overline{H}$	6.08	9.21	12.70	16.57	19.91	22.35	22.00	19.47	15.40	11.30	6.75	4.92
$\overline{K}_T$	.42	.47	.47	.48	.50	.53	.54	.54	.52	.52	.43	.38
$\overline{T}$	−6	−3	2	10	16	22	24	23	18	12	4	−3
DD	747	611	505	255	116	33	19	25	77	207	433	661
Springfield				Latitude			39.8					
$\overline{H}$	6.63	9.77	12.97	17.20	21.17	23.80	23.36	20.50	16.50	12.13	7.68	5.57
$\overline{K}_T$	.43	.47	.47	.50	.53	.57	.57	.56	.55	.54	.46	.40
$\overline{T}$	−3	−1	4	12	17	23	25	24	20	14	6	−1
DD	657	538	443	212	85	18	10	14	47	167	387	592
Indiana												
Evansville				Latitude			38.1					
$\overline{H}$	6.52	9.35	13.06	17.04	20.24	22.51	21.80	19.70	15.92	12.34	7.75	5.66
$\overline{K}_T$	.40	.43	.46	.49	.51	.54	.54	.53	.52	.52	.44	.38
$\overline{T}$	0	2	7	14	19	24	25	25	21	15	7	2
DD	560	451	362	159	60	11	6	9	33	143	337	513
Fort Wayne				Latitude			41					
$\overline{H}$	5.17	7.92	11.14	15.45	18.98	20.91	20.29	18.09	14.46	10.50	5.86	4.19
$\overline{K}_T$	.35	.40	.41	.45	.48	.50	.50	.50	.49	.48	.37	.32
$\overline{T}$	−4	−2	3	10	15	21	23	22	18	12	5	−2
DD	683	580	492	271	134	38	22	30	74	215	414	627
Indianapolis				Latitude			39.7					
$\overline{H}$	5.62	8.47	11.78	15.88	19.16	21.21	20.50	18.65	15.03	11.09	6.58	4.73
$\overline{K}_T$	.36	.41	.43	.46	.48	.51	.50	.51	.50	.49	.40	.34
$\overline{T}$	−2	−1	4	11	17	22	24	23	19	13	5	−1
DD	639	533	439	227	103	26	16	22	59	184	392	588
South Bend				Latitude			41.7					
$\overline{H}$	4.71	7.49	11.27	15.75	19.55	21.82	21.02	18.91	14.66	10.32	5.65	3.86
$\overline{K}_T$	.33	.38	.42	.46	.49	.52	.52	.52	.50	.48	.37	.30
$\overline{T}$	−4	−3	2	9	15	20	22	22	18	12	4	−2
DD	704	603	514	290	144	41	23	28	77	216	425	633
Iowa												
Burlington				Latitude			40.8					
$\overline{H}$	6.58	9.75	13.22	17.46	21.29	24.08	23.67	20.75	16.08	12.04	7.53	5.45
$\overline{K}_T$	.45	.49	.49	.51	.54	.58	.58	.57	.54	.54	.47	.41
$\overline{T}$	−5	−3	3	11	17	22	24	23	19	13	4	−2
DD	726	586	485	239	102	25	12	17	62	189	423	642
Des Moines				Latitude			41.5					
$\overline{H}$	6.59	9.77	13.40	17.67	21.19	24.11	23.80	20.75	16.28	12.12	7.47	5.53
$\overline{K}_T$	.46	.50	.50	.52	.54	.58	.59	.57	.55	.56	.48	.43
$\overline{T}$	−7	−4	1	10	16	21	24	23	18	12	3	−4
DD	785	633	536	273	125	41	22	30	87	212	457	689
Mason City				Latitude			43.2					
$\overline{H}$	6.28	9.50	13.26	17.24	21.51	23.99	23.66	20.81	15.96	11.47	6.82	5.03
$\overline{K}_T$	.47	.51	.51	.51	.55	.57	.58	.58	.56	.55	.47	.43
$\overline{T}$	−10	−8	−2	8	14	20	22	21	16	10	1	−7
DD	874	722	622	329	165	57	35	41	123	265	525	773

	Jan	Feb	Mar	Apr	May	Jun	Jul	Aug	Sep	Oct	Nov	Dec
	Sioux City				Latitude		42.4					
H	6.45	9.55	13.29	17.91	21.58	24.10	24.09	20.95	16.13	11.78	7.29	5.33
K_T	.47	.50	.50	.53	.55	.58	.59	.58	.56	.55	.49	.43
T	−8	−5	1	10	16	21	24	23	17	12	2	−5
DD	809	647	549	274	127	43	22	29	98	231	482	715

Kansas

	Dodge City				Latitude		37.8					
H	9.38	12.73	16.77	21.41	23.72	26.77	26.06	23.33	19.15	14.77	10.14	8.30
K_T	.57	.59	.59	.61	.60	.64	.64	.63	.62	.62	.57	.55
T	−1	2	5	12	18	23	26	26	21	14	6	1
DD	590	463	412	198	77	14	4	6	35	149	372	544

	Goodland				Latitude		39.4					
H	8.96	11.98	16.16	20.76	23.40	26.75	26.33	23.22	18.65	14.39	9.72	7.88
K_T	.57	.57	.58	.60	.59	.64	.65	.63	.62	.63	.58	.56
T	−2	−0	2	9	15	21	24	23	18	12	4	−1
DD	642	521	495	280	141	38	13	17	76	224	443	603

	Topeka				Latitude		39.1					
H	7.74	10.68	14.27	18.64	21.74	24.14	24.16	21.68	17.22	13.02	8.76	6.62
K_T	.49	.51	.51	.53	.55	.58	.59	.59	.57	.57	.52	.46
T	−2	1	5	13	18	23	26	25	20	14	6	−0
DD	637	492	414	196	79	20	9	11	47	160	371	572

	Wichita				Latitude		37.7					
H	8.89	12.02	15.96	20.23	23.11	25.70	25.41	23.06	18.34	14.19	9.88	7.84
K_T	.53	.55	.56	.57	.58	.62	.62	.62	.59	.60	.55	.52
T	−0	2	6	14	19	24	27	26	21	15	7	1
DD	584	451	384	182	88	27	14	16	52	155	350	531

Kentucky

	Lexington				Latitude		38					
H	6.20	8.85	12.48	16.79	19.83	21.54	21.00	19.13	15.46	11.86	7.46	5.51
K_T	.38	.41	.44	.48	.50	.52	.52	.52	.50	.50	.42	.37
T	1	2	6	13	18	23	25	24	20	14	7	2
DD	553	464	374	183	73	19	11	14	41	155	344	511

	Louisville				Latitude		38.2					
H	6.19	8.96	12.51	16.65	19.53	21.60	20.85	19.07	15.45	11.84	7.41	5.54
K_T	.38	.42	.44	.47	.49	.52	.51	.52	.50	.50	.42	.37
T	1	2	7	13	18	23	25	24	21	15	7	2
DD	549	457	370	184	89	30	18	21	51	163	343	510

Louisiana

	Baton Rouge				Latitude		30.5					
H	8.92	11.96	15.66	19.08	21.24	21.87	19.82	19.04	16.62	14.77	10.45	8.36
K_T	.42	.47	.50	.52	.53	.53	.49	.50	.50	.54	.47	.43
T	11	12	15	20	24	27	28	28	25	20	15	12
DD	243	176	110	23	4	0	0	0	1	24	121	213

	Lake Charles				Latitude		30.1					
H	8.27	11.46	14.91	17.82	20.99	22.36	20.30	18.81	16.86	15.67	10.40	8.01
K_T	.39	.44	.47	.48	.52	.54	.50	.50	.51	.57	.47	.40
T	11	13	16	21	24	27	28	28	26	21	16	12
DD	234	172	122	40	14	4	3	3	7	35	117	204

	Jan	Feb	Mar	Apr	May	Jun	Jul	Aug	Sep	Oct	Nov	Dec
	New Orleans				Latitude		30					
$\overline{H}$	9.47	12.62	16.06	20.21	22.33	22.75	20.58	19.49	17.19	15.15	11.04	8.85
$\overline{K_T}$	.45	.49	.51	.55	.56	.55	.51	.51	.51	.55	.49	.44
$\overline{T}$	12	13	16	20	24	27	28	28	26	21	16	13
DD	229	169	126	49	20	7	5	5	10	43	126	200
	Shreveport				Latitude		32.5					
$\overline{H}$	8.66	11.79	15.23	18.31	21.41	23.44	22.86	21.31	17.64	14.80	10.54	8.29
$\overline{K_T}$	.44	.48	.50	.50	.53	.57	.56	.57	.54	.56	.50	.45
$\overline{T}$	8	10	14	19	23	27	28	28	25	20	13	10
DD	317	236	170	60	22	5	3	3	10	54	173	282

Maine

	Jan	Feb	Mar	Apr	May	Jun	Jul	Aug	Sep	Oct	Nov	Dec
	Caribou				Latitude		46.9					
$\overline{H}$	4.76	8.22	12.86	16.05	17.91	19.95	20.00	17.04	12.52	7.81	4.16	3.52
$\overline{K_T}$	.43	.50	.53	.49	.46	.48	.50	.49	.46	.42	.34	.37
$\overline{T}$	−12	−11	−5	3	10	15	18	17	12	7	−0	−9
DD	932	809	714	473	277	130	77	104	201	369	559	839
	Portland				Latitude		43.7					
$\overline{H}$	5.11	7.74	11.01	14.80	17.79	19.43	18.83	16.58	13.14	9.34	5.21	4.12
$\overline{K_T}$	.39	.42	.43	.44	.45	.47	.46	.46	.46	.46	.37	.36
$\overline{T}$	−6	−5	−0	6	12	17	20	19	15	9	4	−3
DD	747	656	573	379	233	104	57	69	146	288	442	677

Maryland

	Jan	Feb	Mar	Apr	May	Jun	Jul	Aug	Sep	Oct	Nov	Dec
	Baltimore				Latitude		39.2					
$\overline{H}$	6.67	9.53	13.20	16.89	19.46	21.33	20.70	18.16	15.10	11.32	7.50	5.67
$\overline{K_T}$	.42	.45	.47	.48	.49	.51	.51	.49	.50	.49	.44	.40
$\overline{T}$	1	2	6	12	18	22	25	24	20	14	8	2
DD	548	471	392	215	102	36	21	27	58	174	329	517

Massachusetts

	Jan	Feb	Mar	Apr	May	Jun	Jul	Aug	Sep	Oct	Nov	Dec
	Boston				Latitude		42.4					
$\overline{H}$	5.40	8.05	11.54	15.05	18.40	20.63	19.85	16.88	14.30	10.10	5.71	4.58
$\overline{K_T}$	.39	.42	.44	.44	.47	.49	.49	.47	.49	.48	.38	.37
$\overline{T}$	−2	−1	3	9	15	20	23	22	18	13	7	1
DD	618	538	464	279	139	40	17	24	67	184	335	549

Michigan

	Jan	Feb	Mar	Apr	May	Jun	Jul	Aug	Sep	Oct	Nov	Dec
	Alpena				Latitude		45.1					
$\overline{H}$	4.11	7.00	11.68	15.97	19.53	21.33	21.39	17.97	13.12	8.44	4.34	3.07
$\overline{K_T}$	.34	.40	.47	.48	.50	.51	.53	.51	.47	.43	.32	.29
$\overline{T}$	−8	−8	−3	4	10	16	19	18	14	9	2	−5
DD	812	725	668	417	261	109	67	79	168	312	502	716
	Detroit				Latitude		42.4					
$\overline{H}$	4.74	7.72	11.36	15.88	19.48	21.18	20.83	17.89	14.22	9.95	5.42	3.90
$\overline{K_T}$	.34	.40	.43	.47	.49	.51	.51	.50	.49	.47	.36	.32
$\overline{T}$	−4	−3	2	9	15	21	23	22	18	12	5	−1
DD	680	591	511	293	149	42	23	29	76	206	400	608
	Flint				Latitude		43					
$\overline{H}$	4.35	7.22	10.86	15.20	18.82	20.58	20.40	17.65	13.57	9.40	4.87	3.51
$\overline{K_T}$	.33	.38	.42	.45	.48	.49	.50	.49	.47	.45	.33	.29
$\overline{T}$	−5	−5	0	8	13	19	21	20	16	11	3	−3
DD	734	641	559	322	176	54	30	38	102	245	445	657

	Jan	Feb	Mar	Apr	May	Jun	Jul	Aug	Sep	Oct	Nov	Dec
	Grand Rapids				Latitude		42.9					
$\overline{H}$	4.19	7.36	11.52	16.03	19.92	22.21	21.73	19.03	14.32	9.73	5.05	3.52
K_T	.31	.39	.44	.47	.51	.53	.54	.53	.50	.46	.34	.29
T	−5	−4	1	8	14	20	22	21	17	11	4	−3
DD	719	631	550	313	165	49	28	35	94	238	440	649
	Sault Ste. Marie				Latitude		46.5					
$\overline{H}$	3.69	6.85	11.68	15.70	19.16	20.56	20.83	17.29	11.90	7.64	3.77	2.87
K_T	.33	.41	.48	.48	.49	.49	.52	.49	.44	.40	.30	.29
T	−10	−9	−4	3	9	15	18	17	13	8	0	−7
DD	874	772	705	453	295	153	100	108	196	337	541	773
	Traverse City				Latitude		44.7					
$\overline{H}$	3.53	6.44	11.36	15.95	19.63	21.71	21.68	18.26	13.23	8.55	4.28	2.92
K_T	.29	.36	.45	.48	.50	.52	.54	.51	.47	.43	.31	.27
T	−6	−6	−2	6	12	18	20	20	15	10	3	−3
DD	761	688	624	378	227	84	46	55	131	274	470	673

Minnesota

	Jan	Feb	Mar	Apr	May	Jun	Jul	Aug	Sep	Oct	Nov	Dec
	Duluth				Latitude		46.8					
$\overline{H}$	4.41	7.63	11.74	15.58	18.65	20.06	21.05	17.56	12.43	8.22	4.33	3.32
K_T	.40	.46	.49	.48	.48	.48	.52	.50	.46	.44	.35	.34
T	−13	−11	−5	4	10	15	19	18	12	7	−2	−10
DD	974	823	714	440	278	134	67	83	197	346	610	870
	International Falls				Latitude		48.6					
$\overline{H}$	4.03	7.52	11.87	16.39	19.48	21.04	21.81	18.37	12.73	7.99	3.92	3.09
K_T	.41	.49	.51	.51	.50	.51	.54	.53	.48	.45	.35	.36
T	−17	−14	−6	3	10	16	19	17	12	6	−4	−13
DD	1085	901	762	451	269	121	71	96	217	376	666	967
	Minneapolis				Latitude		44.9					
$\overline{H}$	5.27	8.67	12.53	16.37	19.72	21.88	22.36	19.15	14.24	9.76	5.45	4.01
K_T	.43	.49	.50	.49	.50	.52	.55	.54	.51	.49	.40	.37
T	−11	−9	−2	7	14	19	22	21	16	10	0	−7
DD	908	753	634	338	167	56	28	37	122	271	544	797
	Rochester				Latitude		43.9					
$\overline{H}$	5.42	8.54	12.28	16.00	19.25	21.59	21.67	18.87	14.19	9.87	5.61	4.20
K_T	.42	.47	.48	.48	.49	.52	.53	.53	.50	.48	.40	.37
T	−11	−8	−2	7	13	19	21	20	15	10	0	−7
DD	896	747	639	350	183	67	40	51	134	279	543	793

Mississippi

	Jan	Feb	Mar	Apr	May	Jun	Jul	Aug	Sep	Oct	Nov	Dec
	Jackson				Latitude		32.3					
$\overline{H}$	8.55	11.65	15.54	19.39	22.04	22.98	21.67	20.21	17.13	14.44	10.23	8.04
K_T	.43	.47	.51	.53	.55	.56	.53	.54	.52	.55	.49	.43
T	8	10	13	19	23	26	28	27	24	19	13	9
DD	312	240	169	51	14	2	1	1	6	53	176	282
	Meridian				Latitude		32.3					
$\overline{H}$	8.45	11.49	15.07	18.87	21.12	22.29	20.70	19.74	16.50	14.28	10.18	7.94
K_T	.42	.46	.49	.52	.53	.54	.51	.52	.50	.54	.48	.43
T	8	10	13	19	22	26	27	27	24	18	12	9
DD	318	243	176	62	22	5	3	3	11	71	198	303

Missouri

	Jan	Feb	Mar	Apr	May	Jun	Jul	Aug	Sep	Oct	Nov	Dec
	Columbia				Latitude		38.8					
$\overline{H}$	6.94	9.93	13.38	17.32	21.34	23.72	24.02	21.32	16.47	12.49	7.97	5.93
K_T	.43	.47	.48	.50	.54	.57	.59	.58	.54	.54	.46	.41
T	−2	1	5	13	18	23	25	24	20	14	7	0
DD	615	488	404	185	75	18	8	11	40	151	355	556

	Jan	Feb	Mar	Apr	May	Jun	Jul	Aug	Sep	Oct	Nov	Dec
	Kansas city				Latitude		39.3					
$\overline{H}$	7.36	10.15	13.65	17.88	21.25	23.60	23.86	21.14	16.48	12.40	8.37	6.37
K_T	.47	.49	.49	.51	.53	.57	.59	.58	.54	.54	.50	.45
$\overline{T}$	−3	0	5	12	18	23	25	25	20	14	6	−0
DD	652	507	423	199	82	20	9	11	45	157	382	581
	St. Louis				Latitude		38.8					
$\overline{H}$	7.12	10.05	13.68	17.75	21.24	23.75	23.26	20.62	16.56	12.48	8.16	6.02
K_T	.45	.47	.49	.51	.53	.57	.57	.56	.54	.54	.47	.41
$\overline{T}$	−0	2	6	14	19	24	26	25	21	15	7	1
DD	581	466	378	167	67	15	8	10	37	137	339	526
	Springfield				Latitude		37.2					
$\overline{H}$	7.76	10.51	14.02	18.21	21.37	23.56	23.42	21.26	16.81	12.98	8.80	6.84
K_T	.46	.48	.49	.52	.54	.57	.58	.57	.54	.54	.49	.44
$\overline{T}$	1	3	7	14	18	23	25	25	21	15	8	2
DD	553	436	364	164	69	16	7	8	35	136	330	500

Montana

	Jan	Feb	Mar	Apr	May	Jun	Jul	Aug	Sep	Oct	Nov	Dec
	Billings				Latitude		45.8					
$\overline{H}$	5.52	8.67	13.51	17.32	21.72	24.67	27.06	22.96	16.69	11.20	6.37	4.78
K_T	.47	.51	.55	.53	.56	.59	.67	.65	.60	.58	.49	.47
$\overline{T}$	−6	−3	0	7	13	17	22	21	15	10	2	−3
DD	741	585	559	341	190	75	13	19	122	274	486	657
	Cut Bank				Latitude		48.6					
$\overline{H}$	4.57	7.80	12.80	16.86	21.37	23.22	25.97	21.52	15.35	9.89	5.45	3.79
K_T	.46	.51	.55	.53	.55	.56	.65	.62	.58	.56	.49	.44
$\overline{T}$	−9	−5	−3	4	10	14	18	17	12	7	−1	−6
DD	840	661	658	428	282	174	89	107	219	369	589	751
	Dillon				Latitude		45.3					
$\overline{H}$	5.97	9.61	14.52	18.61	22.58	24.33	27.16	22.97	17.26	11.62	6.84	5.11
K_T	.50	.55	.58	.56	.58	.58	.67	.65	.62	.59	.52	.49
$\overline{T}$	−7	−4	−1	5	10	14	19	18	13	7	−0	−5
DD	772	614	608	399	263	151	60	76	191	350	553	707
	Glasgow				Latitude		48.2					
$\overline{H}$	4.41	7.62	12.54	16.89	20.75	23.24	24.90	21.15	15.22	9.96	5.43	3.79
K_T	.43	.49	.54	.53	.54	.56	.62	.61	.57	.55	.47	.43
$\overline{T}$	−13	−9	−4	6	12	17	21	21	14	8	−2	−8
DD	962	772	686	377	216	110	45	54	167	333	603	825
	Great Falls				Latitude		47.5					
$\overline{H}$	4.77	8.18	13.29	16.90	20.97	23.85	26.44	21.94	15.65	10.50	5.65	3.82
K_T	.45	.51	.56	.52	.54	.57	.66	.63	.58	.57	.47	.41
$\overline{T}$	−6	−3	−1	6	12	16	21	20	14	9	1	−3
DD	765	596	592	365	218	109	36	48	151	293	509	664
	Helena				Latitude		46.6					
$\overline{H}$	4.76	8.04	13.01	16.88	21.12	23.16	26.49	21.91	16.04	10.52	5.92	4.13
K_T	.43	.48	.54	.52	.54	.56	.66	.62	.59	.56	.48	.42
$\overline{T}$	−8	−4	−1	6	11	15	20	19	13	7	−0	−5
DD	807	617	592	376	232	126	44	56	174	343	556	715
	Lewistown				Latitude		47.1					
$\overline{H}$	4.77	7.86	12.81	16.39	20.51	23.38	25.97	21.58	15.57	10.27	5.70	4.12
K_T	.44	.48	.53	.50	.53	.56	.64	.62	.58	.55	.47	.44
$\overline{T}$	−7	−5	−3	4	10	14	19	18	12	8	0	−4
DD	792	642	646	420	281	171	79	89	208	346	548	699

	Jan	Feb	Mar	Apr	May	Jun	Jul	Aug	Sep	Oct	Nov	Dec
	Miles City				Latitude		46.4					
H	5.19	8.46	13.45	17.50	21.52	24.36	26.02	22.44	16.39	10.90	6.26	4.53
K_T	.46	.50	.55	.54	.55	.58	.65	.64	.60	.57	.50	.46
T	−9	−6	−1	7	14	18	24	23	16	9	0	−6
DD	854	675	602	341	192	90	31	40	140	300	546	743
	Missoula				Latitude		46.9					
H	3.54	6.52	11.14	15.69	20.23	21.94	26.42	21.35	15.41	9.22	4.66	3.03
K_T	.32	.40	.46	.48	.52	.53	.66	.61	.57	.49	.38	.32
T	−6	−3	1	7	11	15	19	18	13	7	0	−4
DD	760	588	545	353	230	127	48	62	177	362	543	695

Nebraska

	Jan	Feb	Mar	Apr	May	Jun	Jul	Aug	Sep	Oct	Nov	Dec
	Grand Island				Latitude		41					
H	7.51	10.40	14.36	19.21	22.39	25.45	25.15	22.01	17.13	12.91	8.38	6.46
K_T	.51	.52	.53	.56	.57	.61	.62	.61	.58	.59	.53	.49
T	−5	−2	2	10	16	21	25	24	18	12	3	−3
DD	735	580	511	263	119	30	12	15	75	211	450	654
	North Omaha				Latitude		41.4					
H	7.20	10.13	13.88	17.69	21.25	24.09	23.91	21.09	15.58	11.92	7.31	5.80
K_T	.50	.51	.52	.52	.54	.58	.59	.58	.53	.55	.47	.45
T	−7	−4	1	10	16	21	24	23	18	12	3	−3
DD	772	614	526	260	115	33	15	19	75	204	453	676
	North Platte				Latitude		41.1					
H	7.86	10.88	15.13	19.57	22.56	25.73	25.85	22.58	17.76	13.36	8.62	6.87
K_T	.54	.55	.56	.57	.57	.62	.64	.62	.60	.61	.55	.52
T	−5	−2	1	9	15	20	24	23	17	11	2	−3
DD	716	575	529	294	147	44	16	20	95	251	482	657
	Scottsbluff				Latitude		41.9					
H	7.67	10.79	14.85	18.93	21.94	25.39	25.92	22.69	18.15	12.99	8.21	6.53
K_T	.55	.56	.56	.56	.56	.61	.64	.63	.62	.60	.54	.52
T	−4	−1	1	8	14	19	23	22	16	10	2	−2
DD	689	552	531	323	180	71	25	35	116	271	484	643

Nevada

	Jan	Feb	Mar	Apr	May	Jun	Jul	Aug	Sep	Oct	Nov	Dec
	Elko				Latitude		40.8					
H	7.81	11.74	16.61	21.57	26.15	28.76	29.77	26.28	21.48	15.02	9.22	7.01
K_T	.53	.59	.61	.63	.66	.69	.73	.72	.72	.68	.58	.53
T	−5	−2	2	6	11	15	21	19	14	8	2	−3
DD	719	557	515	360	232	117	28	44	142	315	501	672
	Ely				Latitude		39.3					
H	9.30	12.95	18.23	22.81	26.23	28.52	27.78	25.32	21.97	15.98	10.52	8.20
K_T	.59	.62	.66	.66	.66	.68	.68	.69	.73	.70	.62	.58
T	−5	−2	0	5	10	14	20	19	14	8	1	−3
DD	714	576	557	396	269	150	53	67	163	334	517	668
	Las Vegas				Latitude		36.1					
H	11.10	15.21	20.70	26.33	30.04	31.53	29.38	26.73	23.13	17.48	12.32	10.00
K_T	.63	.67	.71	.74	.75	.76	.72	.72	.74	.71	.66	.62
T	7	9	13	18	23	28	32	31	27	20	12	7
DD	371	263	211	99	37	9	2	4	13	76	226	357
	Lovelock				Latitude		40.1					
H	9.12	13.23	18.80	24.58	29.00	31.21	31.60	28.19	23.01	16.47	10.55	8.11
K_T	.60	.65	.69	.71	.73	.75	.78	.77	.77	.73	.64	.59
T	−2	2	4	9	14	19	24	22	17	11	4	−1
DD	623	465	434	286	165	72	23	37	99	255	445	592

	Jan	Feb	Mar	Apr	May	Jun	Jul	Aug	Sep	Oct	Nov	Dec
Reno					Latitude		39.5					
$\overline{H}$	9.09	13.05	18.72	24.51	28.65	30.67	30.55	27.31	22.67	16.24	10.36	8.01
K_T	.59	.63	.68	.71	.72	.73	.75	.74	.75	.71	.62	.57
$\overline{T}$	–0	3	5	8	13	16	21	19	16	10	4	1
DD	572	436	428	309	196	102	37	54	115	262	417	550
Tonopah					Latitude		38.1					
$\overline{H}$	10.42	14.46	20.17	25.55	29.26	31.64	30.68	27.67	23.18	17.26	11.70	9.38
K_T	.64	.67	.71	.73	.73	.76	.75	.75	.76	.73	.66	.63
$\overline{T}$	–1	1	4	9	14	19	23	21	18	11	4	–0
DD	600	475	441	293	172	71	25	36	88	238	425	573
Winnemucca					Latitude		40.9					
$\overline{H}$	7.84	11.67	16.71	22.33	26.81	29.17	30.40	26.66	21.65	15.00	9.19	7.02
K_T	.53	.58	.62	.65	.68	.70	.75	.73	.73	.68	.58	.53
$\overline{T}$	–2	1	3	7	12	17	22	20	15	9	3	–1
DD	634	480	474	338	214	105	33	53	134	296	465	596

New Hampshire

	Jan	Feb	Mar	Apr	May	Jun	Jul	Aug	Sep	Oct	Nov	Dec
Concord					Latitude		43.2					
$\overline{H}$	5.21	7.79	11.05	14.95	17.96	19.35	19.01	16.52	12.94	9.28	5.25	4.11
K_T	.39	.42	.43	.44	.46	.46	.47	.46	.45	.45	.36	.35
$\overline{T}$	–6	–5	0	7	13	18	21	20	15	10	3	–4
DD	762	659	562	352	196	74	41	57	130	282	454	692

New Jersey

	Jan	Feb	Mar	Apr	May	Jun	Jul	Aug	Sep	Oct	Nov	Dec
Newark					Latitude		40.7					
$\overline{H}$	6.26	9.01	12.59	16.45	19.15	20.38	19.98	17.76	14.45	10.79	6.77	5.16
K_T	.42	.45	.46	.48	.48	.49	.49	.49	.49	.49	.42	.39
$\overline{T}$	–0	0	5	11	17	22	25	24	20	14	8	1
DD	579	507	428	245	120	40	21	27	62	171	326	529

New Mexico

	Jan	Feb	Mar	Apr	May	Jun	Jul	Aug	Sep	Oct	Nov	Dec
Albuquerque					Latitude		35.1					
$\overline{H}$	11.54	15.23	20.06	25.30	28.82	30.41	28.25	26.00	22.39	17.56	12.87	10.53
K_T	.63	.66	.68	.71	.72	.73	.69	.70	.70	.70	.66	.63
$\overline{T}$	2	4	8	13	19	24	26	25	21	15	7	2
DD	513	391	333	171	63	11	5	8	27	142	346	498
Clayton					Latitude		36.5					
$\overline{H}$	10.92	14.08	18.75	23.15	25.23	27.44	25.92	23.81	20.46	16.28	11.67	9.77
K_T	.63	.63	.65	.65	.63	.66	.64	.64	.65	.67	.63	.61
$\overline{T}$	1	2	5	10	16	21	23	22	18	13	6	2
DD	550	450	425	249	125	37	19	24	69	195	382	516
Farmington					Latitude		36.8					
$\overline{H}$	10.72	14.54	19.22	24.22	27.83	30.26	28.13	25.57	21.96	16.79	11.89	9.51
K_T	.62	.65	.67	.68	.70	.73	.69	.69	.70	.69	.65	.60
$\overline{T}$	–2	2	5	10	15	20	24	23	18	12	4	–1
DD	626	464	421	261	120	33	7	12	58	217	430	602
Roswell					Latitude		33.4					
$\overline{H}$	11.88	15.58	20.51	25.17	27.92	29.63	27.70	25.44	21.72	17.33	12.85	10.80
K_T	.62	.64	.68	.70	.70	.71	.68	.68	.67	.67	.63	.60
$\overline{T}$	3	6	10	15	20	25	26	26	21	15	8	4
DD	464	343	277	117	36	5	3	5	24	124	305	442
Truth or Consequences					Latitude		33.2					
$\overline{H}$	12.69	16.48	21.41	26.53	29.02	30.08	26.84	25.16	22.02	17.92	13.81	11.38
K_T	.66	.68	.71	.73	.73	.73	.66	.67	.68	.69	.67	.63
$\overline{T}$	4	7	10	15	20	25	26	25	22	16	9	5
DD	434	314	264	123	42	7	4	7	22	106	278	419

	Jan	Feb	Mar	Apr	May	Jun	Jul	Aug	Sep	Oct	Nov	Dec
	Tucumcari			Latitude			35.2					
$\overline{H}$	11.45	14.72	19.43	23.82	26.26	28.20	26.66	24.57	20.76	16.38	12.18	10.32
K_T	.63	.63	.66	.67	.66	.68	.66	.66	.65	.65	.63	.62
$\overline{T}$	3	5	8	14	19	24	26	25	21	15	8	4
DD	486	374	325	170	77	20	12	17	45	154	323	457
	Zuni			Latitude			35.1					
$\overline{H}$	11.20	14.72	19.15	24.60	28.07	29.53	25.70	23.59	21.51	16.98	12.36	10.13
K_T	.61	.63	.65	.69	.70	.71	.63	.63	.68	.68	.64	.60
$\overline{T}$	−1	1	4	9	14	19	22	21	17	11	4	0
DD	597	476	442	295	176	73	35	47	93	236	419	570

New York

	Jan	Feb	Mar	Apr	May	Jun	Jul	Aug	Sep	Oct	Nov	Dec
	Albany			Latitude			42.8					
$\overline{H}$	5.18	7.81	11.19	15.15	17.82	19.64	19.58	17.01	13.29	9.28	5.19	4.04
K_T	.38	.41	.43	.45	.45	.47	.48	.47	.46	.44	.35	.34
$\overline{T}$	−6	−5	1	8	14	20	22	21	17	11	4	−3
DD	746	645	544	307	152	44	22	33	95	245	425	673
	Binghamton			Latitude			42.2					
$\overline{H}$	4.38	6.53	9.78	14.10	16.98	19.08	18.83	16.17	12.84	8.85	4.70	3.37
K_T	.32	.34	.37	.41	.43	.46	.46	.45	.44	.41	.31	.27
$\overline{T}$	−6	−5	−0	7	13	18	21	20	16	10	3	−4
DD	741	655	579	338	183	55	26	38	104	256	448	682
	Buffalo			Latitude			42.9					
$\overline{H}$	3.96	6.20	10.09	14.93	18.13	20.48	20.16	17.17	13.07	8.91	4.58	3.21
K_T	.30	.33	.39	.44	.46	.49	.50	.48	.45	.42	.31	.27
$\overline{T}$	−5	−4	0	7	13	19	21	20	16	11	4	−2
DD	711	631	565	338	193	62	33	44	103	247	424	639
	Massena			Latitude			44.9					
$\overline{H}$	4.44	7.04	11.10	15.24	18.31	20.20	19.88	16.84	12.76	8.36	4.41	3.34
K_T	.36	.40	.44	.46	.47	.48	.49	.47	.46	.42	.33	.31
$\overline{T}$	−10	−9	−2	6	12	18	21	19	15	9	2	−7
DD	869	750	642	382	207	76	41	58	131	292	484	772
	New York (Central Park)			Latitude			40.8					
$\overline{H}$	5.68	8.19	11.77	15.48	18.57	19.41	19.16	16.83	13.78	10.17	6.05	4.59
K_T	.39	.41	.43	.45	.47	.46	.47	.46	.46	.46	.38	.35
$\overline{T}$	0	1	5	11	17	22	25	24	20	15	9	2
DD	568	493	418	236	116	38	20	26	58	158	306	514
	New York (La Guardia)			Latitude			40.8					
$\overline{H}$	6.21	9.02	12.69	16.54	19.18	20.46	20.25	17.97	14.53	10.79	6.74	5.19
K_T	.42	.45	.47	.48	.48	.49	.50	.49	.49	.49	.42	.39
$\overline{T}$	0	1	5	11	17	22	25	24	20	15	9	2
DD	567	497	426	241	114	34	16	22	53	159	306	509
	Rochester			Latitude			43.1					
$\overline{H}$	4.13	6.35	10.26	15.20	18.23	20.63	20.22	17.24	13.16	8.87	4.59	3.19
K_T	.31	.34	.39	.45	.46	.49	.50	.48	.46	.43	.31	.27
$\overline{T}$	−4	−4	1	8	14	19	22	21	17	11	5	−2
DD	704	624	550	322	170	51	27	37	93	232	410	633
	Syracuse			Latitude			43.1					
$\overline{H}$	4.37	6.49	10.11	15.03	17.91	20.18	19.96	17.07	13.23	8.83	4.52	3.24
K_T	.33	.35	.39	.45	.45	.48	.49	.48	.46	.42	.31	.27
$\overline{T}$	−5	−4	1	8	14	19	22	21	17	11	5	−2
DD	714	627	545	309	159	41	18	27	78	224	401	636

	Jan	Feb	Mar	Apr	May	Jun	Jul	Aug	Sep	Oct	Nov	Dec
North Carolina												
Asheville					Latitude		35.4					
$\bar{H}$	8.19	11.03	14.82	18.93	20.48	21.05	20.16	18.47	15.45	13.03	9.63	7.46
$\bar{K}_T$	.45	.48	.50	.53	.51	.51	.50	.49	.49	.52	.50	.45
$\bar{T}$	3	4	8	13	18	21	23	23	19	14	8	4
DD	470	401	338	178	91	36	24	26	61	173	321	456
Cape Hatteras					Latitude		35.3					
$\bar{H}$	7.78	10.81	15.06	20.14	22.27	23.11	21.80	19.35	16.70	12.90	9.90	7.47
$\bar{K}_T$	.43	.47	.51	.56	.56	.56	.54	.52	.53	.52	.51	.45
$\bar{T}$	7	8	10	15	19	24	26	25	23	18	13	9
DD	342	299	257	127	45	10	4	5	11	60	167	302
Charlotte					Latitude		35.2					
$\bar{H}$	8.17	11.02	14.96	19.24	21.06	21.81	20.79	19.24	16.07	13.32	9.83	7.63
$\bar{K}_T$	.45	.48	.51	.54	.53	.52	.51	.52	.51	.53	.51	.46
$\bar{T}$	6	7	10	16	20	24	26	25	22	17	11	6
DD	399	329	263	114	46	13	8	10	26	109	245	394
Greensboro					Latitude		36.1					
$\bar{H}$	8.12	11.01	14.90	19.10	21.21	22.17	21.15	19.26	16.09	12.96	9.53	7.47
$\bar{K}_T$	.46	.48	.51	.54	.53	.53	.52	.52	.51	.53	.51	.46
$\bar{T}$	4	5	9	15	20	24	25	24	21	15	9	4
DD	453	380	301	130	44	9	5	7	27	127	281	438
Raleigh-Durham					Latitude		35.9					
$\bar{H}$	7.87	10.71	14.48	18.66	20.52	21.16	20.15	18.29	15.63	12.55	9.22	7.21
$\bar{K}_T$	.44	.47	.50	.52	.51	.51	.50	.49	.50	.51	.49	.44
$\bar{T}$	5	6	10	15	20	24	25	25	21	16	10	5
DD	424	355	280	126	50	14	8	10	29	121	259	413
North Dakota												
Bismarck					Latitude		46.8					
$\bar{H}$	5.30	8.80	13.26	16.56	20.98	23.38	24.78	21.31	15.38	10.30	5.76	4.24
$\bar{K}_T$	.48	.53	.55	.51	.54	.56	.61	.61	.57	.55	.47	.44
$\bar{T}$	−13	−10	−4	6	12	18	22	21	14	8	−2	−9
DD	977	800	685	371	205	81	32	41	153	321	602	850
Fargo					Latitude		46.9					
$\bar{H}$	4.71	8.01	12.46	16.75	20.83	22.64	24.07	20.72	14.80	9.92	5.19	3.83
$\bar{K}_T$	.43	.49	.52	.52	.53	.54	.60	.59	.55	.53	.42	.40
$\bar{T}$	−15	−12	−4	6	13	18	21	21	14	8	−2	−11
DD	1016	842	700	380	193	64	26	33	142	317	606	896
Minot					Latitude		48.3					
$\bar{H}$	4.36	7.45	11.86	16.58	20.96	22.42	23.81	20.43	14.49	9.64	4.98	3.52
$\bar{K}_T$	.43	.48	.51	.52	.54	.54	.59	.59	.55	.54	.44	.40
$\bar{T}$	−13	−11	−5	5	12	17	20	20	13	8	−2	−10
DD	982	812	714	399	224	96	42	52	170	333	618	865
Ohio												
Akron-Canton					Latitude		40.9					
$\bar{H}$	4.86	7.37	10.95	15.40	18.93	20.88	20.29	18.12	14.44	10.30	5.72	4.01
$\bar{K}_T$	.33	.37	.41	.45	.48	.50	.50	.50	.49	.47	.36	.30
$\bar{T}$	−3	−2	2	9	15	20	22	21	18	12	5	−1
DD	667	579	498	279	137	36	20	27	74	216	408	611
Cincinnati					Latitude		39.1					
$\bar{H}$	5.68	8.38	11.67	15.88	18.98	20.85	20.10	18.55	14.89	11.23	6.68	4.91
$\bar{K}_T$	.36	.40	.42	.46	.48	.50	.49	.51	.49	.49	.39	.34
$\bar{T}$	−0	1	5	12	17	22	24	24	20	14	7	1
DD	584	493	405	200	90	23	13	16	46	168	356	541

	Jan	Feb	Mar	Apr	May	Jun	Jul	Aug	Sep	Oct	Nov	Dec
	Cleveland				Latitude		41.4					
$\overline{H}$	4.41	6.83	10.47	15.32	19.08	20.92	20.75	17.97	14.07	9.84	5.29	3.61
K_T	.31	.35	.39	.45	.48	.50	.51	.50	.48	.45	.34	.28
T	−3	−2	2	9	15	20	22	21	18	12	5	−1
DD	654	577	499	285	148	47	29	36	79	212	395	596
	Columbus				Latitude		40					
$\overline{H}$	5.21	7.68	11.12	15.36	18.70	20.58	19.92	18.63	14.55	10.73	6.10	4.40
K_T	.34	.37	.41	.44	.47	.49	.49	.51	.49	.48	.37	.32
T	−2	−1	4	11	16	21	23	22	18	12	5	−1
DD	631	538	448	244	119	38	24	31	75	211	393	592
	Dayton				Latitude		39.9					
$\overline{H}$	5.55	8.23	11.64	15.92	19.29	21.27	20.54	18.68	14.96	11.00	6.41	4.62
K_T	.36	.40	.42	.46	.49	.51	.50	.51	.50	.49	.39	.33
T	−2	−1	4	11	16	22	24	23	19	13	5	−1
DD	638	539	453	247	122	39	25	32	71	196	395	589
	Toledo				Latitude		41.6					
$\overline{H}$	4.94	7.72	11.31	15.71	19.49	21.32	20.99	18.34	14.48	10.35	5.65	4.03
K_T	.35	.39	.42	.46	.49	.51	.52	.51	.49	.48	.36	.31
T	−4	−3	2	9	15	21	22	22	18	12	4	−2
DD	693	590	507	292	155	54	37	44	94	232	429	639
	Youngstown				Latitude		41.3					
$\overline{H}$	4.37	6.66	10.11	14.50	18.00	19.97	19.68	17.09	13.55	9.67	5.18	3.58
K_T	.30	.34	.38	.42	.45	.48	.48	.47	.46	.44	.33	.28
T	−3	−3	2	9	14	19	21	21	17	11	5	−2
DD	676	593	513	293	151	45	25	32	82	226	413	623
Oklahoma												
	Oklahoma City				Latitude		35.4					
$\overline{H}$	9.09	11.97	15.89	19.58	21.77	24.34	24.16	22.14	17.64	13.99	10.22	8.23
K_T	.50	.52	.54	.55	.54	.59	.59	.59	.56	.56	.53	.50
T	3	5	9	16	20	25	28	27	23	17	10	4
DD	486	369	298	117	46	10	4	4	20	99	271	435
	Tulsa				Latitude		36.2					
$\overline{H}$	8.30	11.11	14.82	18.20	20.68	22.93	23.05	21.17	16.72	13.21	9.39	7.49
K_T	.47	.49	.51	.51	.52	.55	.57	.57	.53	.54	.50	.46
T	3	5	9	16	20	25	28	27	23	17	10	4
DD	491	374	299	120	53	13	6	7	26	102	272	440
Oregon												
	Astoria				Latitude		46.2					
$\overline{H}$	3.58	6.19	9.83	14.22	18.25	18.46	19.82	17.01	13.43	8.10	4.40	2.95
K_T	.31	.37	.40	.43	.47	.44	.49	.48	.49	.42	.35	.30
T	5	6	7	9	11	14	16	16	15	12	8	6
DD	424	338	361	295	235	167	127	125	141	226	314	387
	Burns				Latitude		43.6					
$\overline{H}$	5.57	8.98	13.47	18.72	23.30	25.87	27.92	23.64	18.39	11.84	6.74	4.88
K_T	.43	.49	.52	.56	.59	.62	.69	.66	.64	.58	.47	.42
T	−4	−1	2	7	11	15	20	19	15	9	2	−2
DD	684	530	498	349	231	127	38	55	136	310	487	638
	Medford				Latitude		42.4					
$\overline{H}$	4.62	8.37	12.86	18.61	23.08	25.85	28.10	24.07	18.04	11.14	5.72	3.82
K_T	.34	.44	.49	.55	.58	.62	.69	.67	.62	.52	.38	.31
T	3	5	7	10	14	18	22	21	18	12	6	3
DD	486	366	348	249	144	55	11	16	53	206	358	468

	Jan	Feb	Mar	Apr	May	Jun	Jul	Aug	Sep	Oct	Nov	Dec
	North Bend				Latitude		43.4					
$\overline{H}$	4.98	8.00	12.01	17.14	21.08	22.64	23.92	20.27	15.64	10.13	5.95	4.33
K_T	.38	.43	.46	.51	.54	.54	.59	.57	.55	.49	.41	.37
T	7	8	8	9	12	14	15	15	15	13	10	8
DD	354	289	316	271	217	155	131	122	132	189	253	321
	Pendleton				Latitude		45.7					
$\overline{H}$	3.95	6.96	11.85	17.06	21.85	24.34	27.19	22.64	17.05	10.31	4.98	3.33
K_T	.34	.41	.48	.52	.56	.58	.67	.64	.62	.53	.38	.32
T	0	4	7	11	15	19	23	22	18	11	5	2
DD	570	411	372	253	156	75	28	39	90	238	399	506
	Portland				Latitude		45.6					
$\overline{H}$	3.52	6.29	10.15	14.85	18.88	20.12	23.13	19.00	13.81	8.21	4.40	2.95
K_T	.30	.37	.41	.45	.48	.48	.57	.54	.50	.42	.34	.29
T	3	6	8	10	14	17	20	19	17	12	7	5
DD	464	347	337	251	167	94	51	55	93	209	332	422
	Redmond				Latitude		44.3					
$\overline{H}$	5.58	8.79	13.51	19.10	23.60	25.97	27.77	23.48	17.98	11.35	6.50	4.82
K_T	.44	.49	.53	.57	.60	.62	.69	.66	.64	.56	.47	.43
T	−1	2	4	7	11	15	19	18	14	9	4	1
DD	600	455	456	349	251	145	70	86	153	296	436	545
	Salem				Latitude		44.9					
$\overline{H}$	3.77	6.68	10.75	15.56	19.73	20.98	24.32	20.15	15.08	8.73	4.66	3.15
K_T	.31	.38	.43	.47	.50	.50	.60	.57	.54	.44	.35	.29
T	4	6	7	10	13	16	19	19	17	12	7	5
DD	454	345	349	264	183	108	60	65	100	221	337	420

Pennsylvania

	Jan	Feb	Mar	Apr	May	Jun	Jul	Aug	Sep	Oct	Nov	Dec
	Allentown				Latitude		40.7					
$\overline{H}$	5.99	8.67	12.24	16.00	18.58	20.17	20.04	17.55	14.05	10.51	6.45	4.88
K_T	.40	.43	.45	.46	.47	.48	.49	.48	.47	.47	.40	.37
T	−2	−1	3	10	16	21	23	22	18	12	6	−1
DD	639	552	465	263	124	36	17	26	70	206	381	591
	Erie				Latitude		42.1					
$\overline{H}$	3.92	6.54	10.45	15.42	18.68	20.96	20.81	16.52	13.64	9.39	4.73	3.15
K_T	.28	.34	.39	.45	.47	.50	.51	.46	.47	.44	.31	.25
T	−4	−4	1	7	13	18	20	20	16	11	4	−2
DD	685	618	553	340	195	68	39	47	101	242	416	618
	Harrisburg				Latitude		40.2					
$\overline{H}$	6.08	8.76	12.29	16.02	18.75	20.48	20.02	17.61	14.38	10.61	6.57	5.08
K_T	.40	.43	.45	.46	.47	.49	.49	.48	.48	.47	.40	.37
T	−1	0	5	12	17	22	25	23	19	13	7	0
DD	602	506	414	210	79	15	6	11	42	174	353	559
	Philadelphia				Latitude		39.9					
$\overline{H}$	6.30	9.02	12.59	16.28	18.84	20.56	19.96	17.88	14.55	10.88	7.03	5.34
K_T	.41	.44	.46	.47	.47	.49	.49	.49	.48	.48	.43	.39
T	0	1	6	12	17	22	25	24	20	14	8	2
DD	563	483	403	221	98	28	14	20	51	166	322	515
	Pittsburgh				Latitude		40.5					
$\overline{H}$	4.82	7.10	10.70	14.95	18.18	19.99	19.17	17.14	13.72	10.15	5.72	3.94
K_T	.32	.35	.39	.43	.46	.48	.47	.47	.46	.46	.35	.29
T	−2	−2	3	10	15	20	22	21	18	12	5	−1
DD	637	555	466	259	133	46	29	38	83	223	397	593

	Jan	Feb	Mar	Apr	May	Jun	Jul	Aug	Sep	Oct	Nov	Dec
	Wilkes-Barre				Latitude		41.3					
$\overline{H}$	5.17	7.81	11.26	15.20	18.06	19.98	19.82	17.17	13.61	10.18	5.55	4.18
K_T	.36	.40	.42	.44	.46	.48	.49	.47	.46	.47	.35	.32
$\overline{T}$	−3	−3	2	9	15	20	22	21	17	11	5	−2
DD	671	587	504	289	156	62	39	51	104	241	410	620

Rhode Island

	Jan	Feb	Mar	Apr	May	Jun	Jul	Aug	Sep	Oct	Nov	Dec
	Providence				Latitude		41.7					
$\overline{H}$	5.75	8.38	11.71	15.60	18.79	20.15	19.24	17.01	13.72	10.29	6.10	4.75
K_T	.41	.43	.44	.46	.48	.48	.47	.47	.47	.48	.40	.37
$\overline{T}$	−2	−1	3	9	14	19	22	21	17	12	6	−0
DD	630	552	486	302	169	58	26	35	86	212	366	577

South Carolina

	Jan	Feb	Mar	Apr	May	Jun	Jul	Aug	Sep	Oct	Nov	Dec
	Charleston				Latitude		32.9					
$\overline{H}$	8.45	11.30	15.20	19.66	21.12	20.93	20.42	17.99	15.82	13.54	10.61	8.18
K_T	.43	.46	.50	.54	.53	.51	.50	.48	.49	.52	.51	.45
$\overline{T}$	9	10	14	18	22	26	27	26	24	19	14	10
DD	288	231	165	63	18	4	2	3	8	54	162	276
	Columbia				Latitude		34					
$\overline{H}$	8.64	11.59	15.38	19.83	21.51	22.10	20.91	19.33	16.33	13.76	10.46	8.20
K_T	.46	.49	.51	.55	.54	.53	.51	.52	.51	.54	.52	.47
$\overline{T}$	7	9	12	18	22	26	27	27	24	18	12	8
DD	344	274	205	76	23	5	3	4	14	77	202	333
	Greenville				Latitude		34.9					
$\overline{H}$	8.28	11.14	15.08	19.26	20.88	21.77	20.77	19.29	15.96	13.39	10.00	7.61
K_T	.45	.48	.51	.54	.52	.52	.51	.52	.50	.53	.51	.45
$\overline{T}$	6	7	11	16	21	24	26	25	22	17	11	6
DD	394	323	254	106	38	10	6	7	22	103	242	382

South Dakota

	Jan	Feb	Mar	Apr	May	Jun	Jul	Aug	Sep	Oct	Nov	Dec
	Huron				Latitude		44.4					
$\overline{H}$	5.54	8.45	12.64	17.37	21.24	23.85	24.77	21.48	16.09	11.22	6.55	4.60
K_T	.44	.47	.50	.52	.54	.57	.61	.60	.57	.56	.48	.42
$\overline{T}$	−11	−8	−2	8	14	20	23	22	16	10	0	−7
DD	902	731	623	329	178	66	29	37	127	283	545	789
	Pierre				Latitude		44.4					
$\overline{H}$	6.02	9.03	13.70	18.32	22.32	24.91	25.86	22.61	16.98	11.94	7.06	5.02
K_T	.48	.50	.54	.55	.57	.60	.64	.63	.60	.59	.51	.45
$\overline{T}$	−9	−6	−1	8	14	20	24	23	17	10	1	−6
DD	850	692	608	326	175	65	26	31	114	268	523	748
	Rapid City				Latitude		44.1					
$\overline{H}$	6.16	9.38	13.95	18.04	21.42	24.19	25.24	22.29	17.23	12.07	7.34	5.41
K_T	.48	.52	.55	.54	.55	.58	.62	.62	.61	.60	.53	.48
$\overline{T}$	−6	−3	−0	7	13	18	23	22	16	10	2	−3
DD	742	608	581	346	194	81	27	32	121	272	495	665
	Sioux Falls				Latitude		43.6					
$\overline{H}$	6.04	9.11	13.07	17.51	21.50	23.84	24.40	20.93	16.00	11.42	6.89	5.01
K_T	.46	.49	.51	.52	.55	.57	.60	.58	.56	.56	.48	.43
$\overline{T}$	−10	−7	−1	8	14	20	23	22	16	10	1	−7
DD	874	708	603	325	162	55	26	32	116	270	533	776

	Jan	Feb	Mar	Apr	May	Jun	Jul	Aug	Sep	Oct	Nov	Dec
Tennessee												
Chattanooga					Latitude		35.0					
$\overline{H}$	7.16	9.75	13.36	17.59	19.66	20.79	19.70	18.50	15.16	12.59	8.77	6.59
K_T	.39	.42	.45	.49	.49	.50	.48	.49	.48	.50	.45	.39
T	5	6	10	16	20	24	26	26	22	16	9	5
DD	430	346	276	122	51	15	9	11	29	123	280	415
Knoxville					Latitude		35.8					
$\overline{H}$	7.04	9.80	13.52	18.15	20.47	21.58	20.48	18.91	15.70	12.72	8.61	6.46
K_T	.40	.43	.46	.51	.51	.52	.50	.51	.50	.51	.45	.39
T	5	6	10	16	20	24	26	25	22	16	10	5
DD	421	346	267	106	33	6	3	4	16	102	266	405
Memphis					Latitude		35.1					
$\overline{H}$	7.75	10.72	14.50	18.61	21.40	23.21	22.39	20.71	16.70	13.68	9.27	7.13
K_T	.43	.46	.49	.52	.54	.56	.55	.55	.53	.55	.48	.42
T	5	7	11	17	22	26	28	27	23	17	11	6
DD	423	329	247	83	21	3	1	2	11	80	241	387
Nashville					Latitude		36.1					
$\overline{H}$	6.58	9.35	12.82	17.53	20.72	22.29	21.47	19.72	15.87	12.64	8.08	5.91
K_T	.37	.41	.44	.49	.52	.54	.53	.53	.50	.51	.43	.36
T	3	5	9	16	20	25	26	26	22	16	9	5
DD	462	376	291	121	46	11	6	8	25	116	286	425
Texas												
Abilene					Latitude		32.4					
$\overline{H}$	10.48	13.43	17.89	20.92	23.13	25.07	24.28	22.21	18.14	14.94	11.44	9.80
K_T	.53	.54	.58	.58	.58	.61	.60	.59	.56	.57	.55	.53
T	7	9	13	18	22	27	29	29	25	19	12	8
DD	367	268	188	47	10	0	0	0	3	42	188	321
Amarillo					Latitude		35.2					
$\overline{H}$	10.90	14.12	18.51	22.92	25.10	27.17	25.89	23.88	19.98	15.94	11.72	9.89
K_T	.60	.61	.63	.64	.63	.65	.64	.64	.63	.64	.61	.59
T	2	4	8	14	19	24	26	25	21	15	8	4
DD	500	394	336	160	58	10	4	6	28	126	317	457
Austin					Latitude		30.3					
$\overline{H}$	9.81	12.77	16.22	18.22	20.81	23.52	23.90	21.92	18.23	15.13	11.20	9.37
K_T	.47	.50	.52	.50	.52	.57	.59	.58	.55	.55	.51	.47
T	10	12	15	20	24	28	29	29	26	21	15	11
DD	280	202	141	51	20	6	3	3	9	43	139	239
Brownsville					Latitude		25.9					
$\overline{H}$	10.36	12.89	16.55	19.72	21.88	24.01	25.11	23.01	19.23	16.33	11.97	9.79
K_T	.44	.46	.50	.53	.55	.59	.63	.60	.56	.56	.49	.44
T	16	17	20	24	26	28	29	29	28	24	20	17
DD	115	70	43	10	3	1	1	1	1	8	36	86
Corpus Christi					Latitude		27.8					
$\overline{H}$	10.20	13.03	16.23	18.64	21.18	23.76	24.82	22.60	19.15	16.07	11.84	9.59
K_T	.45	.48	.50	.50	.53	.58	.62	.59	.56	.56	.50	.45
T	14	15	18	23	26	28	29	30	27	23	18	15
DD	170	112	67	16	6	1	1	1	2	14	64	131
Dallas					Latitude		32.9					
$\overline{H}$	9.33	12.15	16.14	18.47	21.43	24.24	24.09	22.14	18.01	14.48	10.63	8.86
K_T	.48	.50	.53	.51	.54	.59	.59	.59	.55	.55	.51	.49
T	7	10	13	19	23	28	30	30	26	20	13	9
DD	348	252	188	64	24	5	2	2	10	55	179	301

	Jan	Feb	Mar	Apr	May	Jun	Jul	Aug	Sep	Oct	Nov	Dec
Del Rio				Latitude			29.4					
$\overline{H}$	10.88	13.69	17.94	19.29	20.74	22.98	23.32	21.98	17.98	15.44	12.03	10.25
$\overline{K}_T$	.50	.52	.56	.52	.52	.56	.58	.58	.54	.55	.53	.50
$\overline{T}$	10	13	17	22	26	29	30	30	27	22	15	11
DD	267	171	111	35	14	4	3	3	10	42	140	243
El Paso				Latitude			31.8					
$\overline{H}$	12.77	16.80	21.67	26.83	29.52	30.45	27.82	25.93	22.56	18.61	14.12	11.70
$\overline{K}_T$	.63	.67	.70	.74	.74	.74	.69	.69	.69	.70	.66	.62
$\overline{T}$	6	9	13	18	22	27	28	27	23	18	11	7
DD	372	261	190	67	16	1	1	2	9	69	230	357
Fort Worth				Latitude			32.8					
$\overline{H}$	9.14	12.14	15.99	18.34	21.46	24.44	24.47	22.51	18.40	14.68	10.64	8.69
$\overline{K}_T$	.47	.50	.52	.51	.54	.59	.60	.60	.57	.56	.51	.48
$\overline{T}$	7	9	13	18	23	27	29	29	25	20	13	9
DD	353	.260	192	65	21	3	1	1	7	48	174	303
Houston				Latitude			30					
$\overline{H}$	8.77	11.74	14.73	17.28	20.15	21.55	20.75	19.14	16.70	14.48	10.48	8.28
$\overline{K}_T$	.41	.45	.47	.47	.50	.52	.51	.50	.50	.53	.47	.42
$\overline{T}$	11	13	16	21	24	27	29	29	26	22	16	13
DD	242	176	126	46	19	7	5	5	10	40	117	205
Laredo				Latitude			27.5					
$\overline{H}$	10.88	13.57	17.21	19.60	22.16	23.53	24.19	22.81	19.35	15.98	11.81	10.10
$\overline{K}_T$	.48	.50	.53	.53	.55	.58	.60	.60	.57	.56	.50	.47
$\overline{T}$	14	16	20	25	27	30	31	31	28	24	18	15
DD	180	108	60	16	7	2	1	2	4	19	77	152
Lubbock				Latitude			33.7					
$\overline{H}$	11.70	15.12	20.00	24.60	27.19	28.88	27.37	25.07	20.66	16.66	12.66	10.61
$\overline{K}_T$	.61	.63	.66	.68	.68	.70	.67	.67	.64	.65	.63	.60
$\overline{T}$	4	6	9	16	20	25	26	26	22	16	9	5
DD	451	352	288	123	48	10	7	9	30	117	281	411
Lufkin				Latitude			31.2					
$\overline{H}$	9.01	12.14	15.62	18.43	21.19	23.33	22.77	21.16	17.38	15.31	10.94	8.71
$\overline{K}_T$	.44	.48	.50	.50	.53	.57	.56	.56	.53	.57	.51	.45
$\overline{T}$	9	11	14	20	23	27	28	28	25	20	14	10
DD	292	213	157	55	21	6	4	4	10	52	160	261
Midland-Odessa				Latitude			31.9					
$\overline{H}$	12.28	15.70	20.88	24.89	27.59	29.09	27.12	25.09	20.93	17.28	13.35	11.35
$\overline{K}_T$	.61	.63	.68	.68	.69	.70	.67	.66	.64	.65	.63	.60
$\overline{T}$	6	9	12	18	22	27	28	28	24	19	12	8
DD	381	281	217	94	39	12	9	9	24	84	224	343
Port Arthur				Latitude			30					
$\overline{H}$	9.08	12.15	15.36	18.28	21.24	22.83	20.96	19.71	17.33	15.00	10.81	8.56
$\overline{K}_T$	.43	.47	.49	.50	.53	.56	.52	.52	.52	.54	.48	.43
$\overline{T}$	11	13	16	21	24	27	28	28	26	21	16	12
DD	235	169	119	34	11	2	1	1	4	30	112	203
San Angelo				Latitude			31.4					
$\overline{H}$	10.92	13.72	18.23	21.01	23.05	24.82	24.09	22.32	18.24	15.17	11.85	10.15
$\overline{K}_T$	.53	.54	.59	.57	.58	.60	.59	.59	.55	.56	.55	.53
$\overline{T}$	8	10	14	20	24	28	29	29	25	20	13	9
DD	327	235	164	49	16	3	1	1	9	52	177	294

	Jan	Feb	Mar	Apr	May	Jun	Jul	Aug	Sep	Oct	Nov	Dec
	San Antonio				Latitude		29.5					
$\overline{H}$	10.17	13.10	16.46	18.30	21.50	23.49	24.08	22.10	18.59	15.32	11.45	9.62
K_T	.47	.50	.52	.50	.54	.57	.60	.58	.55	.55	.51	.48
T	10	13	16	21	24	28	29	29	26	21	15	12
DD	262	183	123	42	17	4	3	3	8	39	131	225
	Sherman				Latitude		33.7					
$\overline{H}$	9.01	11.78	15.50	18.28	21.02	24.00	23.57	21.92	17.94	14.39	10.43	8.44
K_T	.47	.49	.51	.51	.53	.58	.58	.58	.56	.56	.52	.48
T	5	8	11	18	22	26	29	29	24	19	12	7
DD	404	300	231	76	25	4	1	1	9	60	206	352
	Waco				Latitude		31.6					
$\overline{H}$	9.45	12.45	16.21	18.30	20.13	23.98	24.18	22.23	18.17	14.77	10.86	9.11
K_T	.46	.50	.52	.50	.50	.58	.60	.59	.55	.55	.51	.48
T	8	11	14	20	24	28	30	30	26	21	14	10
DD	323	233	170	59	22	5	3	3	9	50	158	277
	Wichita Falls				Latitude		34					
$\overline{H}$	9.78	12.74	16.71	20.01	22.90	25.22	24.59	22.35	18.18	14.66	10.87	9.06
K_T	.52	.53	.56	.56	.57	.61	.60	.60	.57	.57	.54	.52
T	5	8	11	18	22	27	30	30	25	19	12	7
DD	413	307	241	91	37	8	4	4	17	79	226	369

Utah

	Jan	Feb	Mar	Apr	May	Jun	Jul	Aug	Sep	Oct	Nov	Dec
	Bryce Canyon				Latitude		37.7					
$\overline{H}$	10.37	14.03	19.13	24.22	27.85	30.13	27.51	24.48	21.80	16.63	11.53	9.29
K_T	.62	.64	.67	.69	.70	.72	.68	.66	.71	.70	.65	.61
T	−7	−5	−2	3	8	12	16	16	12	6	−1	−5
DD	778	650	624	457	335	206	119	137	223	390	573	732
	Cedar City				Latitude		37.7					
$\overline{H}$	10.02	13.39	18.57	23.75	28.01	30.71	28.42	25.44	22.34	16.57	11.27	8.92
K_T	.60	.61	.65	.67	.70	.74	.70	.69	.72	.70	.63	.59
T	−2	1	4	8	13	18	23	22	17	11	4	−1
DD	623	496	457	302	171	61	15	22	78	243	437	590
	Salt Lake City				Latitude		40.8					
$\overline{H}$	7.26	11.22	16.50	21.50	26.82	29.07	29.41	25.58	20.92	14.69	8.94	6.46
K_T	.49	.56	.61	.63	.68	.70	.72	.70	.71	.66	.56	.49
T	−2	1	4	10	15	19	25	24	18	11	4	−1
DD	637	492	440	272	150	61	12	18	73	235	436	596

Vermont

	Jan	Feb	Mar	Apr	May	Jun	Jul	Aug	Sep	Oct	Nov	Dec
	Burlington				Latitude		44.5					
$\overline{H}$	4.37	6.88	10.68	14.71	17.87	19.63	19.54	16.74	12.74	8.41	4.25	3.21
K_T	.35	.38	.42	.44	.46	.47	.48	.47	.45	.42	.31	.29
T	−8	−7	−2	6	13	18	21	20	15	9	3	−5
DD	827	719	618	370	194	67	35	50	127	290	468	729

Virginia

	Jan	Feb	Mar	Apr	May	Jun	Jul	Aug	Sep	Oct	Nov	Dec
	Norfolk				Latitude		36.9					
$\overline{H}$	7.70	10.57	14.54	19.04	21.42	22.71	21.04	19.07	15.85	12.29	9.21	7.08
K_T	.45	.48	.50	.54	.54	.54	.52	.51	.51	.51	.50	.45
T	5	5	9	14	19	24	26	25	22	17	11	6
DD	422	367	296	139	44	7	3	4	15	93	230	392

	Jan	Feb	Mar	Apr	May	Jun	Jul	Aug	Sep	Oct	Nov	Dec
Richmond					Latitude		37.5					
$\bar{H}$	7.17	9.96	13.74	17.78	20.00	21.25	20.14	18.17	15.30	11.72	8.33	6.43
K_T	.43	.45	.48	.50	.50	.51	.49	.49	.49	.49	.46	.42
$\bar{T}$	3	4	8	14	19	23	26	25	21	15	9	4
DD	473	399	318	147	56	14	7	10	31	131	276	449
Roanoke					Latitude		37.3					
$\bar{H}$	7.50	10.21	14.03	17.96	20.02	21.37	20.39	18.39	15.41	12.27	8.68	6.70
K_T	.44	.46	.49	.51	.50	.51	.50	.50	.50	.51	.48	.43
$\bar{T}$	2	3	7	13	18	22	24	23	20	14	8	3
DD	496	420	345	175	79	25	15	18	46	157	310	477

Washington

	Jan	Feb	Mar	Apr	May	Jun	Jul	Aug	Sep	Oct	Nov	Dec
Olympia					Latitude		47					
$\bar{H}$	3.06	5.71	9.59	14.24	18.53	19.22	21.72	17.58	13.13	7.22	3.85	2.51
K_T	.28	.35	.40	.44	.48	.46	.54	.50	.49	.39	.32	.26
$\bar{T}$	3	5	6	9	12	15	18	17	15	10	6	4
DD	481	376	380	289	209	137	86	96	140	262	366	440
Seattle-Tacoma					Latitude		47.5					
$\bar{H}$	2.98	5.62	9.64	14.69	19.46	20.46	25.52	18.34	13.03	7.45	3.83	2.40
K_T	.28	.35	.41	.45	.50	.49	.63	.53	.49	.41	.32	.26
$\bar{T}$	3	6	7	9	13	15	18	18	15	11	7	5
DD	465	355	364	279	194	123	74	81	126	234	344	424
Spokane					Latitude		47.6					
$\bar{H}$	3.58	6.88	11.81	16.97	21.77	23.65	26.76	22.05	16.29	9.54	4.52	2.90
K_T	.34	.43	.50	.53	.56	.57	.66	.63	.61	.52	.38	.32
$\bar{T}$	−4	0	3	8	13	16	21	20	15	9	2	−2
DD	683	510	473	321	195	100	34	44	123	303	494	621
Yakima					Latitude		46.6					
$\bar{H}$	4.15	7.56	12.74	18.14	22.80	24.63	26.77	22.42	16.83	10.11	5.04	3.35
K_T	.37	.45	.53	.56	.58	.59	.66	.64	.62	.54	.41	.34
$\bar{T}$	−3	2	5	10	14	18	21	20	16	10	4	−0
DD	646	455	405	268	154	73	32	45	106	267	443	581

West Virginia

	Jan	Feb	Mar	Apr	May	Jun	Jul	Aug	Sep	Oct	Nov	Dec
Charleston					Latitude		38.4					
$\bar{H}$	5.66	8.02	11.46	15.39	18.61	20.16	19.10	17.19	14.44	11.04	6.96	5.00
K_T	.35	.37	.41	.44	.47	.48	.47	.47	.47	.47	.40	.34
$\bar{T}$	1	3	7	13	18	22	24	23	20	14	7	2
DD	527	444	361	176	79	26	17	22	52	168	335	499
Huntington					Latitude		38.4					
$\bar{H}$	5.97	8.60	12.11	16.44	19.41	20.93	20.08	17.94	14.82	11.39	7.24	5.30
K_T	.37	.40	.43	.47	.49	.50	.49	.49	.48	.49	.42	.36
$\bar{T}$	1	2	7	13	18	22	24	23	20	14	8	2
DD	532	453	370	188	94	36	25	31	65	180	337	504

Wisconsin

	Jan	Feb	Mar	Apr	May	Jun	Jul	Aug	Sep	Oct	Nov	Dec
Eau Claire					Latitude		44.9					
$\bar{H}$	5.12	8.47	12.38	16.19	19.08	21.25	21.41	18.40	13.57	9.38	5.11	3.87
K_T	.42	.48	.49	.49	.49	.51	.53	.52	.49	.47	.38	.36
$\bar{T}$	−11	−9	−3	7	13	19	21	20	15	9	0	−8
DD	917	770	649	346	173	54	27	39	133	288	550	808
Green Bay					Latitude		44.5					
$\bar{H}$	5.12	8.22	12.54	16.33	19.51	21.66	21.43	18.41	13.82	9.31	5.28	3.98
K_T	.41	.46	.50	.49	.50	.52	.53	.52	.49	.46	.39	.36
$\bar{T}$	−9	−8	−2	7	13	18	21	20	15	10	1	−6
DD	853	730	626	355	198	69	36	47	133	279	513	760

	Jan	Feb	Mar	Apr	May	Jun	Jul	Aug	Sep	Oct	Nov	Dec
	La Crosse				Latitude		43.9					
$\overline{H}$	5.46	8.68	12.49	16.19	19.45	21.63	21.57	18.91	14.10	9.80	5.61	4.19
$\overline{K}_T$	.43	.48	.49	.48	.49	.52	.53	.53	.50	.48	.40	.37
$\overline{T}$	−9	−7	−0	9	15	20	23	22	17	11	2	−6
DD	840	701	586	302	152	56	33	41	113	250	496	746
	Madison				Latitude		43.1					
$\overline{H}$	5.85	9.13	12.89	15.88	19.79	22.11	21.96	19.39	14.75	10.34	5.72	4.42
$\overline{K}_T$	.44	.49	.50	.47	.50	.53	.54	.54	.51	.50	.39	.37
$\overline{T}$	−8	−7	−1	7	13	19	21	20	15	10	2	−6
DD	828	695	602	340	194	79	50	60	140	282	508	743
	Milwaukee				Latitude		43					
$\overline{H}$	5.44	8.36	12.36	16.38	20.07	22.44	22.27	19.51	14.87	10.30	5.95	4.29
$\overline{K}_T$	.41	.45	.47	.49	.51	.54	.55	.54	.52	.49	.41	.36
$\overline{T}$	−7	−5	−0	7	12	18	21	21	16	11	3	−4
DD	785	660	578	344	211	79	41	46	113	256	477	701

Wyoming

	Jan	Feb	Mar	Apr	May	Jun	Jul	Aug	Sep	Oct	Nov	Dec
	Casper				Latitude		42.9					
$\overline{H}$	7.76	11.51	16.36	20.97	25.01	28.40	28.77	25.26	19.85	13.83	8.69	6.75
$\overline{K}_T$	.58	.61	.63	.62	.63	.68	.71	.70	.69	.66	.59	.56
$\overline{T}$	−5	−3	−1	6	12	17	22	21	15	9	1	−3
DD	719	593	587	375	224	93	24	32	133	304	516	667
	Cheyenne				Latitude		41.2					
$\overline{H}$	8.69	12.12	16.27	20.09	22.64	25.64	25.32	22.31	18.92	14.10	9.34	7.62
$\overline{K}_T$	.60	.61	.60	.59	.57	.61	.62	.61	.64	.64	.59	.58
$\overline{T}$	−3	−2	−0	6	11	16	21	20	15	9	2	−2
DD	662	562	576	379	239	114	49	59	149	308	495	619
	Rock Springs				Latitude		41.6					
$\overline{H}$	8.35	12.37	17.37	22.07	26.61	29.23	28.92	25.43	20.80	14.82	9.38	7.38
$\overline{K}_T$	.59	.63	.65	.65	.67	.70	.71	.70	.71	.68	.61	.58
$\overline{T}$	−7	−5	−2	4	10	15	20	19	14	7	−1	−5
DD	788	646	621	417	261	132	43	59	163	352	571	729
	Sheridan				Latitude		44.8					
$\overline{H}$	5.87	8.95	13.68	17.45	21.38	24.48	26.44	22.77	17.05	11.42	6.70	5.01
$\overline{K}_T$	.48	.51	.54	.53	.55	.59	.65	.64	.61	.57	.50	.46
$\overline{T}$	−6	−3	−1	6	12	16	21	21	14	9	1	−4
DD	757	608	587	363	222	108	35	40	148	305	527	680

Pacific

Australia

	Jan	Feb	Mar	Apr	May	Jun	Jul	Aug	Sep	Oct	Nov	Dec
	Aspendale				Latitude		−38.0					
$\overline{H}$	24.74	21.91	17.06	11.31	7.67	6.40	7.12	9.45	13.36	18.07	22.09	24.31
$\overline{K}_T$	.57	.57	.54	.48	.45	.46	.47	.46	.48	.50	.53	.55
$\overline{T}$	20	20	18	15	13	10	10	11	12	14	16	18
DD	19	18	39	107	184	244	271	244	181	133	80	39
	Darwin				Latitude		−12.4					
$\overline{H}$	17.54	18.73	18.37	20.84	19.27	19.86	21.07	22.55	23.64	23.23	24.62	19.52
$\overline{K}_T$	.44	.47	.49	.61	.64	.71	.73	.70	.66	.60	.62	.49
$\overline{T}$	28	28	28	28	27	25	25	26	28	29	29	29
DD	0	0	0	0	0	0	0	0	0	0	0	0

	Jan	Feb	Mar	Apr	May	Jun	Jul	Aug	Sep	Oct	Nov	Dec
	Perth				Latitude		−31.9					
$\bar{H}$	25.05	24.09	19.55	13.91	10.58	8.83	10.21	13.45	17.89	21.72	23.43	25.89
K_T	.58	.61	.58	.53	.52	.50	.55	.57	.59	.58	.56	.59
T	24	25	22	19	16	14	13	13	15	16	19	21
DD	1	0	6	29	89	137	170	170	110	89	29	11

Fiji

	Jan	Feb	Mar	Apr	May	Jun	Jul	Aug	Sep	Oct	Nov	Dec
	Nandi				Latitude		−17.8					
$\bar{H}$	19.62	19.22	17.21	15.93	14.24	13.12	13.72	15.98	18.00	19.73	20.94	21.16
K_T	.47	.48	.47	.49	.51	.52	.52	.53	.52	.51	.51	.51
T	27	27	27	26	25	24	23	24	24	25	26	27
DD	0	0	0	0	0	0	1	0	0	0	0	0

New Zealand

	Jan	Feb	Mar	Apr	May	Jun	Jul	Aug	Sep	Oct	Nov	Dec
	Wellington				Latitude		−41.3					
$\bar{H}$	22.26	19.49	14.57	10.04	6.27	5.10	5.48	7.73	12.16	17.10	21.24	22.40
K_T	.52	.52	.48	.46	.42	.42	.41	.42	.46	.49	.51	.50
T	16	17	16	14	11	9	8	9	11	12	14	15
DD	87	56	87	135	228	280	320	289	221	198	135	113

Phillipines

	Jan	Feb	Mar	Apr	May	Jun	Jul	Aug	Sep	Oct	Nov	Dec
	Quezon City				Latitude		14.6					
$\bar{H}$	13.84	16.27	18.77	20.58	18.94	17.43	15.63	14.16	15.31	14.09	13.64	12.83
K_T	.47	.50	.52	.54	.49	.46	.41	.37	.42	.42	.45	.45
T	25	26	27	29	29	29	28	27	27	27	26	25
DD	0	0	0	0	0	0	0	0	0	0	0	0

United States Pacific Territory (also, see Hawaii)

	Jan	Feb	Mar	Apr	May	Jun	Jul	Aug	Sep	Oct	Nov	Dec
	Koror Island				Latitude		7.3					
$\bar{H}$	15.86	17.70	18.51	19.21	17.80	16.53	16.37	16.83	17.33	17.11	16.45	15.20
K_T	.48	.50	.50	.51	.48	.46	.45	.45	.47	.48	.49	.47
T	27	27	27	28	28	28	27	27	28	28	28	27
DD	1	0	1	0	0	0	1	1	0	0	0	1
	Kwajalein Island				Latitude		8.7					
$\bar{H}$	17.87	19.89	20.39	19.75	18.47	18.28	18.28	19.15	18.33	17.31	16.55	16.47
K_T	.55	.57	.55	.52	.50	.50	.50	.51	.49	.49	.50	.52
T	27	27	28	28	28	28	28	28	28	28	28	28
DD	0	0	0	0	0	0	0	0	0	0	0	0
	Wake Island				Latitude		19.3					
$\bar{H}$	15.31	17.84	20.56	22.18	23.32	23.23	21.84	21.25	19.73	17.83	16.31	14.88
K_T	.56	.58	.59	.58	.59	.59	.56	.56	.55	.56	.58	.57
T	25	25	25	26	27	28	28	28	28	28	27	26
DD	16	12	14	12	10	7	7	6	5	7	8	12

South America

Argentina

	Jan	Feb	Mar	Apr	May	Jun	Jul	Aug	Sep	Oct	Nov	Dec
	Buenos Aires				Latitude		−34.6					
$\bar{H}$	25.20	22.91	18.51	13.44	9.69	7.40	8.18	11.49	15.00	18.93	23.95	24.73
K_T	.58	.59	.57	.54	.51	.46	.48	.52	.51	.52	.57	.56
T	24	23	21	16	13	11	11	12	14	17	20	22
DD	2	3	12	87	170	222	229	200	137	68	19	7

Chile

	Jan	Feb	Mar	Apr	May	Jun	Jul	Aug	Sep	Oct	Nov	Dec
	Valparaiso				Latitude		−33.0					
$\bar{H}$	21.91	18.02	14.45	9.83	6.46	5.23	6.25	9.11	12.31	15.46	19.39	21.32
K_T	.51	.46	.44	.38	.33	.31	.34	.40	.41	.42	.46	.48
T	18	18	17	14	13	12	12	12	13	14	15	17
DD	44	42	62	134	168	191	198	198	162	154	108	67

	Jan	Feb	Mar	Apr	May	Jun	Jul	Aug	Sep	Oct	Nov	Dec
Ecuador												
Izobamba					Latitude		−0.4					
$\overline{H}$	15.23	14.81	14.60	13.50	14.78	14.57	15.17	16.03	16.11	14.97	15.02	15.16
$\overline{K}_T$	.42	.39	.39	.37	.43	.44	.45	.45	.43	.40	.41	.42
$\overline{T}$	13	14	14	14	14	13	13	14	14	13	13	14
DD	155	136	149	147	146	149	161	149	141	164	152	146
Peru												
Huancayo					Latitude		−12.1					
$\overline{H}$	26.56	24.17	23.56	23.56	22.23	22.60	22.68	24.28	25.83	26.88	27.91	25.97
$\overline{K}_T$	.66	.61	.63	.69	.74	.80	.78	.76	.72	.70	.70	.65
$\overline{T}$	12	12	12	12	11	10	10	11	12	12	13	13
DD	197	177	197	190	227	249	258	227	190	197	161	167
Venezuela												
Caracas					Latitude		10.5					
$\overline{H}$	14.65	16.15	16.94	16.24	15.92	16.05	16.86	17.10	16.82	15.11	14.22	13.50
$\overline{K}_T$	.46	.47	.46	.43	.42	.43	.45	.45	.45	.43	.44	.44
$\overline{T}$	19	20	21	22	22	22	21	22	22	22	21	20
DD	24	13	8	3	3	4	6	4	3	4	7	14
Maracibo					Latitude		10.6					
$\overline{H}$	15.63	17.11	17.81	16.50	15.48	16.64	17.62	17.43	16.47	15.08	13.93	14.41
$\overline{K}_T$	.49	.50	.48	.43	.41	.45	.47	.46	.45	.43	.43	.47
$\overline{T}$	27	27	27	28	28	29	29	29	29	28	28	27
DD	0	0	0	0	0	0	0	0	0	0	0	0

Appendix H

SOLAR POSITION DIAGRAMS (See Section 1.9)

The following solar position diagrams are described in Section 1.9. They are for latitudes of $0°$ to $70°$ in increments of $5°$.

Months and times are not shown on these diagrams in order to improve their legibility. Months and times are shown in Figure 1.9.1. These are correct for the northern hemisphere; for the southern hemisphere the months should be interchanged as shown on Figure 1.8.2. The order of months does not change from one latitude to another (except in a transition from one hemisphere to the other).

The top-to-bottom orders of the months, by hemispheres, are as follows.

Northern	Southern
June	December
July	January
May	November
August	February
April	October
September	March
March	September
October	April
February	August
November	May
January	July
December	June

Solar time is noon for the central time line; lines are shown for each hour before noon (on the left) and after noon (on the right).

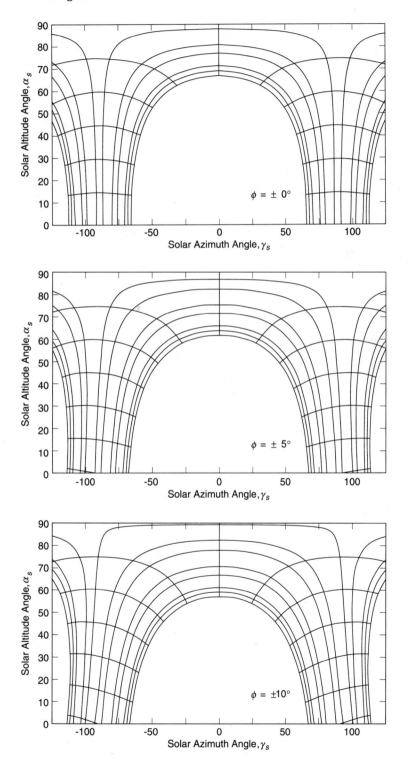

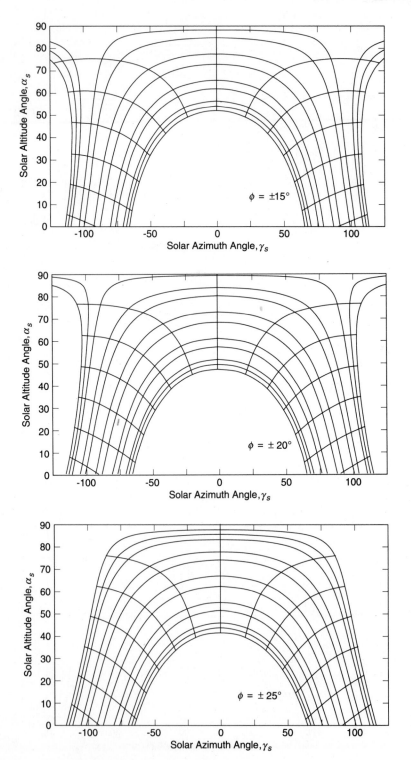

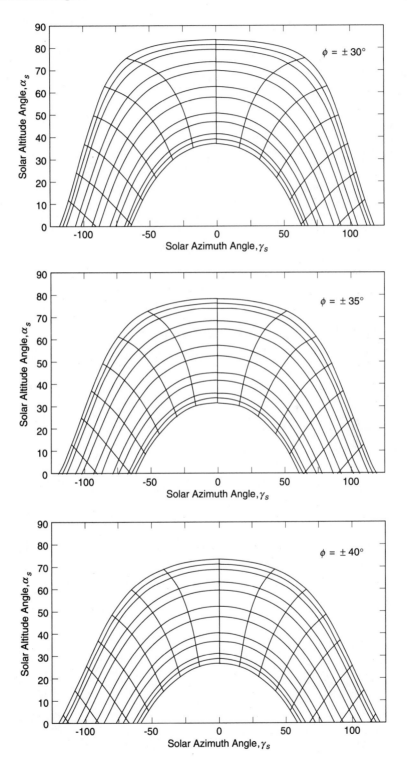

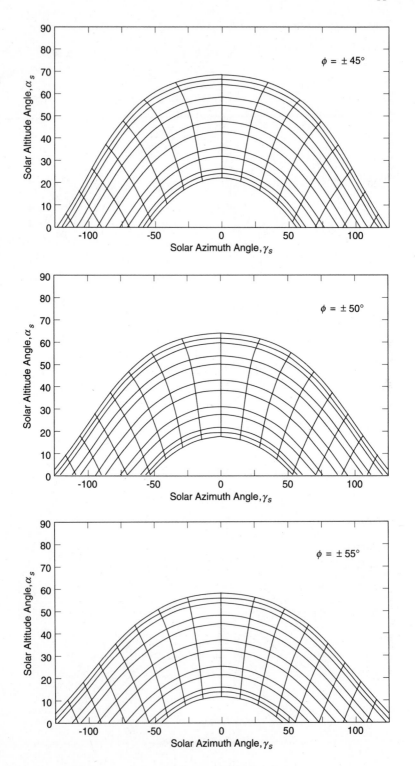

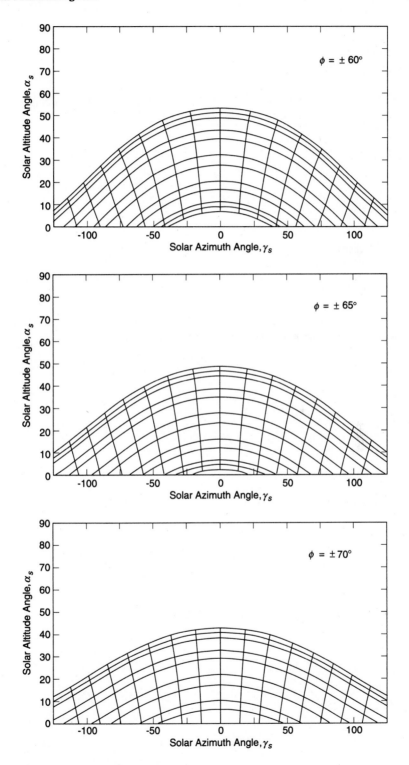

Appendix I

AVERAGE SHADING FACTORS FOR OVERHANGS

The following charts show monthly average fraction of receiver area receiving beam radiation as a function of relative overhang dimensions, for latitudes 35°, 45°, and 55°. For use in the southern hemisphere, interchange months as shown in Figure 1.8.2. See Section 14.4 for methods of use of these factors and interpolation routines.

Utzinger, D. M., M.S. Thesis, Mechanical Engineering, U. of Wisconsin-Madison (1979). "Analysis of Building Components Related to Direct Solar Heating of Buildings."

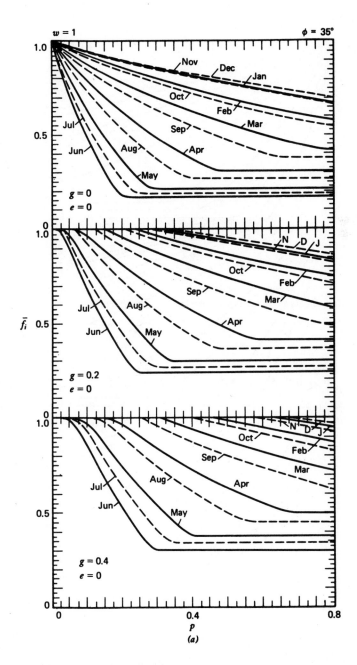

Figure I-1(a-e) Monthly mean fraction of vertical receiver area receiving beam radiation as a function of relative overhang dimensions, for latitude 35°. For the southern hemisphere interchange months as shown in Figure 1.8.2.

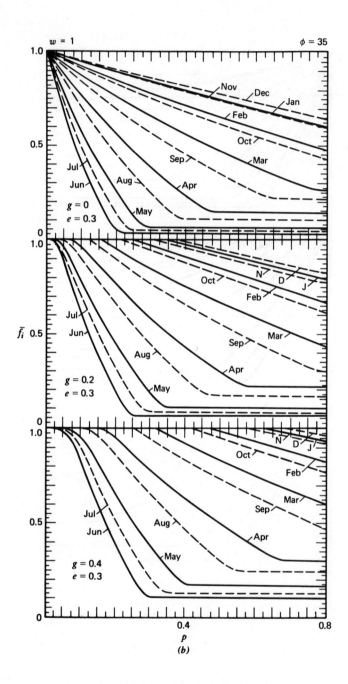

(b)

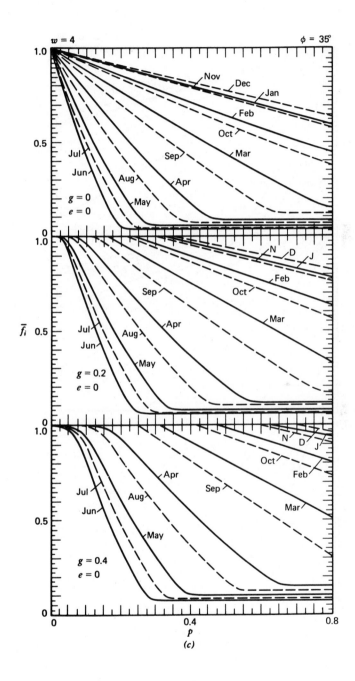

(c)

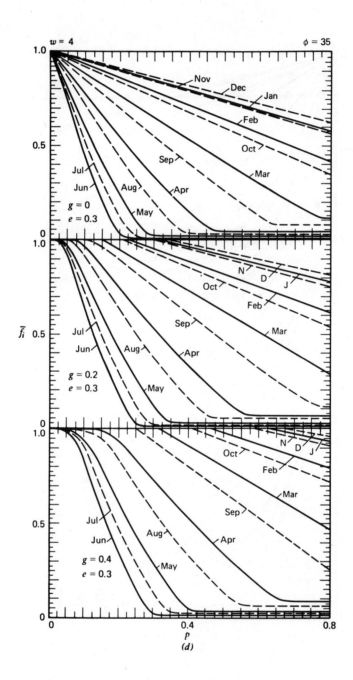

(d)

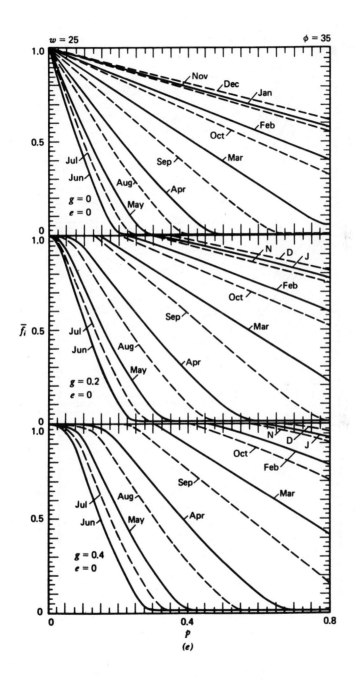

(e)

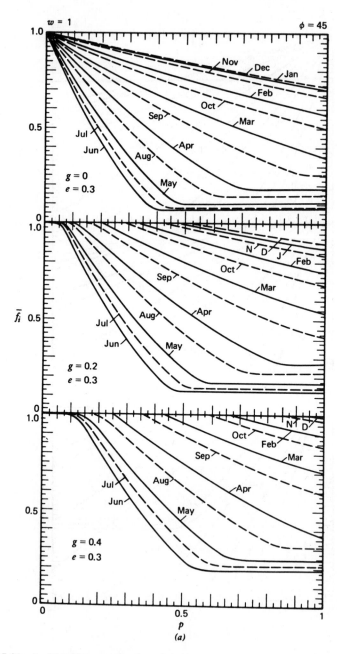

Figure I-2(a-e)   Monthly mean fraction of vertical receiver area receiving beam radiation as a function of relative overhang dimensions, for latitude 45°. For the southern hemisphere interchange months as shown in Figure 1.8.2.

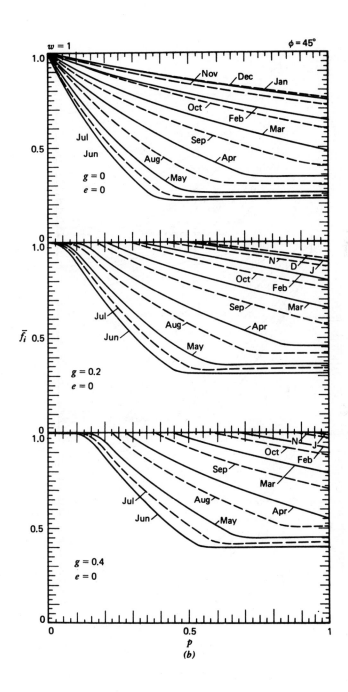

(b)

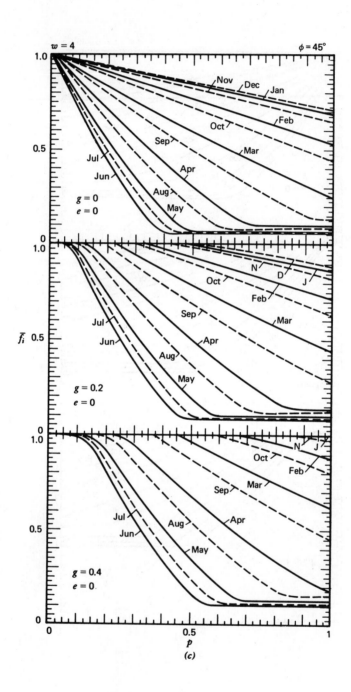

(c)

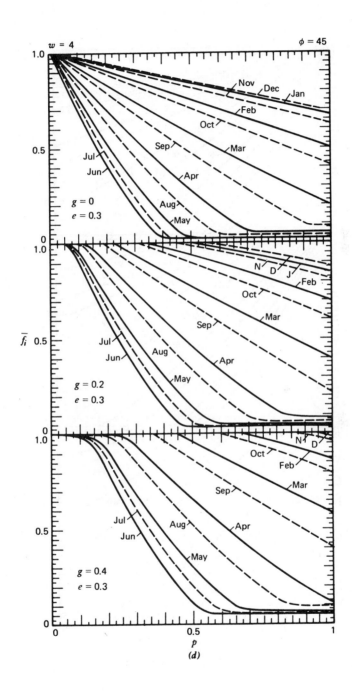

$$p$$
$$(d)$$

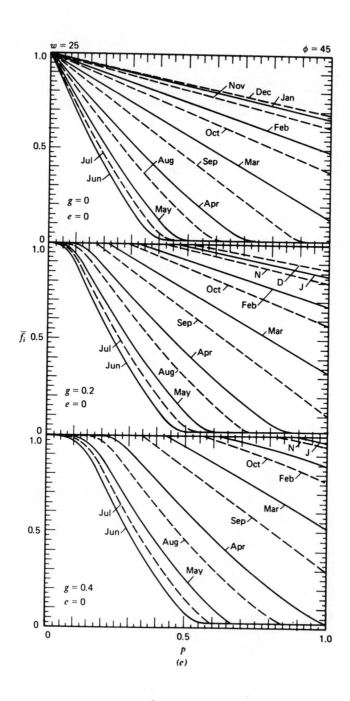

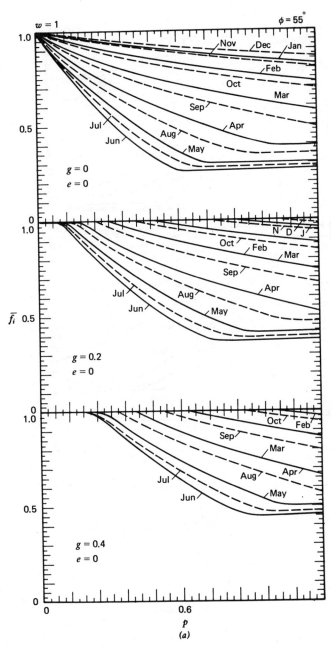

Figure I-3(a-e)   Monthly mean fraction of vertical receiver area receiving beam radiation as a function of relative overhang dimensions, for latitude 45°. For the southern hemisphere interchange months as shown in Figure 1.8.2.

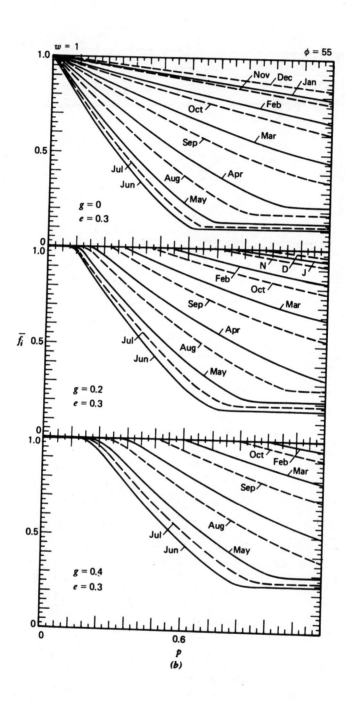

(b)

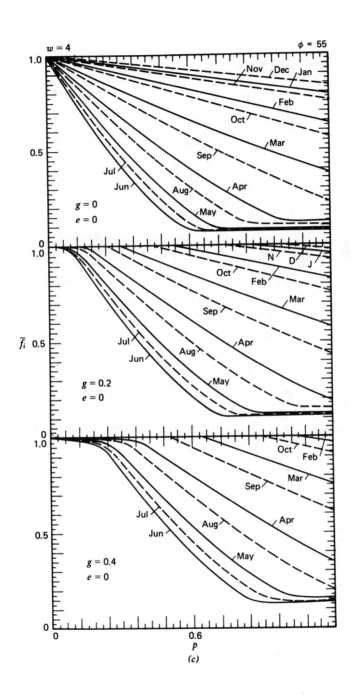

(c)

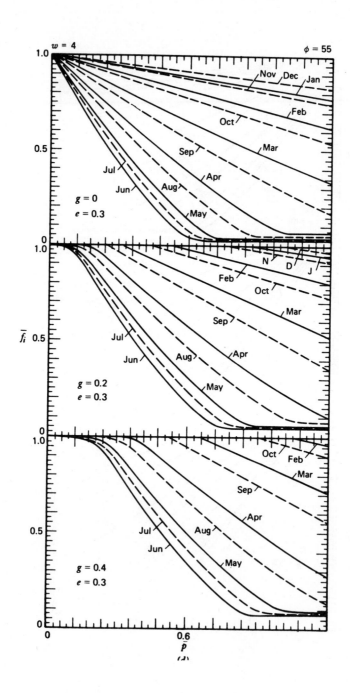

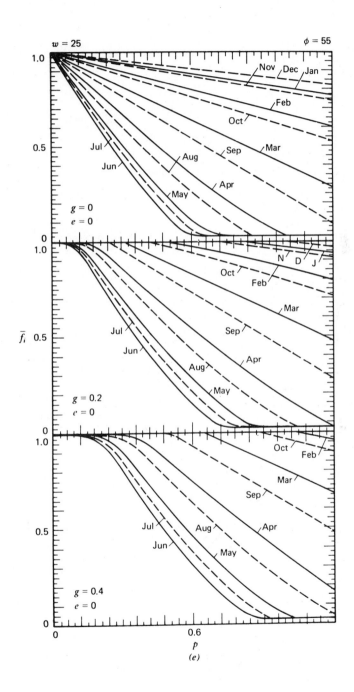

$w = 25$ $\phi = 55$

$\bar{f}_i$

(e)

Author Index

Subject Index